PROBABILITY

Combinations: $C(n, r) = \dfrac{n!}{r!(n-r)!}$

Permutations: $P(n, r) = \dfrac{n!}{(n-r)!}$

Addition Rule: $P(E \cup F) = P(E) + P(F) - P(E \cap F)$
  Mutually Exclusive Events: $P(E \cup F) = P(E) + P(F)$
Multiplication Rule: $P(E \cap F) = P(F)P(E|F)$
  Independent Events: $P(E \cap F) = P(F)P(E)$
Complementary Events: $P(E) + P(E') = 1$
Binomial Probability: $C(n, r)p^r(1-p)^{n-r}$
Bayes' Rule:

$$P(A_i|B) = \dfrac{P(A_i)P(B|A_i)}{P(A_1)P(B|A_1) + P(A_2)P(B|A_2) + \cdots + P(A_n)P(B|A_n)}$$

Expected Value: $E(X) = x_1P(x_1) + x_2P(x_2) + \cdots + x_nP(x_n)$

FINANCE

GEOMETRIC PROGRESSION

$a_n = r^{n-1}a_1$

$S_n = \dfrac{a_1(1 - r^n)}{1 - r}$

SIMPLE INTEREST

Interest: $I = PRT$

Present Value: $P = \dfrac{S}{1 + RT}$

COMPOUND INTEREST

Compound Amount: $S = P(1 + i)^n$

Present Value: $P = S(1 + i)^{-n}$

ORDINARY ANNUITY

Amount: $S = R\left[\dfrac{(1+i)^n - 1}{i}\right]$

Periodic Payment (Sinking Fund): $R = S\left[\dfrac{i}{(1+i)^n - 1}\right]$

Present Value: $A = R\left[\dfrac{1 - (1+i)^{-n}}{i}\right]$

Periodic Payment (Amortization): $R = A\left[\dfrac{i}{1 - (1+i)^{-n}}\right]$

THE PENNSYLVANIA
STATE UNIVERSITY
LIBRARIES

APPLIED FINITE
MATHEMATICS
AND CALCULUS

APPLIED FINITE MATHEMATICS AND CALCULUS

GEORGE J. KERTZ
The University of Toledo

Mathematics is the gate and key of the sciences.... Neglect of mathematics works injury to all knowledge, since he who is ignorant of it cannot know the other sciences or the things of this world.

Roger Bacon (1214?–1294)

WEST PUBLISHING COMPANY
St. Paul   New York   Los Angeles   San Francisco

**TO
SUSAN,
ERIC, MARK, AND JOAN,
FOR WHOM THIS BOOK WAS WRITTEN**

Copy editing: Linda Thompson
Design: Janet Bollow
Artwork: Carl Brown
Composition: Syntax International
Production management: Phyllis Niklas
Cover: No. 9, New York, July 1940, by Charles Biederman. Reproduced with permission of the artist and the owner, Mrs. Raymond F. Hedin. Design by Taly Design Group.

COPYRIGHT © 1985 By WEST PUBLISHING CO.
50 West Kellogg Boulevard
P. O. Box 43526
St. Paul, Minnesota 55164

All rights reserved

Printed in the United States of America

LIBRARY OF CONGRESS CATALOGING IN PUBLICATION DATA

Kertz, George J.
   Applied finite mathematics and calculus.

   Includes index.
   1. Mathematics—1961– . I. Title.
QA37.2.K47 1985 510 84-17389
ISBN 0-314-85317-0

CONTENTS

Preface   xi

CHAPTER 1 INTRODUCTION   1

1.1   A Physical Problem   2
1.2   A Mathematical Solution   6
1.3   The Application of Mathematics   11

CHAPTER 2 EQUATIONS AND THEIR GRAPHS   17

2.1   Equations as Mathematical Models   18
2.2   The Real Number Line and Cartesian Plane   24
2.3   Graphs of Equations   28
2.4   Equation of a Straight Line   31
2.5   Linear Equations   37
2.6   Linear Equations as Mathematical Models   40
2.7   Quadratic Equations   46
2.8   Quadratic Equations as Mathematical Models   49
2.9   Chapter Summary   52
      Review Exercises   54

CHAPTER 3 SYSTEMS OF LINEAR EQUATIONS — MATRICES   57

3.1   Systems of Two Linear Equations   58
3.2   Systems of Equations and Matrices   66
3.3   Gauss-Jordan Elimination   75
3.4   Matrix Addition and Scalar Multiplication   80
3.5   Matrix Multiplication   87
3.6   Identity Matrices and Inverses   95
*3.7   Leontief Input-Output Models   101
3.8   Chapter Summary   105
      Review Exercises   107

* Sections marked with an asterisk can be omitted without loss of continuity.

CHAPTER 4 LINEAR PROGRAMMING — 113

- 4.1 Graphing Linear Inequalities — 114
- 4.2 Linear Programming—Geometric Method — 119
- 4.3 Basic Solutions of Systems of Linear Equations — 127
- 4.4 Linear Programming—Simplex Method — 132
- *4.5 Minimization — 145
- 4.6 Chapter Summary — 151
- Review Exercises — 154

CHAPTER 5 PROBABILITY — 159

- 5.1 Sample Spaces and Events — 160
- 5.2 Calculation of Probabilities — 165
- 5.3 Counting Techniques — 173
- 5.4 Combinations — 179
- 5.5 The Addition Rule — 186
- 5.6 The Multiplication Rule — 193
- 5.7 Chapter Summary — 201
- Review Exercises — 203

CHAPTER 6 ADDITIONAL TOPICS IN PROBABILITY — 207

- *6.1 Binomial Probabilities — 208
- *6.2 Bayes' Rule — 215
- 6.3 Markov Chains — 221
- *6.4 Regular Markov Chains — 230
- 6.5 Expected Value — 236
- 6.6 Game Theory — 243
- *6.7 Games with Mixed Strategies — 248
- 6.8 Chapter Summary — 256
- Review Exercises — 258

CHAPTER 7 INTRODUCTORY STATISTICS — 261

- 7.1 Organizing Data — 262
- 7.2 Measures of Central Tendency — 266
- 7.3 Measures of Dispersion — 272
- 7.4 The Sign Test — 281
- 7.5 The Normal Curve — 289
- *7.6 A Test of a Claimed Mean — 299
- 7.7 Chapter Summary — 303
- Review Exercises — 304

CHAPTER 8 EXPONENTS AND LOGARITHMS — 309

- 8.1 Graphs of Exponential Equations — 310
- 8.2 Logarithms — 317
- 8.3 Graphs of Logarithmic Equations — 325
- 8.4 Chapter Summary — 329
- Review Exercises — 330

CHAPTER 9 MATHEMATICS OF FINANCE — 333

- 9.1 Geometric Progressions — 334
- 9.2 Simple Interest — 341
- *9.3 Average Daily Balance — 346
- 9.4 Compound Interest — 350
- 9.5 Ordinary Annuities — 355
- 9.6 Present Value of an Annuity — 360
- 9.7 Chapter Summary — 364
- Review Exercises — 366

CHAPTER 10 FUNCTIONS AND LIMITS — 371

- 10.1 Functions — 372
- 10.2 Graphs of Functions — 378
- 10.3 Limits — 383
- 10.4 Continuity — 390
- *10.5 Rational Functions — 395
- 10.6 Chapter Summary — 403
- Review Exercises — 404

CHAPTER 11 THE DERIVATIVE — 409

- 11.1 Average Rate of Change — 410
- 11.2 The Derivative — 414
- 11.3 Interpretation of the Derivative — 419
- 11.4 Rules for Derivatives — 424
- 11.5 The Chain Rule — 432
- 11.6 The Product and Quotient Rules — 436
- 11.7 Derivatives of Exponential and Logarithmic Functions — 440
- 11.8 Differentiable Functions — 446
- 11.9 Chapter Summary — 450
- Review Exercises — 451

CONTENTS

CHAPTER 12 ADDITIONAL TOPICS ON THE DERIVATIVE — 455

- 12.1 Derivatives of Higher Order — 456
- 12.2 Increasing and Decreasing Functions — 458
- 12.3 Relative Maxima and Minima — 466
- 12.4 Absolute Maxima and Minima — 472
- 12.5 Maxima–Minima Applications — 475
- *12.6 Inflection Points — 481
- 12.7 Chapter Summary — 484
- Review Exercises — 486

CHAPTER 13 INTEGRATION — 491

- 13.1 Indefinite Integrals — 492
- 13.2 Some Applications of Indefinite Integrals — 498
- 13.3 Definite Integrals — 502
- 13.4 Evaluating Definite Integrals — 507
- 13.5 Properties of Definite Integrals — 510
- 13.6 Some Applications of Definite Integrals — 514
- 13.7 Chapter Summary — 520
- Review Exercises — 522

CHAPTER 14 ADDITIONAL TOPICS IN INTEGRATION — 525

- 14.1 Integration by Substitution — 526
- 14.2 Use of Integration Tables — 532
- 14.3 Integration by Parts — 537
- 14.4 Differential Equations — 540
- 14.5 Improper Integrals — 548
- *14.6 Probability Density Functions — 551
- 14.7 Chapter Summary — 556
- Review Exercises — 558

CHAPTER 15 MULTIVARIABLE CALCULUS — 563

- 15.1 Functions of Several Variables — 564
- 15.2 Graphs in Three Dimensions — 566
- 15.3 Partial Derivatives — 570
- 15.4 Interpretation of Partial Derivatives — 573
- 15.5 Second-Order Partial Derivatives — 576
- 15.6 Maxima and Minima — 580

15.7	Lagrange Multipliers	586
15.8	Chapter Summary	593
	Review Exercises	595

APPENDIX A ALGEBRA REVIEW 599

A.1	Evaluating Expressions	600
A.2	Algebraic Expressions	603
A.3	Factoring	608
A.4	Solving Linear and Quadratic Equations	613
A.5	Algebraic Fractions	618
A.6	Solving Additional Types of Equations	627
A.7	Exponents	630
A.8	Sets	634

APPENDIX B TABLES 641

TABLE I	Binomial Probabilities	642
TABLE II	Areas Under the Standard Normal Curve	644
TABLE III	Values of e^x and e^{-x}	645
TABLE IV	Natural Logarithms	650
TABLE V	Common Logarithms	652
TABLE VI	Values of $(1+i)^n$	654
TABLE VII	Values of $(1+i)^{-n}$	655

APPENDIX C ANSWERS TO ODD-NUMBERED EXERCISES AND TO REVIEW EXERCISES 657

Index   707

PREFACE

OBJECTIVES

What should be the objectives of a finite mathematics and calculus sequence beyond the presentation of the appropriate mathematics? The appropriate mathematics having become more or less standardized, this was the question I attempted to answer several years ago when I served as a committee of one to choose a text for our sequence. In discussing this issue with another faculty member, I was particularly impressed with his belief that, beyond enabling students to "do mathematics," the primary objective of the sequence should be to enable students "to express themselves mathematically." And so the philosophy of this book was born. I have attempted to write a text that will enable students to express themselves mathematically by developing:

1. Some insight into how mathematics is used in solving nonmathematical problems
2. A knowledge of some of the major areas of mathematics used in solving such problems
3. An ability to express nonmathematical problems mathematically
4. An ability to verbalize mathematically

The first of these objectives is the primary purpose of Chapter 1, especially the final section in which the use of mathematical models is discussed. This section becomes the theme for the rest of the text and lays the foundation for the third of the above objectives. The use of models is discussed frequently in conjunction with the illustrations and examples throughout the text, and students are asked to analyze their use in many of the exercises. For several reasons, a few elementary ideas from graph theory were chosen as the vehicle for discussing mathematical models in Section 1.3. They provide a light and easy introduction for the course, they provide a psychological relief for those students who are weary of manipulating x's and y's in previous mathematics courses, and (most importantly) their interesting applications provide a perfect setting for examining the use of mathematical models and the application of mathematics.

    The greatest portion, by far, of the text is devoted to the second of the above objectives. The usual introductory areas of mathematics used in solving problems in the management, biological, and social sciences are presented. Great care was taken to present the topics in a manner that is readable to students in courses for which algebra is the only mathematical prerequisite.

The fourth objective is probably the most challenging. I have found that one of the most serious deficiencies in the mathematical background of even some of the best students is their inability to verbalize mathematically. To help students to develop this ability and to understand the material better, they are asked in the exercises to explain the meaning of the various mathematical concepts introduced. In some instances they are asked to discuss some aspects of the material of the section.

Hopefully, the final result will be that the students will be able to express themselves mathematically, both verbally and in solving problems—mathematical and otherwise—with some insight into the process of using mathematics and into the nature of the mathematics involved.

FEATURES

Chapter Introductions

Each chapter begins with a chapter outline previewing the contents of the chapter and an introduction to motivate the material or an application to illustrate a type of problem the students will be able to solve after completing the chapter.

Chapter Summaries and Review Exercises

Chapters 2–15 each conclude with a chapter summary and a set of review exercises. The major concept of Chapter 1, the use of mathematical models in the application of mathematics, is reviewed throughout the text.

Margin Practice Problems

Margin practice problems are provided throughout each chapter to enable students to check their understanding of the material as they progress through the text. These practice problems can also be used as additional examples in class, if desired.

Step-by-Step Descriptions

In order to help students develop an organized technique for problem-solving procedures, step-by-step descriptions of these procedures are provided when appropriate. These are accompanied by one or more examples illustrating the steps described.

Exercises

The exercises in the text are of three types:

1. *Drill Exercises.* The large number of these exercises, generally given in increasing difficulty, provide students with the opportunity to develop efficiency in solving routine mathematical problems.

2. *Application Exercises.* There are numerous exercises of this type, carefully prepared in the belief that they are one of the major reasons for the existence of the courses for which this text is intended.

3. *Discussion Exercises.* In these exercises students are asked to explain the meaning of mathematical concepts, to discuss some aspect of the material of the corresponding section, or to analyze the use of mathematical models and the four-step procedure involved in the application of mathematics in a particular example or exercise.

The answers for odd-numbered exercises are given in Appendix C. Answers for both odd- and even-numbered review exercises are also given in the appendix.

Analytic Geometry — Probably more than the usual emphasis is placed on analytic geometry in the belief that as much geometric insight as possible is a major factor, when applicable, in understanding the mathematics involved. The introductory sections on analytic geometry in Chapter 2 can be covered quickly in classes with well-prepared students.

Algebra Review — An algebra review based on Appendix A can be undertaken at the beginning of the course, or individual topics can be reviewed, either by students individually or in class, as they are needed in the course. Reference is made in the text to the review topics when these topics are first required. For example, a review of the first four sections is suggested at the beginning of Chapter 2; a review of sets is suggested at the beginning of Chapter 5.

Student Solutions Manual — A solutions manual is available as a study aid for students using this text. This manual contains worked-out solutions for most of the odd-numbered exercises.

Optional Sections — Sections that are marked with an asterisk can be omitted without loss of continuity.

ACKNOWLEDGMENTS

My sincere appreciation to all the individuals involved in the development and production of this text. To Kay Locke and Sheila Lee for converting handwritten copies into beautifully typed manuscripts. To all the students who were subjected to a preliminary version of this text and who influenced this final form more than they realize. To the following reviewers for their constructive criticisms:

Douglas B. Crawford
College of San Mateo

Joseph S. Evans
Middle Tennessee State University

Herbert Gindler
San Diego State University

Bruce A. Jensen
Portland State University

Steve Marsden
Glendale Community College

Walter Roth
University of North Carolina, Charlotte

James R. Senft
University of Wisconsin, River Falls

Robin G. Symonds
Indiana University, Kokomo

Lowell Doerder
Black Hawk Community College

Judy Fine
North Hennepin Community College

Bernard Hoerbelt
Genesee Community College

Alonzo F. Johnson
West Virginia University

Robert A. Moreland
Texas Tech University

Wes Sanders
Sam Houston State University

John Spellman
Southwest Texas State University

To Susan Vayo for preparing the student solutions manual, the answers for the appendix, and the solutions for the instructor's manual. To Barbara Fuller, Senior Production Editor, for coordinating the transformation from manuscript to published text. To Peter Marshall, Executive Editor, for years of assistance and encouragement from the time this text was only an idea until its present realization. To my family who made it all worth the effort.

G.J.K.

CHAPTER 1

INTRODUCTION

OUTLINE

1.1 A Physical Problem
1.2 A Mathematical Solution
1.3 The Application of Mathematics

Most people would probably agree that mathematics is used to solve many different types of problems in many different areas, such as business administration, finance, sociology, psychology, chemistry, and engineering. Surprising, perhaps, is the fact that there is a common procedure underlying these many different applications of mathematics. Understanding this procedure is a major factor in understanding how mathematics is and can be used to solve problems in these other areas.

The major purpose of this opening chapter is to develop some insight into this common procedure. In order to provide the mathematical context in which this is to be done, we shall also develop some elementary concepts of a branch of mathematics called *graph theory*. Born of almost trivial circumstances, graph theory is now used to describe and solve problems in a variety of areas, some of which were just mentioned. However, our primary purpose in discussing graph theory, although interesting in itself, at this time is to provide the context in which to discuss the processes involved in the application of mathematics. Consequently, we limit ourselves to the minimal amount of graph theory required for this purpose.

1.1 A PHYSICAL PROBLEM

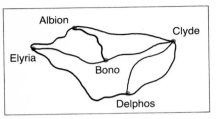

FIGURE 1.1

The delivery truck of a company makes stops along all the rural highways included in the map shown in Figure 1.1. The company's transportation director, desiring the most economical and efficient route, is trying to determine a route that travels each of the highways exactly once. Is such a route possible? At first it might seem that the easiest way to determine if such a route exists would be by trial and error, tracing possible routes along the map with pencil. All possible routes could be tried until—perhaps—the desired type of route could be found. But, as we shall see, there is a much simpler and faster mathematical method for solving the transportation director's problem.

We start with a simple representation of the truck route in which the lines representing the highways are labeled for easier reference (Figure 1.2). The question is then whether or not a point can move continuously along the figure in such a way that it moves along each line once and only once. Our objective is to answer this question for Figure 1.2 (without resorting to trial and error) and, at the same time, for all figures of this type.

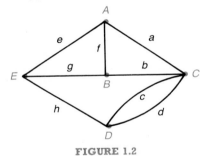

FIGURE 1.2

A figure made up of a set of points and of lines connecting the points, such as Figure 1.2, is called a **graph.** The points connected by lines are called **vertices,** and the connecting lines are called **edges.** In Figure 1.2, A, B, C, D, and E are vertices; a, b, c, $\ldots$, h are edges.

Whether or not the desired type of route exists in a graph depends on the number of edges that meet at the various vertices. This number—the number of edges that meet at a vertex—is called the **degree** of the vertex. In Figure 1.3, A is a vertex of degree 1, B and D are of degree 2, and C is of degree 3. A vertex is **odd** if it is of odd degree and is **even** otherwise. As depicted in Figure 1.3, A and C are odd vertices; B and D are even. ▮1

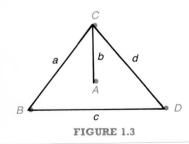

FIGURE 1.3

▮1 **PRACTICE PROBLEM 1**

Determine the degree of each vertex in the graph at the right. Indicate whether each vertex is odd or even.

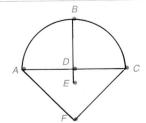

Answer

VERTEX	DEGREE	
A	3	Odd
B	3	Odd
C	3	Odd
D	4	Even
E	1	Odd
F	2	Even

A ***path*** is a sequence of edges traversed consecutively in such a way that no edge is traversed more than once. (Note that *path* is defined so that it describes the type of action desired in the route problem—that is, edges are traversed consecutively and no edge is traversed twice.)

In order to facilitate discussing paths, we shall use the following notation. (The letters refer to Figure 1.3.)

BcD	Denotes the path that begins at B and traverses edge c to arrive at D.
BcDDdC	Denotes the path that begins at B, traverses c to arrive at D, and then traverses d to arrive at C.
BcDDdCCbA	Denotes the same path as the last, except that it continues on from C to A via edge b.

If the number of vertices or edges in a figure could be infinite, the figure could never be described by a single path. Therefore, we restrict our attention to graphs that have only a finite number of vertices and of edges.

Also, it is impossible to traverse a graph such as that shown in Figure 1.4 by a single path, as there is no way to move from one part of the graph to the other. In order to distinguish this type of graph from those given earlier, we say that a graph is **connected** if, for any two vertices A and B in the graph, there is a path that begins at A and ends at B. Since there is no path from A to E, for example, the graph in Figure 1.4 is *not* connected. Each of the graphs shown in Figure 1.5 is connected. The degree of each vertex is indicated. In parts (a) and (c), note that there is no vertex at the intersection of two edges unless indicated by a heavy dot.

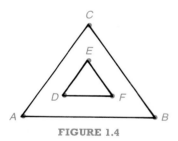

FIGURE 1.4

Note also that the sum of the degrees in each of these graphs is twice the number of edges. In the first graph there are seven edges, and the sum of the degrees is $7 \times 2 = 14$. In the second graph there are six edges and the sum of the degrees is $6 \times 2 = 12$. A similar statement is

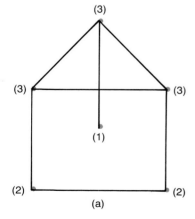 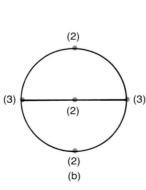 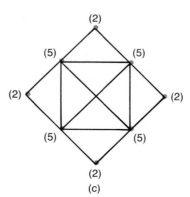 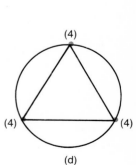

FIGURE 1.5

true for the third and fourth graphs. This must always be true because each edge contributes 2 to the sum of the degrees—1 to the degree of each of the vertices at its two endpoints.

As a result, the sum of degrees of the vertices in a graph is always an even number—twice the number of its edges. This tells us something about the number of odd vertices in a graph. Since the sum of an odd number of odd numbers is again odd, there must always be an even number of odd vertices in a graph. There will never be one, or three, or five odd vertices in a graph.

2 PRACTICE PROBLEM 2

Determine the number of edges and the sum of the degrees of the vertices in the following graph:

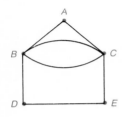

Answer

7 edges; the sum of degrees is 14.

EXERCISES

Determine the degree of each of the vertices in the graphs.

1.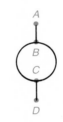

2.

3.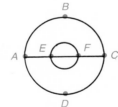

4.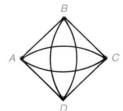

5.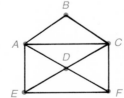

6.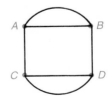

7.

8.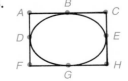

In each graph indicated, which vertices are even and which are odd? What is the sum of the degrees of the vertices in each graph? How many edges are there in each graph?

9. Exercise 1
10. Exercise 2
11. Exercise 3
12. Exercise 4
13. Exercise 5
14. Exercise 6
15. Exercise 7
16. Exercise 8

Construct a graph containing at least four vertices satisfying the given conditions.

17. All the vertices are even.
18. All the vertices are odd.
19. Two of the vertices are even and the rest odd.
20. The sum of the degrees of all the vertices is 20.

If possible, construct a graph containing five vertices satisfying the given conditions.

21. Two of the vertices are odd and the rest are even.
22. Two of the vertices are even and the rest are odd.
23. The sum of the degrees of all of the vertices is 15.
24. The sum of the degrees of all of the vertices is 12.

State whether you can draw the specified graph without lifting your pencil from the paper and without retracing any edge.

25. Exercise 1
26. Exercise 2
27. Exercise 3
28. Exercise 4
29. Exercise 5
30. Exercise 6
31. Exercise 7
32. Exercise 8

33. *Family tree* Albert had two children, Betty and Charles; Betty had three children, named Donald, Evelyn, and Francis. Charles had two children, Gerald and Harriett. Construct a graph (family tree) indicating the descendants of Albert.

34. *Acquaintance graphs* The vertices in the two given graphs represent six different people. Two vertices are connected by an edge only if the two people they represent are acquainted. In the graph on the left, B, C, and E all know each other (they are connected by a triangle). In the graph on the right, none of A, C, or E know each other (there is no edge between any two of them). Draw at least four more possible acquaintance graphs to convince yourself that in any group of six people, there must

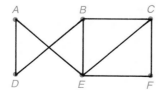 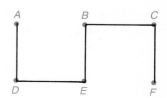

be three who know each other or three who are complete strangers to each other (no two are acquainted). The graphs need not be connected; for example:

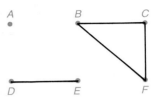

In this case B, C, and F all know each other; A, E, and C are complete strangers, as are A, D, and F, and so forth.

Explain the meaning of each term.

35. A graph
36. The degree of a vertex
37. An odd vertex
38. A path
39. A connected graph.
40. Explain why there must be an even number of odd vertices in a graph.

1.2 A MATHEMATICAL SOLUTION

Our central question is whether or not a graph can be completely described by a single path. Since we shall want to refer often to such paths, we ought to give them a name. This type of path is called an *Euler path* (after Leonhard Euler, 1707–1783, the founder of graph theory). An **Euler path** is a single path that completely traverses a figure, or a path that contains each edge in a figure exactly once.

Graphs for which there are Euler paths are easily constructed. For the graph shown in Figure 1.6, *CaAAcBBdCCbA* is one such path. For the graph shown in Figure 1.7, *CfDDeCCdDDcBBaAAbC* is one of many Euler paths. Examine the number of times each capital letter appears in the Euler paths given for these two graphs. The number of times each such letter appears is equal to the degree of the corresponding vertex. For example, in Figure 1.7, vertex C is of degree 4 and C appears 4 times in the given path; vertex B is of degree 2 and B appears 2 times in the path. The number of times each letter appears must always equal the degree of the vertex because an Euler path must enter or leave each vertex once for each edge at that vertex.

A vertex that apppears *only* in the interior of an Euler path must be even because such vertices appear only in groups of two. Consequently, if a graph contains odd vertices, these must appear at the beginning and at the end (and perhaps an even number of times in the

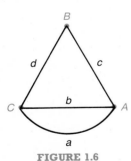

FIGURE 1.6

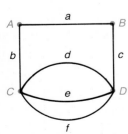

FIGURE 1.7

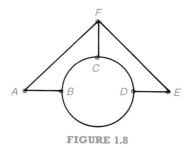

FIGURE 1.8

interior). This is the case with vertices C and A in the Euler path given above for the graph in Figure 1.6. In the graph shown in Figure 1.8, no Euler path is possible because there are more than two odd vertices; B, C, D, and F are all odd vertices, and it would be impossible for an Euler path either to begin or end at each of these four vertices. In general, if there are more than two odd vertices, no Euler path is possible.

We now have an important fact that must be true of every graph as a consequence of the definitions. Such facts are called **theorems** and are usually stated quite formally.

THEOREM 1

In a graph with more than two odd vertices, no Euler path is possible.

Theorem 1 also solves the transportation director's problem. In Figure 1.2, there are four odd vertices. Therefore, no Euler path is possible, which means that it would be impossible to travel each of the highways exactly once!

With Theorem 1 we have accomplished our objective of solving the route problem. Theorems 2 and 3, which are given for the sake of completeness, can also be shown to be valid. Their proofs require a somewhat deeper analysis than that of the first theorem.

THEOREM 2

In a connected graph with only even vertices, an Euler path (which must begin and end at the same vertex) is always possible. Any vertex in the graph can be chosen to be the beginning and ending vertex.

THEOREM 3

In a connected graph with just two odd vertices, an Euler path that begins at one odd vertex and ends at the other is always possible.

Theorems 1, 2, and 3 give us enough information to determine whether or not *any* given graph has an Euler path and some information about constructing such a path when one exists.

1 INTRODUCTION

EXAMPLE 1 Find an Euler path for the following graph, if there is one:

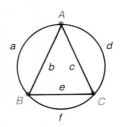

Solution The graph is connected and all the vertices are even. Theorem 2 guarantees that there is a path of the required type. One such path is

AaBBfCCeBBbAAcCCdA

Note that it begins and ends at the same vertex, vertex A, as the theorem requires. ■

EXAMPLE 2 Find an Euler path for the following graph, if there is one:

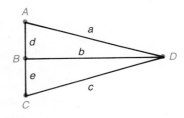

Solution The graph is connected and there are two odd vertices. Theorem 3 guarantees that there is a path of the required type. However, the path must begin and end at the odd vertices; two of several such paths are

BbDDaAAdBBeCCcD   and   *DaAAdBBbDDcCCeB*   ■

In order to simplify path notation, when the edges traversed in a path are either obvious or immaterial, we shall denote the path simply

---

3 PRACTICE PROBLEM 3

Find an Euler path for each of the following graphs, if there is one:

(a)

(b)

Answer

(a) All the vertices are even. By Theorem 2 there is an Euler path. One such path is *AaBBbCCcAAdCCfDDeA*.

(b) There are more than two odd vertices. By Theorem 1, no Euler path is possible.

1.2 A MATHEMATICAL SOLUTION

by indicating the order in which the vertices occur. The notation for the paths given in Example 2 then would be simply *BDABCD* and *DABDCB*.

EXAMPLE 3 A mail carrier's route is given in the following map:

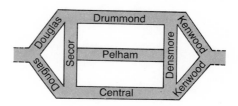

She delivers mail on each side of the street on Secor, Pelham, and Densmore streets but only on the insides of the remaining streets. The post office is at the corner of Secor and Pelham.

The best route for her would be one that begins and ends at the post office and contains the sides of every block on which she delivers mail just once. Is such a route possible? If so, find one.

Solution The map of the mail route can be represented by the graph shown. The vertices represent the intersections and the edges represent the sides of the blocks on which she delivers mail. For example, the two edges between *C* and *B* represent the two sides of Secor between Pelham and Drummond and the edge between *B* and *E* represents the side of Drummond on which she delivers mail.

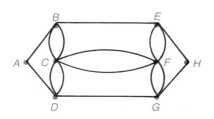

The desired route corresponds to an Euler path that begins and ends at vertex *C*. Since all of the vertices are even, Theorem 2 guarantees that there is such a path. *CBCDABEFCFEHGFGDC* is one possibility. The corresponding route would be Secor from Pelham to Drummond to Pelham to Central, and so forth. ∎

EXERCISES

On the basis of the theorems of this section, determine whether or not each of the following has an Euler path. Find an Euler path, if there is one.

1.
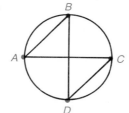

2.
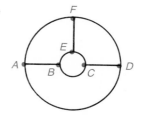

3.
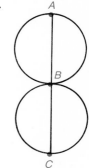

4.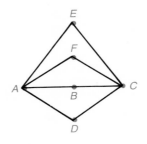

5.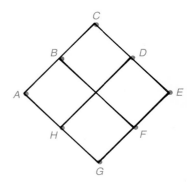

6.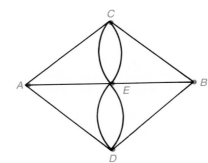

7.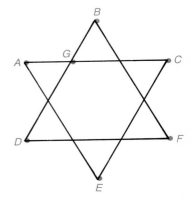

8. Would Theorems 1, 2, and 3 have helped in Exercises 25–32 of Section 1.1?

9. *Inspection route* A street inspector plans to inspect the streets in the neighborhood represented by the graph. The vertices represent intersections and the edges represent the streets. The most efficient route would be one in which he goes down each street exactly once. Does such a route exist? Find one if it does.

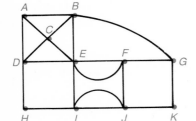

10. *Delivery route* The delivery truck for your company makes deliveries along all the highways indicated in the graph. The vertices represent towns and the edges represent highways. Your company is located at D. The most efficient route would be represented by an Euler path that begins and ends at D. Does such a route exist? Find one if it does.

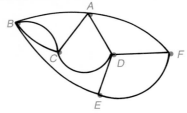

11. *Tour route* The graph represents the floor plan of an exhibition hall for an automobile show. The edges represent the aisles for the visitors to the show, and the vertices indicate where the aisles meet. Is it possible for the show manager to route the visitors so that they enter at the door at H, go down each aisle exactly once, and exit at the door at J? If so, find such a route.

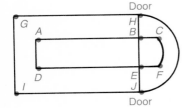

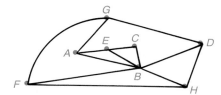

12. *Sales route* A salesperson must call on customers in eight different cities, denoted by vertices *A* through *H* in the graph. The edges correspond to highways connecting the cities. She decides that the most interesting and efficient route would be one that begins and ends at *A* (where she lives), takes her to each of the other cities exactly once, and never travels any highway more than once.
    (a) Does the route she seeks correspond to an Euler path?
    (b) Determine a route of the type she seeks, if there is one.

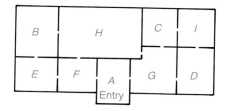

13. *Tour route* The floor plan for a museum is pictured here.
    (a) Construct a graph in which the vertices represent the rooms and any two vertices are connected by an edge if there is a door between the two rooms.
    (b) A conducted tour of the museum would begin and end at the entry and, preferably, visit each of the other rooms exactly once. Would such a tour correspond to an Euler path in the graph?
    (c) Find a route for such a tour, if possible.

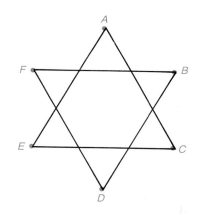

14. The figure shown in the margin has only even vertices but has no Euler path. Why not? (*Hint:* Consider the statement of Theorem 2 carefully.)

Explain the meaning of each term.

15. An Euler path
16. A theorem

State the circumstances under which a graph has the indicated type of path.

17. An Euler path
18. No Euler path
19. An Euler path that begins and ends at the same vertex

1.3 THE APPLICATION OF MATHEMATICS

By what device does a physical problem become a mathematical problem when mathematics is used to solve physical problems? Here the term *physical* is used in a very broad sense, including any real-life situation—be it financial, biological, administrative, sociological, or whatever. The answer to the question is a "mathematical model."

A **mathematical model** is a mathematical description of a physical situation. A mathematical model is the means by which the transition is made from the physical to the mathematical world. Every application of mathematics requires that this transition be made, and—consequently—every application of mathematics requires a mathematical model.

For the sake of illustration, we examine the process by which we solved the transportation director's problem in the last section:

1. There was a physical problem—whether or not a route that traveled each of the highways exactly once was possible.

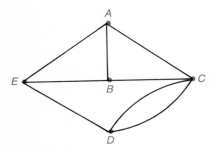

2. The graph in the margin was constructed to describe the situation mathematically—that is, a mathematical model was constructed. The physical problem then became the mathematical problem of whether or not the graph had an Euler path.
3. The solution of the mathematical problem—that no Euler path was possible because the graph had more than two odd vertices—was found by using graph theory.
4. The mathematical solution was interpreted within the physical situation. A route of the desired type was not possible.

Every application of mathematics to solving a physical problem follows a similar four-step process.

> 1. A question is asked concerning a physical situation.
> 2. A mathematical model is constructed to describe the physical situation. The question in the first step now becomes a mathematical question.
> 3. The mathematical solution is found; that is, the mathematical question of Step 2 is answered.
> 4. The mathematical solution is interpreted within the physical situation; that is, the question of Step 1 is answered.

The transition to the mathematical world is made in Step 2 by means of a mathematical model. The transition back to the physical world occurs in Step 4, where the mathematical solution is interpreted. The process is summarized in the following diagram:

Physical situation (physical question) → Mathematical model (mathematical question) → Solution of mathematical question → Interpretation of solution in physical situation

EXAMPLE 1 Identify each aspect of the four-step problem-solving procedure in Example 3, Section 1.2.

Solution
1. The physical situation was the mail carrier's route (shown in the map). The physical question was whether there was a route that included each block just once and began and ended at the post office.
2. The mathematical model was the graph given in the solution. The mathematical question was whether an Euler path exists that begins and ends at vertex C in that graph.

3. The answer to the mathematical question was yes because all of the vertices in the graph are even. The path found was

   CBCDABEFCFEHGFGDC

4. The interpretation of the mathematical solution was that the desired type of route indicated in Step 1 was possible and that Secor from Pelham to Drummond and so on was such a route. ∎

In the exercises you are asked to do analyses similar to that of Example 1 for several exercises in Section 1.2. In addition, we shall see this four-step procedure used again and again throughout the text.

If an appropriate mathematical model is not readily available, then a great amount of experimental work and testing of possible models (which is not reflected in the diagram on page 12) is usually involved. The testing of a possible model amounts to determining if the solution given by the model to the physical problem agrees with the result of the experimental work. If not, the model must be discarded and a new model sought.

Suppose, for example, that you as a social scientist were interested in predicting the annual income of an individual on the basis of the number of years of his or her education and the number of years of his or her employment. You collect this information on 1,000 individuals with varying educational and employment backgrounds. You then construct a mathematical model to relate annual income to years of education and years of employment and test the model with your data. If the model gives the appropriate annual income (with some reasonable degree of accuracy) for only 225 of the individuals that you studied, it would not be considered very reliable in predicting annual income. If you are persistent, you then attempt to construct another model. You may or may not be successful in the end. Furthermore, once a satisfactory model is obtained it must be reviewed and updated periodically if it is to be used again and is to remain realistic.

The advances of physics and engineering are due, to a great extent, to the availability of good mathematical models. Because the mathematical approach has served the sciences so well, other areas of study have begun to adapt the mathematical-model method to their own problems. One result has been the establishment of new branches of study within these other areas, such as econometrics in economics, biomathematics in biology, and operations research in business administration.

You must have encountered the use of mathematical models many times—perhaps several times in the course of a single day—but you probably have not thought of the process in terms of models. The next several examples also illustrate the use of mathematical models. Do not be concerned about the technical details of Examples 2 and 3; these examples are intended primarily to indicate the variety of uses of mathematical models. Their technical details are not our concern here.

EXAMPLE 2 What effect does government spending have on the gross national product?

Solution First we have to construct a mathematical model, or mathematical description, of the gross national product. The mathematical models used by economists consist of up to 400 equations involving as many as 1,000 variables. These, of course, take years to develop and are much too lengthy to present here. Instead, we shall construct our own simplified version, appreciating the fact that it probably is not very reliable.

The gross national product is the total production of goods and services in this country for one year. We consider the production of goods and services to be of four types:

Consumer spending, C

Investments, I

Government spending, G

Net exports (the difference between exports and imports), X_n

Then the gross national product, Y, is given by

$$Y = C + I + G + X_n$$

In addition, we consider consumer spending to be approximately $\frac{9}{10}$ of the gross national product, and we agree that some consumer spending will still occur even if the gross national product is 0. This gives

$$C = a + 0.9Y$$

where a is the amount of consumer spending that would occur when the gross national product, Y, is 0.

Our model then consists of the two equations

$$Y = C + I + G + X_n$$
$$C = a + 0.9Y$$

The mathematical question is, How does Y change as G changes?

If we substitute the value of C from the second equation into the first, we get

$$Y = (a + 0.9Y) + I + G + X_n$$

which gives

$$Y - 0.9Y = a + I + G + X_n$$

Combining the two terms on the left gives

$$\tfrac{1}{10}Y = a + I + G + X_n$$

or

$$Y = 10(a + I + G + X_n) = 10a + 10I + 10G + 10X_n$$

The answer to our mathematical question is that for every increase or decrease of one unit in G, there results a corresponding increase or decrease of 10 units in Y, as indicated by the $10G$ in the last equation.

The interpretation is that every $1 increase or decrease in government spending (that is, G changes by 1) results in a corresponding $10 increase or decrease in the gross national product.

Our model is not entirely unrealistic, as the equation

$$Y = C + I + G + X_n$$

is the basic equation in most models actually used by economists. ■

EXAMPLE 3 A certain bacterium reproduces in such a way that there are about 10% more bacteria present at the end of each hour than there were at the beginning. If a sample of the bacteria contains 300 bacteria, how many will it contain 10 hours later?

Solution The bacteria described are said to have an **exponential growth rate**. The mathematical model that gives the number present (y) at any time (t) is

$$y = ke^{0.1t}$$

where k is the number present when the counting process begins and e is a constant approximately equal to 2.718. (The number e, like π, is an infinite, nonrepeating decimal that can be approximated to as many decimal places as desired; models such as this will be discussed in more detail later in the text.)

The mathematical question is, What is the value of y when t is 10? If $t = 10$ and $k = 300$, the above equation gives

$$y = 300e^{(0.1)(10)} = 300e \approx 300(2.718) = 815.4$$

There will be about 815 bacteria in the sample after 10 hours. ■

EXAMPLE 4 Word problems in an algebra course all entail the use of a mathematical model. A common type of problem of this sort is the following:

If Betty is twice as old as Paul and in 3 years Paul will be 21 years old, how old is Betty?

Solution The mathematical model of the situation is as follows: Let y be Betty's age and x be Paul's age. Then

$$y = 2x$$
$$x + 3 = 21$$

The mathematical question is, What is the value of y? If we subtract 3 from each side of the second equation, we get $x = 18$. Letting $x = 18$ in the first equation gives $y = 36$.

The physical interpretation is that Betty is 36 years old. ■

You may have noticed a common process in the manner in which the problems were solved in the last three examples. The four-step procedure discussed earlier was followed in each case. You are asked to verify this in the exercises that follow.

EXERCISES

Identify each aspect of the four-step procedure involved in the application of mathematics in each of the following:

1. *Inspection route*   Exercise 9, Section 1.2
2. *Delivery route*   Exercise 10, Section 1.2
3. *Tour route*   Exercise 11, Section 1.2
4. *Sales route*   Exercise 12(b), Section 1.2
5. *Gross national product*   Example 2, Section 1.3
6. *Bacteriology*   Example 3, Section 1.3
7. *Age problem*   Example 4, Section 1.3
8. Explain the four-step procedure involved in the application of mathematics in solving physical problems.
9. Explain what is meant by a mathematical model.

CHAPTER 2

EQUATIONS AND THEIR GRAPHS

OUTLINE

2.1 Equations as Mathematical Models
2.2 The Real Number Line and Cartesian Plane
2.3 Graphs of Equations
2.4 Equation of a Straight Line
2.5 Linear Equations
2.6 Linear Equations as Mathematical Models
2.7 Quadratic Equations
2.8 Quadratic Equations as Mathematical Models
2.9 Chapter Summary

Review Exercises

Mathematical models that describe physical problems can be in any one of a variety of forms. In this chapter we consider some mathematical models in the form of equations and then consider methods of obtaining graphs of these equations.* The resulting graphs give a second mathematical model—a geometric model—for the problem.

The setting within which these graphs are obtained is generally referred to as *analytic geometry*. Historically, the development of analytic geometry is due to the French philosopher and mathematician Rene Descartes (1596–1650). His discovery resulted in a union of algebra and geometry that has been beneficial to both. The process by which this union is accomplished is, as we shall see, so simple that one must wonder why it was not discovered earlier.

---

* The use of the word *graph* in this chapter is entirely different from its use in Chapter 1.

2.1 EQUATIONS AS MATHEMATICAL MODELS

As just indicated, we shall be interested in constructing equations to describe physical relationships and eventually in obtaining graphs of these equations. Equations are appropriate as mathematical models for stating equality relationships between two or more physical quantities. They are among the most commonly used methods of describing physical situations mathematically.

EXAMPLE 1 Manufacturing costs for a particular type of electrical fuse consists of a setup cost of $500 for preparing the machines for production plus 7¢ per fuse for material and labor.

(a) Construct an equation to describe the total manufacturing costs for any number of fuses.

(b) What would be the total cost if 5,000 fuses were manufactured? If 9,750 were manufactured?

Solution (a) The $500 setup cost and the 7¢ unit cost for materials and labor are constant; the variable quantities are the *number of fuses* manufactured and the corresponding *total costs*. Let x and y denote these variable quantities, where x is the number of fuses manufactured and y is the total cost (in dollars) of manufacturing them. Then

$$y = 500 + 0.07x$$

is the desired equation.

(b) If we set x (the number of fuses manufactured) equal to 5,000, we get

$$y = 500 + 0.07(5,000)$$

or

$$y = 850$$

Thus the cost of 5,000 fuses is $850.
If we set x equal to 9,750, we get

$$y = 1,182.50$$

Hence, 9,750 fuses cost $1,182.50.

In Example 1(a), the verbal expression of the desired relationship is:

Total cost equals setup cost plus 7¢ per fuse.

Note how closely the mathematical expression parallels the verbal expression:

Total cost equals setup cost plus 7¢ per fuse.
$\quad\quad y \quad\quad = \quad\quad 500 \quad + \quad 0.07 \quad x$

EXAMPLE 2 (a) Suppose the annual rate of inflation is 12%, so that the cost of living next year will increase over this year's cost by 12%. Write an equation to express the cost next year of maintaining a family of four at the same standard at which they are living this year.

(b) If a family is spending $15,000 this year, what amount will they have to spend next year to maintain the same standard of living?

Solution (a) The verbal expression of the relationship between the cost of living for the two years is:

Cost next year is this year's cost plus 12% of this year's cost.

The variable quantities are the *cost next year* and the *cost this year*. If we denote the first of these by y and the second by x, then we can convert the verbal sentence to a mathematical sentence as follows:

Cost next year is this year's cost plus 12% of this year's cost.
$$y \quad = \quad x \quad + \quad 0.12 \quad \times \quad x$$

or

$$y = x + 0.12x$$

This simplifies to

$$y = (1 + 0.12)x = 1.12x$$

(b) Since x represents this year's cost, let $x = 15{,}000$ in the equation of part (a). Then

$$y = 1.12(15{,}000)$$
$$= 16{,}800$$

Thus it will take $16,800 to maintain the family next year. **1**

EXAMPLE 3 An apartment manager estimates that it will require 5,000 worker-hours to refurbish one of her apartment houses.

---

1 PRACTICE PROBLEM 1

A book publisher estimates that printing costs next year will be 10% higher than this year.

(a) Write a verbal expression of the printing cost next year in terms of the cost this year.

(b) Convert the verbal expression to an equation that gives the cost next year (y) in terms of the cost this year (x).

(c) If it costs $16.00 this year to print a particular mathematics text, how much will it cost next year?

Answer

(a) Cost next year is this year's cost plus 10% of this year's cost.

(b) $y = x + 0.10x = 1.10x$

(c) $17.60

(a) Write an equation to express the number of hours required to complete the refurbishment in terms of the number of workers she hires to do the job.

(b) Using the equation obtained in part (a), determine the number of workers she hired if the refurbishment took 200 hours.

Solution (a) The variable quantities are the *total number of hours required* (y) and the *number of individuals working* (x). Consequently, we get:

$$\underbrace{\text{Hours required}}_{y} \underbrace{\text{ equals}}_{=} \underbrace{\text{total hours}}_{5{,}000} \underbrace{\text{ divided by}}_{\div} \underbrace{\text{number of workers.}}_{x}$$

or

$$y = \frac{5{,}000}{x}$$

(b) If we set y (the number of hours required) equal to 200 in the equation of part (a), we get

$$200 = \frac{5{,}000}{x}$$

$$200x = 5{,}000$$

or

$$x = \frac{5{,}000}{200} = 25$$

She hired 25 workers.

EXAMPLE 4 According to his contract, the yearly salary (y), in dollars, of a basketball player will be

$$y = 25{,}000 + 2{,}000x$$

where x is the number of games the team wins.

(a) How much does the player receive for each game the team wins?

(b) How much will the player receive if the team wins no games? If the team wins 60 games?

(c) What does the 25,000 represent in the given equation?

PRACTICE PROBLEM 2

Suppose that you rented a garden tiller by the hour and that you paid a total of $36.00 in rent for using the tiller.

(a) Express a verbal relationship for the hourly rental rate in terms of the number of hours that you had the machine.

(b) Convert the verbal expression to an equation that gives the hourly rate (y) in terms of the number of hours (x) that you had the machine.

(c) If the hourly rate was $4.50, how many hours did you have the machine?

Answer

(a) The hourly rate is equal to 36 divided by the number of hours you have the machine.

(b) $y = \dfrac{36}{x}$ (c) 8 hours

Solution (a) The term $2{,}000x$ indicates that the player receives \$2,000 for each game won in addition to his base salary of \$25,000.

(b) If $x = 0$ in the given equation, we get
$$y = 25{,}000$$
If $x = 60$, we get
$$y = 145{,}000$$

So the player would get \$25,000 for no games won and \$145,000 for 60 games won.

(c) The player's salary starts at \$25,000 and increases with each game won. Consequently, it can be considered his base salary, independent of the number of games won. ∎

EXERCISES

Exercises 1–17 describe situations that involve equality relationships between various quantities. Construct equations that express these relationships, using the notation indicated. Then use the models to answer any additional questions. (Many of the equations constructed here will be used again later in this chapter.)

1. *Computer component manufacturing* It costs a manufacturing company \$2.50 in parts and labor to produce a particular computer component. In addition, the cost for setting up the machinery to produce the parts amounts to \$575.
   (a) Write an equation to express the total manufacturing costs (y) in terms of the number of components (x) manufactured.
   (b) Determine the total manufacturing costs if 2,000 components are manufactured.

2. *Furniture manufacturing* The costs in material and labor for a company to make a particular type of chair amount to \$63. The setup costs are negligible and can be ignored. However, an initial survey costing \$500 was conducted to determine the number of chairs the company could expect to sell. The survey cost (not included in the \$63 unit cost) is to be considered as part of the manufacturing costs.
   (a) Write an equation to express the total cost (y) in terms of the number of chairs (x) produced.
   (b) If the total manufacturing costs amounted to \$38,300, determine the number of chairs manufactured.

3. *Paneling sales* A company charges \$17.00 per sheet for its plywood paneling.
   (a) Write an equation to express its total revenue (y) in terms of the number of sheets (x) that it sells.
   (b) During one week the company sold 3,000 sheets of paneling. Determine its total revenue that week from its paneling sales.

4. *Consultant fee* Dan, a statistics consultant, charges \$250 per day for his services.

(a) Write an equation to express Dan's total fee (y) in terms of the number of days worked (x).
(b) If Dan's consulting fee totals $1,250 for his work with one company, determine the number of days he worked for the company.

5. *Shoplifting* The owner of a bookstore estimates that one-tenth of the books she puts on the shelves of her store "disappear" due to shoplifters.
  (a) Write an equation to express the number of books stolen (y) in terms of the number of books put on the shelf (x).
  (b) During a particular month the owner shelved 500 new books. How many of these will disappear?

6. *High school dropouts* The principal of a high school has found that about 7% of each class of entering freshmen become dropouts before their class graduates.
  (a) Write an equation to express the number of dropouts (y) in terms of the number of entering freshmen (x).
  (b) This year there were 250 entering freshmen. How many of these will become dropouts?

7. *Quality control* The quality control manager of an electrical supplies firm estimates that 11% of the light bulbs manufactured by the firm are defective.
  (a) Write an equation to express the number of defective bulbs (y) in terms of the total number of bulbs manufactured (x).
  (b) Last week the company manufactured 10,000 light bulbs. How many of these were defective?

8. *Police officer authorization* The city council of a certain city has authorized one police officer for each 10,000 residents.
  (a) Write an equation to express the number of police officers authorized (y) in terms of the number of residents (x).
  (b) How many police officers are authorized if there are 763,429 residents? (Round any fraction to the nearest whole number.)

9. *Credit reminders* Past experience indicates that 2 out of every 25 bills mailed each month by the credit manager of a certain department store will not be paid without a reminder.
  (a) Write an equation to express the number of reminders that will have to be sent (y) in terms of the number of bills mailed (x).
  (b) During March the credit manager sent out 250 reminder notices. How many bills were sent out?

10. *Income tax refund* In anticipation of his income tax refund, Jim decides that he will save $200 plus $\frac{1}{10}$ of the total amount he receives as a refund.
  (a) Write an equation to express the total amount that he will save (y) in terms of the amount he receives (x).
  (b) If Jim receives a refund of $450.50, how much will he save?

11. *Sales mark-down* A clothing store has marked all of its prices down 20%, or $\frac{1}{5}$, for its annual spring sale.
  (a) Write an equation to express the original price (y) in terms of the sale price (x).
  (b) If the sale price of a coat is $40.00, determine the original price of the coat.

12. *Population growth* By analyzing the population behavior of a certain country over a number of years, sociologists have determined that the

number of people living in the country at the end of a year is equal to the number of immigrants plus 110% of the number of people living in the country at the beginning of the year. If the number of immigrants during a particular year is 2,000, write an equation to express the total population of the country (y) at the end of that year in terms of the population of the country (x) at the beginning of the year.

13. *Chair sales* Joan manufactures and sells a modernistic chair that she has designed. Her profit per chair is $7.00, not taking into account her annual overhead expenses of $2,300. Write an equation to express her annual profit (y) in terms of the number of chairs sold during the year (x).

14. *Airline revenue* An airline company has scheduled a 100-seat airplane for a charter flight to the Super Bowl. The price per person is $140 plus $5.00 additional for each empty seat.
    (a) Write an equation to express the number of people (w) on the flight in terms of the number of empty seats (x).
    (b) Write an equation to express the price per person (z) in terms of the number of empty seats (x).
    (c) Using parts (a) and (b), write an equation to express the income of the airline company (y) from the flight in terms of the number of empty seats (x).
    (d) If there are 83 people on the flight, how much did each passenger pay?

15. *Agribusiness* A farmer has an apple orchard with 90 trees. He estimates that his annual profit per tree is $100.00 and that for each additional tree planted in the orchard, his profit per tree will decrease by $1.00.
    (a) Write an equation to express the total number of trees (w) in terms of the number of additional trees planted (x).
    (b) Write an equation to express the profit per tree (z) in terms of the number of additional trees planted (x).
    (c) Using parts (a) and (b), write an equation to express the farmer's total profit (y) in terms of the number of additional trees planted.
    (d) If he plants 20 additional trees, what is his total profit?

16. *Depot dimensions* The owners of a company intend to convert a rectangular part of its property into a small storage depot and have 60 feet of fencing with which to surround the depot. At first they plan to allow 15 feet of fencing per side, but they suspect that if they vary the length of the sides, they might possibly obtain a larger area.
    (a) Denote the length of the storage depot by x. Write an equation to express the width (w) in terms of x, keeping in mind that the depot must be able to be surrounded by 60 feet of fencing.
    (b) Using part (a), write an equation to express the area (y) of the depot in terms of the length (x). (*Note:* The area of a rectangle is length times width.)
    (c) What is the area of the depot if the width is 10 feet?

17. *Apartment rental* A real estate management company manages an apartment complex with 20 units. The manager estimates that he can rent all 20 units if he charges $150 per unit per month and that for each increase of $25 in rent, one apartment will be vacated.
    (a) Considering each vacant apartment to correspond to a $25 increase in rent, write an equation to express the rent per apartment (r) in terms of the number of empty apartments (x).

(b) Write an equation to express the number of apartments occupied (z) in terms of the number of apartments empty (x).
(c) Using parts (a) and (b), write an equation to express the total income per month (y) in terms of the number of empty apartments (x).
(d) If there are four empty apartments, what rent is being charged?

18. *Salesperson salary* As a television salesperson, Tom's monthly salary (y), in dollars, in terms of the number of television sets (x) that he sells is given by

$$y = 200 + 75x$$

(a) What is his commission on each television set that he sells?
(b) How much does he make each month in which he sells no television sets? In which he sells 20 television sets?
(c) What does the 200 represent in the given equation?

19. *Questionnaire return* A research sociologist has observed that of each 500 questionnaires that she sends out for her experiments, the number returned (y) in terms of the number of minutes it takes her to complete the questionnaire (x) is approximately

$$y = 300 - 5x$$

(a) If it takes her 20 minutes to complete the form, how many returns can she expect for each 500 questionnaires she sends?
(b) Regardless of the time involved, what is the maximum number of returns that she can expect for each 500 questionnaires sent?
(c) For every increase of 1 minute in time, how many fewer questionnaires can she expect to be returned for each 500 sent?
(d) For what amount of time involvement can she expect no returns?

20. *Effective drug time* The time required (y), in minutes, for a drug to be effective after it is administered in terms of the temperature (x), in degrees Fahrenheit, of the patient is given by

$$y = 2 - 0.10(98.6 - x)$$

(a) Determine the time required for the drug to be effective in a person with normal temperature (98.6°F).
(b) Does the required time increase or decrease as the patient's temperature increases?

21. When are equations appropriate mathematical models?

2.2 THE REAL NUMBER LINE AND CARTESIAN PLANE

Analytic geometry, the setting in which graphs of equations are obtained, begins with the construction of the real number line, on which the points on a line are put into one-to-one correspondence with the real numbers.

On a horizontal straight line pick two points. Label the point on the left 0 and the point on the right 1, as shown in Figure 2.1. The point

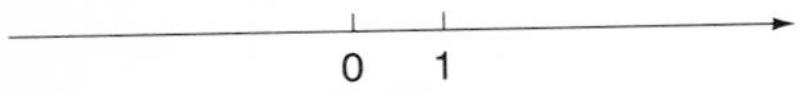

FIGURE 2.1

2.2 THE REAL NUMBER LINE AND CARTESIAN PLANE

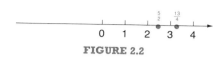

FIGURE 2.2

FIGURE 2.3

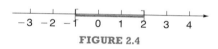

FIGURE 2.4

labeled 0 is called the **origin** and the distance between the two points is the **unit of distance.**

Every point on the line to the right of the origin is labeled with the positive real number that is equal to the number of units distance that the point is from the origin (Figure 2.2). The arrow indicates the positive direction.

Every point on the line to the left of the origin is labeled with the negative of the number that is equal to the number of units distance that the point is from the origin (Figure 2.3).

In this way every point on the line is labeled with a real number, positive if the point is to the right of the origin and negative if the point is to the left of the origin. Conversely, every real number is the label of some point on the line.

The number associated with any point on the line is called the **coordinate** of that point. While the number 3 is not the same as the point whose coordinate is 3, standard mathematical terminology does not distinguish between the two. "The point whose coordinate is 3" is shortened to "the point 3." In general, "the point whose coordinate is x" is replaced by the more simple phrase "the point x" for any real number x.

The color line segment on the real number line in Figure 2.4 represents *all* numbers between -1 and 2, including both -1 and 2. A set of numbers such as this, which can be represented by a line segment on the number line, is called an **interval.** Intervals in general may include both endpoints (as in Figure 2.4, which included both -1 and 2), may include just one of the endpoints, or may include neither of the two endpoints. The classification of intervals and their representations on number lines depend upon which, if any, of the two endpoints are included.

Examples of some of the more common types of intervals are given in Table 2.1.

TABLE 2.1 TYPES OF INTERVALS

DESCRIPTION	TYPE	NOTATION	GEOMETRIC REPRESENTATION
All numbers between 1 and 3, including both 1 and 3	Closed	[1, 3]	
All numbers between 1 and 3, including 1 and excluding 3	Half-open	[1, 3)	
All numbers between 1 and 3, including 3 and excluding 1	Half-open	(1, 3]	
All numbers between 1 and 3, excluding both 1 and 3	Open	(1, 3)	

1 PRACTICE PROBLEM 1

Give the geometric representations of the following intervals:

(a) (2, 4) (b) (−3, 1] (c) [0, 2]

Answer

(a)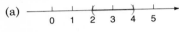
(b)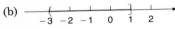
(c)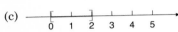

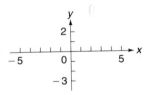

FIGURE 2.5

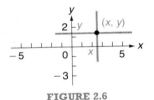

FIGURE 2.6

2 PRACTICE PROBLEM 2

Locate the points with coordinates (1, 4) and (−3, −2) in a coordinate system.

Answer

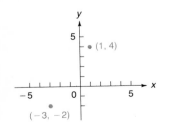

In both the notation and the geometric representation, a square bracket indicates that the corresponding endpoint is included; a parenthesis indicates that the corresponding endpoint is excluded. All numbers in between are always included.

Just as the points of a line can be put into a one-to-one correspondence with the real numbers, the points of a plane can be put into a one-to-one correspondence with pairs of real numbers. A plane is similar to the surface of a sheet of paper or of a blackboard that extends indefinitely in all directions.

To establish this correspondence, construct two real number lines at right angles to each other and intersecting at their origins, as shown in Figure 2.5. The two number lines are called the **coordinate axes;** the horizontal axis is called the **x-axis** and the vertical axis is called the **y-axis.** The arrows indicate the positive directions.

Every point in the plane is associated with a pair of numbers, (x, y) in the following way. Through any given point, draw the line perpendicular to the x-axis; the coordinate of the point at which the line intersects the x-axis is the **x-coordinate** of the point (Figure 2.6). Through the same point draw the line perpendicular to the y-axis; the coordinate of the point of intersection with the y-axis is the **y-coordinate** of the point. The pair of numbers (x, y) determined in this way are the **coordinates** of the point.

Conversely, if we start with a pair of real numbers (x, y), the point with these coordinates can be determined by reversing the above process. Through the point x on the x-axis draw the line perpendicular to the x-axis; through the point y on the y-axis draw the line perpendicular to the y-axis. The point of intersection of these two lines is the point with coordinates (x, y).

EXAMPLE 1 Locate the point with coordinates $(-2, 3)$ in a coordinate system.

Solution First draw the line perpendicular to the x-axis at the coordinate -2; then draw the line perpendicular to the y-axis at the coordinate 3. The point of intersection of these two lines is the desired point, as shown in the figure.

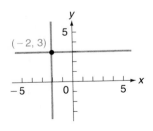

2.2 THE REAL NUMBER LINE AND CARTESIAN PLANE

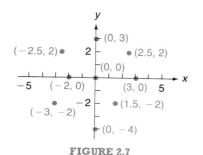

FIGURE 2.7

A plane with coordinates assigned to each point in the manner we have described is called a **Cartesian plane,** or a **coordinate system.** The point (0, 0) at the intersection of the two axes is called the **origin** of the system.

The locations of some representative points together with their coordinates are given in the coordinate system shown in Figure 2.7.

EXAMPLE 2 Where are all the points whose x-coordinates are 4 located in a coordinate system?

Solution Since the value of the y-coordinate is immaterial, these would be all the points on the line drawn through the point (4, 0) and perpendicular to the x-axis, as shown in the figure.

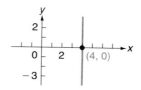

3 PRACTICE PROBLEM 3

Where are all the points whose y-coordinates are 3 located in a coordinate system?

Answer

All points on the line drawn through the point (0, 3) and parallel to the x-axis.

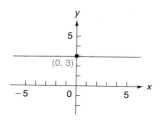

Just as with the real number line, the point with coordinates (x, y) is distinct from the pair of real numbers (x, y). This distinction is usually ignored, and the phrase "the point with coordinates (x, y)" is usually shortened to "the point (x, y)."

EXERCISES

1. Construct a real number line, indicating the location of the following points: $\frac{1}{2}$, 2.3, 3.5, 5, π, -0.7, -1.5, -3, $-\sqrt{2}$, -5. (Note: π is approximately 3.1; $\sqrt{2}$ is approximately 1.4.)

Give the geometric representation of each of the following intervals:

2. $[-2, 2]$ 3. $(-2, 2]$ 4. $(-2, 2)$ 5. $[-2, 2)$
6. $(-3, 0]$ 7. $(-3, -1)$ 8. $[1, 4)$ 9. $[1, 4]$
10. $(1, 4)$ 11. $[0, 3)$ 12. $(-3, 4]$ 13. $[-5, 0]$
14. $[-2, 0)$ 15. $[8, 11]$ 16. $[3, 6]$ 17. $(-9, -5]$

18. Locate the following points in a coordinate system:

$(1, -1)$, $(-3, -2)$, $(0.5, 0)$, $(0, -0.5)$, $(0, 0)$, $(2, 8)$, $(-5, 3)$, $(2.3, -6.4)$

19. Locate the following points in a coordinate system:

    $(1, 1)$, $(2, -3)$, $(-3, 2)$, $(-1, -1.5)$, $(0, 2)$, $(2, 0)$, $(0, -\frac{1}{2})$, $(-\frac{1}{2}, 0)$

Where are the following points located in a coordinate system?

20. All the points whose y-coordinates are -1
21. All the points whose x-coordinates are 5
22. All the points whose x-coordinates are 2
23. All the points whose y-coordinates are -3
24. All the points whose y-coordinates are 0
25. All the points whose x-coordinates are 0

26. (a) Name three types of intervals considered in this section.
    (b) What determines the type of a particular interval?

27. Where are all the points whose x-coordinates are equal to their y-coordinates located in a coordinate system?

Explain the meaning of each term.

28. The origin on a number line
29. A (real) number line
30. The coordinate of a point on a number line
31. The unit distance on a number line
32. The coordinates of a point in a plane
33. Coordinate axes
34. The origin of a coordinate system
35. A Cartesian plane

2.3 GRAPHS OF EQUATIONS

The **graph** of an equation is the set of all points (x, y) whose coordinates satisfy the equation. For example, the point $(2, 5)$ is in the graph of the equation $y = 2x + 1$ because when x is given the value 2 and y is given the value 5 in the equation, equality holds: $5 = 2(2) + 1$. On the other hand, the point $(-2, 0)$ is *not* in the graph of $y = 2x + 1$ because when x and y are given the values -2 and 0, respectively, equality does not hold: $0 \neq 2(-2) + 1$.

To **graph an equation** means to locate all the points (x, y) whose coordinates satisfy the equation. Generally, this means locating an infinite number of points in the plane. Obviously, it is impossible to locate every point individually. Instead, a few points are located, and these are then connected by a curve. The following examples illustrate this procedure.

1 PRACTICE PROBLEM 1

Which of the following points are in the graph of the equation $2y = x + 1$?

(a) $(1, 2)$ (b) $(-3, -1)$
(c) $(0, \frac{1}{2})$

Answer

(a) Not in the graph
(b) In the graph   (c) In the graph

EXAMPLE 1   Graph the equation $y = x^2 + 1$.

Solution   We pick several values for x, say $x = 0, 1, 2, -1$, and -2. For each of these values of x, we determine the corresponding value of y from the equation $y = x^2 + 1$. This pairing of values of x and y can be represented in tabular form, as shown in the margin. Five points in the graph of $y = x^2 + 1$ are easily read from the table: $(0, 1), (1, 2), (2, 5), (-1, 2)$, and $(-2, 5)$. These points are then located in a coordinate system, as shown. Then, to complete the graph of the equation, these points are connected by a smooth curve.

x	y
0	1
1	2
2	5
-1	2
-2	5

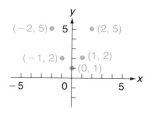
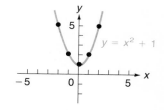

EXAMPLE 2   Graph the equation $x + 2y = 3$.

Solution   We pick several values of x, say $x = -1, 0, 1, 2$, and 3. The corresponding values of y are determined from the equation and written in tabular form in the margin. The points $(-1, 2), (0, \frac{3}{2}), (1, 1), (2, \frac{1}{2})$, and $(3, 0)$ are then located and connected by a curve, which in this case is a straight line.

x	y
-1	2
0	$\frac{3}{2}$
1	1
2	$\frac{1}{2}$
3	0

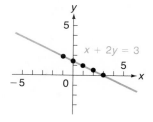

PRACTICE PROBLEM 2

Graph the equation $2y = x + 1$, choosing $x = -3, -1, 0$, and 1.

Answer

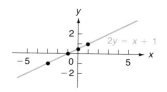

It is important to appreciate the relationship between an equation and its graph. The graph is the set of all those points—and only those points—whose coordinates satisfy the equation. By means of the graph, a geometric picture of the equation is obtained.

The method of obtaining the graph of an equation by the method considered in this section cannot always be expected to be completely accurate. After all, only a few points are used to determine the location of an infinite number of points. More-sophisticated methods of graphing will be introduced in subsequent sections. Generally, these methods will be dependent upon an analysis of the *form* of the equation before any points are located.

EXERCISES

Without constructing the graph, determine which of the following points are in the graph of the equation $y = 3x - 1$:

1. $(0, -1)$
2. $(\frac{1}{3}, 0)$
3. $(1, 1)$
4. $(2, -4)$
5. $(-1, -2)$
6. $(-3, -10)$

Without constructing the graph, determine which of the following points are in the graph of the equation $3y + x = 0$:

7. $(0, 0)$
8. $(-3, 1)$
9. $(-1, \frac{1}{3})$
10. $(6, 2)$
11. $(-2, 0)$
12. $(\frac{1}{6}, \frac{1}{4})$

Without constructing the graph, determine which of the following points are in the graph of the equation $y = 2x^2 + 2$:

13. $(1, 2)$
14. $(-1, 4)$
15. $(\frac{1}{2}, 0)$
16. $(-2, 10)$
17. $(2, -10)$
18. $(0, -2)$

Without constructing the graph, determine which of the following points are in the graph of the equation $y = \sqrt{x}$:

19. $(0, 0)$
20. $(1, 1)$
21. $(2, 2)$
22. $(4, 2)$
23. $(2, 4)$
24. $(-9, 3)$

Choosing the x-coordinate to be $x = -2, -1, 0, 1,$ and 2, determine five points in the graph of each of the following equations:

25. $y = x^2$
26. $y = 2x + 1$
27. $y = -x^2 + 1$
28. $2y = 2x - 1$
29. $x - y = 3$
30. $y - 1 = x^2$
31. $3x + y = 0$
32. $y = -3x + 1$
33. $y = x^3$
34. $y = (x - 1)^3$

Graph each of the equations in the stated exercises by first locating the five points determined in the exercises and then connecting the points with a curve.

35. Exercise 25
36. Exercise 26
37. Exercise 27
38. Exercise 28
39. Exercise 29
40. Exercise 30
41. Exercise 31
42. Exercise 32
43. Exercise 33
44. Exercise 34

45. *Employee errors* The graph in the margin is a graph of the equation that gives the number of errors (y) per hour of a new employee in terms of the number of days (x) that the employee has been on the job.
   (a) Estimate the y-coordinate of the point in the graph whose x-coordinate is 1.
   (b) How many errors are committed per hour by a new employee after 1 day on the job?
   (c) Estimate the y-coordinate of the point in the graph whose x-coordinate is 5.
   (d) How many errors are committed per hour by a new employee after 5 days on the job?

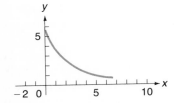

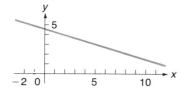

46. *Chemical reaction time* The graph gives the reaction time (y) in minutes for a chemical experiment conducted at various temperatures (x) in degrees Celsius.
    (a) Estimate the y-coordinate of the point in the graph whose x-coordinate is 0.
    (b) What is the reaction time if the experiment is performed at 0°C?
    (c) Estimate the y-coordinate of the point in the graph whose x-coordinate is 10.
    (d) What is the reaction time if the experiment is performed at 10°C?

Explain the meaning of each term.

47. The graph of an equation
48. To graph an equation

2.4 EQUATION OF A STRAIGHT LINE

In the previous section we started with an equation and then obtained the graph of the equation. The reverse approach is just as natural—given a graph, we can try to determine the equation of the graph. This latter approach will be used in this section to determine the equation when the graph is a straight line.

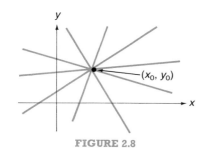

FIGURE 2.8

The basic concept in our approach is the *slope* of a line. Intuitively, the slope of a line indicates the manner in which the line is slanted. Figure 2.8 shows several lines, all passing through the point (x_0, y_0). It is the direction in which each line is slanted that distinguishes it from the other lines.

In order to define the slope of a line (not parallel to the y-axis) mathematically, choose any two points—denoted by (x_1, y_1) and (x_2, y_2)—on the line. If we complete the right triangle determined by the points (x_1, y_1) and (x_2, y_2) in the manner indicated in Figure 2.9, the coordinates of the third vertex of the triangle will be (x_2, y_1).

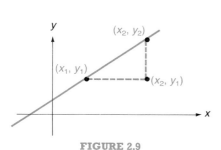

FIGURE 2.9

The length of the vertical side of the triangle is $y_2 - y_1$; the length of the horizontal side is $x_2 - x_1$. The **slope** of the given line is the quotient of these two lengths,

$$\frac{y_2 - y_1}{x_2 - x_1}$$

Think of the point (x_2, y_2) as being obtained by moving from the point (x_1, y_1) along the line to (x_2, y_2). This quotient represents the ratio of the change in the y-coordinates to the change in x-coordinates as the point is moved along the line from (x_1, y_1) to (x_2, y_2).

Slope is used by engineers and architects to measure steepness. The pitch of a roof or steepness of a hill are measured in this way. In these contexts, the vertical change ($y_2 - y_1$) is referred to as the *rise* and the horizontal change ($x_2 - x_1$) is referred to as the *run*. The pitch of a roof, for instance, is then the rise divided by the run.

The standard notation for the slope of a line is the letter m; that is,

$$m = \frac{y_2 - y_1}{x_2 - x_1}$$

EXAMPLE 1 Determine the slope of the line passing through the points $(-4, -1)$ and $(1, 3)$.

Solution Let $(x_2, y_2) = (1, 3)$ and $(x_1, y_1) = (-4, -1)$. Then

$$m = \frac{y_2 - y_1}{x_2 - x_1} = \frac{3 - (-1)}{1 - (-4)} = \frac{4}{5}$$

Note that

$$\frac{y_2 - y_1}{x_2 - x_1} = \frac{y_1 - y_2}{x_1 - x_2}$$

As a consequence of this equality, it is immaterial which of two given points is chosen to be (x_1, y_1) and which is chosen to be (x_2, y_2) in calculating the slope of a line. The value of the resulting quotient will be the same. For instance, in Example 1, if we let $(x_2, y_2) = (-4, -1)$ and $(x_1, y_1) = (1, 3)$, the value of m is the same:

$$\frac{y_2 - y_1}{x_2 - x_1} = \frac{-1 - 3}{-4 - 1} = \frac{-4}{-5} = \frac{4}{5}$$

If a line is slanted upward to the right (as the one in Figure 2.9), both $y_2 - y_1$ and $x_2 - x_1$ are positive numbers. Their quotient is also a positive number. Consequently, such lines have a positive slope.

If a line is parallel to the x-axis, the y-coordinates of any two of its points are the same; that is, $y_2 - y_1 = 0$. Such lines have a slope of 0.

If a line slants downward to the right, as shown in Figure 2.10, $y_2 - y_1$ is negative, but $x_2 - x_1$ is positive. The quotient $(y_2 - y_1) \div (x_2 - x_1)$ is therefore, negative, and such lines have a negative slope.

If a line is parallel to the y-axis, $x_2 - x_1 = 0$. Since division by zero has no meaning, *the slope of such a line is undefined*. That is, lines parallel to the y-axis have no slope (see Figure 2.11).

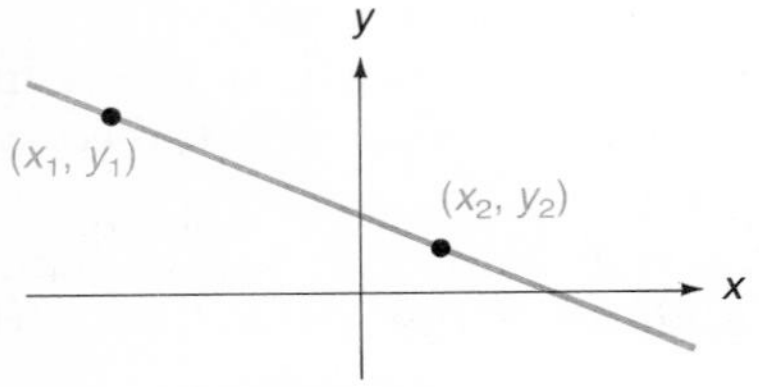

PRACTICE PROBLEM 1

Determine the slope of the line passing through the points $(6, 5)$ and $(3, -2)$.

Answer

$\frac{7}{3}$

FIGURE 2.10

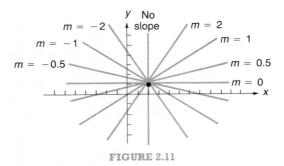

FIGURE 2.11

2.4 EQUATION OF A STRAIGHT LINE

Some representative lines, all passing through the point (2, 1), are given along with their slopes in Figure 2.11.

EXAMPLE 2 In a coordinate system draw the line passing through the point $(-1, 2)$ with slope $-\frac{3}{2}$.

Solution Starting at the point $(-1, 2)$, we let x increase by 2 units. Since the slope is $-\frac{3}{2}$, the corresponding change in y is a decrease of 3 units. In this way we have located another point on the line, and we can easily draw the line, as shown.

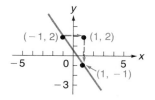

PRACTICE PROBLEM 2

In a coordinate system, draw the line passing through the point $(-1, 2)$ with slope $\frac{2}{3}$.

Answer

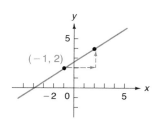

The chief feature of the concept of slope that makes it so useful is the fact that the value of m does not change with the choice of the points (x_1, y_1) and (x_2, y_2) on a line l. In order to see that this is the case, pick any other two points on the line l, such as (x_3, y_3) and (x_4, y_4), as shown in Figure 2.12. The angles of the triangle determined by the points (x_3, y_3) and (x_4, y_4) are the same as the angles of the triangle determined by (x_1, y_1) and (x_2, y_2). Because of this, the triangles are called **similar triangles,** and a theorem of geometry states that the ratios of the lengths of their corresponding sides are equal. Hence,

$$\frac{y_4 - y_3}{x_4 - x_3} = \frac{y_2 - y_1}{x_2 - x_1}$$

That is, either pair of points can be used to determine the slope.

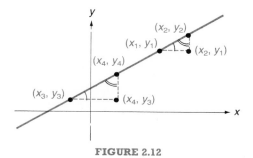

FIGURE 2.12

Suppose now that we start with a given line, and suppose also that we know the coordinates of one point (x_1, y_1) on the line and its slope m. How can we determine the equation whose graph is the given line?

2 EQUATIONS AND THEIR GRAPHS

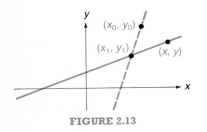

FIGURE 2.13

The problem is to determine an equation such that the coordinates of each point on the line satisfy the equation. At the same time, the equation must be chosen so that the coordinates of no other point satisfy the equation.

Since any two points on the line determine the slope, each point (x, y) on the line must satisfy the equation

$$\frac{y - y_1}{x - x_1} = m$$

At the same time, the coordinates of any point (x_0, y_0) not on the line do not satisfy the equation, as the points (x_1, y_1) and (x_0, y_0) determine a line with a different slope. (See Figure 2.13.)

The desired equation is therefore

$$\frac{y - y_1}{x - x_1} = m$$

Multiplying each side of this equation by the quantity $x - x_1$ gives

$$\boxed{y - y_1 = m(x - x_1)}$$

which is usually referred to as the **point-slope form** of the equation of a straight line.

EXAMPLE 3 Determine the equation of the line passing through the point $(2, 3)$ with slope 5.

Solution Since $(x_1, y_1) = (2, 3)$ and $m = 5$, substituting into the point-slope form gives

$$y - 3 = 5(x - 2)$$

Performing the indicated multiplication on the right side of this last equation gives

$$y - 3 = 5x - 10$$

or

$$y = 5x - 7$$

3 PRACTICE PROBLEM 3

Determine the equation of the line passing through the point $(1, 2)$ with slope -5.

Answer

$y - 2 = -5(x - 1)$, or $y = -5x + 7$

EXAMPLE 4 Determine the equation of the line that passes through the two points $(3, -1)$ and $(4, 3)$.

Solution The slope of the line is

$$m = \frac{3 - (-1)}{4 - 3} = \frac{4}{1} = 4$$

It does not matter which of the two points $(3, -1)$ or $(4, 3)$ we choose to be the point (x_1, y_1), since both are points on the line. Using the point-

slope form with $(x_1, y_1) = (3, -1)$, the equation is
$$y + 1 = 4(x - 3)$$
or
$$y = 4x - 13$$
In Exercise 21 you are asked to verify that the same equation is obtained if the point (x_1, y_1) is chosen to be $(4, 3)$.

▣ PRACTICE PROBLEM 4

Determine the equation of the line passing through the points $(3, -1)$ and $(-5, 2)$.

Answer

$y + 1 = -\frac{3}{8}(x - 3)$, or $8y = -3x + 1$

EXAMPLE 5

Determine the equation of the line that passes through the points $(5, -1)$ and $(3, -1)$.

Solution The slope of the line is
$$m = \frac{-1 + 1}{5 - 3} = 0$$
Choosing the point (x_1, y_1) to be $(5, -1)$, the equation is
$$y + 1 = 0(x - 5) = 0$$
or
$$y = -1$$

EXAMPLE 6

Determine the equation of the line that passes through the points $(2, 1)$ and $(2, -1)$.

Solution
$$x_1 - x_2 = 2 - 2 = 0$$
Therefore, the line has no slope. From the graph of the line shown in the figure, an equation that is satisfied by the coordinates of all points on the line and by the coordinates of no other point is $x = 2$.

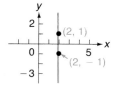

▣ PRACTICE PROBLEM 5

Determine the equation of the line passing through each pair of points.

(a) $(-3, 2)$ and $(4, 2)$
(b) $(-3, 2)$ and $(-3, 0)$

Answer

(a) $y = 2$ (b) $x = -3$

As in most of the examples, the point-slope form can be rewritten
$$y - y_1 = mx - mx_1$$
or
$$y = mx - mx_1 + y_1$$

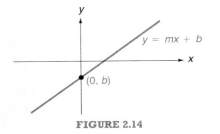

FIGURE 2.14

If b denotes the quantity $-mx_1 + y_1$, this last equation can be written

$$y = mx + b$$

which is the **slope-intercept form** of the equation of a straight line. If x is 0, then $y = b$; that is, the line intersects the y-axis at the point $(0, b)$, as shown in Figure 2.14.

In order to determine the equation of a line, both the point-slope and slope-intercept forms require that the slope of the line and the coordinates of a point on the line be known. However, for the point-slope form, any point on the line can be used; for the slope-intercept form, the point at which the line crosses the y-axis must be known.

EXAMPLE 7 Determine the equation of the line that has slope 4 and intersects the y-axis at the point $(0, -5)$.

Solution From the point $(0, -5)$ we have $b = -5$. The equation is therefore,

$$y = 4x - 5$$

Could we have used the point-slope form in this last example as well? (See Exercise 36.)

6 PRACTICE PROBLEM 6

Use the slope-intercept form to determine the equation of the line that has slope -3 and intersects the y-axis at $(0, 6)$.

Answer

$$y = -3x + 6$$

EXERCISES

Locate each of the following pairs of points in a coordinate system and draw the line that contains the two points. Then determine the slope (if there is one) of the line.

1. $(2, 3)$ and $(-1, 1)$
2. $(2, -3)$ and $(1, -1)$
3. $(\frac{1}{2}, 0)$ and $(\frac{1}{4}, 1)$
4. $(4, 3)$ and $(-1, 3)$
5. $(-3, 4)$ and $(-3, 5)$
6. $(0.5, 0.1)$ and $(0.6, 2)$
7. $(17, 14)$ and $(-18, 14)$
8. $(\frac{1}{3}, -\frac{1}{3})$ and $(\frac{5}{6}, \frac{7}{12})$
9. $(0, 0)$ and $(-0.5, -0.5)$
10. $(-1, 1)$ and $(1, -1)$

In a coordinate system draw the line through the given point with the indicated slope.

11. Through $(2, -3)$ with slope $\frac{3}{2}$
12. Through $(4, 1)$ with slope $-\frac{1}{3}$
13. Through $(0, 0)$ with slope 1
14. Through $(1, -1)$ with slope -2
15. Through $(-1, -2)$ with no slope
16. Through $(-3, 4)$ with slope 0
17. Through $(4, -5)$ with 0 slope
18. Through $(4, -5)$ with no slope
19. Through $(-2, -3)$ with slope 5
20. Through $(-3, 6)$ with slope $-\frac{13}{3}$

21. Verify that the same equation can be obtained in Example 4 if the point (x_1, y_1) is taken to be $(4, 3)$.

Determine the equation of each line.

22. Through (2, 3) with slope -1
23. Through (2, 3) with slope 2
24. Through $(-4, 3)$ with slope 0
25. Through $(\frac{1}{4}, 1)$ with slope -4
26. Through (1, 4) and $(-1, 8)$
27. Through $(-3, 3)$ and $(-3, 5)$
28. Through (2.2, -1.1) and (3.2, 5.9)
29. Through (7, 14) and $(-8, 14)$
30. Through (6, -4) with no slope
31. Through $(\frac{1}{2}, \frac{2}{3})$ and $(\frac{1}{4}, \frac{5}{6})$

Using the slope-intercept form, determine the equation of each line.

32. Has slope -2 and intersects the y-axis at the point (0, 0)
33. Has slope 3 and intersects the y-axis at the point (0, -2)
34. Passes through the point (0, -4) with slope 1
35. Passes through the point (0, 1) with slope -4
36. Use the point-slope form to determine the equation in Example 7.
37. Give the general point-slope and slope-intercept forms of the equation of a straight line.
38. Show that the point-slope form becomes the slope-intercept form when the point (x_1, y_1) is taken to be (0, b).
39. Explain the meaning of the term *slope*.

2.5 LINEAR EQUATIONS

A *linear equation* (in x and y) is any equation that can be written in the form

$$Ax + By + C = 0$$

where A, B, and C are real numbers with A and B not both zero

If A and B were both zero, the equation would become $C = 0$.
  The following are examples of linear equations:

$3x + 2y - 4 = 0$ This equation can be written as

$$3x + 2y + (-4) = 0$$

where $A = 3$, $B = 2$, and $C = -4$.

$-2x = 4y + 1$ This equation can be written as

$$-2x - 4y - 1 = 0$$

by subtracting $4y + 1$ from each side; $A = -2$, $B = -4$, and $C = -1$.

$0.5x + \dfrac{y}{2} - \dfrac{1}{4} = 0$ Here, $A = 0.5$, $B = \frac{1}{2}$, $C = -\frac{1}{4}$.

$3y = -2$ What are A, B, and C?

The important thing to note about a linear equation in the form $Ax + By + C = 0$ is that the highest power of x and y is the first power; there are no terms involving x^2, y^2, or any higher power. There are no denominators involving x or y, nor are there any products involving both x and y, such as xy or xy^2. The following equations, for example, are not linear equations:

$$y = x^2 \qquad 2y^2 - x = 3$$

$$xy + 2x = 0 \qquad x + y = \frac{2}{x}$$

Since the equation of every straight line can be written in the form $y = mx + b$ or in the form $x = c$ (if there is no slope), it follows that the equation of every straight line is a linear equation. Conversely, the graph of every linear equation is a straight line—hence the term *linear equation*. Observe that the equations of Exercises 26, 28, 29, 31, and 32 in Section 2.3 are linear and that their graphs are straight lines.

In order to verify the statement that the graph of *every* linear equation is a straight line, suppose first that B is not zero in $Ax + By + C = 0$. Then division by B is possible, and the equation can be written

$$y = -\frac{A}{B}x - \frac{C}{B}$$

The graph of this equation is the line passing through $(0, -C/B)$ with slope $-A/B$.

If B is zero, then A cannot be zero, and the general equation is $Ax + C = 0$, which can be rewritten

$$x = -\frac{C}{A}$$

This has as its graph the line parallel to the y-axis and passing through the point $(-C/A, 0)$. Consequently, *the graph of a linear equation is always a straight line,* whether or not B is zero.

Knowing this fact makes it particularly easy to graph any linear equation. The location of two points on any line determines completely the location of the line. Therefore, it is sufficient to locate only two points and then to draw the line through them to obtain the graph.

EXAMPLE 1   Graph the equation $3x - y = 4$.

Solution   Choose any two values of x, say $x = 0$ and $x = 1$, and determine the corresponding values of y:

x	y
0	-4
1	-1

2.5 LINEAR EQUATIONS

1 PRACTICE PROBLEM 1

Graph the equation $2x - 3y = 1$.

Answer

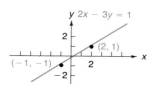

Therefore, $(0, -4)$ and $(1, -1)$ are two points in the graph, as shown.

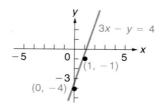

EXAMPLE 2 Determine the slope of the line whose equation is $2x + 3y = 4$.

Solution By rewriting the given equation in the form $y = mx + b$, we can obtain the slope from the coefficient of x. Subtracting $2x$ from each side of the given equation, we have

$$3y = -2x + 4$$

Dividing both sides by 3 gives

$$y = -\tfrac{2}{3}x + \tfrac{4}{3}$$

which is in the form $y = mx + b$. The slope is therefore $-\tfrac{2}{3}$.

2 PRACTICE PROBLEM 2

Determine the slope of the line whose equation is $2x - 3y = 1$.

Answer

$\tfrac{2}{3}$

EXERCISES

Graph each equation.

1. $x - 3y = -2$
2. $2x + y = 0$
3. $y = 2x + 3$
4. $-3x - y + 1 = 0$
5. $x + y = 0$
6. $x = y$
7. $x = 1.5$
8. $y = -1.5$
9. $0.5x - 0.2y = 0$
10. $4y = 7$

Determine the slope of each of the lines whose equations are given in the stated exercise.

11. Exercise 1
12. Exercise 2
13. Exercise 3
14. Exercise 4
15. Exercise 5
16. Exercise 6
17. Exercise 7
18. Exercise 8
19. Exercise 9
20. Exercise 10

21. (a) Graph the following equations in the same coordinate system:

$$y = 3x + 2$$
$$y = 3x - 1$$

   (b) What does part (a) suggest concerning the relationship between the slopes of parallel lines?

22. (a) Graph the following equations in the same coordinate system:

$$y = 2x - 3$$
$$y = -\tfrac{1}{2}x + 2$$

(b) What does part (a) suggest concerning the relationship between the slopes of perpendicular lines?

23. Explain the meaning of the expression *linear equation in x and y*.

2.6 LINEAR EQUATIONS AS MATHEMATICAL MODELS

When a physical problem involves a relationship between two quantities and the mathematical model that expresses this relationship is an equation, the graph of the equation gives a second mathematical model—a geometric model. When the equation is a linear equation, the geometric model is, of course, a straight line. Linear equations are appropriate mathematical models when one quantity (y) increases or decreases with another quantity (x) at a constant rate.

By way of illustration, a small company has determined that its operating expenses each week amount to $700 plus 13% of sales. In dollars, the operating expenses (y) in terms of sales (x) are $y = 700 + 0.13x$. In this case, expenses increase 13¢ ($0.13) for each increase of $1 in sales. The graph of the expense equation shown in Figure 2.15 gives a geometric model. Note that the slope of the line is also 0.13. The relatively small slope of the line indicates that operating expenses increase slowly with sales.

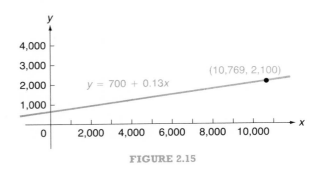

FIGURE 2.15

If the operating expenses for a particular month amount to $2,100.00, then the total sales for that month can be determined by solving the equation

$$2,100 = 700 + 0.13x$$

to obtain $x = 10{,}769$ (approximately). The corresponding point is located on the graph of the expense equation in Figure 2.15.

Note that the points of the graph with negative x-coordinates are meaningless from the point of view of the expense problem; after all, how can one have negative sales!

The unit distance in the last coordinate system was chosen to be very small. This was done for the sake of convenience, as the numbers under consideration were relatively large. The visual effect of a graph can be distorted by using a different unit distance on each of the two axes, a trick sometimes used to obtain a more impressive effect. Compare Figure 2.16, which also shows the graph of the operating expense equation, with Figure 2.15. In Figure 2.16, the line seems to have a greater slope, giving the impression that expenses grow more quickly with sales than is actually the case. In general, caution is warranted when graphs are used to prove a point if different units of distance are used on the two axes.

1 PRACTICE PROBLEM 1

The number of cases (y) of glasses that a manufacturer will supply each day in terms of the price (x), in dollars, is indicated by the graph.

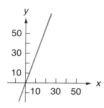

(a) Determine the y-coordinate of the point in the graph whose x-coordinate is 15.

(b) How many cases a day will the manufacturer supply each day if the price is $15.00 a case?

Answer

(a) About 45   (b) About 45 cases

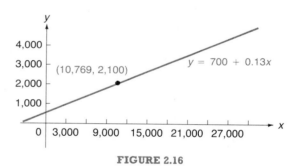

FIGURE 2.16

Another common example of the use of linear equations occurs in the determination of **linear** (or **straight-line**) **depreciation** because the total depreciation increases at a constant rate with time. This type of depreciation occurs when a piece of equipment is considered to depreciate at an equal amount each year over the life of the machine. As a consequence, the value of the equipment also decreases each year at a constant rate. A general equation to describe this relationship is easy to construct.

If a piece of equipment costing c dollars has an expected life of n years and the equipment is considered to depreciate the same amount each year, then the yearly depreciation will be $(1/n)c$, or c/n. After t years the total depreciation (d) will be

$$d = \left(\frac{c}{n}\right)t$$

The value (v) at that time will be the original cost less the depreciation, or

$$v = c - \left(\frac{c}{n}\right)t$$

Note that d increases at a constant rate with t and v decreases at a constant rate with t.

EXAMPLE 1 A piece of heavy construction equipment costing $80,000 has an expected life of 10 years. Using linear depreciation, write equations to express the total depreciation at any time t and the value of the equipment at any time t. Graph both equations.

Solution Using the above relationships with $c = 80,000$ and $n = 10$, the desired equations are

$$d = 8{,}000t$$

and

$$v = 80{,}000 - 8{,}000t$$

Their graphs are shown below. Note that the depreciation equation is linear in t and d and that the value equation is linear in t and v. Consequently, the corresponding changes are made in labeling the axes for their graphs. Also note that only those parts of the graphs corresponding to values of t from 0 to 10 are relevant to the physical situation. Why?

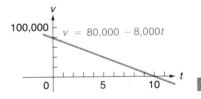

2 PRACTICE PROBLEM 2

A company buys a machine, which has an expected life of 12 years, for $10,800. Using linear depreciation, write equations to express the total depreciation at any time t and the value of the machine at any time t.

Answer

$d = 900t$; $v = 10{,}800 - 900t$

When two quantities are either known or assumed to be linearly related, the values of these quantities at only two points are sufficient to write an equation describing their relationship and/or to graph such an equation. The next example illustrates this type of situation.

EXAMPLE 2 It is known that the rate of blood flow of a type of seal is linearly related to how far below the surface of the water the seal is located. Experiments have determined that the blood flow rate is 25 milliliters per second when the seal is 10 meters below the surface and 50 milliliters per second on the surface (0 meters below the surface). Write a

2.6 LINEAR EQUATIONS AS MATHEMATICAL MODELS

linear equation that describes the blood flow rate (r) in terms of the depth of the seal (d), and graph the equation.

Solution From the given information, we know that (10, 25) and (0, 50) are two points in the graph and that the graph is the line through these points.

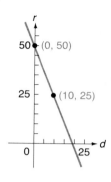

The slope of the line is

$$\frac{50 - 25}{0 - 10} = \frac{25}{-10} = -2.5$$

Using the slope-intercept form, the equation is

$$r = -2.5d + 50$$

Many of the equations that you constructed in Section 2.1 are linear equations. You are asked to graph some of these equations in the following exercises.

PRACTICE PROBLEM 3

Taxi fare depends linearly on the number of miles traveled. If it costs $3.00 to ride 2 miles and $4.50 to ride 4 miles, construct an equation to express taxi fare (y) in terms of the number of miles traveled (x).

Answer

$$y = 1.50 + 0.75x$$

EXERCISES

1. *Computer component manufacturing*
   (a) Graph the equation of Exercise 1, Section 2.1, which gives the manufacturing costs in terms of the number of computer components produced.
   (b) If 2,000 components are produced, locate the corresponding point in the graph.
   (c) Which points in the graph are relevant to this particular situation? (*Hint:* Consider realistic values of x.)

2. *Furniture manufacturing*
   (a) Graph the equation of Exercise 2, Section 2.1, which gives the manufacturing costs in terms of the number of chairs produced.
   (b) If the total manufacturing costs amounted to $38,300, locate the corresponding point in the graph.
   (c) Which points in the graph are relevant to this particular situation?

3. *Shoplifting*
   (a) Graph the equation of Exercise 5, Section 2.1, which gives the number of books stolen in terms of the number of books shelved.
   (b) If 500 new books were shelved during a particular month, locate the corresponding point on the graph.

4. *Sales mark-down*
   (a) Graph the equation of Exercise 11, Section 2.1, which gives the original price of merchandise in terms of the sale price.
   (b) If the sale price of an item is $40.00, locate the corresponding point in the graph.

5. *Population growth* Graph the equation of Exercise 12, Section 2.1, which gives the population of a country at the end of a year in terms of the population at the beginning.

6. *Airline revenue*
   (a) Graph the equation that expresses the price per person in terms of the number of empty seats in the chartered flight of Exercise 14, Section 2.1. (For consistent notation, let the y-axis be the z-axis here.)
   (b) If there are 83 people on the flight, locate the corresponding point in the graph.

7. *Agribusiness*
   (a) Graph the equation of Exercise 15, Section 2.1, which gives the total number of trees in terms of the number of additional trees planted.
   (b) If 20 additional trees are planted, locate the corresponding point in the graph.

8. *Typewriter depreciation* A typewriter costing $1,200 is to be linearly depreciated over a period of 8 years.
   (a) Construct an equation that gives the total depreciation at any time.
   (b) Construct an equation that gives the value of the typewriter at any time.
   (c) Graph the two equations obtained in parts (a) and (b).
   (d) Which points in the graphs of the equations are relevant to this particular situation?

9. *Microscope depreciation* An electron microscope costing $15,000 is to be linearly depreciated over a period of 12 years.
   (a) Construct an equation that gives the total depreciation at any time.
   (b) Construct an equation that gives the value of the microscope at any time.
   (c) Graph the two equations obtained in parts (a) and (b).
   (d) Which points in the graphs of the equations are relevant to this particular situation?

The equations for linear depreciation constructed in this section assumed that the items being depreciated had no salvage value. Assume now that the items do have a salvage value (s) and that the difference between the cost and the salvage value ($c - s$) is to be linearly depreciated over the life of the item (n).

10. Construct equations that give the total depreciation (d) and the value (v) of an item at any time (t).

11. *Typewriter depreciation* Rework Exercise 8 if the typewriter has a salvage value of $200.

12. *Microscope depreciation* Rework Exercise 9 if the microscope has a salvage value of $3,000.

13. *Simple interest* When money is borrowed at simple interest, the amount of interest to be paid is given by the expression

    Principal × Rate of interest × Time

    (a) Construct an equation that gives the simple interest (I) to be paid in terms of the principal (P), rate of interest (R), and time (T).
    (b) Construct an equation that gives the total amount (A) to be repaid (principal plus interest) using the notation of part (a).
    (c) Using the equations constructed in parts (a) and (b), write the equations that give the interest and the total amount to be repaid on a loan of $500 at 11% simple interest in terms of the duration of the loan.
    (d) Graph the equations obtained in part (c).

14. *Disease propagation* Experiments have indicated that, with no outside interference, the number of diseased mice will increase linearly each day after one of the mice in their cage is infected with a particular type of disease-causing germ. In one experiment, there were 7 diseased mice 3 days after the first exposure and 13 diseased mice after 6 days.
    (a) Construct an equation that will give the number of diseased mice after any number of days.
    (b) Graph the equation obtained in part (a).
    (c) If there are 35 mice in the cage, how long will it take until they are all diseased?

15. *Product demand* Company officials believe that the demand (number of units sold) for one of their products decreases just about linearly as the price increases. Their marketing researchers found that in a particular city, an average of 125 units were sold each day when the unit price was $5.00 and 5 units were sold each day when the price was $15.00.
    (a) Construct an equation that will approximate the demand in terms of the price. (Use the letter q to denote demand and p to denote price.)
    (b) Graph the equation obtained in part (a).
    (c) At what price can the company expect to sell no units at all?
    (d) Rewrite the equation obtained in part (a) to obtain price in terms of demand.
    (e) Graph the equation obtained in part (d).

16. *Production cost* Given the data in the table in the margin, construct an equation that will give the total cost (y) in terms of the production level (x), assuming that the cost increases linearly with the production level.

LEVEL OF PRODUCTION	0	10	20
TOTAL COST	100	120	140

17. *Temperature conversion* The relationship between degrees Celsius and degrees Fahrenheit for measuring temperature is a linear relationship. If

    $$0°C = 32°F \quad \text{and} \quad 100°C = 212°F$$

    write an equation to express degrees Fahrenheit (F) in terms of degrees Celsius (C).

18. *Weight–height relation* Suppose the weights of male college freshmen are linearly related to their heights. If a freshman 60 inches tall weighs 115 pounds and a freshman 72 inches tall weighs 180 pounds, write an equation to express weight (w) in terms of height (h).

2.7 QUADRATIC EQUATIONS

The second general type of equation that we shall consider is the quadratic equation. Formally, any equation that can be written in the form

$$y = Ax^2 + Bx + C$$
where A, B, and C are real numbers with $A \neq 0$

is said to be an equation **quadratic in x** (or, for the sake of simplicity, a **quadratic equation**).

The following are examples of quadratic equations:

$2x^2 - 3x + y = 0.6$ This equation can be written as
$$y = -2x^2 + 3x + 0.6$$
where $A = -2$, $B = 3$, and $C = 0.6$.

$y + 3 = x^2$ This equation can be written as
$$y = x^2 - 3$$
where $A = 1$, $B = 0$, and $C = -3$.

$2y = 3x^2 + 4x - 1$ This equation can be written as
$$y = (\tfrac{3}{2})x^2 + 2x - \tfrac{1}{2}$$
where $A = \tfrac{3}{2}$, $B = 2$, and $C = -\tfrac{1}{2}$.

$y = x^2$ What are A, B, and C?

In an equation quadratic in x, y always appears to the first power and x always appears to the second power, regardless of whether or not the first power of x appears. There are no higher powers of x and y nor are there any terms involving both x and y, such as xy or x^2y.

The graph of a quadratic equation is called a **parabola**. In order to graph a given quadratic equation, we shall rewrite it (by a procedure discussed shortly) to get it into the form

$$q(y - k) = (x - h)^2$$

Whether the graph of such an equation opens upward or downward depends on whether q is positive or negative, as shown in Figure 2.17.

Note that the coordinates of the point (h, k) at which the parabola reaches its highest or lowest point are also given in the equation. This point, called the **vertex,** will be the most important point in the parabola when applications involving quadratic equations are discussed later in this chapter.

The procedure whereby a quadratic equation is put into the above form is called *completing the square in x,* a process sometimes used in solving quadratic equations. This procedure is explained in the box on the next page; for the purpose of illustration, the various steps are performed on the equation $y = 2x^2 - 8x + 12$.

2.7 QUADRATIC EQUATIONS

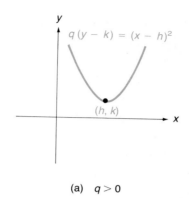

(a) $q > 0$

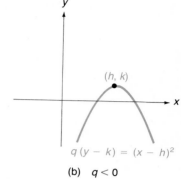

(b) $q < 0$

FIGURE 2.17

Starting with the equation in the form:

| $y = Ax^2 + Bx + C$ | $y = 2x^2 - 8x + 12$ |

1. Divide both sides of the equation by the coefficient of x^2.

1. Dividing both sides of the equation by 2 gives

$$\frac{y}{2} = x^2 - 4x + 6$$

2. Subtract the constant term from each side of the resulting equation.

2. Subtracting 6 from each side gives

$$\frac{y}{2} - 6 = x^2 - 4x$$

3. To each side of the resulting equation, add the square of half the coefficient of x.

3. $\left(-\dfrac{4}{2}\right)^2 = (-2)^2 = 4$

$$\frac{y}{2} - 2 = x^2 - 4x + 4$$

4. Factor the right side into the form $(x - h)^2$. (*Note:* $-h$ will always be the quantity that was squared in Step 3.)

4. $\dfrac{y}{2} - 2 = (x - 2)^2$

5. Put the left side into the form $q(y - k)$ by factoring the coefficient of y from *both* terms on the left.

5. $\dfrac{1}{2}(y - 4) = (x - 2)^2$

This is the equation of a parabola with vertex at $(2, 4)$ and opening upward. Its graph is given in Figure 2.18.

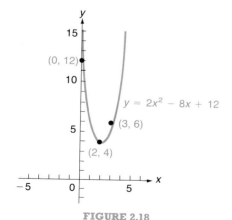

FIGURE 2.18

In order to determine how wide the parabola in Figure 2.18 opens, two additional points, (0, 12) and (3, 6), were determined by choosing two arbitrary values of x, one from each side of the vertex.

EXAMPLE 1 Graph the equation $-2y = x^2 - 6x + 11$.

Solution First the equation must be written in the standard form by completing the square in x. The steps are numbered to conform with the given description.

Step 1. The coefficient of x^2 is 1; this step can be omitted.
Step 2. $-2y - 11 = x^2 - 6x$
Step 3. $\quad (-\frac{6}{2})^2 = (-3)^2 = 9$
$\quad -2y - 2 = x^2 - 6x + 9$
Step 4. $-2y - 2 = (x - 3)^2$
Step 5. $-2(y + 1) = (x - 3)^2$

This is the equation of a parabola with vertex at $(3, -1)$ and opening downward, as shown in the graph.

PRACTICE PROBLEM 1

Graph the equation $3y = x^2 - 2x + 7$.

Answer

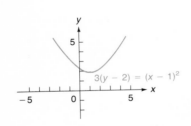

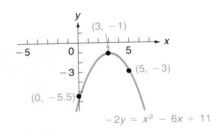

EXAMPLE 2 Without graphing the equation, determine the coordinates of the vertex of the parabola with equation

$$2y = 3x^2 + 6x + 3$$

Solution The equation must be put into the standard form by completing the square. The steps are again numbered in conformity with the given description.

Step 1. $\quad \frac{2}{3}y = x^2 + 2x + 1$
Step 2. $\quad \frac{2}{3}y - 1 = x^2 + 2x$
Step 3. $\quad (\frac{2}{2})^2 = 1^2 = 1$
$\quad \frac{2}{3}y = x^2 + 2x + 1$
Step 4. $\quad \frac{2}{3}y = (x + 1)^2$
Step 5. $\quad \frac{2}{3}(y - 0) = (x + 1)^2$

PRACTICE PROBLEM 2

Without graphing the equation, determine the coordinates of the vertex of the parabola with the equation $y = -x^2 - 4x - 2$. Indicate whether the graph opens upward or downward.

Answer

$(-2, 2)$; graph opens downward

The vertex is therefore at $(-1, 0)$. The parabola opens upward since $q = \frac{2}{3}$ is positive.

EXERCISES

Graph each equation.

1. $6y = 3x^2 - 6x + 9$
2. $2y = -2x^2 + 4x + 1$
3. $y = -\dfrac{x^2}{8} + x - 3$
4. $y = \dfrac{x^2}{4} - \dfrac{x}{2} + \dfrac{1}{4}$
5. $y = 6x + x^2$
6. $y = 2 - x^2$
7. $y = -x^2 - 5x - 4.75$
8. $y = x^2 + 0.6x + 0.09$
9. $4y = -x^2 + 2x - 9$
10. $-3y = x^2 + 6x + 4.5$
11. $y = x^2$
12. $y = -x^2$
13. $y = x^2 + 3$
14. $y = -x^2 - 4$
15. $y + 2 = x^2$

Without graphing the equations, determine the coordinates of the vertex of each parabola and determine whether it opens upward or downward.

16. $y = x^2 - 1$
17. $3y = x^2 + 2x + 7$
18. $y = -x^2 + 8x - 16$
19. $y = -x^2 + 10x - 29$
20. $x^2 + 2y + 2\sqrt{3}\,x = -1$
21. $y = 8x^2$
22. $y = x^2 - 5$
23. $y = (x - 4)^2$
24. $y - 3 = x^2$

Explain the meaning of each term.

25. The vertex of a parabola
26. A quadratic equation

2.8 QUADRATIC EQUATIONS AS MATHEMATICAL MODELS

When the mathematical model for a physical problem is a quadratic equation, the corresponding geometric model is a parabola. This type of model is particularly advantageous because of the role played by the vertex.

In the case of a parabola opening downward, its vertex, (h, k), is the *highest point* on the graph; k is therefore the largest value acquired by y and h is the value of x at which this largest value is acquired. In the case of a parabola opening upward, its vertex, (h, k), is the *lowest point* on the graph; k is therefore the smallest value acquired by y and h is the value of x at which this smallest value is acquired.

This information can often be used to great advantage, as the following examples illustrate.

EXAMPLE 1 The cost to the Style Clothing Company for manufacturing shirts of a particular type is given by

$$y = x^2 - 100x + 2{,}750$$

where x is the number of shirts produced and y is the cost, in cents, per shirt. The company decides to manufacture the number of shirts for which the cost is minimal. How many shirts should the company produce? What will be the total cost? What will be the cost per shirt?

Solution Completing the square in the cost equation gives

$$y - 250 = (x - 50)^2$$

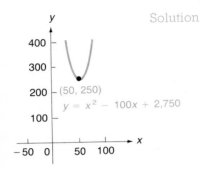

The graph is a parabola opening upward with vertex at (50, 250). The minimal cost per shirt occurs if the company produces 50 shirts, and at this level the cost per shirt is $2.50. The total cost is $2.50 × 50 = $125.00.

In the graph of the equation only those points with integral x-coordinates greater than or equal to zero are relevant. Partial shirts or negative numbers of shirts are not manufactured!

EXAMPLE 2 A company ships its products in cardboard boxes 2 inches deep. The cost of these containers is based upon the perimeters of the bases of the boxes. The company decides that its most economical size is a box with a perimeter of 36 inches. For what width will the box meeting these specifications have the maximum volume?

Solution The volume of a rectangular container is length times width times depth. Let x denote the width (in inches); in order that the perimeter be 36 inches, the length must be $18 - x$ (inches). The volume (y) is, therefore,

$$y = 2(18 - x)x = -2x^2 + 36x$$

cubic inches.

Completing the square in x gives

$$-\tfrac{1}{2}(y - 162) = (x - 9)^2$$

which is the equation of a parabola opening downward with vertex (9, 162). The maximum volume is 162 cubic inches, which is attained when the width is 9 inches.

Those points of the graph with x-coordinate less that 0 or greater than 18 are irrelevant for this particular situation. Not only must the width be positive ($x > 0$), but it must also be less than 18 inches ($x < 18$), since the sum of the length and width can be only 18 inches.

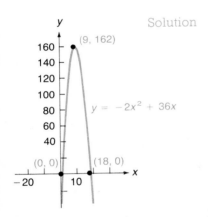

PRACTICE PROBLEM 1

Suppose that the price y, in dollars, of an item in terms of demand x, where x is measured in hundreds of items, is given by $y = -x^2 + 6x + 5$. Determine the demand that gives the maximum price and the corresponding maximum price.

Answer

A demand of 300 units gives a price of $14.00.

The four-step procedure involved in the application of mathematics (Section 1.3) was again used in Example 2:

1. The physical situation was the size of the shipping containers and the physical question was, what size box meeting the required specifications will have the maximum volume?

2. The mathematical model was the equation $y = -2x^2 + 36x$ along with its graph; the mathematical question was, what are the coordinates of the vertex of the parabola?

3. The mathematical solution was (9, 162).
4. The interpretation of the mathematical solution was that the maximum volume of 162 cubic inches was attained when the width was 9 inches. (The full dimensions were 2 by 9 by 9 inches.)

EXAMPLE 3 The velocity v, in centimeters per second, of the blood flowing through a small section of capillary at any point is given by

$$v = 1.2 - 19{,}000r^2$$

where r is the distance of the point P from the center C of the capillary indicated in the drawing of an enlarged cross section. At what points would the velocity be maximum?

Solution Completing the square in r gives

$$-\frac{1}{19{,}000}(v - 1.2) = r^2 = (r - 0)^2$$

The vertex of the parabola is at (0, 1.2), indicating that the maximum velocity occurs at $r = 0$—that is, at the center. At such points the velocity is 1.2 centimeters per second.

2 PRACTICE PROBLEM 2

A manufacturer of computer terminal keyboards estimates that the cost per unit y (in dollars) in terms of the size x of a production run to be given by $y = \frac{1}{2}x^2 - 40x + 890$. Determine the size of a production run that gives the minimum unit cost and the corresponding minimum unit cost.

Answer

40 keyboards per run give a unit cost of $90.00.

EXERCISES

1. *Employee errors* A psychologist has noted that the number of errors (y) committed per day by an employee on a particular job is approximately

   $$y = 12x - x^2$$

   where x is the number of days that the employee has been on that job.
   (a) Complete the square in x and graph the given equation.
   (b) At what day will the employee commit the most errors? How many errors will he or she commit on that day?
   (c) What does the graph indicate about the number of errors committed after the twelfth day? How would you interpret this result?
   (d) In view of the graph of the equation, how can the number of errors be related to the employee's learning process?

2. *Annual profit* The total annual profit (y), in dollars, of a company manufacturing automobile accessories from the manufacture and sale of x rearview mirrors is given by

   $$y = -\frac{x^2}{400} + 2x - 75$$

   (a) Complete the square in x and graph the equation.
   (b) How many mirrors must the company sell per year in order to achieve maximum profit?
   (c) What is the maximum annual profit that the company can achieve from the sale of the mirrors?

(d) What is the profit per mirror when the maximum is achieved?
(e) Which points on the graph are relevant to this particular situation?

3. *Personnel increment*   A large department store believes that it loses sales in a particular department due to a shortage of personnel. A management consultant advises that if an additional part-time employee were to be hired to work x hours per week in the department, the net loss (y) in sales would be

$$y = 400 + x(x - 20)$$

where net loss equals increase in sales less the employee's salary. How many hours should the part-time employee work to attain minimum loss of sales?

4. *Disease propagation*   The number of new cases (y), in hundreds, of diseased individuals from an exposed population is given by

$$y = 12x - x^2 + 1$$

where x is the number of days from the time the first case is observed. At what day will the maximum number of new cases occur? How many cases will occur on that day?

5. *Airline revenue*
   (a) What is the maximum income that the airline company of Exercise 14, Section 2.1, can receive from its charter flight?
   (b) How many empty seats are there when the maximum income is attained?
   (c) What is the price per person when the maximum is attained?

6. *Agribusiness*
   (a) What is the maximum profit that the farmer of Exercise 15, Section 2.1, can achieve per year?
   (b) How many additional trees must he plant in order to achieve the maximum profit?
   (c) What is the profit per tree when the maximum is achieved?

7. *Depot dimensions*   What are the dimensions of the largest storage area that can be fenced in Exercise 16, Section 2.1?

8. *Apartment rental*   What is the maximum total income that can be realized from the apartment complex of Exercise 17, Section 2.1?

Identify the steps of the four-step process involved in the application of mathematics (Section 1.3) in each case.

9. *Blood flow*   Example 3
10. *Personnel increment*   Exercise 3
11. *Apartment rental*   Exercise 8
12. *Depot dimensions*   Exercise 7

2.9 CHAPTER SUMMARY

The **graph** of an equation is the set of all points whose coordinates satisfy the equation. Whether or not a point is in the graph of a particular equation can be determined by direct substitution of its coordinates into the equation.

To **graph an equation** means to locate, in a coordinate system, all the points whose coordinates satisfy the equation. Unless the shape of the graph can be determined from the form of the equation, a rough estimate of the graph can be determined by choosing a set of x-coordinates and calculating the corresponding y-coordinates to obtain a set of points in the graph. These points are located in a coordinate system and then connected by a curve in what seems to be the most reasonable manner. (This procedure can be greatly improved using the methods of calculus.)

Two specific types of equations were considered, linear equations and quadratic equations. A **linear equation** is an equation that can be written in the form

$$Ax + By + C = 0$$

where A and B are not both zero; its graph is a straight line. A **quadratic equation** is an equation that can be written in the form

$$y = Ax^2 + Bx + C$$

with A not zero. Its graph is a parabola, the vertex of which can be determined by completing the square.

Instead of starting with an equation and determining its graph, the reverse procedure is just as natural—given a graph, to determine the equation of the graph. This procedure was considered for straight lines. The basic concept in developing the equation of a line is its **slope,** defined to be

$$m = \frac{y_2 - y_1}{x_2 - x_1}$$

where (x_1, y_1) and (x_2, y_2) are two points on the line. This definition holds if the line is not parallel to the y-axis. Lines parallel to the y-axis (vertical lines) have no slope.

The equation of every nonvertical line can be written as

$$y - y_1 = m(x - x_1) \quad \textbf{\textit{Point-slope form}}$$

which simplifies to the alternate form

$$y = mx + b \quad \textbf{\textit{Slope-intercept form}}$$

The equation of a vertical line is always in the form

$$x = c$$

where c is the x-coordinate of any point on the line.

Finally, if the mathematical model for a physical problem is an equation, the graph of the equation gives a second mathematical model. The graph gives a geometric representation of the relationship between the quantities involved.

REVIEW EXERCISES

Without constructing the graphs, determine whether each of the following points is in the graph of the given equation:

1. $(3, -2)$; $y = 4 - 2x$
2. $(0, 15)$; $4x + 3y = 5$
3. $(0, -3)$; $\dfrac{x+5}{2} = y + 1$
4. $(6, 3)$; $y = \sqrt{15 - x}$
5. $(2, 7)$; $-x^2 + 3 = y$
6. $(2, -1)$; $(-x)^2 + 3 = y$
7. $(-0.6, 1.8)$; $y = \frac{1}{3}(x + 6)$

Graph each of the following equations. When possible, construct the graph on the basis of the form of the equation. Otherwise, in determining individual points, choose the x-coordinates to be $x = -2, -1, 0, 1,$ and 2, unless otherwise indicated.

8. $2x - 3y = 7$
9. $y = 4x^2 - 24x + 35$
10. $x = -2$
11. $2y = 9$
12. $y = x^3 - 3x$
13. $y = \sqrt{x}$ (Pick the x-coordinates to be $x = 0, 1, 4,$ and 9.)
14. $0.2y - 0.3x = 1$
15. $x^2 + \frac{2}{3}x + \frac{11}{18} - y = 0$
16. $y - 2x^2 + 8x - 4 = 0$
17. $y = \dfrac{1}{x - 1}$ (Pick the x-coordinates to be $x = -5, 0, 0.5, 1.5, 2,$ and 5.)

Indicate whether the equation in the given exercise is linear or quadratic. For each linear equation, give the slope of the corresponding line.

18. Exercise 8
19. Exercise 9
20. Exercise 10
21. Exercise 11
22. Exercise 12
23. Exercise 13
24. Exercise 14
25. Exercise 15
26. Exercise 16
27. Exercise 17

In a coordinate system, draw the line with the given characteristics.

28. Passes through $(3, 1)$ with slope $\frac{4}{3}$
29. Passes through $(-2, 0)$ with slope -4
30. Passes through $(1, -1)$ with no slope
31. Passes through $(-1, -2)$ with slope 0

Determine the equation of the line with the given characteristics.

32. Passes through $(3, 1)$ with slope $\frac{4}{3}$
33. Passes through $(-2, 0)$ with slope -4
34. Passes through $(1, -1)$ with no slope
35. Passes through $(-1, 2)$ with slope 0

36. Passes through (4, 6) and (−1, 3)
37. Passes through ($\frac{1}{2}$, −1) and (0, $\frac{1}{3}$)
38. Passes through (1, 0) and is parallel to the x-axis
39. Passes through (1, 0) and is parallel to the y-axis
40. Crosses the y-axis at (0, −$\frac{2}{3}$) with slope −2

41. *Manufacturing costs* It costs an electronics company $9.25 in parts and labor to manufacture one of its hand calculators. In addition, fixed costs (such as machinery depreciation, building maintenance, insurance) amount to $450 for each lot manufactured.
    (a) Write an equation to express the total manufacturing costs (y) per lot in terms of the number of calculators (x) manufactured per lot.
    (b) Graph the equation obtained in part (a).
    (c) Which points in the graph obtained in part (b) are relevant to the physical situation?
    (d) Use the graph obtained in part (b) to estimate the total manufacturing costs for a lot consisting of 200 calculators.
    (e) Use the equation obtained in part (a) to determine the total manufacturing costs for a lot consisting of 350 calculators.

42. *Laboratory animals* A biologist estimates that 17% of the rats she acquires are not satisfactory for experimental purposes.
    (a) Write an equation to express the number of unsatisfactory rats (y) in terms of the number of rats acquired (x).
    (b) Graph the equation obtained in part (a).
    (c) Which points in the graph obtained in part (b) are relevant to the physical situation?
    (d) Use the graph obtained in part (b) to estimate the number of unsatisfactory rats in a shipment of 50.
    (e) Use the equation obtained in part (a) to estimate the number of unsatisfactory rats in a shipment of 75.

43. *Tractor depreciation* A tractor costing $35,000 has an expected life of 14 years. Using linear depreciation, write equations to express the total depreciation at any time t and the value of the tractor at any time t. Graph both equations.

44. *Car rental* As the result of several weeks of experiments and intensive advertising, a car rental agency has acquired the data given in the table.
    (a) Write an equation to relate the number of cars that will be rented to the daily rental charge, assuming the relationship to be linear.
    (b) Graph the equation obtained in part (a).
    (c) At what rental charge will the number of daily rentals become zero?

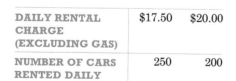

DAILY RENTAL CHARGE (EXCLUDING GAS)	$17.50	$20.00
NUMBER OF CARS RENTED DAILY	250	200

45. *Annual profit* A company has determined that its annual profit y (in hundreds of dollars) from the manufacture of x riding mowers to be given by $y = 4x - 0.01x^2$.
    (a) Graph the profit equation.
    (b) How many mowers should the company manufacture to obtain the maximum profit?
    (c) What is the maximum profit?

46. *Store revenue*   A furniture store can sell 10 sofas per week at $450 each. It can sell an additional 2 sofas for each $10 reduction in price. At what price should the sofas be sold to maximize revenue? What is the maximum revenue? (*Hint:* Let x be the number of $10 reductions. First determine the value of x for which revenue is maximum.)

Identify the steps of the four-step procedure involved in the application of mathematics (Section 1.3) in each case.

47. *Laboratory animals*   Exercise 42

48. *Store revenue*   Exercise 46

CHAPTER 3

SYSTEMS OF LINEAR EQUATIONS—MATRICES

OUTLINE

- **3.1** Systems of Two Linear Equations
- **3.2** Systems of Equations and Matrices
- **3.3** Gauss-Jordan Elimination
- **3.4** Matrix Addition and Scalar Multiplication
- **3.5** Matrix Multiplication
- **3.6** Identity Matrices and Inverses
- ***3.7** Leontief Input-Output Models
- **3.8** Chapter Summary

    Review Exercises

In this chapter we consider methods of obtaining solutions of systems of linear equations and the application of systems of linear equations in solving physical problems. We conclude the chapter with operations with matrices, which are a natural outgrowth of the study of linear equations.

* This section can be omitted without loss of continuity.

3.1 SYSTEMS OF TWO LINEAR EQUATIONS

At the beginning of each month, the owner of a large record store replenishes her stock of records by ordering as many records as she sold the previous month. During May she sold 1,452 records, so she will therefore order 1,452 additional records in June.

A large portion of the customers are classical-music devotees. In order to provide a good selection for these customers, the number of classical records that she orders is in proportion to the number of classical records sold. She estimates that during May the number of classical records sold was twice the number of nonclassical. When she orders, how many of the 1,452 records should be classical and how many should be nonclassical?

For the construction of a mathematical model describing the record problem, let x denote the number of classical and y the number of nonclassical records the owner buys. Since the total number is to be 1,452, x and y must satisfy the relation

$$x + y = 1,452$$

In addition, since the number of classical records must be twice the number of nonclassical records, x and y must also satisfy the equation

$$x = 2y, \quad \text{or} \quad x - 2y = 0$$

These two equations,

$$x + y = 1,452 \quad \text{and} \quad x - 2y = 0$$

together form the mathematical model for solving the record problem. Their graphs are given in Figure 3.1.

Since the number of records to be bought, x and y, must be the coordinates of a point on both of these graphs, they must be the coordinates of the point of intersection of these two lines.

The coordinates of the point of intersection can be approximated from the graph, or they can be obtained exactly by solving the pair of linear equations

$$x + y = 1,452$$
$$x - 2y = 0$$

To obtain this solution, subtract the first equation from the second to eliminate x from the second equation:

$$x + y = 1,452$$
$$-3y = -1,452$$

Multiply both sides of the second equation by $-\frac{1}{3}$ to obtain

$$x + y = 1,452$$
$$y = 484$$

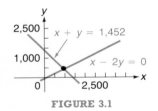

FIGURE 3.1

Then to eliminate y from the first equation, subtract the second from the first to reveal the solution:

$$x = 968$$
$$y = 484$$

The owner should buy 968 classical and 484 nonclassical records.

A collection of two or more equations such as those occurring in the illustration just given, is called a **system of equations.** In this chapter we will be concerned only with systems of linear equations. A **solution** of such a system is a set of values for the variables that satisfies each of the equations. Geometrically, the solution of a system consists of the point(s) that the graphs of the equations have in common.

The method of solution used in the above illustration is an **elimination method.** The object is to eliminate all but one of the variables from each of the equations in order to obtain an **equivalent system** (that is, a system with the same solution as the original system) in which the solution is obvious. You may be familiar with other methods of obtaining solutions; however, we concentrate on the elimination method here because this method will be extended to larger systems in the next section.

There are three legitimate operations that we can perform in the elimination method in order to be sure that the solution of the final system is a solution of the original system:

> 1. Both sides of any equation can be multiplied by a constant other than zero.
> 2. Both sides of any equation can be multiplied by a constant and the result added to one of the other equations.
> 3. The order in which the equations are written can be changed.

EXAMPLE 1 Solve the system of equations:

$$x + 2y = 1$$
$$-2x + y = -2$$

Solution The graphs of the two equations intersect in one point, the coordinates of which give the solution of the system. We want to eliminate one of the variables from the first equation and the other variable from the second. Suppose we choose to eliminate y from the first. To do so we multiply the second equation by -2 and add the result to the first:

$$5x = 5$$
$$-2x + y = -2$$

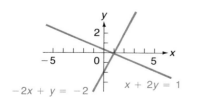

Note that this operation *causes no change in the second equation.* To obtain a coefficient of 1 for x in the first equation, we multiply that equation by $\frac{1}{5}$:

$$x = 1$$
$$-2x + y = -2$$

Finally, to eliminate x from the second equation, we multiply the first equation by 2 and add the result to the second equation:

$$x = 1$$
$$y = 0$$

This final system reveals the solution, which can be verified by substituting these values into both of the original equations.

In Example 1 and the first illustration, the graphs of the two equations intersected in one point, and the coordinates of that point were the solution of the system. The graphs in Figure 3.2 give two other possible arrangements of the graphs of two linear equations. In the coordinate system in Figure 3.2(a), the two lines are parallel and consequently never intersect. The equations of these two lines form a system that has *no* solution.

1 PRACTICE PROBLEM 1

Solve the system of equations

$$x + 2y = 1$$
$$2x - 3y = -12$$

Answer

$x = -3, y = 2$

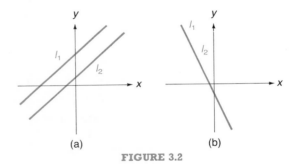

FIGURE 3.2

In the coordinate system in Figure 3.2(b), the two lines coincide, and the coordinates of any point on the line are a solution of the system made up of the equations of the two lines. In this case there are *infinitely many* solutions.

The procedure we have outlined for solving a system of linear equations will reveal whether a system has one, none, or infinitely many solutions. In case of no solutions, the elimination procedure results in an equation that can never be valid. In case of infinitely many solutions, the procedure results in an equation that is always valid. Examples 2 and 3 illustrate these two cases.

EXAMPLE 2 Solve the system of equations:

$$2x - 6y = 4$$
$$-3x + 9y = 4$$

Solution   In order to get a coefficient of 1 for x in the first equation, we multiply that equation by $\frac{1}{2}$:

$$x - 3y = 2$$
$$-3x + 9y = 4$$

To eliminate x from the second equation, we add three times the first equation to the second:

$$x - 3y = 2$$
$$0y = 10$$

The resulting equation, $0y = 10$, has no solution. Consequently, the original system—which is equivalent to the final system—has no solution either.

EXAMPLE 3   Solve the system of equations:

$$0.2x - 0.3y = 1$$
$$1.2x - 1.8y = 6$$

Solution   In order to clear the equations of decimals, we multiply both sides of each equation by 10 to obtain

$$2x - 3y = 10$$
$$12x - 18y = 60$$

Multiplying both sides of the first equation by $\frac{1}{2}$, we get

$$x - \tfrac{3}{2}y = 5$$
$$12x - 18y = 60$$

Multiplying the first equation by -12 and adding it to the second, we get

$$x - \tfrac{3}{2}y = 5$$
$$0y = 0$$

This last equation is true for all values of y, and the first equation gives

$$x = 5 + \tfrac{3}{2}y$$

which has *infinitely many solutions*. For example, if $y = 4$, $x = 11$; if $y = -6$, $x = -4$. The solutions of the system can be expressed by the equation

$$x = 5 + \tfrac{3}{2}y$$

PRACTICE PROBLEM 2

Solve the following systems of equations:

(a) $0.4x + 0.3y = 0.1$
    $1.6x + 1.2y = -2$

(b) $-2x + 6y = -8$
    $x - 3y = 4$

Answer

(a) No solution

(b) $x = 4 + 3y$ or $y = \dfrac{x-4}{3}$

EXAMPLE 4 Solve the system of equations:

$$\frac{x}{4} + \frac{2y}{5} = 4$$

$$\frac{y}{2} - \frac{x}{3} = 5$$

Solution In order to clear the equations of fractions, multiply both sides of each equation by the least common denominator of the fractions involved. For the first equation the least common denominator is 20; multiplying each side by 20 gives

$$\left(20 \cdot \frac{x}{4}\right) + \left(20 \cdot \frac{2y}{5}\right) = 20 \cdot 4$$

or

$$5x + 8y = 80$$

Similarly, multiply each side of the second equation by 6 to obtain

$$3y - 2x = 30$$

The pair of equations

$$5x + 8y = 80$$
$$3y - 2x = 30$$

can then be solved by the elimination method to obtain the solution $x = 0$ and $y = 10$. ∎

The four-step procedure involved in the application of mathematics (Section 1.3) was again used in the illustration at the beginning of this section:

1. The physical situation was the reordering of records; the physical question was, How many of the 1,452 records should be classical and how many nonclassical?
2. The mathematical model was the given system of equations, together with their graphs. The mathematical question was, What is the solution of the system? or What are the coordinates of the point of intersection of the graphs?
3. The mathematical solution was $x = 968$ and $y = 484$.
4. The interpretation was that the owner should buy 968 classical and 484 nonclassical records.

We shall see this process again in the exercises that follow and shall extend it to larger systems in the next section.

EXERCISES

Solve the following systems of equations by the elimination method. (Graph each pair of equations if you find that such a graph is helpful.)

1. $x + y = 1$
   $3x - y = 7$

2. $2x + y = 10$
   $x - 2y = -5$

3. $3x + 4y = 1$
   $2x + y = -1$

4. $\dfrac{x}{2} + \dfrac{y}{3} = 2$
   $\dfrac{x}{4} - \dfrac{y}{6} = 0$

5. $0.3x + 0.9y = 0.3$
   $-1.2x - 3.6y = 0$

6. $\dfrac{y}{2} - \dfrac{x}{2} = 1$
   $x - y = 1$

7. $y - x = 4$
   $2y - 2x = 8$

8. $2x + 2y = 0$
   $0.5x + 0.5y = 0$

9. $3x + y = 0.1$
   $x - 2y = 0.5$

10. Verify that the solution given in Example 4 is correct by verifying that the solution satisfies both of the original equations.

11. *Purchasing*   Eric owns a bicycle shop and wants to order 50 bicycles for the spring rush. The three-speed bicycles cost $60 each and the ten-speeds cost $70 each. He has $3,200 with which to buy the bicycles.
    (a) Let x denote the number of three-speed bicycles he buys and y the number of ten-speed bicycles. Write an equation in x and y to express the fact that he wishes to buy a total of 50 bicycles.
    (b) Write an equation in x and y to express the fact that he plans to spend $3,200 on the bicycles.
    (c) How many of each type of bicycle should Eric buy?

12. *Air pollution*   A certain manufacturing company produces two items, A and B. The manufacturing process emits 1.5 cubic feet of carbon monoxide and 3 cubic feet of sulfur dioxide per unit of item A and 2 cubic feet each of carbon monoxide and sulfur dioxide per unit of item B. Government pollution standards allow the company to emit a maximum of 4,775 cubic feet of carbon monoxide and 5,500 cubic feet of sulfur dioxide per week.
    (a) Let x denote the number of units of item A produced per week and y the number of units of item B. Write an equation to express the number of units of item A and of item B that will cause a total emission of 4,775 cubic feet of carbon monoxide.
    (b) Using the same notation as in part (a), write an equation to express the number of units of item A and of item B that will cause a total emission of 5,500 cubic feet of sulfur dioxide.
    (c) How many units of items A and B can be produced at the maximum allowable government standards?

13. *Diet mix*   The diet of some experimental animals is to consist of exactly 19 grams of protein and 5 grams of fat. The lab technician has available two food mixes, whose protein and fat contents are given in the table. How many grams of each mix should the technician give each animal in order to provide the exact amount of protein and fat?

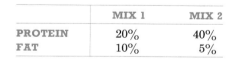

	MIX 1	MIX 2
PROTEIN	20%	40%
FAT	10%	5%

14. *Police personnel*   A city has 104 people on its police force. The number of male police officers is four more than nine times the number of female

police officers. How many male and how many female police officers are on the force?

15. *Economic purchasing* Your company has the following options for buying a particular product:

Option 1: $400 plus $2.00 per item

Option 2: $75 plus $3.00 per item

For what quantities is option 1 the more economical? For what quantities is option 2 the more economical?

16. *Utility rates* The present monthly rate structure and a proposed monthly rate structure of an electric company are given below. Determine the range of kilowatt-hours for which the proposed structure is more economical.

Present: First 60 kilowatt-hours, $6.00
Each additional kilowatt-hour, 6¢ per kilowatt-hour

Proposed: First 40 kilowatt-hours, $4.00
Each additional kilowatt-hour, 8¢ per kilowatt-hour

17. *Break-even point* The cost, in dollars, of manufacturing stereo equipment of a particular type is given by

$$C = 12{,}000 + 120x$$

where x denotes the number of units produced. These units are sold for $200 each; the revenue from x units is therefore

$$R = 200x$$

(a) In a coordinate system, graph the two equations

$$y = 12{,}000 + 120x$$

and

$$y = 200x$$

to represent cost and revenue.

(b) From your graph in part (a), determine the values of x for which cost exceeds revenue and the values of x for which revenue exceeds cost.

(c) The point of intersection of the two graphs in part (a) is called the **break-even point,** as it represents the point at which revenue equals cost. Solve the system of equations to determine the coordinates of the break-even point.

(d) How many units must the company sell in order to break even?

(e) What must be the revenue received in order for the company to break even?

18. *Break-even point* A publishing company is planning to publish a new paperback. Their accountants estimate that it will cost $145,500 to print 500,000 copies and $270,500 to print 1,000,000 copies. They plan to sell the paperback at $2.95 per copy.

(a) Construct an equation that will give the costs in terms of the number of copies printed, assuming a linear cost equation.

(b) Construct an equation that will give the revenue to the company in terms of the number of copies sold.

(c) Graph the equations obtained in parts (a) and (b) in the same coordinate system.

(d) From your graphs in part (c), determine for what number of copies costs exceed revenue and for what number revenue exceeds costs.
(e) Solve the system of equations to determine the break-even point.
(f) How many copies must the company sell just to break even? What will be their revenue at that point?

19. *Break-even point* It costs a company $18,000 plus $15.00 per unit to produce its best-selling product. (The $18,000 is called the **fixed cost** and is constant regardless of the number of units produced; the $15.00 is called the **variable cost**.) The product sells for $30.00 per unit.
    (a) Construct an equation that gives the total cost (y) for producing x units of the product.
    (b) Construct an equation that gives the revenue from the sale of x units of the product.
    (c) Determine the break-even point.
    (d) How many units must the company sell to break even? What will be their revenue at that point?

20. *Break-even point* A company has fixed costs of $10,000 and a variable cost of $6.00 per unit for manufacturing an item that sells for $16.00 per unit. Determine the break-even point, the number of units the company must sell to break even, and the corresponding revenue for the company.

21. *Equilibrium point* In a free marketplace, supply and demand are affected by price, and vice versa. When demand is high, manufacturers will raise the price and increase the supply for greater revenue; when the price is too high, the demand will decrease and manufacturers may find themselves with an excessive supply. The price at which supply and demand are equal is called the **equilibrium price** and the corresponding quantity is called the **equilibrium quantity**. The point at which the graphs of the supply and demand equations intersect is called the **equilibrium point**. Suppose the supply (S) and the demand (D), in hundreds, are given by the linear equations

$$D = -\frac{2}{3}x + 20$$
$$S = 2x - 20$$

where x denotes the price in dollars.
(a) In a coordinate system, graph the two equations

$$y = -\frac{2}{3}x + 20$$
$$y = 2x - 20$$

to represent supply and demand.
(b) Solve the system of equations.
(c) What is the equilibrium price?
(d) What is the equilibrium quantity?

22. *Equilibrium point*
    (a) In Exercise 18, suppose that company officials agree to produce 500,000 copies of the book if the price per copy is to be $2.00 and 681,000 copies if the price is to be $3.00 per copy. At any other price, the number of copies to be produced will vary linearly with these amounts. Construct the supply equation.

(b) Marketing research indicates that 750,000 copies can be sold if the price is $2.00 per copy, and 382,000 copies can be sold if the price is $4.00. Assuming a linear demand equation, construct the demand equation.

(c) Find the equilibrium price. How many copies can be sold at the equilibrium price?

23. *Equilibrium point* The daily demand for y units of a product at x dollars per unit is given by
$$y = 40 - 2x$$
and the supply by
$$y = -240 + 26x$$

(a) Determine the coordinates of the equilibrium point.
(b) What is the equilibrium price?
(c) How many units will be sold at that price?

24. *Equilibrium point* The demand for y units of a product at x dollars per unit is given by
$$y = 100 - 5x$$
and the supply is given by
$$y = 45x$$

(a) Determine the coordinates of the equilibrium point.
(b) How many units must be sold for supply to equal demand?
(c) At what price does supply equal demand?

Identify the steps of the four-step procedure involved in the application of mathematics (Section 1.3) in each case.

25. *Purchasing*   Exercise 11
26. *Air pollution*   Exercise 12
27. *Police personnel*   Exercise 14
28. *Equilibrium point*   Exercise 21

Explain the meaning of each term.

29. A solution for a system of equations
30. Equivalent systems of equations
31. Equilibrium price
32. Equilibrium quantity
33. Break-even point
34. Equilibrium point

35. State the three operations on equations that are allowed in the elimination method.

3.2 SYSTEMS OF EQUATIONS AND MATRICES

In this section and the next, we extend the method of solving a system of two linear equations to larger systems of equations—larger in the sense that they have more than two equations and more than two variables.

In general, a linear equation in the n variables $x_1, x_2, \ldots, x_n$ is an equation that can be written in the form

$$a_1 x_1 + a_2 x_2 + \cdots + a_n x_n = b$$

where $a_1, a_2, \ldots, a_n$ and b are constants. Subscript notation, such as a_1, a_2, x_1, and x_2, is simply a convenience for representing the general form of a series of related variables such as those in the general form of a linear equation. In specific instances the subscripted a's are numbers and the subscripted x's may be replaced by unsubscripted variables, as the following examples of linear equations in several variables illustrate:

$$3x + 4y + 7z = 4$$
$$w + x - 4y + 2z = 3$$
$$2x_1 + 4x_2 + 3x_3 - 2 = 0$$

The important thing to note in a linear equation is that each variable appears only to the first power and no products of the variables, such as xy or $x_1 x_2 x_3$, appear.

A **system** of linear equations is a collection of two or more linear equations, and a **solution** for a system of equations is a set of values for the variables that satisfy each of the equations. The procedure we consider first for solving systems of equations is an extension of the method considered for two equations in the last section. Using the three legitimate operations of the last section, we attempt to eliminate all but one of the variables from each equation.

EXAMPLE 1 Solve the system of equations:

$$2x - y + z = 3$$
$$x + y - 2z = -3$$
$$3x + 2y - 2z = 1$$

Solution We want to use the elimination method of the previous section to obtain a system of equations of the form

$$x = a$$
$$ y = b$$
$$ z = c$$

where a, b, and c are constants.

Since the coefficient of the first variable, x, is 1 in the second equation, we interchange the first and second equations:

$$x + y - 2z = -3$$
$$2x - y + z = 3$$
$$3x + 2y - 2z = 1$$

To eliminate x from the second equation, we multiply the first equation by -2 and add the result to the second equation:

$$x + y - 2z = -3$$
$$-3y + 5z = 9$$
$$3x + 2y - 2z = 1$$

To eliminate x from the third equation, we multiply the first equation by -3 and add the result to the third equation:

$$x + y - 2z = -3$$
$$-3y + 5z = 9$$
$$-y + 4z = 10$$

To obtain a coefficient of 1 for y in the second equation, we multiply that equation by $-\frac{1}{3}$:

$$x + y - 2z = -3$$
$$y - \tfrac{5}{3}z = -3$$
$$-y + 4z = 10$$

To eliminate y from the first equation, multiply the second equation by -1 and add the result to the first equation:

$$x - \tfrac{1}{3}z = 0$$
$$y - \tfrac{5}{3}z = -3$$
$$-y + 4z = 10$$

Then add the second equation to the third to eliminate y from the last equation:

$$x - \tfrac{1}{3}z = 0$$
$$y - \tfrac{5}{3}z = -3$$
$$\tfrac{7}{3}z = 7$$

To obtain a coefficient of 1 for z in the third equation, multiply that equation by $\frac{3}{7}$:

$$x - \tfrac{1}{3}z = 0$$
$$y - \tfrac{5}{3}z = -3$$
$$z = 3$$

Then, using the third equation, eliminate z from the first equation:

$$x = 1$$
$$y - \tfrac{5}{3}z = -3$$
$$z = 3$$

Finally, eliminate z from the second:

$$x = 1$$
$$y = 2$$
$$z = 3$$

This reveals the solution $x = 1$, $y = 2$, and $z = 3$. The fact that this is a solution of the original system can be verified by substituting these values for the variables in each of those equations. ∎

In solving the system of equations in Example 1, the variables never enter into the calculations. All the calculations were done with the coefficients and the constants on the right side of the equations. Consequently, there really is no need to carry the variables along in the calculations. We illustrate how this can be done by again solving the system in Example 1.

We form the matrix, or rectangular array of numbers,

$$\begin{bmatrix} 2 & -1 & 1 & | & 3 \\ 1 & 1 & -2 & | & -3 \\ 3 & 2 & -2 & | & 1 \end{bmatrix}$$

The entries in each row are the coefficients and the constant from the corresponding equation. The operations on the rows of this matrix will be exactly the same as the operations on the equations of the system. This matrix is called the **augmented coefficient matrix** for the system.

For the sake of comparison, we indicate the parallel equation operations along with the operations on the rows of the matrix:

GIVEN SYSTEM OF EQUATIONS   **AUGMENTED COEFFICIENT MATRIX**

$$\begin{aligned} 2x - y + z &= 3 \\ x + y - 2z &= -3 \\ 3x + 2y - 2z &= 1 \end{aligned} \qquad \begin{bmatrix} 2 & -1 & 1 & | & 3 \\ 1 & 1 & -2 & | & -3 \\ 3 & 2 & -2 & | & 1 \end{bmatrix}$$

Since the coefficient of x in the second equation is 1, we interchange the first and second equations (rows):

$$\begin{aligned} x + y - 2z &= -3 \\ 2x - y + z &= 3 \\ 3x + 2y - 2z &= 1 \end{aligned} \qquad \begin{bmatrix} 1 & 1 & -2 & | & -3 \\ 2 & -1 & 1 & | & 3 \\ 3 & 2 & -2 & | & 1 \end{bmatrix}$$

To eliminate x from the second equation, we multiply the first equation (row) by -2 and add the result to the second equation (row):

$$\begin{aligned} x + y - 2z &= -3 \\ -3y + 5z &= 9 \\ 3x + 2y - 2z &= 1 \end{aligned} \qquad \begin{bmatrix} 1 & 1 & -2 & | & -3 \\ 0 & -3 & 5 & | & 9 \\ 3 & 2 & -2 & | & 1 \end{bmatrix}$$

To eliminate x from the third equation, we multiply the first equation (row) by -3 and add the result to the third equation (row):

$$\begin{aligned} x + y - 2z &= -3 \\ -3y + 5z &= 9 \\ -y + 4z &= 10 \end{aligned} \qquad \begin{bmatrix} 1 & 1 & -2 & | & -3 \\ 0 & -3 & 5 & | & 9 \\ 0 & -1 & 4 & | & 10 \end{bmatrix}$$

To obtain a coefficient of 1 for y in the second equation, we multiply the second equation (row) by $-\frac{1}{3}$:

$$\begin{aligned} x + y - 2z &= -3 \\ y - \tfrac{5}{3}z &= -3 \\ -y + 4z &= 10 \end{aligned} \qquad \begin{bmatrix} 1 & 1 & -2 & | & -3 \\ 0 & 1 & -\tfrac{5}{3} & | & -3 \\ 0 & -1 & 4 & | & 10 \end{bmatrix}$$

To eliminate y from the first equation, multiply the second equation (row) by -1 and add the result to the first equation (row):

$$\begin{aligned} x \phantom{{}+{}} - \tfrac{1}{3}z &= 0 \\ y - \tfrac{5}{3}z &= -3 \\ -y + 4z &= 10 \end{aligned} \qquad \begin{bmatrix} 1 & 0 & -\tfrac{1}{3} & | & 0 \\ 0 & 1 & -\tfrac{5}{3} & | & -3 \\ 0 & -1 & 4 & | & 10 \end{bmatrix}$$

Then add the second equation (row) to the third to eliminate y from the last equation (row):

$$\begin{aligned} x \phantom{{}+{}} - \tfrac{1}{3}z &= 0 \\ y - \tfrac{5}{3}z &= -3 \\ \tfrac{7}{3}z &= 7 \end{aligned} \qquad \begin{bmatrix} 1 & 0 & -\tfrac{1}{3} & | & 0 \\ 0 & 1 & -\tfrac{5}{3} & | & -3 \\ 0 & 0 & \tfrac{7}{3} & | & 7 \end{bmatrix}$$

To obtain a coefficient of 1 for z in the third equation, multiply that equation (row) by $\tfrac{3}{7}$:

$$\begin{aligned} x \phantom{{}+{}} - \tfrac{1}{3}z &= 0 \\ y - \tfrac{5}{3}z &= -3 \\ z &= 3 \end{aligned} \qquad \begin{bmatrix} 1 & 0 & -\tfrac{1}{3} & | & 0 \\ 0 & 1 & -\tfrac{5}{3} & | & -3 \\ 0 & 0 & 1 & | & 3 \end{bmatrix}$$

Then, using the third equation (row), eliminate z from the first equation:

$$\begin{aligned} x &= 1 \\ y - \tfrac{5}{3}z &= -3 \\ z &= 3 \end{aligned} \qquad \begin{bmatrix} 1 & 0 & 0 & | & 1 \\ 0 & 1 & -\tfrac{5}{3} & | & -3 \\ 0 & 0 & 1 & | & 3 \end{bmatrix}$$

Finally, eliminate z from the second:

$$\begin{aligned} x &= 1 \\ y &= 2 \\ z &= 3 \end{aligned} \qquad \begin{bmatrix} 1 & 0 & 0 & | & 1 \\ 0 & 1 & 0 & | & 2 \\ 0 & 0 & 1 & | & 3 \end{bmatrix}$$

This last matrix is the augmented coefficient matrix for the final system of equations. The solution for the system can read from the final matrix by constructing the corresponding final system of equations. We will see the procedure used in further examples.

The final matrix in this illustration is said to be in *echelon form*, from which—as just indicated—the solution is easily read. The object of the matrix approach to solving systems of equations is to start with the augmented coefficient matrix for the system and, using row operations similar to the three equation operations, reduce the matrix to echelon form.

3.2 SYSTEMS OF EQUATIONS AND MATRICES

A matrix is in **echelon form** provided each of the following conditions is satisfied:

1. All the rows having only 0 entries, if there are any, appear at the bottom of the matrix.
2. If a row has any nonzero entries, the leading nonzero entry in that row is a 1.
3. In any column containing the leading 1 of a particular row, all the other entries are 0.
4. For any two consecutive nonzero rows, the leading 1 in the lower row is farther to the right than the leading 1 of the upper row.

EXAMPLE 2 Which, if any, of the following matrices are in echelon form?

(a) $\begin{bmatrix} 1 & 2 & 3 & 4 \\ 0 & 1 & 0 & 0 \\ 0 & 0 & 1 & 2 \end{bmatrix}$ (b) $\begin{bmatrix} 1 & 0 & 0 & 2 \\ 0 & 0 & 1 & 0 \\ 0 & 0 & 1 & 1 \end{bmatrix}$ (c) $\begin{bmatrix} 1 & 0 & 0 & 2 \\ 0 & 1 & 0 & 3 \\ 0 & 0 & 2 & 1 \\ 0 & 0 & 0 & 0 \end{bmatrix}$

(d) $\begin{bmatrix} 1 & 0 & 0 \\ 0 & 0 & 0 \\ 0 & 0 & 1 \end{bmatrix}$ (e) $\begin{bmatrix} 0 & 1 & 0 \\ 0 & 0 & 1 \\ 0 & 0 & 0 \\ 0 & 0 & 0 \end{bmatrix}$ (f) $\begin{bmatrix} 2 & 0 & 1 & -2 \\ 0 & 0 & 0 & 0 \\ 0 & 0 & 1 & 3 \\ 0 & 1 & 0 & 4 \end{bmatrix}$

Solution Only the matrix in (e) is in echelon form. The matrix in (a) does not satisfy the third condition. The matrix in (b) does not satisfy the third or fourth condition. The matrix in (c) does not satisfy the second condition. The matrix in (d) does not satisfy the first condition. The matrix in (f) does not satisfy any of the conditions. **1**

PRACTICE PROBLEM 1

(a) Verify that the following matrix satisfies the four conditions required for a matrix to be in echelon form:

$\begin{bmatrix} 0 & 1 & 0 & 2 \\ 0 & 0 & 1 & 0 \\ 0 & 0 & 0 & 0 \end{bmatrix}$

(b) Which of the four conditions are not satisfied by the following matrix?

$\begin{bmatrix} 1 & 2 & 3 & 4 \\ 0 & 0 & 1 & 2 \\ 0 & 1 & 2 & 3 \end{bmatrix}$

Answer

(b) Conditions 3 and 4

As we have just indicated, the objective in solving systems of equations is to convert (or to reduce) the corresponding augmented coefficient matrix to echelon form. To do so we will use row operations corresponding to the three equation operations. These row operations are as follows:

1. Any two rows of a matrix may be interchanged.
2. All entries of any row may be multiplied by a constant other than zero.
3. All entries of any row may be multiplied by a constant and the result added to another row.

Before using the matrix method to solve systems of equations, we will consider the use of the row operations to reduce matrices to echelon form. The general plan of attack will be to obtain the leading 1's, starting with the first row. After a leading 1 is obtained, it is used to obtain 0 entries in the rest of its column. If a row of all 0's is obtained in the process, that row is moved to the bottom.

EXAMPLE 3 Reduce the following matrix to echelon form:
$$\begin{bmatrix} 2 & -1 & 3 \\ 1 & 2 & 3 \\ 4 & -2 & 6 \end{bmatrix}$$

Solution In order to get a 1 in the first position of row 1, interchange rows 1 and 2.
$$\begin{bmatrix} 1 & 2 & 3 \\ 2 & -1 & 3 \\ 4 & -2 & 6 \end{bmatrix}$$

In order to get 0's in the rest of column 1, multiply the first row by -2 and add the result to the second row.
$$\begin{bmatrix} 1 & 2 & 3 \\ 0 & -5 & -3 \\ 4 & -2 & 6 \end{bmatrix}$$

Then multiply the first row by -4 and add the result to the third row.
$$\begin{bmatrix} 1 & 2 & 3 \\ 0 & -5 & -3 \\ 0 & -10 & -6 \end{bmatrix}$$

In order to obtain a leading 1 in the second row, multiply that row by $-\tfrac{1}{5}$.
$$\begin{bmatrix} 1 & 2 & 3 \\ 0 & 1 & \tfrac{3}{5} \\ 0 & -10 & -6 \end{bmatrix}$$

In order to get 0's in the rest of column 2, multiply row 2 by -2 and add the result to the first row.
$$\begin{bmatrix} 1 & 0 & \tfrac{9}{5} \\ 0 & 1 & \tfrac{3}{5} \\ 0 & -10 & -6 \end{bmatrix}$$

Then multiply the second row by 10 and add the result to the third row.
$$\begin{bmatrix} 1 & 0 & \tfrac{9}{5} \\ 0 & 1 & \tfrac{3}{5} \\ 0 & 0 & 0 \end{bmatrix}$$

This last matrix is in echelon form. ■

EXAMPLE 4 Reduce the following matrix to echelon form:
$$\begin{bmatrix} 3 & 0 & 6 & -3 & 3 \\ 6 & 1 & 9 & -6 & 1 \\ 3 & 1 & 3 & -3 & -2 \\ 0 & 2 & 1 & 0 & 1 \end{bmatrix}$$

Solution We follow the general plan of attack discussed before Example 3. The notation on the right describes the row operations stated on the left.

3.2 SYSTEMS OF EQUATIONS AND MATRICES

You may find that using this notation is helpful in doing the exercises; it provides a convenient way of keeping record of the operations performed.

Multiply the first row by $\frac{1}{3}$.
$$\begin{bmatrix} 1 & 0 & 2 & -1 & 1 \\ 6 & 1 & 9 & -6 & 1 \\ 3 & 1 & 3 & -3 & -2 \\ 0 & 2 & 1 & 0 & 1 \end{bmatrix} \quad \frac{1}{3}(R1)$$

Add -6 times the first row to the second row.
$$\begin{bmatrix} 1 & 0 & 2 & -1 & 1 \\ 0 & 1 & -3 & 0 & -5 \\ 3 & 1 & 3 & -3 & -2 \\ 0 & 2 & 1 & 0 & 1 \end{bmatrix} \quad (-6)(R1) + (R2)$$

Add -3 times the first row to the third row.
$$\begin{bmatrix} 1 & 0 & 2 & -1 & 1 \\ 0 & 1 & -3 & 0 & -5 \\ 0 & 1 & -3 & 0 & -5 \\ 0 & 2 & 1 & 0 & 1 \end{bmatrix} \quad (-3)(R1) + (R3)$$

The leading entry in the second row is already 1; multiply the second row by -1 and add the result to the third row.
$$\begin{bmatrix} 1 & 0 & 2 & -1 & 1 \\ 0 & 1 & -3 & 0 & -5 \\ 0 & 0 & 0 & 0 & 0 \\ 0 & 2 & 1 & 0 & 1 \end{bmatrix} \quad (-1)(R2) + (R3)$$

Interchange the third and fourth rows.
$$\begin{bmatrix} 1 & 0 & 2 & -1 & 1 \\ 0 & 1 & -3 & 0 & -5 \\ 0 & 2 & 1 & 0 & 1 \\ 0 & 0 & 0 & 0 & 0 \end{bmatrix} \quad (R3) \leftrightarrow (R4)$$

Multiply the second row by -2 and add the result to the third.
$$\begin{bmatrix} 1 & 0 & 2 & -1 & 1 \\ 0 & 1 & -3 & 0 & -5 \\ 0 & 0 & 7 & 0 & 11 \\ 0 & 0 & 0 & 0 & 0 \end{bmatrix} \quad (-2)(R2) + (R3)$$

Multiply the third row by $\frac{1}{7}$.
$$\begin{bmatrix} 1 & 0 & 2 & -1 & 1 \\ 0 & 1 & -3 & 0 & -5 \\ 0 & 0 & 1 & 0 & \frac{11}{7} \\ 0 & 0 & 0 & 0 & 0 \end{bmatrix} \quad \frac{1}{7}(R3)$$

Multiply the third row by -2 and add the result to the first.
$$\begin{bmatrix} 1 & 0 & 0 & -1 & -\frac{15}{7} \\ 0 & 1 & -3 & 0 & -5 \\ 0 & 0 & 1 & 0 & \frac{11}{7} \\ 0 & 0 & 0 & 0 & 0 \end{bmatrix} \quad (-2)(R3) + (R1)$$

Multiply the third row by 3 and add the result to the second.

$$\begin{bmatrix} 1 & 0 & 0 & -1 & -\frac{15}{7} \\ 0 & 1 & 0 & 0 & -\frac{2}{7} \\ 0 & 0 & 1 & 0 & \frac{11}{7} \\ 0 & 0 & 0 & 0 & 0 \end{bmatrix} \quad 3(R3) + (R2)$$

This final matrix is in echelon form.

② PRACTICE PROBLEM 2

Reduce each of the following to echelon form:

(a) $\begin{bmatrix} 1 & 2 & 0 \\ 2 & 5 & -2 \\ 0 & 0 & 2 \end{bmatrix}$ (b) $\begin{bmatrix} 2 & 4 & 6 \\ 2 & 7 & 3 \\ 0 & -1 & 1 \end{bmatrix}$

Answer

(a) $\begin{bmatrix} 1 & 0 & 0 \\ 0 & 1 & 0 \\ 0 & 0 & 1 \end{bmatrix}$ (b) $\begin{bmatrix} 1 & 0 & 5 \\ 0 & 1 & -1 \\ 0 & 0 & 0 \end{bmatrix}$

EXERCISES

Which of the following matrices are in echelon form? For those that are not, indicate which of the four required conditions is (are) not satisfied.

1. $\begin{bmatrix} 1 & 0 & -2 \\ 0 & 1 & 3 \\ 0 & 0 & 0 \end{bmatrix}$

2. $\begin{bmatrix} 1 & 0 & 0 & 2 \\ 0 & 1 & 0 & -1 \\ 0 & 0 & 1 & 0 \end{bmatrix}$

3. $\begin{bmatrix} 1 & 0 & 0 & -1 \\ 0 & 1 & 0 & -2 \\ 0 & 1 & 0 & -3 \\ 0 & 0 & 1 & -4 \end{bmatrix}$

4. $\begin{bmatrix} 0 & 1 & 0 & 0 \\ 0 & 0 & 1 & 0 \\ 0 & 0 & 0 & 1 \\ 0 & 0 & 0 & 0 \end{bmatrix}$

5. $\begin{bmatrix} 1 & 2 & -3 & 4 \\ 0 & 1 & 11 & 1 \end{bmatrix}$

6. $\begin{bmatrix} 0 & 1 \\ 0 & 0 \\ 0 & 0 \end{bmatrix}$

7. $\begin{bmatrix} 1 & 0 & 0 \\ 0 & 1 & 0 \\ 0 & 0 & 1 \end{bmatrix}$

8. $\begin{bmatrix} 2 & 0 & 18 & -2 \\ 0 & 0 & 0 & 0 \\ 0 & 0 & 1 & 18 \end{bmatrix}$

9. $\begin{bmatrix} 0.1 & 0 \\ 0 & 0.1 \end{bmatrix}$

Reduce each of the following matrices to echelon form:

10. $\begin{bmatrix} 2 & 4 \\ 3 & -1 \end{bmatrix}$

11. $\begin{bmatrix} 1 & 3 & -1 \\ 2 & 5 & -3 \end{bmatrix}$

12. $\begin{bmatrix} 0.2 & 1.5 \\ 0.1 & 0.3 \end{bmatrix}$

13. $\begin{bmatrix} 24 & -32 \\ -18 & 24 \end{bmatrix}$

14. $\begin{bmatrix} 3 & 8 & 2 & -1 \\ 1 & 2 & 0 & 1 \\ 4 & 6 & 2 & 0 \end{bmatrix}$

15. $\begin{bmatrix} -2 & 8 & 0 \\ -3 & 10 & 2 \\ 2 & -11 & 0 \end{bmatrix}$

16. $\begin{bmatrix} 2 & 0 & -2 & 0 \\ 0 & 1 & 0 & 1 \\ 0 & -1 & 1 & 2 \\ 3 & 0 & -3 & -1 \end{bmatrix}$

17. $\begin{bmatrix} 5 & 6 & -3 \\ 2 & 0 & 1 \\ 9 & 6 & 1 \\ 7 & 6 & -2 \end{bmatrix}$

18. $\begin{bmatrix} 1 & 2 & 3 & 4 \\ 5 & 6 & 7 & 8 \\ 9 & 10 & 11 & 12 \end{bmatrix}$

19. $\begin{bmatrix} 2 & 4 & 6 & 8 \\ 1 & 2 & -2 & -2 \\ 0 & 1 & 1 & 1 \end{bmatrix}$

20. State the three row operations allowed in reducing a matrix to echelon form.

Explain the meaning of each term.

21. A linear equation in the variables $x_1, x_2, \ldots, x_n$
22. A system of linear equations
23. The solution for a system of equations
24. A matrix
25. The augmented coefficient matrix of a system of equations
26. A matrix in echelon form.

3.3 GAUSS-JORDAN ELIMINATION

We return now to using our matrix considerations in solving systems of linear equations and illustrate the method by solving the following system:

$$\begin{aligned} 2x - 3y + z &= 2 \\ x + 6y + 3z &= 1 \\ 6y + z &= 1 \end{aligned}$$

Its augmented coefficient matrix is

$$\begin{bmatrix} 2 & -3 & 1 & | & 2 \\ 1 & 6 & 3 & | & 1 \\ 0 & 6 & 1 & | & 1 \end{bmatrix}$$

which we must reduce to echelon form.

To take advantage of the leading 1 in the second row, interchange the first two rows.

$$\begin{bmatrix} 1 & 6 & 3 & | & 1 \\ 2 & -3 & 1 & | & 2 \\ 0 & 6 & 1 & | & 1 \end{bmatrix} \quad (R1) \leftrightarrow (R2)$$

Multiply row 1 by -2 and add the result to row 2.

$$\begin{bmatrix} 1 & 6 & 3 & | & 1 \\ 0 & -15 & -5 & | & 0 \\ 0 & 6 & 1 & | & 1 \end{bmatrix} \quad (-2)(R1) + (R2)$$

Multiply the second row by $-\frac{1}{15}$.

$$\begin{bmatrix} 1 & 6 & 3 & | & 1 \\ 0 & 1 & \frac{1}{3} & | & 0 \\ 0 & 6 & 1 & | & 1 \end{bmatrix} \quad (-\tfrac{1}{15})(R2)$$

Multiply the second row by -6 and add the result to the first row.

$$\begin{bmatrix} 1 & 0 & 1 & | & 1 \\ 0 & 1 & \frac{1}{3} & | & 0 \\ 0 & 6 & 1 & | & 1 \end{bmatrix} \quad (-6)(R2) + (R1)$$

Add the same multiple of row 2 to row 3.
$$\begin{bmatrix} 1 & 0 & 1 & | & 1 \\ 0 & 1 & \frac{1}{3} & | & 0 \\ 0 & 0 & -1 & | & 1 \end{bmatrix}$$
$(-6)(R2) + (R3)$

Multiply the last row by -1.
$$\begin{bmatrix} 1 & 0 & 1 & | & 1 \\ 0 & 1 & \frac{1}{3} & | & 0 \\ 0 & 0 & 1 & | & -1 \end{bmatrix}$$
$(-1)(R3)$

Multiply the third row by -1 and add the result to the first row.
$$\begin{bmatrix} 1 & 0 & 0 & | & 2 \\ 0 & 1 & \frac{1}{3} & | & 0 \\ 0 & 0 & 1 & | & -1 \end{bmatrix}$$
$(-1)(R3) + (R1)$

Multiply the last row by $-\frac{1}{3}$ and add the result to the second row.
$$\begin{bmatrix} 1 & 0 & 0 & | & 2 \\ 0 & 1 & 0 & | & \frac{1}{3} \\ 0 & 0 & 1 & | & -1 \end{bmatrix}$$
$(-\frac{1}{3})(R3) + (R2)$

This last matrix, in echelon form, is the augmented coefficient matrix of the system

$$\begin{aligned} x &= 2 \\ y &= \tfrac{1}{3} \\ z &= -1 \end{aligned}$$

which has the obvious solution.

1 PRACTICE PROBLEM 1

Solve by reducing the augmented coefficient matrix to echelon form.

$$\begin{aligned} x + y + z &= 2 \\ 2x + 3y - z &= -3 \\ x + 2y + 3z &= 5 \end{aligned}$$

Answer

$x = 1, y = -1, z = 2$

The following examples illustrate how the system of equations corresponding to given augmented coefficient matrices (in echelon form) are constructed and how the solution for the system is determined. In each case, the variables in the equations are $w, x, y,$ and z in that order.

EXAMPLE 1 Determine the solution for the system of equations if the final matrix, in echelon form, is

$$\begin{bmatrix} 1 & 0 & 0 & 2 & | & 3 \\ 0 & 1 & 0 & -1 & | & 4 \\ 0 & 0 & 1 & 4 & | & -1 \\ 0 & 0 & 0 & 0 & | & 0 \end{bmatrix}$$

Solution   The corresponding system of equations is

$$\begin{aligned} w \quad\quad\quad\quad + 2z &= 3 \\ x \quad\quad - z &= 4 \\ y + 4z &= -1 \\ 0z &= 0 \end{aligned}$$

This system has *infinitely many solutions,* one for any given value of z. The solution of the system is, therefore, usually written in terms of z as

$$w = 3 - 2z, \quad x = 4 + z, \quad y = -1 - 4z$$

3.3 GAUSS-JORDAN ELIMINATION

If z is given the value 1, for example, then $w = 1$, $x = 5$, $y = -5$, and $z = 1$ is one solution. ∎

EXAMPLE 2 Determine the solution for the system of equations if the final matrix, in echelon form, is

$$\begin{bmatrix} 1 & 0 & 0 & 4 & | & -3 \\ 0 & 1 & 0 & 2 & | & 4 \\ 0 & 0 & 1 & -1 & | & 1 \\ 0 & 0 & 0 & 0 & | & 1 \end{bmatrix}$$

Solution The corresponding system of equations is

$$\begin{aligned} w + 4z &= -3 \\ x + 2z &= 4 \\ y - z &= 1 \\ 0z &= 1 \end{aligned}$$

Because this last equation has no solution, the system has *no* solution. ∎

The method of solution we have been considering is called **Gauss-Jordan elimination** after the German mathematician Carl F. Gauss (1777–1855) and the French mathematician Camille Jordan (1838–1922).

2 PRACTICE PROBLEM 2

If the variables are x, y, and z, determine the solution of the system when the final matrix in echelon form is as shown.

(a) $\begin{bmatrix} 1 & 0 & 0 & | & -2 \\ 0 & 1 & 2 & | & 3 \\ 0 & 0 & 0 & | & 1 \end{bmatrix}$

(b) $\begin{bmatrix} 1 & 0 & 4 & | & -3 \\ 0 & 1 & 0 & | & 6 \\ 0 & 0 & 0 & | & 0 \end{bmatrix}$

Answer

(a) No solution
(b) $x = -3 - 4z$, $y = 6$

EXERCISES

Determine the solution of each system, given the following echelon matrices for systems of equations in the variables w, x, y, and z.

1. $\begin{bmatrix} 1 & 0 & 0 & 0 & | & -\frac{1}{2} \\ 0 & 1 & 0 & 0 & | & 2 \\ 0 & 0 & 1 & 0 & | & 3 \\ 0 & 0 & 0 & 1 & | & -1 \end{bmatrix}$

2. $\begin{bmatrix} 1 & 0 & 0 & 0 & | & 0 \\ 0 & 0 & 1 & 0 & | & 0 \\ 0 & 0 & 0 & 1 & | & 0 \\ 0 & 0 & 0 & 0 & | & 1 \end{bmatrix}$

3. $\begin{bmatrix} 1 & 0 & 2 & 0 & | & -1 \\ 0 & 1 & -2 & 0 & | & 8 \\ 0 & 0 & 0 & 1 & | & 2 \\ 0 & 0 & 0 & 0 & | & 0 \end{bmatrix}$

4. $\begin{bmatrix} 1 & 2 & -1 & 0 & | & 4 \\ 0 & 0 & 0 & 1 & | & -3 \\ 0 & 0 & 0 & 0 & | & 0 \\ 0 & 0 & 0 & 0 & | & 0 \end{bmatrix}$

Determine the solution of each system, given the following matrices for systems of equations in the variables x_1, x_2, and x_3:

5. $\begin{bmatrix} 1 & 0 & 0 & | & 0.5 \\ 0 & 1 & 0 & | & -\frac{1}{3} \\ 0 & 0 & 1 & | & 0 \end{bmatrix}$

6. $\begin{bmatrix} 1 & -3 & 0 & | & 2 \\ 0 & 0 & 1 & | & -3 \\ 0 & 0 & 0 & | & 0 \end{bmatrix}$

7. $\begin{bmatrix} 1 & 0 & 0 & | & \frac{1}{8} \\ 0 & 1 & 0 & | & 0 \\ 0 & 0 & 1 & | & -\frac{1}{3} \\ 0 & 0 & 0 & | & 0 \end{bmatrix}$

8. $\begin{bmatrix} 1 & 2 & 3 & | & 4 \\ 0 & 0 & 0 & | & 0 \\ 0 & 0 & 0 & | & 0 \end{bmatrix}$

Solve the following systems of equations using the Gauss-Jordan method:

9. $\begin{aligned} 2x - y + z &= 2 \\ -x + 2y + z &= 3 \\ x + y + z &= 4 \end{aligned}$

10. $\begin{aligned} 4x - 3y &= 1 \\ 2x + 4z &= 2 \\ 3y - 8z &= -1 \end{aligned}$

11. $\begin{aligned} 2x - 6y &= -1 \\ 3x + 4y &= 5 \end{aligned}$

12. $\begin{aligned} 2x - y + z &= 0 \\ -x + 2y &= 3 \\ x + y + z &= 3 \end{aligned}$

13. $\begin{aligned} x - 2y + z &= 0 \\ 2x + 3y - 2z &= 0 \\ -x + y + 4z &= 0 \end{aligned}$

14. $\begin{aligned} 2x + 5y &= 0 \\ -3x + 4y &= 0 \end{aligned}$

15. $\begin{aligned} x - 5y &= 3 \\ -2x + 10y &= -6 \end{aligned}$

16. $\begin{aligned} 3x - 6y &= 9 \\ 2x + 4y - 4z &= 1 \\ 7x + 2y - 8z &= 11 \end{aligned}$

17. $\begin{aligned} 5x + 3y + 4z &= 3 \\ -2x + 4y &= 6 \\ 3x + 7y + 4z &= 0 \end{aligned}$

18. $\begin{aligned} 2x - y + z &= -2 \\ 4x + 2y - z &= 10 \\ -3x + 2z &= 1 \end{aligned}$

19. $\begin{aligned} 3x_1 + x_2 - x_3 + x_4 &= 4 \\ 2x_2 + x_3 - 2x_4 &= -5 \\ -3x_1 - 2x_2 + 2x_3 &= -3 \\ 6x_1 + 3x_4 &= 8 \end{aligned}$

20. $\begin{aligned} w - 2y + z &= 4 \\ x - y - 2z &= -3 \\ w + x - 3y - z &= 1 \\ w - 2x + 5z &= 10 \end{aligned}$

21. *Purchasing* Solve Exercise 11(c) of Section 3.1 using the method of this section.

22. *Diet mix* Solve Exercise 13 of Section 3.1 using the method of this section.

23. *Police personnel* Solve Exercise 14 of Section 3.1 using the method of this section.

24. *Break-even point* Solve Exercise 17(c) of Section 3.1 using the method of this section.

25. *Break-even point* Solve Exercise 19(c) of Section 3.1 using the method of this section.

26. *Equilibrium point* Solve Exercise 23(a) of Section 3.1 using the method of this section.

27. *Process utilization* Suppose a company manufactures three products A, B, and C, and that each item requires three processes, assembling, painting, and packaging. The total number of worker-hours per day available for these three processes are assembling, 80 hours, painting, 52 hours, and packaging, 30 hours. The number of worker-hours required for processing each item is indicated in the following array:

	A	B	C
Assembling	2	1	1
Painting	1	1	0.5
Packaging	0.5	0.5	0.5

Determine the number of items of each of the products A, B, and C that should be produced each day if all of the available worker-hours are to be used.

Recall that one has to know the coordinates of only two points on a line to determine the equation of the whole line. In the case of a parabola, one has to know only the coordinates of three points on the parabola in order to determine its equation. The procedure is outlined in Exercise 28.

28. Determine the equation of the parabola passing through the three points (2, 9), (1, 4), and (0, 1). The coordinates of these points will satisfy the quadratic equation $y = Ax^2 + Bx + C$, whose graph is the parabola containing these three points. For example, using the coordinates of the first point gives

$$9 = 4A + 2B + C$$

   (a) Using the coordinates of the two remaining points, determine two more equations in A, B, and C.
   (b) Solve the system made up of these three equations for the values of A, B, and C.
   (c) Substitute the values of A, B, and C into $y = Ax^2 + Bx + C$ to get the equation of the parabola.

29. Determine the equation of the parabola passing through the three points $(1, -1)$, $(0, -2)$ and $(-1, 1)$. (*Hint:* See Exercise 28.)

30. *Fuel cost*   Joan has a small company, which makes clay pots. The process of making the pots requires that they be baked in a heated oven. A batch, or lot, of pots is placed in the oven for several hours; during this time the oven is kept at a constant temperature. Increasing fuel costs have made Joan very conscious of the expense involved in heating the oven. Too much heat is wasted when a lot contains just a few pots. At the other extreme, a considerable amount of additional heat is required to maintain the oven at the correct temperature when the lot size is very large. Somewhere in between should be the ideal lot size at which the cost per pot is minimal. But what is this ideal size? Joan carefully calculated the amount of gas used for lot sizes of 5, 10, and 15 pots. She then calculated the cost per pot and made a chart of these costs from her calculations:

LOT SIZE	COST PER POT
5	$2.60
10	1.35
15	0.60

   (a) Determine the quadratic equation that contains the three corresponding points in its graph.
   (b) What lot size gives the minimum fuel cost per pot?
   (c) What is the minimum fuel cost per pot?
   (*Hint:* See Exercise 28.)

Solve each system of equations.

31. $x + 2y + z = 0$
    $x + 3y - 2z = 1$

32. $x + 3y - z + w = 1$
    $x + 4y + 2z + 3w = -1$
    $2x + 6y - 3z - w = 3$

33. $\begin{aligned} x + 2y + z &= 0 \\ x - 3y - 2z &= 5 \\ 2x - y + 2z &= -10 \\ 4x - 2y + z &= 3 \end{aligned}$

34. $\begin{aligned} 2x - y &= 4 \\ x + 3y &= -3 \\ 4x + 5y &= -2 \end{aligned}$

35. $\begin{aligned} 2x - y + z &= 1 \\ -x + y - 2z &= 2 \\ x - z &= 3 \\ y - 3z &= 5 \end{aligned}$

36. In view of Exercises 33–35, what can be said about the number of solutions when the number of equations is greater than the number of variables?

37. In view of Exercises 31 and 32, what can be said about the number of solutions when the number of equations is less than the number of variables?

Identify the steps of the four-step procedure involved in the application of mathematics (Section 1.3) in each case.

38. *Fuel cost*   Exercise 30

39. *Process utilization*   Exercise 27

3.4 MATRIX ADDITION AND SCALAR MULTIPLICATION

In the last section, we used matrices to solve systems of equations. Matrices are used in a variety of other situations as well. Sometimes they provide a convenient method of presenting information. Team standings, such as those of the National Hockey League or of the American Baseball League, are usually presented in matrix form; the daily report of the activity of the New York Stock Exchange appearing in many newspapers forms a huge matrix. At other times matrices provide a convenient method of constructing mathematical models and/or performing calculations. In Section 3.7 and again in Chapter 6, we consider situations in which matrices are used to solve physical problems. In this section and the next, we consider some of the operations with matrices used in calculations.

These operations are of three types: addition and subtraction of matrices, multiplication of matrices by a constant, and multiplication of two matrices.

First, some notation. When it is convenient to denote a matrix by a single letter, capital letters are used. For example, we might denote the matrix

$$\begin{bmatrix} 2 & 6 \\ 3 & 1 \\ 4 & -7 \end{bmatrix}$$

by a single letter A and write

$$A = \begin{bmatrix} 2 & 6 \\ 3 & 1 \\ 4 & -7 \end{bmatrix}$$

When a capital letter is used to denote a matrix, the corresponding lowercase letter is used to denote the elements of the matrix; subscripts on the lowercase letters are used to denote the row and the column of a particular element. In matrix A,

$$a_{21} = 3$$

because 3 is the element in the second row and first column;

$$a_{32} = -7$$

because -7 is the element in the third row and second column. What are a_{13}, a_{23}, and a_{12} in matrix A? In general, a_{ij} denotes the element in the ith row and jth column.

If

$$B = \begin{bmatrix} 6 & -2 & -1 & 3 \\ 11 & 0 & 4 & -4 \\ 6 & 1 & 2 & 9 \end{bmatrix}$$

then $b_{12} = -2$, $b_{34} = 9$, $b_{21} = 11$, and $b_{23} = 4$.

To emphasize the elements of the matrix, the notation $A = [a_{ij}]$ or $B = [b_{ij}]$ is sometimes used. For our matrices A and B, we could write

$$A = [a_{ij}] = \begin{bmatrix} 2 & 6 \\ 3 & 1 \\ 4 & -7 \end{bmatrix}$$

and

$$B = [b_{ij}] = \begin{bmatrix} 6 & -2 & -1 & 3 \\ 11 & 0 & 4 & -4 \\ 6 & 1 & 2 & 9 \end{bmatrix}$$

Finally, the **dimension** of a matrix indicates the number of rows and number of columns of the matrix. For the two matrices just considered, A is a 3 by 2 (or 3×2) matrix because A has 3 rows and 2 columns. Similarly, B is a 3×4 matrix. In general, an $m \times n$ matrix is a matrix with m rows and n columns. Note that the number of rows is always given first. When the dimension of a matrix is to be emphasized, the notation $A_{m \times n}$ is sometimes used. Examples of this notation are

$$A_{3 \times 2}, \quad B_{3 \times 4}, \quad C_{6 \times 1}$$

where, as always, the first subscript indicates the number of rows in the matrix and the second indicates the number of columns.

If A and B are two matrices each of dimension $m \times n$, their **sum (difference)** is defined to be the $m \times n$ matrix C in which each element is the sum (difference) of the corresponding elements of A and B. For example, if

$$A = \begin{bmatrix} 1 & -2 & 3 \\ 6 & 4 & -7 \\ 9 & -1 & 8 \end{bmatrix} \quad \text{and} \quad B = \begin{bmatrix} 4 & 6 & 9 \\ -2 & 1 & 7 \\ 4 & -1 & 0 \end{bmatrix}$$

then

$$A + B = \begin{bmatrix} 1 & -2 & 3 \\ 6 & 4 & -7 \\ 9 & -1 & 8 \end{bmatrix} + \begin{bmatrix} 4 & 6 & 9 \\ -2 & 1 & 7 \\ 4 & -1 & 0 \end{bmatrix}$$

$$= \begin{bmatrix} 1+4 & -2+6 & 3+9 \\ 6-2 & 4+1 & -7+7 \\ 9+4 & -1-1 & 8+0 \end{bmatrix} = \begin{bmatrix} 5 & 4 & 12 \\ 4 & 5 & 0 \\ 13 & -2 & 8 \end{bmatrix}$$

Also,

$$A - B = \begin{bmatrix} 1 & -2 & 3 \\ 6 & 4 & -7 \\ 9 & -1 & 8 \end{bmatrix} - \begin{bmatrix} 4 & 6 & 9 \\ -2 & 1 & 7 \\ 4 & -1 & 0 \end{bmatrix}$$

$$= \begin{bmatrix} 1-4 & -2-6 & 3-9 \\ 6+2 & 4-1 & -7-7 \\ 9-4 & -1+1 & 8-0 \end{bmatrix} = \begin{bmatrix} -3 & -8 & -6 \\ 8 & 3 & -14 \\ 5 & 0 & 8 \end{bmatrix}$$

In terms of the notation introduced above, the sum and difference of two matrices are defined as follows. If $A_{m \times n} = [a_{ij}]$ and $B_{m \times n} = [b_{ij}]$, then

$$A + B = C \quad \text{and} \quad A - B = D$$

where $C = [c_{ij}]$ and $D = [d_{ij}]$ are $m \times n$ matrices, $c_{ij} = a_{ij} + b_{ij}$, and $d_{ij} = a_{ij} - b_{ij}$. Note that the sum and difference are defined only for matrices of the same dimension. For two matrices of different dimensions, these operations are undefined and consequently have no meaning.

EXAMPLE 1 For

$$A = \begin{bmatrix} 2 & 6 & 1 \\ -4 & 3 & -2 \end{bmatrix}, \quad B = \begin{bmatrix} 1 & -1 \\ 2 & 0 \end{bmatrix}, \quad C = \begin{bmatrix} 1 & 2 & 3 \\ -4 & -5 & -6 \end{bmatrix}$$

compute each of the following, if defined:

(a) $A + B$ (b) $A - C$ (c) $C - B$ (d) $C + A$

Solution (a) $A + B$ is not defined because A and B are not of the same dimension.

(b) $A - C = \begin{bmatrix} 2-1 & 6-2 & 1-3 \\ -4+4 & 3+5 & -2+6 \end{bmatrix} = \begin{bmatrix} 1 & 4 & -2 \\ 0 & 8 & 4 \end{bmatrix}$

(c) $C - B$ is not defined because C and B are not of the same dimension.

3.4 MATRIX ADDITION AND SCALAR MULTIPLICATION

1 PRACTICE PROBLEM 1

For

$$A = \begin{bmatrix} 2 & -1 \\ -3 & 4 \\ 0 & -2 \end{bmatrix}$$

and

$$B = \begin{bmatrix} 1 & 1 \\ 0 & -2 \\ 2 & -3 \end{bmatrix}$$

compute:

(a) $A + B$ (b) $B - A$

Answer

(a) $\begin{bmatrix} 3 & 0 \\ -3 & 2 \\ 2 & -5 \end{bmatrix}$ (b) $\begin{bmatrix} -1 & 2 \\ 3 & -6 \\ 2 & -1 \end{bmatrix}$

(d) $C + A = \begin{bmatrix} 1+2 & 2+6 & 3+1 \\ -4-4 & -5+3 & -6-2 \end{bmatrix} = \begin{bmatrix} 3 & 8 & 4 \\ -8 & -2 & -8 \end{bmatrix}$ ■

The **product** of a number k times the $m \times n$ matrix A is the $m \times n$ matrix B, in which each entry is k times the corresponding entry in A. For example, if $k = 3$ and

$$A = \begin{bmatrix} 3 & 2 & 1 \\ 0 & 1 & 2 \\ 3 & 4 & 5 \end{bmatrix}$$

then

$$3A = \begin{bmatrix} 3 \times 3 & 3 \times 2 & 3 \times 1 \\ 3 \times 0 & 3 \times 1 & 3 \times 2 \\ 3 \times 3 & 3 \times 4 & 3 \times 5 \end{bmatrix} = \begin{bmatrix} 9 & 6 & 3 \\ 0 & 3 & 6 \\ 9 & 12 & 15 \end{bmatrix}$$

Each element in A is multiplied by 3.

In a matrix setting, the numbers that form the entries of the matrices (for us, the real numbers) are called **scalars**. The type of multiplication that we have been discussing is then called **scalar multiplication**. In terms of the notation introduced earlier, we can write the definition of scalar multiplication as follows. If c is a scalar and $A_{m \times n} = [a_{ij}]$, then

$$cA = B$$

where $B = [b_{ij}]$ is the $m \times n$ matrix in which

$$b_{ij} = ca_{ij}$$

EXAMPLE 2 For

$$A = \begin{bmatrix} 1 & -1 & 6 \\ 3 & 2 & 4 \\ 1 & 7 & -1 \end{bmatrix}, \quad B = \begin{bmatrix} 1 & 2 \\ 3 & 4 \\ 5 & 6 \end{bmatrix}, \quad C = \begin{bmatrix} -1 & 2 \\ 4 & -3 \\ 7 & 0 \end{bmatrix}$$

compute each of the following, if defined:

(a) $2A$ (b) $-3B$ (c) $2A - C$ (d) $B + 2C$ (e) $4(B + C)$

Solution (a) $2A = \begin{bmatrix} 2 & -2 & 12 \\ 6 & 4 & 8 \\ 2 & 14 & -2 \end{bmatrix}$ (b) $-3B = \begin{bmatrix} -3 & -6 \\ -9 & -12 \\ -15 & -18 \end{bmatrix}$

(c) $2A$ is a 3×3 matrix and C is a 2×3 matrix. Since they are of different dimensions, $2A - C$ is not defined.

(d) $2C = \begin{bmatrix} -2 & 4 \\ 8 & -6 \\ 14 & 0 \end{bmatrix}$, so $B + 2C = \begin{bmatrix} 1 & 2 \\ 3 & 4 \\ 5 & 6 \end{bmatrix} + \begin{bmatrix} -2 & 4 \\ 8 & -6 \\ 14 & 0 \end{bmatrix}$

$$= \begin{bmatrix} -1 & 6 \\ 11 & -2 \\ 19 & 6 \end{bmatrix}$$

2 PRACTICE PROBLEM 2

For

$$A = \begin{bmatrix} 1 & -2 \\ 3 & 4 \end{bmatrix}$$

and

$$B = \begin{bmatrix} -1 & 0 \\ 2 & 1 \end{bmatrix}$$

compute:

(a) $5A$ (b) $3(A - B)$
(c) $2A + B$

Answer

(a) $\begin{bmatrix} 5 & -10 \\ 15 & 20 \end{bmatrix}$ (b) $\begin{bmatrix} 6 & -6 \\ 3 & 9 \end{bmatrix}$

(c) $\begin{bmatrix} 1 & -4 \\ 8 & 9 \end{bmatrix}$

(e) $4(B + C) = 4 \begin{bmatrix} 0 & 4 \\ 7 & 1 \\ 12 & 6 \end{bmatrix} = \begin{bmatrix} 0 & 16 \\ 28 & 4 \\ 48 & 24 \end{bmatrix}$

We have used an equal sign with matrices either to introduce notation or to apply a definition. We also want to use an equal sign to compare two matrices—for example, does $A = B$? In what sense are two matrices equal? Two matrices A and B of the same dimension are said to be **equal** if each of their corresponding entries are the same: If $A_{m \times n} = [a_{ij}]$ and $B_{m \times n} = [b_{ij}]$, then

$$A = B$$

means $a_{ij} = b_{ij}$ for every i and j.

For example, if

$$\begin{bmatrix} 6 & 1 \\ 2 & -3 \\ -1 & 4 \end{bmatrix} = \begin{bmatrix} 6 & 1 \\ x & y \\ -1 & z \end{bmatrix}$$

then $x = 2$, $y = -3$, and $z = 4$, which can be determined by comparing corresponding entries.

EXAMPLE 3   If

$$\begin{bmatrix} -5 & 0 \\ y & 3z \end{bmatrix} = \begin{bmatrix} 2x + 3 & 0 \\ 4 & z + 1 \end{bmatrix}$$

determine the values of x, y, and z.

Solution   Comparing corresponding entries, we get $2x + 3 = -5$, which gives $x = -4$; $y = 4$; and $3z = z + 1$, which gives $z = \frac{1}{2}$. ■

Some properties of the operations that we have been discussing are given below. We shall not attempt to prove them here; hopefully, they seem reasonable in view of the definitions of the operations and the properties of the real numbers.

If A, B, and C are matrices and c and d are scalars, then:

A1. $A + B = B + A$ Commutative law of addition
A2. $A + (B + C) = (A + B) + C$ Associative law of addition
S1. $(c + d)A = cA + dA$
S2. $c(A + B) = cA + cB$
S3. $c(dA) = (cd)A$

These statements all presume that the indicated operations are defined.

EXAMPLE 4 Verify that property A2 holds for

$$A = \begin{bmatrix} 1 & 2 \\ -2 & 3 \end{bmatrix}, \quad B = \begin{bmatrix} 4 & -5 \\ 1 & 0 \end{bmatrix}, \quad C = \begin{bmatrix} 2 & 4 \\ -6 & -8 \end{bmatrix}$$

Solution   In this case, the left side of A2 is

$$A + (B + C) = \begin{bmatrix} 1 & 2 \\ -2 & 3 \end{bmatrix} + \left(\begin{bmatrix} 4 & -5 \\ 1 & 0 \end{bmatrix} + \begin{bmatrix} 2 & 4 \\ -6 & -8 \end{bmatrix} \right)$$

$$= \begin{bmatrix} 1 & 2 \\ -2 & 3 \end{bmatrix} + \begin{bmatrix} 6 & -1 \\ -5 & -8 \end{bmatrix} = \begin{bmatrix} 7 & 1 \\ -7 & -5 \end{bmatrix}$$

The right side of A2 is

$$(A + B) + C = \left(\begin{bmatrix} 1 & 2 \\ -2 & 3 \end{bmatrix} + \begin{bmatrix} 4 & -5 \\ 1 & 0 \end{bmatrix} \right) + \begin{bmatrix} 2 & 4 \\ -6 & -8 \end{bmatrix}$$

$$= \begin{bmatrix} 5 & -3 \\ -1 & 3 \end{bmatrix} + \begin{bmatrix} 2 & 4 \\ -6 & -8 \end{bmatrix} = \begin{bmatrix} 7 & 1 \\ -7 & -5 \end{bmatrix}$$

Comparing the final results for the two sides, we see that A2 does hold in this case. ∎

EXAMPLE 5 Verify that property S1 holds for scalars $c = 3$ and $d = 4$, if A is the matrix of Example 4.

Solution   In this case, the left side of S1 is

$$(c + d)A = (3 + 4) \begin{bmatrix} 1 & 2 \\ -2 & 3 \end{bmatrix} = 7 \begin{bmatrix} 1 & 2 \\ -2 & 3 \end{bmatrix} = \begin{bmatrix} 7 & 14 \\ -14 & 21 \end{bmatrix}$$

The right side of S1 is

$$cA + dA = 3 \begin{bmatrix} 1 & 2 \\ -2 & 3 \end{bmatrix} + 4 \begin{bmatrix} 1 & 2 \\ -2 & 3 \end{bmatrix}$$

$$= \begin{bmatrix} 3 & 6 \\ -6 & 9 \end{bmatrix} + \begin{bmatrix} 4 & 8 \\ -8 & 12 \end{bmatrix} = \begin{bmatrix} 7 & 14 \\ -14 & 21 \end{bmatrix}$$

Comparing the final results for the two sides, we see that S1 does hold in this case. ∎

EXERCISES

1. *Polling data* A polling company interviewed 100 people from each of four regions of the country, asking them to classify the president's performance as poor, satisfactory, or very good. In the Northeast, 46 indicated poor, 30 indicated satisfactory, and 24 indicated very good. In the South, 21 responded poor, 27 satisfactory, and 52 very good. In the Midwest, 23 thought the president's performance was poor, 56 thought satisfactory, and 21 thought very good. In the West, 39 indicated poor, 41

satisfactory, and 20 very good. Present this information in matrix form by completing the following array:

$$\begin{array}{c} \\ \text{Northeast} \\ \text{South} \\ \text{Midwest} \\ \text{West} \end{array} \begin{array}{ccc} Poor & Satisfactory & Very\ good \end{array} \\ \begin{bmatrix} 46 & 30 & \\ & & \\ & & \\ & & \end{bmatrix}$$

2. *Psychology data* Affectivity scores (percent of 200 words indicated as "pleasant") for 75 men and 75 women participating in a psychological test were as follows. In the 90 to 99% range, there were 15 men and 23 women; in the 80 to 89% range, there were 21 men and 27 women; in the 70 to 79% range, 23 men and 16 women; in the 60 to 69% range, 7 men and 8 women; and in the 0 to 59% range, 9 men and 1 woman. Present this data in matrix form. (Label the columns and rows in a manner similar to that in Exercise 1.)

Use matrices

$$A = \begin{bmatrix} 2 & 1 \\ -1 & 3 \end{bmatrix} \quad and \quad B = \begin{bmatrix} 0 & -1 \\ 2 & 4 \end{bmatrix}$$

to compute each of the following:

3. $A + B$
4. $A - B$
5. $B - A$
6. $B + A$
7. $3A + B$
8. $3(A + B)$
9. $2A - B$
10. $-4A + 2B$

Use matrices

$$C = \begin{bmatrix} 1 & 2 \\ 5 & 3 \\ -4 & 0 \end{bmatrix} \quad and \quad D = \begin{bmatrix} -1 & 1 \\ 2 & -2 \\ 0 & 3 \end{bmatrix}$$

to compute each of the following:

11. $C + D$
12. $D + C$
13. $C - D$
14. $D - C$
15. $C + 2D$
16. $C - 2D$
17. $-4C - 2D$
18. $3(C + 4D)$
19. $3C + 4D$
20. $C + (D - 2C)$

Use matrices

$$A = \begin{bmatrix} 2 & 6 \\ 1 & -3 \end{bmatrix}, \quad B = \begin{bmatrix} 0 & 1 \\ -1 & 0.5 \end{bmatrix}, \quad C = \begin{bmatrix} \frac{1}{3} & -1 \\ 1 & \frac{2}{3} \end{bmatrix}$$

to compute each of the following:

21. $A + B$
22. $3C$
23. $B - A$
24. $2A + C$
25. $6C + (2A - B)$
26. $6C + 2(A - B)$
27. $A - (B - C)$
28. $(A - B) - C$

Use matrices

$$A = \begin{bmatrix} 1 & 2 \\ -1 & -2 \\ 0 & 1 \end{bmatrix}, \quad B = \begin{bmatrix} 1 & 0 & 0 \\ 0 & 1 & 0 \\ 0 & 0 & 1 \end{bmatrix}, \quad C = \begin{bmatrix} 1 & -1 \\ 2 & -2 \\ 3 & 0 \end{bmatrix}, \quad D = \begin{bmatrix} 2 & 4 \\ 6 & 8 \\ -2 & 0 \end{bmatrix}$$

to compute each of the following, if defined:

29. $3(-4A)$
30. $3(C + D)$
31. $2A - 3D$
32. $2A - 3B$
33. $(-2 + 5)B$
34. $-2(C + B)$

35. Determine the values of x, y, and z if

$$\begin{bmatrix} 1 & x \\ y-1 & 0 \\ 6 & 2z \end{bmatrix} = \begin{bmatrix} 1 & -3 \\ 4 & 0 \\ 6 & z+3 \end{bmatrix}$$

36. Determine the values of x and y if

$$\begin{bmatrix} 2x & 5 \\ -4 & 9 \end{bmatrix} = \begin{bmatrix} 6 & \sqrt{y} \\ -4 & x^2 \end{bmatrix}$$

In Exercises 37 and 38 use scalars $c = -2$ and $d = 4$ and matrices

$$A = \begin{bmatrix} 1 & 4 \\ 2 & 5 \\ 3 & 0 \end{bmatrix} \quad \text{and} \quad B = \begin{bmatrix} 2 & -1 \\ 4 & -2 \\ 6 & -3 \end{bmatrix}$$

37. Verify that property S2 holds.
38. Verify that property S3 holds.

39. Which of the properties A1, A2, and S2 do you think hold for matrix subtraction in place of matrix addition? Test your conclusions with one or two examples.

Explain the meaning of each term.

40. The dimension of a matrix
41. Scalars

42. State the meaning of property S1 in words, using no symbols.
43. State the meaning of property A2 in words, using no symbols.

3.5 MATRIX MULTIPLICATION

The remaining operation with matrices to be considered is **matrix multiplication.** The definition of matrix multiplication may seem unnecessarily complicated, but it is the definition that has been found to be the most useful and—in some applications—the most natural. It states that if $A = [a_{ij}]$ is an $m \times p$ matrix and $B = [b_{ij}]$ is a $p \times n$ matrix, then

$$A \times B = C$$

where $C = [c_{ij}]$, is the $m \times n$ matrix in which

$$c_{ij} = a_{i1}b_{1j} + a_{i2}b_{2j} + \cdots + a_{ip}b_{pj}$$

This last equation states that the element c_{ij} in the ith row and jth column of C is formed by multiplying each element in the ith row of A times the corresponding element of the jth column of B and then adding the results. For example, suppose

$$A = \begin{bmatrix} 1 & 6 \\ 2 & 3 \\ -4 & -1 \end{bmatrix} \quad \text{and} \quad B = \begin{bmatrix} 2 & 4 & 1 \\ 5 & -1 & 0 \end{bmatrix}$$

Note that $m = 3$, $p = 2$, and $n = 3$, so that the product C will be a 3×3 matrix. The element c_{23}, for example, in the product will be formed from the second row of A and the third column of B by multiplying corresponding entries and adding the results:

$$(2)(1) + 3(0) = 2 + 0 = 2 = c_{23}$$

Similarly, c_{31} is formed from the third row of A and the first column of B:

$$(-4)(2) + (-1)(5) = -8 - 5 = -13 = c_{31}$$

Also, c_{21} comes from the second row of A and the first column of B:

$$(2)(2) + (3)(5) = 4 + 15 = 19 = c_{21}$$

All nine entries in the product matrix are formed in the same way. The result is

$$A \times B = \begin{bmatrix} 32 & -2 & 1 \\ 19 & 5 & 2 \\ -13 & -15 & -4 \end{bmatrix}$$

Hereafter we will follow the usual convention of indicating multiplication of matrices by writing the letters or matrices adjacent to each other with no operation symbol in between—that is, we will write AB instead of $A \times B$ or

$$\begin{bmatrix} 1 & 6 \\ 2 & 3 \\ -4 & -1 \end{bmatrix} \begin{bmatrix} 2 & 4 & 1 \\ 5 & -1 & 0 \end{bmatrix}$$

to denote multiplication.

The definition of multiplication requires that p, the number of columns of A, must be equal to the number of rows of B in order for the product AB to exist; otherwise, AB remains undefined and therefore has no meaning.

EXAMPLE 1 Calculate the following for

$$A = \begin{bmatrix} 1 & 2 & 6 \\ 3 & -4 & 9 \\ 0 & 1 & -1 \end{bmatrix} \quad B = \begin{bmatrix} 1 & 2 \\ 4 & -1 \\ 3 & -2 \end{bmatrix} \quad C = \begin{bmatrix} 1 & 1 \\ -1 & -2 \end{bmatrix}$$

(a) AB (b) BC (c) CA (d) BA

Solution (a) A is a 3×3 matrix and B is a 3×2 matrix. Its product is defined and will be a 3×2 matrix. The entries in the product are:

3.5 MATRIX MULTIPLICATION

c_{11} (from first row of A, first column of B) $= 1 + 8 + 18 = 27$
c_{12} (from first row of A, second column of B) $= 2 - 2 - 12 = -12$
c_{21} (from second row of A, first column of B) $= 3 - 16 + 27 = 14$
c_{22} (from second row of A, second column of B) $= 6 + 4 - 18$
$\qquad\qquad\qquad\qquad\qquad\qquad\qquad\qquad\qquad\qquad\qquad = -8$
c_{31} (from third row of A, first column of B) $= 0 + 4 - 3 = 1$
c_{32} (from third row of A, second column of B) $= 0 - 1 + 2 = 1$

The product is

$$AB = \begin{bmatrix} 27 & -12 \\ 14 & -8 \\ 1 & 1 \end{bmatrix}$$

(b) B is a 3×2 matrix and C is a 2×2 matrix. The product is defined and is the 3×2 matrix

$$\begin{bmatrix} -1 & -3 \\ 5 & 6 \\ 5 & 7 \end{bmatrix}$$

(c) C is 2×2 and A is 3×3; the product CA is undefined.
(d) B is 3×2 and A is 3×3; the product BA is undefined.

Some properties of matrix multiplication are given below. As with the properties of matrix addition and scalar multiplication, we shall not attempt to prove that they are valid. Rather we shall verify that one or two of them hold for particular matrices.

If A, B, and C are matrices and c is a scalar, then:

M1. $A(BC) = (AB)C$ $\qquad$ Associative law of multiplication
M2. $A(B + C) = AB + AC$
M3. $(A + B)C = AC + BC$ $\qquad$ Distributive laws
M4. $A(cB) = c(AB) = (cA)B$

These statements all presume that the indicated operations are defined.

---

1 PRACTICE PROBLEM 1

For

$$A = \begin{bmatrix} 2 & 1 \\ -1 & 3 \end{bmatrix}, \quad B = \begin{bmatrix} 3 & 2 \\ -2 & 1 \end{bmatrix},$$

$$C = \begin{bmatrix} 1 & 2 & 3 \\ 4 & 5 & 6 \end{bmatrix}$$

compute:

(a) AB $\quad$ (b) BC

Answer

(a) $AB = \begin{bmatrix} 4 & 5 \\ -9 & 1 \end{bmatrix}$

(b) $BC = \begin{bmatrix} 11 & 16 & 21 \\ 2 & 1 & 0 \end{bmatrix}$

EXAMPLE 2 Verify that property M2 holds for

$$A = \begin{bmatrix} 2 & 3 \\ -1 & -2 \end{bmatrix}, \quad B = \begin{bmatrix} 1 & 4 & 2 \\ 0 & -3 & 2 \end{bmatrix}, \quad C = \begin{bmatrix} 2 & -1 & 3 \\ 1 & 4 & 0 \end{bmatrix}$$

Solution After adding matrices B and C, the left side of M2 is

$$\begin{bmatrix} 2 & 3 \\ -1 & -2 \end{bmatrix} \begin{bmatrix} 3 & 3 & 5 \\ 1 & 1 & 2 \end{bmatrix} = \begin{bmatrix} 9 & 9 & 16 \\ -5 & -5 & -9 \end{bmatrix}$$

Performing the two matrix multiplications, AB and AC, indicated on the right side of M2, gives

$$\begin{bmatrix} 2 & -1 & 10 \\ -1 & 2 & -6 \end{bmatrix} + \begin{bmatrix} 7 & 10 & 6 \\ -4 & -7 & -3 \end{bmatrix} = \begin{bmatrix} 9 & 9 & 16 \\ -5 & -5 & -9 \end{bmatrix}$$

Comparing the final result for the two sides, we see that M2 does hold. ∎

EXAMPLE 3 Verify that property M4 holds for $c = -2$, and

$$A = \begin{bmatrix} 2 & -1 & -2 \\ 0 & 4 & 3 \end{bmatrix} \quad \text{and} \quad B = \begin{bmatrix} 1 & 2 & 0 \\ 2 & 1 & 2 \\ 0 & 2 & 1 \end{bmatrix}$$

Solution The left side of M4 indicates that matrix B is first to be multiplied by the scalar $c = -2$; the result is then multiplied by A:

$$\begin{bmatrix} 2 & -1 & -2 \\ 0 & 4 & 3 \end{bmatrix} \begin{bmatrix} -2 & -4 & 0 \\ -4 & -2 & -4 \\ 0 & -4 & -2 \end{bmatrix} = \begin{bmatrix} 0 & 2 & 8 \\ -16 & -20 & -22 \end{bmatrix}$$

The middle form of M4 indicates that the product AB is to be calculated first and the result multiplied by the scalar:

$$-2 \begin{bmatrix} 0 & -1 & -4 \\ 8 & 10 & 11 \end{bmatrix} = \begin{bmatrix} 0 & 2 & 8 \\ -16 & -20 & -22 \end{bmatrix}$$

The right side of M4 indicates that matrix A is first to be multiplied by the scalar and then B is to be multiplied by the result:

$$\begin{bmatrix} -4 & 2 & 4 \\ 0 & -8 & -6 \end{bmatrix} \begin{bmatrix} 1 & 2 & 0 \\ 2 & 1 & 2 \\ 0 & 2 & 1 \end{bmatrix} = \begin{bmatrix} 0 & 2 & 8 \\ -16 & -20 & -22 \end{bmatrix}$$

The final form for each of the three cases indicates that M4 does hold. ∎

You may have noticed that there is no commutative law in the properties of matrix multiplication just given. This is due to the fact that the order in which two matrices are multiplied does, in general, make a difference. For example, if

$$A = \begin{bmatrix} 1 & -2 \\ 2 & 0 \end{bmatrix} \quad \text{and} \quad B = \begin{bmatrix} -1 & 0 \\ 3 & 4 \end{bmatrix}$$

2 PRACTICE PROBLEM 2

For

$$A = \begin{bmatrix} -1 & 2 \\ 0 & 2 \end{bmatrix}, \quad B = \begin{bmatrix} 4 & -3 \\ 1 & -2 \end{bmatrix},$$

$$C = \begin{bmatrix} -1 & 4 & 3 \\ 5 & -6 & 0 \end{bmatrix}$$

compute:

(a) $2AB$ (b) $(A + 2B)C$

Answer

(a) $\begin{bmatrix} -4 & -2 \\ 4 & -8 \end{bmatrix}$ (b) $\begin{bmatrix} -27 & 52 & 21 \\ -12 & 20 & 6 \end{bmatrix}$

then

$$AB = \begin{bmatrix} 1 & -2 \\ 2 & 0 \end{bmatrix} \begin{bmatrix} -1 & 0 \\ 3 & 4 \end{bmatrix} = \begin{bmatrix} -7 & -8 \\ -2 & 0 \end{bmatrix}$$

while

$$BA = \begin{bmatrix} -1 & 0 \\ 3 & 4 \end{bmatrix} \begin{bmatrix} 1 & -2 \\ 2 & 0 \end{bmatrix} = \begin{bmatrix} -1 & 2 \\ 11 & -6 \end{bmatrix}$$

That is, $AB \neq BA$.

Previously in this chapter, we used augmented coefficient matrices for obtaining solutions for systems of linear equations. Another important relationship between these two concepts—matrices and systems of linear equations—is that such systems can also be written as matrix equations. For example, if

$$A = \begin{bmatrix} 2 & 3 & -1 \\ 1 & 1 & 2 \\ 4 & -6 & 1 \end{bmatrix}, \quad X = \begin{bmatrix} x \\ y \\ z \end{bmatrix}, \quad B = \begin{bmatrix} 1 \\ 3 \\ 4 \end{bmatrix}$$

then the equation

$$AX = B$$

is

$$\begin{bmatrix} 2 & 3 & -1 \\ 1 & 1 & 2 \\ 4 & -6 & 1 \end{bmatrix} \begin{bmatrix} x \\ y \\ z \end{bmatrix} = \begin{bmatrix} 1 \\ 3 \\ 4 \end{bmatrix}$$

Performing the indicated multiplication gives

$$\begin{bmatrix} 2x + 3y - z \\ x + y + 2z \\ 4x - 6y + z \end{bmatrix} = \begin{bmatrix} 1 \\ 3 \\ 4 \end{bmatrix}$$

Both sides of this last equation are 3×1 matrices, and the definition of equal matrices requires that corresponding entries be equal, that is,

$$2x + 3y - z = 1$$
$$x + y + 2z = 3$$
$$4x - 6y + z = 4$$

In this case, matrix A is called the **coefficient matrix** for the corresponding system of linear equations. Also, matrices consisting of a single row or a single column are sometimes called **vectors**. Matrices X and B in this illustration are vectors.

EXAMPLE 4 Write the system

$$2x - y + z = 4$$
$$x \quad\quad + 2z = -3$$
$$3x + 6y - 7z = 1$$

as a matrix equation.

$$\begin{bmatrix} 2 & -1 & 1 \\ 1 & 0 & 2 \\ 3 & 6 & -7 \end{bmatrix} \begin{bmatrix} x \\ y \\ z \end{bmatrix} = \begin{bmatrix} 4 \\ -3 \\ 1 \end{bmatrix}$$

PRACTICE PROBLEM 3

Write the system below as a matrix equation.

$$2x - 3y = 7$$
$$x + 4y = -2$$

Answer

$$\begin{bmatrix} 2 & -3 \\ 1 & 4 \end{bmatrix} \begin{bmatrix} x \\ y \end{bmatrix} = \begin{bmatrix} 7 \\ -2 \end{bmatrix}$$

EXERCISES

Form the matrix products.

1. $\begin{bmatrix} 1 & 2 \\ -1 & 1 \end{bmatrix} \begin{bmatrix} 3 & -2 \\ 0 & 4 \end{bmatrix}$

2. $\begin{bmatrix} 2 & 6 \\ -1 & 5 \end{bmatrix} \begin{bmatrix} 1 & 3 & 4 \\ 0 & -2 & 3 \end{bmatrix}$

3. $\begin{bmatrix} 2 & 9 \\ 7 & -6 \end{bmatrix} \begin{bmatrix} 3 \\ 4 \end{bmatrix}$

4. $\begin{bmatrix} 3 & 4 \end{bmatrix} \begin{bmatrix} 2 & 9 \\ 7 & -6 \end{bmatrix}$

5. $\begin{bmatrix} 3 & 2 & 0 \\ 4 & -1 & 1 \\ 6 & -4 & 2 \end{bmatrix} \begin{bmatrix} 1 & 4 \\ 2 & 6 \\ -3 & 7 \end{bmatrix}$

6. $\begin{bmatrix} 4 & 6 & 7 \\ 1 & 2 & -3 \end{bmatrix} \begin{bmatrix} 3 & 2 & 0 \\ 4 & -1 & 1 \\ 6 & -4 & 2 \end{bmatrix}$

7. $\begin{bmatrix} 1 & -2 & 3 \\ 4 & 5 & 6 \\ 7 & -8 & 9 \end{bmatrix} \begin{bmatrix} 2 \\ 3 \\ 5 \end{bmatrix}$

8. $\begin{bmatrix} 5 & -9 \\ 3 & 7 \end{bmatrix} \begin{bmatrix} x \\ y \end{bmatrix}$

Use matrices

$$A = \begin{bmatrix} -1 & 2 \\ 3 & 2 \end{bmatrix}, \quad B = \begin{bmatrix} 1 & 2 & -1 \\ -2 & 3 & 4 \end{bmatrix}, \quad C = \begin{bmatrix} 2 & 0 & 1 \\ -3 & -1 & 0 \\ 0 & 1 & 4 \end{bmatrix}$$

to compute each of the following:

9. AB
10. BC
11. $3AB$
12. $A(3B)$
13. $2B + AB$
14. $2(B + AB)$
15. $3B - BC$
16. $3(B - BC)$
17. AA or A^2
18. CC or C^2

Use matrices

$$A = \begin{bmatrix} 2 & -1 \\ 0 & 1 \\ 3 & -2 \end{bmatrix}, \quad B = \begin{bmatrix} -2 & 4 \\ 3 & 1 \end{bmatrix}, \quad C = \begin{bmatrix} 1 & 0 \\ 2 & -2 \end{bmatrix}$$

to compute each of the following, if defined:

19. AB
20. CB
21. CA
22. $A(B + C)$
23. $B(3C)$
24. $(-2B)A$
25. $A + AB$
26. $AC - 4A$

Use matrices

$$A = \begin{bmatrix} 1 & 0 & 0 \\ 0 & 1 & 0 \\ 0 & 0 & 1 \end{bmatrix}, \quad B = \begin{bmatrix} 3 & 1 & 0 \\ 1 & 1 & 1 \\ 1 & -1 & 2 \end{bmatrix}, \quad C = \begin{bmatrix} \frac{3}{8} & -\frac{2}{8} & \frac{1}{8} \\ -\frac{1}{8} & \frac{6}{8} & -\frac{3}{8} \\ -\frac{2}{8} & \frac{4}{8} & \frac{2}{8} \end{bmatrix}$$

to compute each of the following:

27. AB
28. AC
29. BC
30. CB

31. Determine the value of x and y if
$$\begin{bmatrix} 2 & 1 \\ -1 & 3 \end{bmatrix} \begin{bmatrix} x & -2 \\ y & 0 \end{bmatrix} = \begin{bmatrix} 0 & -4 \\ -7 & 2 \end{bmatrix}$$

32. Determine the value of x, y and z if
$$\begin{bmatrix} 1 & 2 \\ -2 & 3 \end{bmatrix} \begin{bmatrix} 2x & 1 \\ -y & 1 \end{bmatrix} = \begin{bmatrix} 6 & z \\ 2 & 1 \end{bmatrix}$$

33. Use matrices
$$A = \begin{bmatrix} 1 & 2 \\ 3 & 4 \end{bmatrix} \quad \text{and} \quad B = \begin{bmatrix} -1 & 1 \\ 1 & 0 \end{bmatrix}$$

   (a) Compute the products AB and BA.
   (b) What conclusion can you make concerning a commutative law for matrix multiplication?

Use matrices A, B, and C given with Exercises 9–18.

34. Verify that property M1 holds.
35. Verify that property M3 does not hold.

For each system, write the corresponding matrix equation.

36. $\begin{aligned} 2x_1 + x_2 - x_3 &= 0 \\ x_1 - 2x_2 + x_3 &= 1 \\ x_1 - 2x_3 &= 4 \\ -3x_1 + x_2 - 4x_3 &= -3 \end{aligned}$

37. $\begin{aligned} 3w - 2x + y - z &= 4 \\ 2w + x - 2y &= -1 \\ -2w + z &= 0 \\ x + y - 3z &= 2 \end{aligned}$

For each matrix equation, write the corresponding system of linear equations.

38. $\begin{bmatrix} 0.1 & 0.6 & 0.1 & -0.4 \\ 1 & 0 & -0.8 & 0.3 \\ 1.2 & -2.4 & 0.6 & 1 \end{bmatrix} \begin{bmatrix} x_1 \\ x_2 \\ x_3 \\ x_4 \end{bmatrix} = \begin{bmatrix} 0.7 \\ -0.9 \\ 1 \end{bmatrix}$

39. $\begin{bmatrix} 2 & 6 & -1 \\ 4 & 0 & 3 \\ -2 & 1 & 0 \end{bmatrix} \begin{bmatrix} x \\ y \\ z \end{bmatrix} = \begin{bmatrix} 1 \\ 2 \\ -2 \end{bmatrix}$

40. *Forestry growth* The number of blue spruce trees in a controlled-growth forest at the beginning of this year, classified according to height, is given by the population vector:

   $\begin{array}{l} \text{0–1 foot} \\ \text{1–6 feet} \\ \text{6–15 feet} \\ \text{15 feet and over} \end{array} \begin{bmatrix} 500 \\ 1{,}000 \\ 2{,}000 \\ 4{,}000 \end{bmatrix} = P$

   It is estimated that $\tfrac{1}{5}$ of the trees in each classification (except the last) grow into the next higher classification each year.

   (a) Form the product GP, where G is the growth matrix
   $$\begin{bmatrix} \tfrac{4}{5} & 0 & 0 & 0 \\ \tfrac{1}{5} & \tfrac{4}{5} & 0 & 0 \\ 0 & \tfrac{1}{5} & \tfrac{4}{5} & 0 \\ 0 & 0 & \tfrac{1}{5} & 1 \end{bmatrix} = G$$

(b) What do the entries in the product GP represent physically?

(c) Past experience indicates that the number of trees in each classification that can be expected to die or to be removed this year is given by the removal matrix:

$$\begin{array}{c} \textit{0--1 foot} \\ \textit{1--6 feet} \\ \textit{6--15 feet} \\ \textit{15 feet and over} \end{array} \begin{bmatrix} 60 \\ 100 \\ 200 \\ 1{,}000 \end{bmatrix} = R$$

Calculate the difference $GP - R$. What do the entries in this difference represent physically?

(d) Determine the product $G(GP - R)$. What do the entries in this product represent physically?

41. *Material requirements* A company manufactures three types of file cabinets, Models A-4, B-2, and C-2. The amount of material required to manufacture one cabinet of each model is given in the materials matrix:

$$\begin{array}{c} \textit{Sheet metal (square feet)} \\ \textit{Paint (quarts)} \\ \textit{Drawer-assembly kits} \\ \textit{File-assembly kits} \end{array} \begin{matrix} \textit{A-4} & \textit{B-2} & \textit{C-2} \\ \begin{bmatrix} 42 & 24 & 24 \\ 1 & \frac{3}{4} & \frac{3}{4} \\ 4 & 2 & 1 \\ 0 & 0 & 1 \end{bmatrix} = M \end{matrix}$$

The number of each model that the company intends to manufacture next month is given by the supply vector:

$$\begin{array}{c} \textit{A-4} \\ \textit{B-2} \\ \textit{C-2} \end{array} \begin{bmatrix} 300 \\ 160 \\ 100 \end{bmatrix} = S$$

(a) Form the product MS. Considering how the matrix product is formed, what do the entries in this product represent to the company?

(b) The amount of the various required materials expected to be available at the end of this month is given in the available-materials vector:

$$\begin{array}{c} \textit{Sheet metal (square feet)} \\ \textit{Paint (quarts)} \\ \textit{Drawer-assembly kits} \\ \textit{File-assembly kits} \end{array} \begin{bmatrix} 10{,}200 \\ 350 \\ 1{,}000 \\ 100 \end{bmatrix} = A$$

Calculate the difference $MS - A$. What do the entries in this difference represent to the company?

42. *Animal population classification* In a certain animal population the maximum age attainable by the females is 12 years. The number of females currently in the population classified according to age is given by the population vector:

$$\begin{array}{c} \textit{0--4 years} \\ \textit{4--8 years} \\ \textit{8--12 years} \end{array} \begin{bmatrix} 300 \\ 600 \\ 500 \end{bmatrix} = P$$

During a 4-year period, the average number of daughters in each classification born to a female is given by the birth vector:

$$\begin{array}{ccc} 0\text{--}4 & 4\text{--}8 & 8\text{--}12 \\ [0 & 4 & 2] \end{array} = B$$

(a) Form the product BP. What does its entry represent?

(b) Two-thirds of the females in the 0–4 year classification and three-fourths of those in the 4–8 year classification survive for 4 years and consequently move into the next-higher classification. Form the product AP, where

$$A = \begin{bmatrix} 0 & 0 & 0 \\ \frac{2}{3} & 0 & 0 \\ 0 & \frac{3}{4} & 0 \end{bmatrix}$$

What do the entries in the product represent?

(c) Form the product LP, where

$$L = \begin{bmatrix} 0 & 4 & 2 \\ \frac{2}{3} & 0 & 0 \\ 0 & \frac{3}{4} & 0 \end{bmatrix}$$

What do the entries in this product represent?

(d) Form the product $L^2 P = L(LP)$. What do the entries in this product represent?

43. Explain the meaning of property M2 in words without using any symbols.

44. Explain the meaning of property M4 in words without using any symbols.

45. Explain the meaning of the term *vector*.

3.6 IDENTITY MATRICES AND INVERSES

We wish to observe the behavior in multiplication by the two matrices

$$I_2 = \begin{bmatrix} 1 & 0 \\ 0 & 1 \end{bmatrix} \quad \text{and} \quad I_3 = \begin{bmatrix} 1 & 0 & 0 \\ 0 & 1 & 0 \\ 0 & 0 & 1 \end{bmatrix}$$

For example, if

$$A = \begin{bmatrix} 1 & -2 \\ -3 & 4 \end{bmatrix}$$

then

$$I_2 A = \begin{bmatrix} 1 & 0 \\ 0 & 1 \end{bmatrix} \begin{bmatrix} 1 & -2 \\ -3 & 4 \end{bmatrix} = \begin{bmatrix} 1 & -2 \\ -3 & 4 \end{bmatrix} = A$$

That is, $I_2 A = A$. Similarly, if

$$B = \begin{bmatrix} 1 & -2 & 7 \\ 4 & 0 & 9 \\ -3 & 6 & 8 \end{bmatrix}$$

then

$$I_3 B = \begin{bmatrix} 1 & 0 & 0 \\ 0 & 1 & 0 \\ 0 & 0 & 1 \end{bmatrix} \begin{bmatrix} 1 & -2 & 7 \\ 4 & 0 & 9 \\ -3 & 6 & 8 \end{bmatrix} = \begin{bmatrix} 1 & -2 & 7 \\ 4 & 0 & 9 \\ -3 & 6 & 8 \end{bmatrix} = B$$

That is, $I_3 B = B$. For this reason, I_2 and I_3 are called *identity matrices*. In order to define identity matrices correctly, we need some preliminary terminology. First, a **square matrix** is a matrix in which the number of rows is the same as the number of columns. A, B, I_2, and I_3 above are all square matrices. The matrix

$$\begin{bmatrix} 1 & -2 \\ -3 & 4 \\ 5 & -6 \end{bmatrix}$$

is not square.

Secondly, the **main diagonal** of a square matrix is made up of the elements $a_{11}, a_{22}, a_{33}, \ldots, a_{nn}$. That is, the main diagonal of a square matrix consists of the elements on the diagonal from the upper left corner to the lower right corner:

$$\begin{bmatrix} a_{11} & a_{12} & a_{13} & \cdots & a_{1n} \\ a_{21} & a_{22} & a_{23} & \cdots & a_{2n} \\ a_{31} & a_{32} & a_{33} & \cdots & a_{3n} \\ \vdots & \vdots & \vdots & \ddots & \vdots \\ a_{n1} & a_{n2} & a_{n3} & \cdots & a_{nn} \end{bmatrix}$$

For any positive integer n, the n-dimensional **identity matrix,** I_n, is the $n \times n$ matrix with 1's on its main diagonal and 0's everywhere else.

EXAMPLE 1 Construct I_4.

Solution

$$I_4 = \begin{bmatrix} 1 & 0 & 0 & 0 \\ 0 & 1 & 0 & 0 \\ 0 & 0 & 1 & 0 \\ 0 & 0 & 0 & 1 \end{bmatrix}$$ ▪1

When the value of n is either obvious or immaterial, we will drop the subscript n in the notation I_n and simply write I.

If A and B are square matrices of the same dimension such that

$$AB = I$$

then B is called the *inverse* of A. For example, since

$$\begin{bmatrix} 1 & 2 \\ 3 & 4 \end{bmatrix} \begin{bmatrix} -2 & 1 \\ \frac{3}{2} & -\frac{1}{2} \end{bmatrix} = \begin{bmatrix} 1 & 0 \\ 0 & 1 \end{bmatrix} = I_2$$

▪1 **PRACTICE PROBLEM 1**

Construct I_5.

Answer

$$I_5 = \begin{bmatrix} 1 & 0 & 0 & 0 & 0 \\ 0 & 1 & 0 & 0 & 0 \\ 0 & 0 & 1 & 0 & 0 \\ 0 & 0 & 0 & 1 & 0 \\ 0 & 0 & 0 & 0 & 1 \end{bmatrix}$$

3.6 IDENTITY MATRICES AND INVERSES

the matrix

$$\begin{bmatrix} -2 & 1 \\ \frac{3}{2} & -\frac{1}{2} \end{bmatrix}$$

is the inverse of

$$A = \begin{bmatrix} 1 & 2 \\ 3 & 4 \end{bmatrix}$$

We use the notation A^{-1} (read "A inverse") to denote the inverse of a matrix A. In this illustration,

$$A^{-1} = \begin{bmatrix} -2 & 1 \\ \frac{3}{2} & -\frac{1}{2} \end{bmatrix}$$

Also, if A and B are square matrices and $AB = I$, then $BA = I$ as well. Consequently, A and B are then inverses of each other. In this illustration,

$$B^{-1} = \begin{bmatrix} 1 & 2 \\ 3 & 4 \end{bmatrix} = A \quad \text{for} \quad B = \begin{bmatrix} -2 & 1 \\ \frac{3}{2} & \frac{1}{2} \end{bmatrix} = A^{-1}$$

Not all square matrices have inverses. Our immediate concern is how to find the inverse of a given matrix if there is one and how to determine that there is none when such is the case.

For example, if we want to determine the inverse for

$$A = \begin{bmatrix} 1 & 2 & -1 \\ 2 & 1 & 0 \\ -1 & 0 & 3 \end{bmatrix}$$

we must look for a matrix

$$B = \begin{bmatrix} b_{11} & b_{12} & b_{13} \\ b_{21} & b_{22} & b_{23} \\ b_{31} & b_{32} & b_{33} \end{bmatrix}$$

such that

$$\begin{bmatrix} 1 & 2 & -1 \\ 2 & 1 & 0 \\ -1 & 0 & 3 \end{bmatrix} \begin{bmatrix} b_{11} & b_{12} & b_{13} \\ b_{21} & b_{22} & b_{23} \\ b_{31} & b_{32} & b_{33} \end{bmatrix} = \begin{bmatrix} 1 & 0 & 0 \\ 0 & 1 & 0 \\ 0 & 0 & 1 \end{bmatrix}$$

From the definition of matrix multiplication, the first column of the identity matrix on the right is formed from A and the first column of B; that is,

$$\begin{bmatrix} 1 & 2 & -1 \\ 2 & 1 & 0 \\ -1 & 0 & 3 \end{bmatrix} \begin{bmatrix} b_{11} \\ b_{21} \\ b_{31} \end{bmatrix} = \begin{bmatrix} 1 \\ 0 \\ 0 \end{bmatrix} \quad (1)$$

2 PRACTICE PROBLEM 2

For

$$A = \begin{bmatrix} 1 & 2 \\ 1 & 3 \end{bmatrix}$$

and $\quad B = \begin{bmatrix} 3 & -2 \\ -1 & 1 \end{bmatrix}$

form the product AB to verify that B is A^{-1}.

Answer

$$AB = \begin{bmatrix} 1 & 0 \\ 0 & 1 \end{bmatrix} = I_2$$

Similarly, for the second and third columns of the identity,

$$\begin{bmatrix} 1 & 2 & -1 \\ 2 & 1 & 0 \\ -1 & 0 & 3 \end{bmatrix} \begin{bmatrix} b_{12} \\ b_{22} \\ b_{32} \end{bmatrix} = \begin{bmatrix} 0 \\ 1 \\ 0 \end{bmatrix} \quad (2)$$

$$\begin{bmatrix} 1 & 2 & -1 \\ 2 & 1 & 0 \\ -1 & 0 & 3 \end{bmatrix} \begin{bmatrix} b_{13} \\ b_{23} \\ b_{33} \end{bmatrix} = \begin{bmatrix} 0 \\ 0 \\ 1 \end{bmatrix} \quad (3)$$

Seeking the matrix B then amounts to solving the three matrix equations (1), (2), and (3), or the three equivalent linear systems. Since the row operations for solving each of these three systems are exactly alike, we might as well solve all three systems at the same time. We can do this by setting up *one augmented matrix* for all three systems and then row-reduce as usual.

For matrix A, this augmented matrix is

$$\left[\begin{array}{ccc|ccc} 1 & 2 & -1 & 1 & 0 & 0 \\ 2 & 1 & 0 & 0 & 1 & 0 \\ -1 & 0 & 3 & 0 & 0 & 1 \end{array} \right]$$

Reducing this matrix to echelon form by row operations, or row-reducing, leads to

$$\left[\begin{array}{ccc|ccc} 1 & 0 & 0 & -\frac{3}{10} & \frac{6}{10} & -\frac{1}{10} \\ 0 & 1 & 0 & \frac{6}{10} & -\frac{2}{10} & \frac{2}{10} \\ 0 & 0 & 1 & -\frac{1}{10} & \frac{2}{10} & \frac{3}{10} \end{array} \right]$$

The solution for systems (1), (2), and (3), respectively, are then

$$\begin{bmatrix} -\frac{3}{10} \\ \frac{6}{10} \\ -\frac{1}{10} \end{bmatrix}, \quad \begin{bmatrix} \frac{6}{10} \\ -\frac{2}{10} \\ \frac{2}{10} \end{bmatrix}, \quad \begin{bmatrix} -\frac{1}{10} \\ \frac{2}{10} \\ \frac{3}{10} \end{bmatrix}$$

and so

$$A^{-1} = \begin{bmatrix} -\frac{3}{10} & \frac{6}{10} & -\frac{1}{10} \\ \frac{6}{10} & -\frac{2}{10} & \frac{2}{10} \\ -\frac{1}{10} & \frac{2}{10} & \frac{3}{10} \end{bmatrix}$$

The procedure amounts to starting with the augmented matrix consisting of the matrix of coefficients on the left-hand side and the identity matrix on the right; then row reduce until the identity is obtained on the left, in which case the right side is the inverse.

In Exercise 1 you are asked to verify that the final matrix just given is the inverse of the original matrix A.

EXAMPLE 2 Find the inverse of

$$C = \begin{bmatrix} 2 & -4 \\ 1 & 3 \end{bmatrix}$$

3.6 IDENTITY MATRICES AND INVERSES

Solution The augmented matrix is

$$\begin{bmatrix} 2 & -4 & | & 1 & 0 \\ 1 & 3 & | & 0 & 1 \end{bmatrix}$$

Row-reducing leads to

$$\begin{bmatrix} 1 & 0 & | & \frac{3}{10} & \frac{2}{5} \\ 0 & 1 & | & -\frac{1}{10} & \frac{1}{5} \end{bmatrix}$$

Therefore,

$$C^{-1} = \begin{bmatrix} \frac{3}{10} & \frac{2}{5} \\ -\frac{1}{10} & \frac{1}{5} \end{bmatrix}$$

The process we have been discussing will also reveal when a given matrix has no inverse. In this case, one or more of the systems corresponding to those in (1), (2), and (3) will have no solution, and this is indicated by a row of 0's occurring in the row-reduction process. Example 3 illustrates what happens.

EXAMPLE 3
Find the inverse of

$$B = \begin{bmatrix} 1 & -1 & 1 \\ 0 & 2 & -4 \\ 1 & 1 & -3 \end{bmatrix}$$

Solution The augmented matrix is

$$\begin{bmatrix} 1 & -1 & 1 & | & 1 & 0 & 0 \\ 0 & 2 & -4 & | & 0 & 1 & 0 \\ 1 & 1 & -3 & | & 0 & 0 & 1 \end{bmatrix}$$

Row reduction leads to

$$\begin{bmatrix} 1 & -1 & 1 & | & 1 & 0 & 0 \\ 0 & 1 & -2 & | & 0 & \frac{1}{2} & 0 \\ 0 & 0 & 0 & | & -1 & -1 & 1 \end{bmatrix}$$

The row of 0's on the left-hand side indicates that B does not have an inverse.

3 PRACTICE PROBLEM 3
Find the inverse of

$$A = \begin{bmatrix} 1 & 2 \\ 1 & 3 \end{bmatrix}$$

Answer

$$A^{-1} = \begin{bmatrix} 3 & -2 \\ -1 & 1 \end{bmatrix}$$

4 PRACTICE PROBLEM 4
Find the inverse, if there is one, of

$$A = \begin{bmatrix} 1 & 0 & 2 \\ 0 & 1 & 0 \\ 1 & -1 & 2 \end{bmatrix}$$

Answer

No inverse exists.

EXERCISES

1. Verify, by multiplication, that if

$$A = \begin{bmatrix} 1 & 2 & -1 \\ 2 & 1 & 0 \\ -1 & 0 & 3 \end{bmatrix}, \quad \text{then} \quad A^{-1} = \begin{bmatrix} -\frac{3}{10} & \frac{6}{10} & -\frac{1}{10} \\ \frac{6}{10} & -\frac{2}{10} & \frac{2}{10} \\ -\frac{1}{10} & \frac{2}{10} & \frac{3}{10} \end{bmatrix}$$

2. Verify, by multiplication, that if

$$C = \begin{bmatrix} 2 & -4 \\ 1 & 3 \end{bmatrix}, \quad \text{then} \quad C^{-1} = \begin{bmatrix} \frac{3}{10} & \frac{2}{5} \\ -\frac{1}{10} & \frac{1}{5} \end{bmatrix}$$

Find the inverse of each of the following matrices:

3. $\begin{bmatrix} 1 & 2 \\ 1 & -3 \end{bmatrix}$
4. $\begin{bmatrix} 2 & 3 \\ -1 & 5 \end{bmatrix}$
5. $\begin{bmatrix} 1 & 2 & -1 \\ 1 & 3 & -1 \\ 0 & 4 & 1 \end{bmatrix}$

6. $\begin{bmatrix} 2 & 4 & -6 \\ 0 & 1 & 3 \\ -1 & -2 & 5 \end{bmatrix}$
7. $\begin{bmatrix} 1 & -2 & 3 & 4 \\ 0 & 1 & 2 & -3 \\ 0 & 0 & 1 & 2 \\ 0 & 0 & 0 & -1 \end{bmatrix}$
8. $\begin{bmatrix} 1 & 0 & 0 & 0 \\ 0 & 2 & 0 & 0 \\ 0 & 0 & 3 & 0 \\ 0 & 0 & 0 & 4 \end{bmatrix}$

Find the inverse, if it exists, of each of the following matrices:

9. $\begin{bmatrix} 2 & -3 \\ -4 & 6 \end{bmatrix}$
10. $\begin{bmatrix} 0.4 & 0.8 \\ 0.3 & 0.9 \end{bmatrix}$

11. $\begin{bmatrix} 3 & 6 & -9 \\ 1 & 4 & -11 \\ 1 & 1 & 1 \end{bmatrix}$
12. $\begin{bmatrix} \frac{1}{4} & \frac{1}{2} & -\frac{1}{2} \\ 1 & \frac{5}{2} & 2 \\ -1 & -2 & \frac{5}{3} \end{bmatrix}$

13. $\begin{bmatrix} 2 & -4 & 2 & 1 \\ 5 & -10 & 5 & 2.5 \\ 3 & 0 & -2 & 0.5 \\ 0 & 1 & 4 & 1.5 \end{bmatrix}$
14. $\begin{bmatrix} 1 & 6 \\ 4 & -3 \\ 9 & 7 \end{bmatrix}$

For a 2×2 matrix

$$A = \begin{bmatrix} a & b \\ c & d \end{bmatrix}$$

if $ad - bc = 0$, then A has no inverse; if $ad - bc \neq 0$, then

$$A^{-1} = \frac{1}{ad - bc} \begin{bmatrix} d & -b \\ -c & a \end{bmatrix}$$

This usually is an easier method for finding the inverse of a 2×2 matrix than the one involving row-reducing. Use this method to determine the inverse, if it exists, of each of the following:

15. $\begin{bmatrix} 2 & 4 \\ -3 & 1 \end{bmatrix}$
16. $\begin{bmatrix} 10 & -15 \\ 16 & -24 \end{bmatrix}$

17. $\begin{bmatrix} 0.5 & 0.3 \\ -0.2 & 0.7 \end{bmatrix}$
18. $\begin{bmatrix} 1 & \frac{1}{2} \\ \frac{1}{3} & 1 \end{bmatrix}$

Explain the meaning of each term.

19. A square matrix
20. The main diagonal of a square matrix
21. An identity matrix
22. The inverse of a matrix

*3.7 LEONTIEF INPUT-OUTPUT MODELS

We have seen matrices used in solving systems of equations and will see additional applications in later sections on Markov chains and game theory. We conclude this chapter with a discussion of one use of matrices in economics. In particular, we consider Leontief input-output models. Leontief models deal with the question of at what level of output each of the several industries in an economy should produce in order to meet the demand for their products. Because of the interaction between the several industries, this question is not as easy to answer as it might seem at first.

Consider an oversimplified economy consisting of three industries, say R, S, and T. These might be the auto, oil, and transportation industries. Our three industries not only supply the consumer but supply themselves and each other as well. For example, the transportation industry transports gas for the oil industry and cars for the auto industry; the auto industry supplies vehicles for the oil, transportation, and auto industries, and so forth.

During a particular time period, say a year, the activities of our three industries are summarized in Table 3.1. Entries represent millions of dollars. The rows give the production, or output, for each industry. For example, industry S supplies $60 million of its product to industry R, $30 million to itself, $30 million to industry T, and $30 million to the consumer, for a total of $150 million.

TABLE 3.1

PRODUCTION (OUTPUT)	CONSUMPTION (INPUT)			CONSUMER DEMAND	TOTAL
	R	S	T		
R	24	15	60	21	120
S	60	30	30	30	150
T	36	60	30	174	300

The column for each industry gives the value of the products that industry uses (inputs) in supplying its indicated total. Industry R for example, uses $24 million of its own product, $60 million from industry S, and $36 million from industry T in producing its total of $120 million.

All this activity was for the purpose of supplying consumers with the indicated $21 million, $30 million, and $174 million from the various industries. The corresponding vector

$$D = \begin{bmatrix} 21 \\ 30 \\ 174 \end{bmatrix}$$

is called the ***demand vector.***

---

* This section can be omitted without loss of continuity.

Suppose the expected demand vector for next year is

$$D = \begin{bmatrix} 28 \\ 40 \\ 145 \end{bmatrix}$$

Consumer demand will increase for R and S but will decrease for T. However, because of the interaction between the industries, it is not obvious at what levels the industries should produce next year in order to meet the new consumer demand. Our purpose is to determine these levels on the assumption that the production processes implied by the input-output table remain essentially the same; that is, prices are relatively stable and technology changes are at most minor. We want to use past activity as the basis for determining future requirements.

From the table we see that it took $15 million of the product of R to produce $150 million of the product of S. Consequently, it took $\frac{15}{150} = \frac{1}{10} = \$0.10$ to produce \$1 of the product of S. Similarly, it took $\frac{30}{150} = \$0.20$ of the product of S and $\frac{60}{150} = \$0.40$ of the product of T to produce \$1 of S's product.

Consequently, if we construct the matrix

$$A = [a_{ij}] = \begin{bmatrix} \frac{24}{120} & \frac{15}{150} & \frac{60}{300} \\ \frac{60}{120} & \frac{30}{150} & \frac{30}{300} \\ \frac{36}{120} & \frac{60}{150} & \frac{30}{300} \end{bmatrix}$$
$$= \begin{bmatrix} 0.2 & 0.1 & 0.2 \\ 0.5 & 0.2 & 0.1 \\ 0.3 & 0.4 & 0.1 \end{bmatrix}$$

from the input-output table, then a_{ij} is the value of the product of industry i used in producing \$1 of the product of industry j. The matrix A is called the **input-output coefficient matrix**.

Now let x_1 denote the level of production required of industry R to meet the new consumer demands, as indicated in the demand vector D, x_2 be the level required of S, and x_3 the level required of T. Each industry must satisfy

Total production = Industry requirements + Consumer demand

That is,

$$\begin{aligned} x_1 &= 0.2x_1 + 0.1x_2 + 0.2x_3 + 28 \\ x_2 &= 0.5x_1 + 0.2x_2 + 0.1x_3 + 40 \\ x_3 &= 0.3x_1 + 0.4x_2 + 0.1x_3 + 145 \end{aligned} \quad (1)$$

If we let X be the **production vector**,

$$X = \begin{bmatrix} x_1 \\ x_2 \\ x_3 \end{bmatrix}$$

the above system of equations can be written as the matrix equation

$$X = AX + D$$

which we must solve for X. Since $X = IX$, subtracting AX from each side of the equation gives

$$IX - AX = D$$

Then from the distributive laws,

$$(I - A)X = D$$

Multiplying each side by $(I - A)^{-1}$ gives

$$\underbrace{(I - A)^{-1}(I - A)}_{I}X = (I - A)^{-1}D$$
$$X = (I - A)^{-1}D$$

or simply

$$X = (I - A)^{-1}D \qquad (2)$$

Equation (2) gives the production vector for a given input-output coefficient matrix A and demand vector D. In our illustration

$$I - A = \begin{bmatrix} 1 & 0 & 0 \\ 0 & 1 & 0 \\ 0 & 0 & 1 \end{bmatrix} - \begin{bmatrix} 0.2 & 0.1 & 0.2 \\ 0.5 & 0.2 & 0.1 \\ 0.3 & 0.4 & 0.1 \end{bmatrix}$$
$$= \begin{bmatrix} 0.8 & -0.1 & -0.2 \\ -0.5 & 0.8 & -0.1 \\ -0.3 & -0.4 & 0.9 \end{bmatrix}$$

Then by the method discussed in the last section,

$$(I - A)^{-1} = \begin{bmatrix} 1.67 & 0.42 & 0.42 \\ 1.18 & 1.62 & 0.44 \\ 1.08 & 0.86 & 1.45 \end{bmatrix}$$

and, from equation (2),

$$\begin{bmatrix} x_1 \\ x_2 \\ x_3 \end{bmatrix} = \begin{bmatrix} 1.67 & 0.42 & 0.42 \\ 1.18 & 1.62 & 0.44 \\ 1.08 & 0.86 & 1.45 \end{bmatrix} \begin{bmatrix} 28 \\ 40 \\ 145 \end{bmatrix} = \begin{bmatrix} 124.46 \\ 161.64 \\ 274.89 \end{bmatrix}$$

That is, next year R should produce \$124.46 million of its product, S should produce \$161.64 million of its product, and T should produce \$274.89 million of its product.

The advantage in using $(I - A)^{-1}$ and equation (2) to determine new production levels rather than solving the system of linear equations of (1) lies in the fact that once $(I - A)^{-1}$ has been determined, it can be used to calculate new production levels for several different demand vectors. This amounts to a simple matrix multiplication rather than again solving a new system of equations similar to (1).

In our illustration, suppose that later government predictions establish a new demand vector,

$$\begin{bmatrix} 25 \\ 35 \\ 168 \end{bmatrix}$$

The new production levels can then be quickly determined:

$$\begin{bmatrix} x_1 \\ x_2 \\ x_3 \end{bmatrix} = \begin{bmatrix} 1.67 & 0.42 & 0.42 \\ 1.18 & 1.62 & 0.44 \\ 1.08 & 0.86 & 1.45 \end{bmatrix} \begin{bmatrix} 25 \\ 35 \\ 168 \end{bmatrix} = \begin{bmatrix} 127.01 \\ 160.12 \\ 300.7 \end{bmatrix}$$

In this case, note that T's output will increase slightly over last year even though consumer demand for its product will decrease. How can this be?

Leontief input-output models provide a mathematical description of the relationships between the industries in an economy and consumer demand. They show at what levels the various industries can expect to produce to meet predicted demand. The following steps outline the procedure to determine these levels.

1. On the basis of known input, output, and consumer demand for a given time period, construct the input-output coefficient matrix A.
2. Calculate the difference $I - A$.
3. Determine $(I - A)^{-1}$.
4. Form the product $(I - A)^{-1}D$, where D is the new demand vector. The vector X resulting from the multiplication gives the new production levels.

These models were first constructed by the American economist Wassily Leontief. For his work in this area he was awarded the 1973 Nobel prize in economics while at Harvard University.

Our economy was greatly oversimplified in order to make the calculations reasonable without the use of a computer. The U.S. Department of Commerce has published input-output tables for the United States for various years. These, of course, are considerably more complicated and take a great deal of time to compile.

EXERCISES

In Exercises 1–4, you are given the input-output tables for different economies for a single year and the forecast demand vector for the next year. In each case, determine the output levels required to meet the forecast demand. All entries are in millions of dollars.

1.

PRODUCTION (OUTPUT)	CONSUMPTION (INPUT)			TOTAL	DEMAND
	R	S	Consumer		
R	15	20	25	60	$\begin{bmatrix} 35 \\ 30 \end{bmatrix}$
S	30	50	20	100	

2.

PRODUCTION (OUTPUT)	CONSUMPTION (INPUT)			TOTAL	DEMAND
	R	S	Consumer		
R	36	60	24	120	$\begin{bmatrix} 20 \\ 25 \end{bmatrix}$
S	48	75	27	150	

3.

PRODUCTION (OUTPUT)	CONSUMPTION (INPUT)				TOTAL	DEMAND
	R	S	T	Consumer		
R	20	15	40	25	100	$\begin{bmatrix} 20 \\ 45 \\ 100 \end{bmatrix}$
S	50	30	20	50	150	
T	30	60	20	90	200	

(*Hint:* Compare the input-output coefficient matrix with that of the illustration in the text.)

4.

PRODUCTION (OUTPUT)	CONSUMPTION (INPUT)				TOTAL	DEMAND
	R	S	T	Consumer		
R	6	5	14	5	30	$\begin{bmatrix} 5 \\ 12 \\ 20 \end{bmatrix}$
S	9	10	21	10	50	
T	6	25	21	18	70	

5. Determine the output levels in Exercise 1 if the forecast demand for the next year is changed to

$$D = \begin{bmatrix} 32 \\ 28 \end{bmatrix}$$

6. Determine the output levels in Exercise 3 if the forecast demand for the next year is changed to

$$D = \begin{bmatrix} 25 \\ 47 \\ 95 \end{bmatrix}$$

7. Determine the output levels in Exercise 4 if the forecast demand for the next year is changed to

$$D = \begin{bmatrix} 7 \\ 11 \\ 21 \end{bmatrix}$$

3.8 CHAPTER SUMMARY

The two major topics of this chapter were solutions of systems of linear equations and matrix operations. Geometrically, the solutions—if any—of a system of linear equations are the coordinates of the points of intersection of the graphs of the equations. Since these are graphs of linear equations, they will always be lines.

Gauss-Jordan elimination, the principal algebraic method considered for solving a system of linear equations, starts with the matrix of coefficients augmented by the constants. This augmented matrix is reduced to echelon form using the following ***row operations:***

1. Any two rows of the matrix may be interchanged.
2. All entries of any row may be multiplied by a constant other than zero.
3. All entries of any row may be multiplied by a constant and the result added to another row.

From the resulting matrix, which is the augmented coefficient matrix of an equivalent system, the solutions or the fact that there is none can easily be determined. This final matrix is in **echelon form,** that is:

1. All the rows having only 0 entries, if there are any, appear at the bottom.
2. If a row has any nonzero entries, the leading nonzero entry in that row is a 1.
3. In any column containing the leading 1 of a particular row, all the other entries are 0.
4. For any two consecutive nonzero rows, the leading 1 in the lower row is farther to the right than the leading 1 of the upper row.

Definitions of the basic operations with matrices are summarized in Table 3.2. In the table, $A = [a_{ij}]$ and $B = [b_{ij}]$; the requirements on the dimensions of A and B, as well as the dimension of the resulting matrix C, are given in Table 3.2.

TABLE 3.2

OPERATION	DEFINITION	DIMENSIONS
Sum and difference	$A \pm B = C$, where $c_{ij} = a_{ij} \pm b_{ij}$	$A: m \times n$ $B: m \times n$ $C: m \times n$
Scalar multiplication	$kA = C$, where $c_{ij} = ka_{ij}$	$A: m \times n$ $C: m \times n$
Product	$AB = C$, where $c_{ij} = a_{i1}b_{1j} + a_{i2}b_{2j} + \cdots + a_{ip}b_{pj}$	$A: m \times p$ $B: p \times n$ $C: m \times n$

To a great extent, properties of these operations, such as the commutative law of addition and the associative law of multiplication, parallel the properties of operations with the real numbers. The major exception is that there is no commutative law of multiplication; in general, the order in which two matrices are multiplied (when the dimensions allow) does make a difference.

An **identity matrix** I is a square matrix with 1's on its main diagonal and 0's elsewhere. Identity matrices have the property that, whenever the dimensions allow,

$$AI = A \quad \text{and} \quad IA = A$$

for every matrix A.

The **inverse** of a square matrix A is a matrix A^{-1} of the same dimension such that

$$AA^{-1} = I$$

Then $A^{-1}A = I$ also. To determine the inverse of a square matrix A, row-reduce the augmented matrix $[A \mid I]$ to the form $[I \mid A^{-1}]$. If this latter form cannot be obtained—that is, if the left side does not reduce to an identity matrix, then A has no inverse.

One application of inverses of matrices was considered in conjunction with **Leontief input-output models.** These models indicate at what level of output each of several industries in an economy should produce to meet the demand for their product, considering the interaction between the industries as well as consumer demand. The procedure to determine these levels is summarized as follows:

1. On the basis of known input, output, and consumer demand for a given time period, construct the input-output coefficient matrix A.
2. Calculate the difference $I - A$.
3. Determine $(I - A)^{-1}$.
4. Form the product

$$(I - A)^{-1}D$$

where D is the new demand vector. The vector X resulting from this multiplication gives the new production levels.

REVIEW EXERCISES

Solve each system of equations using the Gauss-Jordan method.

1. $2x + 3y = -3$
   $x - y = -4$

2. $2x - 3y = 14$
   $5x + 7y = 6$

3. $\begin{bmatrix} 1 & -2 \\ -3 & 6 \end{bmatrix} \begin{bmatrix} x \\ y \end{bmatrix} = \begin{bmatrix} 4 \\ 6 \end{bmatrix}$

4. $\begin{bmatrix} 1 & -2 \\ -3 & 6 \end{bmatrix} \begin{bmatrix} x \\ y \end{bmatrix} = \begin{bmatrix} 4 \\ -12 \end{bmatrix}$

5. $3x - 5y + z = 7$
$-2x + 3y + 2z = 1$
$x - 2y - z = 0$

6. $x + y + 2z = 0.5$
$4x + 5y - z = 0.3$
$-2x + 3y - 4z = -0.5$

7. $2x + y + z = 4$
$x - 2y + z = 7$
$3x - y + 8z = 11$

8. $\begin{bmatrix} 2 & 4 & -6 \\ -4 & 1 & 2 \\ -2 & 5 & -4 \end{bmatrix} \begin{bmatrix} x \\ y \\ z \end{bmatrix} = \begin{bmatrix} 1 \\ 5 \\ -1 \end{bmatrix}$

9. $5x + 7y = 55$
$-3x + 4y = 8$
$2x + 11y = 63$

10. $-4x + 7y = 12$
$3x - y = -5$
$2x + 5y = 3$

11. $3x - 2y + z = 5$
$-2x + y + 2z = -3$

12. $3w + 2y + z = 5$
$-7w + x - y + 3z = 4$
$4w - 5x + z = 0$

For

$$A = \begin{bmatrix} 2 & -1 \\ -2 & 3 \\ -3 & 2 \end{bmatrix}, \quad B = \begin{bmatrix} 1 & -1 \\ 2 & 0 \end{bmatrix},$$

$$C = \begin{bmatrix} 1 & 2 & 0 \\ 2 & 4 & 1 \end{bmatrix}, \quad D = \begin{bmatrix} 1 & 1 & 1 \\ -1 & -1 & -1 \end{bmatrix}$$

evaluate each of the following or indicate that it is undefined, as appropriate:

13. AC
14. $AC - 3B$
15. $(AC)^{-1}$
16. $3C - 2D$
17. B^{-1}
18. AB^{-1}
19. $B + CA$
20. $2B(C + D)$

For

$$A = \begin{bmatrix} 1 & 0 & -1 \\ 2 & -1 & 0 \end{bmatrix}, \quad B = \begin{bmatrix} 1 & 0 & 1 \\ 2 & 1 & -1 \\ 0 & 1 & -2 \end{bmatrix},$$

$$C = \begin{bmatrix} 1 & 2 \\ 3 & -2 \end{bmatrix}, \quad D = \begin{bmatrix} 1 & 0 \\ 0 & 1 \end{bmatrix}$$

evaluate each of the following or indicate that it is undefined, as appropriate:

21. AB
22. $(AB)^{-1}$
23. $2C + 3D$
24. CD
25. $4D(C - D)$
26. $CA - 2D$
27. B^{-1}
28. C^{-1}

29. **Production—revenue** A company has three factories, each of which produces two items, A and B. The weekly production of the factories is given by the following production matrix:

$$\begin{array}{c} \\ \text{Factory 1} \\ \text{Factory 2} \\ \text{Factory 3} \end{array} \begin{matrix} A & B \\ \begin{bmatrix} 10 & 30 \\ 20 & 20 \\ 40 & 10 \end{bmatrix} \end{matrix} = P$$

Each unit of item A sells for \$1,000 and each unit of item B sells for \$1,500. Form the product PR, where

$$R = \begin{bmatrix} 1{,}000 \\ 1{,}500 \end{bmatrix}$$

What do the entries in the product PR represent to the company?

REVIEW EXERCISES

30. *Agribusiness* The yield, in bushels, from three orchards is given by the following yield matrix:

$$\begin{array}{c} \\ \text{Oranges} \\ \text{Peaches} \end{array} \begin{array}{c} \overbrace{\text{Orchard 1} \quad \text{Orchard 2} \quad \text{Orchard 3}} \\ \begin{bmatrix} 110 & 120 & 130 \\ 90 & 100 & 80 \end{bmatrix} \end{array} = Y$$

Oranges sell for $9 per bushel and peaches sell for $11 per bushel. Form the product RY, where $R = [9 \quad 11]$. What do the entries in the product RY represent physically?

31. *Cholesterol experiment* A biologist, conducting a cholesterol experiment, carefully monitors the weight of an experimental guinea pig and a control guinea pig. The experimental pig weighs 0.4 kilogram less than the control pig. How much does each weigh if their combined weight is 2.33 kilograms?

32. *Calculator sales* A distributor sells two types of hand calculators, model I and model II. The distributor's cost and selling price, in dollars, for these calculators are:

	Cost	Selling price
Model I	10	17
Model II	15	25

(a) During one week the revenue from the calculators amounted to $5,050. If the cost to the distributor for these calculators was $3,000, how many of each model were sold?
(b) On one large order the distributor made a profit of $880 on calculators costing $1,300. How many of each model were ordered?

33. *Economic purchasing* Your company has the following options for buying a particular product:

Option 1: $250 plus $2.00 per item

Option 2: $100 plus $3.00 per item

For what quantities is option 1 the more economical? For what quantities is option 2 the more economical?

34. *Break-even point* The monthly overhead cost for a bicycle manufacturing company is $15,015. In addition, it costs $90 in material and labor to manufacture each bicycle. The company sells the bicycles at $145 each.
(a) Write an equation to express the total monthly manufacturing cost (y) in terms of the number (x) of bicycles manufactured.
(b) Write an equation to express the total monthly revenue (y) in terms of the number (x) of bicycles sold.
(c) Presuming the company sells all the bicycles manufactured, how many bicycles must the company sell each month to break even?
(d) What must be the monthly revenue for the company to break even?
(e) What are the coordinates of the break-even point?

35. *Break-even point* The annual cost, in dollars, for manufacturing a designer chair is given by

$$C = 12{,}875 + 55x$$

where x is the number of chairs produced. These chairs sell for $180 each;

the revenue from x chairs is, therefore,

$$R = 180x$$

(a) How many chairs must the manufacturer sell each year to break even?
(b) What must the annual revenue be for the manufacturer to break even?
(c) What are the coordinates of the break-even point?

36. *Equilibrium point* The supply (S) and demand (D) in hundreds of units for a product are given by

$$S = 40x \quad \text{and} \quad D = 100 - \tfrac{3}{4}x$$

where x denotes the price in dollars. Determine the equilibrium price and the equilibrium quantity.

37. *Chemical additive mix* A contract for marine paint specifies that the paint shall contain the following minimum amounts of rust- and corrosion-resistant chemicals A, B, and C:

CHEMICAL	NUMBER OF UNITS
A	68
B	65
C	77

The paint firm who won the contract has available three additives containing these chemicals. Each gallon of the first additive contains 1 unit of chemical A, 1 of chemical B, and 1 of chemical C. Each gallon of the second contains 2 units of A, none of B, and 0.5 unit of C. Each gallon of the third contains none of A and 1.5 units each of B and C. How many gallons of each additive should the firm use to meet the minimum specifications?

38. The sum of three numbers is 66. The first number is equal to the sum of the second and third. The second number is 3 less than twice the third. What are the three numbers?

Input-output model The given table is the input-output table for an economy for a single year. All entries are in millions of dollars. Determine the output level for each industry if the forecast demand vector for next year is as given.

PRODUCTION (OUTPUT)	CONSUMPTION (INPUT)			TOTAL
	R	S	Consumer	
R	16	40	24	80
S	40	60	20	120

39. $D = \begin{bmatrix} 25 \\ 18 \end{bmatrix}$

40. $D = \begin{bmatrix} 23 \\ 20 \end{bmatrix}$

41. *Agribusiness* Identify clearly each step of the four-step process involved in the application of mathematics in the following problem and its solution: A farmer raises both wheat and corn. It takes 5 pounds of fertilizer and 13 hours of labor to produce 1 ton of wheat, costing a total of $46.30 per ton. It takes 4 pounds of fertilizer and 15 hours of labor to produce 1 ton of corn, costing a total of $51.30 per ton. How much does the farmer pay for a pound of fertilizer and for an hour of labor?

REVIEW EXERCISES

Let x denote the cost of a pound of fertilizer and y the cost of an hour of labor. Then

$$5x + 13y = 46.30$$
$$4x + 15y = 51.30$$

We need to determine the value of x and of y. The augmented coefficient matrix for the system of equations is

$$\begin{bmatrix} 5 & 13 & | & 46.30 \\ 4 & 15 & | & 51.30 \end{bmatrix}$$

which reduces to

$$\begin{bmatrix} 1 & 0 & | & 1.2 \\ 0 & 1 & | & 3.1 \end{bmatrix}$$

Therefore, $x = 1.2$ and $y = 3.1$. The farmer pays \$1.20 for a pound of fertilizer and \$3.10 for an hour of labor.

CHAPTER 4
LINEAR PROGRAMMING

OUTLINE

4.1 Graphing Linear Inequalities
4.2 Linear Programming—Geometric Method
4.3 Basic Solutions of Systems of Linear Equations
4.4 Linear Programming—Simplex Method
*4.5 Minimization
4.6 Chapter Summary

Review Exercises

A trust manager intends to invest up to a total of $30,000 in class A bonds yielding 10% annually and class AA bonds yielding 8% annually. Furthermore, she believes that the amount invested in class A bonds should be at most one-third of the amount invested in class AA bonds. How much should she invest in each type of bond to maximize the income from the trust?

In order to describe this situation mathematically, let x be the amount to be invested in class A bonds, y be the amount to be invested in class AA bonds, and z be the annual income from these investments. Then

$$z = 0.10x + 0.08y$$

The constraints on x and y also require that

$$x + y \leq 30{,}000 \quad \text{and} \quad x \leq \tfrac{1}{3}y$$

In addition, the amounts to be invested cannot be negative, so that

$$x \geq 0 \quad \text{and} \quad y \geq 0$$

The mathematical problem of maximizing

$$z = 0.10x + 0.08y$$

subject to

$$x + y \leq 30{,}000$$
$$x \leq \tfrac{1}{3}y$$
$$x \geq 0$$
$$y \geq 0$$

is an example of a linear programming problem. The expression to be maximized is

$$z = 0.10x + 0.08y$$

It is called the **objective function** and is linear in x, y, and z. Furthermore, the four inequality constraints on x and y are called *linear inequalities*. (Section 4.1 considers linear inequalities in more detail.) In general, a **linear programming problem** seeks to maximize (or

---
* This section can be omitted without loss of continuity.

minimize) a linear objective function subject to a number of linear inequality constraints.

We shall consider two methods of solving a linear programming problem—a geometric method and an algebraic method. The geometric method requires that we be able to graph linear inequalities. Our first consideration will be how such graphs are obtained.

4.1 GRAPHING LINEAR INEQUALITIES

A *linear inequality* in two variables x and y is an expression that can be written in one of the forms

$$ax + by \leq c \quad ax + by \geq c$$
$$ax + by < c \quad ax + by > c$$

where a, b, and c are constants with a and b not both 0

Note that the form of a linear inequality is the same as that of a linear equation except that the equals sign is replaced by one of the four inequality signs.

A *solution of an inequality* is a set of pairs of values (x, y) that satisfy the inequality. For

$$x - 3y \leq 6$$

$x = 4$ and $y = 6$ is a solution because the inequality

$$4 - 3(6) \leq 6$$

holds when these values are substituted for x and y; the left side equals -14, which is less than or equal to 6. Furthermore,

$$7 - 3(20) \leq 6$$

so that $x = 7$ and $y = 20$ is another solution. In general, a linear inequality has infinitely many solutions. On the other hand, $x = 35$ and $y = 7$ is not a solution, since

$$35 - 3(7) \nleq 6$$

The *graph of an inequality* is the set of all points whose coordinates satisfy the inequality, and to *graph an inequality* means to locate all those points in a coordinate system. Before attempting to construct graphs of inequalities, consider how the line

$$x - 3y = 6$$

obtained by replacing $\leq$ by $=$ in our earlier inequality, divides the plane into two parts, or regions, as shown in Figure 4.1.

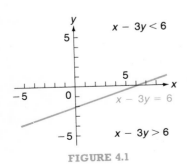

FIGURE 4.1

The coordinates of *all points above* the line satisfy the inequality $x - 3y < 6$. If the coordinates of the point $(-2, 3)$, for example, are substituted, we have $-2 - 3(3) < 6$. Similarly, the coordinates of *all points below* the line satisfy the inequality $x - 3y > 6$. For the point $(4, -4)$, for instance, we have $4 - 3(-4) > 6$. And, of course, *all points on* the line satisfy the equation $x - 3y = 6$.

The graph of the inequality

$$x - 3y \leq 6$$

is all points on and above the line; the graph of

$$x - 3y \geq 6$$

is all points on and below the line.

In general, the graph of a linear inequality will be one of the half-planes determined by the graph (a line) of the corresponding linear equation. The line itself is included only if equality is allowed. To graph a linear inequality, we need only graph the corresponding linear equation and pick a test point on either side of the resulting line to see which inequality sign corresponds to each half-plane. The procedure is illustrated in the next several examples.

EXAMPLE 1 Graph the inequality $2x + 4y \leq -3$.

Solution The corresponding linear equation is $2x + 4y = -3$, whose graph is shown in the given coordinate system.

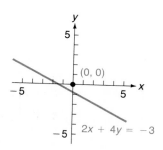

We now pick any point not on the line; $(0, 0)$ is the easiest with which to work. Substituting these values for x and y into the left side of the inequality, we get

$$2(0) + 4(0) > -3$$

Consequently, the whole half-plane above the line satisfies

$$2x + 4y > -3$$

Therefore, the graph of $2x + 4y \leq -3$ is the lower half-plane together with the line itself, as shown below:

EXAMPLE 2 Graph the linear inequality $2x + 3y \geq -6$.

Solution We graph the equation $2x + 3y = -6$ and pick a point—for example, $(-1, -1)$—to determine which inequality corresponds to each half-plane.

$$2(-1) + 3(-1) = -5 > -6$$

Since the coordinates of the point $(-1, -1)$ satisfy the given inequality, the graph consists of the half-plane containing the point $(-1, -1)$.

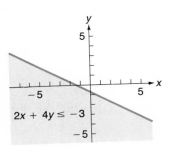

1 PRACTICE PROBLEM 1

Graph the linear inequality $x - 2y \leq 3$.

Answer

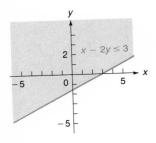

EXAMPLE 3 Graph the system of inequalities

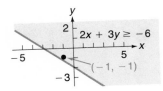

$$-2x + y \leq 2$$
$$4x + 3y \leq 12$$
$$x \geq 0$$

Solution In this case we want to locate all the points whose coordinates satisfy *each* of the given inequalities. Normally, this is done in a single coordinate system. However, in this example we graph each of the inequalities in a separate coordinate system using the procedure explained earlier, and then combine the results in a single coordinate system.

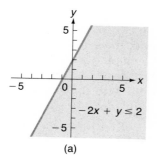
(a)

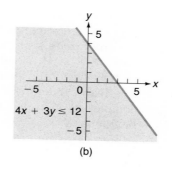
(b)

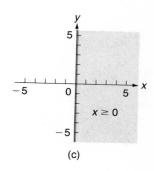
(c)

The coordinates of the points that lie in *all three* of these regions satisfy each of the inequalities. Consequently, the points that lie in the area in the plane where all three regions overlap form the graph of the given system:

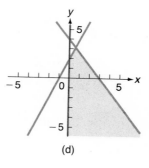
(d)

EXAMPLE 4 Graph the system of inequalities

$$x + 2y \leq 6$$
$$-2x + y \geq -4$$
$$x \geq 0$$
$$y \geq 0$$

Solution Graphing all four of the corresponding linear equations in the same coordinate system and proceeding as before, we obtain the shaded region in the given coordinate system.

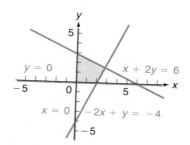

Compare the general shape of the graphs of the systems in Examples 3 and 4. In Example 4 the graph is contained within a finite portion of the coordinate system; we could, if we so desired, draw a circle with its center at the origin that completely encircles the graph of the system. The graph is said to be *bounded*. In Example 3 the graph extends downward indefinitely; no circle with its center at the origin would completely encircle the graph. In this latter case the graph is said to be *unbounded*.

Formally, the graph of a system of linear inequalities is said to be **bounded** if it can be completely encircled by a circle with its center at the origin and is said to be **unbounded** otherwise. This distinction will be important in our discussion of the solution of linear programming problems in the next sections.

2 PRACTICE PROBLEM 2

(a) Graph the system of inequalities.
$$x - 2y \leq 3$$
$$x + y \leq 2$$
$$x \geq 0$$

(b) Is the graph bounded or unbounded?

Answer

(a)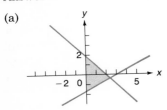

(b) Bounded

EXERCISES

Graph each inequality.

1. $-x + y \leq 2$
2. $3x - 2y \geq -6$
3. $2x + 4y \geq 6$
4. $2x - y \leq 4$
5. $-x - 4y \leq 8$
6. $x + 4y \geq -8$
7. $y \leq -2$
8. $x \geq 3$
9. $y \leq 0$
10. $x \geq 0$
11. $\frac{1}{2}x - \frac{1}{4}y \leq 1$
12. $0.4x + 0.5y \geq 0$

Graph each system of inequalities.

13. $2x + 3y \leq 18$
    $-4x + y \geq -8$

14. $x + y \geq 4$
    $x - y \leq 4$

15. $5y + 4x \leq 15$
    $5y - 3x \geq -20$

16. $3y - 4x \geq 12$
    $x \leq 2$

17. $2x + 3y \leq 18$
    $-4x + y \geq -8$
    $x \geq 0$
    $y \geq 0$

18. $5y + 4x \leq 15$
    $5y - 3x \geq -20$
    $x \geq 0$

19. $8y - 3x \leq 17$
    $3y - 7x \geq -23$
    $5y + 4x \geq -7$

20. $8y - 3x \leq 17$
    $3y - 7x \geq -23$
    $5y + 4x \geq -7$
    $y \geq 0$

21. $x - y \leq 0$
    $-3x + y \leq -2$

22. $x - y \leq 0$
    $-3x + y \leq -2$
    $x \leq 4$

Determine whether the graph in the indicated exercise is bounded or unbounded.

23. Exercise 13
24. Exercise 14
25. Exercise 15
26. Exercise 16
27. Exercise 17
28. Exercise 18
29. Exercise 19
30. Exercise 20
31. Exercise 21
32. Exercise 22

Explain the meaning of each term.

33. A linear inequality in the two variables x and y
34. A linear programming problem

35. A solution of a linear inequality in two variables x and y
36. The graph of a linear inequality
37. The graph of a system of linear inequalities
38. A bounded graph of a system of linear inequalities
39. An unbounded graph of a system of linear inequalities

4.2 LINEAR PROGRAMMING—GEOMETRIC METHOD

Consider the following linear programming problem:
Maximize
$$z = 2x + y$$
subject to
$$2x + 3y \leq 18$$
$$-4x + y \geq -8$$
$$x \geq 0$$
$$y \geq 0$$

The solution of the problem consists of values of x and y that satisfy all four inequalities and *give the largest value of z*. Consequently, the solution will be the coordinates of a point in the graph of the system made up of the four given inequalities. The coordinates of any point in this graph is said to be a **feasible solution** of the problem. Our task, then, is to find the feasible solution that gives the greatest value of z.

In the coordinate system shown in Figure 4.2, we have graphed the system of inequalities for this linear programming problem, along with the objective function
$$z = 2x + y$$

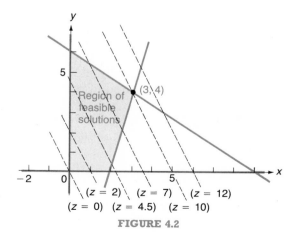

FIGURE 4.2

for several values of z. The line labeled $z = 7$, for example, is the graph of
$$7 = 2x + y$$
For the coordinates of all points on this line, the objective function has the value 7.

We can consider all the lines $z = 2x + y$ to be obtained by moving one of them (say the one corresponding to $z = 2$) parallel to itself across the region of feasible solutions. As the line moves, the value of z increases (or decreases, depending on the direction of movement) until the line corresponding to the maximum (or minimum) value of z intersects the region of feasible solutions at one of the corner points determined by the inequalities.

In this case, the maximum value of z for points in the region of feasible solutions is 10, and the corresponding line intersects the region at the point (3, 4). The solution for the problem is, therefore, $x = 3$, $y = 4$, which yields a maximum value of $z = 10$.

Note also that the minimum value of z—namely, $z = 0$—occurs at (0, 0), another corner point.

The above discussion suggests the following theorem, which is proved in more advanced texts.

THEOREM 1

Suppose that the region of feasible solutions for a linear programming problem is bounded (and contains the line segments making up its boundary, which will always be the case in this text). Then the maximum and minimum values of the objective function occur at corner points.

Because the maximum and minimum values of the objective function occur at corner points, to solve a linear programming problem for which the region of feasible solutions is bounded, we need only determine the corner points of the region and then determine which corner point gives the maximum or minimum value of z. Graphing the region of feasible solutions is not absolutely necessary; however doing so will usually help in identifying the corner points. The procedure is illustrated in the following examples.

EXAMPLE 1 Solve the following linear programming problem:
Minimize
$$z = 2x + 4y$$
subject to
$$5y - 6x \leq 8$$
$$y + 8x \leq 20$$
$$3y + x \geq -9$$

Solution Graphing the system of inequalities gives the region of feasible solutions shown in the figure.

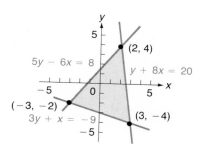

The coordinates for each corner point come from solving the appropriate system of linear equations. For example, (2, 4) is the result of solving the system

$$y + 8x = 20$$
$$5y - 6x = 8$$

We then determine the value of z at each of the corner points. From the table we see that the minimum value of z—namely, -14—occurs when $x = -3$ and $y = -2$.

CORNER POINT	z
(2, 4)	20
(−3, −2)	−14
(3, −4)	−10

EXAMPLE 2 Determine the maximum value of $z = 3x - 2y$ subject to the same constraints as in Example 1.

Solution Since we already know the corner points, all we need to do is construct the table. The maximum value of z—namely, 17—occurs at $x = 3$ and $y = -4$.

CORNER POINT	z
(2, 4)	−2
(−3, −2)	−5
(3, −4)	17

EXAMPLE 3 Solve the linear programming problem of the trust manager in the illustration at the beginning of this chapter.

Solution The problem is:
 Maximize

$$z = 0.10x + 0.08y$$

subject to

$$x + y \leq 30{,}000$$
$$x - \tfrac{1}{3}y \leq 0$$
$$x \geq 0$$
$$y \geq 0$$

First, we graph the region of feasible solutions, and then we set up a table listing the corner points.

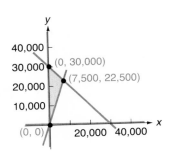

CORNER POINT	z
(7,500, 22,500)	2,550
(0, 30,000)	2,400
(0, 0)	0

The maximum value of z occurs at $x = 7{,}500$ and $y = 22{,}500$. That is, she should invest $7,500 in class A bonds and $22,500 in class AA bonds for a return of $2,550.

The four-step procedure in the application of mathematics, discussed in Section 1.3, was used again in Example 3:

1. The physical situation was described in the first paragraph of this chapter. The question was how much should be invested in each type of bond for the maximum return and what would be the maximum return.
2. The mathematical model was the linear programming problem, along with the graph of Example 3. The mathematical question was to find the values of x and y that maximize z.
3. The mathematical solution was $x = 7{,}500$ and $y = 22{,}500$ for a maximum value of $z = 2{,}550$.
4. The interpretation was given in the last sentence of Example 3.

Each of the previous examples had a single solution. However, two special cases—more than one solution or no solution—also arise. Examples 4 and 5 illustrate how these cases can be detected.

PRACTICE PROBLEM 1

Solve the linear programming problems:

(a) Maximize
$$z = x + 2y$$
subject to
$$3y + 2x \leq 12$$
$$7y - 3x \geq 5$$
$$4y - 5x \leq 16$$

(b) Minimize $z = -2x + y$ subject to the constraints in part (a).

Answer

(a) $z = 8$ at $x = 0, y = 4$
(b) $z = -4$ at $x = 3, y = 2$

EXAMPLE 4 Solve the following linear programming problem:

Maximize
$$z = x + y$$
subject to
$$x + y \leq 3$$
$$x \geq 0$$
$$y \leq 2$$
$$y \geq 0$$

Solution The graph and table of corner points are:

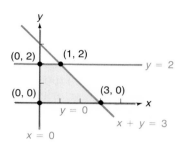

CORNER POINT	z
(0, 0)	0
(0, 2)	2
(1, 2)	3
(3, 0)	3

The table indicates that the maximum value of z is obtained at two of the corner points, and each corner point yields a solution. The reason that there are two solutions in this case is that the graph of the objective function with $z = 3$,

$$3 = x + y$$

lies on one of the boundary lines for the region of feasible solutions. Consequently, every point on that line from (1, 2) to (3, 0) yields a solution; there are *infinitely many* solutions. These solutions can be described collectively as the set of all x and y where

$$x + y = 3 \quad \text{and} \quad 1 \leq x \leq 3$$

In any problem, the indication that there is more than one solution will be the fact that the maximum value of z (or minimum, as the case may be) is attained at two different corner points.

2 PRACTICE PROBLEM 2

Solve the linear programming problem:
Minimize
$$z = y - x$$
subject to
$$2y + x \leq 14$$
$$y - x \geq 1$$
$$x \geq 0$$

Answer
$z = 1$ for $y - x = 1$ and $0 \leq x \leq 4$

EXAMPLE 5 Solve the following linear programming problem:
Maximize
$$z = x + 2y$$
subject to
$$3y + 4x \geq 12$$
$$5y + 2x \geq 10$$

Solution In this case the region of feasible solutions is the unbounded portion of the coordinate system above the two lines. Because the region of interest is unbounded, the possibility of no solution arises. Notice that

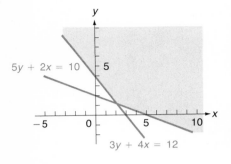

by picking points on the y-axis, $(0, 5)$, $(0, 9)$, $(0, 20)$, $(0, 100)$, and so on, we get successively larger values of z. We can get values of z as large as we wish by picking the point high enough on the y-axis. Consequently, there is no point in the region that yields a maximum value of z. There is no solution in this case.

If the region of feasible solutions is unbounded, there is no general method for determining whether a linear programming problem has a solution or not. We must carefully examine the values of the objective function in the region of feasible solutions. For example, if the problem is to minimize $z = x + 2y$ subject to the same inequalities as Example 5, the minimum value is $z = \frac{31}{7}$, occurring at the corner point $(\frac{15}{7}, \frac{8}{7})$. One way to determine this minimum is to follow the procedure illustrated on page 119, constructing the graph of $z = x + 2y$ for different values of z.

3 PRACTICE PROBLEM 3

Solve the linear programming problem:
Minimize
$$z = 3y + x$$
subject to
$$x \geq 0$$
$$x \leq 4$$
$$2y - x \leq -6$$

Answer

No solution

EXERCISES

Solve each linear programming problem.

1. Maximize
$$z = y - 3x$$
subject to
$$5y + 2x \leq 30$$
$$y - 4x \geq -16$$
$$x \geq 0$$
$$y \geq 0$$

2. Minimize $z = y - 3x$ subject to the constraints in Exercise 1.

3. Maximize $z = y + 3x$ subject to the constraints in Exercise 1.

4. Minimize
$$z = -3x - 4y$$
subject to
$$x + y \leq 5$$
$$y - x \geq -3$$
$$x \geq 1$$
$$y \geq 0$$

5. Maximize $z = 5x + 2y$ subject to the constraints in Exercise 4.

6. Maximize
$$z = 7x - 10y$$
subject to
$$5y + x \leq 20$$
$$3y - 5x \geq -16$$
$$x \geq 0$$
$$y \geq -2$$

7. Minimize $z = \frac{1}{2}y + \frac{3}{4}x$ subject to the constraints in Exercise 6.

8. Maximize
$$z = 4y - 5x$$
subject to
$$x + y \leq 3$$
$$-5x + 4y \leq 30$$
$$x \leq 0$$
$$y \geq 0$$

9. Maximize
$$z = x + 5y$$
subject to
$$4y + 3x \geq 12$$
$$4y - 3x \leq -12$$

10. Minimize
$$z = 3y + 4x$$
subject to
$$7y + 3x \leq -21$$
$$y \leq 0$$

11. Minimize
$$z = y - x$$
subject to
$$3y - 4x \leq -12$$
$$x \geq 0$$

12. Maximize
$$z = 3x + 6y$$
subject to
$$2y + x \leq 10$$
$$4y - 9x \geq -24$$
$$x \geq 0$$

13. *Manufacturing* A company manufactures two items, A and B. Each unit of item A requires $1\frac{1}{2}$ worker-hours for assembling and $\frac{1}{4}$ worker-hour for painting. Each unit of item B requires 1 worker-hour for assembling and $\frac{2}{3}$ worker-hour for painting. The total number of worker-hours available each day is 336 for assembling and 116 for painting. The company makes a profit of $1 on each unit of item A sold and 75¢ on each unit of item B sold. The company can sell all units of both items that it can manufacture and wants to determine how many units of each item it should produce each day to maximize its profit.
   (a) Let x denote the number of units of item A the company should produce each day to maximize profit and y denote the corresponding number of units of item B. Write an equation to express the daily profit (z) in terms of x and y.
   (b) Write an inequality in x and y to express the fact that the maximum number of worker-hours which can be used each day for assembling is 336.
   (c) Write an inequality in x and y to express the fact that the maximum number of worker-hours that can be used each day for painting is 116.
   (d) Write two inequalities to express the fact that the number of units of each item to be produced each day must be zero or greater.
   (e) Determine the number of units of each item the company should produce each day for a maximum daily profit.

14. *Manufacturing* How many units of each item should the company in Exercise 13 produce each day if the profit is $1 on each unit of item A and 50¢ on each unit of item B?

15. *Chemical utilization* A researcher for a pharmaceutical company is experimenting with two new mouthwashes, type I and type II. Each quart of type I requires 1 ounce of chemical A, 3 ounces of chemical B, and 1 ounce of chemical C. Each quart of type II requires 3 ounces of chemical A, 5 ounces of chemical B, and 1 ounce of chemical C. There are 36 ounces of chemical A, 84 ounces of chemical B, and 20 ounces of chemical C available. The researcher wishes to make the maximum number of quarts possible with the available chemicals.
    (a) Let x denote the number of quarts of type I made and y the number of quarts of type II. Write an equation to express the total number (z) of quarts made.
    (b) Write an inequality in x and y to express the fact that the maximum amount of chemical A that can be used is 36 ounces.
    (c) Write an inequality in x and y to express the fact that the maximum amount of chemical B that can be used is 84 ounces.
    (d) Write an inequality in x and y to express the fact that the maximum amount of chemical C that can be used is 20 ounces.
    (e) Write two inequalities to express the fact that the number of quarts of each type must be zero or greater.
    (f) Determine the maximum number of quarts of mouth wash that the researcher can make.

16. *Chemical utilization* If the researcher in Exercise 15 decides that at least 5 quarts of the type II mouth wash are required, what is the maximum number of quarts that can be made?

17. *Service utilization* An hotel is refurbishing its rooms, each of which will accommodate either one or two persons. A one-person room requires 1.5 hours service (cleaning and maintenance) each day and a two-person room requires 2 hours of service each day. The hotel is able to refurbish up to 60 rooms and has available up to 100 hours a day for service. If at least 20 of the rooms must be one-person rooms, how many rooms of each type should be established in order to accommodate the most people? What is this maximum number of people?

18. *Hotel rental* In Exercise 17 if a one-person room rents for $40.00 a day and a two-person room for $50.00 a day, how many rooms of each type should be established to obtain the maximum daily rental? What is this maximum rental?

19. *Air pollution* A manufacturing company produces two items, A and B. The manufacturing process emits 1.5 cubic feet of carbon monoxide and 3 cubic feet of sulfur dioxide per unit of item of A and 2 cubic feet of both carbon monoxide and sulfur dioxide per unit of item B. Government pollution standards allow the company to emit a maximum of 4,775 cubic feet of carbon monoxide and 5,500 cubic feet of sulfur dioxide per week. The company can sell all of items A and B that it can produce and makes a profit of $2.00 per unit of item A and $1.00 per unit of item B. Determine the number of units of each item the company should produce each week to maximize its profit without exceeding government pollution standards.

20. *Diet mix* A hospital dietician is preparing a special diet that is to contain at least 375 units of calcium, 180 units of iron, and 300 units of vitamin A. The table gives the units of calcium, iron, and vitamin A per

	FOOD I	FOOD II
CALCIUM	15	25
IRON	12	10
VITAMIN A	10	30

ounce for the two foods that the dietician intends to use. Each ounce of food I contains 40 calories and each ounce of food II contains 60 calories. How many ounces of each food should the dietician use to minimize the calorie content? What is the minimum calorie content?

21. *Psychology study* A psychologist studying the learning process with the help of two assistants intends to observe one group of children in a mathematics learning environment and another group in a reading learning environment. The psychologist and two assistants will each observe every child in both groups but in different environments and during different time intervals. The time (in minutes) required for each to observe a child is given in the table. The psychologist has 8 hours available for the observations, the first assistant has 5 hours available, and the second assistant has $4\frac{1}{2}$ hours available. How many children should they choose for each group in order to maximize the total number of children observed? What is this maximum total?

	MATH	READING
PSYCHOLOGIST	15	20
ASSISTANT I	12	10
ASSISTANT II	10	10

22. (a) Give the three main steps in the geometric procedure of this section for solving a linear programming problem.
    (b) Under what conditions are we sure that the objective function of a linear programming problem has a maximum or minimum?
    (c) Under what conditions are we not sure that the objective function of a linear programming problem has a maximum or minimum?

Identify the steps of the four-step procedure involved in the application of mathematics (Section 1.3) in each exercise.

23. *Manufacturing* Exercise 13
24. *Chemical utilization* Exercise 15

4.3 BASIC SOLUTIONS OF SYSTEMS OF LINEAR EQUATIONS

In the last section, the objective function z was given in terms of the two variables x and y. Furthermore, the inequality constraints were all in terms of the same two variables. If these were expressed in terms of even three variables, the (two-dimensional) geometric method of the last section would no longer be applicable. Another dimension would be required for each additional variable. In actual applications, the number of variables might be a hundred or more! Our geometric method would quickly be overpowered.

The *simplex* method of the next section presents a method used to solve these larger linear programming problems. Before discussing the simplex method itself, we need to consider a preliminary concept, that of a *basic solution* for a system of linear equations.

Suppose that we wish to solve the system of equations

$$2x + 4y + w + u + 2v = 20$$
$$3x + 7y + 3w + u + v = 18$$
$$x + y + w + 2u + v = 10$$

expressing x, y, and w in terms of u and v. We set up the augmented coefficient matrix

$$\begin{array}{c} \begin{array}{ccccc} x & y & w & u & v \end{array} \\ \begin{bmatrix} 2 & 4 & 1 & 1 & 2 & | & 20 \\ 3 & 7 & 3 & 1 & 1 & | & 18 \\ 1 & 1 & 1 & 2 & 1 & | & 10 \end{bmatrix} \end{array}$$

and, using the usual row-reduction operations, reduce this matrix to

$$\begin{array}{c} \begin{array}{ccccc} & x & y & w & u & v \end{array} \\ \begin{array}{c} x \\ y \\ w \end{array} \begin{bmatrix} 1 & 0 & 0 & \frac{11}{4} & \frac{5}{2} & | & 19 \\ 0 & 1 & 0 & -\frac{5}{4} & -\frac{1}{2} & | & -3 \\ 0 & 0 & 1 & \frac{1}{2} & -1 & | & -6 \end{bmatrix} \end{array}$$

from which it follows that

$$x + \tfrac{11}{4}u + \tfrac{5}{2}v = 19$$
$$y - \tfrac{5}{4}u - \tfrac{1}{2}v = -3$$
$$w + \tfrac{1}{2}u - v = -6$$

or

$$x = 19 - \tfrac{11}{4}u - \tfrac{5}{2}v$$
$$y = -3 + \tfrac{5}{4}u + \tfrac{1}{2}v$$
$$w = -6 - \tfrac{1}{2}u + v$$

The original system has infinitely many solutions, one for every choice of u and v. For example, if $u = 4$ and $v = 2$, we get

$$x = 3 \qquad y = 3 \qquad w = -6$$
$$u = 4 \qquad v = 2$$

If $u = 4$ and $v = -4$, we get

$$x = 18 \qquad y = 0 \qquad w = -12$$
$$u = 4 \qquad v = -4$$

In particular, if we let $u = 0$ and $v = 0$, we get

$$x = 19 \qquad y = -3 \qquad w = -6$$
$$u = 0 \qquad v = 0$$

This last solution with u and v equal to zero is called a **basic solution;** x, y, and w are called the **basic variables;** u and v are called the **nonbasic variables.**

The basic solution was determined as follows: The system of equations was solved by the Gauss-Jordan method to obtain the basic variables in terms of the nonbasic variables; then each nonbasic variable was set equal to 0.

4.3 BASIC SOLUTIONS OF SYSTEMS OF LINEAR EQUATIONS

Several things should be observed in the final matrix:

1. The columns corresponding to the basic variables form an identity matrix.
2. The columns corresponding to the nonbasic variables have no special form.
3. The column on the extreme right gives the basic solution.
4. The variable on the left of each row is the basic variable which that row expresses in terms of the nonbasic variables.

Throughout this section, the number of basic variables will always be equal to the number of equations, and the remaining variables will be nonbasic.

In the illustration above, we arbitrarily chose x, y, and w to be the basic variables. We could, for example, make x, w, and v the basic variables by row-reducing to obtain the identity matrix in the columns corresponding to these variables. To do so, we start with the same augmented coefficient matrix,

$$\begin{array}{c} \\ \\ \\ \end{array} \begin{array}{ccccc} x & y & w & u & v \end{array}$$
$$\begin{bmatrix} 2 & 4 & 1 & 1 & 2 & | & 20 \\ 3 & 7 & 3 & 1 & 1 & | & 18 \\ 1 & 1 & 1 & 2 & 1 & | & 10 \end{bmatrix}$$

and, without interchanging the columns, row-reduce to obtain

$$\begin{array}{c} x \\ w \\ v \end{array} \begin{bmatrix} 1 & 5 & 0 & -\frac{7}{2} & 0 & | & 4 \\ 0 & -2 & 1 & 3 & 0 & | & 0 \\ 0 & -2 & 0 & \frac{5}{2} & 1 & | & 6 \end{bmatrix}$$

Then

$$x = 4 - 5y + \tfrac{7}{2}u$$
$$w = 0 + 2y - 3u$$
$$v = 6 + 2y - \tfrac{5}{2}u$$

The basic variables are x, w, and v; the basic solution is $x = 4$, $w = 0$, $v = 6$, $y = 0$, and $u = 0$.

EXAMPLE 1 For the given row-reduced augmented coefficient matrix, determine the basic variables, the nonbasic variables, and the corresponding basic solution.

$$\begin{array}{ccccc} x & y & w & s & t \end{array}$$
$$\begin{bmatrix} 3 & 0 & 1 & 2 & 0 & | & 1 \\ 4 & 1 & 0 & -1 & 0 & | & -1 \\ -6 & 0 & 0 & 4 & 1 & | & 8 \end{bmatrix}$$

---

1 PRACTICE PROBLEM 1

Given the row-reduced matrix:

$$\begin{array}{ccccc} x & y & z & u & v \end{array}$$
$$\begin{bmatrix} 1 & 0 & 0 & 2 & -3 & | & 3 \\ 0 & 1 & 0 & 0.5 & 0 & | & -5 \\ 0 & 0 & 1 & -1 & 1 & | & 8 \end{bmatrix}$$

(a) What are the basic variables?
(b) What are the nonbasic variables?
(c) What is the basic solution?

Answer

(a) x, y, z (b) u and v
(c) $x = 3$, $y = -5$, $z = 8$, $u = 0$, $v = 0$

4 LINEAR PROGRAMMING

Solution   The columns under y, w, and t form an identity matrix, so these are the basic variables; then x and s are nonbasic. The corresponding basic solution is $x = 0$, $y = -1$, $w = 1$, $s = 0$, and $t = 8$.

EXAMPLE 2   In the following system of equations, let x and t be the basic variables. Write the basic variables in terms of the nonbasic variables and find the corresponding basic solution:

$$2x + 3y + t + u = 10$$
$$x + y + u = 7$$

Solution   The augmented coefficient matrix is

$$\begin{array}{cccc} x & y & t & u \end{array}$$
$$\begin{bmatrix} 2 & 3 & 1 & 1 & | & 10 \\ 1 & 1 & 0 & 1 & | & 7 \end{bmatrix}$$

which row-reduces to

$$\begin{array}{cccc} & x & y & t & u \end{array}$$
$$\begin{array}{c} x \\ t \end{array} \begin{bmatrix} 1 & 1 & 0 & 1 & | & 7 \\ 0 & 1 & 1 & -1 & | & -4 \end{bmatrix}$$

Then

$$x = 7 - y - u$$
$$t = -4 - y + u$$

The corresponding basic solution is $x = 7$, $t = -4$, $y = 0$, and $u = 0$.

EXAMPLE 3   Repeat Example 2, letting y and t be the basic variables.

Solution   The augmented coefficient matrix of Example 2 also row-reduces to

$$\begin{array}{cccc} & x & y & t & u \end{array}$$
$$\begin{array}{c} y \\ t \end{array} \begin{bmatrix} 1 & 1 & 0 & 1 & | & 7 \\ -1 & 0 & 1 & -2 & | & -11 \end{bmatrix}$$

PRACTICE PROBLEM 2

Given the row-reduced matrix:

$$\begin{array}{ccccc} x & y & z & u & v \end{array}$$
$$\begin{bmatrix} 2 & 0 & \frac{1}{2} & 0 & 1 & | & -9 \\ -4 & 1 & 0 & 0 & 0 & | & 12 \\ 3 & 0 & 5 & 1 & 0 & | & 18 \end{bmatrix}$$

(a) What are the basic variables?
(b) What are the nonbasic variables?
(c) What is the basic solution?

Answer

(a) y, u, v (b) x, z
(c) $x = 0, y = 12, z = 0, u = 18, v = -9$

Then

$$y = 7 - x - u$$
$$t = -11 + x + 2u$$

The corresponding basic solution is $y = 7$, $t = -11$, $x = 0$, and $u = 0$.

③ PRACTICE PROBLEM 3

Given

$$x - 3y + 2t - u = 5$$
$$4x - 6y + 2t + 5u = 8$$

Write the basic variables in terms of the nonbasic variables and find the corresponding basic solution for each pair of basic variables.

(a) x and y

(b) x and t

Answer

(a) $x = -1 + t - \frac{7}{2}u$; $y = -2 + t - \frac{3}{2}u$; $x = 1$, $y = -2$, $t = 0$, $u = 0$

(b) $x = 1 + y - 2u$; $t = 2 + y + \frac{3}{2}u$; $x = 1$, $y = 0$, $t = 2$, $u = 0$

EXERCISES

For each of the following row-reduced augmented coefficient matrices, identify the basic variables, the nonbasic variables, and the corresponding basic solution:

1.
$$\begin{matrix} x & y & s & t \\ \begin{bmatrix} 1 & 2 & -1 & 0 \\ 0 & 3 & 4 & 1 \end{bmatrix} & \begin{matrix} 7 \\ -6 \end{matrix} \end{matrix}$$

2.
$$\begin{matrix} x & y & u & v \\ \begin{bmatrix} 17 & 1 & 0 & 8 \\ -9 & 0 & 1 & 0 \end{bmatrix} & \begin{matrix} 14 \\ -12 \end{matrix} \end{matrix}$$

3.
$$\begin{matrix} x & y & s & t & u \\ \begin{bmatrix} 6 & -1 & 0 & 2 & 1 \\ 3 & 0 & 1 & -4 & 0 \end{bmatrix} & \begin{matrix} 3 \\ 12 \end{matrix} \end{matrix}$$

4.
$$\begin{matrix} x & y & u & v \\ \begin{bmatrix} 1 & 0 & 0.4 & 0.9 \\ 0 & 1 & -0.2 & 0.7 \end{bmatrix} & \begin{matrix} -0.5 \\ 0.3 \end{matrix} \end{matrix}$$

5.
$$\begin{matrix} x & y & w & s & t \\ \begin{bmatrix} 1 & 0 & 4 & 7 & 0 \\ 0 & 1 & -3 & 2 & 0 \\ 0 & 0 & -1 & 0 & 1 \end{bmatrix} & \begin{matrix} 4 \\ -3 \\ 1 \end{matrix} \end{matrix}$$

6.
$$\begin{matrix} x & y & w & s & t & u \\ \begin{bmatrix} 2 & 1 & 0 & 6 & 4 & 0 \\ -1 & 0 & 0 & 7 & 10 & 1 \\ 3 & 0 & 1 & -9 & 12 & 0 \end{bmatrix} & \begin{matrix} 0 \\ 7 \\ -9 \end{matrix} \end{matrix}$$

7.
$$\begin{matrix} x & y & w & s & t & u \\ \begin{bmatrix} 1 & 0 & 0 & \frac{1}{2} & \frac{2}{3} & -\frac{1}{2} \\ 0 & 0 & 1 & \frac{3}{4} & -\frac{2}{3} & \frac{5}{8} \\ 0 & 1 & 0 & \frac{9}{5} & 1 & 3 \end{bmatrix} & \begin{matrix} -\frac{1}{3} \\ \frac{3}{5} \\ \frac{4}{9} \end{matrix} \end{matrix}$$

For each of the following systems of equations, find the basic solution corresponding to the indicated basic variables:

8. $2x + 3y + s - 2t = 4$
   $x - 2y + 2s - 3t = 6$
   Basic variables: x and s

9. Use the system in Exercise 8 with basic variables x and y.

10. $x + y - 2s + 2t = 10$
    $-3x - 5y + 8s - 7t = -12$
    Basic variables: y and s

11. Use the system in Exercise 10 with basic variables x and t.

12. $7x - 9y + s = 12$
    $8x + 4y + t = 9$
    Basic variables: s and t

13. $x + 2y + 8w + t = 10$
    $ y + 3w + 2t = 3$
    $-x + y + 9w + s - t = 1$
    Basic variables: x, y, and t

14. Use the system in Exercise 13 with basic variables x, y, and s.

15. $x + y - 3w + s = 4$
    $2x + 3y - 6w + t = 2$
    $x - 2y - 4w + u = -1$
    Basic variables: s, t, and u

16. Use the system in Exercise 15 with basic variables x, y, and w.

For each of the following systems of equations, find at least four different basic solutions by choosing four different pairs of basic variables:

17. $x + y + u + 3v = 4$
    $x - y - 3u - v = 8$

18. $x - y + 2s - t = 0$
    $x + 2y + s + t = 6$

4.4 LINEAR PROGRAMMING— SIMPLEX METHOD

While the simplex method is designed for solving large linear programming problems, we will borrow a smaller problem from the exercises of Section 4.2 to illustrate the procedure.

A company manufactures two items, A and B. Each unit of item A requires $1\frac{1}{2}$ worker-hours for assembling and $\frac{1}{4}$ worker-hour for painting. Each unit of item B requires 1 worker-hour for assembling and $\frac{2}{3}$ worker-hour for painting. The total number of worker-hours available each day is 336 for assembling and 116 for painting.

The company makes a profit of \$1 on each unit of item A sold and 75¢ on each unit of item B sold. The company can sell all units of both items that it manufactures and wants to determine how many units of each item it should produce each day to maximize its profit.

We let x denote the number of units of item A the company should produce each day and y the number of units of item B. The linear programming problem is then:

Maximize
$$z = x + \tfrac{3}{4}y$$

subject to
$$\tfrac{3}{2}x + y \leq 336$$
$$\tfrac{1}{4}x + \tfrac{2}{3}y \leq 116$$
$$x \geq 0$$
$$y \geq 0$$

4.4 LINEAR PROGRAMMING—SIMPLEX METHOD

The **simplex method** first introduces a new variable for each of the inequalities, other than the nonnegativity inequalities, to convert these into equations. We will use s and t. Then the problem can be restated:

Maximize
$$z = x + \tfrac{3}{4}y$$
subject to

$$\left.\begin{array}{r}\tfrac{3}{2}x + y + s = 336 \\ \tfrac{1}{4}x + \tfrac{2}{3}y + t = 116\end{array}\right\} \quad (1)$$

$$\left.\begin{array}{r}x \geq 0 \\ y \geq 0 \\ s \geq 0 \\ t \geq 0\end{array}\right\} \quad (2)$$

The new variables, s and t in this case, are called **slack variables;** they "take up the slack" between the two sides of the original inequalities. In our illustration, s represents the amount of available assembling time unused when x units of item A and y units of item B are produced, and t represents the unused painting time. Both will be nonnegative, as indicated by the last two inequalities. Note that s and t do not enter into the evaluation of the objective function, so that the maximum value of z does not depend on the values of s and t. **1**

The feasible solutions are now the values of x, y, s, and t that satisfy both equations (1) and the four inequalities (2). The problem is to determine the feasible solution that maximizes the value of z. An important theorem of linear programming states that if there is a feasible solution that maximizes z in our problem, then there is a basic feasible solution that maximizes z. *Consequently, we need to examine only the solutions that are both basic and feasible.* The basic feasible solutions now correspond to the corner points in the geometric method of Section 4.2.

Table 4.1 gives all the basic feasible solutions for our problem, along with the value of the objective function for each such solution. The maximum value of z occurs at the last basic feasible solution; the company should produce 144 units of item A and 120 units of item B per day.

1 PRACTICE PROBLEM 1

Rewrite the linear programming problem below, replacing the first two constraints by equations containing slack variables:

Maximize
$$z = 2x + 3y$$
subject to
$$2x + 4y \leq 12$$
$$6x + 5y \leq 20$$
$$x \geq 0$$
$$y \geq 0$$

Answer

Maximize
$$z = 2x + 3y$$
subject to
$$2x + 4y + s = 12$$
$$6x + 5y + t = 20$$
$$x \geq 0$$
$$y \geq 0$$
$$s \geq 0$$
$$t \geq 0$$

TABLE 4.1

BASIC FEASIBLE SOLUTION				$z = x + \tfrac{3}{4}y$
$x = 0$	$y = 0$	$s = 336$	$t = 116$	0
$x = 0$	$y = 174$	$s = 162$	$t = 0$	130.5
$x = 224$	$y = 0$	$s = 0$	$t = 60$	224
$x = 144$	$y = 120$	$s = 0$	$t = 0$	234

For this problem, there are two additional basic solutions:

$$x = 0, \quad y = 336, \quad s = 0, \quad t = -108$$

and

$$x = 464, \quad y = 0, \quad s = -360, \quad t = 0$$

However, these two are not feasible. (Why?)

In larger problems, the number of basic feasible solutions becomes too large to check the value of the objective function at each one. It is not unusual to have several hundred such solutions. What the simplex method does is to move from one basic feasible solution to another, choosing the one that promises the greatest increase in the objective function. This is done in an organized manner, which eliminates the need to consider every possible basic feasible solution, until the maximum value of z is obtained.

In our illustration, suppose we start with the basic feasible solution

$$x = 0 \quad s = 336$$
$$y = 0 \quad t = 116$$

Then the value of the objective function is $z = 0$. Since $z = x + \frac{3}{4}y$, it appears that by increasing x we would obtain a greater increase in z than we would by increasing y. So we first move to a solution that contains x as a basic variable and in which y is still 0.

By how much can we increase x? From equations (1), with $y = 0$, we get

$$s = 336 - \frac{3}{2}x, \quad \text{which equals 0 for} \quad x = \frac{336}{\frac{3}{2}} = 224 \quad (3)$$

$$t = 116 - \frac{1}{4}x, \quad \text{which equals 0 for} \quad x = \frac{116}{\frac{1}{4}} = 464$$

We can increase x to 224 [increasing x to 464 would make s negative, which inequalities (2) prohibit]. Doing so makes $s = 0$ and $t = 60$, and we have obtained a new basic feasible solution,

$$x = 224 \quad s = 0$$
$$y = 0 \quad t = 60$$

from those in Table 4.1. From the equation

$$z = x + \tfrac{3}{4}y$$

we see that the corresponding value of z is 224, clearly an increase over the previous value of $z = 0$. Note that in moving from the previous basic feasible solution to the present one, x was the variable *entering* the set of basic variables and s was the variable *departing* from the set of basic variables.

The current solution has $y = 0$. We next look for a new basic feasible solution that increases the value of y and at the same time increases the value of the objective function. Following a process similar to that above, we obtain the solution

$$x = 144 \quad s = 0$$
$$y = 120 \quad t = 0$$

which gives a value of $z = 234$. Note that in moving from the previous solution to this one, y entered the set of basic variables and t departed.

The simplex method repeats this process of picking entering and departing variables for the set of basic variables until the maximum value of the objective function is obtained. This is usually done with all of the relevant information in table form, called a **simplex tableau.** Using the tableau enables all the calculations to be presented in simpler form. We demonstrate the complete process for the problem we have been considering:

Maximize

$$z = x + \tfrac{3}{4}y$$

subject to

$$\tfrac{3}{2}x + y + s = 336 \atop \tfrac{1}{4}x + \tfrac{2}{3}y + t = 116 \qquad (1)$$

$$\begin{aligned} x &\geq 0 \\ y &\geq 0 \\ s &\geq 0 \\ t &\geq 0 \end{aligned} \qquad (2)$$

The objective function is first rewritten

$$z - x - \tfrac{3}{4}y = 0$$

and the corresponding tableau is:

	x	y	s	t	
s	$\tfrac{3}{2}$	1	1	0	336
t	$\tfrac{1}{4}$	$\tfrac{2}{3}$	0	1	116
z	-1	$-\tfrac{3}{4}$	0	0	0

$336 \div \tfrac{3}{2} = 224$
$116 \div \tfrac{1}{4} = 464$

The first two rows in the tableau form the augmented coefficient matrix for equations (1); the last row gives the coefficients for the (rewritten) objective function. The initial basic variables are s and t; their values in the corresponding basic solution are given in the column on the right. The entry in the lower right-hand corner gives the value of z for the current basic feasible solution, $x = 0$, $y = 0$, $s = 336$, and $t = 116$. (Note that this is the initial solution considered in the discussion above.)

The entering variable is the variable with the greatest (in absolute value) negative coefficient in the objective function—in this case, the

2 PRACTICE PROBLEM 2

(a) Construct the simplex tableau for the linear programming problem in Practice Problem 1.
(b) Identify the departing variable, entering variable, and the pivot.

Answer

(a)

	x	y	s	t	
s	2	4	1	0	12
t	6	5	0	1	20
z	-2	-3	0	0	0

(b) Entering variable: y; departing variable: s; pivot: 4

3 PRACTICE PROBLEM 3

(a) Use pivoting to construct the next tableau for the linear programming problem in Practice Problem 2.
(b) What is the new basic feasible solution and what is the corresponding value of the objective function?

Answer

(a)

	x	y	s	t	
y	$\frac{1}{2}$	1	$\frac{1}{4}$	0	3
t	$\frac{7}{2}$	0	$-\frac{5}{4}$	1	5
z	$-\frac{1}{2}$	0	$\frac{3}{4}$	0	9

(b) $x = 0$, $y = 3$, $s = 0$, $t = 5$; $z = 9$

variable x indicated by the arrow below the tableau. The calculations to determine the departing variable are given on the right and are the same as in (3). The variable corresponding to the smallest nonnegative ratio on the right—in this case, s—is the departing variable. The entry in the row corresponding to the departing variable and the column corresponding to the entering variable is called the **pivot**—$\frac{3}{2}$ in this case.

To obtain the new basic feasible solution, the usual row-reduction techniques are applied to obtain a 1 in the pivot position and 0's everywhere else in the pivot column. This results in the following tableau:

	x	y	s	t	
x	1	$\frac{2}{3}$	$\frac{2}{3}$	0	224
t	0	$\frac{1}{2}$	$-\frac{1}{6}$	1	60
z	0	$-\frac{1}{12}$	$\frac{2}{3}$	0	224

The first two rows form the augmented coefficient matrix for the new basic feasible solution, $x = 224$, $y = 0$, $s = 0$, and $t = 60$; the lower right-hand corner gives the corresponding value of the objective function. These values are the same as those of the second solution in our earlier discussion. This process of obtaining a new tableau is called *pivoting*.

We repeat the process again, using the last tableau, to increase further the value of z.

To determine the entering basic variable, we look for the largest negative coefficient in the z-row—in this case, y.

	x	y	s	t	
x	1	$\frac{2}{3}$	$\frac{2}{3}$	0	224
t	0	$(\frac{1}{2})$	$-\frac{1}{6}$	1	60
z	0	$-\frac{1}{12}$	$\frac{2}{3}$	0	224

$224 \div \frac{2}{3} = 336$
$60 \div \frac{1}{2} = 120$

To determine the departing basic variable, we divide each value of the current basic variables in the column on the right by the corresponding entry in the y-column. The variable corresponding to the smallest nonnegative quotient is the departing variable—in this case, t. The pivot then is the entry in the t-row and y-column. Finally, we row-reduce to obtain a 1 in the pivot position and 0's elsewhere in its column; the result is the following tableau:

	x	y	s	t	
x	1	0	$\frac{8}{9}$	$-\frac{4}{3}$	144
y	0	1	$-\frac{1}{3}$	2	120
z	0	0	$\frac{23}{36}$	$\frac{1}{6}$	234

4.4 LINEAR PROGRAMMING—SIMPLEX METHOD

The new basic feasible solution is $x = 144$, $y = 120$, $s = 0$, and $t = 0$. The corresponding value of z is 234.

Since there are no more negative coefficients in the bottom row, the value of z cannot be increased further; $z = 234$ is maximum. This agrees with the result obtained previously by examining the value of z at all of the basic feasible solutions.

> The procedure involved in working with each tableau of the simplex method is summarized as follows:
>
> 1. Find the entering basic variable by determining the greatest (in absolute value) negative coefficient of the objective function. (If there are two the same, either variable may be chosen.)
> 2. Find the departing basic variable by dividing each value of the current basic variables in the column on the right by the corresponding entry in the column of the entering variable. The departing variable corresponds to the smallest nonnegative quotient obtained.
> 3. Determine the pivot element, which will be in the column of the entering variable and the row of the departing variable.
> 4. Row-reduce to obtain a 1 in the pivot position and 0's elsewhere in its column. The value of the objective function for the new basic feasible solution will appear in the lower right-hand corner of the resulting tableau.
>
> The maximum value of the objective function is attained when there are no more negative coefficients left in the bottom row of the table.

We illustrate the procedure again in Example 1.

EXAMPLE 1 Maximize
$$z = 3x + 5y - w$$
subject to
$$x + 3y + 2w \leq 36$$
$$x + y + 4w \leq 24$$
$$-x + y - w \leq 32$$
$$x \geq 0$$
$$y \geq 0$$
$$w \geq 0$$

Solution We rewrite the objective function and introduce slack variables s, t, and u so that the problem becomes:

Maximize z, where
$$z - 3x - 5y + w = 0$$

subject to
$$x + 3y + 2w + s = 36$$
$$x + y + 4w + t = 24$$
$$-x + y - w + u = 32$$
$$x \geq 0 \quad s \geq 0$$
$$y \geq 0 \quad t \geq 0$$
$$w \geq 0 \quad u \geq 0$$

The simplex tableau is then:

	x	y	w	s	t	u		
s	1	③	2	1	0	0	36	$36 \div 3 = 12$
t	1	1	4	0	1	0	24	$24 \div 1 = 24$
u	-1	1	-1	0	0	1	32	$32 \div 1 = 32$
z	-3	-5	1	0	0	0	0	

↑

The steps in working with each tableau are numbered according to the outline given on the previous page.

Step 1. The entering variable y corresponds to the largest negative coefficient in the z-row, as indicated by the arrow.

Step 2. The smallest nonnegative quotient on the right indicates that s is the departing basic variable.

Step 3. The pivot is then the 3 circled in the tableau.

Step 4. Row-reduction to obtain a 1 in the pivot position and 0's elsewhere in its column leads to the following tableau:

	x	y	w	s	t	u		
y	$\frac{1}{3}$	1	$\frac{2}{3}$	$\frac{1}{3}$	0	0	12	$12 \div \frac{1}{3} = 36$
t	$②/③$	0	$\frac{10}{3}$	$-\frac{1}{3}$	1	0	12	$12 \div \frac{2}{3} = 18$
u	$-\frac{4}{3}$	0	$-\frac{5}{3}$	$-\frac{1}{3}$	0	1	20	$20 \div -\frac{4}{3} = -15$
z	$-\frac{4}{3}$	0	$\frac{13}{3}$	$\frac{5}{3}$	0	0	60	

↑

Step 1. The entering basic variable is x.

Step 2. The departing basic variable is t.

Step 3. The pivot then is the $\frac{2}{3}$ circled in the tableau.

Step 4. Row-reduction leads to the following tableau:

	x	y	w	s	t	u	
y	0	1	−1	$\frac{1}{2}$	$-\frac{1}{2}$	0	6
x	1	0	5	$-\frac{1}{2}$	$\frac{3}{2}$	0	18
u	0	0	5	−1	2	1	44
z	0	0	11	1	2	0	84

Since none of the coefficients for z in the last row are negative, we are done. The solution, in terms of the original variables, is $x = 18$, $y = 6$, and $w = 0$; the corresponding maximum value of z is 84. ◼

In the next to the last tableau of Example 1, there really was no need to perform the division $20 \div -\frac{4}{3} = -15$ for the u-row since the choice of the departing variable is concerned only with the smallest *nonnegative* quotient. In fact, *if all of the entries in the pivot column above the z-row are negative, there is no solution.* The computation can be stopped at that point.

EXAMPLE 2 Maximize

$$z = 4x + y + 3w$$

subject to

$$x + 2y + w \leq 18$$
$$2x + 3y - 4w \leq 24.$$
$$x \geq 0$$
$$y \geq 0$$
$$w \geq 0$$

Solution We rewrite the objective function and introduce slack variables s and t so that the problem can be restated:

Maximize z, where

$$z - 4x - y - 3w = 0$$

subject to

$$x + 2y + w + s = 18$$
$$2x + 3y - 4w + t = 24$$
$$x \geq 0 \qquad s \geq 0$$
$$y \geq 0 \qquad t \geq 0$$
$$w \geq 0$$

We construct the tableau and follow the four-step procedure of this section with each tableau. However, in this case we present only the tableau.

---

4 PRACTICE PROBLEM 4

Complete the linear programming problem of Practice Problem 3.

Answer

The final tableau is

	x	y	s	t	
y	0	1	$\frac{3}{7}$	$-\frac{1}{7}$	$\frac{16}{7}$
x	1	0	$-\frac{5}{14}$	$\frac{2}{7}$	$\frac{10}{7}$
z	0	0	$\frac{4}{7}$	$\frac{1}{7}$	$\frac{68}{7}$

The maximum value of z is $\frac{68}{7}$ for $x = \frac{10}{7}$ and $y = \frac{16}{7}$.

	x	y	w	s	t		
s	1	2	1	1	0	18	$18 \div 1 = 18$
t	②	3	-4	0	1	24	$24 \div 2 = 12$
z	-4	-1	-3	0	0	0	
	↑						
s	0	$\frac{1}{2}$	③	1	$-\frac{1}{2}$	6	$6 \div 3 = 2$
x	1	$\frac{3}{2}$	-2	0	$\frac{1}{2}$	12	
z	0	5	-11	0	2	48	
			↑				
w	0	$\frac{1}{6}$	1	$\frac{1}{3}$	$-\frac{1}{6}$	2	
x	1	$\frac{11}{6}$	0	$\frac{2}{3}$	$\frac{1}{6}$	16	
z	0	$\frac{41}{6}$	0	$\frac{11}{3}$	$\frac{1}{6}$	70	

The maximum value of z is 70, attained when $w = 2$, $x = 16$, and $y = 0$.

The simplex method, as we have been considering it, requires that a linear programming problem has the following three properties:

> 1. The objective function is to be maximized.
> 2. All the variables must be nonnegative: $x \geq 0$, $y \geq 0$, $w \geq 0$, and so forth.
> 3. The remaining inequalities (that is, other than the nonnegativity constraints) are of the type $\leq b$, where b is a constant greater than or equal to 0. The inequality $2x + 3y - 4w \leq 7$ is one of this type; b in this case is 7.

Each example in this section has each of these three properties. The simplex method must be modified in order to solve a linear programming problem that is lacking one or more of these properties. To consider how these modifications are done would require a much more detailed explanation than we can present here. However, in the next section we do consider one method of solving a linear programming problem in which the objective function is to be minimized.

Our purpose here has been to introduce the simplex method. Additional questions that may have occurred to you, such as how the simplex method recognizes multiple solutions for a problem, require a deeper analysis of the method. Such topics are treated in courses that allow more time for the discussion of linear programming.

5 PRACTICE PROBLEM 5

Solve the linear programming problem:

Maximize
$$z = 2x + 3y + w$$

subject to
$$x + 2y + 2w \leq 10$$
$$2x + y - w \leq 18$$
$$x \geq 0$$
$$y \geq 0$$
$$w \geq 0$$

Answer

Maximum value of z is $\frac{58}{3}$ for $x = \frac{26}{3}$, $y = \frac{2}{3}$, and $w = 0$.

EXERCISES

Set up the initial simplex tableau for each of the following:

1. Maximize
$$z = 2x + 3y$$
subject to
$$x - 2y \leq 10$$
$$2x + 4y \leq 18$$
$$x \geq 0$$
$$y \geq 0$$

2. Maximize
$$z = 3x - y$$
subject to
$$x + 2y \leq 12$$
$$2x - y \leq 10$$
$$3x + 2y \leq 20$$
$$x \geq 0$$
$$y \geq 0$$

3. Maximize
$$z = x + 2y + 2w$$
subject to
$$2x + y - w \leq 10$$
$$x + 3y + w \leq 14$$
$$x \geq 0$$
$$y \geq 0$$
$$w \geq 0$$

4. Maximize
$$z = 3x + 5y + 2w$$
subject to
$$x + y + w \leq 100$$
$$2x - y + 3w \leq 120$$
$$3x + 2y + 4w \leq 95$$
$$x \geq 0$$
$$y \geq 0$$
$$w \geq 0$$

For each of the following, determine the next tableau from the given tableau:

5.

	x	y	s	t	
x	1	2	0	4	10
s	0	4	1	-1	15
z	0	-3	0	4	2

6.

	x	y	w	s	t	
w	4	0	1	-8	2	9
y	5	1	0	2	0	12
z	-2	0	0	-1	3	23

7.

	x	y	w	s	t	
s	8	4	1	1	0	10
t	-1	5	3	0	1	20
z	-2	-4	-1	0	0	0

8.

	x	y	w	s	t	u	
s	0	12	9	1	-6	0	3
x	1	2	$\frac{1}{2}$	0	$\frac{1}{4}$	0	7
u	0	0	1	0	3	1	9
z	0	-2	-3	0	$\frac{2}{3}$	0	13

Each of the following is the final tableau for a linear programming problem with the three properties listed in this section. For each, give the maximum value of the objective function and the corresponding value of each of the variables.

9.

	x	y	s	t	
x	1	0	$\frac{1}{2}$	-2	7
y	0	1	1	$\frac{1}{3}$	5
z	0	0	2	1	51

10.

	x	y	s	t	
y	1	1	$\frac{1}{3}$	0	26
t	3	0	$\frac{1}{2}$	1	40
z	4	0	1	0	52

11.

	x	y	w	s	t	
w	0	-1	1	-2	-1	143
x	1	4	0	3	0	67
z	0	3	4	1	2	287

12.

	x	y	w	s	t	u	
y	0	1	0	3	-1	1	7
w	0	0	1	7	4	-1	4
x	1	0	0	9	-2	5	12
z	0	0	0	2	1	4	68

Use the simplex method to solve each problem.

13. Maximize
$$z = x + 2y$$
subject to
$$3x + y \leq 21$$
$$2x + 5y \leq 40$$
$$x \geq 0$$
$$y \geq 0$$

14. Maximize $z = x + 4y$ subject to the constraints in Exercise 13.

15. Maximize
$$z = 2x + 3y + w$$
subject to
$$x + 2y - w \leq 8$$
$$4x - y + 2w \leq 11$$
$$x \geq 0$$
$$y \geq 0$$
$$w \geq 0$$

16. Maximize $z = 2x + 3y - 2w$ subject to the constraints in Exercise 15.

17. Maximize
$$z = -2x + 2y - w$$
subject to
$$x + y + w \leq 12$$
$$2x + y \leq 6$$
$$y - w \leq 2$$
$$x \geq 0$$
$$y \geq 0$$
$$w \geq 0$$

18. Maximize
$$z = 3x + 2y + w$$
subject to
$$x - y \leq 2$$
$$x + 2y + 3w \leq 8$$
$$2x + 3y \leq 19$$
$$x \geq 0$$
$$y \geq 0$$
$$w \geq 0$$

Solve each of the following using the simplex method:

19. *Manufacturing* Exercise 14 of Section 4.2

20. *Air pollution* Exercise 19 of Section 4.2

21. *Personnel utilization* A city has three types of garbage crews—a *full* crew consisting of a truck, a driver, and three pickup people; a *partial* crew consisting of a truck, a driver, and two pickup people; and a *half* crew, consisting of a truck, a driver, and one pickup person. It is estimated that a full crew hauls 12 tons of garbage a day, a partial crew hauls 10 tons, and a half crew hauls 6 tons. One day when the city had 100 trucks in working order, 320 people (each of whom could be a driver or a pickup person) showed up for work. The administration wants to know how many crews of each type should be formed to maximize the amount of garbage to be hauled to the dump that day.
    (a) Let x denote the number of full crews to be formed, y the number of partial crews, and w the number of half crews. Write an equation to express the total number of tons (z) that will be hauled in terms of x, y, and w.
    (b) Write an inequality in x, y, and w to express the fact that the total number of people assigned must be less than or equal to 320.
    (c) Write an inequality in x, y, and w to express the fact that the total number of trucks to be used must be less than or equal to 100.
    (d) Write inequalities to express the fact that the number of crews of each type must be greater than or equal to 0.
    (e) Use the simplex method to determine how many crews of each type should be formed to maximize the amount of garbage to be hauled and the maximum amount of garbage to be hauled.

22. *Process utilization* A clothing company makes shirts, blouses, and jackets, each of which goes through three manufacturing phases—cutting, sewing, and finishing. The number of minutes each type of product requires in each of the three phases is given in the table. There are 4,800

	CUTTING	SEWING	FINISHING
SHIRT	10	18	6
BLOUSE	18	20	6
JACKET	16	20	12

minutes of cutting time, 6,200 minutes of sewing time, and 2,040 minutes of finishing time each day. The company can sell all of the items it manufactures and makes a profit of $1.00 on each shirt, $1.25 on each blouse, and $1.50 on each jacket. Determine the number of shirts, blouses, and jackets the company should manufacture each day in order to obtain the maximum profit.

23. *Material utilization* A small company produces three products each of which requires three different materials, as indicated in the table. At present they have 350 units of material A, 260 units of material B, and 420 units of material C delivered each day. Product I sells for $40.00 per item, product II for $20.00 per item, and product III for $15.00 per item. How many items of each product should the company produce each day to obtain the maximum revenue?

PRODUCT	NUMBER OF UNITS OF MATERIAL REQUIRED PER ITEM		
	A	B	C
I	5	4	6
II	7	1	4
III	2	1	3

24. List the three requirements if a linear programming problem is to be solvable by the simplex method discussed in this section.

Why can the following not be solved by the simplex method discussed in this section?

25. Minimize

$$z = 2x + 5y$$

subject to

$$2x + 7y \leq 13$$
$$6x - 4y \leq 9$$
$$x \geq 0$$
$$y \geq 0$$

26. Maximize

$$z = 4x + 2y + 3w$$

subject to

$$x + 2y + w \leq 15$$
$$3x - y + 2w \geq 9$$
$$x \geq 0$$
$$y \geq 0$$
$$w \geq 0$$

27. Exercise 16 of Section 4.2

28. Exercise 17 of Section 4.2

Identify the steps of the four-step procedure involved in the application of mathematics (Section 1.3) in each case.

29. *Personnel utilization* Exercise 21
30. *Process utilization* Exercise 22

*4.5 MINIMIZATION

Consider the two linear programming problems:

Minimize

$$w = 10y_1 + 4y_2$$

subject to

$$2y_1 + 3y_2 \geq 5$$
$$6y_1 - 4y_2 \geq 7$$
$$y_1 \geq 0$$
$$y_2 \geq 0$$

Maximize

$$z = 5x_1 + 7x_2$$

subject to

$$2x_1 + 6x_2 \leq 10$$
$$3x_1 - 4x_2 \leq 4$$
$$x_1 \geq 0$$
$$x_2 \geq 0$$

For each of these problems, we construct the augmented coefficient matrix, omitting the slack variables but including the coefficients of the objective function in the last row:

$$\begin{bmatrix} 2 & 3 & | & 5 \\ 6 & -4 & | & 7 \\ \hline 10 & 4 & | & 0 \end{bmatrix} \qquad \begin{bmatrix} 2 & 6 & | & 10 \\ 3 & -4 & | & 4 \\ \hline 5 & 7 & | & 0 \end{bmatrix} \qquad (1)$$

The first row in the matrix on the left is the same as the first column of the matrix on the right; the same is true for the second row on the left and second column on the right and for the third row on the left and third column on the right. Each of the two matrices is said to be the *transpose* of the other. Formally, the ***transpose*** of the $m \times n$ matrix A is an $n \times m$ matrix, each of whose columns has the same entries as the corresponding row of A. The transpose of matrix A is denoted by A^t.

EXAMPLE 1 Determine the transpose of

$$A = \begin{bmatrix} 1 & 2 & 6 \\ 3 & -1 & 4 \\ 2 & 0 & 1 \\ 5 & 9 & -8 \end{bmatrix}$$

Solution Using the rows of A to form the columns, we construct

$$A^t = \begin{bmatrix} 1 & 3 & 2 & 5 \\ 2 & -1 & 0 & 9 \\ 6 & 4 & 1 & -8 \end{bmatrix}$$ ▮

The two linear programming problems giving rise to the matrices in (1) are said to be the ***duals*** of each other. We shall see shortly that the solution of either problem gives the solution of the other. Consequently, we can solve the maximization problem using the simplex

1 PRACTICE PROBLEM 1

Determine the transpose of

$$A = \begin{bmatrix} 6 & 7 & -9 \\ -3 & 0 & 4 \end{bmatrix}$$

Answer

$$A^t = \begin{bmatrix} 6 & -3 \\ 7 & 0 \\ -9 & 4 \end{bmatrix}$$

* This section can be omitted without loss of continuity.

method and thereby obtain the solution of the dual minimization problem. But first we consider some examples illustrating the procedure of obtaining the dual problem for a given problem. (In such situations, the given problem is called the ***primal*** problem.)

EXAMPLE 2 Construct the dual problem of the following linear programming problem.

Minimize
$$w = 3y_1 + 2y_2$$

subject to
$$4y_1 + y_2 \geq 5$$
$$6y_1 - y_2 \geq 4$$
$$y_1 \geq 0$$
$$y_2 \geq 0$$

Solution We form the augmented coefficient matrix, as in the illustration at the beginning of this section:

$$\begin{bmatrix} 4 & 1 & | & 5 \\ 6 & -1 & | & 4 \\ \hline 3 & 2 & | & 0 \end{bmatrix}$$

Next, we form its transpose,

$$\begin{bmatrix} 4 & 6 & | & 3 \\ 1 & -1 & | & 2 \\ \hline 5 & 4 & | & 0 \end{bmatrix}$$

and then construct the maximization problem for which the transpose is the coefficient matrix:

Maximize
$$z = 5x_1 + 4x_2$$

subject to
$$4x_1 + 6x_2 \leq 3$$
$$x_1 - x_2 \leq 2$$
$$x_1 \geq 0$$
$$x_2 \geq 0$$

This maximization problem is the dual of the given (primal) problem. ∎

EXAMPLE 3 Construct the dual problem for the following primal problem:

Maximize
$$z = x_1 + 2x_2 + 3x_3$$

4.5 MINIMIZATION

subject to

$$x_1 + 2x_2 \leq 5$$
$$x_1 + 3x_3 \leq 10$$
$$x_2 - x_3 \leq 15$$
$$x_1 \geq 0$$
$$x_2 \geq 0$$
$$x_3 \geq 0$$

Solution We form the augmented coefficient matrix:

$$\left[\begin{array}{ccc|c} 1 & 2 & 0 & 5 \\ 1 & 0 & 3 & 10 \\ 0 & 1 & -1 & 15 \\ \hline 1 & 2 & 3 & 0 \end{array}\right]$$

We then form its transpose,

$$\left[\begin{array}{ccc|c} 1 & 1 & 0 & 1 \\ 2 & 0 & 1 & 2 \\ 0 & 3 & -1 & 3 \\ \hline 5 & 10 & 15 & 0 \end{array}\right]$$

From the transpose we construct the corresponding minimization problem:

Minimize
$$w = 5y_1 + 10y_2 + 15y_3$$
subject to
$$y_1 + y_2 \geq 1$$
$$2y_1 + y_3 \geq 2$$
$$3y_2 - y_3 \geq 3$$
$$y_1 \geq 0$$
$$y_2 \geq 0$$
$$y_3 \geq 0$$

The following three properties of dual problems should be noted:

1. The dual of a minimization problem is a maximization problem, and vice versa.
2. All the variables in both the primal and dual problems are required to be nonnegative (greater than or equal to 0).
3. The remaining inequalities (other than the nonnegativity constraints) are of the form $\leq b$ in the maximization problem and of the form $\geq b$ in the minimization problem, where b is a constant greater than or equal to zero.

The dual problem for other types of linear programming problems requires a deeper analysis of the theory of linear programming than we can consider here.

2 PRACTICE PROBLEM 2

Construct the dual problem for the following primal problem:
Minimize
$$w = 20y_1 + 21y_2$$
subject to
$$y_1 + 3y_2 \geq 4$$
$$5y_1 + 2y_2 \geq 3$$
$$y_1 \geq 0$$
$$y_2 \geq 0$$

Answer
Maximize
$$z = 4x_1 + 3x_2$$
subject to
$$x_1 + 5x_2 \leq 20$$
$$3x_1 + 2x_2 \leq 21$$
$$x_1 \geq 0$$
$$x_2 \geq 0$$

The following theorem of linear programming indicates how the simplex method can be used to solve a minimization problem:

THEOREM 1

If either the primal or dual problem has a solution, then so does the other problem. In this case, the maximum value of z is the same as the minimum value of w.

Consequently, to solve a minimization problem (with the properties listed above), it is sufficient to apply the simplex method to the corresponding dual problem. The following exercise illustrates the procedure.

EXAMPLE 4 Solve the following linear programming problem:
Minimize
$$w = 6y_1 + 4y_2$$
subject to
$$y_1 + 3y_2 \geq 5$$
$$2y_1 - y_2 \geq 8$$
$$y_1 \geq 0$$
$$y_2 \geq 0$$

Solution We form the coefficient matrix,
$$\begin{bmatrix} 1 & 3 & | & 5 \\ 2 & -1 & | & 8 \\ \hline 6 & 4 & | & 0 \end{bmatrix}$$

and its transpose,
$$\begin{bmatrix} 1 & 2 & | & 6 \\ 3 & -1 & | & 4 \\ \hline 5 & 8 & | & 0 \end{bmatrix}$$

The dual problem is:
Maximize
$$z = 5x_1 + 8x_2$$
subject to
$$x_1 + 2x_2 \leq 6$$
$$3x_1 - x_2 \leq 4$$
$$x_1 \geq 0$$
$$x_2 \geq 0$$

We solve this problem using the simplex method. The first tableau (with s_1 and s_2 as slack variables) is:

4.5 MINIMIZATION

	x_1	x_2	s_1	s_2	
s_1	1	2	1	0	6
s_2	3	−1	0	1	4
z	−5	−8	0	0	0

The final tableau is:

	x_1	x_2	s_1	s_2	
x_2	0	1	$\frac{3}{7}$	$-\frac{1}{7}$	2
x_1	1	0	$\frac{1}{7}$	$\frac{2}{7}$	2
z	0	0	$\frac{29}{7}$	$\frac{2}{7}$	26

The maximum value of z is 26; therefore, the minimum value of w in the primal problem is also 26.

The final tableau of the procedure illustrated in Example 4 also gives *the values of y_1 and y_2 for which the minimal value of w is attained in the primal problem. These appear in the z-row under the corresponding slack variable.* In Example 4 these values are $y_1 = \frac{29}{7}$ and $y_2 = \frac{2}{7}$; substituting these values into the objective function $w = 6y_1 + 4y_2$ verifies that they yield the minimum value of 26.

3 PRACTICE PROBLEM 3
Solve the given minimization problem in Practice Problem 2.

Answer
$w = 29$

EXERCISES

Give the transpose for each matrix.

1. $\begin{bmatrix} 1 & 2 & 3 \\ -4 & 4 & 9 \end{bmatrix}$

2. $\begin{bmatrix} -2 & 3 \\ 4 & -8 \end{bmatrix}$

3. $\begin{bmatrix} 4 & -7 \\ 10 & 12 \\ 19 & 0 \end{bmatrix}$

4. $\begin{bmatrix} 1 & 0 & 0 & 1 \\ 0 & 1 & 0 & 0 \\ 0 & 0 & 0 & 1 \\ 1 & 0 & 1 & 1 \end{bmatrix}$

Construct the dual for each of the following:

5. Maximize
$$z = 2x_1 + 3x_2$$
subject to
$$x_1 - x_2 \leq 3$$
$$2x_1 + 3x_2 \leq 10$$
$$4x_1 + 6x_2 \leq 12$$
$$x_1 \geq 0$$
$$x_2 \geq 0$$

6. Maximize
$$z = x_1 + 2x_2 + 4x_3$$
subject to
$$x_1 + x_2 + x_3 \leq 20$$
$$2x_1 + 2x_2 - x_3 \leq 42$$
$$x_1 - x_2 + 4x_3 \leq 50$$
$$x_1 \geq 0$$
$$x_2 \geq 0$$
$$x_3 \geq 0$$

7. Minimize
$$w = 2y_1 + 5y_2$$
subject to
$$2y_1 - y_2 \geq 12$$
$$y_1 + 3y_2 \geq 18$$
$$y_1 \geq 0$$
$$y_2 \geq 0$$

8. Minimize
$$w = 20y_1 + 42y_2 + 50y_3$$
subject to
$$y_1 + 2y_2 + y_3 \geq 1$$
$$y_1 + 2y_2 - y_3 \geq 2$$
$$y_1 - y_2 + 4y_3 \geq 4$$
$$y_1 \geq 0$$
$$y_2 \geq 0$$
$$y_3 \geq 0$$

Solve.

9. Minimize
$$w = 4y_1 + 7y_2$$
subject to
$$2y_1 + y_2 \geq 14$$
$$y_1 + 3y_2 \geq 10$$
$$y_1 \geq 0$$
$$y_2 \geq 0$$

10. Minimize
$$w = y_1 + 2y_2$$
subject to
$$2y_1 + 3y_2 \geq 10$$
$$y_1 + y_2 \geq 12$$
$$y_1 \geq 0$$
$$y_2 \geq 0$$

11. Minimize
$$w = 8y_1 + 11y_2$$
subject to
$$y_1 + 4y_2 \geq 2$$
$$2y_1 - y_2 \geq 3$$
$$-y_1 + 2y_2 \geq 1$$
$$y_1 \geq 0$$
$$y_2 \geq 0$$

12. Minimize
$$w = 2y_1 + y_2$$
subject to
$$-2y_1 + y_2 \geq 2$$
$$4y_1 + 3y_2 \geq 12$$
$$y_1 \geq 0$$
$$y_2 \geq 0$$

13. Minimize
$$w = 8y_1 + 6y_2 + 10y_3$$
subject to
$$y_1 + 3y_2 + y_3 \geq 1$$
$$2y_1 + 4y_2 + 4y_3 \geq 6$$
$$2y_1 - 4y_2 - 2y_3 \geq 4$$
$$y_1 \geq 0$$
$$y_2 \geq 0$$
$$y_3 \geq 0$$

14. Minimize
$$w = 18y_1 + 12y_2 + 16y_3$$
subject to
$$y_1 + y_2 - y_3 \geq 3$$
$$2y_1 + 2y_3 \geq 2$$
$$2y_1 + 4y_2 - y_3 \geq 1$$
$$y_1 \geq 0$$
$$y_2 \geq 0$$
$$y_3 \geq 0$$

SUPPLEMENT	VITAMIN	
	A	C
I	1	2
II	$\frac{3}{2}$	2

15. *Diet mix* A biologist, conducting an experiment with guinea pigs, wants each guinea pig to have at least 8 units of vitamin A each day and 12 units of vitamin C each day. Two food supplements, I and II, are available to provide the vitamins. These provide the number of units per gram listed in the table. Supplement I costs 6¢ per gram and supplement II costs 8¢ per gram.

(a) Let y_1 denote the number of grams of supplement I to be given each guinea pig per day and y_2 the number of grams of supplement II. Write an equation to express the daily cost (w) per guinea pig (in cents) in terms of y_1 and y_2.

(b) Write an inequality in y_1 and y_2 to express the fact that the amount of vitamin A per day must be at least 8 units.

(c) Write an inequality in y_1 and y_2 to express the fact that the amount of vitamin C per day must be at least 12 units.

(d) Write inequalities to express the fact that number of grams of each supplement must be nonnegative.

(e) Determine the minimal cost per guinea pig per day for the supplements satisfying the above requirements. What quantities give this minimal cost?

16. *Product mix* A publishing company intends to publish a second edition of one of its best-selling books. It costs the company $1.50 to produce each paperback copy and $7.50 to produce each hardback copy. Marketing analysts suggest they publish at least 10,000 hardback copies, that the number of paperbacks be at least twice the number of hardbacks, and that the combined number of paperbacks and hardbacks be at least 40,000. Use linear programming to determine how many paperbacks and how many hardbacks the company should produce to meet the above objectives at minimal cost. What is the minimal cost?

17. *Diet mix* Solve the problem in Exercise 20 of Section 4.2 using the method of this section.

4.6 CHAPTER SUMMARY

The central idea in this chapter has been the solution of **linear programming problems**—that is, the determination of the maximum or minimum value of a linear objective function subject to a number of linear inequality constraints.

A **linear inequality** in two variables x and y is an expression that can be written in one of the forms $ax + by \leq c$, $ax + by < c$, $ax + by \geq c$, or $ax + by > c$, where a, b, and c are constants with a and b not both zero. A **solution** of such an equality is a set of pairs of values of x and y that satisfy the inequality. The **graph of an inequality,** consisting of one of the half-planes determined by the graph of the corresponding linear equation, is the set of all points whose coordinates satisfy the inequality.

The **graph of a system of inequalities,** such as those forming the set of constraints for a linear programming problem, is the set of all points whose coordinates satisfy each of the inequalities. This set of points forms the **region of feasible solutions** for the linear programming problem.

The geometric method of solution involves the following steps:

> 1. Determine the coordinates of the corner points of the region of feasible solutions.
> 2. Examine the value of the objective function at each of the corner points to find the maximum or minimum value.

If the region of feasible solutions is bounded, there is always at least one solution. There may be infinitely many solutions, indicated when the maximum (or minimum) value of the objective function occurs at consecutive corner points. If the region of feasible solutions is unbounded, there may or may not be a solution. Whether there is or not is determined by a close examination of the values of the objective function in the set of feasible solutions.

The **simplex method** is used for solving linear programming problems in which the value of the objective function depends on more than two variables. The simplex method begins by introducing a new variable into each of the inequality constraints, other than the nonnegativity constraints. These new variables are called **slack variables.** The simplex method examines the value of the objective function at the basic solutions of the resulting system of equations, considering only those solutions that are also feasible—that is, those that also satisfy the nonnegativity constraints.

The basic solutions of a system of equations are determined by solving for one set of variables (the **basic variables**) in the equations in terms of the remaining variables (the **nonbasic variables**) using the usual row-reduction techniques. The nonbasic variables are then set equal to zero. The resulting values of the variables form a basic solution. A set of equations can have many basic solutions, depending on which variables are chosen to be the basic and which are chosen to be the nonbasic variables.

The simplex method starts with the basic feasible solution in which the slack variables are the basic variables and, using the pivoting procedure, moves from one basic feasible solution to another, determining the values of the objective function at each such solution encountered. The basic feasible solution chosen at each step (after the first) is the one that promises the greatest increase in the value of the objective function.

The important problem data in the simplex method are presented in table form, called the **simplex tableau,** one tableau for each basic feasible solution considered. The pivoting procedure for determining the next tableau from a given tableau is as follows:

> 1. Find the entering basic variable by determining the greatest (in absolute value) negative coefficient of the objective function.
> 2. Find the departing basic variable by dividing each value of the current basic variables in the column on the right by the corresponding entry in the column of the entering variable. The departing variable corresponds to the smallest nonnegative quotient obtained.
> 3. Determine the pivot element, which will be in the column of the entering variable and the row of the departing variable.
> 4. Row-reduce to obtain a 1 in the pivot position and 0's elsewhere in its column. The value of the objective function for the new basic feasible solution will appear in the lower right-hand corner of the new tableau.

The maximum value of the objective function is attained when there are no more negative coefficients left in the bottom row of the table. If all the entries in the pivot column above the z-row are negative at any step, the problem has no solution and the computation can be stopped.

The simplex method discussed above is applicable only if the linear programming problem has the following three properties:

1. The objective function is to be maximized.
2. All variables are nonnegative: $x \geq 0$, $y \geq 0$, $w \geq 0$, and so forth.
3. The inequalities other than the nonnegativity constraints are all of the type $\leq b$, where b is a constant greater than or equal to zero.

Solving linear programming problems that do not have these three properties requires different variations in the simplex technique. One such type of problem considered has the following properties:

1. The objective function is to be minimized.
2. All variables are nonnegative.
3. The inequalities other than the nonnegativity constraints are all of the form $\geq b$, where b is a constant greater than or equal to zero.

The **dual** of such a problem has the three properties required of a linear programming problem if the simplex technique is to be used. Furthermore, if the dual has a solution, the maximum value of the objective function in the dual problem is the same as the minimum value of the objective function in the original problem. Consequently, we can use the simplex method to solve the dual problem, obtaining the solution of the minimization problem at the same time.

REVIEW EXERCISES

Graph the following inequalities:

1. $2x - y \leq 1$
2. $-3x - 2y \geq 6$
3. $-4x + 2y \geq 3$
4. $7x + 9y \leq 12$
5. $5x - 10y \leq 10$
6. $2x \geq 5$
7. $y \leq -4$
8. $y \geq 0$
9. $x \leq 0$
10. $0.2x - 0.3y \leq 0$
11. $7 \geq x - y$
12. $4x + 3 \leq 2y$

Graph the following systems of inequalities:

13. $2x + y \geq 4$
    $x - 2y \leq 6$

14. $x + y \leq 2$
    $-2x + 3y \geq 4$

15. $-4x + 5y \leq 10$
    $2x - 6y \leq 8$
    $y \geq 0$

16. $6x + 8y \leq 12$
    $-3x + 5y \leq 9$
    $y \leq -1$

17. $3x + 4y \leq 6$
    $x - y \leq 3$
    $2y - 3x \leq 9$

18. $2y + 7x \geq 14$
    $9y + 5x \geq 45$
    $x \leq 6$

19. $3y - 2x \leq 18$
    $3y + 2x \geq 6$
    $x \geq 0$
    $y \geq 0$

20. $2y + x \leq 10$
    $2y - 3x \geq -6$
    $x \geq 0$
    $y \geq 0$

Solve the following problems using the geometric method:

21. Maximize
    $$z = 2y - 3x$$
    subject to
    $$3y + x \leq 12$$
    $$2y - x \geq -2$$
    $$y - 3x \leq 14$$

22. Maximize $z = 4x - 3y$ subject to the constraints of Exercise 21.

23. Minimize $z = 2x + 2y$ subject to the constraints of Exercise 21.

24. Maximize
    $$z = 8y - 10x$$
    subject to
    $$3y + 10x \leq 67$$
    $$4y - 5x \leq 16$$
    $$y \geq -1$$

25. Minimize $z = 3x - 3y$ subject to the constraints of Exercise 24.

26. Minimize
    $$z = 7x - 2y$$
    subject to
    $$3y - 2x \leq 15$$
    $$4y + 9x \leq 20$$
    $$y + x \geq 0$$

27. Maximize $z = 6y - 4x$ subject to the constraints of Exercise 26.

28. Minimize
$$z = y - x$$
subject to
$$2y + 3x \geq 16$$
$$2y - x \geq 0$$

29. Maximize
$$z = 2x + 7y$$
subject to
$$y + x \geq 6$$
$$y - x \geq 0$$

For each of the following row-reduced augmented coefficient matrices, identify the basic variables, the nonbasic variables, and the corresponding basic solution:

30.
$$\begin{array}{cccc} x & y & s & t \end{array}$$
$$\begin{bmatrix} 2 & -3 & 1 & 0 & | & 5 \\ 4 & -9 & 0 & 1 & | & 7 \end{bmatrix}$$

31.
$$\begin{array}{ccccc} x & y & t & u & v \end{array}$$
$$\begin{bmatrix} 10 & 0 & 8 & 1 & 8 & | & -3 \\ -9 & 1 & 12 & 0 & 9 & | & 7 \end{bmatrix}$$

32.
$$\begin{array}{ccccc} x & y & w & s & t \end{array}$$
$$\begin{bmatrix} 1 & 0 & 7 & 0 & 4 & | & 10 \\ 0 & 0 & 12 & 1 & -2 & | & 14 \\ 0 & 1 & -8 & 0 & -2 & | & -19 \end{bmatrix}$$

33.
$$\begin{array}{ccccc} x & y & w & s & t & u \end{array}$$
$$\begin{bmatrix} 1 & 4 & 1 & 0 & 0 & 7 & | & -3 \\ 2 & 5 & 0 & 1 & 0 & 8 & | & -2 \\ 3 & 6 & 0 & 0 & 1 & 9 & | & -1 \end{bmatrix}$$

For each of the following systems of equations, find the basic solution corresponding to the indicated basic variables:

34. $-3x - 10y + 4s + 2t = -11$
$\quad\ \ x + 3y - 2s + t = 7$
Basic variables: x and y

35. Use the system in Exercise 34 with basic variables x and t.

36. $-8x + 4y - 2s + t + 2u = 4$
$\ \ \ 4x - 3y + 2s + 2t + u = 5$
Basic variables: y and u

37. Use the system in Exercise 36 with basic variables s and t.

38. $\ \ \ x + 2y - 2w + s \ \ \ \ \ \ \ \ - u = 4$
$-3x - 9y + 9w + 6s + 3t + 4u = -9$
$\ \ 2x + 4y - 5w + 3s + 2t - 2u = 5$
Basic variables: x, y, and w

39. Use the system in Exercise 38 with basic variables x, y and t.

For each of the following, determine the next tableau from the given tableau. Give the basic solution and corresponding value of the objective function displayed in the resulting tableau.

40.

	x	y	s	t	
s	4	3	1	0	12
t	8	2	0	1	9
z	-3	-5	0	0	0

41.

	x	y	s	t	
y	3	1	0	0	15
s	6	0	1	-9	18
z	-8	0	0	4	22

42.

	x	y	w	s	t	
t	4	10	0	7	1	25
w	4	$\frac{3}{4}$	1	$\frac{1}{3}$	0	34
z	-11	-7	0	4	0	46

43.

	x	y	w	s	t	u	
s	7	$-\frac{3}{2}$	0	1	0	$\frac{13}{4}$	68
t	8	14	0	0	1	4	84
w	-4	$\frac{1}{3}$	1	0	0	$\frac{7}{4}$	72
z	-5	-1	0	0	0	4	143

Solve each problem using the simplex method.

44. Maximize

$$z = 3x + y$$

subject to

$$x + 2y \leq 16$$
$$6x - y \leq 18$$
$$x \geq 0$$
$$y \geq 0$$

45. Maximize

$$z = 12x + 3y$$

subject to

$$4x - y \leq 20$$
$$7y - 4x \leq 28$$
$$x \geq 0$$
$$y \geq 0$$

46. Maximize

$$z = x + 4y + 6w$$

subject to

$$x + 2y + 2w \leq 12$$
$$3x + y + 2w \leq 10$$
$$x \geq 0$$
$$y \geq 0$$
$$z \geq 0$$

47. Maximize

$$z = 15x + 19y + 20w$$

subject to

$$\tfrac{1}{2}x + y + 2w \leq 100$$
$$5x + 6y + 2w \leq 360$$
$$3x + 3y + 2w \leq 240$$
$$x \geq 0$$
$$y \geq 0$$
$$w \geq 0$$

48. *Process utilization* A computer company makes two models of microcomputers, an industrial model and a personal model. Each industrial model requires 10 hours of assembly time and 3 hours of testing time. Each personal model requires 8 hours of assembly time and 2 hours of testing time. The company has available 440 hours of assembly time and 120 hours of testing time each day. It makes a profit of $1,000 on each industrial model and $750 on each personal model. How many of each model should the company make each day to maximize its daily profit?
(a) Write this problem as a linear programming problem.
(b) Solve the linear programming problem using the geometric method.
(c) Solve the linear programming problem using the simplex method.

49. *Investment mix* An individual intends to invest up to $24,000, presently in a savings account, in stocks and bonds. The bonds will yield 11% annually and return on the stocks should be about 9% during the first year. Furthermore, the person believes that the amount invested in bonds

should be at most one-half of the amount invested in stocks. How much should be invested in stocks and in bonds to maximize the first year's return?
(a) Write this problem as a linear programming problem.
(b) Solve the linear programming problem using the geometric method.
(c) Solve the linear programming problem using the simplex method.

50. *Coffee mix* A domestic coffee company sells three types of coffee, each of which is a blend of Brazilian, Colombian, and Guatemalan coffee. The amount (in pounds) of each of these coffees going into each pound of its domestic types is indicated in the table. The company makes a profit of 40¢ on each pound of deluxe coffee, 48¢ on each pound of superior coffee, and 36¢ on each pound of special coffee. Next week the company will have available 4,000 pounds of Brazilian, 6,000 pounds of Colombian, and 2,000 pounds of Guatemalan coffee. How many pounds of each type should the company produce to maximize its profit for the week? What is the maximum profit?

TYPE	Brazilian	Colombian	Guatemalan
Deluxe	0.3	0.6	0.1
Superior	0.2	0.5	0.3
Special	0.4	0.4	0.2

For each of the following, construct the dual problem and determine the minimum value of the given objective function.

51. Minimize
$$w = 21y_1 + 40y_2$$
subject to
$$3y_1 + 2y_2 \geq 1$$
$$y_1 + 5y_2 \geq 2$$
$$y_1 \geq 0$$
$$y_2 \geq 0$$

52. Minimize
$$w = 16y_1 + 18y_2$$
subject to
$$y_1 + 6y_2 \geq 3$$
$$2y_1 - y_2 \geq 1$$
$$y_1 \geq 0$$
$$y_2 \geq 0$$

53. Minimize
$$w = 8y_1 + 11y_2$$
subject to
$$y_1 + 4y_2 \geq 2$$
$$2y_1 - y_2 \geq 3$$
$$-y_1 + 2y_2 \geq 1$$
$$y_1 \geq 0$$
$$y_2 \geq 0$$

54. Minimize
$$w = 200y_1 + 360y_2 + 240y_3$$
subject to
$$y_1 + 5y_2 + 3y_3 \geq 15$$
$$2y_1 + 6y_2 + 3y_3 \geq 19$$
$$4y_1 + 2y_2 + 2y_3 \geq 20$$
$$y_1 \geq 0$$
$$y_2 \geq 0$$
$$y_3 \geq 0$$

Identify the steps in the four-step procedure involved in the application of mathematics (Section 1.3) in each case.

55. *Process utilization*   Exercise 48
56. *Investment mix*   Exercise 49

CHAPTER 5

PROBABILITY*

OUTLINE

5.1 Sample Spaces and Events
5.2 Calculation of Probabilities
5.3 Counting Techniques
5.4 Combinations
5.5 The Addition Rule
5.6 The Multiplication Rule
5.7 Chapter Summary

Review Exercises

An automobile parts manufacturer has prepared an order for 25 carburetors for shipment. Before shipment can be made, the company's quality control department will pick a sample of 6 of the carburetors for inspection; if any one of these 6 is found to be defective, the shipment will be rejected and all 25 will be inspected. If there actually is 1 defective in the 25, what is the probability that the shipment will be rejected: $\frac{1}{25}$, $\frac{1}{6}$, $\frac{6}{25}$, or what?

Later in this chapter we shall see that the probability is $\frac{6}{25}$ and also what this number means. For now, note the difference between the above problem and those considered earlier in this text. The mathematical models that we have considered in previous chapters are called *deterministic* in that the behavior of the elements of the physical problems were considered determined or able to be predicted exactly (within a reasonable degree of accuracy). Not all physical problems can be considered to be of this type. No one can say for sure whether the shipment in our illustration will be rejected or not. This is typical when a sample is chosen to represent a whole group; also, in problems involving individuals or groups of individuals, human behavior can be predicted only to within some degree of probability. Models that attempt to predict an outcome in this manner are called *probabilistic* and are based on appropriate probabilities.

The first book that treated probability was *The Book on Games,* by Jerome Cardan (1501–1576). This versatile Italian was a physician, mathematician, philosopher, astrologer, and gambler. Having predicted the day on which he would die, he committed suicide on that day in order that his prediction come true!

Cardan's book did not treat probability as a theory in itself but related probability to problems in gambling. The founders of the theory of probability are generally recognized to be Blaise Pascal (1623–1662) and Pierre de Fermat (1601–1665). These two French mathematicians also became interested in probability in connection with problems related to gambling. Today applications of probability occur in such diverse areas as business administration, industrial engineering, biology, psychology, sociology, and physics.

---

* The use of sets is fundamental in our treatment of probability; it may be helpful to review Section A.8 on sets in the algebra review of Appendix A at this time.

5.1 SAMPLE SPACES AND EVENTS

The term *experiment* in probability is used in the very broad sense of referring to any observable phenomenon from the choice of a marriage partner to the relationship between the planets in the solar system, to the choice of a marketing strategy for a new product, or to the spin of a roulette wheel—as contrasted to its usual meaning of a procedure by a scientist in a laboratory. Before being able to determine the probability of any particular outcome of an experiment, we must *first be able to determine all the possible outcomes* of the experiment. A set that describes all the possible outcomes of an experiment is called a **sample space** for the experiment.

EXAMPLE 1   The experiment is the roll of a single die. The sample space is, therefore, the set

$$S = \{1, 2, 3, 4, 5, 6\}$$

where 1 denotes the outcome of obtaining the side of the die with a single dot, 2 denotes the outcome of obtaining the side of the die with two dots, and so on. The set S describes all the possible outcomes when a die is rolled.

EXAMPLE 2   The experiment consists of tossing 3 coins, a nickel, dime, and quarter. The sample space is

$$S = \{HHH, HHT, HTH, HTT, THH, THT, TTH, TTT\}$$

where the first position in each triple indicates the outcome on the nickel, the second position indicates the outcome on the dime, and the third the outcome on the quarter. Therefore, *HHT* indicates that heads are obtained on the nickel and dime and a tail is obtained on the quarter. The set S describes all of the possible outcomes when the three coins are tossed.

EXAMPLE 3   The experiment consists of a quiz in which 5 questions are given and a student is to answer any 3 of the 5. The sample space of possible choices of the student is

$$S = \{(123), (124), (125), (134), (135), (145), (234), (235), (245), (345)\}$$

where (235), for example, indicates that the student chooses to answer questions 2, 3, and 5.

EXAMPLE 4   The experiment consists of drawing a card from an ordinary deck of 52 playing cards. The sample space is

$$S = \{C2, C3, C4, C5, C6, C7, C8, C9, C10, CJ, CQ, CK, CA,$$
$$D2, D3, D4, D5, D6, D7, D8, D9, D10, DJ, DQ, DK, DA,$$
$$H2, H3, H4, H5, H6, H7, H8, H9, H10, HJ, HQ, HK, HA,$$
$$S2, S3, S4, S5, S6, S7, S8, S9, S10, SJ, SQ, SK, SA\}$$

where C2, for example, indicates that the 2 of clubs is drawn, DJ indicates that the jack of diamonds is drawn, HQ indicates that the queen

5.1 SAMPLE SPACES AND EVENTS

1 PRACTICE PROBLEM 1

The integers from 1 through 10 are written on slips of paper and placed in a box. Then one of the numbers is drawn from the box. Construct the sample space for this experiment.

Answer

$\{1, 2, 3, 4, 5, 6, 7, 8, 9, 10\}$

2 PRACTICE PROBLEM 2

Give the elements in the following events for the experiment in Practice Problem 1:

(a) The number chosen is an even number greater than 3.
(b) The number chosen is divisible by 5.

Answer

(a) $\{4, 6, 8, 10\}$ (b) $\{5, 10\}$

of hearts is drawn, SK indicates that the king of spades is drawn, and so forth.

In connection with Example 1, suppose that we are interested in the outcome of obtaining an even number. This outcome would correspond to the subset $\{2, 4, 6\}$ of the sample space. Such subsets are called *events*. Formally, an **event** is a subset of a sample space.

Again in connection with Example 1, if E is the event that a number greater than 2 is obtained, then

$$E = \{3, 4, 5, 6\}$$

In Example 3, if E is the event that exactly 2 heads are obtained, then

$$E = \{HHT, THH, HTH\}$$

In Example 4, if E is the event that a red card is drawn, then

$$E = \{D2, D3, D4, D5, D6, D7, D8, D9, D10, DJ, DQ, DK, DA,$$
$$H2, H3, H4, H5, H6, H7, H8, H9, H10, HJ, HQ, HK, HA\}$$

Note that sample spaces and events are the mathematical descriptions of all possible outcomes and of particular outcomes, respectively, for an experiment. Sample spaces and events are sets; hence, any further relevant characteristics of the outcome of an experiment must also be expressed in the terminology of sets. In particular, we shall be interested in the manner in which combinations of events are expressed in the terminology of sets.

In Example 1, let E_1 be the event "an even number is obtained" and E_2 be the event "a number greater than 3 is obtained." Then

$$E_1 = \{2, 4, 6\}$$

and

$$E_2 = \{4, 5, 6\}$$

If F is the event "an even number is obtained *or* a number greater than 3 is obtained," then

$$F = \{2, 4, 5, 6\}$$

which is the same as $E_1 \cup E_2$; that is,

$$F = E_1 \cup E_2$$

In general, the connective *or* between events is expressed as the *union* of the events.

Similarly, the connective *and* between events is expressed as the *intersection* of the events. In connection with the events E_1 and E_2 just given, if G is the event "an even number is obtained *and* a number greater than 3 is obtained," then

$$G = \{4, 6\}$$

which is the same as $E_1 \cap E_2$.

Note that *or* indicates one or the other or possibly both events occur; *and* indicates that both events occur.

Finally, the event E *not* happening is expressed by E', the *complement* of E, with the sample space being the universal set. If H is the event "a number greater than 3 is not obtained," then

$$H = \{1, 2, 3\}$$

Note that H is the complement of E_2, where E_2 is the event "a number greater than 3 is obtained" considered earlier; that is, $H = E'_2$.

EXAMPLE 5 In the sample space for the toss of 3 coins given previously, let E_1 be the event "a head is obtained on the nickel" and E_2 be the event "more heads than tails are obtained." If F is the event "a head is obtained on the nickel *and* more heads than tails are obtained," then

$$F = E_1 \cap E_2$$

If G is the event "more heads than tails are *not* obtained," then

$$G = E'_2$$

And if H is the event "more heads than tails are obtained *or* a head is obtained on the nickel," then

$$H = E_2 \cup E_1$$

3 PRACTICE PROBLEM 3

Write each of the following in terms of the events E and F of Practice Problem 2, where E is the event "The number chosen is an even number greater than 3" and F the event "The number chosen is divisible by 5." Then list the elements in each event.

(a) The number chosen is an even number greater than 3 or is divisible by 5.

(b) The number chosen is an even number greater than 3 and is divisible by 5.

(c) The number chosen is not divisible by 5.

Answer

(a) $E \cup F = \{4, 5, 6, 8, 10\}$
(b) $E \cap F = \{10\}$
(c) $F' = \{1, 2, 3, 4, 6, 7, 8, 9\}$

EXERCISES

Give the elements of the following events in the sample space of Example 1:

1. An odd number is obtained.
2. A number less than 4 is obtained.
3. The number 4 is not obtained.

Give the elements of the following events in the sample space of Example 2:

4. One tail is obtained.
5. At least two heads are obtained.
6. No heads are obtained.

Give the elements of the following events in the sample space of Example 3:

7. Question 1 is one of the questions answered.
8. Questions 2 and 3 are among the questions answered.
9. Question 1 is not one of the questions answered.

Give the elements of the following events in the sample space of Example 4:

10. A black card is drawn.
11. A picture card (jack, queen, or king) is drawn.
12. A red ace is drawn.

5.1 SAMPLE SPACES AND EVENTS

The experiment consists of the roll of a pair of dice, one red and one green.

13. Give the sample space where the elements of the sample space are of the form (2, 3), which indicates that a 2 was obtained on the red die and a 3 on the green. That is, (2, 3) is one of the elements; you are to determine the rest.

14. List the elements in each event.
    (a) A sum of 5 is obtained.   (b) A sum less than 5 is obtained.
    (c) The sum obtained is an odd number less than 7.

Six horses, with numbers 1 through 6, run in a race.

15. Give a sample space that gives the possible ways the horses can finish in first and second positions (assuming no ties, that these are the only two positions of interest, and that at least two horses finish the race).

16. List the elements in each event.
    (a) Horse 1 finished in first position.
    (b) Horse 1 finished in first or second position.
    (c) Horse 1 did not finish in first or second position.
    (d) No horse finished in first or second position.

Wine classification   A wine manufacturer classifies each wine made as white (W) or red (R), dry (D) or sweet (S), and carbonated (C) or noncarbonated (N).

17. If a federal examiner chooses a bottle of wine at random, construct a sample space with eight elements to describe the possible choices. The elements should be of the form *WSC*, for example, which would indicate that a white, sweet, carbonated wine is chosen.

18. Give the elements in each event.
    (a) A white wine is chosen.   (b) A carbonated wine is chosen.
    (c) A red, noncarbonated wine is chosen.

Family composition   A study is to be made of families with 4 children and the sex of each child from the oldest to the youngest is to be recorded.

19. Construct a sample space describing the possible sexes by age, where the elements are of the form *GBGB*, for example, which would indicate that the oldest child is a girl, the second oldest, a boy, the third oldest, a girl, and the youngest, a boy. (There should be sixteen elements.)

20. Give the elements in each event.
    (a) The oldest child in the family is a girl.
    (b) The oldest and the second oldest children are boys.
    (c) All the children are of the same sex.

Union holidays   In preparing for its contract negotiations, a union asks its members to rank, by preference, three holidays: Martin Luther King Day (K), Presidents' Day (P), and Veterans' Day (V).

21. Construct a sample space describing an individual's possible rankings where the elements are of the form *VKP*, for example, which would indicate that an individual's first choice is Veterans' Day, second choice is Martin Luther King Day, and third choice is Presidents' Day.

22. Give the elements in each event.
    (a) The first choice is Martin Luther King Day.

(b) The third choice is Presidents' Day.
(c) The choices are in alphabetical order.

Racial interaction A sociologist studying racial interaction between adults and children is dividing the volunteers for her study into groups of three—an adult male, an adult female, and a child—any one of whom may be black (B) or white (W).

23. Construct a sample space describing the possible racial makeup of the groups where the elements are of the form BBW, for example, which would indicate that the adult male (the first B) is black, the adult female (the second B) is black, and the child is white.

24. Give the elements in each event.
    (a) All three in the group are of the same race.
    (b) The adults in the group are of different races.
    (c) The adult female is black.

25. Let the sample space be that of Example 3. Let E_1 be the event "Question 3 is answered" and E_2 be the event "Question 5 is answered." Write the following events in terms of one or both of the events E_1 and E_2; then list the elements in each event:
    (a) Question 3 is answered and question 5 is answered.
    (b) Question 3 is answered or question 5 is answered.
    (c) Question 5 is not answered.

26. Let the sample space be that of Exercises 13–14. Let E_1 be the event "A 1 is obtained on the red die" and E_2 be the event "an even sum is obtained." Write the following events in terms of one of or both of the events E_1 and E_2; then list the elements in each event:
    (a) A 1 is obtained on the red die and an even sum is obtained.
    (b) A 1 is obtained on the red die or an even sum is obtained.
    (c) An odd sum is obtained.
    (d) A 1 is obtained on the red die and an odd sum is obtained.

27. Let the sample space be that of Exercises 15–16. Let E_1 be the event "Horse 3 finished in first place," E_2 be the event "Horse 4 finished in first place," and E_3 be the event "Horse 5 finished in second place." Write the following events in terms of one or more of the events E_1, E_2, and E_3; then list the elements in each event.
    (a) Horse 3 finished in first place and horse 5 finished in second place.
    (b) Either horse 3 or 4 finished in first place.
    (c) Horse 4 did not finish in first place.
    (d) Horse 3 and horse 4 finished in first place.
    (e) Horse 3 finished in first place or horse 5 did not finish in second place.

28. *Wine classification* Let the sample space be that of Exercises 17–18. List the elements in the events in parts (a), (b), and (c).
    (a) A white carbonated wine is chosen.
    (b) A sweet wine or a red wine is chosen.
    (c) A dry wine is not chosen.
    (d) Write the event in part (a) as the union or intersection (whichever is appropriate) of two events E_1 and E_2, specifying what the events E_1 and E_2 are.
    (e) Repeat part (d) for the event in part (b).

(f) Express the event in part (c) in terms of the event E, "A sweet wine is chosen."

29. *Racial interaction* Let the sample space be that of Exercises 23–24. List the elements in the events (a), (b), and (c).
    (a) The adult male and the child are black.
    (b) The adult male is white or the adult female is black.
    (c) The child is white.
    (d) Write the event in part (a) as the union or intersection (whichever is appropriate) of two events E_1 and E_2, specifying what the events E_1 and E_2 are.
    (e) Repeat part (d) for the event in part (b).
    (f) Express the event in part (c) above in terms of the event E, "The child is black."

30. *Family composition* Let the sample space be that of Exercises 19–20. List the elements in the events in parts (a), (b), and (c).
    (a) The oldest or the youngest child is a girl.
    (b) The oldest child is a boy and the third oldest is a girl.
    (c) The second oldest child is a boy.
    (d) Write the event in part (a) as the union or intersection (whichever is appropriate) of two events E_1 and E_2, specifying what the events E_1 and E_2 are.
    (e) Repeat part (d) for the event in part (b).
    (f) Express the event in part (c) in terms of the event E, "The second oldest child is a girl."

Explain the meaning of each term.

31. Sample space for an experiment   32. Event

In the following, E and F are events. Complete each sentence with a proper expression in terms of the occurrence of E and F.

33. $E \cap F$ is the event that _____.

34. $E \cup F$ is the event that _____.

35. $E' \cup F$ is the event that _____.

36. $E \cap F'$ is the event that _____.

5.2 CALCULATION OF PROBABILITIES

The calculation of probabilities, as developed here, is appropriate only for sample spaces with a finite number of elements, which has been the case for all sample spaces considered thus far. Other sample spaces may contain an infinite number of elements. For example, if we are interested in the exact amount of time (in hours) that a space satellite stays in orbit, the sample space would consist of all numbers between 0 and the maximum amount of time a satellite could conceivably stay in orbit. No finite sample space could list all the numbers in any such interval. In such situations probabilities must be calculated in a manner different from that considered here.

There are two essential ingredients for calculating the probabilities for different outcomes of an experiment with a finite number of possible outcomes: a sample space and a probability assignment. Sample spaces were the subject of the previous section, which leaves the probability assignment concept for this section. A ***probability assignment*** for a sample space S assigns to each element x in S a number [denoted by $P(x)$ and called the ***probability of x***] in such a way that the following two requirements are satisfied:

1. The probability assigned to each element x in S is a positive number; that is, $P(x) > 0$ for each $x \in S$.

2. The sum of all the probabilities assigned to the elements of S is 1.

For example, if the sample space for the roll of a single die is $S = \{1, 2, 3, 4, 5, 6\}$ and the assignment of probabilities is

$$P(1) = \tfrac{1}{6} \qquad P(2) = \tfrac{1}{6} \qquad P(3) = \tfrac{1}{6}$$
$$P(4) = \tfrac{1}{6} \qquad P(5) = \tfrac{1}{6} \qquad P(6) = \tfrac{1}{6}$$

the two given requirements are satisfied.

The probability of the occurrence of any event in the sample space is then defined in terms of the probabilities of the elements in S as follows: for any event $E \subset S$, ***the probability of E*** is the sum of probabilities assigned to the elements of E. The probability of an event E is denoted by $P(E)$. **To calculate the probability of an event E, add up the probabilities assigned to the elements of E.** For example, in the roll of a single die, the event of obtaining an even number of dots less than 5 is $E = \{2, 4\}$, so $P(E)$ is given by

$$P(2) + P(4) = \tfrac{1}{6} + \tfrac{1}{6} = \tfrac{1}{3}$$

What interpretation should be given to the number $\tfrac{1}{3}$ just obtained for the probability of E? To what extent does it indicate the likelihood of obtaining an even number of dots less than 5 on the next roll of a die?

The probability of an event E indicates the fraction of times that E can be *expected* to occur if the experiment is performed a great many times. Since the probability that an even number less than 5 is obtained is $\tfrac{1}{3}$, one would expect the fraction of times that this event occurs in, say 600 rolls of the die, to be approximately $\tfrac{200}{600}$ because

$$\tfrac{200}{600} = \tfrac{1}{3}$$

Only to this extent does $\tfrac{1}{3}$ indicate the likelihood of obtaining a 2 or 4 on any one roll.

This definition for the probability of an event E does not apply to the empty set, since it does not contain any elements whose probabilities could be added. Since the sample space describes *all* the possible outcomes, the empty set corresponds to impossibility. It is given probability zero; that is, $P(\emptyset) = 0$.

Notice that the probability of any event E is determined by the manner in which probabilities are assigned to the elements in the sample space. For this reason, care must be given to the manner in which this assignment is made in order to give realistic probabilities.

5.2 CALCULATIONS OF PROBABILITIES

We could assign probabilities to the elements in the sample space for the roll of a single die as follows:

$P(1) = \frac{1}{12}$ $\qquad$ $P(3) = \frac{1}{4} = \frac{3}{12}$ $\qquad$ $P(5) = \frac{1}{6} = \frac{2}{12}$
$P(2) = \frac{1}{3} = \frac{4}{12}$ $\qquad$ $P(4) = \frac{1}{12}$ $\qquad$ $P(6) = \frac{1}{12}$

The two requirements for a probability assignment are satisfied. However, if the die is a balanced die so that the likelihood of any number appearing is the same as every other number, this assignment would lead to unrealistic probabilities; the model constructed to give the likelihood of particular outcomes for the experiment would be faulty!

In general, if each element in the sample space is as likely to occur as any other, each should be given the same probability. In the case of the die discussed earlier (presuming the die to be a balanced die), each of the six elements in the sample space was given a probability of $\frac{1}{6}$. The general rule is that *if a sample space contains n equilikely elements, each element is given a probability of 1/n.*

Other types of probability assignments are discussed below.

EXAMPLE 1 Three fair coins are tossed. What is the probability of getting two or more tails?

Solution The sample space is

$$S = \{HHH, HHT, HTH, HTT, THH, THT, TTH, TTT\}$$

Since S contains 8 equilikely elements, each is given a probability of $\frac{1}{8}$.

The event of getting two or more tails is

$$E = \{HTT, THT, TTH, TTT\}$$

Therefore, the probability of E is $\frac{1}{8} + \frac{1}{8} + \frac{1}{8} + \frac{1}{8} = \frac{1}{2}$. ∎

EXAMPLE 2 An ordinary deck of playing cards is shuffled well and a card is chosen at random. What is the probability that the card chosen is a king?

Solution The sample space, consisting of 52 elements, was given in the last section. Since the cards are well shuffled and the card is chosen at random, each element in the sample space is as likely to occur as any other. Each is given a probability of $\frac{1}{52}$. The event E of interest is

$$E = \{CK, DK, HK, SK\}$$

Therefore, $P(E) = \frac{1}{52} + \frac{1}{52} + \frac{1}{52} + \frac{1}{52} = \frac{4}{52} = \frac{1}{13}$. ∎

---

1 PRACTICE PROBLEM 1

The integers from 1 through 10 are written on slips of paper and placed in a box. Then one of the numbers is drawn at random from the box. What is the probability of each outcome?

(a) The number chosen is an even number greater than 3.

(b) The number chosen is divisible by 5.

Answer

(a) $\frac{4}{10} = \frac{2}{5}$ $\qquad$ (b) $\frac{2}{10} = \frac{1}{5}$

If the elements of the sample space are not equilikely, the method of determining probabilities described in the last several examples is not realistic. Another method of assigning probabilities is to study outcomes of past occurrences of the experiment.

Suppose an experiment consists of a student taking a particular mathematics course and the interest is in whether the student, randomly chosen, will pass or fail the course. The sample space is $\{P, F\}$. What probabilities should be assigned to P and to F?

Suppose also that during the past year 500 students took the same course from the same instructor and that of the 500, 450 passed and 50 failed the course. On this basis, a probability of $\frac{450}{500} = \frac{9}{10}$ would be assigned P and $\frac{50}{500} = \frac{1}{10}$ would be assigned F. Note that each element in the sample space is assigned a probability equal to the fraction of the times that element occurred during the past observations. In this way each element in the sample space is assigned a probability in such a way that the two requirements for a probability assignment are satisfied.

EXAMPLE 3 Your company receives frequent shipments from one of its suppliers. Many of these shipments are late. A study of the arrival dates for 30 shipments resulted in the following data:

ARRIVAL DATE	NUMBER OF SHIPMENTS
3 days late	4
2 days late	6
1 day late	5
0 days late	10
1 day early	5
Total:	30

The sample space is chosen to be $S = \{3, 2, 1, 0, -1\}$, where 0 indicates arrival on time and -1 indicates arrival 1 day early. On the basis of these results, the probabilities are assigned as follows:

$$P(3) = \tfrac{4}{30}, \qquad P(2) = \tfrac{6}{30}, \qquad P(1) = \tfrac{5}{30},$$
$$P(0) = \tfrac{10}{30}, \qquad P(-1) = \tfrac{5}{30}$$

What is the probability that any particular future shipment will be late? The event of interest is $E = \{3, 2, 1\}$; then

$$P(E) = \tfrac{4}{30} + \tfrac{6}{30} + \tfrac{5}{30} = \tfrac{15}{30} = \tfrac{1}{2}$$

EXAMPLE 4 A fish hatchery has an experimental pond containing perch. During one week, a record of the sizes of the fish removed from the pond revealed the following percentage for the various sizes, where the smaller length is included in each range but the greater is not:

5 to 6 inches	30%
6 to 7 inches	40%
7 to 8 inches	20%
8 to 9 inches	10%

5.2 CALCULATIONS OF PROBABILITIES

Although the exact number of fish of each size is not known, the percentages give the fraction, or proportion, of fish of each size and can be used to assign probabilities in the same manner as in Example 3. If the sample space is $S = \{5-6, 6-7, 7-8, 8-9\}$, the probabilities are

$$P(5-6) = 0.30, \quad P(6-7) = 0.40, \quad P(7-8) = 0.20, \quad P(8-9) = 0.10$$

What is the probability that a fish chosen at random from the pond is at least 7 inches long? The event is $E = \{7-8, 8-9\}$; therefore, $P(E) = 0.20 + 0.10 = 0.30$.

Assigning probabilities in the manner indicated in Examples 3 and 4 may not always be satisfactory. One problem is the question, To what extent does the past indicate future behavior? In Example 3, for example, if the transportation director for the supplier initiates a program for more prompt shipments, the probability of $\frac{1}{2}$ that a future shipment will be late may no longer be realistic.

A similarly unrealistic result would be obtained in Example 4 if the fish used to determine the probabilities for the elements in the sample space were not truly representative of the total population of fish in the pond. In both of these cases, the results would be mathematically correct, but the assignment of probabilities to the elements in the sample space result in faulty models for predicting future behavior. However, in some instances this may be the best that can be achieved.

Finally, if the equilikely method is not applicable and there is no past experience upon which to rely, a remaining possibility is to assign probabilities to the elements of the sample space on the basis of the *subjective opinion* of someone familiar with the details of the experiment—essentially an educated guess.

Suppose 5 horses run in a race; the sample space for the winning of the race is $\{1, 2, 3, 4, 5\}$. It is not likely that each horse has the same chance of winning. Supposing also that no 2 of these 5 horses ever appeared together in a previous race; there is no past experience to rely upon. Someone who is familiar with the 5 horses would be in a position to give an opinion about what probabilities to assign to each horse's winning. Probabilities determined in this manner fall into the class of subjective probability assignments.

Or suppose a company is about to market a new product and that it hopes to sell 150,000 units during the first year. The company can make a profit from the product if it charges at least $2 per unit but wants to make a larger profit if it can still accomplish its sales objective of 150,000 units. Company executives who are familiar with the market behavior are asked to give their estimate of the probability of attaining the sales objective if the price is set at $2.00, $2.25, and $2.45. The sample space is then $\{2.00, 2.25, 2.45\}$. Probabilities are assigned on the basis of the subjective opinions of the executives, and the final decision is made accordingly.

2 PRACTICE PROBLEM 2

A coin suspected of being unfair is tossed 150 times. A head occurs 90 times and a tail, 60 times. On this basis, what probability should be assigned each element in the sample space $S = \{H, T\}$?

Answer

$P(H) = \frac{3}{5}; \quad P(T) = \frac{2}{5}$

EXAMPLE 5   Each ticket in a state lottery contains two four-digit numbers (from 0000 to 9999). The drawing consists of randomly choosing one number between 0000 and 9999. Each person who holds a ticket with one of the numbers the same as that drawn is a winner.

Susan has a ticket with the numbers 0531 and 2507. What is the probability that she will be a winner?

Solution   The sample space has 10,000 elements—the numbers from 0000 to 9999. How should probabilities be assigned to these elements? Are they equilikely so that each element should be given the same probability? Are they not equilikely so that probabilities should be assigned based on past drawings? Or should we attempt to find an educated guess?

As long as the lottery is honest, the first type of probability should be assigned; each element is as likely to occur as any other. Two of these, 0531 and 2507, are in the event "Susan wins." The corresponding probability is therefore $2 \times \frac{1}{10,000} = \frac{1}{5,000}$. ∎

EXAMPLE 6   In the given Venn diagram, the numbers indicate the total probability assigned to the elements of the various subsets. Determine $P(E_1')$ and $P(E_2 \cup E_3)$.

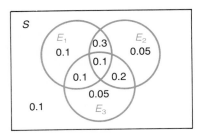

Solution   If we add the probabilities assigned to the points of E_1', we get

$$P(E_1') = 0.05 + 0.2 + 0.05 + 0.1 = 0.4$$

If we add the probabilities assigned to the points of $E_2 \cup E_3$, we get

$$P(E_2 \cup E_3) = 0.3 + 0.1 + 0.2 + 0.05 + 0.1 + 0.05 = 0.8$$ ∎

Every application of probability involves the four-step procedure of Section 1.3 on the application of mathematics. For example, consider the solution of the problem, What is the likelihood of obtaining a sum of 5 in a roll of a pair of dice?

1. A question is asked concerning a physical situation.
2. A mathematical model is constructed—the sample space with subsets as events and the probability assignment, which indicates the relative likelihoods that the various elements in the sample space occur. The original question becomes the mathematical question, What is the probability of $E = \{(1, 4), (4, 1), (3, 2), (2, 3)\}$?

3 PRACTICE PROBLEM 3

In the Venn diagram, the numbers indicate the total probability assigned to the elements of the various subsets. Determine $P(F')$ and $P(E \cap G)$.

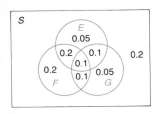

Answer

$P(F') = 0.4$; $P(E \cap G) = 0.2$

3. A mathematical solution is found: $P(E) = \frac{1}{9}$.
4. The mathematical solution is interpreted in the physical situation: In the long run, a sum of 5 will be obtained on the roll of a pair of dice $\frac{1}{9}$ of the time, and in this sense the solution indicates the degree of certainty of obtaining a sum of 5 on any future roll.

EXERCISES

On the roll of a single die, what is the probability of each outcome? (See Exercises 1–3, Section 5.1.)

1. An odd number is obtained.
2. A number less than 4 is obtained.
3. The number 4 is not obtained.

Three coins are tossed. What is the probability of each outcome? (See Exercises 4–6, Section 5.1.)

4. One tail is obtained.
5. At least two heads are obtained.
6. No heads are obtained.

A student is given a quiz in which 5 questions are given, of which any 3 are to be answered. The student is not well prepared and chooses 3 questions completely at random. What is the probability of each outcome? (See Exercises 7–9, Section 5.1.)

7. Question 1 is answered.
8. Questions 2 and 3 are answered.
9. Question 1 is not answered.

A card is chosen at random from a well-shuffled deck of ordinary playing cards. What is the likelihood of each outcome? (See Exercises 10–12, Section 5.1.)

10. A black card is drawn.
11. A picture card is drawn.
12. A red ace is drawn.

On the roll of a pair of dice, what is the probability of each outcome? (See Exercises 13–14, Section 5.1.)

13. A sum of 5 is obtained.
14. A sum less than 5 is obtained.
15. The sum obtained is an odd number less than 7.
16. Eric and Mark each roll a die. What is the probability that Mark rolls a higher number than Eric? (*Hint:* Use the same sample space as that for Exercises 13–15.)
17. *Company shipments* What is the probability that the shipment in Example 3 will not be late?
18. *State lottery* In the state lottery of Example 5, Mark has a ticket with the numbers 0483 and 5361.
    (a) What is the probability that either Susan or Mark will win at the drawing?

(b) What is the probability that both Susan and Mark will win at the drawing?

19. *Company switchboard calls* Over a 6-month period, a company has observed that during the first hour each morning, its switchboard receives 0, 1, 2, 3, 4, 5 or more calls on 10%, 25%, 25%, 15%, 15%, and 10% of the days, respectively. What is the probability that on a randomly chosen day the company will receive less than 3 calls during the first hour?

20. Eric has gotten into the habit of counting the number of cars on the commuter train that he rides to work each day. During 8 weeks (40 workdays), there were 12 cars on 14 days, 13 cars on 12 days, 14 cars on 10 days, and 15 cars on 4 days. On this basis, what is the probability that the train will have more than 12 cars next Monday?

21. The letters of the word *MATHEMATICS* are written on 11 cards, one letter per card. The cards are placed in a hat and one is chosen at random. What is the probability that the card drawn contains a letter that precedes n in the alphabet?

22. *Smoking survey* Of 2,000 randomly chosen adults in a small city, 832 have never smoked, 618 have quit smoking, and 550 still smoke. On this basis, what is the probability that an adult in the city does not smoke?

23. *Tire tread depth* Stretch Rubber Company tested 1,000 tires of a particular type. Of these, 750 maintained a certain depth of tread for more than 40,000 miles and 250 did not. What is the probability that one of this type of tire bought by a customer will maintain the same depth of tread?

24. You have two pairs of shoes in a dark closet. You randomly pick two of the shoes. What is the probability that you have a pair?

25. *Learning experiment* In a learning experiment, food for a mouse was placed at the end of a maze. On 12 attempts, the mouse found its way to the food in less than 5 minutes. On 3 attempts, the mouse either did not find the food or failed to execute the maze in less than 5 minutes. On this basis, what is the probability that the mouse will fail to find the food in less than 5 minutes on the next trial?

Suppose that the sample space is $S = \{a, b, c, d\}$ and $P(a) = 0.5$, $P(b) = 0.3$, $P(c) = 0.1$, and $P(d) = 0.1$. Determine the probability of each event.

26. $E_1 = \{a, c, d\}$ 27. $E_2 = \{a, b, d\}$ 28. $E_1 \cup E_2$
29. $E_1 \cap E_2$ 30. S 31. $\emptyset$
32. Determine $P(E')$ if $E = \{c, d\}$.

If the probability assigned to the points of S is as indicated in the Venn diagram, determine each probability.

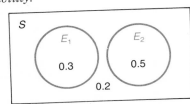

33. $P(E'_1)$ 34. $P(E_1 \cup E_2)$ 35. $P(E_1 \cap E_2)$ 36. $P(E_1 \cap E'_2)$

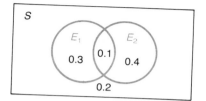

If the probability assigned to the points of S is as indicated in the Venn diagram, determine each probability.

37. $P(E_1 \cup E_2)$ 38. $P(E_1 \cap E_2)$ 39. $P(E_1' \cap E_2)$ 40. $P[(E_1 \cap E_2)']$

Explain each of the following:

41. The role of a probability assignment
42. Three different methods of assigning probabilities to the elements in a sample space
43. The meaning of the probability of an event
44. The physical interpretation of the probability of an event

Identify the steps of the four-step procedure involved in the application of mathematics (Section 1.3) in each case.

45. Exercise 11   46. *Company switchboard calls*   Exercise 19

5.3 COUNTING TECHNIQUES

Often the elements in a sample space, although finite in number, are too numerous to list individually. For instance, there are over 2 million possible 5-card hands that can be dealt from an ordinary deck of 52 playing cards. It would be virtually impossible to list all the elements of the sample space for this experiment by hand.

However, when the elements are equilikely, it is sufficient to know only *how many* elements are in a sample space and in the events in which we are interested. There is no need to construct the actual sample space. The counting techniques of this section are used to calculate probabilities by determining the number of elements in sample spaces and in events without constructing them.

There are two basic techniques to be considered: the *fundamental counting principle* and *combinations*. To introduce the fundamental counting principle, suppose that a salesperson's route requires that he travel from Toledo to Detroit and then to Chicago. Suppose, further, that he can choose one of two highways (A or B) from Toledo to Detroit and one of three highways (R1, R2, or R3) from Detroit to Chicago. In how many ways can he pick his route? We illustrate the possible routes by means of the *tree graph* shown in Figure 5.1.

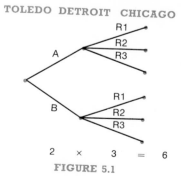

FIGURE 5.1

If he picks route A to get to Detroit, he has a choice of 3 highways to get to Chicago. The same is true, if he chooses route B. His possible routes correspond to the paths through the tree starting at the left vertex and ending at one of the vertices on the right. His first choice can be made in 2 ways, his second in 3. Altogether, he has $2 \times 3 = 6$ different possible choices.

This is a special case of the *fundamental counting principle*, which states: **If one thing can happen in n_1 ways and for each of these n_1 ways a second thing can then happen in n_2 ways, the two things can occur in order in $n_1 \times n_2$ ways.**

EXAMPLE 1   A student must take one of three mathematics courses at 9:00 and one of four English courses at 10:00. In how many ways can she arrange her 9:00–11:00 schedule?

Solution   Since the first thing (choosing a mathematics course) can happen in 3 ways and the second (choosing an English course) can happen in 4 ways, together they can happen in $3 \times 4 = 12$ ways.

The fundamental counting principle can be extended to counting the number of ways more than two things can happen. Suppose the salesperson to whom we referred earlier could return from Chicago to Toledo by two different highways and we are to determine the number of routes at his disposal for the round trip. The final 12 branches in Figure 5.2 correspond to the number of paths through the tree, or the number of possible routes that the salesperson can choose.

1 PRACTICE PROBLEM 1

A man has 3 coats and 5 ties, each of which can be worn with any of the coats.

(a) What are the two steps involved in picking a coat-tie combination each morning?

(b) How many coat-tie combinations can he choose?

Answer

(a) (1) Picking a coat and (2) Picking a tie

(b) $3 \times 5 = 15$

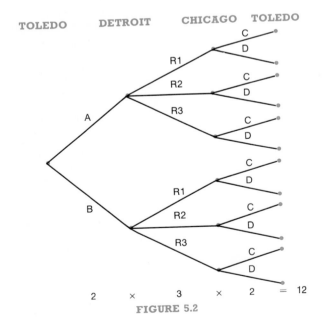

FIGURE 5.2

To determine the number of ways that three events can occur in order, the ***fundamental counting principle*** states: **If one thing can happen in n_1 ways, if for each of these n_1 ways, a second thing can then happen in n_2 ways, and if for each of the $n_1 \times n_2$ ways in which these two things can happen in order, a third thing can then happen in n_3 ways, the three things can happen in order $n_1 \times n_2 \times n_3$ ways.**

EXAMPLE 2 How many three-digit numbers can be formed using the ten digits 0, 1, 2, 3, 4, 5, 6, 7, 8, 9 if the leading digits may be zero?

Solution The construction of a three-digit number involves three things—the choosing of the first digit (which can happen in any one of 10 ways), the choosing of the second digit (which can also happen in one of 10 ways) and the choosing of the third digit (which can also happen in one of 10 ways). Hence the total number of such digits is, not unexpectedly,

$$10 \times 10 \times 10 = 1{,}000$$

EXAMPLE 3 Suppose that no digit can be used twice in the construction of the three-digit numbers of Example 2. Then how many such numbers can be formed?

Solution The first choice can still be made in one of 10 ways, but after the choice of the first digit, there are only 9 ways to choose the second digit, and then 8 ways to choose the third digit. Hence, there are

$$10 \times 9 \times 8 = 720$$

three-digit numbers without repetition of any one digit.

EXAMPLE 4 A particular type of combination lock has 12 numbers on it. How many combinations of 4 different numbers can be formed to open the lock?

Solution The first number can be chosen in 12 ways; the second, in 11; the third, in 10; and the fourth, in 9. Therefore, there are

$$12 \times 11 \times 10 \times 9 = 11{,}880$$

different combinations.

PRACTICE PROBLEM 2

"Words" consisting of any four letters are to be formed from the first seven letters of the alphabet. (For example, *adcf*, *dbgg*, and *fade* are words.)

(a) What are the four steps involved in forming such a word?

(b) How many such words can be formed?

(c) How many such words can be formed if no letter can be used twice in the same word?

Answer

(a) (1) Picking the first letter; (2) Picking the second letter; (3) Picking the third letter; (4) Picking the fourth letter

(b) $7 \times 7 \times 7 \times 7 = 2{,}401$

(c) $7 \times 6 \times 5 \times 4 = 840$

EXAMPLE 5 A mathematics honorary fraternity consists of 15 students, 8 girls and 7 boys. In how many ways can the fraternity choose a president, vice-president, treasurer, and secretary?

Solution The president can be chosen in 15 ways, the vice-president in 14 ways, and so on, so the officers can be chosen in a total of

$$15 \times 14 \times 13 \times 12 = 32{,}760$$

different ways. ∎

EXAMPLE 6 In how many ways can the fraternity of Example 5 pick their officers if they agree that the president and secretary must be girls and that the vice-president and treasurer must be boys?

Solution In this case there would be

$$8 \times 7 \times 7 \times 6 = 2{,}352$$

different slates of officers.

Recall that we are interested in the counting techniques from the point of view of calculating probabilities. The next two examples illustrate the use of the fundamental counting principle in probability.

EXAMPLE 7 If a three-digit number is chosen at random from the numbers 000 to 999, what is the probability that the number contains three different digits?

Solution The sample space consists of 1,000 equilikely elements—the numbers from 000 to 999, each with a probability of $\frac{1}{1{,}000}$. From Example 3, the event of interest has 720 elements; its probability is, therefore,

$$720 \times \left(\tfrac{1}{1{,}000}\right) = \tfrac{18}{25}$$

∎

EXAMPLE 8 Kay, Susan, and Joan are three of the girls in the fraternity of Example 6. If the choice of officers is made by lot with the distinction by sex as indicated, what is the probability that two of these three girls will be officers?

Solution The sample space consists of the 2,352 different slates of officers determined in Example 6, so we assign the probability of $\frac{1}{2{,}352}$ to each slate.

By the fundamental counting principle, the event that two of Kay, Susan, or Joan fill the positions of president and secretary contains

$$3 \times 2 \times 7 \times 6 = 252$$

PRACTICE PROBLEM 3

(a) In how many ways can a committee of 12 people choose a chairperson, vice-chairperson, and a secretary if no person can hold more than one office?

(b) If the committee consists of 7 women and 5 men, in how many ways can the offices be filled if the chairperson and secretary are to be women and the vice-chairperson a man?

Answer

(a) $12 \times 11 \times 10 = 1{,}320$

(b) $7 \times 5 \times 6 = 210$

4 PRACTICE PROBLEM 4

(a) If in Practice Problem 2, a four-letter word made up from the first seven letters is chosen at random, how many elements are there in the sample space? How many elements are there in the event "The word chosen is made up of four different letters"? What is the probability that the word chosen is made up of four different letters?

(b) If in Practice Problem 3, the officers are to be chosen by lot, how many elements are there in the sample space? How many elements are there in the event "The chairperson and secretary chosen are female and the vice-chairperson male"? What is the probability that the chairperson and secretary chosen are female and the vice-chairperson male?

Answer

(a) 2,401; 840; $\frac{840}{2,401} = \frac{120}{343}$

(b) 1,320; 210; $\frac{210}{1,320} = \frac{7}{44}$

elements. The probability of this event is, therefore,

$$252 \times \left(\frac{1}{2,352}\right) = \frac{3}{28}$$

In Examples 7 and 8, the fundamental counting principle is used twice, once to determine the number of elements in the sample space and once to determine the number of elements in the event of interest.

Before doing the exercises, note that the fundamental counting principle is concerned with *order*—counting the number of ways things can happen in *order*. This includes the number of *ordered* arrangements that can be made from a specified number of objects chosen from some given set. The 4 numbers to be chosen from 12 in Example 4 were to be in a specific order. Any attempt to open a combination lock with the correct numbers in the wrong order is doomed to failure!

Also, the 4 officers to be chosen from 15 in Example 5 were to be in a specific order—president, vice-president, secretary, and treasurer. If the 4 persons chosen interchanged offices, the result would be a different slate of officers.

However, if the sequence of happenings are independent in the sense that the outcome of each in no way depends on the outcome of any other, then the total number of outcomes is the same whether they are performed simultaneously or in order. Consequently, they can always be considered to be performed in order and the fundamental counting principle can be used to count the number of possible outcomes. For example, if a coin is tossed and a die is rolled, there are $2 \times 6 = 12$ possible outcomes whether the two are done simultaneously or in sequence. Observe that the three things happening in Example 2 are independent and could occur simultaneously with the same number of total outcomes. This is not true for the happenings of Example 3. Why not?

EXERCISES

1. *Television set classification* Every television set manufactured by a particular manufacturing company is classified by picture, style and quality in the following manner:

   Picture   Black-white or color
   Style   Floor, table, or portable
   Quality   Good, medium, or deluxe

   (a) How many different types of sets are manufactured by this company?
   (b) Draw a tree graph to verify your answer to part (a).

2. *Account classification* A credit card company classifies its delinquent accounts as follows:

   Frequency   First offense or several offenses
   Amount   Less than $500 or more than $500
   Time   Up to 60 days or over 60 days

(a) How many classifications of delinquent accounts are there?
(b) Draw a tree graph to verify your answer to part (a).

3. A mathematics club has 12 members. In how many ways can the club select a president and vice-president?

4. A gourmet club has 12 members, 7 women and 5 men. In how many ways can the club select a president and vice-president if the president must be a woman and the vice-president must be a man?

5. The Greek alphabet consists of 24 letters. How many names of fraternities can be formed if each name consists of three letters and the given condition holds?
   (a) No repetitions of letters are allowed.
   (b) Repetitions of letters are allowed.

6. How many batting orders are possible for players on a baseball team in each case? (There are 9 players on a baseball team and a tenth player may be designated a pinch hitter to bat instead of the pitcher.)
   (a) There is no designated pinch hitter and the pitcher must bat last.
   (b) There is a designated pinch hitter.

7. How many four-digit integers less than 5,000 can be formed using the digits 2, 3, 4, 5, and 6 in each case?
   (a) Repetitions of digits are allowed.
   (b) Repetitions of digits are not allowed.

8. In a particular state, each license plate contains 2 letters followed by 4 digits. If no digit can be used twice in any one license plate, how many different license plates can be made?

9. *Family composition* A sociologist is making a study of families with 3 children and is interested in the arrangement of the children by sex; for example, whether the sex order is *BBG* or *GBB*, where the first letter indicates the sex of the oldest child, the second letter indicates the sex of the middle child, and the third letter that of the youngest. How many such arrangements by sex are possible?

10. *Albinism research* A research project classifies its subjects by sex and whether or not the subject has albinism or not. How many different classifications of subjects are there?

11. An experiment consists in rolling 4 dice (white, red, green, and yellow).
    (a) How many elements are there in the sample space? Assume the elements are of the form (2431), which indicates that a 2 is obtained on the white, a 4 on the red, a 3 on the green, and a 1 on the yellow.
    (b) How many elements in the above sample space are in the event "A 2 is obtained on the white die"?
    (c) What is the probability of obtaining a 2 on the white die?
    (d) How many elements in the above sample space are in the event "A 2 is obtained on the white die and a 1 on the yellow"?
    (e) What is the probability of obtaining a 2 on the white die and a 1 on the yellow?

12. The 4 aces, 4 kings, and 4 queens are removed from an ordinary deck of playing cards and shuffled (the remaining cards are discarded). An experiment consists of drawing 4 cards with each card being replaced after it is drawn.

(a) How many elements are there in the sample space?
(b) How many elements are there in the event "The first 3 cards drawn are aces but the fourth is not an ace"?
(c) What is the probability that the first 3 cards drawn (but not the fourth) are aces?
(d) How many elements are there in the event "The first 3 or the first 4 cards drawn are aces"?
(e) What is the probability that the first three or first four cards drawn are aces?

13. *Personnel assignment*
    (a) In how many ways can a company assign its 5 salespeople (A, B, C, D, and E) to the five territories Northwest, Southwest, Midwest, South, and East?
    (b) In how many ways can the assignment be made if B must be assigned to the Northwest territory and D to the Midwest?
    (c) If the assignment in part (a) is made by lot, what is the probability that C is assigned to the Southwest territory?
    (d) If the remaining assignments in part (b) are made by lot, what is the probability that C is assigned to the Southwest territory?

14. *Memory test*   A psychologist is preparing a memory test by forming words, a word being any arrangement of three letters in which the first letter is from the set $\{b, s, t, m\}$, the second letter is from the set $\{a, e, i, o, u\}$, and the third letter is from the set $\{g, x, w\}$.
    (a) How many such words are possible?
    (b) If one of the words is chosen at random from the set of all possible such words, what is the probability that the word chosen contains either an a or an e?

15. Use the fundamental principle to verify that there are 2^n subsets of a set that contains n elements. (*Hint:* Consider a subset to be constructed by making a yes or no decision about each of the elements.)

5.4 COMBINATIONS

If an organization has 12 members and wants to send a committee of 2 to the national convention, how many different committees can be formed? Committees do not have structure, so that we are no longer interested in ordered arrangements—the committee made up of Maria and Larry is the same as the committee made up of Larry and Maria. Consequently, we now want to know how many *subsets* of 2 members can be formed *without regard to order*. Such unordered subsets are called **combinations**.

Our next objective is to establish a method for counting the number of possible combinations. In order to do so, we shall have to consider counting ordered arrangements from another point of view.

In Example 3 of Section 5.3, we were interested in how many three-digit numbers could be constructed using the digits 0 through 9 without repetition. There were $10 \times 9 \times 8 = 720$ different such numbers.

(Ordered arrangements of this type without repetitions are sometimes called **permutations**.)

Since

$$10 \times 9 \times 8 = 10 \times 9 \times 8 \times \frac{7 \times 6 \times 5 \times 4 \times 3 \times 2 \times 1}{7 \times 6 \times 5 \times 4 \times 3 \times 2 \times 1} \qquad (1)$$

either the right side of this last equation or the left side could be used for the calculation. The merit of this second form is that the notation can be simplified by use of **factorials**.

The product $n \times (n-1) \times (n-2) \times \cdots \times 3 \times 2 \times 1$, for any natural number n, is written $n!$ (read, "n factorial"). It follows, for example, that

$$5! = 5 \times 4 \times 3 \times 2 \times 1 = 120$$
$$4! = 4 \times 3 \times 2 \times 1 = 24$$
$$7! = 7 \times 6 \times 5 \times 4 \times 3 \times 2 \times 1 = 5{,}040$$

Now we can write the right-hand side of equation (1) as

$$\frac{10!}{7!}$$

Note that the numerator corresponds to the number of objects available and the denominator corresponds to the number *not* used.

In general, the number of ordered arrangements without repetitions that can be made using r objects when there are n from which to choose is given by

$$\frac{n!}{(n-r)!}$$

In the illustration just given, n is 10 and r is 3, so that $n - r$ is 7. If all n objects are to be used, then $r = n$. In this case, the denominator in the above expression becomes $0!$ Since 0 is not a natural number, $0!$ has no meaning from the definition just given for factorials.

By the fundamental counting principle, the number of ordered arrangements of n objects when all are used is $n!$. If $0!$ has the value 1, then

$$\frac{n!}{0!} = n!$$

Therefore, we define $0!$ to be equal to 1. Then the expression

$$\frac{n!}{(n-r)!} \qquad (2)$$

gives the number of ordered arrangements of r objects that can be made from n objects without repetitions

5.4 COMBINATIONS

whether or not n equals r. For example, if $n = 7$ and $r = 4$, then equation (2) becomes

$$\frac{7!}{(7-4)!} = \frac{7!}{3!} = \frac{7 \times 6 \times 5 \times 4 \times 3 \times 2 \times 1}{3 \times 2 \times 1} = 840$$

Or, if $n = 4$ and $r = 4$, we get

$$\frac{4!}{(4-4)!} = \frac{4!}{0!} = \frac{4 \times 3 \times 2 \times 1}{1} = 24$$

We return now to the method of counting combinations. We can expect there to be fewer combinations than ordered arrangements. This is because to one subset, there are usually many corresponding ordered arrangements. For a specific example, consider how many ordered arrangements of the letters in the set $\{a, b, c\}$ are possible. By the fundamental counting principle, there should be $3 \times 2 \times 1 = 6$; these six are

$$\begin{array}{ccc} abc & bca & cab \\ acb & bac & cba \end{array}$$

To the one set of 3 elements, there are 3! corresponding ordered arrangements.

Therefore, in order to obtain the number of (unordered) subsets of 3 digits that can be formed from all 10 digits, we divide the number of ordered arrangements by 3!, obtaining

$$\frac{10!}{7!} \div 3! = \frac{10!}{7!3!} = 120$$

In the same way, to any one set of r elements, there correspond $r!$ ordered arrangements. In order, therefore, to determine the number of subsets of r objects (without regard to order) that can be chosen from a given set of n objects, it is sufficient to divide the number of ordered arrangements of n objects chosen r at a time by $r!$; that is, using the expression in equation (2) for the number of ordered arrangements, the number of such subsets of r objects is

$$\frac{n!}{(n-r)!} \div r! \quad \text{or} \quad \frac{n!}{r!(n-r)!}$$

The notation used for this last expression is $C(n, r)$ (read, "the number of combinations of n objects chosen r at a time").

The number of unordered subsets (combinations) of r objects that can be formed from n objects is given by

$$C(n, r) = \frac{n!}{r!(n-r)!}$$

For example, the number of combinations of 4 objects that can be made from 7 objects is

$$C(7, 4) = \frac{7!}{4!3!} = \frac{7 \times 6 \times 5 \times 4 \times 3 \times 2 \times 1}{4 \times 3 \times 2 \times 1 \times 3 \times 2 \times 1} = 35$$

The number of combinations of 6 objects that can be made from 6 objects is

$$C(6, 6) = \frac{6!}{6!0!} = \frac{6 \times 5 \times 4 \times 3 \times 2 \times 1}{6 \times 5 \times 4 \times 3 \times 2 \times 1 \times 1} = 1$$

1 PRACTICE PROBLEM 1

Evaluate $C(8, 5)$.

Answer

56

EXAMPLE 1 Returning to the illustration given earlier, how many committees of 2 can be formed from the membership of an organization if there are a total of 12 members?

Solution In this case $n = 12$ and $r = 2$. There are

$$C(12, 2) = \frac{12!}{2!10!} = \frac{12 \times 11 \times 10!}{2 \times 1 \times 10!} = 66$$

possible committees.

EXAMPLE 2 How many different 5-card hands can be dealt from an ordinary deck of 52 playing cards?

Solution Since we are not concerned with the order in which the cards are dealt, the number of such possible hands is

$$C(52, 5) = \frac{52!}{5!47!} = 2{,}598{,}960$$

Why should we not use the fundamental counting principle in Example 2? If we did, we would allow the first thing to happen 52 ways, the second, 51, and so on, to obtain

$$52 \times 51 \times 50 \times 49 \times 48 = 311{,}875{,}200$$

This is the number of ways the 5 cards could be dealt *in order* or could be held in your hand *in order*. Normally, in playing cards the important thing is just the 5 cards you have; there is no concern about order. Consequently, we are interested only in the number of sets of 5 cards.

2 PRACTICE PROBLEM 2

A mathematics honors class consists of 14 students. How many groups of three can be chosen to participate in the state mathematics contest?

Answer

$$C(14, 3) = 364$$

EXAMPLE 3 A bag contains 14 balls, 8 red and 6 white. Suppose an experiment consists of drawing out 3 of the balls. How many elements are there in the sample space?

Solution As long as the order in which the balls are drawn is of no concern, the possible outcomes of the experiment are all combinations of 3 balls. The number of such combinations is

$$C(14, 3) = \frac{14!}{3!11!} = 364$$

5.4 COMBINATIONS

In Example 3, we did not distinguish between the red and the white; all 14 balls were treated on an equal basis. If we had been interested in obtaining the number of ways of obtaining 2 red balls and 1 white ball, then we would have had to count within the red balls and within the white balls separately. In general, if counting combinations involves more than one group of objects, *all of which are not treated on exactly the same basis,* then combinations must be first counted *within each group* and then the results combined by the fundamental principle. The fundamental counting principle applies here because counting within one group is independent from counting within any other group—the set of red balls can be chosen first and then the set of white balls.

EXAMPLE 4 In the sample space of Example 3, how many elements are there in the event "two red balls and one white ball are drawn"?

Solution The number of ways that two red balls can be drawn is $C(8, 2) = 28$; the number of ways that one white ball can be drawn is $C(6, 1) = 6$. Together these two outcomes can occur in $28 \times 6 = 168$ ways.

EXAMPLE 5 What is the probability of drawing two red balls and one white ball in Example 3?

Solution There are 364 elements in the sample space (Example 3). Of these, 168 are in the event of two red and one white (Example 4). The probability is, therefore, $\frac{168}{364} = \frac{6}{13}$.

As Example 5 illustrates, the role of the method of counting combinations in probability is much the same as that of the fundamental counting principle—to count the number of elements in sample spaces and in events. When the elements in the sample space are equilikely, these counts are all that is required to calculate probabilities; the actual construction of the sample space and events is not required.

---

3 PRACTICE PROBLEM 3

A biologist has 9 individuals willing to participate in an experiment, 5 with blood type A and 4 with blood type O. How many different groups can be formed consisting of 2 people with blood type A and 2 with blood type O?

Answer

$C(5, 2) \times C(4, 2) = 60$

---

4 PRACTICE PROBLEM 4

Answer the following questions for the illustration at the beginning of this chapter:

(a) How many different samples of 6 can the quality control department choose?

(b) How many different samples consist of 1 defective and 5 good?

(c) Verify that the probability of rejection is $\frac{6}{25}$.

Answer

(a) $C(25, 6) = 177{,}100$
(b) $C(1, 1) \times C(24, 5) = 42{,}504$

EXERCISES

Evaluate each of the following:

1. 3!
2. 6!
3. 11!
4. 1!
5. 0!
6. $C(12, 3)$
7. $C(7, 7)$
8. $C(9, 1)$

9. *Automobile display* An automobile dealer receives 12 new cars. He wants to display 3 in his showroom. How many groups of 3 can he choose to show?

10. *Pizza toppings* Peano's Pizza Parlor makes sausage, anchovie, mushroom, pepper, onion, pepperoni, and hamburger pizzas. Peano's Columbus Day special consists of combinations of any three toppings for the price of one. In how many ways can the Columbus Day special be made?

11. A quiz consists of 5 questions, of which the student is to answer 3. In how many ways can the student pick the 3 questions to be answered?

12. *Factory sites* In how many ways can a company pick 3 of 7 possible sites for new factories?

13. From a class of 12 trumpet players, a group of 4 is to be chosen to play in a jazz band. How many different groups of 4 can be chosen?

14. How many 5-card hands consisting of 3 hearts and 2 clubs can be dealt from an ordinary deck of playing cards?

15. *Patient groups* Seven of a doctor's patients, 3 men and 4 women, have agreed to an experimental treatment. How many groups of 2 men and 2 women can the doctor pick for the treatment?

16. *Automobile safety check* There are 14 cars in a parking lot, 9 domestic and 5 foreign. An experiment consists in picking out 3 of the cars at random for a safety check.
    (a) How many elements are there in the sample space?
    (b) How many elements are there in the event "All three cars chosen are domestic"?
    (c) What is the probability that all three chosen are domestic?

17. *Personnel choices*
    (a) A department store manager has 7 qualified applicants available for 2 saleswomen positions in a certain department and 5 qualified applicants available for 3 salesmen positions in the same department. In how many ways can he fill the 5 positions?
    (b) If Eric, Mark, and Joan are three of the qualified applicants, what is the probability that they all will be hired, assuming that the possible combination of 5 applicants are equilikely?

In Exercises 18–26, use the method of combinations and/or the fundamental counting principle, as appropriate.

18. Ten men on a football team are all able to play each of the 4 backfield positions. How many different backfield lineups are available?

19. *Congressional committees* In the United States Congress, a conference committee is to be composed of 3 senators and 3 representatives. In how

many ways can this be done? (There are 435 representatives and 100 senators.)

20. *Botany experiment* A botanist wants to plant each of 3 fields with a different type of hybrid corn. If he has 10 different types of seed available, how many different groups of three can he choose to plant?

21. A class consists of 7 boys and 7 girls. If the boys must sit in the odd-numbered seats and there are 14 seats in the room, how many different seating arrangements are possible?

22. *Political committees* A PAC (political action committee) consists of 12 Democrats and 10 Republicans. A committee of 5 is to be chosen to investigate the presidential candidates' histories of support for women's rights.
    (a) How many committees of 5 can be chosen?
    (b) How many of these committees are made up of 3 Democrats and 2 Republicans?
    (c) If the committee is chosen by lot, what is the probability that the committee chosen will consist of 3 Democrats and 2 Republicans?

23. *Quality control* A carton contains 12 light bulbs. The quality control director will pick a sample of 3 bulbs for inspection.
    (a) How many different samples can be chosen?
    (b) If 3 of the 12 bulbs are defective and the sample is chosen at random, what is the probability that the sample will contain 1 defective and 2 good bulbs?
    (c) If there are 3 defective bulbs and the sample is chosen at random, what is the probability that the sample contains no defective bulbs?

24. *Personnel choices*
    (a) How many 4-salesperson teams can be formed from a sales force of 15 people?
    (b) If Kay and Ben are two of the 15 salespeople, how many of the teams include Kay and Ben?
    (c) If the team is chosen at random, what is the probability that the team chosen includes Kay and Ben?

25. *Television programming*
    (a) A television executive must schedule 3 half-hour programs in three open time slots of a particular evening's program. If she has 7 such half-hour programs from which to choose, in how many ways can she design the evening's program?
    (b) Suppose that the 7 available programs are all of about the same quality, so that the choice of programs is essentially a random choice. If 1 of the 7 is a quiz program, what is the probability that the quiz program is chosen to be shown? (*Hint:* First consider the number of time slots in which the quiz program can be shown.)

26. What is the probability of obtaining exactly 3 queens in a 5-card hand dealt from an ordinary deck of playing cards?

27. Three balls are drawn from a bag containing 4 red, 4 blue, and 4 green balls. What is the probability of obtaining 2 blue balls and 1 green ball?

28. What is the probability of obtaining three of a kind (3 matching cards and 2 unmatched cards) in a 5-card hand dealt from an ordinary deck of playing cards?

29. Recall that ordered arrangements all taken from the same set are called *permutations* if repetitions are not allowed. The notation for the number of such arrangements of n objects using r at a time is $P(n, r)$. Equation (2) in this section gives the value of $P(n, r)$:

$$P(n, r) = \frac{n!}{(n-r)!}$$

(a) Rework Exercise 3 of Section 5.3 by counting permutations.
(b) Rework Exercise 5(a) of Section 5.3 by counting permutations.

30. Explain the role of the counting methods in probability.

31. Explain the meaning of the term *combination*.

5.5 THE ADDITION RULE

Up to now we have calculated the probability of an event by setting up a sample space with a probability assignment and then adding the probabilities assigned to the elements in the event. In the case of large sample spaces and events with equilikely elements, the counting methods enable us to do the calculations without actually constructing the sample spaces and events. In this section we develop a few relationships that provide additional methods for calculating probabilities. In particular, we shall be interested in the relationships between:

1. The probability of $E \cup F$ and that of E and of F
2. The probability of E and of its complement E'

Consider first the two events E and F depicted in Figure 5.3, where S is the universal set. What can we say about the probability of $E \cup F$? We know that it is the sum of the probabilities assigned to the elements in $E \cup F$. To calculate this sum, suppose we first add the probabilities assigned to the elements in E and then we add to that the probabilities assigned to the elements in F, giving a sum equal to $P(E) + P(F)$. Since the elements in $E \cap F$ are in both E and F, their probabilities have been added in twice—once in $P(E)$ and once in $P(F)$. To compensate for this double addition, we subtract $P(E \cap F)$ once and obtain

$$P(E \cup F) = P(E) + P(F) - P(E \cap F)$$

We state this result formally as Theorem 1.

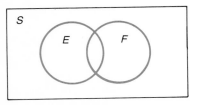

FIGURE 5.3

THEOREM 1 THE ADDITION RULE

If E and F are events in S, then

$$P(E \cup F) = P(E) + P(F) - P(E \cap F)$$

5.5 THE ADDITION RULE

Recall that $E \cup F$ is the event that either E or F or both occur. Consequently, the addition rule gives a method for calculating the probability that either E or F or both occur.

For example, suppose an experiment consists in choosing a number at random from the integers between 1 and 15 (including both 1 and 15). Then

$$S = \{1, 2, 3, 4, 5, 6, 7, 8, 9, 10, 11, 12, 13, 14, 15\}$$

and each element in S has a probability of $\frac{1}{15}$. Let E be the event "The number chosen is divisible by 2" and F be the event "The number chosen is divisible by 3." Then

$$E = \{2, 4, 6, 8, 10, 12, 14\} \qquad F = \{3, 6, 9, 12, 15\}$$

and

$$E \cap F = \{6, 12\}$$

By the addition rule, the probability that the number chosen is divisible by 2 or by 3 (or both) is

$$P(E \cup F) = P(E) + P(F) - P(E \cap F) = \tfrac{7}{15} + \tfrac{5}{15} - \tfrac{2}{15} = \tfrac{10}{15} = \tfrac{2}{3}$$

This is the same result that would be obtained by applying the definition of the probability of an event directly to

$$E \cup F = \{2, 3, 4, 6, 8, 9, 10, 12, 14, 15\}$$

1 PRACTICE PROBLEM 1

If $P(E) = 0.25$, $P(F) = 0.47$, and $P(E \cap F) = 0.14$, determine $P(E \cup F)$.

Answer
0.58

EXAMPLE 1 On the roll of a pair of dice, one red and one green, what is the probability of getting at least one 2?

Solution Getting at least one 2 means getting a 2 on the red die or getting a 2 on the green die (or both). For events

$$E: \text{ A 2 is obtained on the red die}$$

and

$$F: \text{ A 2 is obtained on the green die}$$

$$P(E \cup F) = P(E) + P(F) - P(E \cap F) = \tfrac{6}{36} + \tfrac{6}{36} - \tfrac{1}{36} = \tfrac{11}{36}$$

EXAMPLE 2 A survey indicates that in a particular city, 60% of the patients visiting a doctor have physical disorders, 35% have emotional disorders, and 25% have both physical and emotional disorders. What is the probability that a randomly chosen person visiting a doctor in that city has either a physical or emotional disorder (or both)?

Solution For events

$$E: \text{ The person has a physical disorder}$$

2 PRACTICE PROBLEM 2

A manufacturer of table linens has found that the following usually occur in each lot of its table cloths:

3% are defective because of faulty material.
4% are defective because of faulty sewing.
1% are defective because of both faulty material and faulty sewing.

On this basis, determine the probability that one of the table cloths manufactured will be defective.

Answer
0.06

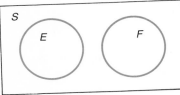

FIGURE 5.4

and

F: The person has an emotional disorder

$$P(E \cup F) = P(E) + P(F) - P(E \cap F) = 0.60 + 0.35 - 0.25 = 0.70$$ 2

If E and F have no points in common, as shown in Figure 5.4, then $E \cap F = \emptyset$ and $P(E \cap F) = 0$. In this special case,

$$P(E \cup F) = P(E) + P(F) - P(E \cap F)$$

becomes simply

$$P(E \cup F) = P(E) + P(F)$$

As a result we have Theorem 2.

THEOREM 2

If E and F are events in S such that $E \cap F = \emptyset$, then

$$P(E \cup F) = P(E) + P(F)$$

If events E and F have no elements in common (that is, $E \cap F = \emptyset$), they are said to be ***mutually exclusive.*** Mutually exclusive events represent outcomes that cannot happen together. For example, getting a 2 and getting a 3 on a single roll of a die or a company winning a contract and not winning the contract are mutually exclusive. Owning a sports car and running out of gas are *not* mutually exclusive—they can happen together. Theorem 2 states that the probability of either of two mutually exclusive events occurring is the sum of their individual probabilities.

For example, if the experiment consists of the roll of a single die, E is the event "A number less than 3 is obtained," and F is the event "A number 4 or greater is obtained," then E and F are mutually exclusive—they cannot occur at the same time. In this case,

$$S = \{1, 2, 3, 4, 5, 6\} \qquad E = \{1, 2\} \qquad F = \{4, 5, 6\}$$
$$E \cup F = \{1, 2, 4, 5, 6\} \qquad E \cap F = \emptyset$$

If each element in the sample space is given a probability of $\frac{1}{6}$, then

$$P(E) = \tfrac{2}{6} = \tfrac{1}{3} \qquad P(F) = \tfrac{3}{6} = \tfrac{1}{2}$$

and $P(E \cup F) = \tfrac{5}{6}$ which, as Theorem 2 requires, is the same as $P(E) + P(F)$.

EXAMPLE 3 In a human population, 33% have type A blood; 45%, type B; 19%, type AB; and 3%, type O. What is the probability that a person randomly chosen from the population will have type A or type O blood?

5.5 THE ADDITION RULE

Solution: Since having type A and having type O are mutually exclusive, the probability of having one or the other is $0.33 + 0.03 = 0.36$.

EXAMPLE 4 A box contains 12 ball point pens, 8 red and 4 blue. Two pens are chosen at random. What is the probability that they are the same color?

Solution: Picking two pens of the same color can be written as $E \cup F$, where E is the event "the two pens chosen are red" and F is the event "the two pens chosen are blue." Since only 2 pens are chosen, E and F are mutually exclusive. Therefore, $P(E \cup F) = P(E) + P(F)$. We use the counting methods to calculate $P(E)$ and $P(F)$. There are $C(12, 2) = 66$ possible pairs of pens from which to choose. Of these, $C(8, 2) = 28$ are both red and $C(4, 2) = 6$ are both blue. Therefore, $P(E) = \frac{28}{66} = \frac{14}{33}$ and $P(F) = \frac{6}{66} = \frac{1}{11}$, so that the desired probability is

$$P(E \cup F) = \frac{14}{33} + \frac{1}{11} = \frac{17}{33}$$

Theorem 2 extends to any number of events, all of which are mutually exclusive; that is, no two of them can occur at the same time.

> If E, F, and G are mutually exclusive, then
> $$P(E \cup F \cup G) = P(E) + P(F) + P(G)$$

This last equation states that the probability that at least one of three mutually exclusive events occurs is the sum of their individual probabilities. In Example 3, the probability that the person chosen has type A, B, or O blood is $0.33 + 0.45 + 0.03 = 0.81$.

Our next theorem is another result of the addition rule. In this case we take F to be E'. Since $E \cap E' = \varnothing$, Theorem 2 states that

$$P(E \cup E') = P(E) + P(E') \quad (1)$$

But $E \cup E' = S$ and the sum of the probabilities assigned to all of the elements in S is 1. Therefore,

$$P(E \cup E') = P(S) = 1$$

and equation (1) becomes

$$1 = P(E) + P(E')$$

This gives us the next theorem.

> **THEOREM 3**
> For any event E in S,
> $$P(E) + P(E') = 1$$

PRACTICE PROBLEM 3

A manufacturer of table linens has found that the following usually occur in each lot of its table cloths:

2% are oversized.
3% are undersized.

If an inspector randomly selects a table cloth from a lot, what is the probability that it is the incorrect size?

Answer

0.05

In calculating probabilities, Theorem 3 is most often used in the alternate forms:

$$P(E) = 1 - P(E') \quad \text{or} \quad P(E') = 1 - P(E)$$

These alternate forms state that the probability of an event occurring is equal to 1 minus the probability that the event does not occur, and vice versa. If it is known, for instance, that $P(E) = \frac{1}{3}$, then $P(E') = 1 - \frac{1}{3} = \frac{2}{3}$. Likewise, if it is known that $P(E') = 0.65$, then $P(E) = 1 - 0.65 = 0.35$.

4 PRACTICE PROBLEM 4

(a) If $P(E) = 0.79$, determine $P(E')$.
(b) If $P(E') = \frac{5}{8}$, determine $P(E)$.

Answer
(a) 0.21 (b) $\frac{3}{8}$

EXAMPLE 5 On a roll of a pair of dice, what is the probability of obtaining a sum greater than 3?

Solution For

$$E: \text{ The sum is greater than 3}$$

it is easier to calculate $P(E')$ than $P(E)$. So we proceed in that direction. In the sample space for the roll of a pair of dice,

$$E' = \{(1, 1), (1, 2), (2, 1)\}$$

and $P(E') = \frac{3}{36} = \frac{1}{12}$. Therefore,

$$P(E) = 1 - \frac{1}{12} = \frac{11}{12}$$

EXAMPLE 6 In Example 2 of this section, what is the probability of a person having neither a physical nor an emotional disorder visiting a doctor?

Solution In Example 2 we found that the probability of having at least one type of disorder is 0.70. Therefore, the probability of having neither is $1 - 0.70 = 0.30$.

EXAMPLE 7 Determine in two ways the probability of getting 2 heads and 1 tail on the toss of 3 coins.

Solution As before, the sample space is

$$S = \{HHH, HHT, HTH, HTT, THH, THT, TTH, TTT\}$$

in which each element is given a probability of $\frac{1}{8}$. First let E be the event of getting any combination *other than* 2 heads and 1 tail. Then

$$E = \{HHH, HTT, THT, TTH, TTT\}$$

and $P(E) = \frac{5}{8}$. The event we are interested in is E'; by Theorem 3,

$$P(E') = 1 - \frac{5}{8} = \frac{3}{8}$$

The second method, using the definition of the probability of an event, is already familiar. The event of getting two heads and one tail is the subset

$$\{HHT, HTH, THH\}$$

PRACTICE PROBLEM 5

Use Theorem 3 to determine the probability of obtaining at least one head when three coins are tossed.

Answer

$1 - \frac{1}{8} = \frac{7}{8}$

Its probability is the sum of the probabilities assigned to the three elements of this subset, which is again $\frac{3}{8}$.

As Example 7 illustrates, the properties of probability given by the three theorems of this section give alternate methods for calculating probabilities. There are still the basic methods developed in earlier sections. Which method is used in a particular situation depends on what information is available or simply which method is easier. However, the three theorems are also important for theoretical considerations and will be used again in the further development of probability in the next chapter.

EXERCISES

1. A number is chosen at random from the integers between 10 and 25, inclusive. Determine the probability that the number chosen is divisible by 3 or by 5. (Use Theorem 1.)

2. *Botany experiment* In an experiment with garden peas, 25% of the plants have wrinkled peas, 60% have yellow peas, and 20% have both yellow and wrinkled peas. What is the probability that a randomly chosen plant has either yellow or wrinkled peas (or both)? (Use Theorem 1.)

3. *Voting behavior* In a survey of 1,000 registered voters in a particular city, 240 usually do not vote, 170 usually vote a straight Republican ticket, 110 vote a straight Democratic ticket, 430 vote a split ticket, and 50 vote for independent candidates. On this basis, what is the probability that a randomly chosen registered voter will vote a straight Republican or Democratic ticket? (Use Theorem 2.)

4. An urn contains 5 white, 6 red, and 8 green balls. If 2 balls are randomly chosen from the urn, what is the probability that they are the same color? (Use Theorem 2.)

5. On the roll of a pair of dice, what is the probability of obtaining a sum less than 10? (Use Theorem 3.)

6. In Exercise 4 what is the probability that the 2 balls chosen will be of different colors? (Use Theorem 3.)

7. Suppose the sample space is $S = \{a, b, c, d\}$, $P(a) = 0.5$, $P(b) = 0.3$, $P(c) = 0.1$, and $P(d) = 0.1$. If $E = \{a, d\}$, find $P(E')$ by two different methods.

8. You ask a cashier for change for a half a dollar and request no pennies. Assuming that the manner in which he makes the change is purely arbitrary, determine in two ways the probability that the change he gives you will include just one quarter.

9. In the game Over-and-Under, a pair of dice is rolled and one can bet whether the sum of dots showing on the two dice is *over* 7, *under* 7, or *exactly* 7. Determine in three different ways the probability that the house wins when you bet that over 7 wins.

10. If a card is randomly chosen from an ordinary deck of playing cards, determine in two different ways the probability that the card chosen is as

follows:
(a) Either a king or a heart   (b) Either a queen or a black card

11. *Transportation* A company makes 42% of its shipments by truck, 37% by train, 15% by plane, and 6% by mail. What is the probability that one of its shipments, randomly chosen, will be made as follows?
(a) By train or by plane   (b) By truck, train, or mail

12. *Customer purchasing* A survey of 200 customers during a department store sale indicated that 44 bought sale items and 26 bought items not on sale. Of those who made purchases, 10 bought both types of items. What is the probability that the next customer who comes in the door will buy something?

13. *Customer purchasing* In Exercise 12, what is the probability that the next customer who comes in the door will buy nothing?

14. *Personnel testing* Each person applying to work on the police force in a particular city must take a written test and a test of physical ability. Based on past experience, the probability that an applicant passes the written test is 0.40, that an applicant passes the physical-ability test is 0.70, and that an applicant passes both tests is 0.35. What is the probability that an applicant will pass either the written or the physical-ability test?

15. *Air pollution* In a particular city the quantity of hydrocarbons in the air exceeds minimum air quality standards on 12% of the days, the quantity of nitrogen oxides exceeds minimum standards on 8% of the days, and both exceed the minimum standards on 5% of the days. What is the probability that the quantity of either will exceed minimum standards on any one day?

16. *Computer customers* The manager of a regional office for a computer company notes that among the 500 companies that she considers to be potential customers, 150 already have one of the company's computers, 200 have a computer manufactured by one of the company's competitors, and 50 have both types of computers. What is the probability that a randomly chosen potential customer already has a computer?

17. *Colorblindness* An experimental group consists of 100 men and 200 women. Two percent of the men and 1% of the women are colorblind. If one of the group is randomly chosen, what is the probability that the person is a man or is colorblind?

Given that $P(E) = 0.3$, $P(F) = 0.5$, and $P(E \cap F) = 0.2$, determine each probability.

18. $P(E')$ 19. $P(E \cup F)$ 20. $P[(E \cup F)']$ 21. $P(E \cup E')$

Given that $P(E) = \frac{1}{4}$, $P(F) = \frac{3}{8}$, and $P(E \cup F) = \frac{1}{2}$, determine each probability.

22. $P(F')$ 23. $P(E \cap F)$ 24. $P[(E \cap F)']$ 25. $P(F \cup F')$

Odds give a mathematical description of the likelihood of an event occurring as compared to the likelihood of it not occurring. Probability enters in a natural way. The **odds in favor of an event E** *are determined by the quotient*

$$\frac{P(E)}{P(E')} = \frac{P(E)}{1 - P(E)}$$

Odds are generally given in terms of positive integers. These are obtained by rewriting the odds as an equivalent quotient of two positive integers with no common factor greater than 1. For example, since the probability of obtaining "snake eyes" (two 1's) is $\frac{1}{36}$, the odds in its favor are

$$\frac{\frac{1}{36}}{\frac{35}{36}} = \frac{1}{35}$$

The odds are then said to be 1 to 35 in favor of the occurrence of snake eyes.

26. In Exercise 9, what are the odds in favor of you winning when you bet that over 7 wins?
27. The probability of an event occurring is 0.40. What are the odds in favor of the event?
28. In Exercise 13, what are the odds that the customer will buy nothing?
29. In Exercise 12, what are the odds that the next customer in the door will buy something?
30. If a card is drawn at random from an ordinary deck of playing cards, what are the odds in favor of it being a picture card (king, queen, or jack)?

If the odds in favor of an event are a to b, the corresponding probability that the event will occur is $a/(a + b)$. For example, if the odds in favor of E occurring are 2 to 3, then $P(E) = 2/(2 + 3) = \frac{2}{5}$.

31. The odds in favor of an event are 3 to 7. What is the probability that the event will occur?
32. A businessman figures that the odds are 3 to 1 that a new store will succeed. What is the subjective probability that he is assigning to the store succeeding?
33. Explain the meaning of *mutually exclusive events*.

5.6 THE MULTIPLICATION RULE

Up to this point we have added probabilities in a variety of situations, but we have never *multiplied* them. Multiplication of probabilities enters into determining the probability of each of several events occurring. In order to see how this is done, we shall need the concept of *conditional probability*.

To illustrate this concept, we suppose that a car is chosen at random from the lot of an auto sales company that handles both foreign and domestic cars. Let F be the event "A foreign car is chosen" and E be the event "A sports car is chosen." Then the probability that E occurs given that F occurs—that is, the probability that a sports car is chosen given that a foreign car is chosen—is an example of a conditional probability. The notation used to indicate this type of probability is $P(E|F)$ (read, "the probability of E given F").

In order to calculate this conditional probability, suppose that the company has on its lot 36 cars classified as shown in Table 5.1. The corresponding sample space has 36 elements representing the 36 cars.

TABLE 5.1

	DOMESTIC	FOREIGN	TOTAL
SEDANS	18	6	24
SPORTS CARS	4	8	12
TOTAL	22	14	36

However, the condition that F occurs reduces the choice to the foreign cars (Table 5.2).

The corresponding reduced sample space has just 14 elements, and the event E of choosing a sports car in this reduced sample space has 8 elements. Therefore,

$$P(E|F) = 8 \times \tfrac{1}{14} = \tfrac{8}{14} = \tfrac{4}{7}$$

TABLE 5.2

	FOREIGN
SEDANS	6
SPORTS CARS	8
TOTAL	14

EXAMPLE 1 A card is drawn from a well-shuffled deck of 52 playing cards. What is the probability that it is a black queen given that it is a picture card?

Solution The condition is the event F, "A picture card is drawn"; the event of interest under this condition is E, "A black queen is drawn." The condition reduces the sample space to 12 equilikely elements, two of which are in E. Therefore, $P(E|F)$ is $\tfrac{2}{12}$, or $\tfrac{1}{6}$.

Returning to the illustration of domestic versus foreign cars, by examining Table 5.1 for all 36 cars, we see that 8 is the number of elements in the event $E \cap F$ (the car chosen is both a sports car and a foreign car). Also, 14 is the number of elements in the event F (a foreign car is chosen). Consequently, we might expect that the conditional probability

$$P(E|F) = \tfrac{8}{14}$$

could just as well have been determined in the sample space for all 36 cars.

In fact, $P(E \cap F) = \tfrac{8}{36}$ and $P(F) = \tfrac{14}{36}$, so that

$$\frac{P(E \cap F)}{P(F)} = \frac{8}{36} \div \frac{14}{36} = \frac{8}{36} \times \frac{36}{14} = \frac{8}{14}$$

that is,

$$\frac{P(E \cap F)}{P(F)} = P(E|F)$$

By multiplying each side of this last equation by $P(F)$, we get the more useful form of the next theorem.

---

1 PRACTICE PROBLEM 1

Use the notation $P(E|F)$ to denote the probability of obtaining a 4 given that an even number is obtained on the roll of a single die.

(a) What are the events E and F?

(b) Determine the value of $P(E|F)$.

Answer

(a) E: A 4 is obtained
  F: An even number is obtained

(b) $\tfrac{1}{3}$

> **THEOREM 4 THE MULTIPLICATION RULE**
>
> For any events E and F,
>
> $$P(E \cap F) = P(F)P(E|F)$$

5.6 THE MULTIPLICATION RULE

The multiplication rule gives a method for determining the probability that *each of two events, E and F*, occurs. (Recall the meaning of the expression $E \cap F$.) It states that the probability that both E and F occur is equal to the probability that F occurs times the probability that E occurs given that F occurs.

Use of the multiplication rule is not the only way of determining the probability that each of two events occurs. However, it often simplifies the sample spaces and/or calculations required.

EXAMPLE 2 Two cards are drawn at random from an ordinary deck of playing cards. The first is not replaced before the second is drawn. What is the probability that both are aces?

Solution Let F be the event "An ace is drawn on the first draw" and E be the event "An ace is drawn on the second draw." We wish to determine $P(E \cap F)$. The sample space for the first draw contains 52 elements, 4 of which are in the event F; hence, $P(F) = \frac{4}{52}$, or $\frac{1}{13}$. The sample space for the second draw contains 51 elements, 3 of which are in the event E, assuming that an ace was drawn on the first try; hence, $P(E|F)$, the probability of an ace on the second draw given that an ace is drawn on the first, is $\frac{3}{51}$, or $\frac{1}{17}$. By the multiplication rule,

$$P(E \cap F) = \tfrac{1}{13} \times \tfrac{1}{17} = \tfrac{1}{221}$$

In Example 2 the sample space relative to the outcome of event F contains only 52 elements, while that of $E|F$ contains 51. This leads to relatively simple calculations in comparison to those in the sample space of all possible 2-card combinations, which consists of $C(52, 2) = 1,326$ elements. Of these 1,326, $C(4, 2) = 6$ consist of 2 aces. The probability of the 2-ace draw in this sample space is again $6 \times \tfrac{1}{1,326} = \tfrac{1}{221}$.

EXAMPLE 3 If the first card in Example 2 is replaced before the second is drawn, what is the probability of drawing 2 aces?

Solution Again, $P(F) = \tfrac{1}{13}$. If the first card is replaced before the second is drawn, there are again 4 aces in the 52 cards, so that $P(E|F)$ is now $\tfrac{1}{13}$. By the multiplication rule,

$$P(E \cap F) = \tfrac{1}{13} \times \tfrac{1}{13} = \tfrac{1}{169}$$

EXAMPLE 4 A coin and die are tossed simultaneously. What is the probability of obtaining a head on the coin and a 5 on the die?

---

2 PRACTICE PROBLEM 2

Two cards are drawn at random from an ordinary deck of playing cards. What is the probability that both cards are clubs in each case?

(a) The first card is not replaced before the second is drawn.

(b) The first card is replaced before the second is drawn.

Answer (a) $\tfrac{13}{52} \times \tfrac{12}{51} = \tfrac{1}{17}$ (b) $\tfrac{13}{52} \times \tfrac{13}{52} = \tfrac{1}{16}$

Solution We are interested in each of the two events

E: A head is obtained on the coin
F: A 5 is obtained on the die

occurring. Since $P(F) = \frac{1}{6}$ and $P(E|F) = \frac{1}{2}$,

$$P(E \cap F) = P(F)P(E|F) = \frac{1}{6} \times \frac{1}{2} = \frac{1}{12}$$

In Example 4, $P(E|F) = P(E)$; this reflects the fact that the outcome of the die does not affect the outcome of the coin. Such events are called independent. Formally, two events E and F are **independent** if $P(E|F) = P(E)$. Events that are not independent are called **dependent**. Example 2 was concerned with dependent events, Example 3 with independent.

In the case of independent events, the multiplication rule takes the simple form of Theorem 5:

THEOREM 5

For independent events E and F,

$$P(E \cap F) = P(F)P(E)$$

That is, the probability of the occurrence of *each of two independent events* is the product of their individual probabilities.

EXAMPLE 5 A company has developed a new product and hopes to start both the manufacturing and advertising of the product on July 1. Management estimates that the probability that the factory will be ready for the manufacturing process by that date is 0.80 and the probability that the advertising campaign will be ready is 0.90. What is the probability that both will be ready by July 1?

Solution If F is the event that the advertising campaign will be ready and E is the event that the factory will be ready, we are interested in the probability of $E \cap F$. Since E and F are independent events,

$$P(E \cap F) = P(F)P(E) = (0.90)(0.80) = 0.72$$

---

3 PRACTICE PROBLEM 3

A university computer center has two new computers, each made by a different manufacturer. The probability that the operating system of the first will fail on any particular day during the first 6 months is $\frac{1}{5}$ and that of the second, $\frac{1}{10}$. What is the probability that both operating systems will fail on any particular day during the first 6 months?

Answer

$\frac{1}{50}$

EXAMPLE 6 Many of the physical characteristics of plants, animals and humans are dependent on a pair of genes, one from the male parent and one from the female parent. The probability foundation for the distribution of genes from one generation to the next is due to a great extent to the work of the Augustinian monk Gregor Mendel (1822–1884). He first reported his work, based on experiments with garden peas, in 1865, but its importance was not appreciated until the beginning of the twentieth century—well after his death.

Consider only one aspect of Mendel's work concerning the color of peas. From its parent plants, each plant has two genes, designated by Y for yellow and y for green, which determine the color of its peas. The possible combinations, called **genotypes,** and the color of the peas with that genotype are given below:

GENOTYPE	COLOR OF PEAS
YY	Y
yY	Y
Yy	Y
yy	y

The first gene in each genotype is from the female parent and the second is from the male parent. For example, the genotype yY received the y gene from its female parent and Y from its male parent.

The right-hand column in the table indicates that as long as the yellow gene, Y, is present, the peas will be yellow; only when the plant is of genotype yy will the peas be green. For this reason the yellow gene is called the **dominant** gene and the green gene the **recessive** gene of the color pair.

Mendel's experiments led to the conclusion that any plant, male or female, will pass on either of its color genes to a plant of the next generation with equal probability. For example, a Yy genotype will pass on its Y gene or its y gene, each with a probability of $\frac{1}{2}$. If the parent plants are of genotype yY in the case of the female parent and Yy in the case of the male, what is the probability that the next generation plant will be of genotype YY? The resulting YY genotype requires two events: a Y gene from its female parent, which has probability $\frac{1}{2}$, and a Y gene from its male parent, which also has a probability of $\frac{1}{2}$. The probability that *both* occur is then $\frac{1}{2} \times \frac{1}{2} = \frac{1}{4}$; the type of gene from the male is independent of the type from the female, and vice versa. ■

EXAMPLE 7 If, in Example 6, the parent plants are of genotype yY in the case of the female and yy in the case of the male, what is the probability that the next generation plant will have yellow peas?

Solution In order for the next generation plant to have yellow peas, it must be of one of the following genotypes:

GENOTYPE	PROBABILITY
YY	$\frac{1}{2} \times 0 = 0$
yY	$\frac{1}{2} \times 0 = 0$
Yy	$\frac{1}{2} \times 1 = \frac{1}{2}$

Since the three genotypes are mutually exclusive, the probability that one or the other occurs is the sum of their probabilities: $0 + 0 + \frac{1}{2}$. The probability is $\frac{1}{2}$ that the peas will be yellow. ∎

Note the use of the multiplication rule (for independent events) and the addition rule (for mutually exclusive events) in Example 7.

The multiplication rule extends to the joint occurrence of three or more events. For three events, by Theorem 4,

$$P(E \cap F \cap G) = P(F \cap G) \times P(E|F \cap G)$$

where the event F of Theorem 4 is taken to be $F \cap G$. Rewriting $P(F \cap G)$, again using Theorem 4, gives

$$P(E \cap F \cap G) = P(G) \times P(F|G) \times P(E|F \cap G)$$

If the three events are independent,

$$P(F|G) = P(F) \quad \text{and} \quad P(E|F \cap G) = P(E)$$

Consequently, we have the following result:

If E, F, and G are independent events,
$$P(E \cap F \cap G) = P(G)P(F)P(E)$$

EXAMPLE 8 Three cards are drawn at random from an ordinary deck of 52 playing cards. What is the probability that all 3 cards are aces?

Solution Let G be the event "An ace is drawn on the first draw," F be the event "An ace is drawn on the second draw," and E be the event "An ace is drawn on the third draw." Then

$$P(G) = \tfrac{4}{52} = \tfrac{1}{13}$$
$$P(F|G) = \tfrac{3}{51} = \tfrac{1}{17}$$
$$P(E|F \cap G) = \tfrac{2}{50} = \tfrac{1}{25}$$

Therefore,
$$P(E \cap F \cap G) = \tfrac{1}{13} \times \tfrac{1}{17} \times \tfrac{1}{25} = \tfrac{1}{5{,}525}$$
∎

EXERCISES

1. Two cards are drawn at random from an ordinary deck of 52 playing cards. What is the probability that the first is a king and the second is a queen in each case?
   (a) The first card is replaced before the second is drawn.
   (b) The first card is not replaced before the second is drawn.

2. Three cards are drawn at random from an ordinary deck of 52 playing cards. What is the probability that all three are 7's in each case?
   (a) Each card is replaced after it is drawn.
   (b) The cards drawn are not replaced.

3. An urn contains 5 white and 7 red balls. Two balls are drawn at random. If the first is replaced before the second is drawn, what is the probability of each outcome?
   (a) Two red balls are drawn.
   (b) The first ball is red and the second is white.
   (c) The first ball is white and the second is red.
   (d) One of the balls is white and one is red.

4. What are the probabilities in Exercise 3 if the first ball is not replaced before the second is drawn?

Are the two events dependent or independent in each case?

5. Exercise 1(a)   6. Exercise 1(b)
7. Exercise 3   8. Exercise 4

9. *Computer personnel* The staff of a computer company consists of 20 salespeople and 30 systems analysts. Of the salespeople, 15 are male and 5 are female. Of the systems analysts, 15 are male and 15 are female. If one of the staff members is chosen at random to discuss a particular computer application, determine the probability that the person is a male systems analyst.
   (a) Use the multiplication rule.   (b) Do not use the multiplication rule.

10. *Senate membership* Assume that the membership of the United States Senate is divided as indicated in the table. A senator is chosen at random to attend a White House luncheon. Determine in three different ways the probability that the one chosen is a Republican woman.

	DEMOCRATS	REPUBLICANS	INDEPENDENTS
MEN	45	28	2
WOMEN	15	9	1

11. *Genetics* Brown eyes (B) is dominant over blue eyes (b). What is the probability that a child of parents both of genotype Bb will have the color eyes indicated?
    (a) Brown   (b) Blue

12. *Genetics* In his experiments Mendel (Example 6) determined that round peas (R) were dominant over wrinkled peas (r). If the parent plants are of genotype YyRr in the case of the male and yYrr in the case of the female, determine the probability that the resulting plant will have round green peas. Assume that color and shape are independent.

13. *Family composition* A couple expects to have 3 children. Assuming that the probability of either sex at each birth is $\frac{1}{2}$, what is the probability of each outcome?
    (a) All 3 children will be girls.
    (b) They will have at least 1 boy. [Use part (a).]

14. *Employment agencies* A small city has two employment agencies, A and B: A handles only accountants and places 80% of their clients; B handles other types of clients as well and places 60% of their accountant applicants. Mark, an accountant looking for a job, intends to apply at agency A first and, failing there, to apply at agency B. What is the probability that he will be placed by agency B?

15. *Defective equipment* An electronic device has two switches, which operate independently of each other. The probability that the first switch is defective is 0.01 and the probability that the second is defective is 0.05. What is the probability of each outcome?
    (a) Both switches are defective.
    (b) Either of the switches is defective.
    (c) None of the switches is defective.

16. Two urns, A and B, are setting on a desk. In urn A there are 30 red and 20 green balls; in urn B, 20 red and 40 green balls. A person enters the room, randomly chooses one of the urns, and randomly picks a ball from the chosen urn. What is the probability that the ball chosen is as specified?
    (a) A green ball from urn A   (b) A green ball   (c) Not a green ball

The game of craps is played with a pair of dice. If the thrower throws a sum of 7 or 11 on the first throw, he wins. If he throws a sum of 2, 3, or 12 on the first throw, he loses. If he throws any other sum on the first throw, that sum becomes his point. He then continues to throw the dice until he obtains his point again or a sum of 7. If he throws the 7 first, he loses. If he throws his point first, he wins. What is the probability of each of the following?

17. The thrower wins on the first throw.
18. The thrower loses on the first throw.
19. The thrower throws a sum of 6 on the first throw and wins on the second throw.
20. The thrower throws a sum of 6 on the first throw and loses on the second throw.
21. The thrower throws a sum of 6 on the first throw and neither wins nor loses on the second throw.
22. The thrower throws a sum of 6 on the first throw and wins on a subsequent throw. (*Hint:* Use a conditional probability with the condition being that a 6 or a 7 was thrown.)
23. *Personnel testing* Each person applying to be a police officer in a particular city must take a written test and a physical-ability test. Based on past experience, the probability that an applicant passes the written test is 0.38 and that an applicant passes the physical-ability test is 0.67. What is the probability that an applicant passes both tests?

24. *Computer security* A bank maintains its computer records on discs and, for security, every evening duplicates its records onto tapes, which are shipped out of the city. The bank's officers estimate that the probability of the disc records being destroyed on any one day is 0.000001 and of the backup tapes being destroyed on any one day is 0.000003. On this basis, what is the probability that both the disc records and the backup tapes are destroyed on any one day?

25. *Air pollution* In a particular city, the quantity of hydrocarbons (due primarily to vehicle exhaust) in the air exceeds minimum air standards on 9% of the days. The quantity of sulfur oxides (due primarily to industrial processes) exceeds minimum standards on 13% of the days. What is the probability that both exceed minimum standards on any one day?

26. What is wrong with the following—or is it wrong?
    The probability of obtaining a 5 on the roll of a die is $\frac{1}{6}$.
    Therefore, the probability of obtaining two 5's on the roll of two dice is $\frac{1}{6} + \frac{1}{6} = \frac{1}{3}$.

27. Explain the meaning of the term *independent events*.

For what type of compound event does each rule give the probability?

28. Addition rule
29. Multiplication rule

What is the special form of each rule?

30. Addition rule for mutually exclusive events
31. Multiplication rule for independent events

5.7 CHAPTER SUMMARY

There are two essential ingredients in the calculation of probabilities, a sample space and an assignment of probabilities to the elements in the sample space. A **sample space** is a set that describes all possible outcomes of an experiment. A **probability assignment** for a sample space S assigns to each element x in S a number, denoted by $P(x)$ and called the **probability of x**, in such a way that:

1. The probability assigned to each element in S is a positive number.
2. The sum of all the probabilities assigned to the elements of S is 1.

Three types of probability assignments were considered. The first was based on equilikely elements, the second on past observations, and the third on subjective opinion.

The **probability of an event** E, $E \subset S$, is the sum of the probabilities assigned to the elements of E. We have considered a number of ways in which probabilities can be calculated and a number of probabilities that have been calculated in different ways. One of the first things to be determined (and quite often is half the battle!) is the *method* to be used to calculate a particular probability. The following methods are available to us at this point:

1. ***Basic definitions*** To determine $P(E)$, it is always possible (at least theoretically) to construct the sample space, determine an appropriate probability assignment, determine the elements in the event E, and—finally—calculate $P(E)$ by adding the probabilities assigned to the elements of E. However, this method becomes impractical if the number of elements in the sample space is very large. The remaining methods are alternatives that may be easier to use when applicable.

2. ***The addition rule*** The probability of the compound event $E \cup F$, that *either E or F occurs*, is given by

$$P(E \cup F) = P(E) + P(F) - P(E \cap F)$$

This is particularly useful when E and F are *mutually exclusive* ($E \cap F = \emptyset$). Then $P(E \cap F) = 0$ and the addition rule becomes

$$P(E \cup F) = P(E) + P(F)$$

Use of the addition rule requires that $P(E)$ and $P(F)$, along with $P(E \cap F)$ when necessary, can be calculated.

3. ***The multiplication rule*** The probability of the compound event $E \cap F$, that *each of E and F occur*, is given by

$$P(E \cap F) = P(F)P(E|F)$$

If E and F are **independent** so that $P(E|F) = P(E)$, the multiplication rule becomes

$$P(E \cap F) = P(F)P(E)$$

Use of the multiplication rule requires that $P(E)$ and $P(E|F)$ or $P(E)$, when appropriate, can be calculated.

4. ***$P(E) + P(E') = 1$*** If $P(E')$ is known or relatively easy to calculate, then $P(E)$ can be determined from the relation

$$P(E) + P(E') = 1$$

This applies regardless of the nature of the event E—for example, regardless of whether E is simple or compound.

5. ***Counting techniques*** If the elements in the sample space are equilikely, then each of the probabilities encountered above can be calculated using the counting techniques. The sample space and events do not have to be constructed. However, at least a verbal description of the sample space is required; otherwise, we cannot determine whether or not the elements are equilikely.

As indicated in the text, the addition rule and the multiplication rule can be extended to any given number of events. This is particularly useful when working with a series of mutually exclusive or independent events.

REVIEW EXERCISES

The given diagram illustrates a mechanical device in which a marble is dropped into the funnel at the top and works its way through the channels into one of the cups at the bottom. At each intersection in the channels, the probability is $\frac{1}{2}$ that the marble will go in either direction. In the following exercises, use the method indicated to determine the probability of the given event. The number of the method corresponds to the number given in the Chapter Summary.

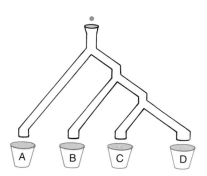

1. Use Method 3 to determine the probability that the marble will end up in cup B; in cup C; in cup D.

2. Use Method 2 [along with Exercise 1] to determine the probability that the marble will end up in cups B, C, or D.

3. Use Method 4 [together with Exercise 2] to determine the probability that the marble will end up in cup A.

4. Set up a sample space that describes the various cups in which the marble can end its trip through the channels. Using Exercises 1 and 3, what probability should be assigned to each element of the sample space?

5. Use Method 1 and Exercise 4 to determine the probability that the marble will end up in cup B or in cup C.

6. Use Method 2 [together with Exercise 4] to determine the probability that the marble will end up in cup B or cup C.

7. Let $S1$ and $S2$ be the indicated sets of cups:

$$S1 = \{A, B\} \qquad S2 = \{B, C\}$$

Use Method 2 to determine the probability that the marble will end up in a cup in either $S1$ or $S2$.

8. If two marbles are dropped into the funnel, one at a time, use Method 3 to determine each probability.
   (a) Both will end up in cup A.
   (b) The first will end up in cup B or D and the second will end up in cup B or D.

A committee of two is to be formed from the five men Larry, Harry, Terry, Jerry, and Barry (denoted by L, H, T, J and B, respectively).

9. Using the counting methods, determine how many different committees can be formed.

10. List the elements in the sample space for this experiment.

11. Given the events

   E_1: Larry is on the committee chosen
   E_2: Terry is on the committee chosen
   E_3: Larry and Terry are on the committee chosen
   E_4: Larry or Terry are on the committee chosen
   E_5: Terry is not on the committee chosen

   write E_3, E_4, and E_5 in terms of E_1 and/or E_2.

12. For the events in Exercise 11, determine $P(E_1)$, $P(E_2)$, $P(E_3)$ in two different ways, $P(E_4)$ in two different ways, and $P(E_5)$ in two different ways, assuming that the committee is chosen at random.

A coin is tossed 4 times and the sequence of heads and tails obtained is observed.

13. Use one of the counting methods to determine the number of possible outcomes.

14. Construct the sample space for this experiment, where the elements in the sample space are in the form *HHTT*, which would indicate a head on the first toss, a head on the second, a tail on the third, and a tail on the fourth.

15. Given the events

   E_1: More tails than heads occur
   E_2: The first 2 tosses are heads
   E_3: More tails than heads occur or the first 2 tosses are heads
   E_4: More tails than heads occur and the first 2 tosses are heads
   E_5: The first 2 tosses are not both heads

   write E_3, E_4, and E_5 in terms of E_1 and/or E_2.

16. For the events in Exercise 15, determine $P(E_1)$, $P(E_2)$, $P(E_3)$ in two different ways, $P(E_4)$ in two different ways, and $P(E_5)$ in two different ways, assuming the coin is fair.

If $S = \{v, w, x, y, z\}$, $P(v) = 0.2$, $P(w) = 0.3$, $P(x) = 0.1$, $P(y) = 0.2$, $E = \{v, w, x\}$, and $F = \{w, x, z\}$, determine each probability.

17. $P(F)$ 18. $P(E)$ 19. $P(E')$
20. $P(E \cap F)$ 21. $P(E \cup F)$ 22. $P(E' \cup F')$

If the probability assigned to the points in S is as indicated in the given Venn diagram, determine each probability.

23. $P(E)$ 24. $P(F)$ 25. $P(E \cup F)$
26. $P(E \cap F)$ 27. $P(E')$ 28. $P[(E \cap F)']$
29. $P(E \cap G)$ 30. $P(F \cup G)$ 31. $P[(E \cap G)']$

Use $P(E) = \frac{3}{4}$, $P(F) = \frac{4}{5}$, and $P(E \cap F) = \frac{3}{5}$.

32. Determine $P(F')$.   33. Determine $P(E \cup F)$.
34. Determine $P(E|F)$.
35. Are E and F mutually exclusive? Explain.
36. Are E and F independent? Explain.

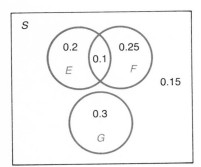

Use $P(E) = 0.4$, $P(F) = 0.5$, and $P(E|F) = 0.1$.

37. Determine $P(E \cap F)$.
38. Determine $P(E \cup F)$.
39. Are E and F mutually exclusive? Explain.
40. Are E and F independent? Explain.
41. *Customer queue* A bank manager has observed that the number of customers in the waiting queue at closing time over a period of 40 days was distributed as shown in the table. On this basis find the probability that the number of customers in the queue when the bank closes tomorrow will be:
    (a) Exactly two (b) At least two (c) At most two

NUMBER OF CUSTOMERS	NUMBER OF DAYS
0	4
1	8
2	11
3	13
4 or more	4

42. *Computer models* A computer company makes a personal computer with either a 8K or 12K byte memory. For each model the optional equipment includes either a 5 megabyte or a 10 megabyte disk drive and a printer. How many different configurations are available to a person buying one of the company's computers?

43. *Political candidates* Twelve candidates are running for the 5 county commissioner positions.
    (a) In how many ways can the 5 positions be filled?
    (b) If 7 of the candidates are Republicans and 5 are Democrats, in how many ways can the 5 positions be filled with all Republicans?
    (c) If 7 of the candidates are Republicans and 5 are Democrats, in how many ways can the five positions be filled with 3 Republicans and 2 Democrats?

44. *Marketing research* A marketing research organization is asking individuals to rank 6 different brands of cheddar cheese by preference.
    (a) How many rankings are possible?
    (b) If the organization asked individuals simply to pick the 3 brands they preferred most rather than ranking them, how many different groups of 3 are possible?

45. A box contains 10 marbles. Three marbles are chosen at random.
    (a) How many different choices are possible?
    (b) If 6 of the marbles are green and 4 are white, what is the probability that the 3 chosen contain (exactly) 2 green?
    (c) If 6 of the marbles are green and 4 are white, what is the probability that at least 2 are green?

46. Two boxes of marbles are sitting on a table. The first contains 6 green and 4 white marbles. The second contains 7 green and 5 white marbles. A box is chosen at random and three marbles are picked at random from the box chosen. What is each probability?
    (a) (Exactly) 2 green marbles are chosen.
    (b) At least 2 green marbles are chosen.

47. *Salary survey* An economist making a salary survey in a small city interviewed 200 families. In these she found that 120 of the husbands and 90 of the wives made an annual salary of over $10,000. Among these were 40 instances in which both the husband and wife made over $10,000 annually. If one of the 200 families is chosen at random, what is the probability that the family chosen has a wage earner with an annual salary of over $10,000?

48. *Salary survey* In Exercise 47, if a family is chosen at random, what are the odds in favor of each of the following?
    (a) The wife makes an annual salary of over $10,000.
    (b) The husband makes an annual salary of over $10,000.

49. *Sales calls* Experience shows that each month the probability that a salesperson for a company will call on any one of his or her customers is $\frac{1}{2}$, and if a salesperson calls, the probability that he or she will receive an order is $\frac{1}{3}$.
    (a) What is wrong with the following—or is it wrong?
        The probability that a salesperson will receive an order from a randomly chosen customer next month is $\frac{1}{6}$.
    (b) What is the probability that, for a randomly chosen customer, a salesperson will call and not receive an order?

50. If two cards are drawn from an ordinary deck of playing cards and the first card is not replaced before the second is drawn, what is the probability of each of the following?
    (a) The first card is a king and the second is a queen.
    (b) The first card is a queen and the second is a king.
    (c) One king and one queen are drawn.
    (d) The first card is a king and the second is a spade.

51. What are the probabilities in Exercise 50 if the first card is replaced before the second is drawn?

CHAPTER 6

ADDITIONAL TOPICS IN PROBABILITY

OUTLINE

*6.1 Binomial Probabilities
*6.2 Bayes' Rule
 6.3 Markov Chains
*6.4 Regular Markov Chains
 6.5 Expected Value
 6.6 Game Theory
*6.7 Games with Mixed Strategies
 6.8 Chapter Summary
   Review Exercises

In Chapter 5 we considered the concepts of probability that are basic for any reasonable discussion of probability for finite sample spaces. In this chapter we consider some additional and special topics in probability. The five general areas—binomial probabilities, Bayes' rule, Markov chains, expected value, and game theory—can be studied independently of each other, except that Section 6.5 on expected value is required for Section 6.7 on games with mixed strategies. We continue to assume that only finite sample spaces are involved.

---

* Sections marked with an asterisk can be omitted without loss of continuity.

*6.1 BINOMIAL PROBABILITIES

Suppose a die is rolled 5 times. What is the probability that a 4 is obtained exactly twice? The sample space for this experiment contains 6^5, or 7,776, elements. Of these, 1,250 are in the event of obtaining exactly two 4's.[†] The probability is, therefore, $\frac{1,250}{7,776}$, or approximately 0.16.

In this section we shall consider an alternate method of calculating probabilities for experiments, such as the one just described, which are repeated a number of times. First, we shall do a detailed analysis of the method applied to calculating the probability of exactly two 4's in 5 rolls of a die.

We shall denote the event of obtaining a 4 by S (for a success) and of obtaining a 1, 2, 3, 5, or 6 by F (for a failure). Then, on any roll the probability of a success is $\frac{1}{6}$; the probability of a failure is $\frac{5}{6}$.

When the die is rolled 5 times, the success could occur on the first 2 rolls, followed by 3 failures. This succession of events will be denoted by $SSFFF$. This outcome requires that each of the following events occurs:

Success on the first roll

Success on the second roll

Failure on the third roll

Failure on the fourth roll

Failure on the fifth roll

The outcome on any roll is independent of the outcome on any of the other rolls. Therefore, the probability that *each* of the above events occurs is the *product* of their individual probabilities. That is, the probability that $SSFFF$ occurs is

$$\tfrac{1}{6} \times \tfrac{1}{6} \times \tfrac{5}{6} \times \tfrac{5}{6} \times \tfrac{5}{6} = (\tfrac{1}{6})^2 (\tfrac{5}{6})^3$$

If the successes occur on the third and fifth rolls, we have $FFSFS$. As before, the probability of this outcome is

$$\tfrac{5}{6} \times \tfrac{5}{6} \times \tfrac{1}{6} \times \tfrac{5}{6} \times \tfrac{1}{6} = (\tfrac{1}{6})^2 (\tfrac{5}{6})^3$$

Or, the successes could occur on the first and fifth rolls: $SFFFS$. The probability of this outcome is also

$$\tfrac{1}{6} \times \tfrac{5}{6} \times \tfrac{5}{6} \times \tfrac{5}{6} \times \tfrac{1}{6} = (\tfrac{1}{6})^2 (\tfrac{5}{6})^3$$

The total number of outcomes that result in exactly two successes is $C(5, 2) = 10$. Other than the three just considered, these are

SFSFF	*FSFSF*	*FFSSF*
SFFSF	*FSFFS*	*FFFSS*
FSSFF		

---

* This section can be omitted without loss of continuity.

† The rolls on which the 4's appear can occur in $C(5, 2) = 10$ ways; for each of these, the other three rolls can result in any of the other five numbers. Total: $10 \times 5^3 = 1,250$.

Since no two of these ten outcomes can occur together (that is, the ten are mutually exclusive) the probability that any *one* of them occurs is the *sum* of their individual probabilities. Then, $(\frac{1}{6})^2(\frac{5}{6})^3$ added to itself ten times gives

$$10 \times (\tfrac{1}{6})^2(\tfrac{5}{6})^3$$

This product simplifies to $\frac{1,250}{7,776}$, the result obtained in the first paragraph of this section.

In terms of the number of trials, the number of successes, and the probability of success on each trial, the probability at which we arrived in the last paragraph can be written

$$C(5, 2)(\tfrac{1}{6})^2(1 - \tfrac{1}{6})^3$$

A similar analysis for the probability of just one 4 in 5 rolls gives

$$C(5, 1)(\tfrac{1}{6})^1(1 - \tfrac{1}{6})^4$$

For obtaining a 4 or a 5 in 3 out of 7 rolls, we would obtain

$$C(7, 3)(\tfrac{2}{6})^3(1 - \tfrac{2}{6})^4$$

In this last case, the probability of a success on any trial is now $\frac{2}{6}$, or $\frac{1}{3}$.

Generalizing, we would expect that the probability of obtaining r successes in n trials, when the probability of a success in each trial is p, would be given by

$$C(n, r)p^r(1 - p)^{n-r}$$

But under what circumstances is the calculation of probabilities in this way valid?

There are two features of this illustration that were essential in our calculations. First, the outcome on each of the 5 rolls was independent of the other rolls. Second, the probability of a success ($\frac{1}{6}$) was the same on each of the rolls. When either or both of these two conditions are not present, the general form just given is not valid. (The mutual exclusiveness of the possible sequences of successes and failures is always automatic.)

In summary, we have Theorem 1.

THEOREM 1

The probability of exactly r successes in n independent trials is given by

$$C(n, r)p^r(1 - p)^{n-r}$$

where p is the probability of a success in each trial.

The term *success* does not refer to any special type of desirable outcome. It simply refers to the outcome of interest in a particular experiment and is therefore a relative term. In the computation of mortality rates, a death would be considered a success!

EXAMPLE 1 It is estimated that 1 out of 5 people have type B blood. If there are 10 people in an emergency room, what is the probability that 3 (exactly) have type B blood, assuming that none of the 10 people are related?

Solution   A success in this case is a person having type B blood. The probability of a success is $p = \frac{1}{5}$. The number of trials is $n = 10$. The probability of $r = 3$ successes is, therefore,

$$C(10, 3)(\tfrac{1}{5})^3(\tfrac{4}{5})^7 = 120(0.20)^3(0.80)^7 = 120(0.008)(0.20972)$$

or approximately 0.201.

Note that the conditions for using the formula of Theorem 1 are satisfied. The experiment of checking a person for blood type is considered to be repeated 10 times. Since none of the 10 people are related, the outcome on any trial is independent of the outcome on the other trials. Also, the probability of a success ($\tfrac{1}{5}$) remains the same on each trial.

A hand calculator with a button for evaluating powers is especially helpful for evaluating the expression

$$C(n, r)p^r(1 - p)^{n-r}$$

in particular problems. A common designation for such buttons is $\boxed{x^y}$.

In algebra an expression such as $p + q$, representing the sum of two quantities, is called a **binomial**. The quantities

$$C(n, r)p^r(1 - p)^{n-r}$$

are called **binomial probabilities** because they are the various terms in the product

$$(p + q)^n$$

where $q = 1 - p$.

For example, if we multiplied $p + q$ by itself three times, we would get

$$(p + q)^3 = 1p^3q^0 + 3p^2q + 3pq^2 + 1p^0q^3$$

or

$$(p + q)^3 = C(3, 0)p^3q^0 + C(3, 1)p^2q + C(3, 2)pq^2 + C(3, 3)p^0q^3$$

The terms on the right are the probabilities of the number of successes when $n = 3$, p is the probability of a success, and $q = 1 - p$. The number r of successes is then said to have a **binomial distribution**. Our only concern is the calculation of binomial probabilities. We mention these facts only to explain the terminology.

---

1 PRACTICE PROBLEM 1

In Example 1, what is the probability that exactly 4 of the people in the emergency room have type B blood?

Answer

$C(10, 4)(\tfrac{1}{5})^4(\tfrac{4}{5})^6$, or about 0.088

6.1 BINOMIAL PROBABILITIES

EXAMPLE 2 Suppose that you are given a 7-question multiple-choice quiz in which there are four choices for each question. Since you have not studied, you decide to pick an answer randomly for each question. What is the probability that you answer (exactly) 5 correctly?

Solution Here the number of trials is $n = 7$. The probability of a success (answering correctly) on each trial is $\frac{1}{4}$. The number of successes in which we are interested is $r = 5$. The probability is, therefore,

$$C(7, 5)(\tfrac{1}{4})^5(\tfrac{3}{4})^2 = 21(0.25)^5(0.75)^2 \approx 21(0.00098)(0.5625)$$

or approximately 0.01.

EXAMPLE 3 Of the records produced by a certain recording company, 5% are defective. The quality control inspector randomly picks 15 records for inspection. What is the probability that none will be defective?

Solution A success in this case is finding a record to be defective. The probability of a success on each of the 15 trials is 0.05, or $\frac{1}{20}$. The probability of zero successes is

$$C(15, 0)(\tfrac{1}{20})^0(\tfrac{19}{20})^{15} = 1(0.05)^0(0.95)^{15} \approx 1(1)(0.46329)$$

or approximately 0.463.

Theorem 1 gives a method for calculating probabilities in situations satisfying the conditions indicated in the theorem without having actually to construct the sample space of all possible sequences of successes and failures. The probabilities calculated in this manner can be used in conjunction with the theorems of the previous chapter, as illustrated in the following example.

EXAMPLE 4 What is the probability that the inspector of Example 3 finds 2 or fewer defectives records?

PRACTICE PROBLEM 2

In Example 2, what would be the probability of answering exactly 5 questions correctly if there were 8 questions instead of 7 on the quiz?

Answer

$C(8, 5)(\tfrac{1}{4})^5(\tfrac{3}{4})^3$, or about 0.023

PRACTICE PROBLEM 3

A biologist plants 6 seeds of a new variety of tomato plant that she is studying. If the germination rate is 0.60, what is the probability she will obtain exactly 5 plants?

Answer

$C(6, 5)(0.60)^5(0.40)^1$, or about 0.187

4 PRACTICE PROBLEM 4

In Practice Problem 3, what is the probability of obtaining either 5 or 6 plants?

Answer

$C(6, 5)(0.60)^5(0.40)^1$
  $+ C(6, 6)(0.60)^6(0.40)^0 \approx 0.234$

Solution Since finding no defective, exactly one defective and exactly two defective records are mutually exclusive, the probability of any one occurring is the sum of their individual probabilities:

$$C(15, 0)(\tfrac{1}{20})^0(\tfrac{19}{20})^{15} + C(15, 1)(\tfrac{1}{20})^1(\tfrac{19}{20})^{14} + C(15, 2)(\tfrac{1}{20})^2(\tfrac{19}{20})^{13}$$
$$\approx 0.463 + 0.366 + 0.135 = 0.964$$

4

In Examples 1, 2, and 3, (and in each of the three mutually exclusive events of Example 4), we were interested in the outcome of each trial in a sequence of independent trials. Furthermore, in each trial we were interested in only two events, success and failure, and the probability of a success on each trial was the same. Such trials are called **Bernoulli trials,** after the Swiss mathematician Jacob Bernoulli (1654–1705). In this setting, the number of combinations $C(n, r)$ is often written $\binom{n}{r}$, read, "n over r"; both $C(n, r)$ and $\binom{n}{r}$ denote the same quantity, the number of combinations of n things chosen r at a time. With this terminology and notation Theorem 1 can be written as follows:

THEOREM 1′

The probability of exactly r successes in n Bernoulli trials is given by

$$\binom{n}{r}p^r(1-p)^{n-r}$$

where p is the probability of a success in each trial.

Table I in Appendix B gives the binomial probabilities

$$\binom{n}{r}p^r(1-p)^{n-r} = C(n, r)p^r(1-p)^{n-r}$$

for some values of n, r, and p if a hand calculator with a button for evaluating powers is not available.

EXAMPLE 5 A sales manager has just hired a new salesperson. The manager knows from experience that a new salesperson will make 1 sale for every 10 customers on whom he or she calls. What is the probability that the new salesperson will make 2 (exactly) sales to the first 7 customers visited?

Solution We consider the outcomes from the calls to the 7 customers all to be independent. Consequently, we have a sequence of 7 Bernoulli trials, where the probability of a success (making a sale) at each trial is 0.10; from Table I (Appendix B), part of which is reproduced here, we see that for $n = 7$, $r = 2$, and $p = 0.10$, the probability is 0.124.

PORTION OF TABLE I: BINOMIAL PROBABILITIES

n	r	0.05	0.1	0.2	0.3	0.4	*p* 0.5	0.6	0.7	0.8	0.9	0.95
5	0	0.774	0.590	0.328	0.168	0.078	0.031	0.010	0.002			
	1	0.204	0.328	0.410	0.360	0.259	0.156	0.077	0.028	0.006		
	2	0.021	0.073	0.205	0.309	0.346	0.312	0.230	0.132	0.051	0.008	0.001
	3	0.001	0.008	0.051	0.132	0.230	0.312	0.346	0.309	0.205	0.073	0.021
	4			0.006	0.028	0.077	0.156	0.259	0.360	0.410	0.328	0.204
	5				0.002	0.010	0.031	0.078	0.168	0.328	0.590	0.774
6	0	0.735	0.531	0.262	0.118	0.047	0.016	0.004	0.001			
	1	0.232	0.354	0.393	0.303	0.187	0.094	0.037	0.010	0.002		
	2	0.031	0.098	0.246	0.324	0.311	0.234	0.138	0.060	0.015	0.001	
	3	0.002	0.015	0.082	0.185	0.276	0.312	0.276	0.185	0.082	0.015	0.002
	4		0.001	0.015	0.060	0.138	0.234	0.311	0.324	0.246	0.098	0.031
	5			0.002	0.010	0.037	0.094	0.187	0.303	0.393	0.354	0.232
	6				0.001	0.004	0.016	0.047	0.118	0.262	0.531	0.735
7	0	0.698	0.478	0.210	0.082	0.028	0.008	0.002				
	1	0.257	0.372	0.367	0.247	0.131	0.055	0.017	0.004			
	2	0.041	0.124	0.275	0.318	0.261	0.164	0.077	0.025	0.004		
	3	0.004	0.023	0.115	0.227	0.290	0.273	0.194	0.097	0.029	0.003	

EXAMPLE 6 In Example 5, what is the probability that the salesperson will make 1 or more sales on the first 7 customer visits?

Solution We are interested in the event E of 1, 2, 3, ..., 7 sales. It would be easier to calculate the probability of E', that no sales are made in the 7 Bernoulli trials. From Table I, this probability is 0.478. Therefore, the probability of E is $1 - 0.478 = 0.522$.

EXERCISES

Whenever possible, use a hand calculator rather than Table I in these exercises.

Determine the probability of obtaining the following results when a coin is tossed 6 times. (Hint: In each case first determine the value of n, r, and p.)

1. Four heads
2. One tail
3. Six heads
4. No heads
5. Two heads
6. Four tails

Determine the probability of obtaining the following results when a die is rolled 4 times. (Hint: In each case first determine the value of n, r, and p.)

7. Two 1's
8. Three 1's
9. No 1's
10. Four 1's
11. Two 5's
12. No 6's

Family life-style It is estimated that one-third of the families in the United States have two cars. If 8 families are chosen at random, what is the probability of each outcome?

13. Four of them are two-car families.
14. None of them are two-car families.
15. All of them are two-car families.
16. Three of them are two-car families.

Genetics If two green parakeets are mated, the probability that an offspring is green is approximately 0.5. What is the probability that among three randomly chosen offspring of green parakeets, each of the following occurs?

17. One green   18. Two green   19. All green   20. No green

21. A dart thrower in an English pub hits the bull's eye on 80% of his shots. What is the probability that he misses the bull's eye on 2 of his next 6 shots?

22. A baseball player's batting average against the opposition's pitcher in the current game is 0.400. If he bats 4 times during the game, what is the probability that he will get 3 hits?

23. *Television advertising* Experience indicates that 2 out of 10 viewers of a particular television advertisement will buy the product at least once. Out of 12 randomly chosen viewers who have seen the ad, what is the probability that 7 will buy the product at least once?

Six coins are tossed. What is the probability of each outcome?

24. Four heads are obtained.
25. Four or more heads are obtained.
26. Fewer than 4 heads are obtained.
27. At most 4 heads are obtained.
28. Fewer than 2 or more than 4 heads are obtained.

Disease propagation The probability of contracting a particular disease after having been exposed to it is 0.60. If 15 students are all exposed to the disease, what is the probability of each outcome?

29. Twelve students contract the disease.
30. Twelve or more of the students contract the disease.
31. Less than 12 of the students contract the disease.
32. At most 12 of the students contract the disease.
33. Less than 2 or more than 10 of the students contract the disease.

Employee fatigue In a particular factory it is estimated that 3 out of 10 accidents are due to fatigue.

34. On this basis what is the probability that 2 out of 4 accidents are due to fatigue?
35. What is the probability that at most 2 out of 4 accidents are due to fatigue?

Political poll A newspaper poll indicates that 60% of the voters in a city prefer the Republican nominee for their Congressional district. If 7 of the voters are chosen at random, what is the probability of each outcome?

36. Three of them prefer the Republican nominee.
37. Less than half prefer the Republican nominee.
38. Less than half prefer the Democratic nominee, assuming that there are only 2 nominees for the office and the newspaper poll was based on which of the 2 the voters preferred.

Cancer treatment A new treatment for cancer causes serious side effects in 10 out of 50 patients. What is the probability of each outcome, based on 15 patients receiving the treatment?

39. Three experience serious side effects.
40. Less than 4 experience serious side effects.
41. More than 10 do not experience serious side effects.

42. *Quality control* Of the tires produced by a certain company, 10% are defective. The quality control inspector randomly picks 10 from each shipment. If 2 or more of these 10 are found to be defective, the entire shipment must be examined. What is the probability of each outcome?
    (a) An entire shipment will not be examined.
    (b) An entire shipment will be examined.

43. In a sequence of independent Bernoulli trials, the probability of a success is 0.40.
    (a) What is the probability of at least three successes if $n = 5$?
    (b) What is the probability of at least three successes if $n = 6$?
    (c) What is the minimum value of n for which the probability of at least three successes is 0.60 or higher?

44. *Inoculation experiment* Experiments have indicated that 20% of the diseased mice inoculated with a particular serum are cured. What is the minimum number of diseased mice that should be inoculated with the serum in order to be sure with a probability of 0.60 that at least 2 will be cured?

45. List the three conditions required of repeated trials in order that they can be considered to be a sequence of Bernoulli trials.

*6.2 BAYES' RULE

Consider the following situation: A printing company orders its paper from four different suppliers. Thirty percent of their paper shipments are from supplier W, 25% are from X, 20% are from Y, and 25% are from supplier Z. Of the shipments from supplier W, 5% are damaged; of those from X, 4% are damaged; of those from Y, 7% are damaged; and of those from Z, 5% are damaged. If a shipment received is found to be damaged, what is the probability that it came from supplier Y?

* This section can be omitted without loss of continuity.

Note the difference between the problem here and those of the previous sections. In the previous sections we discussed methods for determining the probability of particular outcomes *before* the experiment was performed; in this problem we are to determine a probability *after* the experiment was performed—after the shipment was received. Bayes' rule, named after the English minister and mathematician Reverend Thomas Bayes (1702–1763), gives a method of determining probabilities of this type after the experiment is performed.

In the given problem, we have four mutually exclusive events:

A_1: The shipment received is from W

A_2: The shipment received is from X

A_3: The shipment received is from Y

A_4: The shipment received is from Z

Their union is the whole sample space S, since all shipments to the printing company came from one of these suppliers. We want to determine the conditional probability $P(A_3|B)$, where B is the event that a damaged shipment has been received; the event A_3, as indicated, is that the shipment received is from supplier Y.

More generally, suppose that we have four mutually exclusive events A_1, A_2, A_3, and A_4 whose union is all of S; that is, $A_i \cap A_j = \emptyset$ when $i \neq j$ and $A_1 \cup A_2 \cup A_3 \cup A_4 = S$.

Suppose further that B is also an event in S (as shown in Figure 6.1), that the experiment is performed, and that B actually occurs. Bayes' rule gives a method for finding the probability that any one of the A_i's occurred; that is, $P(A_1|B)$, $P(A_2|B)$, and so forth.

First, using the multiplication rule, for each A_i

$$P(A_i|B) = \frac{P(A_i \cap B)}{P(B)} = \frac{P(B \cap A_i)}{P(B)} = \frac{P(A_i)P(B|A_i)}{P(B)} \qquad (1)$$

Next we obtain an expression for $P(B)$ in the denominator of equation (1).

In the Venn diagram shown in Figure 6.1, note that B is the union of its various shaded parts, which represent $B \cap A_1$, $B \cap A_2$, $B \cap A_3$, and $B \cap A_4$; that is,

$$B = (B \cap A_1) \cup (B \cap A_2) \cup (B \cap A_3) \cup (B \cap A_4)$$

Since $B \cap A_1$, $B \cap A_2$, etc., are mutually exclusive, by the addition rule

$$P(B) = P(B \cap A_1) + P(B \cap A_2) + P(B \cap A_3) + P(B \cap A_4)$$

Using the multiplication rule to rewrite each probability on the right in this last equation gives

$$P(B) = P(A_1)P(B|A_1) + P(A_2)P(B|A_2) + P(A_3)P(B|A_3) + P(A_4)P(B|A_4)$$

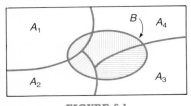

FIGURE 6.1

Substituting this last expression for $P(B)$ in the denominator on the right in equation (1) gives

$$P(A_i|B) = \frac{P(A_i)P(B|A_i)}{P(A_1)P(B|A_1) + P(A_2)P(B|A_2) + P(A_3)P(B|A_3) + P(A_4)P(B|A_4)}$$

which is Bayes' rule for four A_i's. Since we divide by $P(B)$ in (1), we must require that B be an event such that $P(B) \neq 0$.

Returning to the printing company illustration, the event B that has occurred is that a damaged shipment has been received. We are interested in the probability $P(A_3|B)$. According to the information given above, the probabilities required for using Bayes' rule are:

$P(A_1) = 0.30 \qquad P(A_2) = 0.25 \qquad P(A_3) = 0.20 \qquad P(A_4) = 0.25$
$P(B|A_1) = 0.05 \qquad P(B|A_2) = 0.04 \qquad P(B|A_3) = 0.07 \qquad P(B|A_4) = 0.05$

Then Bayes' rule gives

$$P(A_3|B) = \frac{(0.20)(0.07)}{(0.30)(0.05) + (0.25)(0.04) + (0.20)(0.07) + (0.25)(0.05)}$$
$$= \frac{0.0140}{0.0150 + 0.0100 + 0.0140 + 0.0125} = \frac{0.0140}{0.0515} = 0.27$$

EXAMPLE 1 In the printing company illustration, what is the probability that the damaged shipment came from supplier X?

Solution Using the given probabilities in Bayes' rule yields

$$P(A_2|B) = \frac{(0.25)(0.04)}{(0.30)(0.05) + (0.25)(0.04) + (0.20)(0.07) + (0.25)(0.05)}$$
$$= \frac{0.0100}{0.0515} = 0.19$$

EXAMPLE 2 If A_1, A_2, A_3, and A_4 are mutually exclusive events whose union is all of S, and if B is an event in S such that

$P(A_1) = 0.3 \qquad P(A_2) = 0.5 \qquad P(A_3) = 0.1 \qquad P(A_4) = 0.1$
$P(B|A_1) = 0.04 \qquad P(B|A_2) = 0.02 \qquad P(B|A_3) = 0.01 \qquad P(B|A_4) = 0.05$

determine $P(A_4|B)$.

1 PRACTICE PROBLEM 1

In the printing company illustration, determine the probability that the damaged shipment came from supplier W.

Answer

$$\frac{0.0150}{0.0515} = 0.291$$

Solution

Using the given probabilities in Bayes' rule gives

$$P(A_4|B) = \frac{(0.1)(0.05)}{(0.3)(0.04) + (0.5)(0.02) + (0.1)(0.01) + (0.1)(0.05)}$$

$$= \frac{0.005}{0.028} = 0.179$$

In our discussion at the beginning of this chapter, we arbitrarily chose the case that there were four of the A_i's. The same analysis can be made for any number n of A_i's; we state the general rule for arbitrary n in Theorem 2.

2 PRACTICE PROBLEM 2

In Example 2, determine $P(A_2|B)$.

Answer

$$\frac{0.010}{0.028} = 0.357$$

THEOREM 2 BAYES' RULE

If $A_1, A_2, \ldots, A_n$ are mutually exclusive events whose union is the whole sample space S, and if B is any event in S for which $P(B) \neq 0$, then for $i = 1, 2, \ldots, n$,

$$P(A_i|B) = \frac{P(A_i)P(B|A_i)}{P(A_1)P(B|A_1) + P(A_2)P(B|A_2) + \cdots + P(A_n)P(B|A_n)}$$

EXAMPLE 3 In doing a marketing survey for one of its new products, a company has divided the United States into three regions, eastern, central, and western. The company estimates that 40% of the potential customers for their product are in the eastern region, 40% are in the central region, and 20% are in the western region. The survey indicates that 50% of the potential customers in the eastern region, 45% of the potential customers in the central region, and 48% of those in the western region will buy the product.

If a potential customer chosen at random indicates that he will buy the product, what is the probability that the customer is from the central region?

Solution We are asked to calculate $P(A_1|B)$ for

  A_1: The customer is from the central region
  B: The customer will buy the product

Then the other A_i's are

  A_2: The customer is from the eastern region
  A_3: The customer is from the western region

The required probabilities for Bayes' rule are:

$$P(A_1) = 0.40 \qquad P(A_2) = 0.40 \qquad P(A_3) = 0.20$$
$$P(B|A_1) = 0.45 \qquad P(B|A_2) = 0.50 \qquad P(B|A_3) = 0.48$$

Bayes' rule gives

$$P(A_1|B) = \frac{(0.40)(0.45)}{(0.40)(0.45) + (0.40)(0.50) + (0.20)(0.48)}$$

$$= \frac{0.1800}{0.4760} = 0.38$$

EXAMPLE 4 (a) In Example 3, what is the probability that a potential customer chosen at random is from the central region and buys the product?

(b) What is the probability that a potential customer chosen at random buys the product?

Solution (a) We now are interested in the probability that both A_1 and B occur:

$$P(B \cap A_1) = P(A_1)P(B|A_1) = (0.40)(0.45) = 0.1800$$

(b) For this part we want to calculate the probability of B occurring. In view of the information available, we must use the fact that B is the union of the mutually exclusive events $B \cap A_1$, $B \cap A_2$, and $B \cap A_3$. From the given information,

$$P(B \cap A_1) = P(A_1)P(B|A_1) = (0.40)(0.45) = 0.1800$$
$$P(B \cap A_2) = P(A_2)P(B|A_2) = (0.40)(0.50) = 0.2000$$
$$P(B \cap A_3) = P(A_3)P(B|A_3) = (0.20)(0.48) = 0.0960$$

Consequently,

$$P(B) = 0.1800 + 0.2000 + 0.0960 = 0.4760$$

Note how the probabilities 0.1800 and 0.4760 of Example 4 enter into the calculations in Example 3. It is interesting to compare Examples 3 and 4 using a Venn diagram. Example 4(b) considers the probability of B and Example 4(a) considers the probability for the shaded region $B \cap A_1$ in Figure 6.2, both giving probabilities *before* a potential customer is chosen. Example 3 gives the probability for the same

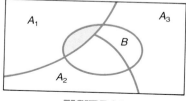

FIGURE 6.2

PRACTICE PROBLEM 3

Of the cars sold by an automobile dealer during the last 5 years, 25% were luxury cars, 35% were midsize, and 40% were compacts. A survey of these customers indicates that 90% of those buying luxury cars, 70% of those buying midsize cars, and 80% of those buying compacts would buy another car from the same dealer. Suppose the dealer sold another car to one of these previous customers. We want to determine the probability that the previous car bought by the customer was a compact.

(a) What are the mutually exclusive events whose union forms the sample space S?

(b) What is the event B for Bayes' rule?

(c) If the dealer sold another car to one of these previous customers, what is the probability that the previous car bought by the customer was a compact?

Answer

(a) A_1: The previous car bought by the customer was a luxury car
A_2: The previous car bought by the customer was a midsize car
A_3: The previous car bought by the customer was a compact

(b) The dealer sold another car to one of these previous customers.

(c) $P(A_3|B) = 0.405$

shaded region *when the sample space has been reduced to B alone;* that is, *after* a potential customer who will buy has been chosen.

Probabilities such as those in Example 4, which are calculated before the experiment is performed, are called ***a priori probabilities*** (Latin for *from before*). Those calculated after the experiment is performed such as in Example 3 are called ***a posteriori probabilities*** (Latin for *from after*).

EXERCISES

1. In the printing company illustration at the beginning of this section, if a damaged shipment is received, what is the probability that it came from supplier Z?

2. In Example 3, if a potential customer who will buy the product is chosen, what is the probability that the customer is from the western region?

Suppose A_1, A_2, A_3, and A_4 are mutually exclusive events whose union is the whole sample space. Also,

$P(A_1) = 0.2 \qquad P(A_2) = 0.4 \qquad P(A_3) = 0.1 \qquad P(A_4) = 0.3$

$P(B|A_1) = 0.12 \qquad P(B|A_2) = 0.22 \qquad P(B|A_3) = 0.08 \qquad P(B|A_4) = 0.19$

3. Determine $P(A_1|B)$.

4. Determine $P(A_4|B)$.

Suppose A_1, A_2, and A_3 are mutually exclusive events whose union is the whole sample space. Also,

$P(A_1) = 0.59 \qquad P(A_2) = 0.13 \qquad P(A_3) = 0.28$

$P(B|A_1) = 0.23 \qquad P(B|A_2) = 0.09 \qquad P(B|A_3) = 0.17$

5. Determine $P(A_2|B)$.

6. Determine $P(A_1|B)$.

7. *Manufacturing* A television company has three plants to manufacture its sets. Plant X manufactures 20% of its sets, plant Y manufactures 30% of the sets, and plant Z manufactures 50% of the sets. Of the sets manufactured by plant X, 7% are defective; of those manufactured by plant Y, 8% are defective; and of those manufactured by plant Z, 5% are defective. If a randomly chosen television set is found to be defective, what is the probability that the set was manufactured at the indicated plant?
   (a) Plant X   (b) Plant Z
   (*Hint:* First determine the A_i events and the event B.)

8. Two urns are sitting on a table. The first urn contains 6 white and 18 blue balls; the second contains 12 white and 24 blue balls. A person enters the room, randomly chooses one of the urns, and randomly picks a ball from the urn. If the ball chosen is blue, what is the probability that it was chosen from the first urn? (*Hint:* First determine the A_i events and the event B.)

9. *Investment* An investment manager classifies stocks that she recommends to clients as good for short term, good for long term, or mixed. Of the stock owned by her clients, 65% is good for short term, 25% is good for long term, and 10% is mixed. At the beginning of a business day she estimates that the good-for-short-term stocks will appreciate in value with a probability of 0.65, that the good-for-long-term stocks will appreci-

ate with a probability of 0.55, and the mixed stocks will appreciate with a probability of 0.50. If a randomly chosen stock owned by her clients does appreciate that day and assuming her estimates are reasonable, determine the probability that the stock was a mixed stock.

10. *Colorblindness* In a human population, half male and half female, 4% of the males and 0.2% of the females are colorblind. If a person randomly chosen from the population is found to be colorblind, what is the probability that the person is a female?

11. *Pregnancy test* An experimental pregnancy test being tested on mice has been found to give a positive result 79% of the time when tested on pregnant mice and 11% of the time when tested on mice that are not pregnant. In a group of 10 mice, 7 are known to be pregnant and 3, not to be pregnant. One of the mice was chosen for the test, which indicated a positive result. What is the probability that the mouse was not pregnant?

12. *Lactose-intolerance* In a group of 60 male children and 60 female children, 10 of the males and 15 of the females have been found to be lactose-intolerant. If one of the children is chosen at random and found to be lactose-intolerant, what is the probability that the child is a female?

13. *Lie-detector accuracy* Suppose that a lie-detector test is 99% accurate for innocent people and 95% accurate for guilty people. One of 10 used-car salespeople is guilty of theft, but all 10 are to be given the test in random order. The detector indicates that the first salesperson tested is guilty.
    (a) What is the probability that the salesperson is actually guilty? (*Hint:* Let the A_i's denote actual guilt or innocence and B denote that test indicates guilt.)
    (b) If the detector indicates that the first salesperson is innocent, what is the probability that he or she is actually innocent?

14. *Manufacturing* In Exercise 7, if a randomly chosen television set is found to be good, what is the probability that it was manufactured at the given plant?
    (a) Plant X   (b) Plant Z

15. *Colorblindness* In Exercise 10, if a randomly chosen person is found not to be colorblind, what is the probability that the person chosen is a male?

16. *Manufacturing* In Exercise 7, if a television set is randomly chosen, what is the probability of each of the following?
    (a) The set is from plant Y and is defective.
    (b) The set is defective.

17. *Colorblindness* In Exercise 10, if a person is randomly chosen, what is the probability of each of the following?
    (a) The person is male and is colorblind.
    (b) The person is not colorblind.

18. Explain the difference between a priori and a posteriori probabilities.

6.3 MARKOV CHAINS

Consider a sequence of experiments that consist of determining, at the end of each business day, whether or not a particular stock has increased in value that day, decreased in value that day, or remained the same in value that day. The sample space would be

$$S = \{I, D, R\}$$

A long-range analysis of the behavior of the stock gives the following matrix of probabilities for the behavior of the stock on any given day in terms of the previous day's behavior:

$$\text{From} \begin{array}{c} \\ I \\ D \\ R \end{array} \overset{\overset{To}{\begin{array}{ccc} I & D & R \end{array}}}{\begin{bmatrix} 0.5 & 0.3 & 0.2 \\ 0.3 & 0.6 & 0.1 \\ 0.4 & 0.4 & 0.2 \end{bmatrix}} = P$$

The entry p_{ij} denotes the probability of obtaining the outcome at the top of the jth column given the outcome on the previous day, which appears at the left of row i. For example, the probability that the price will remain the same (R) on any day given a decrease (D) on the previous day is 0.1; the probability of an increase (I) on any day given that the price remained the same on the previous day is 0.4.

This sequence of experiments of determining the day-to-day behavior of the stock is one example of a Markov chain. In general, a **Markov chain** is a sequence of repeated experiments, or *trials*, having the following three properties:

1. The sample space $S = \{a_1, a_2, \ldots, a_m\}$ remains the same for each experiment, or trial.
2. The probability of obtaining any particular element in the sample space at each trial possibly depends upon the outcome of the preceding trial but upon no other previous trial.
3. For each i and j, the probability p_{ij} that a_j will occur given that a_i occurred on the preceding trial remains constant throughout the sequence.

Markov chains are named after the Russian mathematician Andrei A. Markov (1856–1922).

In the illustration just given, the trials were the day-to-day observations of the behavior of the stock. In the sample space, $m = 3$, with $a_1 = I$, $a_2 = D$, and $a_3 = R$. The probabilities given in the matrix P for each outcome depend only on the outcome of the previous day (on the previous trial) and are presumed to remain constant from one day to the next (or from one trial to the next).

In a Markov chain setting, the physical situation being observed (the behavior of the stock in the illustration) is called the **system** being observed; the elements $a_1, a_2, \ldots, a_m$ in the sample space are called the **states** of the system; the probabilities p_{ij} are called **transition probabilities,** and the matrix $P = [p_{ij}]$ is called the **transition matrix.**

6.3 MARKOV CHAINS

There are two questions concerning Markov chains to be considered:

1. If the system starts in state a_i, what is the probability after n trials that the system is in state a_j?
2. Can the long-term behavior of the system be predicted?

With respect to the first question and our stock illustration, suppose that on a particular day (day 0) the stock has increased in value (that is, the system is in state I). What are the probabilities of I, D, and R for the next day (day 1)? For 2 days afterward (day 2)?

The probabilities for day 1 are the 0.5, 0.3, and 0.2, respectively, given in the first row of the transition matrix P. Calculating the probabilities for day 2 is a bit more complicated. The tree graph in Figure 6.3 gives the possible behavior over the 3 days.

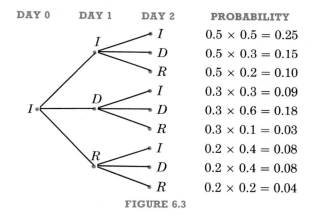

FIGURE 6.3

An increase on day 2 can be the result of one of the three mutually exclusive sequences

$$III, \quad IDI, \quad \text{or} \quad IRI$$

so that the probability of an increase on the second day is the sum of the individual probabilities:

$$0.25 + 0.09 + 0.08 = 0.42$$

Similarly, the probability of a decrease on day 2 is:

$$0.15 + 0.18 + 0.08 = 0.41$$

and the probability of remaining the same on day 2 is

$$0.10 + 0.03 + 0.04 = 0.17$$

If we want to calculate corresponding probabilities for a decrease in value or for remaining the same in value on day 0 and/or extend the number of days to 4, 5, 6, and so forth, the above process becomes tedious and eventually unmanageable.

A more efficient method for calculating these probabilities using matrix operations is available. First, we present the given probabilities in vector form (recall that a vector is a matrix consisting of one row or one column):

$$P_1 = [0.5 \quad 0.3 \quad 0.2] \quad \text{For day 1}$$
$$P_2 = [0.42 \quad 0.41 \quad 0.17] \quad \text{For day 2}$$

where the entries are $P(I)$, $P(D)$, and $P(R)$ for day 1 in the case of P_1 and for day 2 in the case of P_2. Since we know (or presume) that an increase occurred on day 0, we take the probability vector for day 0 to be

$$P_0 = [1 \quad 0 \quad 0]$$

Now note that

$$P_0 P = [1 \quad 0 \quad 0] \begin{bmatrix} 0.5 & 0.3 & 0.2 \\ 0.3 & 0.6 & 0.1 \\ 0.4 & 0.4 & 0.2 \end{bmatrix} = [0.5 \quad 0.3 \quad 0.2] = P_1$$

$$P_0 P^2 = [1 \quad 0 \quad 0] \begin{bmatrix} 0.42 & 0.41 & 0.17 \\ 0.37 & 0.49 & 0.14 \\ 0.40 & 0.44 & 0.16 \end{bmatrix} = [0.42 \quad 0.41 \quad 0.17] = P_2$$

Continuing this process gives

$$P_0 P^3 = [1 \quad 0 \quad 0] \begin{bmatrix} 0.401 & 0.440 & 0.159 \\ 0.388 & 0.461 & 0.151 \\ 0.396 & 0.448 & 0.156 \end{bmatrix}$$
$$= [0.401 \quad 0.440 \quad 0.159] = P_3$$

$$P_0 P^4 = [1 \quad 0 \quad 0] \begin{bmatrix} 0.3961 & 0.4479 & 0.1560 \\ 0.3927 & 0.4534 & 0.1539 \\ 0.3948 & 0.4500 & 0.1552 \end{bmatrix}$$
$$= [0.3961 \quad 0.4479 \quad 0.1560] = P_4$$

and so forth. From P_3, the probabilities of I, D, and R on the third day are, respectively, 0.401, 0.440, and 0.159, given that an increase occurred on day 0. Similar probabilities hold for day 4 from P_4, and so forth.

You may have noticed that:

P_1 is the same as the first row of P.
P_2 is the same as the first row of P^2.
P_3 is the same as the first row of P^3.

And so forth. In general, the first row of P^n gives the probabilities of arriving at the various states on the nth day, given that I occurred on day 0.

Now suppose that D occurred on day 0. What changes would result in the above calculations? P_0 would now be [0 1 0] in each equation and:

P_1 would be the same as the second row of P.

P_2 would be the same as the second row of P^2.

P_3 would be the same as the second row of P^3.

In general, the second row of P^n would give the probabilities of arriving at the various states on the nth day, given that D occurred on day 0.

Similarly, the elements in the third row of P^n, with $P_0 = [0\ 0\ 1]$, would give the probabilities of arriving at the various states on the nth day, given that R occurred on day 0.

The corresponding result for arbitrary Markov chains is stated in Theorem 3.

THEOREM 3

In a Markov chain with states $a_1, a_2, \ldots, a_m$ and transition matrix

$$\begin{array}{c} \\ a_1 \\ a_2 \\ \vdots \\ a_m \end{array} \begin{array}{cccc} a_1 & a_2 & \cdots & a_m \end{array} \\ \begin{bmatrix} p_{11} & p_{12} & \cdots & p_{1m} \\ p_{21} & p_{22} & \cdots & p_{2m} \\ \vdots & \vdots & & \vdots \\ p_{m1} & p_{m2} & \cdots & p_{mm} \end{bmatrix} = P$$

the probability of obtaining state a_j at the nth trial given that the system was in state a_i at the beginning of the first trial is given by the element in the ij-position in P^n.

As a result of Theorem 3, all that is required to determine the probabilities of being in the various states after n trials is to calculate P^n.

EXAMPLE 1 In our stock-value illustration,

$$P^5 = \begin{bmatrix} 0.3948 & 0.4500 & 0.1552 \\ 0.3939 & 0.4514 & 0.1547 \\ 0.3945 & 0.4505 & 0.1550 \end{bmatrix}$$

with the entries rounded to four decimal places. Then the probabilities to which Theorem 3 refers are given in the table on the next page.

DAY 0 System state	DAY 5 State	PROBABILITY OF OCCURRING
Increase	Increase	0.3948
	Decrease	0.4500
	Remaining the same	0.1552
Decrease	Increase	0.3939
	Decrease	0.4514
	Remaining the same	0.1547
Remaining the same	Increase	0.3945
	Decrease	0.4505
	Remaining the same	0.1550

EXAMPLE 2 A manufacturing company leases a machine to make bolts, which are required to meet exact specifications. The output from the machine is watched carefully and a long-term analysis of the output indicates that if the machine produces a good bolt, the probability is 0.9 that the next bolt will also be good. If the machine produces a defective bolt, the probability that the next bolt will be good reduces to 0.8. The system being observed in this case is the machine production with states good (G) and defective (D).

(a) Set up the transition matrix P and calculate P^2, P^3, P^4, and P^5.

(b) If the machine is observed to produce a good bolt, what is the probability that the second bolt after that will also be good? That the third bolt after the observed one will be defective?

(c) If the machine is observed to produce a defective bolt, what is the probability that the fourth bolt after that will be good? That the fifth bolt after the observed bolt will be defective?

Solution (a) The sample space for each observation is $S = \{G, D\}$. Each element in S corresponds to one row and one column in P.

$$\begin{array}{c} \\ G \\ D \end{array} \begin{array}{c} G D \\ \begin{bmatrix} 0.9 & 0.1 \\ 0.8 & 0.2 \end{bmatrix} \end{array} = P$$

1 PRACTICE PROBLEM 1

The transition matrix for a Markov chain is

$$\begin{array}{c} \\ a_1 \\ a_2 \end{array} \begin{array}{c} a_1 a_2 \\ \begin{bmatrix} 0.3 & 0.7 \\ 0.6 & 0.4 \end{bmatrix} \end{array} = P$$

(a) Calculate P^2 and P^3.

(b) If a_1 occurs on a trial, what is the probability that a_2 occurs on the next trial? On the second trial? On the third trial?

(c) If a_2 occurs on a trial, what is the probability that a_1 occurs on the second following trial? That a_2 occurs on the third following trial?

Answer

(a) $P^2 = \begin{bmatrix} 0.51 & 0.49 \\ 0.42 & 0.58 \end{bmatrix}$;

$P^3 = \begin{bmatrix} 0.447 & 0.553 \\ 0.474 & 0.526 \end{bmatrix}$

(b) 0.7, 0.49, 0.553

(c) 0.42, 0.526

$$P^2 = \begin{bmatrix} 0.89 & 0.11 \\ 0.88 & 0.12 \end{bmatrix} \qquad P^3 = \begin{bmatrix} 0.889 & 0.111 \\ 0.888 & 0.112 \end{bmatrix}$$

$$P^4 = \begin{bmatrix} 0.8889 & 0.1111 \\ 0.8888 & 0.1112 \end{bmatrix} \qquad P^5 = \begin{bmatrix} 0.88889 & 0.11111 \\ 0.88888 & 0.11112 \end{bmatrix}$$

(b) From P^2, the probability that the second will be good is 0.89. From P^3, the probability that the third will be defective is 0.111.

(c) From P^4, the probability that the fourth will be good is 0.8888. From P^5, the probability that the fifth will be defective is 0.11112.

2 PRACTICE PROBLEM 2

A communications corporation regularly checks its communication satellite and will adjust its flight path if it is out of orbit. Several months of observations indicate that if the satellite is in orbit at one observation, the probability that it will be in orbit at the next observation is 0.99 and that it will be out of orbit at the next observation is 0.01. If it is out of orbit at an observation, the corresponding probabilities are 0.90 and 0.10, respectively.

(a) If the situation is to be modeled as a Markov chain, what are the states of the system? Determine the transition matrix P.

(b) Calculate P^2 and P^3.

(c) If the satellite is in orbit at an observation, what is the probability that it will be out of orbit at the second following observation? At the third following observation?

Answer

(a) $\begin{array}{c} \\ In \\ Out \end{array} \begin{array}{cc} In & Out \\ \begin{bmatrix} 0.99 & 0.01 \\ 0.90 & 0.10 \end{bmatrix} \end{array} = P$

(b) $P^2 = \begin{bmatrix} 0.9891 & 0.0109 \\ 0.9810 & 0.0190 \end{bmatrix}$;

$P^3 = \begin{bmatrix} 0.9890 & 0.0110 \\ 0.9883 & 0.0117 \end{bmatrix}$

(c) 0.0109; 0.0110

EXERCISES

For Exercises 1–3, use the transition matrix

$$\begin{array}{c} \\ A \\ B \end{array} \begin{array}{cc} A & B \\ \begin{bmatrix} 0.3 & 0.7 \\ 0.5 & 0.5 \end{bmatrix} \end{array} = P$$

1. Calculate P^2 and P^3.

2. If A occurs on a particular trial, what is the probability of each event?
   (a) A occurs on the next trial. (b) B occurs on the second trial.
   (c) A occurs on the third trial.

3. If B occurs on a particular trial, what is the probability of each event?
   (a) A occurs on the next trial. (b) A occurs on the second trial.
   (c) B occurs on the third trial.

For Exercises 4–6, use the transition matrix

$$\begin{array}{c} \\ A \\ B \end{array} \begin{array}{cc} A & B \\ \begin{bmatrix} 0.20 & 0.80 \\ 0.35 & 0.65 \end{bmatrix} \end{array} = P$$

4. Calculate P^2 and P^3.

5. If A occurs on a particular trial, what is the probability of each event?
   (a) B occurs on the next trial. (b) A occurs on the second trial.
   (c) B occurs on the third trial.

6. If B occurs on a particular trial, what is the probability of each event?
   (a) B occurs on the next trial. (b) B occurs on the second trial.
   (c) A occurs on the third trial.

For Exercises 7–9, use the transition matrix

$$\begin{array}{c} \\ A \\ B \\ C \end{array} \begin{array}{ccc} A & B & C \\ \begin{bmatrix} 0.3 & 0 & 0.7 \\ 0.1 & 0.7 & 0.2 \\ 0.0 & 0.5 & 0.5 \end{bmatrix} \end{array} = P$$

7. Calculate P^2 and P^3.

8. If B occurs on a particular trial, what is the probability of each event?
   (a) C occurs on the next trial.   (b) A occurs on the second trial.
   (c) B occurs on the third trial.

9. If C occurs on a particular trial, what is the probability that B occurs as follows?
   (a) On the next trial   (b) On the second trial
   (c) On the third trial

For Exercises 10–12, use the transition matrix

$$\begin{array}{c} \\ A \\ B \\ C \end{array} \begin{array}{ccc} A & B & C \end{array} \\ \left[\begin{array}{ccc} 0.4 & 0.6 & 0 \\ 0.1 & 0.1 & 0.8 \\ 0 & 0 & 1 \end{array} \right] = P$$

10. Calculate P^2 and P^3.

11. If A occurs on a particular trial, what is the probability of each event?
    (a) A occurs on the next trial.   (b) B occurs on the second trial.
    (c) C occurs on the third trial.

12. If C occurs on a particular trial, what is the probability that C occurs as follows?
    (a) On the next trial   (b) On the second trial
    (c) On the third trial

13. *Grocery purchasing* At the only grocery store in a small town, the grocer has observed that 70% of the people who buy frozen orange juice at his store will buy the same the next time they buy orange juice and that 40% of those who buy unfrozen orange juice will buy the unfrozen again the next time.
    (a) Choosing the states from the sample space $S = \{F, U\}$, construct the transition matrix P and calculate P^2, P^3, and P^4. (*Hint:* Each element in S corresponds to one row and one column of P.)
    (b) If on a visit to the store a customer buys frozen orange juice, what is the probability that she will buy unfrozen the next time that she buys orange juice? The third time? The fourth time?
    (c) If on a visit to the store, a customer buys unfrozen orange juice, what is the probability that she will buy unfrozen the next time that she buys orange juice? The second time? The third time?

14. *Voting patterns* A study of the history of the voting results of a particular precinct for its congressional representative reveals the following. If the precinct votes Republican at one election, the probabilities are 0.6 that it will vote Republican, 0.3 that it will vote Democratic, and 0.1 that they will vote other (for an independent candidate or a candidate from the minor parties) on the next election. If it votes Democratic at one election, the probabilities are 0.4 that it will vote Republican, 0.5 that it will vote Democratic, and 0.1 that it will vote other on the next election. If it votes other at one election, the corresponding probabilities are 0.6 for Republican, 0.4 for Democratic, and 0 for other on the next election.
    (a) Choosing the states from the sample space $S = \{R, D, O\}$, construct the transition matrix P and calculate P^2 and P^3.
    (b) If the precinct voted Republican at the last congressional election, what is the probability that it will vote Democratic on the second future election? On the third?

(c) If the precinct voted Democratic at the last congressional election, what is the probability that it will vote other on the second future election? On the third?

15. *Genetics* The table gives the probability of a child being one of the four possible genotypes for each possible genotype of its mother when the mother marries an aA genotype, where A is the dominant gene and a the recessive. If a female child should also marry an aA genotype, construct an appropriate transition matrix and use the transition matrix to determine the probability that a grandchild will be an AA genotype given that its grandmother is Aa, its grandfather being aA.

		CHILD			
		AA	Aa	aA	aa
	AA	$\frac{1}{2}$	$\frac{1}{2}$	0	0
FEMALE	Aa	$\frac{1}{4}$	$\frac{1}{4}$	$\frac{1}{4}$	$\frac{1}{4}$
PARENT	aA	$\frac{1}{4}$	$\frac{1}{4}$	$\frac{1}{4}$	$\frac{1}{4}$
	aa	0	0	$\frac{1}{2}$	$\frac{1}{2}$

16. *Car rental* A car rental company that rents cars in three different cities (A, B, and C) has found that a car rented in city A is turned in at city A with probability 0.6, is turned in at city B with probability 0.3, and is turned in at city C with probability 0.1. A car rented in city B is turned in at city A with probability 0.4, at city B with probability 0.5, and at city C with probability 0.1. A car rented in city C is turned in at city A with probability 0, at city B with probability 0.4, and at city C with probability 0.6. Construct the transition matrix P and, using P, determine the probability that a car rented in city B will be turned in at city A at the second following rental.

17. *Water pollution* A factory dumps its waste products into a river, which empties into the ocean. The administration maintains that on any given day each particle of waste product in the river is carried to the ocean with a probability of 0.99. If a particle of waste is dumped into the river on Monday, use the transition matrix to determine the probability that the particle is still in the river on the following Monday. Assume that any waste carried to the ocean will not return to the river.

The following exercises do not require powers of the transition matrices for their solutions; however, enough information is available for their solutions. (Hint: Consider whether the addition rule or multiplication rule is applicable.)

18. *Voting patterns* In Exercise 14 if the precinct votes Democratic at one election, what is the probability that the precinct votes Republican on each of the next three elections?

19. *Manufacturing* In Example 2 if a good bolt is produced, what is the probability that each of the next three bolts is good?

The following exercises do not require powers of the transition matrices for their solution; however, you will have to use both the addition rule and the multiplication rule.

20. *Voting patterns* In Exercise 14 suppose that at any congressional election, the probability that the precinct will vote Republican is 0.70, Democratic is 0.25, and other is 0.05. Of the next two elections, what is the probability that at the second election the precinct will vote Democratic?

21. *Manufacturing* In Example 2 suppose that at any time, the probability that a bolt will be good is 0.97. Of the next two bolts manufactured, what is the probability that the second one will be good?

22. Why will the sum of the entries in each row of a transition matrix always be 1?

Explain the meaning of each term.

23. Markov chain
24. The states of a system in a Markov chain
25. Transition probabilities

*6.4 REGULAR MARKOV CHAINS

We turn now to the second question to be considered concerning Markov chains, "Can the long-term behavior of the system be predicted?" Consider the information developed in the beginning of Section 6.3 relative to the stock-value illustration:

$$P_0 = [1 \quad 0 \quad 0]$$

$$P = \begin{bmatrix} 0.5 & 0.3 & 0.2 \\ 0.3 & 0.6 & 0.1 \\ 0.4 & 0.4 & 0.2 \end{bmatrix} \qquad P_1 = [0.5 \quad 0.3 \quad 0.2]$$

$$P^2 = \begin{bmatrix} 0.42 & 0.41 & 0.17 \\ 0.37 & 0.49 & 0.14 \\ 0.40 & 0.44 & 0.16 \end{bmatrix} \qquad P_2 = [0.42 \quad 0.41 \quad 0.17]$$

$$P^3 = \begin{bmatrix} 0.401 & 0.440 & 0.159 \\ 0.388 & 0.461 & 0.151 \\ 0.396 & 0.448 & 0.156 \end{bmatrix} \qquad P_3 = [0.401 \quad 0.440 \quad 0.159]$$

$$P^4 = \begin{bmatrix} 0.3961 & 0.4479 & 0.1560 \\ 0.3927 & 0.4534 & 0.1539 \\ 0.3948 & 0.4500 & 0.1552 \end{bmatrix} \qquad P_4 = [0.3961 \quad 0.4479 \quad 0.1560]$$

$$P^5 = \begin{bmatrix} 0.3948 & 0.4500 & 0.1552 \\ 0.3939 & 0.4514 & 0.1547 \\ 0.3945 & 0.4505 & 0.1550 \end{bmatrix} \qquad P_5 = [0.3948 \quad 0.4500 \quad 0.1552]$$

$$P^6 = \begin{bmatrix} 0.3945 & 0.4505 & 0.1550 \\ 0.3942 & 0.4509 & 0.1549 \\ 0.3944 & 0.4507 & 0.1550 \end{bmatrix} \qquad P_6 = [0.3945 \quad 0.4505 \quad 0.1550]$$

In P^5 and P^6 the entries have been rounded to four decimal places. Vectors such as P_0, P_1, P_2, and so forth, with nonnegative entries whose sum is 1, are called **probability vectors**.

Note the behavior of the entries in $P_1, P_2, \ldots, P_6$. The first entries in these vectors seem to be getting closer and closer to 0.394, the second entries seem to be getting close to 0.450, and the third entries seem to be getting close to 0.155. Furthermore, the entries in each row of $P^1, P^2, \ldots, P^6$ seem to be getting closer and closer to the same numbers. That is, as n gets larger and larger, the entries in P_n and in

---

* This section can be omitted without loss of continuity.

each row of P^n seem to be converging toward some particular numbers. If we could determine these numbers, they would give us an indication of the long-term behavior of the system, as "long-term" corresponds to large n.

We can determine such numbers for a large class of Markov chains, those for which the transition matrix is regular. A transition matrix P is **regular** if P or some power of P contains only positive numbers; in this case the Markov chain is also said to be regular.

$$P = \begin{bmatrix} 0.2 & 0.7 & 0.1 \\ 0.3 & 0.6 & 0.1 \\ 0.4 & 0.5 & 0.1 \end{bmatrix} \quad \text{Regular}$$

$$P = \begin{bmatrix} \frac{1}{2} & \frac{1}{2} \\ 1 & 0 \end{bmatrix} \quad \text{Regular because } P^2 = \begin{bmatrix} \frac{3}{4} & \frac{1}{4} \\ \frac{1}{2} & \frac{1}{2} \end{bmatrix}$$

$$P = \begin{bmatrix} 1 & 0 \\ \frac{1}{2} & \frac{1}{2} \end{bmatrix} \quad \text{Not regular because } P^2 = \begin{bmatrix} 1 & 0 \\ \frac{3}{4} & \frac{1}{4} \end{bmatrix}, P^3 = \begin{bmatrix} 1 & 0 \\ \frac{7}{8} & \frac{1}{8} \end{bmatrix},$$

and all higher powers of P will contain a zero in the upper right-hand corner.

Theorem 4 states that the properties suggested by the behavior of the entries in the matrices and probability vectors above do hold for Markov chains with regular transition matrices. We shall not attempt its proof.

THEOREM 4

If P is a regular transition matrix for a Markov chain and A is a probability vector that satisfies the equation

$$AP = A \qquad (1)$$

then as n increases:

(i) The entries in each row of P^n approach the corresponding entry in A.

(ii) For any initial probability vector P_0, the entries in

$$P_n = P_0 P^n$$

approach the corresponding entries in A.

Part (i) of Theorem 4 indicates that the vector A, called the ***fixed probability vector*** of P, is the key for determining the long-term behavior of the system. Part (ii) indicates that this long-term behavior is the same regardless of the initial state of the system! Before considering several examples, note that we can rewrite equation (1) as

$$AP - A = 0, \text{ or}$$

$$A(P - I) = 0 \qquad (2)$$

It is generally more convenient to use equation (2) rather than equation (1) to determine A.

EXAMPLE 1 Verify that

$$P = \begin{bmatrix} 0.2 & 0.8 \\ 0.6 & 0.4 \end{bmatrix}$$

is regular and determine the fixed probability vector of P.

Solution P is regular because it contains only positive entries. For $A = [a_1 \ a_2]$, equation (2) becomes

$$[a_1 \ a_2] \begin{bmatrix} -0.8 & 0.8 \\ 0.6 & -0.6 \end{bmatrix} = [0 \ 0]$$

or

$$-0.8a_1 + 0.6a_2 = 0$$
$$0.8a_1 - 0.6a_2 = 0$$

This system has infinitely many solutions: $a_1 = \frac{3}{4}a_2$; however, if A is to be a probability vector, then $a_1 + a_2 = 1$, so that

$$\tfrac{3}{4}a_2 + a_2 = 1 \quad \text{or} \quad \tfrac{7}{4}a_2 = 1$$

Consequently, a_2 must equal $\frac{4}{7}$ and $a_1 = \frac{3}{4}a_2 = \frac{3}{7}$. That is, $A = [\frac{3}{7} \ \frac{4}{7}]$ is the fixed probability vector of P. ∎

EXAMPLE 2 Verify that

$$P = \begin{bmatrix} 0.2 & 0.3 & 0.5 \\ 1 & 0 & 0 \\ 0 & 0.5 & 0.5 \end{bmatrix}$$

is regular and find the fixed probability vector of P.

Solution Since

$$P^2 = \begin{bmatrix} 0.34 & 0.31 & 0.35 \\ 0.20 & 0.30 & 0.50 \\ 0.50 & 0.25 & 0.25 \end{bmatrix}$$

has only positive entries, P is regular. Equation (2), with $A = [a_1 \ a_2 \ a_3]$, becomes

$$[a_1 \ a_2 \ a_3] \begin{bmatrix} -0.8 & 0.3 & 0.5 \\ 1 & -1 & 0 \\ 0 & 0.5 & -0.5 \end{bmatrix} = [0 \ 0 \ 0]$$

which has infinitely many solutions: $a_1 = a_3$, $a_2 = \frac{4}{5}a_3$. In order for A to be a probability vector, $a_1 + a_2 + a_3$ must equal 1. In terms of a_3,

$$a_3 + \tfrac{4}{5}a_3 + a_3 = 1$$

Solving this equation for a_3 gives $a_3 = \frac{5}{14}$. Consequently,

$$A = [\tfrac{5}{14} \quad \tfrac{2}{7} \quad \tfrac{5}{14}]$$

For the stock-value illustration, the transition matrix is

$$P = \begin{bmatrix} 0.5 & 0.3 & 0.2 \\ 0.3 & 0.6 & 0.1 \\ 0.4 & 0.4 & 0.2 \end{bmatrix}$$

which is obviously regular. To find the fixed probability vector of P, equation (2) becomes

$$[a_1 \quad a_2 \quad a_3] \begin{bmatrix} -0.5 & 0.3 & 0.2 \\ 0.3 & -0.4 & 0.1 \\ 0.4 & 0.4 & -0.8 \end{bmatrix} = [0 \quad 0 \quad 0]$$

which has infinitely many solutions: $a_1 = \frac{28}{11}a_3$, $a_2 = \frac{32}{11}a_3$. But $a_1 + a_2 + a_3 = 1$ requires a_3 to equal $\frac{11}{71}$, so that

$$A = [\tfrac{28}{71} \quad \tfrac{32}{71} \quad \tfrac{11}{71}] = [0.3944 \quad 0.4507 \quad 0.1549]$$

rounded to four decimal places. Note that the entries in P_n and in the rows in P^n do, in fact, seem to be approaching the corresponding entries in A, as Theorem 4 requires.

Furthermore, after a reasonable period of time (in this example about 5 or 6 days), the probability that the stock will increase in value on any day is about 0.3944; that it will decrease in value, 0.4507; and that it will remain the same in value, 0.1549—regardless of any initial status of the stock.

EXAMPLE 3 Determine the long-term production behavior of the machine in Example 2 of Section 6.3.

Solution Examining the entries in the powers of

$$P = \begin{bmatrix} 0.9 & 0.1 \\ 0.8 & 0.2 \end{bmatrix}$$

in Example 2 indicates that P is regular and what entries (rounded to four decimal places) to expect in the fixed probability vector A. However, following the procedure described above, we get

$$[a_1 \quad a_2] \begin{bmatrix} -0.1 & 0.1 \\ 0.8 & -0.8 \end{bmatrix} = [0 \quad 0]$$

for equation (2), with solution $a_1 = 8a_2$. Then $a_1 + a_2 = 1$ requires that $a_2 = \frac{1}{9}$, so that $A = [\tfrac{8}{9} \quad \tfrac{1}{9}] = [0.8889 \quad 0.1111]$. One can therefore expect

---

1 PRACTICE PROBLEM 1

Verify that

$$P = \begin{bmatrix} 0.3 & 0.7 \\ 1 & 0 \end{bmatrix}$$

is regular and find the fixed probability vector of P.

Answer

$$A = [\tfrac{10}{17} \quad \tfrac{7}{17}]$$

② PRACTICE PROBLEM 2

Determine the fixed probability vector and the long-term behavior of the satellite in Practice Problem 2 of Section 6.3.

Answer

$$A = [0.989 \quad 0.011]$$

The satellite will be in orbit with a probability of 0.989 and out of orbit with a probability of 0.011.

a good bolt with probability 0.8889 and a defective bolt with probability 0.1111 at any time.

Observe the four-step process in the application of mathematics (Section 1.3) in the stock-value illustration.

1. The physical situation was the behavior of the stock; the physical problem was to determine the long-term behavior of the stock when the day-to-day behavior was known.
2. The mathematical model was the transition matrix; the mathematical question was, What is the fixed probability vector of the transition matrix?
3. Using Theorem 4, the mathematical solution was found to be [0.3944 0.4507 0.1549].
4. The interpretation was that on a long-term basis, the probability of an increase on any particular day is 0.3944, of a decrease, 0.4507, and for remaining the same, 0.1459.

The model for the stock-value problem, as for any Markov chain, is good only while the probabilities in the transition matrix are realistic. When changing conditions require new probabilities, the results obtained from the old probabilities (although mathematically correct) are no longer realistic and a new model must be constructed.

In discussing both Bernoulli trials and Markov chains, we were interested in a sequence of repeated experiments, or trials. However, there are significant differences in the trials themselves. For each experiment in Bernoulli trials, there are only two outcomes of interest, success and failure; in Markov chain trials there can be 2, 3, 4, or more outcomes (states) of interest at each trial. Furthermore, in the Bernoulli trials the outcome on each trial is independent of the outcome of any of the previous trials, whereas in Markov chain trials the outcome on each trial can be dependent on the outcome of the previous trial. Is it correct to say that Bernoulli trials are special cases of Markov chains; that is, do Bernoulli trials satisfy the three conditions required for a sequence of repeated trials to be a Markov chain? (See Exercise 16.)

EXERCISES

Determine which of the following matrices are regular. For each regular matrix, determine the fixed probability vector.

1. $\begin{bmatrix} 0.4 & 0.6 \\ 0.8 & 0.2 \end{bmatrix}$

2. $\begin{bmatrix} 0.3 & 0.7 \\ 1 & 0 \end{bmatrix}$

3. $\begin{bmatrix} 0.25 & 0.75 \\ 0.65 & 0.35 \end{bmatrix}$

4. $\begin{bmatrix} 1 & 0 \\ 0.5 & 0.5 \end{bmatrix}$

5. $\begin{bmatrix} 0.2 & 0.4 & 0.4 \\ 0 & 1 & 0 \\ 0.1 & 0.4 & 0.5 \end{bmatrix}$
6. $\begin{bmatrix} 0.3 & 0.2 & 0.5 \\ 1 & 0 & 0 \\ 0.6 & 0.2 & 0.2 \end{bmatrix}$

7. $\begin{bmatrix} 0.2 & 0.2 & 0.6 \\ 0.4 & 0.3 & 0.3 \\ 0.1 & 0.7 & 0.2 \end{bmatrix}$
8. $\begin{bmatrix} 0.1 & 0.1 & 0.8 & 0 \\ 0.2 & 0.2 & 0 & 0.6 \\ 0.3 & 0 & 0.4 & 0.3 \\ 0.5 & 0 & 0 & 0.5 \end{bmatrix}$

9. *Grocery purchasing*
   (a) Determine the fixed probability vector of the transition matrix in Exercise 13 of Section 6.3.
   (b) In view of the result of part (a), what fraction of the orange juice stocked by the grocer should be frozen? What portion unfrozen?

10. *Voting patterns*
    (a) Determine the fixed probability vector of the transition matrix in Exercise 14 of Section 6.3.
    (b) In view of the results of part (a), should the Democratic or Republican party work harder to obtain votes in this precinct?

11. *Car rental*
    (a) Determine the fixed probability vector of the transition matrix in Exercise 16 of Section 6.3.
    (b) The company wants to build a maintenance garage in one of the three cities. In view of the results of part (a), in which city should they locate the garage?

12. *Water pollution*
    (a) Determine whether or not the transition matrix in Exercise 17 of Section 6.3 is regular.
    (b) Verify that [0 1] is a fixed probability vector for the transition matrix in part (a). Note that Theorem 4 does not say that a matrix that is not regular cannot have a fixed probability vector.

13. *Machine comparison* Suppose a company leases two machines to make bolts, one being the machine of Example 3. The second, which is exactly the same model as the first and leases for the same number of dollars as the first, has transition matrix

$$\begin{array}{c c} & \begin{array}{cc} A & D \end{array} \\ \begin{array}{c} A \\ D \end{array} & \begin{bmatrix} 0.95 & 0.05 \\ 0.75 & 0.25 \end{bmatrix} \end{array}$$

The company has determined to cancel its lease of one of the machines. Which of the two machines should the company give up on the basis of the available information?

14. *Genetics* In Exercise 15 of Section 6.3, if each offspring continues to marry an aA genotype, what fraction of future generations can be expected to be of each of the four possible genotypes AA, Aa, aA, aa?

15. (a) If

$$P = \begin{bmatrix} p_{11} & p_{12} \\ p_{21} & p_{22} \end{bmatrix}$$

is a regular transition matrix with p_{21} and p_{12} not both zero, verify that

$$A = \begin{bmatrix} \dfrac{p_{21}}{p_{12}+p_{21}} & \dfrac{p_{12}}{p_{12}+p_{21}} \end{bmatrix}$$

is the fixed probability vector of P. [*Hint:* Substitute into equation (1) or equation (2) and use the relations $p_{11} = 1 - p_{12}$ and $p_{22} = 1 - p_{21}$.]

(b) Part (a) gives an alternate method for determining the fixed probability vector of a 2×2 regular transition matrix. Use the method of part (a) to verify your solution in Exercises 1 and 9(a).

16. (a) Verify that Bernoulli trials are special cases of Markov chains by verifying that the three conditions required for a sequence of repeated trials (page 222) are satisfied by Bernoulli trials. (Recall that for Bernoulli trials the sample space is $\{S, F\}$.)
    (b) How can the questions considered regarding Markov chains (page 223) be answered in the case of Bernoulli trials?
    (c) What was the central question asked in Bernoulli trials that was not considered in Markov chains?

Explain the meaning of each term.

17. Probability vector
18. Transition matrix
19. Regular transition matrix
20. Fixed probability vector
21. *Grocery purchasing* Identify the steps of the four-step process involved in the application of mathematics (Section 1.3) in Exercise 9.

6.5 EXPECTED VALUE

If you repeatedly tossed 4 coins a large number of times and kept a record of the number of heads obtained, intuitively you might expect the average number of heads per toss to be 2. The purpose of this section is to make this intuitive concept of the expected average mathematically precise. First we examine the process as it applies to tossing 4 coins.

The sample space for tossing 4 coins would be the 16 elements of the form *HHTH, TTHH,* and so forth, where the first letter indicates the result obtained on the first coin, the second letter indicates the result for the second coin, and so forth. Also, the number of heads that can be obtained is 0, 1, 2, 3 or 4. Each of these numbers corresponds to an event in the sample space, as illustrated in Table 6.1. The probabilities of obtaining each possible number of heads are listed below the table.

We also get the expected average by multiplying the possible number of heads by their respective probabilities and adding the result:

$$0(\tfrac{1}{16}) + 1(\tfrac{1}{4}) + 2(\tfrac{3}{8}) + 3(\tfrac{1}{4}) + 4(\tfrac{1}{16}) = \tfrac{32}{16} = 2$$

This latter process is the one to be discussed in this section.

We can consider Table 6.1 to be obtained by assigning a number (the number of heads) to each element in the sample space and then collecting all the elements with the same number into an event. For example,

6.5 EXPECTED VALUE

TABLE 6.1

0	1	2	3	4
TTTT	HTTT THTT TTHT TTTH	HHTT HTHT HTTH THHT THTH TTHH	THHH HTHH HHTH HHHT	HHHH

NUMBER OF HEADS	PROBABILITY
0	$\frac{1}{16}$
1	$\frac{4}{16} = \frac{1}{4}$
2	$\frac{6}{16} = \frac{3}{8}$
3	$\frac{4}{16} = \frac{1}{4}$
4	$\frac{1}{16}$

the elements *HTTT*, *THTT*, *TTHT*, and *TTTH* are all assigned the number 1 and consequently together form the event "1."

Such an assignment of a number to each element in a sample space is called a *random variable*. Formally, a **random variable** on a sample space S is a function (or rule) that assigns a unique number to each element in S. Random variables are usually denoted by capital letters chosen from the latter part of the alphabet—W, X, Y, Z. The numbers assigned by a random variable to elements in S are called the **values** of the random variable. Each value of a random variable determines an event in S, the elements in the event being all those assigned that particular value.

If we denote the random variable that assigns the number of heads to each element in the sample space for tossing four coins by X, then the values of X are 0, 1, 2, 3, and 4. Each of these four numbers determine an event in S, as indicated in Table 6.1.

For a given random variable X, if x denotes a value of X, we use the notation $P(X = x)$, or simply $P(x)$, to denote the probability of the event corresponding to x. For the random variable X in our coin illustration,

$$P(X = 0) = P(0) = \tfrac{1}{16}$$
$$P(X = 1) = P(1) = \tfrac{1}{4}$$
$$P(X = 2) = P(2) = \tfrac{3}{8}$$
$$P(X = 3) = P(3) = \tfrac{1}{4}$$
$$P(X = 4) = P(4) = \tfrac{1}{16}$$

in view of the probabilities given in Table 6.1. The expression $P(X = 4)$, for example, is read, "the probability that X is equal to 4."

With this notation the expected average number of heads can also be written

$$0P(0) + 1P(1) + 2P(2) + 3P(3) + 4P(4)$$

This expected average is called the *expected value* of X.

If X is a random variable with values $x_1, x_2, x_3, \ldots, x_n$, the **expected value** of X is the sum

$$x_1 P(x_1) + x_2 P(x_2) + x_3 P(x_3) + \cdots + x_n P(x_n)$$

denoted by $E(X)$.

EXAMPLE 1 Find the expected value of the sum of dots obtained on the roll of a pair of dice.

Solution The random variable X here assigns to each of the 36 elements in the sample space for the roll of a pair of dice a number equal to the sum obtained. The values of X are 2, 3, 4, 5, 6, 7, 8, 9, 10, 11, and 12. The event corresponding to 2 is $\{(1, 1)\}$, to 3 is $\{(1, 2), (2, 1)\}$, to 4 is $\{(1, 3), (2, 2), (3, 1)\}$, and so forth, so that $P(2) = \frac{1}{36}$, $P(3) = \frac{2}{36}$, $P(4) = \frac{3}{36}, \ldots$. Consequently,

$$E(X) = 2(\tfrac{1}{36}) + 3(\tfrac{2}{36}) + 4(\tfrac{3}{36}) + 5(\tfrac{4}{36}) + 6(\tfrac{5}{36}) + 7(\tfrac{6}{36}) + 8(\tfrac{5}{36})$$
$$+ 9(\tfrac{4}{36}) + 10(\tfrac{3}{36}) + 11(\tfrac{2}{36}) + 12(\tfrac{1}{36}) = \tfrac{252}{36} = 7$$

$E(X)$ indicates that if the dice are rolled a large number of times, the average of the sums obtained can be expected to be 7 (or close to 7).

EXAMPLE 2 Suppose you bought three $1 tickets of 300 such tickets sold in a raffle in which the prize is $100. What are your expected net winnings?

Solution The sample space is $S = \{W, L\}$, W for win and L for lose. $P(W) = \frac{3}{300}$ and $P(L) = \frac{297}{300}$. If the random variable Y denotes your net winnings, the value of Y for W is 97 and for L is -3, so that $P(97) = \frac{3}{300}$,

---

1 PRACTICE PROBLEM 1

If the random variable X takes the values 1, 2, 4, and 6, with probabilities as given by the table, determine the expected value of X.

VALUE OF X	PROBABILITY
1	0.4
2	0.1
4	0.3
6	0.2

Answer

3

---

2 PRACTICE PROBLEM 2

A couple getting married will have 3 children and the random variable X denotes the number of boys they will have.

(a) What are the possible values of X?

(b) What are the values of $P(0)$, $P(1)$, $P(2)$, and $P(3)$?

(c) Determine $E(X)$.

Answer

(a) 0, 1, 2, and 3   (b) $\tfrac{1}{8}, \tfrac{3}{8}, \tfrac{3}{8}, \tfrac{1}{8}$

(c) 1.5

$P(-3) = \frac{297}{300}$, and

$$E(Y) = 97(\tfrac{3}{300}) + (-3)(\tfrac{297}{300}) = -\tfrac{600}{300} = -2$$

If you participate in this manner in a large number of raffles of this type, you can expect to lose an average of $2 per raffle. ∎

Example 2 illustrates two important ideas. First, the expected value of a random variable does not have to be one of the values of the random variable. In Example 2 the values of Y were 97 and -3, but $E(Y) = -2$.

Second, quantities such as payments or losses are usually represented by negative values of the random variable.

3 PRACTICE PROBLEM 3

Suppose you bought 5 tickets instead of 3 in Example 2. What are your expected net winnings?

Answer

$-\$10/3$

EXAMPLE 3 Every morning a grocer stocks four units of a perishable item. For each unit she sells, she makes a profit of $2.00; each unit she does not sell that day she must throw out for a loss of $3.00 per unit. Past customer behavior indicates that she could sell 2, 3, 4, 5, or 6 units per day with the probabilities given in the table. She believes that each customer requesting the item after her daily supply is exhausted buys elsewhere and therefore represents a loss of $2.00 in potential profit. This information, along with the net profit for each number of requested units, is summarized in the table. Determine her expected daily profit from this item.

NUMBER UNITS REQUESTED	PROBABILITY	LOSS IN GOODS	PROFIT	LOSS IN SALES	NET PROFIT
2	0.1	-6	4	0	-2
3	0.3	-3	6	0	3
4	0.4	0	8	0	8
5	0.1	0	8	-2	6
6	0.1	0	8	-4	4

Solution The random variable X denotes net profit and has values of -2, 3, 8, 6, and 4, with probabilities as indicated in the table. Consequently,

$$E(X) = (-2)(0.1) + (3)(0.3) + 8(0.4) + 6(0.1) + 4(0.1) = 4.90$$

Over a period of many days, she could expect her average daily profit to be about $4.90 from this item. ∎

In Example 3 we can consider the sample space for the experiment to be $S = \{2, 3, 4, 5, 6\}$, representing the possible number of requests. The random variable X then assigns the values -2 to 2, 3 to 3, 8 to 4, 6 to 5, and 4 to 6. In other situations when the elements in a sample space are numbers, the values of a random variable on that sample space might agree exactly with the elements themselves. This situation is illustrated in Example 4.

EXAMPLE 4 On the roll of a balanced die, what is the expected number of dots to be obtained?

Solution In this case $S = \{1, 2, 3, 4, 5, 6\}$, and the value of the random variable Y (which denotes the number of dots obtained) at each element in S is the same as that element—the value of Y for 1 is 1, for 2 is 2, and so forth. Here

$$E(Y) = 1(\tfrac{1}{6}) + 2(\tfrac{1}{6}) + 3(\tfrac{1}{6}) + 4(\tfrac{1}{6}) + 5(\tfrac{1}{6}) + 6(\tfrac{1}{6}) = \tfrac{21}{6} = 3.5$$

The following steps outline the general procedure involved in determining the expected value of a random variable:

> 1. Determine the sample space S, which describes the possible outcomes when the experiment is performed, and assign an appropriate probability to each element in S.
> 2. Determine the values $x_1, x_2, \ldots, x_n$ of the random variable X, and for each value of X determine the corresponding event in S.
> 3. Determine $P(x_i)$ for each value x_i of X.
> 4. Calculate $E(X) = x_1 P(x_1) + x_2 P(x_2) + \cdots + x_n P(x_n)$.

The procedure is illustrated again in the following example. Each step in the solution is numbered according to the above outline.

EXAMPLE 5 From a well-shuffled deck of ordinary playing cards, you pull a single card. If the card is a picture card (king, queen, or jack), you pay your friend $5.00; otherwise your friend pays you $2.00. What are your expected winnings?

Solution **Step 1.** Since we are not interested in the suit of the card drawn, the sample space is

$$S = \{A, 2, 3, 4, 5, 6, 7, 8, 9, 10, J, Q, K\}$$

Each element in S is given a probability of $\tfrac{1}{13}$.

Step 2. Since the random variable X here denotes your winnings, the possible values of X are -5 and 2. The corresponding events are:

For -5: $\{J, Q, K\}$
For 2: $\{A, 2, 3, 4, 5, 6, 7, 8, 9, 10\}$

Step 3. $P(-5) = \tfrac{3}{13}$ and $P(2) = \tfrac{10}{13}$

Step 4. $E(X) = (-5)\tfrac{3}{13} + (2)\tfrac{10}{13} = \tfrac{5}{13}$

If you played this game a great number of times, your average winnings per game would be $\$\tfrac{5}{13}$, or about 38¢.

4 PRACTICE PROBLEM 4

Suppose you and a friend are playing a game in which one of you rolls a die. If a 1 or 6 shows up, you pay the friend $2.00; but if any other number shows up, the friend pays you $1.00.

(a) What are your expected winnings?
(b) Is this a fair game?

Answer

(a) $0.00   (b) Yes

In Example 5 your friend's expected winnings would be $-38¢$ per game; if your friend is smart, he will not play very long. In order that a game or bet be *fair*, the expected winnings for each player should be 0.

With probability models it is impossible to predict exactly what the outcome will be when a single experiment is performed; if such a prediction can be made in a physical situation, probability models should not be used in any case. What the expected value does is to provide the average value of the random variable that can be expected when the experiment is performed many times.

Note the four-step procedure involved in the application of mathematics (Section 1.3) as it applies to Example 5:

1. The physical question is, What are your expected winnings?
2. The mathematical model consists of the sample space S with each element assigned a probability of $\frac{1}{13}$, along with the random variable X. The mathematical question is, What is the expected value of X?
3. The mathematical solution is $\frac{5}{13}$.
4. The interpretation is that you can expect to win an average of about 38¢ per game in the long run.

EXERCISES

X	P(X)
1	0.4
2	0.25
5	0.2
8	0.15

1. Determine the expected value of the random variable X if the values of X and the corresponding probabilities are as given in the table in the margin.

2. Determine the expected value of the random variable Y whose values and corresponding probabilities are as given in the table.

Y	P(Y)
0	0.32
1.5	0.24
3	0.19
4.5	0.16
6	0.09

3. Determine the expected number of heads when 3 coins are tossed.

4. A die is weighted so that the probability of obtaining the various number of dots is as given in the table. Determine the expected number of dots obtained when the die is rolled.

NUMBER OF DOTS	PROBABILITY
1	$\frac{1}{6}$
2	$\frac{1}{12}$
3	$\frac{1}{3}$
4	$\frac{1}{12}$
5	$\frac{1}{12}$
6	$\frac{1}{4}$

5. Determine the expected value of each of the following when a pair of dice are rolled:
   (a) The difference between the two numbers showing, subtracting the smaller number from the larger
   (b) The larger of the two numbers showing (If both dice show the same number, that number is the larger.)

6. A box contains 2 quarters, 3 dimes, 4 nickels, and 1 penny. If a coin is chosen at random, what is the expected value of the draw?

7. Three thousand tickets are to be sold for $5.00 each in a raffle for a car worth $6,005. What will be your expected net winnings in each case?
   (a) If you buy one ticket   (b) If you buy five tickets

8. If it costs you $1.00 to play Over-and-Under (Exercise 9, Section 5.5), over 7 and under 7 each pay $2.00 (your dollar plus an additional dollar), and exactly 7 pays $5.00 (your dollar plus four additional dollars), determine your expected net winnings if you play as indicated.
   (a) Over 7   (b) Under 7   (c) Exactly 7

9. On American roulette wheels there are the 38 numbers 0, 00, and 1–36. If you bet $1.00 on *odd* and the ball lands on any one of the 18 odd numbers, you win $1.00 (and you get your dollar back). Otherwise you lose your $1.00. What are your expected winnings?

10. On some roulette wheels there are the 37 numbers 0 and 1–36. If you bet $1.00 on *odd* as in Exercise 9, what are your expected winnings?

11. Is each game fair?
    (a) Exercise 8(a)   (b) Exercise 8(c)
    (c) Exercise 9   (d) Exercise 10

12. *Grocery profit* Determine the expected daily profit of the grocer in Example 3 if she would stock 5 units instead of 4 each morning.

UNITS DEMANDED	PROBABILITY
100	0.10
150	0.20
200	0.40
250	0.20
300	0.10

13. *Marketing survey* A company is introducing a new product and marketing surveys indicate the possible monthly demands and corresponding probabilities given in the table. The company makes a profit of $10 on each item sold. Determine the expected monthly profit for the company from this product.

14. *Defective parts* Past experience indicates that in each lot of parts of a particular type, 1% are defective with probability 0.50; 2% are defective with probability 0.30; and 3% are defective with probability 0.20.
    (a) What is the expected number of defective parts in a lot of 10,000 parts?
    (b) If it costs 10¢ to rework each defective part, what is the expected cost for reworking the defectives in a lot of 10,000 parts?

15. *Learning experiment* The number of errors committed in a learning experiment with 200 individuals ranged from 0 to 4. The number of individuals committing each of the various numbers of errors is given in the table. Determine the expected number of errors per individual.

NUMBER OF ERRORS	INDIVIDUALS
0	20
1	70
2	50
3	40
4	20

16. *Dye injection* The effects of a dye injected into the bloodstream for purposes of kidney X rays were monitored in 100 cases. For 50 patients it took 8 hours for the dye to be dispelled from the kidneys, for 40 patients it took 9 hours, and for 10 patients it took 10 hours. Determine the expected dissipation time for the dye.

Explain the meaning of each term.

17. Random variable

18. Values of a random variable

19. Expected value of a random variable

20. Identify the steps of the four-step procedure involved in the application of mathematics (Section 1.3) in Exercise 9.

6.6 GAME THEORY

Imagine that you and a friend are playing a game in which each of you, at the same time, secretly writes down a number from 1 to 3. Then, depending upon the combination of numbers written down, one of you pay the other a sum of money as indicated in the following matrix:

$$\begin{array}{c} & & \textit{Friend} \\ & & 1 \quad\; 2 \quad\; 3 \\ \textit{You} & \begin{array}{c}1\\2\\3\end{array} & \left[\begin{array}{rrr} 2 & 1 & -1 \\ 3 & 2 & 1 \\ -2 & 4 & -1 \end{array}\right] \end{array}$$

If you write 1 and the friend writes 2, she pays you $1.00, as indicated in the row corresponding to 1 and the column corresponding to 2. Negative entries indicate that you pay your friend. If you write 3 and she writes 1, for example, you pay her $2.00.

For easier notation, we shall write r_1 to denote your play corresponding to the first row in the matrix (writing a 1); r_2 and r_3 are defined similarly. In the same way, we shall write c_1 to denote your friend's play corresponding to column 1, c_2 for the play corresponding to column 2, and c_3 for the play corresponding to column 3. Then if you play r_2, for example, and she plays c_1, she pays you $3.00.

You might start the game by playing r_3 because it has the highest potential payoff, but once your friend realizes your strategy, she recognizes that c_1 would be her best play because with your r_3, c_1 has the best payoff (-2) for her.

But once your friend consistently plays c_1, you would change to r_2 because with her c_1, r_2 has the best payoff (3) for you. But when you consistently play r_2, your friend would move to c_3. At this point it would no longer be beneficial for either of you to change plays. As long as she plays c_3, you could not better your payoff by changing to r_1 or r_3, so you continue playing r_2. Similarly, as long as you continue to play r_2, it would not be advantageous for her to change to c_1 or c_2. She would continue to pay you $1.00 on each play until the game ceases.

The above is one example of a **two-person game**, that is, a competition between two entities (such as persons, corporations, or forces). The matrix giving the payoffs for various combinations of plays in a two-person game is called the **payoff matrix**. Actually, any $m \times n$ matrix can be the payoff matrix for a two-person game. The row player would have a choice of m plays and the column player would have a choice of n plays.

The entries in a payoff matrix can be given from the point of view of either player; however, the usual practice is to express them from the point of view of the row player. In the game just described, we could write the payoff matrix from the point of view of your friend as the row player as follows:

$$\begin{array}{c} & \text{You} \\ & \begin{array}{ccc} 1 & 2 & 3 \end{array} \\ \text{Friend} \begin{array}{c} 1 \\ 2 \\ 3 \end{array} & \left[\begin{array}{ccc} -2 & -3 & 2 \\ -1 & -2 & -4 \\ 1 & -1 & 1 \end{array} \right] \end{array}$$

The payoff for any combination of plays is exactly the same as that given by the first matrix; the matrices are just written from a different point of view. In either case, the row player plays to obtain a maximum entry and the column player plays to obtain a minimum entry; each is seeking the most advantage for himself or herself. Any combination of appropriate plays for a player is called a **strategy** for that player.

The final strategies in the game above (using the first form of the playoff matrix) was for you to play r_2 consistently and for your friend to play c_3. These final strategies could be obtained immediately as follows. By observing the payoff matrix, you see that your worst payoff

For r_1 is -1
For r_2 is 1
For r_3 is -1

To guarantee for yourself the best of these possibilities, you choose r_2. Your friend, making similar observations, sees that her worst payoff

For c_1 is 3
For c_2 is 4
For c_3 is 1

To guarantee for herself the best of these possibilities, she chooses c_3. Choosing your strategies independently of each other, you immediately arrive at the final payoff, 1, obtained earlier.

The entry 1 in the payoff matrix is called a **saddle point,** that is, an entry that is the minimum in its row and the maximum in its column. The best strategy is for a row player to play a row with a saddle point and for a column player to play a column with a saddle point—best in the sense that each is pursuing the strategy that guarantees for himself or herself the most advantageous from among the worst payoffs.

EXAMPLE 1   Find the saddle point in the payoff matrix

$$\left[\begin{array}{ccc} -1 & 0 & 2 \\ 3 & -2 & 1 \\ 2 & 0 & 3 \end{array} \right]$$

Solution   The minimum in row 1 is -1, in row 2 is -2, and in row 3 is 0; the only one that is the maximum in its column is 0. The saddle point is 0. ■

6.6 GAME THEORY

EXAMPLE 2 Find the saddle point in the payoff matrix

$$\begin{bmatrix} 2 & 5 \\ 4 & 3 \end{bmatrix}$$

Solution The minimum in row 1 is 2 and in row 2 is 3; however, neither is the maximum in its column. There is no saddle point.

The best policy in Example 1 is for the row player to play r_3 and for the column player to play c_2. If one of the players follows this strategy consistently, the other player gains no advantage by changing.

In Example 2 there is no saddle point; consequently, there is no best strategy in the sense we have been discussing. Games with this type of payoff matrix are considered in the next section.

A two-person game with a payoff matrix in which there is a saddle point is called a **strictly determined** game; in such games the strategy for each player is determined by the saddle point. The game in the illustration at the beginning of this section and the game in Example 1 are both strictly determined games. In a strictly determined game, the value of the saddle point is called the **value** of the game, and a game is called **fair** if its value is 0. The game in Example 1 is a fair game.

EXAMPLE 3 Two retail companies, R and C, each intend to build a new store in the same city. Feasible locations are either downtown (D) or in the suburbs (S). Company R generally has a larger volume of business. The percent of the potential business that R can expect depending on the locations of the two stores is indicated in the following payoff matrix:

$$R \begin{array}{c} \\ D \\ S \end{array} \overset{\begin{array}{cc} C \\ D \quad S \end{array}}{\begin{bmatrix} 60 & 75 \\ 65 & 70 \end{bmatrix}}$$

If R locates downtown and C locates in the suburbs, for example, R would get 75% of the potential business (with C getting 25%). Where should the stores locate?

---

1 PRACTICE PROBLEM 1

Determine the saddle point, if there is one, in each matrix.

(a) $\begin{bmatrix} 1 & -1 & 2 \\ 3 & 4 & 5 \end{bmatrix}$

(b) $\begin{bmatrix} 1 & 2 & 4 \\ 6 & 5 & 3 \end{bmatrix}$

Answer

(a) 3   (b) None

2 PRACTICE PROBLEM 2

Determine the best strategy for the row player and for the column player if the payoff matrix is

$$\begin{bmatrix} 0 & -2 & 1 \\ 1 & 3 & 2 \\ -1 & 1 & -2 \end{bmatrix}$$

Answer

r_2 and c_1

Solution We look for a saddle point in the payoff matrix. In row 1 the minimum is 60, and in row 2 the minimum is 65. The latter is also the maximum in its column. The saddle point in the matrix is 65; R should locate in the suburbs and C should locate downtown. R would then get 65% of the business and C would get 35%. ∎

The game in Example 3 is still called a two-person game, even though the competitors are institutions and not two people; the value of the game is 65.

A payoff matrix can have more than one saddle point. However in such a case, although we shall not attempt to prove the fact, the values of all of the saddle points must be the same.

EXAMPLE 4 Two individuals, R and C, are candidates for the Democratic nomination for representative of a congressional district. The two major cities, A and B, in the district have approximately the same number of Democratic voters. Candidate R's staff believes that if R concentrates on city A and Candidate C does the same, Candidate R will get 40% of the total votes; with R in A and C in B, R will get 65% of the votes; with R in B and C in A, R will get 50% of the votes; and with both concentrating on B, R will get 60% of the votes. Formulate this as a two-person game and determine the best strategy for each candidate.

Solution The payoff matrix in terms of the percent of votes is

$$R \begin{array}{c} \\ A \\ B \end{array} \begin{array}{c} C \\ \begin{array}{cc} A & B \end{array} \\ \begin{bmatrix} 40 & 65 \\ 50 & 60 \end{bmatrix} \end{array}$$

The saddle point being 50, R should concentrate the campaign on B and C should concentrate on A. ∎

---

3 PRACTICE PROBLEM 3

If the payoff matrix in Example 3 were

$$\begin{array}{c} \\ D \\ S \end{array} \begin{array}{c} \begin{array}{cc} D & S \end{array} \\ \begin{bmatrix} 55 & 65 \\ 50 & 75 \end{bmatrix} \end{array}$$

where should each store locate? What percent of the business would each then receive?

Answer

Both downtown; 55% for R and 45% for C

EXERCISES

For each of the following payoff matrices, determine if there is a saddle point. If there is, determine the best strategy for the row player and for the column player; also determine the value of the game.

1. $\begin{bmatrix} 7 & 6 \\ 4 & -3 \end{bmatrix}$

2. $\begin{bmatrix} 12 & 15 \\ 17 & 9 \end{bmatrix}$

3. $\begin{bmatrix} 4 & 0 & 9 \\ 7 & 6 & 3 \end{bmatrix}$
4. $\begin{bmatrix} 5 & -2 & 1 \\ 0 & -3 & -4 \end{bmatrix}$

5. $\begin{bmatrix} 5 & -3 \\ -2 & 0 \\ 4 & 4 \end{bmatrix}$
6. $\begin{bmatrix} 7 & -1 \\ 4 & 0 \\ 3 & -2 \end{bmatrix}$

7. $\begin{bmatrix} 2 & -1 & -3 \\ 4 & 3 & 4 \\ 1 & 2 & -2 \end{bmatrix}$
8. $\begin{bmatrix} 4 & -3 & -2 \\ 1 & 0 & 4 \\ -2 & 1 & 5 \end{bmatrix}$

9. $\begin{bmatrix} 6 & -4 & 3 & -1 \\ -1 & 4 & 2 & -2 \\ 3 & 2 & 1 & 0 \\ 0 & 1 & -1 & -1 \end{bmatrix}$

Which of the games corresponding to the payoff matrices in the indicated exercises are strictly determined? Which of the strictly determined games are fair?

10. Exercise 1 11. Exercise 2 12. Exercise 3
13. Exercise 4 14. Exercise 5 15. Exercise 6
16. Exercise 7 17. Exercise 8 18. Exercise 9

19. *Personnel negotiation* It is contract negotiation time at a particular company. The union and management have agreed on a team of three people from union and three from management. Each day a committee consisting of one member from each team is to meet to discuss the new contract. The members of each team can be classified as nice guy (N), persuasive (P), and outspoken (O). The likely increase in wages (in cents per hour) depending on the classification of the individuals arriving at the final agreement is given in the matrix below. Determine the best strategy for each side and the corresponding wage increase.

$$\begin{array}{c} \\ \\ \text{Union} \end{array} \begin{array}{c} \\ N \\ P \\ O \end{array} \overset{\begin{array}{ccc} \multicolumn{3}{c}{Management} \\ N & P & O \end{array}}{\begin{bmatrix} 30 & 45 & 25 \\ 50 & 40 & 35 \\ 40 & 55 & 30 \end{bmatrix}}$$

		INVADERS		
		A	M	W
	R	-5	-2	3
DEFENDERS	N	-3	3	5
	I	-4	-3	2

20. *Military strategy* A country has been invaded and considers its three choices to be to repel the invaders (R), negotiate a settlement (N), or invade the offending country (I). The choices for the invaders are to advance their army (A), maintain their positions (M), or withdraw their army (W). The relative merits to the two sides for the various combinations of choices is given in the table. Determine the best strategy for each side.

21. Two people are playing a game in which each secretly writes down a number from the set {1, 2, 3}. If the numbers written are 1 and 2, the person writing 1 receives a chip from the other player; if they are 2 and 3, the player writing 2 receives a chip; if the numbers are 1 and 3, the player writing 3 receives a chip. If both write the same number, the result is a draw (neither wins). Construct the payoff matrix and determine whether or not this is a strictly determined game.

22. *Advertising strategy* Two competing companies, R and C, are each intending to launch an advertising campaign to improve sales. Both are

considering television campaigns versus magazine campaigns. Marketing analysts for company R believe that if both companies use television, R will obtain 40% of the potential new business; if R uses television and C uses magazines, R will obtain 60% of the potential new business; if R uses magazines, R will obtain 45% regardless of which method C uses. Formulate this situation as a two-person game by constructing the payoff matrix. Determine the best strategy for each company.

23. *Guarantee strategy* A television manufacturing firm uses a particular type of transistor in its sets. It can buy guaranteed transistors for $5.00 or not guaranteed for $3.00 each. If the transistor goes bad within their 3-year warranty period, it costs the television manufacturer $15.00 for service, regardless of the type of transistor; it costs the manufacturer an additional $3.00 to replace the transistor if it is not guaranteed. Formulate this as a two-person game and determine whether or not the company should use the guaranteed type of transistor. (*Note:* In a conflict such as this in which no particular person or institution is competing against the manufacturer but only the state of the transistors, the opponent is considered to be "nature.")

Explain the meaning of each term.

24. Two-person game
25. Payoff matrix for a two-person game
26. Saddle point in a payoff matrix
27. Strategy for a player
28. Strictly determined game
29. Value of a strictly determined game

*6.7 GAMES WITH MIXED STRATEGIES

In the last section we considered the best strategy only for games whose payoff matrices have saddle points. These were strictly determined games in the sense that the saddle point determined the best strategy for each player. We now consider some two-person games for which there is no saddle point. For the sake of illustration, consider the game with the payoff matrix

$$A = \begin{bmatrix} 2 & -1 \\ -1 & 3 \end{bmatrix}$$

in which there is no saddle point. If the row player played r_1 consistently and the column player became aware of this fact, he or she would consistently play c_2. If the row player played r_2 consistently, the column player would play c_1. Consequently, the row player would have to use a mixed strategy involving both r_1 and r_2. A similar analysis would indicate that the column player would have to use a mixed

---

* This section can be omitted without loss of continuity.

6.7 GAMES WITH MIXED STRATEGIES

PROBABILITY	$\frac{1}{2}$	$\frac{1}{2}$
$\frac{1}{4}$	2	-1
$\frac{3}{4}$	-2	3

strategy involving both c_1 and c_2. We want to develop the best mixed strategy for each using an expected-value approach.

Suppose that the column player decides to play each column an equal number of times but in a random fashion, so that the probability of playing c_1 is $\frac{1}{2}$ and of playing c_2 is also $\frac{1}{2}$. Similarly, suppose that the row player decides to play r_1 an average of one out of four times and r_2 three out of four times in a random fashion so the probability of playing r_1 is $\frac{1}{4}$ and of r_2 is $\frac{3}{4}$. These probabilities are listed along with the payoff matrix in the margin. The probability of an $r_1:c_1$ combination is then $\frac{1}{4} \times \frac{1}{2} = \frac{1}{8}$. Note that we are using the multiplication rule for independent events, since we suppose that the players each choose their strategies without knowledge of the other's choice. Similarly, the probability of an $r_2:c_1$ combination is $\frac{3}{4} \times \frac{1}{2} = \frac{3}{8}$, and so forth.

Using this procedure, we can assign a probability to each element in the sample space of possible combinations:

$$S = \{r_1:c_1, r_2:c_1, r_1:c_2, r_2:c_2\}$$
$$\text{Probability:} \quad \tfrac{1}{8} \quad \tfrac{3}{8} \quad \tfrac{1}{8} \quad \tfrac{3}{8}$$

Next we construct a random variable X on S, which assigns to each element the value of the corresponding payoff.

$$S = \{r_1:c_1, r_2:c_1, r_1:c_2, r_2:c_2\}$$
$$X: \quad 2 \quad -2 \quad -1 \quad 3$$

Then the expected value of X, or the expected payoff, is

$$2(\tfrac{1}{8}) + (-2)(\tfrac{3}{8}) + (-1)(\tfrac{1}{8}) + 3(\tfrac{3}{8}) = \tfrac{4}{8} = \tfrac{1}{2}$$

If the probabilities are changed by either player, then of course the expected payoff changes also. Our objective is to determine the probabilities for each that yield the best payoff.

But note first that if we construct probability vectors

$$P = [\tfrac{1}{4} \quad \tfrac{3}{4}] \quad \text{For the row player}$$

$$Q = \begin{bmatrix} \tfrac{1}{2} \\ \tfrac{1}{2} \end{bmatrix} \quad \text{For the column player}$$

then the expected payoff can be obtained more easily by the matrix product

$$[\tfrac{1}{4} \quad \tfrac{3}{4}] \begin{bmatrix} 2 & -1 \\ -2 & 3 \end{bmatrix} \begin{bmatrix} \tfrac{1}{2} \\ \tfrac{1}{2} \end{bmatrix} = \tfrac{1}{2} \qquad (1)$$

That is, the expected payoff is given by the matrix product PAQ, where A is the payoff matrix. You should verify that the product in equation (1) is, in fact, equal to $\frac{1}{2}$.

EXAMPLE 1 Determine the expected payoff in the illustration just given if the column player decides to play c_1 with probability $\frac{1}{3}$ and c_2 with probability

$\frac{2}{3}$, while the row player decides to play r_1 with probability $\frac{1}{6}$ and r_2 with probability $\frac{5}{6}$.

Solution The probability vectors are now

$$P = [\tfrac{1}{6} \ \tfrac{5}{6}] \quad \text{and} \quad Q = \begin{bmatrix} \tfrac{1}{3} \\ \tfrac{2}{3} \end{bmatrix}$$

The expected payoff is

$$PAQ = [\tfrac{1}{6} \ \tfrac{5}{6}] \begin{bmatrix} 2 & -1 \\ -2 & 3 \end{bmatrix} \begin{bmatrix} \tfrac{1}{3} \\ \tfrac{2}{3} \end{bmatrix} = [-\tfrac{8}{6} \ \tfrac{14}{6}] \begin{bmatrix} \tfrac{1}{3} \\ \tfrac{2}{3} \end{bmatrix} = \tfrac{20}{18} = \tfrac{10}{9}$$

We turn now to the question of the best probabilities for each player. First consider the row player. Continuing with the illustration discussed above Example 1, suppose that the row player plays r_1 with probability p and, therefore, plays r_2 with probability $1 - p$. If the column player plays c_1, then the expected payoff for the row player is

$$E_R = 2p + (-2)(1 - p) = 4p - 2$$

If the column player plays c_2, the expected payoff is

$$E_R = (-1)p + 3(1 - p) = -4p + 3$$

We graph these two linear equations in a coordinate system in Figure 6.4.

The second coordinate for each point on these lines represents the expected payoff for all possible values of p, one line if the column player plays c_1 and the second line if the column player plays c_2. The worst expected payoff for each value of p is the second coordinate for whichever line is lower—the portions corresponding to the line segments shown in color in the graph.

Recall that for strictly determined games, the best policy for a player was determined by choosing the strategy that guarantees the best result from among the worst possible outcomes. Adopting the same policy in this case, the row player would be interested in the value of p corresponding to the best of the worst possible expected values—that is, the value of p corresponding to the highest point on the colored line segments. In this case, that value of p would be $\frac{5}{8}$, determined by solving the system of equations

$$\begin{aligned} E_R &= 4p - 2 \\ E_R &= -4p + 3 \end{aligned} \qquad (2)$$

The row player would play r_1 with probability $\frac{5}{8}$ and r_2 with probability $\frac{3}{8}$. The corresponding probability vector is $P = [\tfrac{5}{8} \ \tfrac{3}{8}]$.

Now consider the column player. Suppose c_1 is played with probability q and, therefore, c_2 is played with probability $1 - q$. If the row player plays r_1, the expected payoff is

$$E_C = 2q + (-1)(1 - q) = 3q - 1$$

PRACTICE PROBLEM 1

Determine the expected payoff in Example 1 if the column player decides to play c_1 and c_2 with probabilities $\frac{1}{5}$ and $\frac{4}{5}$, respectively, and the row player decides to play both r_1 and r_2 each with probability $\frac{1}{2}$.

Answer

$\frac{4}{5}$

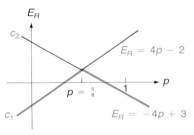

FIGURE 6.4

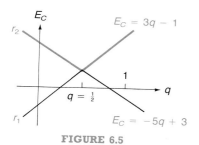

FIGURE 6.5

If the row player plays r_2, the expected payoff is

$$E_C = (-2)q + (3)(1-q) = -5q + 3$$

We graph these two linear equations in a coordinate system in Figure 6.5.

The worst expected payoff for the column player for each value of q is the second coordinate of the higher line corresponding to that value of q. (Recall that the payoff matrix is written in terms of the payoff to the row player—so the higher the payoff in the matrix, the more disadvantageous it is to the column player.)

To seek the best from among the worst possible expected values, the column player would be interested in the value of q corresponding to the lowest point on the colored line segments. In this case, that value of q is $\frac{1}{2}$, determined by solving the system of equations

$$\begin{aligned} E_C &= 3q - 1 \\ E_C &= -5q + 3 \end{aligned} \quad (3)$$

The column player would play each column with probability $\frac{1}{2}$. The corresponding probability vector is

$$Q = \begin{bmatrix} \frac{1}{2} \\ \frac{1}{2} \end{bmatrix}$$

The expected payoff corresponding to vector P, determined above, and Q is

$$PAQ = \begin{bmatrix} \frac{5}{8} & \frac{3}{8} \end{bmatrix} \begin{bmatrix} 2 & -1 \\ -2 & 3 \end{bmatrix} \begin{bmatrix} \frac{1}{2} \\ \frac{1}{2} \end{bmatrix} = \frac{1}{2}$$

In a game not strictly determined, the **value** of the game is the expected payoff corresponding to the probability vectors P and Q that give the best strategies for the players. If the value is positive, the game is to the advantage of the row player; if it is negative, the game is to the advantage of the column player. If the value of the game is zero, the game is considered **fair**.

It is not necessary to graph the equations for E_R and E_C in order to determine the probability vectors P and Q. All that is required is to solve the two systems of equations (2) and (3). The graphs in Figures 6.4 and 6.5 were used to describe the best strategy for each player.

EXAMPLE 2 Determine the probability vectors P and Q, the best strategy for each player, and the value of the game with payoff matrix

$$A = \begin{bmatrix} 4 & -3 \\ -2 & 6 \end{bmatrix}$$

Solution The equations for E_R are

$$E_R = 4p + (-2)(1-p) = 6p - 2$$

and
$$E_R = -3p + 6(1-p) = -9p + 6$$

Solving this system gives $p = \frac{8}{15}$. Therefore, $P = [\frac{8}{15} \quad \frac{7}{15}]$.

The equations for E_C are
$$E_C = 4q + (-3)(1-q) = 7q - 3$$
and
$$E_C = -2q + 6(1-q) = -8q + 6$$

Solving this system gives $q = \frac{3}{5}$. Therefore,
$$Q = \begin{bmatrix} \frac{3}{5} \\ \frac{2}{5} \end{bmatrix}$$

The row player should play r_1 with probability $\frac{8}{15}$ and r_2 with probability $\frac{7}{15}$. The column player should play c_1 with probability $\frac{3}{5}$ and c_2 with probability $\frac{2}{5}$. The value of the game is

$$PAQ = [\tfrac{8}{15} \quad \tfrac{7}{15}] \begin{bmatrix} 4 & -3 \\ -2 & 6 \end{bmatrix} \begin{bmatrix} \frac{3}{5} \\ \frac{2}{5} \end{bmatrix} = \frac{18}{15} = \frac{6}{5}$$

The method for determining best strategies in games not strictly determined is applicable only when the payoff matrix is 2×2. It can be extended to games with larger payoff matrices; furthermore, games with larger payoff matrices can at times be reduced to games with 2×2 matrices. Consider the game with payoff matrix

$$\begin{bmatrix} -1 & 3 & 4 \\ 2 & -2 & 3 \\ 1 & -3 & 2 \end{bmatrix}$$

The row player would never play r_3 because each entry in r_2 is larger than the corresponding entry in r_3; he would always play r_2 when given a choice between r_2 and r_3. Consequently, r_3 can be removed from the payoff matrix to obtain

$$\begin{bmatrix} -1 & 3 & 4 \\ 2 & -2 & 3 \end{bmatrix}$$

The column player would never play c_3 because each entry in c_1 (or c_2) is less than the corresponding entry in c_3. Therefore, c_3 can also be removed:

$$\begin{bmatrix} -1 & 3 \\ 2 & -2 \end{bmatrix}$$

Now that the game has been reduced to a 2×2 matrix, the previous methods are applicable. In this case,

$$E_R = (-1)p + 2(1-p)$$

PRACTICE PROBLEM 2

Determine the probability vectors P and Q, the best strategy for each player, and the value of the game with payoff matrix

$$\begin{bmatrix} 7 & -3 \\ -2 & 3 \end{bmatrix}$$

Answer

$$P = [\tfrac{1}{3} \quad \tfrac{2}{3}]; \quad Q = \begin{bmatrix} \frac{2}{5} \\ \frac{3}{5} \end{bmatrix}$$

The row player should play r_1 with probability $\frac{1}{3}$ and r_2 with probability $\frac{2}{3}$. The column player should play c_1 with probability $\frac{2}{5}$ and c_2 with probability $\frac{3}{5}$. The value of the game is 1.

6.7 GAMES WITH MIXED STRATEGIES

and

$$E_R = 3p + (-2)(1 - p)$$

give a value of $p = \frac{1}{2}$, so that $P = [\frac{1}{2} \; \frac{1}{2}]$. Also,

$$E_C = (-1)q + 3(1 - q)$$

and

$$E_C = 2q + (-2)(1 - q)$$

give $q = \frac{5}{8}$, so that

$$Q = \begin{bmatrix} \frac{5}{8} \\ \frac{3}{8} \end{bmatrix}$$

The value of the game, which is the same as the expected payoff for strategies determine by P and Q, is

$$PAQ = [\tfrac{1}{2} \; \tfrac{1}{2}] \begin{bmatrix} -1 & 3 \\ 2 & -2 \end{bmatrix} \begin{bmatrix} \frac{5}{8} \\ \frac{3}{8} \end{bmatrix} = \tfrac{1}{2}$$

In general, a **row r_i is dominated by row r_j** in a payoff matrix if each entry in r_j is greater than the corresponding entry in r_i. **Column c_i is dominated by column c_j** if each entry in c_j is less than the corresponding entry in c_i. Since dominated rows and columns would never be played, they can be removed from the payoff matrix without affecting the strategies or the payoff.

EXAMPLE 3 Remove the dominated row and column; then determine the probability vectors and the value of the game with payoff matrix

$$\begin{bmatrix} -1 & -4 & -3 \\ 0 & 5 & -2 \\ -3 & 6 & 3 \end{bmatrix}$$

Solution Comparing corresponding entries shows that r_1 is dominated by r_2, so r_1 can be removed:

$$\begin{bmatrix} 0 & 5 & -2 \\ -3 & 6 & 3 \end{bmatrix}$$

Then c_2 is dominated by c_3, so c_2 can be removed:

$$\begin{bmatrix} 0 & -2 \\ -3 & 3 \end{bmatrix}$$

The result is a 2×2 game that is not strictly determined. The method discussed earlier gives probability vectors

$$P = [\tfrac{3}{4} \; \tfrac{1}{4}] \quad \text{and} \quad Q = \begin{bmatrix} \frac{5}{8} \\ \frac{3}{8} \end{bmatrix}$$

3 PRACTICE PROBLEM 3

Remove the dominated row; then determine the probability vectors and the value of the game with payoff matrix

$$\begin{bmatrix} 10 & -20 \\ -6 & 4 \\ -8 & 2 \end{bmatrix}$$

Answer

$P = \begin{bmatrix} \frac{1}{4} & \frac{3}{4} \end{bmatrix}; \quad Q = \begin{bmatrix} \frac{3}{5} \\ \frac{2}{5} \end{bmatrix}$

The value of the game is -2.

The value of the game is

$$PAQ = \begin{bmatrix} \frac{3}{4} & \frac{1}{4} \end{bmatrix} \begin{bmatrix} 0 & -2 \\ -3 & 3 \end{bmatrix} \begin{bmatrix} \frac{5}{8} \\ \frac{3}{8} \end{bmatrix} = -\frac{3}{4} \qquad \boxed{3}$$

The following outline suggests an organized approach using the methods we have discussed to solving two-person games:

> 1. First determine whether the payoff matrix has a saddle point. If so, the location of the saddle point indicates the best strategy for each player. The value of the saddle point is the value of the game.
> 2. If the payoff matrix is a 2 × 2 matrix without a saddle point, determine the probability vector for each player. These vectors indicate the long-term ratio or percent of times each row or column should be played. The corresponding expected payoff is the value of the game.
> 3. If the payoff matrix is larger than a 2 × 2 matrix, remove any dominated rows and columns. If the resulting payoff matrix is 2 × 2, proceed as in Step 2.

All the games we have considered are **zero-sum** games; that is, what one player gains, the other loses, and vice versa. Techniques other than those we have discussed are used to solve games that are not zero-sum games. In fact, our discussion has been just an introduction to game theory. Other than games that are not zero-sum, there are games involving more than two players, as well as two-person games with payoff matrices larger than 2 × 2 without dominated rows and/or columns. None of the techniques used to solve these types of games are discussed here.

Game theory, as we have considered it, is appropriate for modeling competitive situations between two players when the problem is to determine the best strategy for one or both players. The players can be individuals, a group of individuals, an institution, or just the flow of events referred to as nature. Our techniques are applicable only for zero-sum games and assume that each player wants to choose for himself or herself the best of the worst possible results. Two other assumptions are that neither player has any advance knowledge of the play of his or her opponent and, in the case of games that are not strictly determined, that the game is played many times because of the probabilities involved.

It is probably fair to say that not many physical situations fit into all the classifications and assumptions just listed. However, game theory can still be useful for analyzing competitive situations and determining possible strategy, while the final decision must be based on other factors as well.

EXERCISES

For the game with payoff matrix

$$\begin{bmatrix} 2 & -1 \\ -3 & 3 \end{bmatrix}$$

determine the expected payoff in each case.

1. If the row player plays r_1 with probability $\frac{1}{3}$ and r_2 with probability $\frac{2}{3}$, while the column player plays c_1 with probability $\frac{3}{4}$ and c_2 with probability $\frac{1}{4}$.

2. If the row player plays r_1 with probability $\frac{3}{5}$ and the column player plays c_1 with probability $\frac{5}{8}$.

Find the probability vectors, the best strategy for each player, and the expected payoff for the game with the indicated payoff matrix.

3. $\begin{bmatrix} 2 & -1 \\ -3 & 3 \end{bmatrix}$
4. $\begin{bmatrix} -4 & 2 \\ 3 & 0 \end{bmatrix}$
5. $\begin{bmatrix} -8 & 7 \\ 6 & -5 \end{bmatrix}$

6. $\begin{bmatrix} 1 & -1 \\ -1 & 0 \end{bmatrix}$
7. $\begin{bmatrix} -18 & 30 \\ 20 & 15 \end{bmatrix}$
8. $\begin{bmatrix} -1 & 1 \\ 1 & -1 \end{bmatrix}$

Rewrite the following payoff matrices, eliminating the dominated rows and columns:

9. $\begin{bmatrix} -2 & 3 & 1 \\ 3 & 4 & 2 \\ -1 & 5 & 3 \end{bmatrix}$
10. $\begin{bmatrix} 2 & -3 & 3 \\ -1 & 4 & 0 \end{bmatrix}$

11. $\begin{bmatrix} 2 & -1 \\ 0 & -2 \\ -3 & 2 \end{bmatrix}$
12. $\begin{bmatrix} 1 & 4 & 3 & 0 \\ 0 & 3 & 2 & -1 \\ 1 & -2 & 4 & -2 \\ 2 & -1 & 5 & 3 \end{bmatrix}$

Determine the best strategy for each player and the value of the game with each payoff matrix.

13. $\begin{bmatrix} 1 & -2 & 3 \\ 0 & -3 & 2 \\ -2 & 1 & 0 \end{bmatrix}$
14. $\begin{bmatrix} -2 & 3 & -1 \\ 3 & 2 & 1 \\ 4 & -1 & 0 \end{bmatrix}$

15. $\begin{bmatrix} 0 & 1 \\ -1 & 2 \end{bmatrix}$
16. $\begin{bmatrix} 2 & -3 & -2 \\ 1 & 3 & 0 \end{bmatrix}$

17. $\begin{bmatrix} -4 & 3 \\ 1 & -2 \\ -3 & 4 \\ 0 & -3 \end{bmatrix}$
18. $\begin{bmatrix} -4 & 1 & -3 & 0 \\ 3 & -2 & 4 & -3 \end{bmatrix}$

19. *Investment strategy* A financial adviser has $30,000 to invest for 1 year for one of her clients. She prefers to invest the money conservatively and has decided to put part of the money into a money market fund and the rest into stocks of utility companies. She believes if interests rates go up during the next year, she will realize 17% on the money market investment and 11% on the stocks, but if interest rates go down next year, she will realize 9% on the money market investment and 12% on the stock. She wants to decide how much of the $30,000 to invest in each type. Formulate this as a two-person game by constructing the payoff matrix with nature as her competitor. Determine how much money she should invest in each type of investment and how much she can expect to make during the year.

20. A game consists of one player secretly choosing heads or tails on a coin, while the second player secretly choose heads or tails on two different coins. If the result is two heads and one tail, the first player pays the second $1; if the result is two tails and one head, the second player pays the first $1; if there are three heads or three tails, no money is exchanged.
    (a) Formulate this as a two-person game and determine the best strategy for each player.
    (b) What is the value of the game? Is this a fair game?
    (c) If the second player pays $\$\frac{1}{3}$ to the first to play each game, would the game be fair? Verify your decision by constructing a new payoff matrix and determining the new expected payoff.

Explain the meaning of each term.

21. Value of a game that is not strictly determined
22. Row r_i being dominated by row r_j
23. Column c_i being dominated by column c_j
24. Zero-sum game
25. *Investment strategy* Identify the steps of the four-step process involved in the application of mathematics (Section 1.3) in Exercise 19.

6.8 CHAPTER SUMMARY

In this chapter we considered some additional topics in probability. The first topic considered was **binomial probabilities**. The requirements for calculating probabilities in this manner are:

1. We have a sequence of independent trials of an experiment.
2. On each trial we are interested in only two outcomes, a success or a failure.
3. The probability of a success on each trial remains the same.

Then the probability of exactly r successes in n such trials is

$$C(n, r)p^r(1-p)^{n-r}$$

where p is the probability of a success on each trial. Trials satisfying these three conditions are called **Bernoulli trials.**

Bayes' rule gives a method for calculating probabilities of a particular type after an experiment has been performed. It states that if $A_1, A_2, \ldots, A_n$ are mutually exclusive events whose union is the whole sample space S and if B is any event in S for which $P(B) \neq 0$, then for any $i = 1, 2, \ldots, n$,

$$P(A_i|B) = \frac{P(A_i)P(B|A_i)}{P(A_1)P(B|A_1) + P(A_2)P(B|A_2) + \cdots + P(A_n)P(B|A_n)}$$

Markov chains are also concerned with a sequence of trials. In this case, the physical situation being observed is called the **system,** and trial. The elements a_i in S are called the **states** of the system.

1. The sample space $S = \{a_1, a_2, \ldots, a_m\}$ remains the same for each trial. The elements a_i in S are called the **states** of the system.
2. The probability of obtaining any particular element in the sample space at each trial possibly depends upon the outcome of the preceding trial but upon no other previous trial.
3. For each i and j, the probability p_{ij} that a_j will occur given that a_i occurred on the preceding trial remains constant throughout the sequence.

Then if $P = (p_{ij})$, the probability of obtaining state a_j at the nth trial given that the system was in state a_i at the beginning of the first trial is the element in the ij-position of P^n. The matrix P is called the **transition matrix.**

If the transition matrix P is **regular**—that is, P or some power of P contains only positive numbers—and vector A satisfies the equation

$$AP = A \quad \text{or} \quad A(P - I) = 0$$

the entries in A indicate the long-term behavior of the system.

A **random variable** X on a sample space S is a function, or rule, which assigns a unique number to each element in the sample space. The numbers assigned are called the **values** of X. The **expected value** of X then is

$$E(X) = x_1 P(x_1) + x_2 P(x_2) + \cdots + x_n P(x_n)$$

where $x_1, x_2, \ldots, x_n$ are the values of X. $E(X)$ gives the expected average value of X if the experiment is performed many times.

A **two-person game** is a competition between two entities. A **strictly determined** game is a game in which the payoff matrix has a **saddle point,** that is, an entry that is the minimum in its row and the maximum in its column. The saddle point indicates the best strategy for both the row and the column players and is the **value** of the game. The indicated strategy is best in the sense that each player is pursuing the strategy that guarantees the most advantageous from among the worst payoffs.

In a game not strictly determined with a 2 × 2 payoff matrix A, the best strategy for each player is to seek the best from among the worst possible expected payoffs. The object is to determine the probabilities p and $1 - p$ with which the row player should play row 1 and row 2, respectively, as well as the probabilities q and $1 - q$ with which the column player should play column 1 and column 2, respectively, to obtain the best strategy as indicated above.

If p and q are the values so obtained and if

$$P = [p \quad 1 - p] \quad \text{and} \quad Q = \begin{bmatrix} q \\ 1 - q \end{bmatrix}$$

the corresponding expected payoff of the game, given by the product PAQ, is the **value** of the game. If the value of a game is zero, it is said to be a **fair** game.

A game not strictly determined with a larger than 2 × 2 payoff matrix can sometimes be reduced to a 2 × 2 game by removing dominated rows and columns.

REVIEW EXERCISES

A coin is tossed five times. What is the probability of each outcome?

1. Three heads are obtained.
2. More than three heads are obtained.
3. Less than three heads are obtained.
4. At most three heads are obtained.
5. Less than three or more than three heads are obtained. Evaluate this probability in two different ways.
6. *Voting patterns* A poll indicates that 60% of the voters in a city will vote for Kay for mayor. On this basis what is the probability of each outcome, based on ten arbitrarily chosen voters?
   (a) Seven will vote for Kay.   (b) Seven will not vote for Kay.

In a sequence of independent Bernoulli trials the probability of a success is 0.5.

7. What is the probability of at least 4 successes if $n = 9$?
8. What is the probability of at least 4 successes if $n = 8$?
9. What is the minimum value of n for which the probability of at least 4 successes is at least $\frac{1}{2}$?
10. *Test completion* A mathematics teacher noted 80% of her class completed a particular test. Of those who completed the test, 90% passed; of those who did not complete the test, 65% passed.
    (a) What is the probability that a student who passed the test completed the test?

(b) What is the probability that a student who passed the test did not complete the test?

11. *Automobile purchasing* In a small town, 55% of the cars are purchased from dealer A, 35% are purchased from dealer B, and 10% are bought out of town. Eighty percent of the customers buying a car from dealer A are satisfied with their purchase, 65% of the customers of dealer B are satisfied, and 70% of the customers of out-of-town dealers are satisfied. If a randomly chosen car owner is satisfied with his purchase, what is the probability that he bought the car from dealer B?

12. *Family income* The table gives the distribution of income levels and of two-television families within income levels in a small suburb. If a randomly chosen family has two or more television sets, what is the probability that the family has an income of at least the amount indicated?
    (a) $35,000   (b) $25,000

YEARLY FAMILY INCOME	PERCENT OF FAMILIES	PERCENT HAVING TWO OR MORE TELEVISIONS
14,999 or less	20%	30%
15,000–24,999	50%	60%
25,000–34,999	25%	90%
35,000 or greater	5%	100%

For Exercises 13–15, use the transition matrix

$$\begin{array}{c} \\ A \\ B \end{array} \begin{array}{cc} A & B \\ \begin{bmatrix} 0.5 & 0.5 \\ 0.4 & 0.6 \end{bmatrix} \end{array} = P$$

13. If A occurs on a particular trial, what is each probability?
    (a) B occurs on the next trial.   (b) A occurs on the second trial.
    (c) B occurs on the third trial.

14. Explain why P is regular.

15. Determine the fixed probability vector of P.

16. *Advertising results* An intensive advertising campaign is being conducted to promote brand X toothpaste. Surveys during the first few months indicate that 70% of the people using brand X at the beginning of the month continue to use brand X during the month and that 10% of the people not using brand X at the beginning of the month switch to brand X during the month.
    (a) Construct the transition matrix, assuming the transition probabilities continue to hold.
    (b) If a person does not use brand X, what is the probability that he or she will not use brand X 2 months later?
    (c) What portion of the market will be using brand X at the end of the advertising campaign 1 year from now, assuming the transition matrix will remain the same?

17. *Economic levels* An economist has found that in her county the probability of changing from one of the three economic levels [lower income (L), middle income (M), and upper income (U)] to another or remaining at

the same level during a 5-year period are as indicated in the following matrix:

$$\begin{array}{c} \\ \text{From} \end{array} \begin{array}{c} \\ L \\ M \\ U \end{array} \overset{\begin{array}{ccc} & To & \\ L & M & U \end{array}}{\begin{bmatrix} 0.80 & 0.20 & 0 \\ 0.10 & 0.70 & 0.20 \\ 0 & 0.10 & 0.90 \end{bmatrix}}$$

If these probabilities remain constant, what do they indicate concerning the long-range distribution of the population with respect to the three economic levels?

18. The values of a random variable X and their respective probabilities are given in the table. Determine $E(X)$.

VALUE OF X	PROBABILITY
−3	0.05
0	0.27
7	0.38
9	0.19
13	0.11

19. *Insurance profit* A 50-year-old man buys a 1-year, $5,000 insurance policy for $225. If mortality tables indicate that the probability of a 50-year-old man living one more year is 0.992, determine the expected profit of the insurance company.

20. *Learning experiment* In a learning experiment laboratory, rats were introduced to a difficult maze and a record was kept of the number of trials it took each rat to pass through the maze successfully. The results are given in the table. On this basis, determine the expected number of trials it will take a rat, once introduced to the maze, to pass through the maze successfully.

NUMBER OF TRIALS	NUMBER OF RATS
1	1
2	1
3	2
4	1
5	3
6	4
7	5
8	4
9	3
10	1

Following the procedure outlined on page 254 determine the best strategy for each player and the value of the game with the given payoff matrix.

21. $\begin{bmatrix} 1 & -2 & 3 \\ 2 & 0 & 4 \\ -2 & -1 & 3 \end{bmatrix}$

22. $\begin{bmatrix} 1 & 2 \\ 2 & -4 \end{bmatrix}$

23. $\begin{bmatrix} 1 & -5 & 2 \\ 2 & -4 & 3 \\ 3 & 2 & -1 \end{bmatrix}$

24. $\begin{bmatrix} -3 & 4 \\ 2 & -1 \end{bmatrix}$

25. *Advertising strategy* Two companies are each planning to introduce a new personal computer by means of a television and magazine advertising campaign. The payoff matrix indicates the corresponding new sales (in millions of dollars) for each company. What is the best strategy for Company R; that is, what percent of their advertising should be on television and what percent in magazines?

$$\text{Company } R \begin{array}{c} \\ TV \\ Magazine \end{array} \overset{\begin{array}{c} Company \ C \\ TV \quad\quad Magazine \end{array}}{\begin{bmatrix} 2.5 & -1 \\ -1.5 & 2 \end{bmatrix}}$$

26. *Advertising strategy* Identify the steps of the four-step procedure involved in the application of mathematics (Section 1.3) in Exercise 25.

CHAPTER 7

INTRODUCTORY STATISTICS

OUTLINE

7.1 Organizing Data
7.2 Measures of Central Tendency
7.3 Measures of Dispersion
7.4 The Sign Test
7.5 The Normal Curve
*7.6 A Test of a Claimed Mean
7.7 Chapter Summary

Review Exercises

Often an experiment is repeated a number of times and the results are recorded and analyzed. One of the most familiar occurrences of this procedure is that of a class of students taking a test. If there are 20 students taking the test, the experiment of a student taking the test can be considered to be repeated 20 times. The 20 scores form the resulting data, sometimes referred to as the **observations.** The central problem in statistics lies in determining what valid conclusions one can infer from a given set of data. For example, the 20 scores might be testing the effectiveness of a new method of teaching reading to slow learners, and we are interested in whether the scores indicate a significant improvement over the old method or indicate no improvement at all when compared with the scores of students taught by the old method. Just because some of the scores are higher does not necessarily mean that the new method is better.

In general, statistics is concerned with collecting, describing, and interpreting data. In this chapter we shall examine some topics related to these three aspects of statistics.

---

* This section can be omitted without loss of continuity.

7.1 ORGANIZING DATA

Usually the "raw" data collected in an experiment are members of an unordered set of numbers, such as:

86, 88, 93, 91, 83, 84, 84, 86, 86, 86, 89,
87, 90, 85, 91, 83, 86, 87, 93, 88, 88, 91,
90, 86, 88, 88, 90, 84, 85, 87, 85, 90, 85,
84, 85, 88, 90, 88, 87, 86, 86, 85, 84, 86,
83, 88, 86

In this form the numbers are practically meaningless; they have to be put into a more structured arrangement for any type of analysis. There are many ways in which this can be done; we shall discuss two of them.

One way is to list each number (measurement) from the smallest to the largest and to indicate the number of times that each of these appears. The number of times that each appears is called its *frequency* and the resulting table is called a *frequency table.* For the data just given, the frequency table is shown in Table 7.1.

Another form in which data can be presented is a type of bar graph called a *histogram.* A histogram for the given data is shown in Figure 7.1. The horizontal scale of the histogram lists a range of numbers that includes the smallest to the largest in the set of data, spaced an equal distance apart. The height of the bar centered over each measurement is equal to its frequency, as found in the scale to the left.

TABLE 7.1

MEASUREMENT	FREQUENCY
83	3
84	5
85	6
86	10
87	4
88	8
89	1
90	5
91	3
92	0
93	2

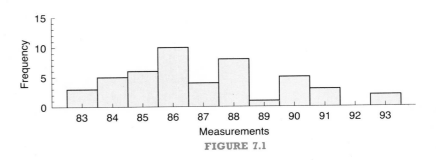

FIGURE 7.1

EXAMPLE 1 Construct a frequency table and a histogram for the following set of data:

21.5, 23.0, 24.0, 22.5, 22.5, 23.0, 24.0, 22.0, 25.0,
22.0, 24.0, 21.5, 23.0, 22.0, 22.0, 23.5, 24.0, 25.0,
24.0, 23.5, 24.0, 22.5, 23.0, 24.0, 23.0, 22.0, 22.0,
25.0

7.1 ORGANIZING DATA

Solution The frequency table is given first. Then the histogram is constructed from the table.

MEASUREMENT	FREQUENCY
21.5	2
22.0	6
22.5	3
23.0	5
23.5	2
24.0	7
24.5	0
25.0	3

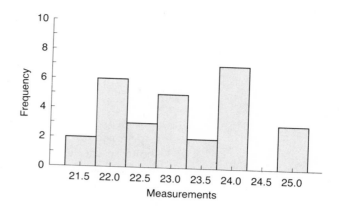

The scales on the bottom and the left of a histogram are chosen as appropriate for each given set of data.

Another method of organizing a large set of data is to determine the number of measurements falling into particular intervals rather than the number of measurements with a specific value. The process is illustrated with the set of data

65, 57, 73, 74, 66, 66, 51, 78, 76, 52, 64,
76, 82, 55, 70, 58, 72, 68, 69, 64, 61

The measurements range in size from 51 to 82. We (arbitrarily) choose intervals of length 5 from 50 to 84:

50–54, 55–59, 60–64, 65–69, 70–74, 75–79, 80–84

There are two measurements, 51 and 52, in the interval 50–54; there are three measurements in the interval 55–59, and so forth. These counts give the frequencies in the frequency table given in Table 7.2.

Data that have been grouped into intervals in this manner are called **grouped data**. Histograms for grouped data are similar to those for

TABLE 7.2

INTERVAL	FREQUENCY
50–54	2
55–59	3
60–64	3
65–69	5
70–74	4
75–79	3
80–84	1

PRACTICE PROBLEM 1

The number of employees absent from a factory on 15 consecutive work days were:

6, 1, 3, 6, 5, 6, 3, 2,
1, 6, 6, 3, 6, 1, 5

Construct a frequency table and histogram for the data.

Answer

NUMBER ABSENT	FREQUENCY
6	6
5	2
4	0
3	3
2	1
1	3

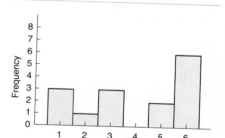

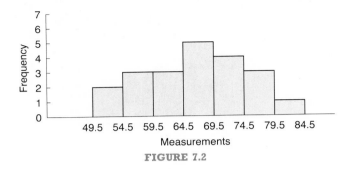

FIGURE 7.2

ungrouped data except that the widths of the bases of the bars correspond to the lengths of the intervals. The histogram in Figure 7.2 illustrates the process for the last set of data. Note that on the horizontal axis, the points of subdivision are halfway between the endpoints of the successive intervals in the frequency table given in Table 7.2.

When the data are presented in a frequency table or histogram, one can tell quickly both the size of the numbers involved and their frequencies. As a consequence, the data are in a form for easier analysis. Frequency tables are generally more useful for calculations and histograms for their visual effects. Grouping data makes it easier to work with large sets of data when the measurements take on many values.

In the next sections, we shall consider two types of models for describing data, as well as some uses of the two models. The two types of models are those that indicate the central tendency of the data and those that indicate the dispersion of the data. Roughly speaking, the central tendency is indicated by the middle measurement, or middle number, of the data, while the dispersion of the data is concerned with how the data cluster around this middle measurement.

② PRACTICE PROBLEM 2

The number of customer accounts opened each week by a new business during its first several months were:

5, 7, 7, 11, 13, 3, 4,
16, 11, 18, 10, 3, 9, 9,
8, 11, 16, 19, 12, 13, 3

Form the data into grouped data using four intervals of length 5, starting with 0: 0–4, 5–9, and so forth. Construct a frequency table and histogram for the grouped data.

Answer

INTERVAL	FREQUENCY
0–4	4
5–9	6
10–14	7
15–19	4

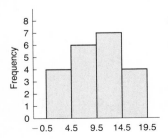

EXERCISES

Construct a frequency table and a histogram for each of the following sets of data:

1. 10, 12, 12, 13, 15, 10, 13, 11, 10, 11, 10, 11,
   12, 14, 10, 13, 11, 11, 15, 10, 15, 12, 10, 15,
   11, 15

2. 66, 69, 69, 67, 70, 71, 72, 73, 70, 72, 66, 70,
   73, 75, 73, 75, 70, 73, 71, 69, 66, 69, 72, 69,
   67, 69, 70, 73, 69, 71, 66, 73, 72, 73

3. 9.2, 9.4, 9.0, 9.4, 9.2, 9.6, 9.8, 9.4, 9.6, 9.4, 9.6,
   9.4, 9.2, 9.6, 9.8, 9.6, 9.6, 10.0, 9.2, 9.4, 9.0, 9.4,
   9.8, 10.0, 9.6, 9.8, 9.6

4. 21.3, 21.5, 21.8, 21.4, 21.3, 21.5, 21.6, 21.5, 21.6,
   21.6, 21.9, 21.8, 21.4, 21.5, 21.1, 21.5, 21.8, 21.2,
   21.2, 21.7, 21.6, 21.2, 21.3, 21.1, 21.6, 21.7, 21.3,
   22.0, 21.9, 21.9, 21.8, 21.6, 21.7, 21.7, 21.5, 22.0,
   21.4, 21.4, 21.4, 21.7

5. *Automobile inventory* The following data give the new-car inventory of an automobile dealer on 50 consecutive business days. Form the data into grouped data using five intervals of length 5 each, starting with 15: 15–19, 20–24, 25–29, and so forth. Then construct a frequency table and histogram for the grouped data.

   38, 34, 33, 29, 28, 32, 29, 27, 25, 20,
   26, 21, 19, 36, 34, 16, 36, 32, 30, 28,
   30, 29, 25, 20, 18, 26, 21, 19, 35, 34,
   16, 37, 34, 33, 30, 32, 30, 29, 27, 26,
   27, 25, 20, 18, 35, 25, 21, 20, 18, 16

6. *Memory test scores* The following data give the scores on a memory test given to 40 individuals. Form the data into grouped data using eight intervals of length 5 each, starting with 11: 11–15, 16–20, 21–25, and so forth. Then construct a frequency table and histogram for the grouped data.

   30, 34, 44, 39, 46, 22, 19, 37, 33, 35,
   36, 45, 42, 37, 25, 31, 46, 24, 14, 40,
   39, 28, 29, 44, 41, 20, 32, 33, 42, 30,
   26, 20, 47, 43, 13, 25, 32, 34, 36, 36

7. *Test scores* The following data are the scores received by 37 students on a math test. Form the data into grouped data using six intervals of length 10 each, starting with 40: 40–49, 50–59, and so forth. Then construct a frequency table and histogram for the grouped data.

   88, 83, 65, 72, 72, 75, 70, 89, 84, 74,
   95, 74, 81, 77, 83, 75, 83, 61, 49, 92,
   46, 70, 60, 73, 98, 61, 78, 72, 71, 98,
   85, 85, 63, 75, 66, 66, 81

8. *Pollution study* Environmentalists are studying the effect of pollution on the fish in Lake Erie. Each week they take a sample of fish from the lake to examine. The following data give the number of fish taken per week over a 6-month period. Form the data into grouped data. Construct a frequency table and histogram for the grouped data.

   25, 36, 25, 37, 37, 24, 23, 30, 23, 22,
   35, 32, 31, 23, 34, 22, 29, 30, 26, 27,
   27, 21, 38, 25, 26, 31

9. *Fermentation experiment* The following data give the number of parts per million of hydrogen sulfide obtained in 45 tests of an experimental fermentation process. Form the data into grouped data using six intervals

of length 2 each, starting with 125: 125.0–126.9, 127.0–128.9, and so forth. Then construct a frequency table and histogram for the grouped data.

133.3,	135.0,	128.3,	131.0,	131.3,	129.1,	130.8,	126.4,
128.7,	131.9,	131.1,	132.3,	132.3,	130.6,	129.0,	127.0,
125.3,	133.0,	134.0,	134.8,	132.4,	132.6,	130.6,	130.1,
130.1,	127.2,	129.3,	125.1,	133.3,	134.1,	135.8,	135.2,
131.1,	130.9,	132.6,	135.1,	125.3,	129.4,	131.4,	127.4,
134.0,	135.9,	133.1,	133.2,	134.1			

10. *Computer CPU time* The following data give the amount of computer CPU (central processing unit) time, in seconds, required to execute a computer programming assignment given to a class of 39 students. Form the data into grouped data. Construct a frequency table and histogram for the grouped data.

1.82,	2.07,	1.74,	2.01,	2.11,	1.95,	2.03,	1.93,	1.89,	1.73,
2.02,	1.85,	1.82,	1.97,	1.84,	1.77,	2.09,	2.05,	1.91,	1.84,
1.97,	1.93,	1.82,	1.99,	2.16,	1.77,	1.76,	2.15,	1.91,	2.02,
1.72,	1.99,	1.80,	2.10,	2.17,	2.16,	1.88,	2.04,	2.07	

7.2 MEASURES OF CENTRAL TENDENCY

Numbers that serve as measures of central tendency of a set of data are called **averages**. We shall consider the two most common types of averages—the mean and the median.

The **mean** of a set of n numbers $\{x_1, x_2, \ldots, x_n\}$ is the quotient

$$\frac{x_1 + x_2 + \cdots + x_n}{n}$$

The **median** of the set of numbers $\{x_1, x_2, \ldots, x_n\}$ is the middle measurement when the numbers are arranged in order of magnitude.

EXAMPLE 1 The numbers of customer calls made by a salesperson during 9 consecutive weeks are 44, 47, 52, 57, 63, 48, 60, 61, and 54. The mean number of calls per week during the 9-week period is

$$\frac{44 + 47 + 52 + 57 + 63 + 48 + 60 + 61 + 54}{9} = \frac{486}{9} = 54$$

If the number of calls are arranged in increasing order, we have

44,   47,   48,   52,   ⑤4,   57,   60,   61,   63

The median number of calls is also 54. ∎

EXAMPLE 2 Suppose that the numbers of calls made by the salesperson of Example 1 during the next 9 weeks are

21, 35, 37, 48, (54), 55, 59, 61, 62

The median is again 54, but the mean is

$$\frac{21 + 35 + 37 + 48 + 54 + 55 + 59 + 61 + 62}{9} = 48$$

The medians in these two examples are the same, but the means differ. The smaller mean in Example 2 is due to the fact that extreme values, like the relatively small values 21, 35, and 37, influence a mean strongly. Example 3 again illustrates this.

EXAMPLE 3 Suppose that the years of service of the 24 teachers in a particular high school are 1, 1, 2, 2, 2, 2, 3, 3, 3, 3, 3, 3, 4, 4, 5, 5, 6, 6, 6, 6, 9, 10, 25, and 30. The mean number of years of service of these teachers is 6. Actually 16, or two-thirds, of the teachers had less than 6 years service. The higher mean is due to the relatively high number of years of service of the two teachers with 25 and 30 years.

In this situation, the median of 3.5 years service indicates that half of the teachers had less than 3.5 years service, despite the fact that the mean number of years is 6.

Note that the number of teachers in Example 3 is an even number; there is no "middle measurement." The median of such a set of numbers is the mean of the two middle numbers. In this example the twelfth-largest number is 3 and the thirteenth-largest number is 4. The median is therefore

$$\frac{3 + 4}{2} = 3.5$$

---

1 PRACTICE PROBLEM 1

Calculate the mean and median:

(a) 7, 9, 10, 13, 14, 15, 16
(b) 22, 28, 26, 20, 21, 30

Answer

(a) 12; 13   (b) 24.5; 24

Both the mean and the median are mathematical descriptions (models) of sets of data. Each, although in different ways, answers the question, What is the central tendency of the data, and the result is interpreted in the physical situation as the middle measure of the number of salesperson calls (Example 1), the middle measure of years of service (Example 3), and so on. Several uses of these middle measures are discussed shortly.

One common use of averages is to represent a set of data by a single number. In using averages in this way, it is preferable to indicate *both* the median and the mean. If only one is to be used, it should be remembered that the mean divides the total evenly among the number of entries; the median indicates the middle position of the entries.

If both the mean and the median are indicated, a mean that is much greater than the median suggests the possibility of a few numbers that are considerably larger than the others. Conversely, a mean that is much smaller than the median suggests a few numbers considerably smaller. Remember that the mean and median are single numbers representing a whole set of numbers. Detail is sacrificed for the sake of simplicity. Whether or not this is misleading depends on the particular purpose. In some cases it may be more desirable to examine the whole set of numbers.

Besides representing sets of data by single numbers, averages are used to compare the relative size of one of the numbers to the total set. The expressions *above average* and *below average* are often used in this connection.

One of the most common occurrences of this second use of averages is in connection with a class taking a test. After the results are known, a student compares his or her score with the class mean to determine how he or she did relative to the rest of the class. Or the student compares his or her score with the class median to determine whether or not he or she scored in the top half of the class.

A third use of averages is in the comparison of two sets of data. The average monthly maintenance costs of several automobiles might be compared to determine which is the most economical. The average score of two classes taking the same exam might be compared in order to determine which class is superior. However, the use of averages for the sake of comparison for groups of people or objects can be misleading. The possibility of error in this use of averages is considered further in Section 7.4.

If a set of ungrouped data is large and the data entries are repeated several times, the calculation of the mean and median can be simplified by using the frequencies. For the mean, instead of adding each measurement each time it appears, the measurements are multiplied by their frequencies and these products are added. The resulting sum is the same, and the calculations are generally simplified. The frequencies are also used to locate the median. The procedure is illustrated in Example 4.

EXAMPLE 4 Calculate the mean and the median of the set of data given at the beginning of Section 7.1.

Solution The measurements, their frequencies, and the required products are given in the table:

MEASUREMENT	FREQUENCY	FREQUENCY × MEASUREMENT
83	3	249
84	5	420
85	6	510
86	10	860
87	4	348
88	8	704
89	1	89
90	5	450
91	3	273
93	2	186
Total:	47	4,089

2 PRACTICE PROBLEM 2

Calculate the mean and median number of absent employees in Practice Problem 1 of Section 7.1.

Answer

4; 5

The mean is then $\frac{4,089}{47} = 87$. Since there are 47 measurements, the median is the 24th number from the bottom. If we start totaling the frequencies from the bottom, we see that the 24th will be one of the 86's.

②

The calculation of the mean for grouped data is similar to the procedure illustrated in Example 4, except that all of the measurements in an interval are considered to be concentrated at the midpoint of that interval. The midpoints of the intervals are then used in place of the measurements. The procedure is illustrated in Example 5.

EXAMPLE 5 Calculate the mean of the set of grouped data given in Section 7.1.

Solution The intervals, their corresponding frequencies, and other required information are given in the table:

INTERVAL	FREQUENCY	INTERVAL MIDPOINT	FREQUENCY × MIDPOINT
50–54	2	52	104
55–59	3	57	171
60–64	3	62	186
65–69	5	67	335
70–74	4	72	288
75–79	3	77	231
80–84	1	82	82
Total:	21		1,397

3 PRACTICE PROBLEM 3

Calculate the mean number of customer accounts opened each week in Practice Problem 2 of Section 7.1.

Answer

9.6

INTERVAL	FREQUENCY
12–15	2
16–19	3
20–23	4
24–27	3
28–31	4

The mean is then $\frac{1,397}{21} = 66.5$.

③

The calculation of the median of grouped data does not consider all of the entries in an interval to be concentrated at its midpoint. Rather, as we shall see, the entries are considered to be spaced an equal distance apart throughout the interval. Consider, for example, the grouped data listed in the table in the margin. There are 16 measurements; the median is the mean of the eighth and ninth, both of which occur in the third interval. The four measurements in the third interval are considered to be spaced evenly throughout the corresponding interval, 19.5–23.5, on the horizontal scale of the histogram for the data:

The four points divide the interval into five subintervals of equal length. The length of each subinterval is therefore the total length divided by 5, or

$$\frac{23.5 - 19.5}{5} = \frac{4}{5} = 0.8$$

The measurements are therefore considered to be

First: $19.5 + (0.8) = 20.3$
Second: $19.5 + 2(0.8) = 21.1$
Third: $19.5 + 3(0.8) = 21.9$
Fourth: $19.5 + 4(0.8) = 22.7$

The eighth and ninth measurements in the set of data are the third and fourth in this interval. The median is therefore

$$\frac{21.9 + 22.7}{2} = \frac{44.6}{2} = 22.3$$

EXAMPLE 6 Calculate the median of the grouped data of Example 5.

Solution There is an odd number of measurements. The median is the eleventh measurement, which is the third of the five measurements in the interval 65–69. The corresponding interval on the histogram scale is 64.5–69.5. The five measurements divide this interval into six subintervals, each of length

$$\frac{69.5 - 64.5}{6} = \frac{5}{6}$$

The third measurement, which is also the median, is

$$64.5 + 3(\tfrac{5}{6}) = 64.5 + 2.5 = 67$$

■ **4 PRACTICE PROBLEM 4**

Calculate the median number of customer accounts opened each week in Practice Problem 2 of Section 7.1.

Answer

$\frac{81}{8} = 10.125$

Two other measurements of central tendency that are sometimes used—but not as frequently as the two we have considered—are the mode and the midrange. The **mode** is the measurement that occurs most often. In Example 4 the mode is 86, since it has the highest frequency.

The **midrange** is the mean of the largest and smallest measurements. In Example 4 the midrange is

$$\frac{83 + 93}{2} = \frac{176}{2} = 88$$

Neither the mode nor the midrange will be considered in detail here.

EXERCISES

Because of the amount of arithmetic involved in statistics, use of a hand calculator is recommended for the exercises of this chapter, particularly in this and the following sections.

Determine the mean and median for each of the following sets of numbers:

1. 19, 23, 20, 34, 31, 35, 30, 26, 25
2. 75, 83, 96, 102, 87, 105, 92, 98

7.2 MEASURES OF CENTRAL TENDENCY 271

3. 199, 204, 185, 180, 190, 210, 220, 235, 216, 221
4. 23.1, 24.2, 24.7, 24.9, 25.0, 25.3, 25.7, 25.8, 26.1, 26.3, 26.4

TYPE A	TYPE B	TYPE A	TYPE B
46.3	47.9	43.7	42.4
42.1	40.3	42.0	41.3
40.1	39.6	44.0	45.1
38.3	42.6	47.2	42.2
39.2	29.9	44.3	44.1
41.9	40.6	39.8	40.2
39.2	38.1	41.9	41.9
		40.1	41.9

5. *Fertilizer comparison* Two types of fertilizer are tested for yield of corn on 15 pairs of plots of soil. The yield in bushels is given in the table.
   (a) Determine the mean yield for each type of fertilizer.
   (b) On the basis of the means, which type of fertilizer is better?
   (c) Determine the median yield for each type of fertilizer.
   (d) On the basis of the medians, which type of fertilizer is better?

6. *Tire comparison* Two brands of tires were tested simultaneously on the rear wheels of 11 cars. The table indicates (in thousands) the number of miles each car traveled before the tread was reduced to a specified level.
   (a) Determine the mean number of miles traveled for each type of tire.
   (b) On the basis of the means, which type of tire is better?
   (c) Determine the median number of miles traveled for each type of tire.
   (d) On the basis of the medians, which type of tire is better?

TYPE I	TYPE II	TYPE I	TYPE II
40.2	40.1	42.4	41.4
42.3	41.0	39.9	39.7
41.6	42.1	42.8	42.7
43.7	42.5	43.1	42.5
41.9	41.1	44.1	43.2
		43.8	42.8

Calculate the mean and the median of each set of data.

7. Exercise 1 of Section 7.1
8. Exercise 2 of Section 7.1
9. Exercise 3 of Section 7.1
10. Exercise 4 of Section 7.1
11. *Automobile inventory* Calculate the mean and the median number of new cars in Exercise 5 of Section 7.1.
12. *Memory test scores* Calculate the mean and the median scores in Exercise 6 of Section 7.1.
13. *Test scores* Calculate the mean and the median scores in Exercise 7 of Section 7.1.
14. *Pollution study* Calculate the mean and the median number of fish in Exercise 8 of Section 7.1.
15. *Fermentation experiment* Calculate the mean and median number of parts per million of hydrogen sulfide in Exercise 9 of Section 7.1.
16. *Computer CPU time* Calculate the mean and median CPU times in Exercise 10 of Section 7.1.
17. Interpret each of the following statements:
    (a) The median family income in the United States in 1979 was $19,684.
    (b) The mean cost of a 1981 car with typical equipment from one American manufacturer was $9,450.
18. Construct a set of nine numbers that has the same number for its mean, median, and mode. (*Hint:* Start in the "middle.")

Explain the meaning of each term.

19. Averages
20. Mean of a set of numbers
21. Median of a set of numbers
22. Mode of a set of numbers
23. Midrange of a set of numbers

7.3 MEASURES OF DISPERSION

Methods of analyzing the dispersion of data are used to indicate how closely the data centers about an average. Compare the histogram in Figure 7.3 with that given in Figure 7.1. Both represent 47 measurements; the corresponding sets of data each have a mean of 87 and a median of 86. In Figure 7.3, the data cluster more closely about the averages. Correspondingly, the measures of dispersion (or of spread) would be smaller in this case than in the former.

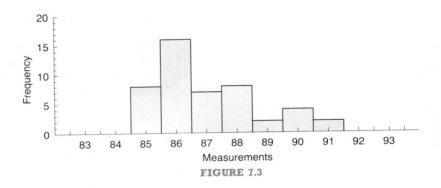

FIGURE 7.3

The standard deviation measures the concentration, or clustering, of a set of numbers about their mean. If one wanted to measure this concentration, a natural approach would be to calculate the difference between each number and the mean and then determine the mean of these differences. But there is a problem involved in this approach! Consider the set

$$\{17, 18, 21, 25, 28, 29, 30, 32\}$$

with mean 25. If we calculate the differences and add them, we get:

$$
\begin{aligned}
17 - 25 &= -8 \\
18 - 25 &= -7 \\
21 - 25 &= -4 \\
25 - 25 &= 0 \\
28 - 25 &= 3 \\
29 - 25 &= 4 \\
30 - 25 &= 5 \\
32 - 25 &= 7 \\
\hline
&0
\end{aligned}
$$

The mean of the difference is $\frac{0}{8} = 0$. For any set of numbers, we would always get the same result. Because of the nature of the mean, the positive differences would always be offset by corresponding negative differences.

7.3 MEASURES OF DISPERSION

Instead, the standard deviation is based on the mean of the *squares* of the differences. The square root of this mean gives the standard deviation. Formally, if $\bar{x}$ denotes the mean of the set of numbers $\{x_1, x_2, \ldots, x_n\}$, **the standard deviation** is the square root of the quotient

$$\frac{(x_1 - \bar{x})^2 + (x_2 - \bar{x})^2 + \cdots + (x_n - \bar{x})^2}{n}$$

The notation for the numerator of this quotient is

$$\sum_{i=1}^{n} (x_i - \bar{x})^2$$

Here, i is called the *index of summation;* the notation indicates that i should take on the values $1, 2, 3, \ldots, n$ in the expression $(x_i - \bar{x})^2$ and the results should be added. With this notation, the standard deviation can be written

$$\sqrt{\frac{\sum_{i=1}^{n} (x_i - \bar{x})^2}{n}}$$

EXAMPLE 1 Determine the standard deviation of the set of numbers

$$\{17, 18, 21, 25, 28, 29, 30, 32\}$$

Solution The mean of the set of numbers is 25.

$$(17 - 25)^2 = (-8)^2 = 64$$
$$(18 - 25)^2 = (-7)^2 = 49$$
$$(21 - 25)^2 = (-4)^2 = 16$$
$$(25 - 25)^2 = 0^2 = 0$$
$$(28 - 25)^2 = 3^2 = 9$$
$$(29 - 25)^2 = 4^2 = 16$$
$$(30 - 25)^2 = 5^2 = 25$$
$$(32 - 25)^2 = 7^2 = 49$$

$$\sum_{i=1}^{8} (x_i - 25)^2 = 64 + 49 + 16 + 0 + 9 + 16 + 25 + 49 = 228$$

The standard deviation is, therefore,

$$\sqrt{\tfrac{228}{8}} = \sqrt{28.5}$$

which is approximately 5.3. ∎

A hand calculator is particularly helpful for determining square roots, such as the value of $\sqrt{28.5}$ in Example 1. On a calculator, $\boxed{\sqrt{x}}$ is a common notation for indicating a square root button.

EXAMPLE 2 The set of numbers below has the same mean as does the set in Example 1:

$$\{1, 7, 12, 25, 28, 35, 43, 49\}$$

Determine the standard deviation and compare with that of Example 1.

Solution

$$(1 - 25)^2 = (-24)^2 = 576$$
$$(7 - 25)^2 = (-18)^2 = 324$$
$$(12 - 25)^2 = (-13)^2 = 169$$
$$(25 - 25)^2 = 0^2 = 0$$
$$(28 - 25)^2 = 3^2 = 9$$
$$(35 - 25)^2 = 10^2 = 100$$
$$(43 - 25)^2 = 18^2 = 324$$
$$(49 - 25)^2 = 24^2 = 576$$
$$\sum_{i=1}^{8}(x_i - 25)^2 = 2{,}078$$

The standard deviation is, therefore,

$$\sqrt{\tfrac{2{,}078}{8}} = \sqrt{259.75}$$

which is approximately 16.1.

The greater standard deviation of this example as compared with that of Example 1 is accounted for by the greater distances of the numbers from the mean. ∎

The position of one of the numbers in a set of data relative to the whole set can be indicated not only by whether the number is above or below the mean but more exactly by determining how many standard deviations above or below.

1 PRACTICE PROBLEM 1

Determine the standard deviation: 22, 24, 26, 29, 32, 34, 36, 37.

Answer

$$\sqrt{\tfrac{222}{8}} \approx 5.27$$

EXAMPLE 3 Suppose the numbers of the set of data of Example 1 are the scores made by a class of eight students on a 35-question, true-false examination. Susan, one of the eight students, had the score of 30. How did she do relative to the rest of the class?

Solution Susan's score of 30 indicates that she scored 5 points above the mean. Her score is $\tfrac{5}{5.3}$, or approximately 1, standard deviation above the mean. ∎

EXAMPLE 4 Suppose that the numbers in the set of data of Example 2 are the scores that the same class of eight students (Example 3) scored on a subsequent true-false examination of 50 questions. On this test Susan scored 35. Did she do better on the second test than she did on the first, relative to the whole class?

7.3 MEASURES OF DISPERSION

Solution Since she scored 10 points above the mean on the second test, she scored $\frac{10}{16.1}$, or approximately $\frac{5}{8}$, standard deviation above the mean. As compared to the 1 standard deviation above the mean of the first test, $\frac{5}{8}$ standard deviation above the mean of the second test indicates that Susan did worse on the second test relative to the entire class. ∎

Just as a set of data is divided into halves by the median, it is divided into fourths by the quartiles. These, like the median, are positions determined by counting. The ***first quartile*** (Q_1) of the set of data $\{x_1, x_2, \ldots, x_n\}$ is a number such that one-fourth of the measurements are less than or equal to this number. The ***third quartile*** (Q_3) is a number such that one-fourth of the measurements are greater than or equal to this number. The ***second quartile*** is the median itself.

The difference between the third and first quartiles, $Q_3 - Q_1$, is called the ***interquartile range*** and measures the spread of the data about the median.

EXAMPLE 5 Determine the interquartile range of the set of numbers of Example 1:

$$\{17, 18, 21, 25, 28, 29, 30, 32\}$$

Solution Since there are eight numbers in the set and $\frac{1}{4} \times 8 = 2$, the first quartile should be the second number (since they are already given in increasing order): Q_1 is therefore 18. Similarly, Q_3 should be the second-to-the-last number, or 30. The interquartile range is $Q_3 - Q_1 = 30 - 18$, or 12. ∎

EXAMPLE 6 The set of numbers of Example 2,

$$\{1, 7, 12, 25, 28, 35, 43, 49\}$$

has the same median (26.5) as does that of Example 5. Determine its interquartile range and compare it with that of Example 5.

Solution $Q_1 = 7$ and $Q_3 = 43$, so that the interquartile range is $43 - 7 = 36$. The much larger result here as compared with that of 12 in Example 5 reflects the fact that the numbers in this case are much more spread out than in the former. ∎

2 PRACTICE PROBLEM 2

Determine the interquartile range of the data in Practice Problem 1.

Answer

$$36 - 24 = 12$$

When the median is used as an average, the position of a number in a set of data can be indicated not only by determining whether the number is above the median but also by whether the number is above or below the third quartile. If the number is below the median, it is more informative to know also whether the number is above or below the first quartile.

Susan (Examples 3 and 4) scored above the median in each test—and at the third quartile on the first and below it on the second. On this basis as well, she did better on the first test relative to the entire class.

Some common uses of the standard deviation and quartiles have been indicated in the examples of this section:

1. The spread of data about the mean or median, as in Examples 1 and 5.
2. A comparison of the spread of the data about the mean or median of two or more sets of data, as in Examples 2 and 6.
3. The relative position of one member of a set of data, as in Example 3.
4. The relative positions of corresponding members of two or more sets of data are compared, as in Example 4 and in the discussion of the last paragraph.

As an illustration of the application of mathematics, the measures of dispersion provide mathematical models to answer questions of how widely the data is spread about the mean or median. The results are interpreted in the manner indicated in the previous examples.

Just as in the calculation of the mean, the calculation of the standard deviation and quartiles can be simplified by use of frequencies when the set of data is large and/or the measurements are repeated a number of times. In calculating the standard deviation, the square of the difference for each measurement is multiplied by its frequency and the results are added. This procedure for ungrouped data is illustrated in Example 7.

EXAMPLE 7 Calculate the standard deviation of the set of data given in Example 4 of Section 7.2.

Solution   The mean of the set of data is 87. The quantities required for the calculation of the standard deviation are given in tabular form:

MEASUREMENT	FREQUENCY	DISTANCE FROM MEAN	FREQUENCY × SQUARE OF DISTANCE FROM MEAN
83	3	−4	3 × 16 = 48
84	5	−3	5 × 9 = 45
85	6	−2	6 × 4 = 24
86	10	−1	10 × 1 = 10
87	4	0	4 × 0 = 0
88	8	1	8 × 1 = 8
89	1	2	1 × 4 = 4
90	5	3	5 × 9 = 45
91	3	4	3 × 16 = 48
93	2	6	2 × 36 = 72
Total:	47		Sum: 304

The standard deviation is $\sqrt{\frac{304}{47}}$ or approximately 2.5. ■

Since there are 47 measurements in the set of data of Example 7, the determination of the quartiles results in the number

$$\tfrac{1}{4} \times 47 = 11.75$$

In such situations the number is rounded to the nearest integer. In this case, Q_1 is the twelfth number from the bottom, or 85, and Q_3 is the twelfth from the top, or 88. The interquartile range is, therefore, $88 - 85 = 3$.

The one exception to the rule just given for calculating quartiles occurs when the decimal involved is 0.5. In this case, the mean of the numbers above and below is used. For example, the set

$$\{2, 7, 8, 12, 15, 20, 22, 25, 27, 30\}$$

has 10 measurements. Since $\tfrac{1}{4} \times 10 = 2.5$, the first quartile is given by the mean of the second and third measurements:

$$Q_1 = \frac{7 + 8}{2} = 7.5$$

Similarly, the third quartile is given by the mean of the second and third largest measurements:

$$Q_3 = \frac{25 + 27}{2} = 26$$

The method of calculating the standard deviation for grouped data is essentially the same as that for a large set of ungrouped data (Example 7), except that the interval midpoints are used in place of the measurements. The procedure is illustrated in Example 8.

3 PRACTICE PROBLEM 3

Calculate the standard deviation and interquartile range for the data of Practice Problem 1 of Section 7.1.

Answer

2; 4

EXAMPLE 8 Calculate the standard deviation for the set of data in Example 5 of Section 7.2.

Solution In Example 5, the mean was found to be 66.5. The quantities required to calculate the standard deviation are given in tabular form:

INTERVAL	FREQUENCY	INTERVAL MIDPOINT	DISTANCE OF MIDPOINT FROM MEAN	FREQUENCY × SQUARE OF DISTANCE FROM MEAN
50–54	2	52	−14.5	2 × 210.25 = 420.50
55–59	3	57	−9.5	3 × 90.25 = 270.75
60–64	3	62	−4.5	3 × 20.25 = 60.75
65–69	5	67	0.5	5 × 0.25 = 1.25
70–74	4	72	5.5	4 × 30.25 = 121.00
75–79	3	77	10.5	3 × 110.25 = 330.75
80–84	1	82	15.5	1 × 240.25 = 240.25
Total:	21			Sum: 1,445.25

The standard deviation is

$$\sqrt{\frac{1{,}445.25}{21}} \approx \sqrt{68.82}$$

or approximately 8.3. ∎

When determining the quartiles for grouped data, the entries are considered to be spaced evenly throughout the intervals, as they were in the determination of the median. For the data in Example 8, $\frac{1}{4} \times 21 = 5.25$, so that Q_1 is the third measurement in the interval 55–59. In this interval the measurements are considered to be spaced

$$\frac{59.5 - 54.5}{4} = \frac{5}{4} = 1.25$$

units apart. The third measurement is $54.5 + 3(1.25) = 58.25$.

Similarly, Q_3—the fifth measurement from the top—is the fourth measurement in the interval 70–74. In this interval the measurements are considered to be spaced

$$\frac{74.5 - 69.5}{5} = \frac{5}{5} = 1$$

unit part. The fourth measurement is $69.5 + 4(1) = 73.5$. The interquartile range is

$$Q_3 - Q_1 = 73.5 - 58.25 = 15.25$$

You are probably more familiar with the term *percentile* than with *quartile*. Percentiles divide a set of data into hundredths instead of fourths. One can define an interpercentile range and locate a particular number in a set of data using percentiles in much the same way we did with quartiles. You have probably seen this done in the interpretation of standardized tests. However, the sets of data with which we are concerned in this text are not large enough to be divided into hundredths.

A final measure of dispersion for a set of data is the difference between its largest and smallest numbers. This difference is called its **range**. For the set

$$\{2, 7, 8, 12, 15, 20, 22, 25, 27, 30\}$$

the range is $30 - 2 = 28$. The range measures the spread of the data independent of any measure of central tendency or average.

4 PRACTICE PROBLEM 4

Calculate the standard deviation and interquartile range for the data of Practice Problem 2 of Section 7.1.

Answer

4.85; 8.66

EXERCISES

Determine the standard deviation and the interquartile range of the following data from Exercises 1–4 of Section 7.2:

1. 19, 23, 20, 34, 31, 35, 30, 26, 25
2. 199, 204, 185, 180, 190, 210, 220, 235, 216, 221
3. 75, 83, 96, 102, 87, 105, 92, 98
4. 23.1, 24.2, 24.7, 24.9, 25.0, 25.3, 25.7, 25.8, 26.1, 26.3, 26.4
5. Determine the standard deviation, the interquartile range, and the range of the measurements

    82, 62, 81, 63, 80, 64, 66, 78, 68, 75, 73, 72
6. Determine the standard deviation, the interquartile range, and the range of the set of numbers

    87, 86, 85, 85, 84, 81, 78, 77, 75, 75, 74, 73
7. Suppose that the data of Exercises 5 and 6 are the scores that a class of 12 students made on two different tests. Suppose Joan scored 66 on the first test (Exercise 5) and 74 on the second.
    (a) Was Joan's score within 1 standard deviation of the mean on the first test? On the second test?
    (b) On the basis of part (a), did Joan do better on the first or on the second test, relative to the entire class?
8. Suppose that grades are given on the following basis:

    A: Scores greater than 2 standard deviations above the mean
    B: Scores between 1 and 2 standard deviations above the mean
    C: Scores in the range from 1 standard deviation above to 1 standard deviation below the mean
    D: Scores between 1 and 2 standard deviations below the mean
    F: Scores less than 2 standard deviations below the mean

    On this basis, what grade did Joan (Exercise 7) get on the first test? On the second test?

Calculate the standard deviation, the interquartile range, and the range of the set of data given in the indicated exercise.

9. Exercise 1 of Section 7.1
10. Exercise 2 of Section 7.1
11. Exercise 3 of Section 7.1
12. Exercise 4 of Section 7.1

Calculate the standard deviation and the interquartile range for the set of grouped data of the indicated exercise.

13. *Automobile inventory* Exercise 5 of Section 7.1
14. *Memory test scores* Exercise 6 of Section 7.1
15. *Test scores* Exercise 7 of Section 7.1
16. *Pollution study* Exercise 8 of Section 7.1

17. *Fermentation experiment*   Exercise 9 of Section 7.1
18. *Computer CPU time*   Exercise 10 of Section 7.1
19. Verify that addition of the difference of the numbers in Exercise 1 from their mean gives a sum of 0.

It can be shown algebraically (see Exercise 30) that

$$\frac{\sum_{i=1}^{n}(x_i - \bar{x})^2}{n} = \frac{\sum_{i=1}^{n} x_i^2 - n\bar{x}^2}{n}$$

Therefore, an alternate formula for calculating the standard deviation is

$$\sqrt{\frac{\sum_{i=1}^{n} x_i^2 - n\bar{x}^2}{n}}$$

For example, for the set of data $\{2, 7, 8, 10, 12\}, \bar{x} = 8$ *and*

$$\sum_{i=1}^{5} x_i^2 = 9 + 49 + 64 + 100 + 144 = 366$$

The standard deviation is, therefore,

$$\sqrt{\frac{366 - 5(64)}{5}} = \sqrt{\frac{46}{5}}$$

Use the above alternate formula to calculate the standard deviation of the indicated set of data.

20. $\{1, 5, 7, 10, 12\}$ 21. Exercise 1   22. Exercise 2

Explain the meaning of each term.

23. Standard deviation of a set of data
24. First quartile of a set of data
25. Third quartile of a set of data
26. Interquartile range of a set of data
27. Range of a set of data

Explain the purpose of each of the following:

28. Measures of central tendency
29. Measure of dispersion
30. *Bonus problem*   The following exercises in the use of summation notation lead to the verification of the formula for the standard deviation given before Exercise 20.
    (a) Verify that

$$\bar{x} = \frac{\sum_{i=1}^{n} x_i}{n}$$

(b) Verify that $\sum_{i=1}^{n}(cx_i) = c\left(\sum_{i=1}^{n}x_i\right)$, where c is a constant or an expression whose value does not depend on i.

(c) Verify that $\sum_{i=1}^{n} c = nc$, where c is the same as in part (b).

(d) Using parts (a), (b), and (c), verify that the formula for the standard deviation given before Exercise 20 is valid. [*Hint:* Start by expanding the expression $(x_i - \bar{x})^2$.]

7.4 THE SIGN TEST

The material of the previous sections is part of descriptive statistics; the object was to describe a set of collected data. As was pointed out at the beginning of this chapter, the central problem in statistics is the determination of the conclusions that are valid from the data. This aspect of statistics is called *inferential statistics*. The purpose of this section is to examine one of the methods used in establishing a conclusion based on a set of data.

In inferential statistics, the set of data is usually a sample drawn from a much larger collection, called the *population*. The object is to arrive at a conclusion concerning the total population on the basis of the sample. One of the most familiar examples of this procedure is the forecasting of the results on election nights by the television networks. Precincts representative of all the voters in a state are selected. On the basis of the voting results in the selected precincts, the outcome is forecast for the whole state. In this case, the voters in the selected precincts form the sample; all the voters in the state form the population.

One of the first requirements, of course, is that the sample be truly representative of the whole population. The study of correct procedures for choosing a sample is the subject of a branch of statistics called *sampling theory*. We shall not go into this aspect of statistics here.

Suppose now that a company plans to buy a new computer and has reduced its choices down to a final two. It is the company's intention to purchase the faster of these two computers; both cost about the same. In order to compare the speeds of the two, their computer operator runs 15 fairly substantial programs on each machine. The time, in seconds, to run each program is given in Table 7.3. (The given data is the sample; the run times of all programs during the lives of the respective machines is the population.)

The lower mean and median of computer B would indicate that computer B is to be preferred. But is this difference due to chance? If 15 different programs were chosen to compare the machines, it is conceivable that the results would have indicated that computer A is preferable to B.

The operator maintains that computer B is, in fact, the faster of the two and recommends it for purchase. The supervisor, who has been

TABLE 7.3

COMPUTER A	COMPUTER B
930	880
940	930
890	920
890	870
900	850
880	870
930	950
910	890
940	910
930	890
910	900
900	920
920	890
930	900
880	870
Mean: 912	895
Median: 910	890

wined and dined by the sales representative for computer A, maintains otherwise. The supervisor maintains that the difference in the trial runs is due to chance and that if the trial runs could be extended, the results might well be reversed.

However, the agreement with each manufacturer was to run 15 trial programs. The management consults a statistician to determine whether or not the data in the table indicate a real difference in the long range performance of the two machines. The statistician bases his decision on the following *sign test*.

The difference in the time for the two machines for each program is determined:

$$
\begin{aligned}
930 - 880 &= 50 \quad + \\
940 - 930 &= 10 \quad + \\
890 - 920 &= -30 \quad - \\
890 - 870 &= 20 \quad + \\
900 - 850 &= 50 \quad + \\
880 - 870 &= 10 \quad + \\
930 - 950 &= -20 \quad - \\
910 - 890 &= 20 \quad + \\
940 - 910 &= 30 \quad + \\
930 - 890 &= 40 \quad + \\
910 - 900 &= 10 \quad + \\
900 - 920 &= -20 \quad - \\
920 - 890 &= 30 \quad + \\
930 - 900 &= 30 \quad + \\
880 - 870 &= 10 \quad +
\end{aligned}
$$

Here, a plus sign indicates that the corresponding difference is greater than 0; a minus sign indicates that the corresponding difference is less than 0. If there is no real difference in the speed of the two machines, the probability of a plus or a minus for any program ought to be the same; that is, each should have a probability of $\frac{1}{2}$. In this case, there are 12 pluses and 3 minuses. The sign test considers the larger of the two — in this case the 12 plus signs. What is the probability of 12 or more plus signs when the probability of a plus for any program is $\frac{1}{2}$?

For 15 programs, the following probabilities hold:*

The probability of 12 plus signs is 0.0139.

The probability of 13 plus signs is 0.0032.

The probability of 14 plus signs is 0.005.

The probability of 15 plus signs is 0.0001.

---

* These probabilities are calculated using binomial probabilities of Section 6.1. However, we do not have to calculate these probabilities to use the sign test.

Since getting (exactly) 12, 13, 14, or 15 plus signs are mutually exclusive events, the probability of getting 12 or more is their sum:

$$0.0139 + 0.0032 + 0.0005 + 0.0001 = 0.0177$$

Since the probability of obtaining a value in this range (12 to 15) is so small when plus has a probability of $\frac{1}{2}$, we can conclude that plus actually has a probability greater than $\frac{1}{2}$. But if plus, in fact, does have a probability greater than $\frac{1}{2}$, then computer A can be expected to have the longer run times over the lives of the two machines!

In general, the sign test begins with the assumption that there is no difference in the objects being compared (the run times of the two machines in the above illustration), so that plus and minus each have a probability of $\frac{1}{2}$. Then the decision to accept or reject this assumption depends upon the number of plus and minus signs obtained in the sample. If the greater of these two numbers lies in a range with very small probability, the assumption of no difference is rejected; otherwise, the assumption is accepted. If the assumption is rejected, then one of the objects is considered to be better than the other.

Two questions arise. How small is a "small probability," and how much confidence can we have in our decision? We shall return to these questions after some further illustrations.

EXAMPLE 1 A company has developed an additive for its insect repellent and tested the repellent both with and without the additive in 14 different houses. A record was kept of the number of days each was effective; this record is given in the table. Use the sign test to determine whether or not the data indicate that the repellent with the additive is better.

WITH	WITHOUT	WITH	WITHOUT
139	130	158	151
164	160	148	144
145	138	160	157
160	170	130	133
183	183	148	142
151	159	150	160
149	145	164	171

Solution Taking the differences of the data for each of the 14 houses above yields the following list of plus and minus signs:

```
  +       +
  +       +
  +       +
  −       −
  0       +
  −       −
  +       −
```

There are 8 pluses and 5 minuses; *a difference of 0 is ignored*. What significance should we attach to the 8 pluses, the greater of the two numbers?

We assume first that there is no difference between the repellent with the additive and the repellent without the additive. Then the probability of plus or minus at any house is $\frac{1}{2}$. Suppose that we will reject the assumption of no difference *only if the probability of 8 or more pluses is less than 0.05*.

In a total of 13 houses, the following probabilities holds:

The probability of 13 pluses is 0.0001.
The probability of 12 pluses is 0.0016.
The probability of 11 pluses is 0.0095.
The probability of 10 pluses is 0.0349.
The probability of 9 pluses is 0.0873.
The probability of 8 pluses is 0.1571.

Then:

The probability of 12 or 13 pluses is 0.0016 + 0.0001 = 0.0017.

The probability of 11, 12, or 13 pluses is 0.0095 + 0.0016 + 0.0001 = 0.0112.

The probability of 10, 11, 12, or 13 pluses is 0.0349 + 0.0095 + 0.0016 + 0.0001 = 0.0461.

The probability of 9, 10, 11, 12, or 13 pluses is 0.0873 + 0.0349 + 0.0095 + 0.0016 + 0.0001 = 0.1334.

In the same way, the probability of 8, 9, 10, 11, 12, or 13 pluses is the sum of their individual probabilities, or 0.2905.

Since we decided to *reject* the assumption of no difference only if this last probability was less than 0.05, we *accept* the assumption of no difference and conclude that the additive does not improve the effectiveness of the repellent. ∎

In Example 1, observe that the probability of 10 or more pluses is less than 0.05 and the probability of 9 or more is greater than 0.05—that is, the transition from less than 0.05 to greater than 0.05 is made when going from 10 or more to 9 or more. The smallest number of pluses for which we would have rejected the assumption of no difference is 10. The diagram indicates the number of pluses (or of minuses, as the case may be) for which the assumption of no difference is rejected and those for which it is accepted when 13 comparisons are counted.

The smallest number in a rejection region is called the ***critical value.*** The critical value varies with the number of comparisons counted; in the last case the critical value was 10.

In Table 7.4, n denotes the number of comparisons counted—that is, the sum of the number of pluses and the number of minuses. (Remember that 0's are ignored.) The table gives the critical value for n up to 47.

TABLE 7.4 CRITICAL VALUES FOR THE SIGN TEST (0.05 LEVEL)

n	CRITICAL VALUE	n	CRITICAL VALUE	n	CRITICAL VALUE
6	6	20	15	34	23
7	7	21	15	35	23
8	7	22	16	36	24
9	8	23	16	37	24
10	9	24	17	38	25
11	9	25	18	39	26
12	10	26	18	40	26
13	10	27	19	41	27
14	11	28	19	42	27
15	12	29	20	43	28
16	12	30	20	44	28
17	13	31	21	45	29
18	13	32	22	46	30
19	14	33	22	47	30

All that is required to use the sign test is to determine whether the number of pluses or the number of minuses (whichever is greater) is less than the appropriate critical value for acceptance or greater than or equal to the critical value for rejection.

In Example 1, $n = 8 + 5 = 13$. The greater number of signs is the 8 pluses and the critical value is 10. Since 8 is less than 10, the assumption of no difference is accepted.

EXAMPLE 2 Two classes of 34 students each are taught mathematics by one of two different methods. The students are paired by their scores on a mathematics aptitude test. Taking difference of scores yields 9 minuses and 25 pluses. Should the assumption of no difference in the methods be accepted or rejected?

Solution The greater number of signs is the 25 pluses; $n = 9 + 25 = 34$. For $n = 34$, the critical value in the table is 23. Since there are more than 23 pluses, we reject the assumption that there is no difference in the results of the two methods. The data indicate that one method is superior to the other. ∎

EXAMPLE 3 In Example 2, suppose that taking differences of scores yields 18 minuses and 16 pluses. Should the assumption of no difference in the methods be accepted or rejected?

Solution Since the greater number of signs is 18, which is less than the critical value of 23, the assumption of no difference in methods is accepted. **1**

The sign test is applicable only when the entries being compared in the sample would be expected to be the same when there is no difference in the objects being evaluated. In Example 2, the students are paired by mathematical aptitude; if they were paired by chance, the difference in scores would indicate nothing as far as the superiority of either method is concerned. Similarly, the two computers in the illustration were treated as equally as possible. The same programs were run on each computer; if different programs were used on each, the results would not indicate anything concerning their relative speeds.

We return now to the question of what confidence we can have in a decision made in this way. Note that when using the table, the probability of obtaining a value in the critical (or rejection) region when the assumption of no difference is true is 0.05. (The *size* of the critical region is said to be 0.05.) Consequently, the probability of rejecting this assumption when, in fact, it is true is 0.05. That is, an error of this type can be expected to occur about 5% of the time in the long run.

EXAMPLE 4 Thirty-five students each perform the following experiment with the same coin in order to determine whether or not the coin is biased. (The coin is known to be unbiased, but the students are not told this.)

The coin is tossed 40 times by each student. Each keeps a record of his or her results by recording a 1 under heads and a 0 under tails whenever a head is obtained, and vice-versa when a tail is obtained. The record shown in the margin indicates two heads followed by a tail. When the differences are taken, a plus corresponds to obtaining a head; a minus corresponds to obtaining a tail.

Thirty-four of the students obtain a greater number of pluses or minuses in the 20 to 25 range. Since the critical value for $n = 40$ is 26, each of these 34 students accept the assumption of no difference in the number of heads and tails to be expected in the long run.

HEADS	TAILS
1	0
1	0
0	1
⋮	⋮

---

1 **PRACTICE PROBLEM 1**

A police department is testing new .38-caliber pistols made by two manufacturers. Twelve officers fired each of the two; their scores are given in the table.

(a) Do the data indicate that one pistol is more accurate than the other?

(b) What would have been the conclusion if there had been 9 pluses and 2 minuses?

PISTOL A	PISTOL B
41	38
37	39
43	40
37	35
42	40
38	38
39	41
37	39
40	38
41	38
39	36
39	43

Answer

(a) No

(b) One pistol is more accurate.

The 35th student obtains 27 minuses and 13 pluses. He therefore rejects the above assumption!

But note that all 35 students have made the *statistically correct* decision. The last one happens to be erroneous. He rejected an assumption which happened to be true. ∎

We can vary the likelihood of this type of error by varying the size of the critical region. Some common sizes are 0.01, 0.025, 0.05, and 0.10. The risk of the type of error we have been discussing changes accordingly. For the table, the size 0.05 was chosen arbitrarily, except for the fact that it is one of those commonly used.

There is also the possibility of *accepting the assumption of no difference when, in fact, this assumption is false*. If the coin in Example 4 had been known to be biased, the 34 students would have made this type of error—even though their decision was *statistically correct*.

An analysis of this second type of error is beyond the scope of this text. In practice, the statistician chooses the size of the critical region according to the risk he or she is willing to take of committing each of the two types of errors indicated and the relative importance of either type of error in a particular application.

EXERCISES

In each of the following exercises, use the sign test with critical region of size 0.05.

1. *Mileage comparison* The table gives the miles per gallon obtained by 15 drivers using two types of spark plugs. Determine if the results indicate either type of plug to be superior.

TYPE A	TYPE B	TYPE A	TYPE B	TYPE A	TYPE B
14.3	13.9	15.4	16.0	15.8	15.3
16.7	16.5	17.3	16.9	16.2	16.2
15.3	14.9	20.3	20.1	13.8	13.1
16.0	16.9	19.3	18.8	20.1	18.9
14.9	14.3	14.9	14.7	18.3	17.5

2. *Program effectiveness* The weights of 17 people before and after they participated in a 1-month weight reducing program are listed in the table. Determine whether or not the data indicate that the plan is effective.

BEFORE	AFTER	BEFORE	AFTER	BEFORE	AFTER
144	138	166	180	151	148
198	190	222	212	158	161
139	130	150	146	138	141
193	193	176	174	132	130
160	164	120	130	211	202
149	145	170	166		

3. *Fertilizer comparison* Determine if either of the fertilizers in Exercise 5, Section 7.2, is superior.

4. *Tire comparison* Determine if either type of tire in Exercise 6, Section 7.2, is superior.

5. *Gasoline comparison* A car rental company tested the mileage obtained by 20 of its medium-size cars, randomly chosen, using high octane and normal octane gasoline. The miles per gallon using each type of gasoline are given, by car, in the table. Do the results indicate that either type of gasoline gives higher mileage for their medium-size cars?

HIGH	NORMAL	HIGH	NORMAL	HIGH	NORMAL
27.4	26.8	26.8	28.2	31.5	29.9
29.3	30.1	30.1	29.2	29.8	29.1
28.7	27.8	27.3	25.8	28.4	30.0
26.4	25.9	26.3	27.7	27.4	26.9
28.4	24.0	24.2	26.0	25.9	25.3
29.7	29.2	31.5	30.3	32.1	30.8
32.0	31.5	28.2	27.8		

6. Use the Super Bowl victories from 1971 (after the formation of the present National Football League in 1970) to 1984 to test the claim that the National and American Football Conferences are equal in their abilities to win.

SUPER BOWL WINNERS

1971	AFC	1976	AFC	1981	AFC
1972	NFC	1977	AFC	1982	NFC
1973	AFC	1978	NFC	1983	NFC
1974	AFC	1979	AFC	1984	AFC
1975	AFC	1980	AFC		

7. *Diet effectiveness* A diet designed to lower a person's cholesterol level was given to 18 individuals. The table gives the cholesterol concentration (in milligrams of cholesterol per 100 cubic centimeters of blood) for each person before and after the diet. Use the sign test to determine whether the results indicate that the diet is effective.

BEFORE	AFTER	BEFORE	AFTER	BEFORE	AFTER
210	180	169	155	204	183
193	176	198	174	207	193
178	183	174	157	188	168
185	169	220	198	234	207
221	202	169	189	196	210
165	193	183	165	232	218

8. *Wine preference*   Forty-five people chosen at random tasted two different brands of Chianti wine. Of the 45:

   27 preferred the domestic brand.
   15 preferred the imported brand.
   3 had no preference.

   Determine whether the results indicate that the domestic brand is more appealing to the general public.

9. *Political preference*   Of 52 randomly chosen voters in a small city:

   32 preferred the Republican candidate for mayor.
   15 preferred the Democratic candidate.
   5 had no preference.

   Do the results indicate that either candidate has the majority of the votes of all the voters in the city?

10. *Program effectiveness*   Forty people took a speed-reading course. Comparing reading speeds at the beginning and end of the course indicates that 35 improved their reading speeds; for the rest, their reading speeds actually decreased. Do the results indicate that the course is effective?

11. *Flavor preference*   A pharmaceutical company is about to introduce a new drug to the market. The company is trying to decide whether to use mint or cinnamon flavoring to disguise the taste of the drug. Of 60 randomly chosen individuals:

    26 preferred the mint flavor.
    18 preferred the cinnamon flavor.
    16 had no preference.

    What do the results indicate concerning which flavoring to use?

12. *Advertising*   "Homemakers prefer brand ABC dishwashing soap." If 35 randomly chosen homemakers were consulted concerning their dishwashing soap preferences, how many would have to indicate their preference for brand ABC in order to justify the claim?

7.5 THE NORMAL CURVE

While there are many curves used in statistics, the bell-shaped, or normal, curve shown in Figure 7.4 is by far the most important. Its use as a model for the distribution of so many various quantities is consistent with its title, the *normal* curve.

The determination of probabilities using the normal curve is based on the area under the curve. The total area under the curve is 1 unit and the probability of obtaining a value between two numbers a and b

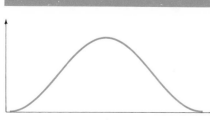

FIGURE 7.4

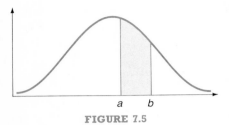

FIGURE 7.5

on the axis at the base of the curve is the same as the area under the curve indicated in Figure 7.5.

Quantities for which the probabilities are determined in this manner are said to have a **normal distribution.** Examples of quantities having a normal distribution will be given in the illustrations and exercises of this section. Our immediate objective is to examine the characteristics of the normal curve and the process of determining probabilities using this curve.

Distributions such as the normal also have a mean and standard deviation. The manner in which these are defined and calculated is entirely different from that for sets of data. However, their purposes are similar—to indicate the center of the distribution and the spread of the distribution about the center. We shall use the standard notation: the Greek letters μ (mu) for the mean and σ (sigma) for the standard deviation.

We shall not examine the manner in which the mean and standard deviation of a normal distribution are calculated, as these are fairly complicated. It is sufficient for our purposes to realize these two characteristics:

1. The mean is the balance point of the distribution:

Half of the area under the curve lies to the right of the mean and half to the left. It is also that value at which the normal curve peaks.

2. In terms of the standard deviation, the area under the curve is distributed approximately in the manner indicated in the diagram.

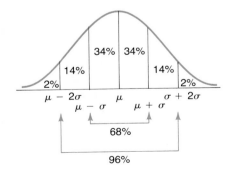

7.5 THE NORMAL CURVE

If the possible outcomes of an experiment have a normal distribution, the probability of obtaining an outcome within 1 standard deviation of the mean is, therefore, 0.68; the probability of an outcome within 2 standard deviations is 0.96. For example, suppose that the life of a particular type of machine is normally distributed with a mean of 5 years and standard deviation of 6 months. Then 68% of these machines have a life of from $4\frac{1}{2}$ to $5\frac{1}{2}$ years, and the probability that any one of these machines will have a life span from $4\frac{1}{2}$ to $5\frac{1}{2}$ years is 0.68.

Figure 7.6(a) depicts several normal curves with the *same mean* but *different standard deviations*. In Figure 7.6(b), these relationships are reversed; the curves have the *same standard* deviation but *different means*. In Figure 7.6(a), the area of the higher curve is closer to the center; of the three curves, it has the smallest standard deviation. The normal curve with mean 0 and standard deviation 1 is called the **standard normal curve.** The reasons that this particular curve is given special attention will become obvious shortly.

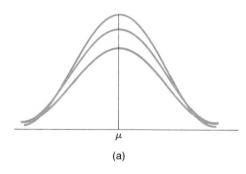
(a)

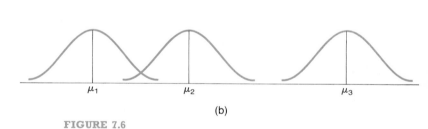
(b)

FIGURE 7.6

Suppose now that z denotes a quantity having the standard normal distribution with mean 0 and standard deviation 1. In order to calculate the probability of z taking a value between 0 and, say 1.31, it is necessary to determine the area indicated in Figure 7.7.

FIGURE 7.7

1 PRACTICE PROBLEM 1

If z has the standard normal distribution, what is the probability that z takes a value between 0 and 1.43?

Answer

0.4236

Areas such as this have been calculated for various values of z, some of which are given in Table II of Appendix B. From the table, part of which is reproduced on page 292, we see that the desired probability for $z = 1.31$ is 0.4049.

PORTION OF TABLE II: AREAS UNDER THE STANDARD NORMAL CURVE

z	0.00	0.01	0.02	0.03	0.04	0.05	0.06	0.07	0.08	0.09
0.0	0.0000	0.0040	0.0080	0.0120	0.0160	0.0199	0.0239	0.0279	0.0319	0.0359
0.1	0.0398	0.0438	0.0478	0.0517	0.0557	0.0596	0.0636	0.0675	0.0714	0.0753
0.2	0.0793	0.0832	0.0871	0.0910	0.0948	0.0987	0.1026	0.1064	0.1103	0.1141
0.3	0.1179	0.1217	0.1255	0.1293	0.1331	0.1368	0.1406	0.1443	0.1480	0.1517
0.4	0.1554	0.1591	0.1628	0.1664	0.1700	0.1736	0.1772	0.1808	0.1844	0.1879
0.5	0.1915	0.1950	0.1985	0.2019	0.2054	0.2088	0.2123	0.2157	0.2190	0.2224
0.6	0.2257	0.2291	0.2324	0.2357	0.2389	0.2422	0.2454	0.2486	0.2517	0.2549
0.7	0.2580	0.2611	0.2642	0.2673	0.2704	0.2734	0.2764	0.2794	0.2823	0.2852
0.8	0.2881	0.2910	0.2939	0.2967	0.2995	0.3023	0.3051	0.3078	0.3106	0.3133
0.9	0.3159	0.3186	0.3212	0.3238	0.3264	0.3289	0.3315	0.3340	0.3365	0.3389
1.0	0.3413	0.3438	0.3461	0.3485	0.3508	0.3531	0.3554	0.3577	0.3599	0.3621
1.1	0.3643	0.3665	0.3686	0.3708	0.3729	0.3749	0.3770	0.3790	0.3810	0.3830
1.2	0.3849	0.3869	0.3888	0.3907	0.3925	0.3944	0.3962	0.3980	0.3997	0.4015
1.3	0.4032	(0.4049)	0.4066	0.4082	0.4099	0.4115	0.4131	0.4147	0.4162	0.4177
1.4	0.4192	0.4207	0.4222	0.4236	0.4251	0.4265	0.4279	0.4292	0.4306	0.4319
1.5	0.4332	0.4345	0.4357	0.4370	0.4382	0.4394	0.4406	0.4418	0.4429	0.4441

When using Table II, it is essential to keep in mind that the area given is *always the area under the curve from 0 to z,* as indicated in Figure 7.8.

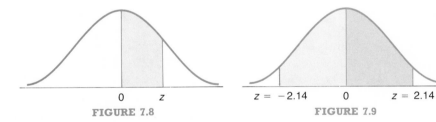

FIGURE 7.8   FIGURE 7.9

Only positive values of z are given in Table II. However, the probabilities for negative values of z can also be determined from the table because the curve is symmetric about 0. For example, the area under the standard normal curve from -2.14 to 0 is the same as that from 0 to 2.14 (see Figure 7.9). From the table, then, the probability of z taking a value between -2.14 and 0 is 0.4838.

Other uses of Table II in the determination of probabilities are illustrated in the examples.

2 PRACTICE PROBLEM 2

If z has the standard normal distribution, what is the probability that z takes a value between -1.43 and 0?

Answer

0.4236

EXAMPLE 1 If z has the standard normal distribution, what is the probability that z takes a value between 0.80 and 1.53?

Solution We have to determine the area indicated in the figure. From Table II, the area from 0 to 1.53 is 0.4370 and the area from 0 to 0.80 is 0.2881. The area from 0.80 to 1.53 is, therefore, $0.4370 - 0.2881$, or 0.1489.

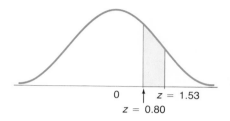

EXAMPLE 2 If z has the standard normal distribution, what is the probability that z takes a value between -1.39 and 2.1?

Solution Again we are interested in the area indicated in the figure. From Table II, the area between 0 and 2.1 is 0.4821 and the area between -1.39 and 0 is 0.4177. The desired probability is $0.4821 + 0.4177 = 0.8998$.

3 PRACTICE PROBLEM 3

If z has the standard normal distribution, what is the probability that z takes a value between each pair of numbers?

(a) 1.43 and 2.11

(b) -1.43 and 2.11

Answer

(a) 0.0590 (b) 0.9062

EXAMPLE 3 If z has the standard normal distribution, what is the probability that z takes a value less than -0.72?

Solution The total area under the curve to the left of 0 is 0.5; from Table II the area between -0.72 and 0 is 0.2642. The area to the left of -0.72 is, therefore, $0.5 - 0.2642 = 0.2358$.

4 PRACTICE PROBLEM 4

If z has the standard normal distribution, what is the probability that z takes a value in the indicated interval?

(a) Less than 1.43

(b) Greater than 1.43

Answer

(a) 0.9236 (b) 0.0764

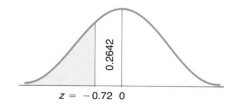

EXAMPLE 4 If z has the standard normal distribution, find the value of z such that 2.5% of the distribution lies to the left of that value.

Solution   If z_0 denotes the desired value, then the area to the left of z_0 is 0.025 and the area between z_0 and 0 is 0.475. The *area* 0.475 in the table corresponds to $z = 1.96$; that is, $z_0 = -1.96$.

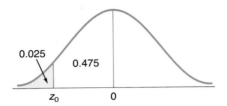

Now let x be a quantity having an arbitrary normal distribution with mean μ and standard deviation σ. If z is a quantity such that

$$z = \frac{x - \mu}{\sigma}$$

then z has the standard normal distribution. Furthermore, the area under the standard normal curve between 0 and z is the same as the area between μ and x under the normal curve for x (see Figure 7.10). Because of this relationship, Table II for the standard normal distribution can be used to determine probabilities for arbitrary normal distributions. For example, suppose x has a normal distribution with mean 20 and standard deviation 2. In order to determine the probability of x taking a value between 20 (the mean) and 23, we first find the *z-value* for $x = 23$:

$$z = \frac{23 - 20}{2} = 1.50$$

From the table for the standard normal distribution, we find the probability to be 0.4332.

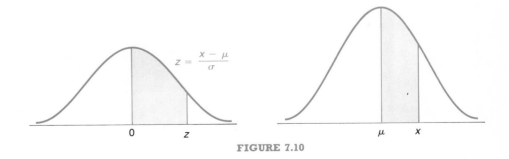

FIGURE 7.10

EXAMPLE 5   If x has a normal distribution with mean 16 and standard deviation 0.5, what is the probability that x takes a value less than 14.82?

7.5 THE NORMAL CURVE

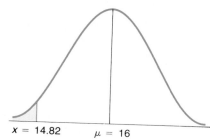

Solution   The z-value for $x = 14.82$ is

$$z = \frac{14.82 - 16}{0.5} = \frac{-1.18}{0.5} = -2.36$$

Since the area of interest is to the left of 14.82, the probability is $0.5 - 0.4909 = 0.0091$.

⑤

EXAMPLE 6   If x has a normal distribution with mean 100 and standard deviation 10, what is the probability that x takes a value between 85 and 105?

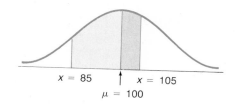

Solution   The corresponding z-values are:

$$z = \frac{105 - 100}{10} = 0.5 \quad \text{and} \quad z = \frac{85 - 100}{10} = -1.5$$

The probability is, therefore, $0.1915 + 0.4332 = 0.6247$.

∎

EXAMPLE 7   If the scores on an IQ test are normally distributed with mean 100 and standard deviation 15, what is the likelihood that a person chosen at random will score higher than 120?

Solution   The z-value for $x = 120$ is

$$z = \frac{120 - 100}{15} = \frac{4}{3}$$

---

⑤ PRACTICE PROBLEM 5

Suppose x has a normal distribution with mean 16 and standard deviation 0.5.

(a) What is the z-value corresponding to $x = 17.1$?

(b) What is the probability that x takes a value between the mean 16 and 17.1?

(c) What is the probability that x takes a value less than 17.1?

Answer   (a) 2.2   (b) 0.4861   (c) 0.9861

or approximately 1.33. The probability of obtaining a value of x greater than 120 is approximately $0.5 - 0.4082 = 0.0918$. This result indicates that in the long run 9.18% of the people taking the test will score higher than 120. Only to this extent does the result indicate the likelihood of a particular person scoring that high.

With respect to the four-step procedure involved in the application of mathematics, the physical situation in Example 7 is taking the IQ test. The physical question is, What is the likelihood that a person taking the test will score higher than 120? The mathematical description of the test scores is the normal distribution with mean 100 and standard deviation 15. The mathematical question is, What is the probability of obtaining a value of x greater than 120?

The answer to the mathematical question is 0.0918. The interpretation of this result is that in the long run 9.18% of the persons taking the test will score higher than 120. Only to that extent does the result indicate the likelihood of any particular person scoring higher than 120.

6 PRACTICE PROBLEM 6

In Example 7, what is the probability that the person chosen will score between 90 and 120?

Answer

$$0.2486 + 0.4082 = 0.6568$$

EXAMPLE 8 The quantity x has a normal distribution with mean 50 and standard deviation 5. Find the value of x such that 5% of the distribution is to the left of that value.

Solution From Table II, the value of z for which 5% of the distribution lies to the left is -1.64. The corresponding value of x satisfies the equation

$$-1.64 = \frac{x - 50}{5}$$

Solving this equation for x gives $x = 41.8$.

EXERCISES

Sketching a graph for each of these exercises will probably be helpful.

If z has the standard normal distribution, find the probability that z takes a value in the indicated interval.

1. Between 0 and 1.24
2. Between 0 and 1.87

3. Between −0.63 and 0
4. Between −1.54 and 0
5. Between 1.50 and 2.65
6. Between 0.44 and 2.01
7. Between −2.41 and −0.58
8. Between −2.38 and −1.4
9. Between −1 and 2
10. Between −2 and 1
11. Greater than 1.80
12. Greater than 2.16
13. Less than 1.5
14. Less than 1.93
15. Less than −0.88
16. Less than −1.70
17. Greater than 2 or less than −2
18. Greater than 1.5 or less than −2.5

If x has a normal distribution with mean 50 and standard deviation 5, find the probability that x takes a value in the indicated interval.

19. Between 50 and 60
20. Between 50 and 57
21. Between 40 and 50
22. Between 38 and 50
23. Between 55 and 60
24. Between 52 and 62
25. Between 37 and 45
26. Between 35 and 46
27. Between 43 and 60
28. Between 46 and 62
29. Greater than 65
30. Greater than 58
31. Greater than 44
32. Greater than 48
33. Less than 65
34. Less than 56
35. Less than 47
36. Less than 39
37. Less than 43 or greater than 61

If x has a normal distribution with mean 2.3 and standard deviation 0.5, find the probability that x takes a value in the indicated interval.

38. Between 2.3 and 2.39
39. Between 2.3 and 2.47
40. Between 1.5 and 2.3
41. Between 0.81 and 2.3
42. Between 2.75 and 3.5
43. Between 3 and 3.5
44. Between 1.2 and 1.8
45. Between 1.2 and 2.28
46. Between 1.11 and 2.41
47. Between 2 and 2.55
48. Greater than 1.5
49. Greater than 2.10
50. Greater than 2.85
51. Greater than 3
52. Less than 1
53. Less than 1.95
54. Less than 2.65
55. Less than 3.02
56. Less than 1.46 or greater than 3.19

57. *Student heights* Suppose the height of first-grade boys is normally distributed with a mean of 40 inches and standard deviation of 2 inches.
    (a) What is the probability that a first-grade boy chosen at random is between 39 and 41 inches tall?
    (b) What is the probability that a first-grade boy chosen at random is between 38.5 and 42.5 inches tall?

58. *Chemical reaction time* A scientist has found that the reaction time for a chemical experiment is normally distributed with a mean of 33.45 seconds and a standard deviation of 0.3 second.
   (a) What is the probability that the reaction time for one of the experiments is between 33.75 and 33.90 seconds?
   (b) What is the probability that the reaction time for one of the experiments is between 32.94 and 34.05 seconds?

59. *Grade-point averages* The quality grade-point averages of college students in a particular college are approximately normally distributed with a mean of 2.4 and a standard deviation of 0.6.
   (a) What is the likelihood that a student randomly chosen has a grade-point average greater than 3.0?
   (b) For what percent of the students is the grade-point average less than 1.5?

60. *Political survey* Suppose that surveys of people intending to vote in the next election are to be taken, 350 people to be interviewed in each survey. Suppose also that 44% of all people intending to vote in the next election favor a particular candidate. The number of people in a survey who favor this candidate has a normal distribution with mean 154 and standard deviation 9.3.
   (a) In what percent of the surveys will the candidate be favored by a majority of those surveyed?
   (b) In what percent of the surveys will the candidate be favored by less than 250 of those surveyed?

61. *Machine performance* The amount of orange juice put into containers by a machine follows a normal distribution with a mean of 12 ounces and a standard deviation of $\frac{1}{3}$ ounce. What percentage of the time will the machine overflow the 13-ounce containers?

62. *Quality control*
   (a) Close examination over a number of years has revealed that a precision manufacturing machine produces bolts with a mean diameter of 0.25 inch and standard deviation of 0.0001 inch. What is the probability that a bolt chosen at random from the machine's output has a diameter greater than 0.2502 inch?
   (b) The quality control inspector occasionally measures one of the bolts produced by the machine. If the diameter lies outside the range 0.2498 to 0.2502 inch, the machine is adjusted. What is the likelihood that the machine is adjusted at inspection time?

If z has the standard normal distribution, find the value of z that satisfies the indicated conditions.

63. 2.5% of the distribution lies to the right of that value.
64. 5% of the distribution lies to the right of that value.
65. 5% of the distribution lies to the left of that value.
66. 10% of the distribution lies to the left of that value.

If x has a normal distribution with mean 60 and standard deviation 7, find the value of x that satisfies the indicated conditions.

67. 2.5% of the distribution lies to the right of that value.
68. 2.5% of the distribution lies to the left of that value.

69. 10% of the distribution lies to the right of that value.

70. 5% of the distribution lies to the left of that value.

In a graph of the normal curve, what percent (approximately) of the area under the curve lies in each range?

71. Between the mean and 1 standard deviation above the mean

72. Between 1 standard deviation and 2 standard deviations above the mean

73. Between the mean and 2 standard deviations below the mean

74. To the right of 1 standard deviation below the mean

Identify the steps of the four-step procedure involved in the application of mathematics (Section 1.3) in each case.

75. *Grade-point averages* Exercise 59(a)

76. *Quality control* Exercise 62(b)

*7.6 A TEST OF A CLAIMED MEAN

We conclude this chapter by examining one use of the normal curve in inferential statistics. The normal curve involved represents the distribution of sample means. Suppose, for example, that you are interested in the annual cost of attending college. In order to get more information on the topic, you send questionnaires to students on every campus in the country. When (and if!) the questionnaires return, you divide them into piles of 20 each. For each pile of 20, you calculate mean cost. These means will obviously not be the same for each possible group of 20 students but rather will follow some sort of distribution of their own—a distribution of sample means.

In this illustration, the cost to every student in the country is the population; the costs to the 20 students in the various piles are the samples. The *size* of each sample in this case is 20.

We are interested in the distribution of the sample means. The following theorem, one of the more important theorems of statistics, describes the distribution:

CENTRAL LIMIT THEOREM

If samples of size n are randomly chosen and if the population has mean μ and standard deviation σ', then the sample means are approximately normally distributed with mean μ and standard deviation $\sigma = \sigma'/\sqrt{n}$.

---

* This section can be omitted without loss of continuity.

On the basis of this theorem, a test has been devised whether or not to accept the claim that a population has a certain mean. (For example, the claim might be that the mean annual cost of attending college in the United States is $3,500.) The test takes advantage of the fact, as stated in the theorem, that the sample means and the population distributions have the same mean. A sample is taken and, on the basis of its mean, a decision is made whether to accept or reject the claimed population mean.

The test begins with the assumption that the mean of the population is, in fact, the claimed mean μ. Then the normal distribution of sample means is as shown in Figure 7.11. The probability of a sample mean taking a value in the shaded regions is 0.05.

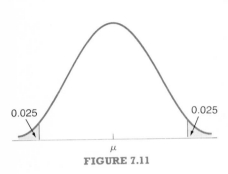

FIGURE 7.11

A random sample of size n is taken and the mean $\bar{x}$ of the sample is calculated. If the value of $\bar{x}$ lies in the shaded region (the *rejection region*), the claim that the population mean is μ is rejected. The reason for rejecting the claim is that the probability of obtaining a sample mean value in the rejection region is so small that the sample means must have a different distribution. The assumption on which this distribution is founded—namely, that the population mean is μ—must, therefore, be rejected.

Before examining the use of the test in several examples, consider the standard normal curve (Figure 7.12) corresponding to the distribution of the sample means of Figure 7.11. Here $z = (\bar{x} - \mu)/\sigma$ and $\sigma = \sigma'/\sqrt{n}$, where μ is the claimed population mean, σ' is the population standard deviation, and n is the sample size. Note that the z-values for the rejection region are always $z > 1.96$ and $z < -1.96$. Consequently, we need calculate only the appropriate z-values and determine whether or not they are in this rejection region.

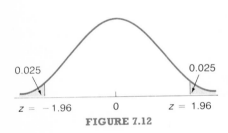

FIGURE 7.12

EXAMPLE 1 The manufacturer of a particular type of television picture tube claims that the mean life of its picture tubes is 7 years. A consumer group randomly picked 36 television sets that use this tube. The mean life of these 36 picture tubes was 6 years and the standard deviation was 6 months. On this basis, should the claim of the manufacturer be accepted or rejected?

Solution Nothing is known about σ', the standard deviation of the population (the life span of all picture tubes of this type). The standard deviation of the sample (6 months) is used as an estimate for σ'. The resulting procedure is valid as long as the sample size is 30 or more.

Calculation of the z-value gives

$$z = \frac{\bar{x} - \mu}{\sigma'/\sqrt{n}} = \frac{6 - 7}{0.5/\sqrt{36}} = -12$$

Since $-12 < -1.96$, the 7-year mean must be rejected.

Note that all measurements must be in terms of the same units—months, years, and so forth. Years were used in this example.

7.6 A TEST OF A CLAIMED MEAN

EXAMPLE 2 If the average age of people actually voting in a particular state is 32.5, then the most effective political campaigner would probably cater to the people in their midtwenties to midthirties. A random sample of 100 actual voters has a mean age of 32 and standard deviation of 3 years. Should the assumption of a mean age of 32.5 be accepted or rejected?

Solution The z-value is

$$z = \frac{32 - 32.5}{3/\sqrt{100}} = \frac{-5}{3}$$

or approximately -1.67. Since $-1.96 < -1.67 < 1.96$, the mean age of 32.5 is accepted.

The test does not prove that the mean age is *exactly* 32.5. It simply indicates that the sample data are compatible with a mean age of 32.5. The true mean is probably not exactly 32.5, but for practical purposes, whether the true mean is 32.5 or close to 32.5 is not of great importance.

EXAMPLE 3 A certain drug was developed to slow down a person's heartbeat. The manufacturer claims that 10 minutes after the drug is administered, the patient's heartbeat will be slowed by 12 beats per minute on the average.

The drug was administered at various times to 49 patients. Their heartbeats had slowed down 10 minutes after receiving the drug by an average (mean) of 12.4 beats with a standard deviation of 2 beats. On this basis, should the manufacturer's claim be accepted or rejected?

Solution The z-value is

$$z = \frac{12.4 - 12}{2/\sqrt{49}} = 1.4$$

Since $-1.96 < 1.4 < 1.96$, the manufacturer's claim is accepted. The claim is accepted in the same sense as in Example 2; the claimed mean is compatible with the sample data.

As with the sign test, the size 0.05 of the rejection region was chosen arbitrarily here and denotes the probability of rejecting the claimed mean when it is correct. Some other common choices for this probability, also denoted by α, are 0.01, 0.025, and 0.10. In fact, all the comments at the end of the section on the sign test regarding the size of the rejection region and the possibility of the two types of errors (even though the decision is statistically correct) are applicable here as well, but now the assumption to be accepted or rejected is that the claimed mean is the true mean.

1 PRACTICE PROBLEM 1

The systems analyst for a university computer claims that the mean CPU time for the programs run on the computer is 2.5 seconds. A check of 64 programs run on the computer gives a mean CPU time of 2 seconds with a standard deviation of 0.25 second. Should the systems analyst's claim be accepted or rejected?

Answer

$z = -16$; reject the claim

EXERCISES

1. *Family finances* Test the claim that a family of four spends an average of $500 on its vacation, if the mean spent by a random sample of 36 such families is $550 and the standard deviation is $150.

2. *Quality control* A company receives a shipment of bolts that are supposed to be 0.25 inch in diameter. The receiving department picks a random sample of 49 bolts for inspection. The sample has a mean of 0.255 inch and a standard deviation of 0.01 inch. The bolts are not required to be precisely 0.25 inch in diameter, only to be "reasonably close." Should the company keep the shipment?

3. *Textbook cost* The mean cost of a random sample of 36 mathematics textbooks is $18.50 and the standard deviation is $2.00. Use this information to test the claim that the average cost of mathematics textbooks is $19.00.

4. *Telephone calls* The lengths of 100 telephone calls placed through a company switchboard were recorded. The calls were randomly chosen, had a mean length of 3 minutes, and had a standard deviation of 15 seconds. Test the claim that the average length of telephone calls placed through the switchboard is 4 minutes.

5. *Chemical process* A chemical process is designed so that the impurities in the output have a mean of $\mu = 0.10$ gram per liter. The impurities in 100 sample liters randomly chosen from a batch have a mean of 0.12 gram per liter and a standard deviation of 0.03 gram. Do the results indicate that the batch meets the designed impurity specification?

6. *Quality control* A supplier for an electric company sells the company reels of wiring that are supposed to contain (a mean of) 1,000 feet per reel. A supervisor at the electric company measures the length of wire on 50 of these reels. She obtains a mean length of 990 feet per reel and a standard deviation of 5 feet. What do the results indicate concerning the supplier's claim of 1,000 feet per reel?

7. *EPA rating* The EPA rating for a particular model of automobile is 29.3 miles per gallon. A test of 60 cars of this model gives a mean of 28.1 miles per gallon with a standard deviation of 0.3. Is the statement "Mileage may vary with driving conditions" particularly significant?

8. *Course evaluation* A college dean maintains that the mean student rating for the courses given by the mathematics department is 3.2 on a scale of 4.0. To test the claim, the mean rating given to 44 randomly selected mathematics courses was 3.26 with a standard deviation of 0.51. Should the dean's claim be accepted or rejected?

9. *Nicotine content* The manufacturer of a new brand of cigarettes claims the mean nicotine content to be 1.5 milligrams per cigarette. A random sample of 144 of the cigarettes had a mean nicotine content of 1.65 milligrams and a standard deviation of 1.2 milligrams. Should the manufacturer's claim be accepted or rejected?

10. *Label versus contents* The label on a soft drink bottle claims the contents to be 2 liters. A random sample of 50 bottles had a mean content of

2.2 liters and a standard deviation of 750 milliliters. Does the label understate the contents?

Test the claim that $\mu = 123$ if a random sample of size 64 gives $\bar{x} = 126$ and the stated standard deviation.

11. $\sigma = 15$ 	 12. $\sigma = 10$

Test the claim that $\mu = 36.5$ if a random sample of size 50 has a standard deviation of 2.6 and the given value of $\bar{x}$.

13. $\bar{x} = 35$ 	 14. $\bar{x} = 35.7$

15. State the central limit theorem.

7.7 CHAPTER SUMMARY

In general, statistics is concerned with collecting, describing, and interpreting data. In this chapter we examined some topics related to these three aspects of statistics.

A large set of statistical data is usually put into an organized form for easier computation and analysis. Two methods of organizing data are **frequency tables** and **histograms.** A frequency table lists the measurements, from the smallest to the largest, along with the number of times each measurement occurred. A histogram is a bar graph; the height of each bar is equal to the frequency of the corresponding measurements.

The same two methods are also used for representing grouped data. In this case, the frequency table gives the number of measurements in each group. Similarly, the height of each bar in a histogram is equal to the number of measurements in the corresponding group.

One of the major objectives of descriptive statistics is to describe the center of a set of data and the dispersion, or spread, of the data about the center. The principal measures of central tendency are the **mean** and the **median.**

The mean of the set of numbers $\{x_1, x_2, \ldots, x_n\}$ is the quotient

$$\frac{x_1 + x_2 + \cdots + x_n}{n}$$

The median is the middle measurement when the numbers are arranged in order of magnitude. For large sets of data, grouped or ungrouped, the use of frequency tables reduces the number of calculations required to determine the mean and the median. The **mode** and the **midrange** are other measures of central tendency, although they are not used as often as the first two.

The principal measure of dispersion of a set of data is the **standard deviation,** which measures the spread of the data about the mean. If

$\bar{x}$ denotes the mean of the set of numbers $\{x_1, x_2, \ldots, x_n\}$, the standard deviation is the square root of the quotient

$$\frac{(x_1 - \bar{x})^2 + (x_2 - \bar{x})^2 + \cdots + (x_n - \bar{x})^2}{n}$$

The **interquartile range** measures the spread of the data about the median. It is defined to be the difference between the third and first quartiles, $Q_3 - Q_1$, where Q_3 is a number such that one-fourth of the measurements are greater or equal to this number and Q_1 is a number such that one-fourth of the measurements are less than or equal to this number. A less commonly used measure of dispersion is the **range** of the data—that is, the difference between the largest and smallest measurements.

The object in inferential statistics is to arrive at a conclusion concerning a population on the basis of a sample from the population. Two examples of the process were considered. The sign test was used to determine which of two comparable populations was "better" in some sense. The second example, based on the normal curve, was used to determine whether a claimed mean for a population should be accepted or rejected. In neither case does the statistical test prove that the resulting conclusion is correct. The tests verify only that the conclusion is compatible with the sample data.

Finally, we considered how to determine probabilities using the normal curve. If a variable has a normal distribution, the probability of that variable taking a value in an interval is equal to the corresponding area under the normal curve. Some of these areas are given in table form for the standard normal curve (mean equal to 0 and standard deviation equal to 1). Then if x has an arbitrary normal distribution with mean μ and standard deviation σ and

$$z = \frac{x - \mu}{\sigma}$$

then z has the standard normal distribution. This relationship allows one to use the tables for the standard normal distribution to calculate probabilities for arbitrary normal distributions.

REVIEW EXERCISES

For each of the following sets of data, determine the mean, the median, the standard deviation, and the interquartile range:

1. 12, 14, 17, 20, 21, 19, 10, 23
2. 100, 95, 103, 107, 92, 97, 110, 112, 90, 89, 94
3. 51.1, 53.7, 49.6, 62.3, 51.4, 48.5, 57.4, 59.2, 56, 54.8
4. 100.9, 100.5, 100.3, 100.5, 100.8, 101, 100.1, 100.4, 100

REVIEW EXERCISES

Use the set of data below for Exercises 5–8.

23, 27, 24, 25, 25, 23, 28, 24, 23, 25, 27, 24, 24, 28, 23, 24, 28

5. Construct the frequency table.
6. Construct the histogram.
7. Determine the mean and median.
8. Determine the standard deviation and the interquartile range.

Use the set of data below for Exercises 9–12.

110.4, 110.0, 110.0, 110.4, 110.1, 110.4, 110.1, 110.0, 110.3, 110.4, 110.1, 110.0, 110.4

9. Construct the frequency table.
10. Construct the histogram.
11. Determine the mean and median.
12. Determine the standard deviation and the interquartile range.

Use the set of data below for Exercises 13–16.

0, 16, 5, 15, 2, 20, 4, 10, 10, 24, 17, 13, 6, 21, 3, 18, 19, 6, 21, 3, 19, 23, 4, 12, 9, 23, 2, 18, 11, 15, 13, 20, 7, 15

13. Form the data into grouped data using five intervals of length 5 each, starting with 0: 0–4, 5–9, 10–14, and so forth.
14. Construct the frequency table and histogram for the grouped data.
15. Using the grouped data, determine the mean and median.
16. Using the grouped data, determine the standard deviation and the interquartile range.

Use the set of data below for Exercises 17–20.

16.3, 17.2, 11.5, 14.7, 15.3, 20.0, 15.5, 21.5, 17.9, 12.3, 10.6, 16.4, 12.2, 17.5, 11.0, 14.5, 18.1, 13.9, 19.3, 13.7, 19.9, 16.9, 20.2, 16.6, 21.6

17. Form the data into grouped data using four intervals of length 3 each, starting with 10: 10.0–12.9, 13.0–15.9, and so forth.
18. Construct the frequency table and histogram for the grouped data.
19. Using the grouped data, determine the mean and median.
20. Using the grouped data, determine the standard deviation and the interquartile range.
21. *Monthly sales* A jewelry company has five stores. The sales for the stores last month were $43,500, $38,200, $36,700, $35,100, and $33,900. Determine the average (mean) sales for the stores.
22. *Computer times* A computer systems analyst compared the running times of four different programs written to perform the same computer task. The times were 2.37, 1.98, 2.60, and 1.83 seconds. Determine the mean running time.
23. *Infant temperatures* The temperature of newborn infants drops below the normal 98.6°F at birth. The temperatures of six newborns in one hospital were found to be 97.8°, 97.6°, 98.2°, 98.0°, 97.5°, and 98.2°F. Determine the mean temperature.

24. *Incentive plan* A billfold-manufacturing firm has the following incentive plan for their assemblers: Each week the mean number of billfolds assembled is determined. Those assemblers who complete from 0 to 1 standard deviation above the mean receive a bonus of $25; those who complete from 1 to 2 standard deviations above the mean receive a bonus of $50, and those who complete above 2 standard deviations above the mean receive a bonus of $100. Those below the mean receive no bonus. The numbers completed by 12 assemblers during one week are listed in the table. Determine which employees receive a bonus and how much.

ASSEMBLER	NUMBER COMPLETED	ASSEMBLER	NUMBER COMPLETED
A	180	G	210
B	193	H	178
C	204	I	206
D	207	J	202
E	183	K	197
F	191	L	189

25. *Battery comparison* A toy manufacturer tested two different brands of batteries on 10 of their electronic football games. The hours of use of each type of battery on the 10 machines are given in the table.

MACHINE	BRAND X	BRAND Y	MACHINE	BRAND X	BRAND Y
1	3.5	3.7	6	3.7	3.7
2	4.2	3.9	7	3.9	3.6
3	4.3	4.0	8	4.1	3.8
4	3.6	3.8	9	4.1	4.3
5	4.5	4.1	10	3.5	3.8

(a) Determine the mean number of hours for each brand. On this basis, which brand seems to be better?
(b) Determine the standard deviation for each brand. Which brand seems to have a more uniform number of hours of use?
(c) Use the sign test to determine whether the results indicate either brand is superior.

26. *Smoking study* A researcher is studying the effect of smoking on pulse rate. She compared the change in the rate between two different brands of cigarettes on 15 different individuals. The results are given in the table, where a positive number indicates an increase in pulse rate and a negative number indicates a decrease.

INDIVIDUAL	BRAND A	BRAND B	INDIVIDUAL	BRAND A	BRAND B
1	+3	+2	9	+1	+2
2	+4	0	10	+3	+1
3	+2	+1	11	−1	−1
4	+1	−2	12	0	−2
5	−2	−1	13	+2	+2
6	+2	+4	14	+3	+4
7	+3	+3	15	−2	−1
8	0	+2			

(a) Determine the mean change in pulse rate for each brand. On this basis, which brand seems to have the least effect on pulse rate?
(b) Determine the standard deviation for each brand. Which brand seems to have a more uniform effect on pulse rate?
(c) Use the sign test to determine whether the results indicate either brand has the least effect on pulse rate. If so, which brand?

27. *Placement exam* Forty-five high school students, randomly chosen, are given a college mathematics placement exam. After an intensive remedial mathematics program, they are given a second, similar test. Of the 45, 2 students received the same grades, 27 received higher grades, and 16 received lower grades. Use the sign test to determine whether the results indicate that the program is effective.

28. *Reaction times* The reaction times of 36 randomly chosen individuals are tested before and after having two drinks. Of the 36, 24 had slower reaction times after the drinks, 10 had faster reaction times, and 2 had the same reaction scores. Use the sign test to determine whether drinking has an effect on reaction times.

If z has the standard normal distribution, find the probability that z takes a value in the indicated interval.

29. Between 0 and 2.35
30. Between -1.4 and 0
31. Between 1.27 and 2.30
32. Between -1.27 and 2.30
33. Greater than -1
34. Greater than 1
35. Less than 1
36. Less than -1

If x has a normal distribution with mean 15 and standard deviation 2, find the probability that x takes a value in the indicated interval.

37. Between 15 and 18
38. Between 10 and 15
39. Less than 13
40. Between 13 and 17
41. Greater than 13
42. Greater than 16
43. Between 11 and 13
44. Less than 19
45. Less than 12 or greater than 16.5
46. Less than 12 and greater than 16.5

If x has a normal distribution with mean 12.3 and standard deviation 0.4, find the probability that x takes a value in the indicated interval.

47. Between 11.9 and 12.7
48. Between 12 and 12.2
49. Less than 11.5
50. Less than 12.9
51. Between 12.3 and 12.8
52. Greater than 12.2
53. Less than 11.7 or greater than 12.5
54. Less than 12.4 or greater than 13.1

Teller efficiency The time required for a bank teller to wait on a customer is believed to be approximately normally distributed with a mean of 5 minutes and a standard deviation of 90 seconds. What is the probability that a customer will be waited on in the indicated time?

55. Less than 2 minutes
56. Between $3\frac{1}{2}$ and $6\frac{1}{2}$ minutes
57. More than 8 minutes
58. Between 5 and 8 minutes

Dissolution time The dissolution time for a pain-reliever tablet is normally distributed with a mean of 2 minutes and a standard deviation of 30 seconds. What is the probability that one of these tablets, randomly chosen, will dissolve in the indicated time?

59. In less than 2 minutes
60. Between 1 and 3 minutes
61. In more than $3\frac{1}{2}$ minutes
62. In less than 30 seconds

63. *Label versus contents* The label on a brand of beans claims the net weight of the contents to be 16 ounces. The state inspector weighs the contents of 100 cans and find their mean weight to be 15.75 ounces with a standard deviation of 1.5 ounces. What do the results indicate concerning the manufacturer's claim of 16 ounces?

64. *Salary claim* A labor union president maintains that the average (mean) annual salary of the union members is $22,500. A survey by an economist of 225 union members finds that their mean salary is $22,650 with a standard deviation of $1,350. What conclusion should the economist make concerning the president's claim?

65. *Water contamination* The water department of a city draws its water from a nearby lake. The department will switch to its reservoir when the concentration of a particular bacteria present in raw sewage becomes too high. In particular, they will switch when the mean contamination-index reading for the lake exceeds 150 per liter. A sample of 40 liters of lake water gives a mean reading of 155 per liter with a standard deviation of 9. What action, if any, should the water department take?

Test the claim that $\mu = 53$ for each random sample.

66. Size 49, $\bar{x} = 54.1$, standard deviation of 1.3
67. Size 42, $\bar{x} = 52.5$, standard deviation of 1.9

CHAPTER 8

EXPONENTS AND LOGARITHMS*

OUTLINE

8.1 Graphs of Exponential Equations
8.2 Logarithms
8.3 Graphs of Logarithmic Equations
8.4 Chapter Summary

Review Exercises

One does not have to be a professional economist to realize that the purchasing power of money decreases as a result of inflation. For example, if the inflation rate is 8% a year, the purchasing power of $100 at any time x (in years) is

$$y = 100(1 - 0.08)^x = 100(0.92)^x$$

At the end of 2 years, for example, $100 will be worth

$$y = 100(0.92)^2 = 84.64$$

or $84.64, in terms of the purchasing power of today's dollar. Our model, valid as long as the inflation rate remains at 8%, is one example of the use of exponential equations.

In this chapter, we consider exponential equations and a closely related concept—logarithms. We shall encounter other uses of exponents and logarithms in the examples and exercises of this chapter.

* It may be helpful to review the basic definitions and operations with exponents in Section A.7 in the algebra review of Appendix A at this time.

8.1 GRAPHS OF EXPONENTIAL EQUATIONS

We have previously evaluated expressions such as 3^2, $4^{1/2}$, and 10^{-1} and slightly more complicated expressions such as $4(2^3) = 32$, $-3(5^2) = -75$, and so forth. Now we want to consider a related type of equation. Specifically, we shall consider exponential equations of the form

$$y = Ab^{kx}$$

where A, b, and k are constants, $b > 0$. The following are equations of this type:

$y = 3(2^{4x})$	$A = 3$, $b = 2$, and $k = 4$
$y = 2^{-5x}$	$A = 1$, $b = 2$, and $k = -5$
$y = -4(\frac{1}{2})^x$	$A = -4$, $b = \frac{1}{2}$, and $k = 1$
$y = 3^x$	What are A, b, and k?

Determining the value of y corresponding to particular values of x involves evaluating exponential expressions. If $y = 3(2^{4x})$ and $x = 2$, then

$$y = 3(2^8) = 3(256) = 768$$

If $y = 2^{-5x}$ and $x = 1$, then

$$y = 2^{-5} = \frac{1}{32} = 0.03125$$

Many evaluations of this type require the use of a hand calculator. For example, if $y = 4(2^{3x})$ and $x = 0.75$, the value of y (which happens to be 19.027) is very difficult without the use of a calculator. **1**

---

1 PRACTICE PROBLEM 1

Determine the value of y for each of the given values of x.

(a) $y = 3(2^x)$, $x = 2$ and $x = -1$
(b) $y = 2(4^{-2x})$, $x = \frac{1}{4}$ and $x = -\frac{1}{4}$
(c) $y = -\frac{1}{6}(3^{4x})$, $x = 0$ and $x = \frac{1}{4}$
(d) $y = 5^{3x}$, $x = 1.5$ and $x = 3$
(e) $y = -100(3^{2x})$, $x = -5$ and $x = \frac{1}{4}$

Answer

(a) 12; $\frac{3}{2}$ (b) 1; 4 (c) $-\frac{1}{6}$; $-\frac{1}{2}$
(d) 1,397.543; 1,953,125
(e) -0.0017; -173.205

8.1 GRAPHS OF EXPONENTIAL EQUATIONS

The graphs of these equations depend, of course, upon the values of A, b, and k. Except for special values of these constants (see Exercises 32–34), the graph of each will be in one of the four forms shown in Figure 8.1.

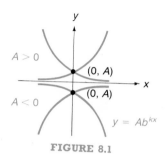

FIGURE 8.1

In order to graph a given equation of this type, it is not necessary to remember each of the possible four forms depicted in the graphs but only to appreciate what type of curve to expect. Determining a few points by picking several positive and negative values of x, along with $x = 0$, is usually sufficient for obtaining a fairly accurate graph.

Note that in the exponential equation $y = Ab^{kx}$, y will have the value A when $x = 0$; consequently, in each of the forms in Figure 8.1, the graph crosses the y-axis at $(0, A)$. Furthermore, no value of x makes b^{kx} equal to zero, so that the graph never crosses the x-axis.

EXAMPLE 1 Graph the equation $y = 4(2^x)$.

Solution Picking the values of x to be -1, -2, 0, 1, and 2, we determine the corresponding values of y, as shown in the table. We locate the corresponding points in a coordinate system and then determine the curve of the type shown in Figure 8.1 passing through them.

x	y
-2	1
-1	2
0	4
1	8
2	16

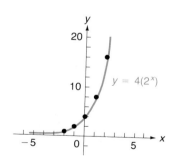

EXAMPLE 2 Graph the equation $y = -(3^{-2x})$.

Solution   Following the procedure in Example 1—constructing a table, locating the corresponding points in a coordinate system, and then determining the curve through these points—gives the following results:

x	y
-1	-9
$-\frac{1}{2}$	-3
0	-1
$\frac{1}{2}$	$-\frac{1}{3}$
1	$-\frac{1}{9}$

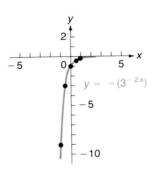

Some of the most important exponential equations are those in which the value of b is a special constant that mathematicians denote by the letter e. This constant e, like the number π, is a number with an infinite number of nonrepeating decimal places, which can be approximated as accurately as desired by using as many decimals as needed. To five decimal places,

$$e = 2.71828$$

Some hand calculators have an $\boxed{e^x}$ button for evaluating e^x for various values of x. Calculators with a $\boxed{y^x}$ button can be used with $y = e = 2.71828$. As a last resort, if your calculator has neither of these capabilities, the values of e^x for some values of x are given in Table III in Appendix B. Using a calculator or the table gives, for example, to five decimal places,

$$e^{3.2} = 24.53253, \qquad e^{0.65} = 1.91554, \qquad e^{-0.65} = 0.52205$$

▣ PRACTICE PROBLEM 2

Graph each of the following equations:

(a) $y = 2^x$ (b) $y = 3(2^{-2x})$

Answer

(a)

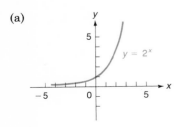

(b)

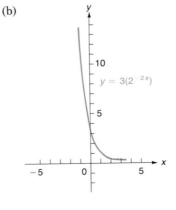

▣ PRACTICE PROBLEM 3

Determine the value of each expression.

(a) e^2 (b) $e^{1.5}$ (c) $e^{-2.37}$

Answer

(a) 7.38906   (b) 4.48169
(c) 0.09348

8.1 GRAPHS OF EXPONENTIAL EQUATIONS

EXAMPLE 3 Graph the equation $y = \frac{1}{2}e^x$.

Solution   We first determine the values of e^x for $x = -1, 0, 1,$ and 2; then construct the table and the graph as before. The values of y are rounded to one decimal place.

x	y
-1	0.2
0	0.5
1	1.4
2	3.7

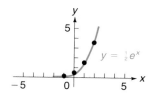

EXAMPLE 4 Graph the equation $y = -e^{-x/2} + 1$.

Solution   This is a variation from our basic equation $y = Ab^{kx}$ because of the added 1. However, for any given value of x, the value of y in $y = -e^{-x/2} + 1$ differs from that in $y = -e^{-x/2}$ by 1 unit, so that the shapes of the graphs of these two equations will be the same. The only difference in the graphs is that the graph of the given equation is shifted 1 unit higher in the coordinate system than that for $y = -e^{-x/2}$. The graphs of both equations are given for the sake of comparison. (The tables for determining points in the graphs are omitted.)

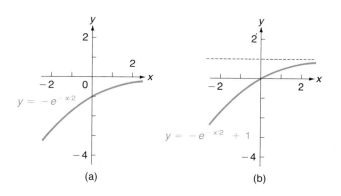

(a)   (b)

PRACTICE PROBLEM 4

Graph the equation $y = -e^x$, using $x = -1, -0.5, 0, 0.5,$ and 1.

Answer

x	y
-1	-0.4
-0.5	-0.6
0	-1
0.5	-1.6
1	-2.7

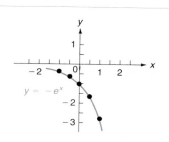

EXAMPLE 5 The balance (y) in a savings account with a $1,000 deposit and drawing 6% interest compounded annually at the end of n years is given by the equation

$$y = 1{,}000(1.06)^n$$

Graph the equation and from the graph estimate the balance in the account at the end of 10 years.

Solution

n	y
0	1,000
1	1,060
3	1,191.02
5	1,338.23

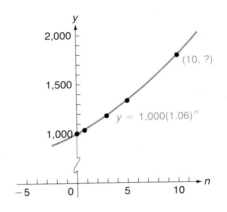

The y-coordinate of the point in the graph corresponding to $x = 10$ is approximately 1,800. (The actual balance is $1,790.85.)

In Example 5, the equation $y = 1{,}000(1.06)^n$ is a mathematical description of the account balance, and the graph of the equation gives a geometric description. Only points in the graph with n-coordinates that are positive integers are relevant in this situation. Since the interest is compounded annually, the interest is added to the account only at the end of each year. Consequently, those points whose n-coordinates represent fraction of years have no meaning here.

5 PRACTICE PROBLEM 5

From the graph in Example 5, estimate the account balance at the end of 6 years. Then determine the exact balance.

Answer

The exact balance is $1,418.52.

EXERCISES

Determine the value of y for the indicated values of x.

1. $y = 3^x$ for $x = 3, 4, -1,$ and 0
2. $y = 5^{-2x}$ for $x = 2, 3, -1,$ and -2
3. $y = -5(2^{3x})$ for $x = -2, \frac{2}{3}, 0,$ and $\sqrt{2}$
4. $y = 7(4^{2x})$ for $x = -1, 1, -\frac{3}{4},$ and 0.25
5. $y = \frac{1}{2}(3^{-2x})$ for $x = -0.5, -2, 0.75,$ and 0.68
6. $y = -0.2(1.5^{-3x})$ for $x = -2, -\frac{2}{3}, 0.45,$ and 1.37
7. $y = e^{4x}$ for $x = -1, -1.3, 0,$ and 2.4
8. $y = 5(e^{-x/2})$ for $x = -4, -3.5, 0.08,$ and 4

Graph each of the following equations:

9. $y = 3(2^x)$
10. $y = -2^x$
11. $y = -2^{-x}$
12. $y = \frac{1}{2}(3^x)$
13. $y = 3^{-x}$
14. $y = 4(\frac{1}{2})^x$
15. $y = e^x$
16. $y = e^{-x}$
17. $y = 2^x + 2$
18. $y = 2^{-x} - 1$
19. $y = 1 - e^{-x}$
20. $y = 1 + 2(3^{x/2})$

Graph the following equations in the same coordinate system and compare their graphs:

21. $y = 2^x$ and $y = 2^{(x+1)}$
22. $y = 2e^x$ and $y = 2e^{(x-1)}$

23. *Compound interest* The balance (y) in a savings account with a $1,000 deposit and drawing 9% interest at the end of n years is given by the equation

$$y = 1{,}000(1.09)^n$$

when the interest is compounded annually.
(a) Graph the given equation.
(b) Which points in the graph are relevant for this situation?
(c) From the graph, estimate the balance in the account at the end of 8 years.
(d) Calculate the exact balance in the account at the end of 8 years.
(e) Calculate the exact balance in the account at the end of 10 years.

24. *Compound interest* The balance (y) at any time (x), in years, in a savings account with a $1,000 deposit and drawing 9% interest compounded continuously is given by the equation

$$y = 1{,}000(e^{0.09x})$$

(a) Graph the given equation.
(b) Which points in the graph are relevant for this situation?
(c) From the graph estimate the balance in the account at the end of 8 years.
(d) Calculate the exact balance in the account at the end of 8 years.
(e) Calculate the exact balance in the account at the end of 10 years.
[Compare the results for parts (d) and (e) with those in Exercise 23.]

25. *Advertising expense* A company spends $1,200 a month on advertising. If their advertising expense increases 1% per month, the expense (y) after n months is given by

$$y = 1{,}200(1 + 0.01)^n$$

(a) Graph the given equation.
(b) From the graph, estimate the company's advertising expense 1 year from now.
(c) Calculate the exact advertising expense 1 year from now.
(d) Calculate the exact advertising expense $1\frac{1}{2}$ years from now.

26. *Inflation* A company employee is retiring with a pension of $1,500 per month. If inflation remains constant at 6% per year (0.5% per month), the purchasing power (y) of her pension after n months will be

$$y = 1{,}500\,(1 - 0.005)^n$$

(a) Graph the given equation.
(b) From the graph estimate her monthly purchasing power after 1 year.

(c) Calculate the exact purchasing power after 1 year.
(d) Calculate the exact purchasing power after 3 years.

27. *Population growth* The population in a rural county was found to be 2,400 during the 1980 census. Its future population (y) is estimated to be

$$y = 2{,}400e^{0.1x}$$

where x denotes number of years after 1980.
(a) Graph the given equation.
(b) From the graph, estimate the population in the county at the end of the year 1990.

28. *Radioactive decay* The amount (y), in grams, of a radioactive substance present at any time t, in years, is given by the equation

$$y = A(\tfrac{1}{2})^{t/5{,}000}$$

where A is the number of grams present when the measurement begins (that is, at $t = 0$). There are 15 grams present when the measurement begins.
(a) Graph the equation that gives the amount present at any future time.
(b) Which points in the graph are relevant?
(c) From the graph, estimate the amount present at the end of 20,000 years.

29. *Supply equation* The number (q) of units of a particular product that a company is willing to produce at the selling price (p), in dollars, is given by the supply equation

$$q = 500(2^p) - 500$$

(a) Graph the supply equation, locating points for $p = 0, 1, 2, 3,$ and 4.
(b) Which points in the graph are relevant to this situation?
(c) From your graph determine how many units the company is willing to produce when the price per unit is $3.50.

30. *Learning curve* Experimental results with psychiatric patients indicate that when the patients are rewarded for making a correct response, their learning curve is given by the graph of the equation

$$y = 1 - 0.5(0.9^x)$$

where y denotes the probability of making a correct response and x denotes the number of trials of the experiment.
(a) Graph the learning curve.
(b) From the graph, estimate the probability of a patient making a correct response at the seventh trial in this experiment.
(c) Calculate the exact probability in part (b).
(d) Calculate the exact probability of a patient making a correct response at the tenth trial in this experiment.

31. *Sales growth* A company expects that its annual sales y, in thousands of units, of a new product to be

$$y = 6 - 5e^{-n} \qquad n \geq 1$$

where n is the number of years after the product has been introduced to the market.
(a) Graph the given equation.

(b) From the graph estimate the number of units expected to be sold during its third year on the market.
(c) Calculate the exact number of units expected to be sold during its third year on the market.
(d) Calculate the exact number of units expected to be sold during its fifth year on the market.

Graph the equation $y = Ab^{kx}$ for each of the given values.

32. $A = 0$ **33.** $A = 2$ and $b = 1$ **34.** $A = 2$ and $k = 0$

8.2 LOGARITHMS

Another concept, which is closely related to the exponent concept and is used to describe physical situations and solve physical problems, is that of a logarithm. Since $3^2 = 9$, for instance, the exponent 2 is called the *logarithm* of 9 to the base 3. In general, if b is a positive number (not equal to 1) and if $x = b^y$, then y is the **logarithm of x to the base b.** For example:

Since $8 = 2^3$, the logarithm of 8 to the base 2 is 3.
Since $100 = 10^2$, the logarithm of 100 to the base 10 is 2.
Since $1 = e^0$, the logarithm of 1 to the base e is 0.
Since $\frac{1}{4} = 2^{-2}$, the logarithm of $\frac{1}{4}$ to the base 2 is -2.

The notation

$$\log_b x$$

is used in place of "the logarithm of x to the base b." With this notation, the above expressions can be rewritten as

$$\log_2 8 = 3, \quad \log_{10} 100 = 2, \quad \log_e 1 = 0, \quad \log_2 \tfrac{1}{4} = -2$$

With this notation, the definition of a logarithm can be written:

> If $b > 0$, $b \neq 1$, then $\log_b x = y$ provided that $x = b^y$.

Note that logarithms are just exponents; it is important not to lose sight of this relationship. In order to determine the value of $\log_3 9$, for example, we must answer the question, To what power should 3 be raised to obtain 9? The answer is 2, and therefore $\log_3 9 = 2$. The exponent 2 in the relationship $3^2 = 9$ is the value of $\log_3 9$.

EXAMPLE 1 Determine the value of each expression.

(a) $\log_5 25$ (b) $\log_3 81$ (c) $\log_4 \tfrac{1}{16}$
(d) $\log_{10} 0.10$ (e) $\log_{10} 1{,}000$ (f) $\log_6 1$

Solution (a) $\log_5 25 = 2$, since $5^2 = 25$

(b) $\log_3 81 = 4$, since $3^4 = 81$

(c) $\log_4 \frac{1}{16} = -2$, since $4^{-2} = \frac{1}{4^2} = \frac{1}{16}$

(d) $\log_{10} 0.10 = -1$, since $10^{-1} = \frac{1}{10} = 0.10$

(e) $\log_{10} 1{,}000 = 3$, since $10^3 = 1{,}000$

(f) $\log_6 1 = 0$, since $6^0 = 1$

1 PRACTICE PROBLEM 1

Determine the value of each expression.

(a) $\log_4 16$ (b) $\log_2 32$
(c) $\log_3 1$ (d) $\log_5 5$
(e) $\log_3 \frac{1}{27}$ (f) $\log_{10} \frac{1}{100}$

Answer

(a) 2   (b) 5   (c) 0
(d) 1   (e) -3 (f) -2

The illustrations in Example 2 might not seem as obvious; however, the solutions all follow from the definition of logarithms.

EXAMPLE 2   Determine the value of each expression.

(a) $\log_7 7^4$ (b) $\log_{12} 12^3$ (c) $\log_e e^5$
(d) $\log_{10} 10^x$ (e) $e^{\log_e 5}$ (f) $10^{\log_{10} 25}$

Solution   (a) $\log_7 7^4 = 4$, since $7^4 = 7^4$

(b) $\log_{12} 12^3 = 3$, since $12^3 = 12^3$

(c) $\log_e e^5 = 5$, since $e^5 = e^5$

(d) $\log_{10} 10^x = x$, since $10^x = 10^x$

(e) $e^{\log_e 5} = 5$, since $\log_e 5$ is the power to which e is raised in order to equal 5

(f) $10^{\log_{10} 25} = 25$, since $\log_{10} 25$ is the power to which 10 is raised in order to equal 25

2 PRACTICE PROBLEM 2

Determine the value of each expression.

(a) $\log_5 5^8$ (b) $\log_4 4^{-3}$
(c) $e^{\log_e 21}$ (d) $10^{\log_{10} 49}$

Answer

(a) 8   (b) -3 (c) 21   (d) 49

Observe carefully in all the examples and illustrations above that the logarithm of x base b is the power to which it is necessary to raise b in order to obtain x.

Logarithms to the base e and base 10 are especially important—base e because of their role in calculus and base 10 because of their role in calculations. Logarithms to the base e are called **natural** logarithms and are usually denoted by ln in place of $\log_e$. For example, $\ln e^2 = 2$ and $\ln e^x = x$. The value of natural logarithms can be determined from your hand calculator if it has an $\boxed{\ln x}$ button. Otherwise, Table IV in Appendix B gives the value of $\ln x$ for some values of x.

8.2 LOGARITHMS

Logarithms to the base 10 are called **common** logarithms and are usually denoted by log in place of $\log_{10}$. For example, $\log 100 = 2$ and $\log \frac{1}{10} = -1$. Some values of common logarithms can be determined from the definition of logarithms, but most values must be obtained from a hand calculator or a table. Your calculator will give you the value of common logarithms if it has a $\boxed{\log x}$ button. Otherwise, Table V in Appendix B gives the value of $\log x$ for some values of x.

EXAMPLE 3 Determine the value of each expression.

(a) ln 1.80   (b) ln 2.74   (c) log 3.27
(d) log 0.01   (e) ln 1   (f) log 1,000

Solution

(a) $\ln 1.80 = 0.58779$ from a calculator or Table IV

(b) $\ln 2.74 = 1.00796$ from a calculator or Table IV

(c) $\log 3.27 = 0.5145$ from a calculator or Table V

(d) $\log 0.01 = -2$, since $10^{-2} = \frac{1}{100} = 0.01$

(e) $\ln 1 = 0$, since $e^0 = 1$

(f) $\log 1,000 = 3$, since $10^3 = 1,000$

③ PRACTICE PROBLEM 3

Determine the value of each expression.

(a) ln 2.43   (b) ln 8.7
(c) log 2.43   (d) log 8.7

Answer

(a) 0.88789   (b) 2.16332
(c) 0.38561   (d) 0.93952

The value of x in each part of Example 4 below depends on the relationship

$$\log_b x = y \quad \text{means} \quad x = b^y$$

EXAMPLE 4 Determine the value of x that satisfies each of the following equations:

(a) $\log_3 x = 2$ (b) $\log_x 64 = 3$ (c) $\ln x = 1.45161$

Solution

(a) $\log_3 x = 2$ means that $x = 3^2 = 9$

(b) $\log_x 64 = 3$ means that $x^3 = 64$; therefore, $x = \sqrt[3]{64} = 4$

(c) $\ln x = 1.45161$ means that $e^{1.45161} = x$; using a hand calculator (or Table IV), $x = e^{1.45161} = 4.27$

④ PRACTICE PROBLEM 4

Determine the value of x.

(a) $\log_x 16 = 2$ (b) $\log_5 x = 3$
(c) $\ln x = 1.22671$

Answer

(a) 4   (b) 125   (c) 3.41

Because logarithms are exponents, it should not be surprising that the properties of logarithms parallel the properties of exponents. Three of the more important properties of logarithms are listed below and are illustrated using the fact that $\log 2 = 0.3010$ and $\log 3 = 0.4771$.

The first property states that the logarithm of a product of two numbers is equal to the sum of the logarithms of the two numbers:

PROPERTY 1
$$\log_b(m \cdot n) = \log_b m + \log_b n$$

Since $6 = 3 \cdot 2$,
$$\log 6 = \log(3 \cdot 2) = \log 3 + \log 2 = 0.4771 + 0.3010 = 0.7781$$

The second property states that the logarithm of the quotient of two numbers is equal to the logarithm of the numerator minus the logarithm of the denominator:

PROPERTY 2
$$\log_b \frac{m}{n} = \log_b m - \log_b n$$

Since $1.5 = \frac{3}{2}$,
$$\log 1.5 = \log \tfrac{3}{2} = \log 3 - \log 2 = 0.4771 - 0.3010 = 0.1761$$

The third property states that the logarithm of a number raised to a power is equal to the power times the logarithm of the number:

PROPERTY 3
$$\log_b m^n = n \log_b m$$

Since $81 = 3^4$,
$$\log 81 = \log 3^4 = 4(\log 3) = 4(0.4771) = 1.9084$$

These properties of logarithms, along with the corresponding properties of exponents, are summarized in Table 8.1.

TABLE 8.1

PROPERTY OF LOGARITHMS	PROPERTY OF EXPONENTS
1. $\log_b(m \cdot n) = \log_b m + \log_b n$	$b^m \cdot b^n = b^{m+n}$
2. $\log_b \dfrac{m}{n} = \log_b m - \log_b n$	$\dfrac{b^m}{b^n} = b^{m-n}$
3. $\log_b m^n = n \log_b m$	$(b^m)^n = b^{m \cdot n}$

For each property in the table, the operation on the right side of each logarithm property is the same as the operation in the exponential position on the right of the corresponding property of exponents!

EXAMPLE 5 Using the three properties of logarithms above, together with the facts that $\ln 4 = 1.38629$ and $\ln 7 = 1.94591$, determine the value of each expression.

(a) $\ln 16$ (b) $\ln 28$ (c) $\ln 1.75$

Solution
(a) $\ln 16 = \ln 4^2 = 2(\ln 4) = 2(1.38629) = 2.77258$
(b) $\ln 28 = \ln(4 \cdot 7) = \ln 4 + \ln 7 = 1.38629 + 1.94591 = 3.3322$
(c) $\ln 1.75 = \ln(\frac{7}{4}) = \ln 7 - \ln 4 = 1.94591 - 1.38629 = 0.55962$

Because we were given the values of $\ln 4$ and $\ln 7$, each of the numbers 16, 28, and 1.75 first had to be written in terms of 4 and 7 before the properties of logarithms could be used. ▫ 5

EXAMPLE 6 Using the three properties of logarithms, together with the facts that $\log 3 = 0.4771$ and $\log 5 = 0.6990$, determine the value of each expression.

(a) $\log 15$ (b) $\log 1.8$ (c) $\log 75$

Solution Note first that $15 = 3 \cdot 5$, $1.8 = 3^2/5$, and $75 = 3 \cdot 5^2$.

(a) $\log 15 = \log(3 \cdot 5) = \log 3 + \log 5 = 0.4771 + 0.6990 = 1.1761$

(b) $\log 1.8 = \log \frac{3^2}{5} = \log 3^2 - \log 5 = 2(\log 3) - \log 5$
$= 2(0.4771) - 0.6990 = 0.9542 - 0.6990 = 0.2552$

(c) $\log 75 = \log(3 \cdot 5^2) = \log 3 + \log 5^2 = \log 3 + 2(\log 5)$
$= 0.4771 + 2(0.6990) = 0.4771 + 1.3980 = 1.8751$ ▫ 6

---

5 PRACTICE PROBLEM 5

Using the three properties of logarithms, determine the value of each expression.

(a) $\ln 49$ (b) $\ln \frac{4}{7}$

Answer

(a) 3.89182 (b) -0.55962

---

6 PRACTICE PROBLEM 6

Using the three properties of logarithms, determine the value of each expression.

(a) $\log 9$ (b) $\log 0.60$ (c) $\log 45$

Answer

(a) 0.9542 (b) -0.2219 (c) 1.6532

The properties of logarithms give an expression for the logarithm of a product, of a quotient, and of a power. There are no corresponding general properties for logarithms of a sum nor of a difference. That is, there is no general form in which the expressions

$$\log_b(m + n) \quad \text{and} \quad \log_b(m - n)$$

can be rewritten.

The following example illustrates the use of properties of logarithms to solve some equations involving logarithms.

EXAMPLE 7 Determine the value of x that satisfies each of the following equations.

(a) $\log_2 x = \log_2 3 + \log_2 5$ (b) $\log_3 x = 2(\log_3 4)$

(c) $\log 5 = \log 30 - \log x$

Solution (a) $\log_2 3 + \log_2 5 = \log_2(3 \cdot 5)$; then, $\log_2 x = \log_2 15$ gives $x = 15$.
(b) $2(\log_3 4) = \log_3 4^2$, so that $x = 16$.
(c) $\log 30 - \log x = \log(30/x)$; then $5 = 30/x$ gives $x = 6$. ∎

The next example illustrates the use of properties of logarithms and the role of logarithms in solving equations when the unknown quantity appears in the exponential position.

EXAMPLE 8 The balance (y) at any time (x), in years, in a savings account with a $1,500 deposit and drawing 8% interest compounded continuously is given by

$$y = 1,500 e^{0.08x}$$

How long will it take for the balance to amount to $2,000?

Solution We want to determine the value of x when $y = 2,000$; that is, we want to solve the equation

$$2,000 = 1,500 e^{0.08x}$$

Dividing each side of this equation by 1,500, we get

$$\frac{2,000}{1,500} = e^{0.08x} \quad \text{or} \quad \frac{4}{3} = e^{0.08x}$$

Taking the natural logarithm of each side gives

$$\ln \frac{4}{3} = \ln e^{0.08x}$$

$$\ln \frac{4}{3} = 0.08x$$

$$\frac{1}{0.08} \ln \frac{4}{3} = x$$

8.2 LOGARITHMS

Using a calculator to determine $\ln \frac{4}{3} = 0.28768$, we get

$$x = \frac{0.28768}{0.08}$$

or approximately 3.596. If we want to use Table IV, we first write $\ln \frac{4}{3} = \ln 4 - \ln 3$. Then, since $\ln 4 = 1.38629$ and $\ln 3 = 1.09861$,

$$x = \frac{1.38629 - 1.09861}{0.08} = 3.596$$

It will take about 3.6 years, or approximately 3 years and 7 months, for the balance to amount to $2,000.

7 PRACTICE PROBLEM 7

How long will it take the balance in Example 8 to amount to $1,800?

Answer

2.279 years

EXERCISES

Evaluate each of the following expressions. (Use of a calculator or the tables should not be necessary for any of these.)

1. $\log_3 9$
2. $\log_3 \frac{1}{3}$
3. $\log_3 27$
4. $\log_2 \frac{1}{16}$
5. $\log_2 16$
6. $\log_2 \frac{1}{8}$
7. $\log_5 25$
8. $\log_5 625$
9. $\log_{1/2} \frac{1}{4}$
10. $\log_{1/2} 8$
11. $\log_{1/2} 1$
12. $\log 100$
13. $\log \frac{1}{10}$
14. $\log 0.1$
15. $\ln e$
16. $\ln e^2$
17. $\log_6 36$
18. $\log_7 \frac{1}{7^2}$
19. $\log 10^5$
20. $\ln e^{-4}$
21. $4^{\log_4 17}$
22. $e^{\ln 2}$
23. $10^{\log 51}$
24. $e^{\ln x}$
25. $e^{\ln e^2}$
26. $\ln e^x$
27. $\log 10^x$
28. $\ln e^{x^2}$

Evaluate each of the following expressions:

29. $\ln 3.12$
30. $\ln 4.97$
31. $\ln 9.4$
32. $\ln e^{-1}$
33. $\log 2.83$
34. $\log 6.49$
35. $\log 4.0$
36. $\log 0.001$
37. $\log 7.55$

Determine the value of x that satisfies each given equation. Use a calculator or tables when necessary.

38. $\log_2 8 = x$
39. $\log_3 \frac{1}{9} = x$
40. $\log_6 36 = x$
41. $\log_2 x = 4$
42. $\log_5 x = -2$
43. $\log_4 x = -1$
44. $\log_9 x = \frac{1}{2}$
45. $\log_8 x = \frac{1}{3}$
46. $\log_x 9 = 2$
47. $\log_x 1,000 = 3$
48. $\log_x \frac{1}{6} = -1$
49. $\log_5 x = -\frac{1}{2}$
50. $\log_x 9 = -2$
51. $\log_x \frac{1}{25} = -2$
52. $\log_x 2 = -1$
53. $\log_x \frac{1}{4} = -1$
54. $\log_x \frac{1}{4} = -2$
55. $\log_x 125 = -3$
56. $\log_x 4 = 2$
57. $\log_x \frac{1}{4} = 2$
58. $\log_x \frac{1}{4} = -\frac{1}{2}$
59. $\ln x = -2$
60. $\log_x \frac{1}{4} = \frac{1}{2}$
61. $\ln x = 2.00148$
62. $\ln x = 1.80336$
63. $\log x = 3$
64. $\log x = 0.7443$

65. (a) Write each of the following numbers in terms of 2 and 5:
   (i) 10   (ii) 25   (iii) 2.5   (iv) 0.80

(b) Using the three properties of logarithms and the facts that ln 2 = 0.69315 and ln 5 = 1.60944, determine the value of each expression.
(i) ln 10 (ii) ln 25 (iii) ln 2.5 (iv) ln 0.80

66. (a) Write each of the following numbers in terms of 4 and 11:
(i) 44 (ii) 2 (iii) 22 (iv) 2.75 (v) $\frac{16}{11}$
(b) Using the three properties of logarithms and the facts that log 4 = 0.6021 and log 11 = 1.0414, determine the value of each expression.
(i) log 44 (ii) log 2 (iii) log 22 (iv) log 2.75 (v) log $\frac{16}{11}$

Using the properties of logarithms and the facts that ln 3 = 1.09861 *and* ln 4 = 1.38629, *determine the value of each expression.*

67. ln 12 68. ln 16 69. ln $\frac{4}{3}$ 70. ln 2.25 71. ln $\frac{2}{3}$

Using the properties of logarithms and the facts that log 2 = 0.30103 *and* log 6 = 0.77815, *determine the value of each expression.*

72. log 12 73. log 3 74. log 72 75. log 48 76. log $\frac{1}{4}$

Use property 2 of logarithms and Table IV to determine the value of each expression.

77. ln $\frac{1}{2}$ 78. ln $\frac{2}{3}$ 79. ln 0.75 80. ln $\frac{1}{3}$

Determine the value of x that satisfies each of the following equations:

81. $\log_3 x = \log_3 4 + \log_3 5$
82. $\log_4 x = 2 \log_4 7$
83. $\log 20 = \log 40 - \log x$
84. $\log x = \log 5 - \log 2$
85. $\ln x = 3 \ln 5$
86. $\ln 14 = \ln 2 + \ln x$
87. $\log_{12} 1{,}000 = \log_{12} x - \log_{12} 2$
88. $\log x = 3 \log 2 + \log 20$

89. *Compound interest* The balance (y) at any time (x), in years, in a savings account with a $2,500 deposit and drawing 9% interest compounded continuously is given by

$$y = 2{,}500 e^{0.09x}$$

How long will it take for the balance to reach each amount?
(a) $2,750 (b) $5,000

90. *Compound interest* The balance (y) at any time (x), in years, in a savings account with a $2,000 deposit and drawing 10% interest compounded continuously is given by

$$y = 2{,}400 e^{0.1x}$$

How long will it take for the balance to reach each amount?
(a) $2,500 (b) $4,000

91. *Population growth* The population in a rural county was found to be 2,400 during the 1980 census. Its future population (y) is estimated to be

$$y = 2{,}400 e^{0.1x}$$

where x denotes the number of years after 1980. How long will it take for the population to reach each number?
(a) 2,750 (b) 3,000

92. *Radioactive decay* The amount (y), in grams, of a radioactive substance present at any time (t), in years, is given by the equation

$$y = A e^{-0.0001t}$$

where A is the amount present initially.

(a) If there are 100 grams present initially, how long will it take for the substance to decay to 75 grams?
(b) For any initial amount A, how long will it take for the substance to decay to one half of its original amount? (*Hint:* Set y equal to $\frac{1}{2}A$.)

93. Explain the meaning of the expression, the logarithm of x to the base b.
94. Why is the restriction $b \neq 1$ included in the definition of logarithms?
95. Why does $\log_2(-4)$ have no value?

Determine whether each of the following equations is true or false. If false, change the right-hand side to obtain a valid equation.

96. $\log_b a + \log_b c = \log_b(a + c)$
97. $\log_b a - \log_b c = \log_b(a - c)$
98. $c \log_b a = \log_b(ca)$

Express each of the following properties of logarithms verbally, without using any mathematical symbols:

99. Property 1
100. Property 2
101. Property 3

8.3 GRAPHS OF LOGARITHMIC EQUATIONS

In this section we shall consider the graphs of logarithmic equations of the form

$$y = A \log_b(kx)$$

where A, b, and k are constants, b is positive, and $b \neq 1$. Some examples of this type of equations are:

$y = 2 \log_3(4x)$	$A = 2, b = 3, k = 4$
$y = -4 \log(6x)$	$A = -4, b = 10, k = 6$
$y = \ln x$	$A = 1, b = e, k = 1$
$y = \log_2 x$	What are A, b, and k in this case?

As long as A is not zero, the graph of this type of equation will always be in one of the forms shown in Figure 8.2, depending on the values of A, of b, and of k.

As with the exponential equations in Section 8.1, it is not necessary to remember the four forms but only to appreciate what type of form to expect. But note that each graph passes through the point $(1/k, 0)$, corresponding to the fact that

$$\log_b\left(k\,\frac{1}{k}\right) = \log_b 1 = 0$$

for each b.

The most important logarithmic equations are probably those with both A and k equal to 1. The graphs will then be one of the forms

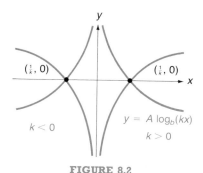

FIGURE 8.2

8 EXPONENTS AND LOGARITHMS

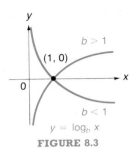

FIGURE 8.3

shown in Figure 8.3, depending on the value of b. These graphs emphasize the fact that $\log_b x$ has no value for $x \leq 0$; the graphs lie entirely to the right of the y-axis.

Because the values of $\log_b x$ change so dramatically with the values of x, it is difficult to choose scales (unit distances) for the axes that make the graphing manageable. Consequently, we will choose only two values of x ($x = 1/k$ and $x = b/k$), locate the corresponding points, connect these points with the proper type of curve, and settle for a rough sketch of the graph. The procedure is illustrated in the next several examples.

EXAMPLE 1 Graph the equation $y = \log_3(2x)$.

Solution As just indicated, we pick $x = 1/k = 1/2$ and $x = b/k = 3/2$ for convenient values, locate the corresponding points, and connect these with the proper type of curve from Figure 8.2.

x	y
$\frac{1}{2}$	0
$\frac{3}{2}$	1

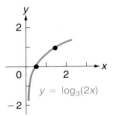

EXAMPLE 2 Graph the equation $y = 3 \log_{1/2}(-4x)$.

Solution In this case, $1/k = 1/(-4)$ and $b/k = -1/8$.

x	y
$-\frac{1}{4}$	0
$-\frac{1}{8}$	3

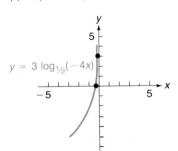

---

1 PRACTICE PROBLEM 1

Graph the equation $y = -3 \log_{1/2}(4x)$.

Answer

x	y
$\frac{1}{4}$	0
$\frac{1}{8}$	-3

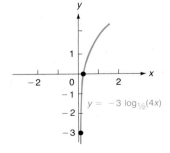

8.3 GRAPHS OF LOGARITHMIC EQUATIONS

EXAMPLE 3 Graph the equation $y = 2 + \ln x$.

Solution In this case, $1/k = 1$ and $b/k = e \approx 2.7$. The 2 in the equation has the effect of increasing the y-coordinate of each point of the graph of $y = \ln x$ by 2 units; that is, the graph of $y = \ln x$ is shifted 2 units upward.

x	y
1	2
e	3

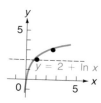

EXERCISES

Graph each of the following equations:

1. $y = \log x$
2. $y = \ln x$
3. $y = 3 \log(2x)$
4. $y = \ln(-3x)$
5. $y = -2 \log_4 x$
6. $y = -\log_{1/2} x$
7. $y = 3 - \log_{1/3}(-x)$
8. $y = 1 + 2 \log x$
9. Graph the following equations in the same coordinate system and compare their graphs:

$$y = \log x \quad \text{and} \quad y = \log(x - 2)$$

(For the first equation, take $x = 1$ and 10; for the second, take $x = 3$ and 12.)

10. Graph the following equations in the same coordinate system and compare their graphs:

$$y = \ln x \quad \text{and} \quad y = \ln(x + 2)$$

(For the first equation, take $x = 1$ and e; for the second, take $x = -1$ and $e - 2$).

11. *Demand equation* The demand equation for a product is given by the equation

$$q = \log_{1/3}(\tfrac{1}{10}p)$$

where q gives the demand in hundreds of units and p denotes the price per unit in dollars.
(a) Graph the demand equation.
(b) Discuss the physical significance of the points on the curve for positive values of p close to 0.
(c) Discuss the physical significance of the point corresponding to $p = 10$.

12. *Richter scale* A seismograph measures the intensity (x) of ground movement during an earthquake. Because of the great variation in the measurements, they are reported on a log scale called the *Richter scale*, where

the Richter number y is given by

$$y = \log\left(\frac{x}{x_0}\right)$$

and x_0 is the least intensity that the seismograph can detect. An earthquake measuring 5 on the Richter scale, for example, has intensity measuring $10^5 x_0$ since

$$5 = \log\left(\frac{x}{x_0}\right)$$

means

$$10^5 = 10^{\log(x/x_0)} = \frac{x}{x_0}$$

or

$$x = 10^5 x_0$$

(a) A 1971 Los Angeles earthquake measured 6.7 on the Richter scale. Determine its intensity.

(b) A 1964 Alaska earthquake measured 8.4 on the Richter scale. How much more intense was this earthquake than the 1971 earthquake in part (a)?

(c) Construct a graph of the equation

$$y = \log\left(\frac{x}{x_0}\right)$$

using the given coordinate system (with different units on the two axes).

(d) In view of the values of x in parts (a) and (b), does the graph constructed in part (c) display much information about the Richter number y of earthquakes?

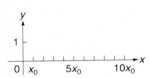

13. *Sound intensity* The intensity of sound is measured in decibels (D) in terms of the energy (I), in watts, that it produces by the equation

$$D = 10 \cdot \log\left(\frac{I}{I_0}\right)$$

I_0 corresponds to the least (approximately) sound the human ear can hear (I_0 equals 10^{-16} watts per square centimeter of the ear). A sound of $100 I_0$ watts, for example, has intensity

$$10 \cdot \log\left(\frac{100 I_0}{I_0}\right) = 10 \cdot \log(100) = 20$$

decibels.

(a) Determine the intensity of a sound of $10{,}000 I_0$ watts.

(b) Ordinary speech transmits about $10^6 I_0$ watts. Determine its intensity.

(c) A jet plane 100 feet away transmits a sound of about $10^{14} I_0$ watts. Determine its intensity.

(d) Determine the number of watts transmitted by a sound of 53 decibels.

14. *Sound intensity* Identify the steps of the four-step procedure involved in the application of mathematics (Section 1.3) in Exercise 13(a).

8.4 CHAPTER SUMMARY

The graph of an exponential equation
$$y = Ab^{kx}$$
where A, b, and k are constants with $b > 0$, is (except for some special values of the constants) one of the following four forms:

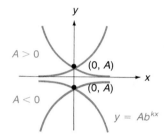

To graph one of these exponential equations, it is sufficient to:

1. Determine the coordinates of a few points in the graph, picking several positive and negative values of x and $x = 0$.
2. Locate the points determined in Step 1 in a coordinate system.
3. Connect these points with the appropriate curve.

One of the most important exponential relationships is that in which the base b is the number $e = 2.71828$ (to five decimal places). The value of e^x for most values of x must be determined using a calculator or from a table.

Closely related to exponents are logarithms. The **logarithm** of x to the base b ($\log_b x$) is defined to be y if $b^y = x$, provided b is a positive number other than 1. Logarithms to the base e and base 10 are especially important. Logarithms to the base e are called **natural** logarithms and are denoted by ln in place of $\log_e$. Logarithms to the base 10 are called **common** logarithms and are denoted by log in place of $\log_{10}$. Most values of natural and common logarithms must be determined using a calculator or from tables.

Three important properties for operating with logarithms are:

1. $\log_b(m \cdot n) = \log_b m + \log_b n$
2. $\log_b \dfrac{m}{n} = \log_b m - \log_b n$
3. $\log_b m^n = n \cdot \log_b m$

The graph of a logarithmic equation
$$y = A \log_b(kx)$$
will be one of the following forms, as long as A is not zero:

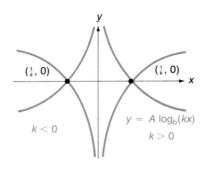

In order to make graphing logarithmic equations manageable, we used the following procedure in locating the graphs:

1. Determine the y-coordinates of the points with x-coordinates $x = 1/k$ and $x = b/k$.
2. Plot the two points determined in Step 1 in a coordinate system.
3. Connect these two points by the appropriate curve.

REVIEW EXERCISES

Determine the value of y for the indicated values of x.

1. $y = 3(4^x)$ for $x = 2, \frac{1}{2}, 0$, and -1
2. $y = -5(3^{2x})$ for $x = 3, 1, -1$, and -2
3. $y = \frac{2}{3}(6^{x/2})$ for $x = 4, 2, -1$, and -2
4. $y = -3(10^{-x})$ for $x = 3, 2, -2$, and -1.5
5. $y = 4(e^{-3x})$ for $x = 2, 0, -0.5$, and -2.3

Graph the following equations:

6. $y = 3^x$
7. $y = 4(2^{-x})$
8. $y = -4(2^x)$
9. $y = -\frac{1}{2}(3^{2x})$
10. $y = 3(4^{-x/2})$
11. $y = -\frac{3}{4}e^{-x}$
12. $y = 1.5e^x$
13. $y = 2^x - 2$

14. *Price increase* A company charges $1,200 for one of its heavy-duty drills. If the price increases at the rate of 6% a year, the cost (y) of the drill in n years will be

$$y = 1,200(1.06)^n$$

(a) Graph the given equation.
(b) From the graph, estimate the cost in 5 years.
(c) Calculate the exact cost in 5 years.
(d) Calculate the exact cost in 7 years.

REVIEW EXERCISES

15. *Compound interest* An individual puts $500 into a savings account that pays 8% compounded continuously. The balance (y) in the account at any time x, in years, is given by

$$y = 500e^{0.08x}$$

(a) Graph the given equation.
(b) From the graph, estimate the balance in the account after 21 months.
(c) Calculate the exact balance in the account after 21 months.
(d) Calculate the exact balance in the account after 4 years.

16. *Inflation* The purchasing power (y) of an amount A when inflation remains at 6% per year will be

$$y = A(1 - 0.06)^n$$

after n years. A foundation sets up a fund that will pay a $1,000 scholarship each year.
(a) Graph the equation that gives the purchasing power of the scholarship after n years.
(b) From the graph, estimate the purchasing power of the scholarship 4 years from now.
(c) Calculate the exact purchasing power of the scholarship 4 years from now.
(d) Calculate the exact purchasing power of the scholarship 10 years from now.

Without using a calculator or tables, give the value of each of the following:

17. $\log_2 4$
18. $\log_5 1$
19. $\log_3 \frac{1}{9}$
20. $\log_{10} 1{,}000$
21. $\log 1{,}000$
22. $\log_5 0.04$
23. $\log_7 7$
24. $\log_4 64$
25. $\ln 1$
26. $\log 0.01$
27. $\log_6 \frac{1}{216}$
28. $\log_3 729$
29. $\log_4 4^5$
30. $\log_6 6^{-3}$
31. $\log_7 \left(\frac{1}{7^3}\right)$
32. $\ln e^3$
33. $\log_2 2^{1.5}$
34. $10^{\log 4 7}$
35. $e^{\ln 3.5}$
36. $7^{\log_7 50}$
37. $6^{-\log_6 2}$
38. $e^{\ln x^2}$
39. $4^{\log_3 9}$
40. $e^{\log_{10} 1{,}000}$
41. $9^{\log_8 2}$
42. $5^{\log_{25} 5}$

Determine the value of x that satisfies each of the following:

43. $\log_6 x = 2$
44. $\log_4 x = 3$
45. $\log_7 x = 0$
46. $\log_5 x = 1$
47. $\log_5 x = -2$
48. $\log_2 x = -1$
49. $\log_4 x = \frac{1}{2}$
50. $\log_{27} x = \frac{1}{3}$
51. $\ln x = 2$
52. $\ln x = 2.11746$
53. $\ln x = 1.80829$
54. $\log x = -2$
55. $\log x = 0.8482$
56. $\log_x 25 = 2$
57. $\log_x 64 = 3$
58. $\log_x \frac{1}{3} = -1$
59. $\log_x \frac{1}{64} = -2$
60. $\log_x \frac{1}{8} = 3$
61. $\log_x 16 = -2$
62. $\log_x \frac{4}{9} = 2$
63. $\log_x 1 = 0$
64. $\log_3 x = \log_3 4 + \log_3 5$
65. $\log_4 7 = \log_4 x - \log_4 6$
66. $\log_5 9 = x \log_5 3$
67. $\log 21 = \log x + \log 7$
68. $\log 16 = 2 \log x$

Use the facts that $\ln 4 = 1.38629$ *and* $\ln 9 = 2.19722$ *to determine the value of each of the following:*

69. $\ln 36$ **70.** $\ln 2.25$ **71.** $\ln 16$ **72.** $\ln 3$ **73.** $\ln 20.25$

Use the facts that $\log 3 = 0.47712$ *and* $\log 12 = 1.0792$ *to determine the value of each of the following:*

74. $\log 4$ **75.** $\log 9$ **76.** $\log 36$ **77.** $\log 6$ **78.** $\log 24$

Compound interest In Exercise 15, how long will it take for the balance to reach the indicated amount?

79. $600   **80.** $700

Salvage value The salvage value (y) of a computer originally costing $42,000 after t years is given by the equation

$$y = 42{,}000 e^{-0.1t}$$

After how many years will the computer be worth the indicated amount?

81. $35,000   **82.** One-half of its original value

Graph the following equations:

83. $y = \log_3 x$ **84.** $y = -3 \log_2 x$ **85.** $y = -3 \log_2 4x$

86. $y = \log_3(-2x)$ **87.** $y = 2 \log 3x$ **88.** $y = 1 - 2 \ln x$

CHAPTER 9

MATHEMATICS OF FINANCE

OUTLINE

9.1 Geometric Progressions
9.2 Simple Interest
*__9.3__ Average Daily Balance
9.4 Compound Interest
9.5 Ordinary Annuities
9.6 Present Value of an Annuity
9.7 Chapter Summary
   Review Exercises

The mathematics of finance is a very technical subject and much too involved to be presented in its entirety in one chapter. The basic concepts considered here are intended as an introduction to the subject and, hopefully, will be useful to you in your personal life as well as in your professional career. The work with geometric progressions in the first section may seem unrelated; however, geometric progressions form one of the basic tools in working with compound interest later in the chapter.

* This section can be omitted without loss of continuity.

9.1 GEOMETRIC PROGRESSIONS

The programming staff of a television network approach you, as an executive of the network, with the following proposal for a new quiz show. In return for answering the questions correctly, the contestant would receive pennies on the red squares of a checkerboard in the following manner. For the first correct answer, the contestant would receive 1 penny on the first red square; for the second correct answer, the contestant would receive an additional 2 pennies on the second red square; for the third correct answer, 4 additional pennies on the third red square; then 8 pennies on the fourth red square, 16 on the fifth, and so forth. What would be your reaction?

The doubling of pennies might sound rather drab. Actually, as we shall see, by the time the contestant reached the 32nd red square, he or she would be a millionaire many times over. You would probably never find a sponsor willing to pay the contestants the money involved.

Or suppose that your parents started a college fund for you on your first birthday by depositing $200 in a savings account that pays 6% interest. Each subsequent year, they deposited another $200. If the interest was allowed to accumulate, how much would be available to you when you started college?

If you emptied your account after the $200 deposit on your eighteenth birthday, you would have about $6,180—quite a bit more than the $3,600 deposited over the years.

Surprising, perhaps, is the fact that the amount of money on the checkerboard squares and the amount of money in your account can be calculated in exactly the same way. The amounts of money on the squares and the amounts of money in your account at the end of each year both form a sequence of numbers called a *geometric progression,* the subject to be studied in this section. One of the topics to be considered is the sum of the numbers in a geometric progression. We shall then be in a position to analyze further the two illustrations discussed above.

The following sequence of numbers also forms a geometric progression:

$$1, 3, 9, 27, 81, 243, 729, \ldots$$

Each number after the first can be obtained by multiplying the preceding number by 3. This is the key feature of a geometric progression, that each number is a constant multiple of the number before it in the sequence.

Formally, a **geometric progression** is a sequence of numbers such that each number in the sequence, after the first, can be obtained from the preceding number by multiplying by some constant. This constant is called the **common ratio** and is usually denoted by the letter r. In the illustration just given, the value of r was 3. In the geometric

9.1 GEOMETRIC PROGRESSIONS

progression

$$2, -4, 8, -16, 32, -64, 128, \ldots$$

$r = -2$; in the progression

$$64, 32, 16, 8, 4, 2, 1, \tfrac{1}{2}, \tfrac{1}{4}, \ldots$$

$r = \tfrac{1}{2}$.

The numbers in a geometric progression are called the **terms.** The first term is denoted by a_1, the second by a_2, the third by a_3, and so forth. In the progression

$$\tfrac{1}{125}, \tfrac{1}{25}, \tfrac{1}{5}, 1, 5, 25, 125, \ldots$$

$a_1 = \tfrac{1}{125}$, $a_2 = \tfrac{1}{25}$, $a_3 = \tfrac{1}{5}$, $a_4 = 1$, and so forth. In the progression

$$1, 2, 4, 8, 16, 32, 64, 128, \ldots$$

$a_1 = 1$, $a_2 = 2$, $a_3 = 4$, $a_4 = 8$, and so forth. What are the values of r in each of these last two cases?

---

1 PRACTICE PROBLEM 1

If $a_1 = 2$ and $r = 3$ in a geometric progression, determine a_2, a_3, and a_4.

Answer

$a_2 = 6$, $a_3 = 18$, $a_4 = 54$

In terms of the a's and r,

$$a_2 = ra_1$$
$$a_3 = ra_2$$
$$a_4 = ra_3$$
$$a_5 = ra_4$$
$$\vdots$$

That is, each term beginning with a_2 can be determined from the preceding term by multiplying by r (as the definition requires). Therefore,

$$a_3 = ra_2 = r(ra_1) = r^2 a_1$$
$$a_4 = ra_3 = r(ra_2) = r^2 a_2$$
$$a_5 = ra_4 = r(ra_3) = r^2 a_3$$
$$a_6 = ra_5 = r(ra_4) = r^2 a_4$$
$$\vdots$$

That is, each term beginning with a_3 can be determined from the second preceding term by multiplying by r^2.

In terms of a_1 and r, we have

$$a_2 = ra_1$$
$$a_3 = ra_2 = r(ra_1) = r^2 a_1$$
$$a_4 = ra_3 = r(r^2 a_1) = r^3 a_1$$
$$a_5 = ra_4 = r(r^3 a_1) = r^4 a_1$$
$$a_6 = ra_5 = r(r^4 a_1) = r^5 a_1$$
$$\vdots$$

That is, the nth term, a_n, is given by

$$a_n = r^{n-1}a_1$$

This relationship enables us to determine any term in a geometric progression without calculating all the preceding terms. For example, if $a_1 = 4$ and $r = 3$, the seventh term is

$$a_7 = 3^6 \times 4 = 2{,}916$$

The 32nd term of the progression

$$1, 2, 4, 8, 16, 32, \ldots$$

is

$$a_{32} = 2^{31} \times 1 = 2{,}147{,}483{,}648$$

which is the number of pennies on the 32nd red square of the quiz program.

PRACTICE PROBLEM 2

If $a_1 = 3$ and $r = 2$, determine a_5 without calculating any other terms.

Answer

$$a_5 = 2^4 \cdot 3 = 48$$

EXAMPLE 1 Determine the first five terms of the geometric progression in which $a_2 = 36$ and $a_3 = 108$.

Solution Since $a_3 = ra_2$, we have $108 = r(36)$, from which it follows that $r = 3$. Once the value of r is known, it is easy to determine the three missing terms.

$$a_4 = 3a_3 = 324$$
$$a_5 = 3a_4 = 972$$
$$a_1 = \frac{a_2}{3} = 12$$

The first five terms are

$$12, 36, 108, 324, 972$$

EXAMPLE 2 Determine the first five terms of the geometric progression in which $a_1 = 2$ and $a_3 = 18$.

Solution Since $a_3 = r^2 a_1$, we have $18 = r^2(2)$, so that $r^2 = 9$. Consequently, r could be 3 or -3. Each case must be treated separately.

If $r = 3$,

$$a_1 = 2$$
$$a_2 = 6$$
$$a_3 = 18$$
$$a_4 = 54$$
$$a_5 = 162$$

9.1 GEOMETRIC PROGRESSIONS

If $r = -3$,

$$a_1 = 2$$
$$a_2 = -6$$
$$a_3 = 18$$
$$a_4 = -54$$
$$a_5 = 162$$

EXAMPLE 3 Can $a_1 = 2$, $a_2 = 4$, and $a_3 = 1$ be the first three terms of a geometric progression?

Solution Comparing a_1 and a_2 requires that r have the value $\frac{4}{2} = 2$. Comparing a_2 and a_3 requires that r have the value $\frac{1}{4}$. Consequently, no geometric progression can begin with the terms 2, 4, and 1, as the value of r must be constant in any one progression.

EXAMPLE 4 Mark is offered a job with a starting salary of $1,000 per month and is told that the average yearly raise is 6%. On this basis, what will be his monthly salary during the fifth year?

Solution His monthly salaries form a geometric progression with $a_1 = 1,000$ and $r = 1.06$. His monthly salary for the fifth year will be

$$a_5 = (1.06)^4 \times 1,000 = 1,262.48$$

or $1,262.48.

Quantities that are positive and form a geometric progression with a common ratio greater than 1 are said to ***increase geometrically.*** The ratio need not be much bigger than 1 for these quantities to increase very rapidly. For example, in the progression of pennies with $a_1 = 1$ and $r = 2$, the 32nd term was found to be

$$a_{32} = 2,147,483,648$$

This represents a huge increase from the first term, $a_1 = 1$.

---

3 PRACTICE PROBLEM 3

Determine the first five terms of each geometric progression.

(a) $a_2 = 4$ and $a_3 = 2$
(b) $a_1 = 4$ and $a_3 = 36$

Answer

(a) $a_1 = 8, a_2 = 4, a_3 = 2, a_4 = 1, a_5 = \frac{1}{2}$
(b) $r = 3$: $a_1 = 4, a_2 = 12,$ $a_3 = 36, a_4 = 108,$ $a_5 = 324$
   $r = -3$: $a_1 = 4, a_2 = -12, a_3 = 36, a_4 = -108, a_5 = 324$

---

4 PRACTICE PROBLEM 4

Can $a_1 = 2$, $a_2 = 4$, and $a_3 = 8$ be the first three terms of a geometric progression?

Answer

Yes ($r = 2$)

On the other hand, quantities that are positive and form a geometric progression with r less than 1 are said to **decrease geometrically**. These quantities generally tend to decrease very rapidly. In Figure 9.1, the various squares are obtained by dividing the next larger square into fourths. The areas of the different-size squares form a geometric progression with $r = \frac{1}{4}$. After only six different sizes, the area of the smallest square is $\frac{1}{1,024}$ times that of the largest.

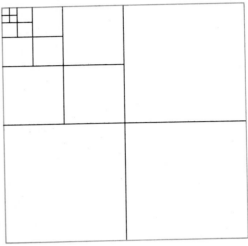

FIGURE 9.1

One of our major objectives in this section is to find a method for determining the sum of the numbers in a geometric progression. We could, of course, calculate the desired terms and add them—but this method becomes cumbersome when the number of terms is very large. There is a shorter method, which requires only the values of a_1, r, and n, the number of terms to be added.

Suppose we wanted to add the first n terms in a geometric progression and we denote their sum by S_n:

$$S_n = a_1 + a_2 + a_3 + \cdots + a_{n-1} + a_n$$

or, in terms of a_1 and r,

$$S_n = a_1 + a_1 r + a_1 r^2 + \cdots + a_1 r^{n-2} + a_1 r^{n-1}$$

If we multiply both sides of this last equation by r, we get

$$rS_n = a_1 r + a_1 r^2 + a_1 r^3 + \cdots + a_1 r^{n-1} + a_1 r^n$$

Subtracting corresponding sides of these last two equations gives

$$\begin{array}{r} S_n = a_1 + a_1 r + a_1 r^2 + \cdots + a_1 r^{n-1} \\ -rS_n = - a_1 r - a_1 r^2 - \cdots - a_1 r^{n-1} - a_1 r^n \\ \hline S_n - rS_n = a_1 - a_1 r^n \end{array}$$

Each side of this last equation can be factored to give
$$S_n(1-r) = a_1(1-r^n)$$
or
$$S_n = \frac{a_1(1-r^n)}{1-r} \qquad (1)$$

Equation (1) gives a method of determining S_n using only the values of a_1, n, and r (except for $r = 1$).

EXAMPLE 5 Calculate the sum of the first six terms in the geometric progression with $a_1 = 2$ and $r = 4$.

Solution In this case n, the number of terms to be added, is 6. We therefore want to calculate
$$S_6 = \frac{2(1-4^6)}{1-4} = \frac{2(1-4{,}096)}{-3} = 2{,}730 \qquad \blacksquare$$

EXAMPLE 6 Calculate the sum of the first seven terms in the geometric progression with $a_2 = 3$ and $a_3 = -6$.

Solution Comparing a_2 and a_3, we see that $r = -2$. Also, $a_1 = a_2/r = -\frac{3}{2} = -1.5$. Since $n = 7$, we calculate
$$S_7 = \frac{-1.5[1-(-2)^7]}{1-(-2)} = \frac{-1.5[1-(-128)]}{1+2} = -64.5 \qquad \boxed{5}$$

Applications of geometric progressions in the mathematics of finance will occur in our discussion of compound interest in later sections.

5 PRACTICE PROBLEM 5

Calculate the sum of the first six terms of each geometric progression.

(a) $a_1 = 4$ and $r = 3$
(b) $a_1 = 8$ and $a_2 = -4$

Answer

(a) 1,456   (b) 5.25

EXERCISES

Which of the following can be the first three terms of a geometric progression?

1. 5, 15, 45
2. $-2, 4, -8$
3. 3, 6, 9
4. 1, 5, 9
5. 4, 2, 1
6. $4, -2, 1$
7. $5, 1, \frac{1}{5}$
8. $-1, -\frac{1}{2}, -\frac{1}{4}$
9. $3, -3, 3$
10. 7, 7, 7
11. $-5, -15, 45$
12. 0.001, 0.005, 0.025
13. 0.1, 1, 100
14. $\frac{2}{3}, \frac{1}{2}, \frac{3}{8}$
15. $\frac{1}{3}, \frac{1}{5}, \frac{1}{15}$
16. $-0.05, 2, -80$

Without calculating any other terms, find the indicated term.

17. a_3 if $a_1 = 4$ and $r = 5$
18. a_3 if $a_1 = 7$ and $r = 4$
19. a_4 if $a_1 = 3$ and $r = -2$
20. a_4 if $a_1 = -4$ and $r = -3$
21. a_7 if $a_1 = 2$ and $r = -1$
22. a_5 if $a_1 = \frac{1}{27}$ and $r = 3$

23. a_6 if $a_1 = -5$ and $r = \frac{1}{2}$
24. a_5 if $a_1 = -2$ and $r = -\frac{1}{3}$
25. a_4 if $a_1 = 7$ and $r = 0.1$
26. a_4 if $a_1 = 10$ and $r = -0.2$
27. a_{27} if $a_1 = 0$ and $r = 9$
28. a_{32} if $a_1 = 17$ and $r = 1$

Determine the first five terms of the geometric progression(s) with the given properties.

29. $a_1 = 3$ and $r = 4$
30. $a_1 = 5$ and $r = -2$
31. $a_1 = -16$ and $r = \frac{1}{4}$
32. $a_1 = 4$ and $r = -\frac{1}{2}$
33. $a_1 = 1$ and $r = 0.01$
34. $a_1 = 10$ and $r = -0.1$
35. $a_1 = -2$ and $a_2 = 4$
36. $a_1 = 6$ and $a_2 = 3$
37. $a_2 = \frac{1}{9}$ and $a_3 = \frac{1}{3}$
38. $a_2 = 1{,}620$ and $a_3 = 1{,}749.6$
39. $a_1 = 15$ and $a_3 = 375$
40. $a_2 = 1.5$ and $a_4 = 13.5$

Find the sum of the indicated number of terms using equation (1).

41. The first five terms, $a_1 = 1$, $r = 2$
42. The first seven terms, $a_1 = -3$, $r = 2$
43. The first four terms, $a_1 = 6$, $r = -\frac{1}{3}$
44. The first eight terms, $a_1 = 4$, $r = \frac{1}{2}$
45. The first five terms, $a_1 = -2$, $a_2 = -1$
46. The first six terms, $a_1 = 1$, $a_2 = 3$
47. The first three terms, $a_2 = 1$, $a_3 = -4$
48. The first four terms, $a_2 = 2$, $a_3 = \frac{1}{2}$
49. The first 1,000 terms, $a_1 = 0$, $r = 5$
50. Use equation (1) twice to find the sum of the seventh, eighth, and ninth terms of the geometric progression in which $a_1 = 2$ and $r = 3$.
51. *Salary increments* If you are offered a job with a starting salary of $1,500 per month and are told that you can expect an annual raise of 11% for each of the next 3 years, what will be your monthly salary during the fourth year?
52. *Inflation rate* If it takes $15,000 this year to maintain a family of four at a reasonable standard of living and the rate of inflation is 10% a year for the next 3 years, how much money will be required to maintain the same family at the same standard of living 3 years hence? Determine the correct amount by finding the fourth term in an appropriate geometric progression. (*Hint:* $r = 1.10$.)
53. *Antibiotic effect* The number of a certain type of bacteria in an organism decreases about $\frac{1}{3}$ every hour for the first 4 hours after an antibiotic is administered. Determine the number of bacteria present at the end of the fourth hour after the antibiotic is administered in a specimen originally containing 9,000 bacteria.
54. *Chemical dilution* A chemical is received in concentrated form in 1-gallon containers. Each time the chemical is used, $\frac{1}{3}$ of the container is removed and replaced by water. After the third time the chemical is used, how much of the original concentrate is left in the container?

55. A water lily that doubles its size every day is planted in a pond. If it fills the pond in 10 days, how many days would it take two such water lilies to fill the pond?

56. *Salary increments* Suppose that a company offers you a starting annual salary of $15,500 with a promise of an 8% raise for each of the next 3 years.
    (a) Determine your salary for the fourth year.
    (b) Use equation (1) to determine your total salary for the first 4 years.

57. *Average salary* Use equation (1) to calculate the average (mean) monthly salary that Mark of Example 4 will make during the first 5 years.

58. Determine the total amount of money on the checkerboard in the illustration at the beginning of this section after the 32nd red square is filled.

Explain the meaning of each term.

59. Geometric progression

60. Common ratio of a geometric progression

61. Quantities that increase geometrically

62. Determine an expression for the sum of the first n terms of a geometric progression if $r = 1$.

9.2 SIMPLE INTEREST

The interest involved in any money-lending situation has two aspects—there are two sides of the coin, so to speak. On the side of the borrower, interest is payment made to someone else for the use of his or her money. In some situations the money itself may be passed from one person to another, as when a bank loan is made, but it need not be. When a purchase is made by using a credit card or on an installment plan, the buyer is using the money that the seller has put into the item purchased.

From the point of view of the lender, interest is payment received for the use of his, her, or its money. This is true whether the lender is a savings and loan association financing a new home or an individual putting his or her money into a savings account so that the bank can use the money for its commercial purposes.

Simple interest is determined only by the amount of money borrowed or loaned, the rate of interest, and the duration of the loan. If you were to borrow $150 from your bank at 12% per year for 2 years, the interest you would pay is

$$150 \times 0.12 \times 2 = 36$$

or $36.00. This $36.00 is *simple interest*.

The $150 in the above transaction is the **principal**, the 12% per year is the **interest rate**, and the loan duration of 2 years is called the **time**. With this terminology, the **simple interest** owed at the end of 2 years is given by

$$\text{Principal} \times \text{Rate} \times \text{Time}$$

or, with the obvious notation,

$$\boxed{I = PRT} \tag{1}$$

Simple interest is always calculated using equation (1).

Should your loan have been arranged in such a way that during the second year you would pay interest on the interest accumulated during the first year, the interest on the loan would have been slightly higher:

For the first year: $I = 150 \times 0.12 \times 1 = 18.00$
For the second year: $I = 168 \times 0.12 \times 1 = \underline{20.16}$
Total interest: 38.16

When interest is paid on interest, as in this last case, the result is compound interest. This section treats only simple interest and considers various situations in which simple interest is applicable.

Note that both the interest rate and the duration of a loan involve a time factor. Both are expressed in terms of some unit, such as days, months, or years. In all interest calculations, *both interest rate and duration of loan must be expressed in terms of the same unit of time.*

EXAMPLE 1   Stephen borrowed $200 from his credit union for 2 months at 9% per year. Find the amount of interest that he will have to pay.

Solution   Since 2 months equals $\frac{1}{6}$ year, the interest he will have to pay is

$$200 \times 0.09 \times \tfrac{1}{6} = 3.00$$

If we choose to do the calculations in terms of months, the interest rate must be expressed as $\frac{9}{12}\%$, or $\frac{3}{4}\%$, per month. The amount of interest is the same:

$$200 \times 0.0075 \times 2 = 3.00$$

1 PRACTICE PROBLEM 1

If you borrow $300 for 6 months at 10% per year, how much interest will you have to pay?

Answer

$15.00

Throughout this chapter, *all interest rates will be understood to be annual rates, unless otherwise indicated.*

Certificates of deposit (C.D.'s) are essentially savings accounts in which a fixed sum of money is deposited for a specified period of time. Simple interest is paid to the owner of the certificate at regular time intervals, typically every 3 months. There may or may not be a minimum amount of money that must be deposited, and the owner pays a penalty in interest received if the money is withdrawn before the time period has elapsed.

9.2 SIMPLE INTEREST

EXAMPLE 2 A $10,000, 12-month C.D. pays 14% interest. The owner is paid simple interest every 3 months. Determine the amount of quarterly interest received.

Solution The quarterly interest is

$$10{,}000 \times 0.14 \times \tfrac{1}{4} = 350$$

or $350.

PRACTICE PROBLEM 2

A $5,000, 12-month C.D. pays 12% interest. Determine the amount of quarterly interest paid the owner.

Answer

$150.00

When a loan is made and the interest is subtracted from the principal before the borrower receives the money, the loan is said to be **discounted**. If an individual arranges for a loan of $300 for 2 years discounted at 10%, the amount of money he or she actually receives is

$$300 - (300 \times 0.10 \times 2) = 240$$

or $240; the interest is subtracted from the principal when the loan is made. The effect is that the individual is paying a rate of interest higher than 10% on the $240 received.

EXAMPLE 3 Jack borrowed $250 for 3 months discounted at 12%. What rate of interest did he pay on the amount of money that he received?

Solution The amount of money that he received was

$$250 - (250 \times 0.12 \times \tfrac{1}{4}) = 242.50$$

or $242.50, for which he is paying $7.50 interest. We want to determine what interest rate this $7.50 represents relative to the $242.50 he receives.

From equation (1), we get

$$\boxed{R = \frac{I}{PT}} \qquad (2)$$

Using the appropriate values in equation (2), Jack's interest rate is

$$R = \frac{7.50}{242.50 \times \tfrac{1}{4}} = 0.1237$$

or 12.37% per year.

PRACTICE PROBLEM 3

If, in Example 3, Jack had borrowed the $250 for 6 months discounted at 12%, what rate of interest would he be paying for the money that he received?

Answer

12.77% per year

EXAMPLE 4 Eric is buying a piece of land for which he can either pay $6,000 now or $6,500 in 3 years. He has the $6,000 invested at $5\tfrac{1}{2}\%$ simple interest. Should he use the $6,000 to pay for the land now or should he wait?

Solution The interest that he would earn on the $6,000 during the 3 years is

$$6{,}000 \times 0.055 \times 3 = 990$$

or $990. Since his $6,000 would be worth $6,990 in 3 years, he would be ahead $490 at that time by waiting.

An alternative method of deciding the question asked in this last example is to find the sum of money which, when invested now at $5\frac{1}{2}\%$, would yield an amount of $6,500 in 3 years. This sum, called the *present value* of $6,500, is easy to determine.

The principal plus the interest of a loan (or investment) is called the *amount*. The *present value* of a particular amount is the principal needed to yield that amount in the specified time. If we designate the amount by S, then

$$S = P + I$$

or, from equation (1),

$$S = P + PRT$$

which becomes

$$S = P(1 + RT)$$

Solving this last equation for P gives

$$P = \frac{S}{1 + RT} \qquad (3)$$

Equation (3) gives the present value of any amount for investments paying simple interest. In Example 4,

$$P = \frac{6,500}{1 + (0.055 \times 3)} = \frac{6,500}{1.165} = 5,579.40$$

which also indicates that Eric should wait. $5,579.40 invested now at $5\frac{1}{2}\%$ will amount to $6,500 in 3 years; $6,000 at the same rate will amount to more.

EXAMPLE 5 What is the present value of $500 due in 6 months at 8%?

Solution From equation (3),

$$P = \frac{500}{1 + (0.08 \times \frac{1}{2})} = \frac{500}{1.04} = 480.77$$

The present value of the $500 is $480.77.

The present value of a future sum of money emphasizes the time value of money. In Example 5, $500 in 6 months is presently worth $480.77 invested at 8% simple interest. These differences become more dramatic as the amount of money, interest rate, and/or time interval increase.

4 PRACTICE PROBLEM 4

What is the present value of $400 due in 1 year at 10%?

Answer

$363.64

EXERCISES

Determine the amount of simple interest earned or paid in each case.

1. $5,000 invested for 2 years at 10%
2. $750 invested for 18 months at 8%
3. $900 invested for 15 months at 9.5%
4. $5,000 borrowed for 2 years at 10%
5. $1,200 borrowed for 16 months at 12%
6. $300 borrowed for 8 months at 11.25%
7. Mark deposits $200 in his credit union account, which pays $6\frac{1}{2}\%$ interest. Determine the amount in his account 4 months later when the interest is added.
8. Joan borrows $350 from her bank for 90 days at 12%. How much interest will she have to pay? What amount will she owe the bank at the end of the 90 days? (Consider a year to have 360 days.)
9. *Certificate of deposit* A $15,000, 12-month C.D. pays 14% interest. The owner is paid simple interest every 6 months. Determine the semiannual interest received.
10. *Certificate of deposit* A $5,000, 30-month C.D. pays 11% interest. The owner is paid simple interest every 3 months. Determine the quarterly interest received.
11. Suppose that your checking account service charge runs about $3.00 each month. If you maintain a minimum balance of $400 each month, there would be no service charge. Would it be to your advantage to move $400 from your savings account (which pays 6%) to your checking account in order to maintain the minimum balance?

Determine the rate of interest involved if you earned or paid interest as indicated.

12. $72 earned on $600 invested for 18 months
13. $140 earned on $1,000 invested for 2 years
14. $77.00 paid for $700 borrowed for 1 year
15. $296.88 earned on $2,500 invested for 15 months
16. $131.25 paid for $750 borrowed for 20 months
17. $33.75 paid for $500 borrowed for 9 months
18. $525 earned on $3,500 invested for 16 months
19. Eric borrowed $500 discounted at 12% for 6 months.
    (a) How much money did he receive when the loan was made?
    (b) What rate of interest is he paying for the money actually received?
20. Catherine borrowed $1,200 for 1 year discounted at 10%.
    (a) How much money did she receive when the loan was made?
    (b) What rate of interest is she paying for the money actually received?
21. *Treasury bill* Sally bought a 90-day (3-month), $10,000 U.S. Treasury bill discounted at 12%; that is, she paid $9,700 for the bill, which the Trea-

sury Department will redeem in 90 days for $10,000. What rate of interest will she be making on her $9,700?

22. *Corporate bonds* Kay buys five $1,000 corporate bonds, which pay interest every 6 months at the rate of 18% per year. Determine the amount of the interest check that she receives every 6 months.

23. Kay (Exercise 22) thought that she could make better use of her money, so she sold her bonds to Susan for $925 each. Susan now receives the semiannual payments. What rate of interest on her investment is Susan receiving from these payments?

24. After Susan (Exercise 23) has the bonds for 10 years, the corporation redeems them (buys them back) for $1,000 each. Find the sum of all her semiannual payments and her profit from the sale of the bonds. Use this sum to determine the annual rate of return on her investment.

25. *Automobile loan* Suppose that you made a $6,000 car loan to be repaid over 3 years in monthly payments of $210.50.
    (a) Determine the total amount to be paid over the 3 years.
    (b) How much of the total paid over the 3 years will be for interest?
    (c) If $135.50 of your first month's payment is applied toward the principal, the rest being interest, what is the monthly rate of interest? The annual rate?

What is the present value of each amount?

26. $1,000 due in 1 year at 14%
27. $350 due in 3 months at 12%
28. $1,078.25 due in 18 months at 9%
29. $832 due in 6 months at 8%

30. How much money invested now at 6% will amount to $450 in 2 years?
31. Gene can settle a debt by either paying $200 now or $208 in 6 months. He has the money invested at $5\frac{1}{2}$%. Which is the better plan for him?

*9.3 AVERAGE DAILY BALANCE

Most charge card accounts (department store, oil company, bank card, and so forth) charge interest each month on the account balance at some specified rate. A rate of 1.5% per month seems fairly typical. The principal on which this interest is calculated is quite often the average daily balance in the account. Our purpose in this section is to examine one method in which interest is calculated in this manner.

Customers with this type of account are usually billed each month on the same day of the month—for example, on January 17, February 17, March 17, and so forth—insofar as possible. To simplify matters, we shall assume that the months considered all have 30 days. The procedure is essentially the same for months with 28, 29, or 31 days.

Suppose that you have an account of this type on which you must pay 1.5% per month interest on the average daily balance. Suppose also that you begin the month (billing period) with a balance of $200.00

---

* This section can be omitted without loss of continuity.

9.3 AVERAGE DAILY BALANCE

owed to the account and that the following activity takes place in your account during the month:

DAY	ACTIVITY	
7	Charge:	$ 35.00
15	Payment:	150.00
20	Charge:	60.00
30	Customer billed	

To calculate your average daily balance we construct Table 9.1. Your average daily balance for the 30 days would be 5,100.00 ÷ 30 = 170.00, or $170.00. At 1.5% per month, the interest charged for the month would be

$$170 \times 0.015 \times 1 = 2.55$$

Note that 1.5% per month is the same as 18% per year!

TABLE 9.1

DAYS	BALANCE	NUMBER OF DAYS × BALANCE	TOTAL
1–6	$200.00	6 × 200.00	= $1,200.00
7–14	235.00	8 × 235.00	= 1,880.00
15–19	85.00	5 × 85.00	= 425.00
20–30	145.00	11 × 145.00	= 1,595.00
			$5,100.00

EXAMPLE 1 Determine the average daily balance and the interest at 1.5% per month based on the average daily balance for an account with the given activity. Opening balance: $160.50.

DAY	ACTIVITY	
4	Charge:	$ 20.25
10	Charge:	16.00
20	Payment:	140.00
25	Charge:	15.15
30	Customer billed	

Solution We construct a table similar to Table 9.1:

DAYS	BALANCE	NUMBERS OF DAYS × BALANCE	TOTAL
1–3	$160.50	3 × 160.50	= $ 481.50
4–9	180.75	6 × 180.75	= 1,084.50
10–19	196.75	10 × 196.75	= 1,967.50
20–24	56.75	5 × 56.75	= 283.75
25–30	71.90	6 × 71.90	= 431.40
			$4,248.65

9 MATHEMATICS OF FINANCE

Average daily balance: $4{,}248.65 \div 30 = 141.62$

Interest: $141.62 \times 0.015 \times 1 = 2.12$

The average daily balance is $141.62, and the interest is $2.12.

1 PRACTICE PROBLEM 1

Determine the average daily balance and interest in Example 1 if the charge on day 25 was for $55.00 instead of $15.15.

Answer

$149.59; $2.24

Another situation in which the average daily balance can be utilized is in determining interest in NOW (negotiable order of withdrawal) accounts. These are essentially checking accounts that pay interest. The details of these accounts vary from one financial institution to another, but they generally require that a minimum balance (for example, $500.00) be maintained in the account and impose a service charge when the balance falls below the required minimum. The method in which the interest is earned by the accounts can also vary among different institutions; we shall consider NOW accounts in which the interest is calculated on the average daily balance.

The process of calculating the average daily balance in these types of accounts is the same as that in charge-card accounts except in this case, checks that clear reduce the balance and deposits made increase the balance.

Suppose you have a NOW account that pays $5\frac{1}{2}\%$ interest on the average daily balance, that you start a month with a balance of $575.00, and that the activity in your account is as follows:

DAY	ACTIVITY	
5	Check cleared:	$ 26.00
9	Check cleared:	10.00
12	Deposit:	270.00
14	Checks cleared:	65.00
		43.00
17	Deposit:	140.00
20	Check cleared:	210.00
30	Interest paid on account at close of business day	

To calculate your average daily balance, we construct Table 9.2, similar to those used for charge-card accounts.

TABLE 9.2

DAYS	BALANCE	NUMBER OF DAYS × BALANCE	TOTAL
1–4	$575.00	4 × 575.00	= $ 2,300.00
5–8	549.00	4 × 549.00	= 2,196.00
9–11	539.00	3 × 539.00	= 1,617.00
12–13	809.00	2 × 809.00	= 1,618.00
14–16	701.00	3 × 701.00	= 2,103.00
17–19	841.00	3 × 841.00	= 2,523.00
20–30	631.00	11 × 631.00	= 6,941.00
			$19,298.00

Then,

Average daily balance: $19{,}298.00 \div 30 = 643.27$

Interest: $643.27 \times 0.055 \times \frac{1}{12} = 2.95$

The average daily balance is $643.27 and the interest is $2.95.

② PRACTICE PROBLEM 2

Determine the average daily balance and interest in the NOW account if the check that cleared on day 9 was for $45.00 instead of $10.00.

Answer

$617.60; $2.83

EXERCISES

1. *Charge-card account* Determine the average daily balance and the interest charged at 1.5% per month on the average daily balance for a charge-card account with the given activity during 1 month. Opening balance: $73.00.

DAY	ACTIVITY	
3	Charge:	$15.00
10	Charge:	22.00
14	Charge:	13.00
17	Payment:	70.00
19	Charge:	14.00
30	Customer billed	

2. *Charge-card account* Determine the average daily balance and the interest charged at 1.5% per month on the average daily balance for a charge-card account with the given activity during 1 month. Opening balance: $116.42.

DAY	ACTIVITY	
4	Charge:	$ 23.00
10	Charge:	17.40
15	Payment:	100.00
20	Charges:	15.49
		41.07
25	Charge:	18.80
30	Customer billed	

3. *Charge-card account* Determine the average daily balance and the interest charged in Exercise 1 if the payment of $70.00 is made on day 5 instead of day 17.

4. *Bank card account* On his trip to the Republican convention, Dan charged a total of $435.00 on his bank card. On the fifteenth day of each of the next 3 months, he intends to pay $115.00 on his account, paying the balance on the fifteenth day of the fourth month. Determine the total amount of interest he will have to pay under the following conditions:
   (i) The charges are entered on his account on day 20.
   (ii) His previous account balance was zero.
   (iii) Interest is charged at the rate of 1.5% per month.
   (iv) Except for these interest charges and payments, no other activity takes place in his account.
   (v) No interest is charged in months in which the account closes the month with a zero balance.
   (vi) For convenience, all months are considered to have 30 days.

5. *NOW account* Determine the average daily balance and the interest earned at the rate of 5.5% on the average daily balance for a NOW account that has the given activity during 1 month. Opening balance: $520.00.

DAY	ACTIVITY	
4	Check cleared:	$ 15.00
10	Deposit:	136.00
12	Check cleared:	25.00
18	Check cleared:	43.00
25	Deposit:	120.00
30	Interest paid on account at close of business day	

6. *NOW account* Determine the average daily balance and the interest earned at the rate of 5.25% on the average daily balance for a NOW account with the given activity during 1 month. Opening balance: $624.30.

DAY	ACTIVITY	
1	Check cleared:	$ 74.68
5	Check cleared:	26.43
9	Deposit:	361.98
17	Checks cleared:	123.42
		64.50
		25.00
23	Check cleared:	14.45
25	Deposit:	79.63
27	Checks cleared:	16.43
		27.19
30	Interest paid on account at close of business day	

7. *NOW account* Determine the average daily balance and the interest earned in the NOW account of Exercise 5 if the check cleared on the twelfth day is for $74.75 instead of for $25.00.

8. *NOW account* Determine the average daily balance and the interest earned in the NOW account of Exercise 6 if the deposit on the ninth day is for $225.50 instead of for $361.98.

9.4 COMPOUND INTEREST

It is probably fair to say that most financial transactions that involve an extended period of time are based on compound interest. In any case, it is an integral element in our financial system. In *The Research Revolution,* Leonard Silk writes, "Explaining compound interest—for a bank or a nation—is the central problem of economics."*

We had a glimpse into the determination of compound interest in Section 9.2; before examining this topic in more detail, let us take a look at the calculations over a longer period of time. In particular, we

* Leonard S. Silk, *The Research Revolution,* New York: McGraw-Hill, 1960, p. 38.

9.4 COMPOUND INTEREST

shall determine the interest for 4 years on $1,000 invested at 8% compounded each year.

FIRST YEAR
Interest: $1,000 \times 0.08 \times 1 = 80$
Compound amount: $1,000 + 80 = 1,080$

SECOND YEAR
Interest: $1,080 \times 0.08 \times 1 = 86.40$
Compound amount: $1,080.00 + 86.40 = 1,166.40$

THIRD YEAR
Interest: $1,166.40 \times 0.08 \times 1 = 93.31$
Compound amount: $1,166.40 + 93.31 = 1,259.71$

FOURTH YEAR
Interest: $1,259.71 \times 0.08 \times 1 = 100.78$
Compound amount: $1,259.71 + 100.78 = 1,360.49$

Note that compound interest is simple interest computed again and again. The difference between compound and simple interest is that (after the first period) compound interest is based on the interest from the previous periods as well as on the original principal; simple interest is based only on the original principal.

To compare the effects of compound and simple interest, Table 9.3 gives the balance in an imaginary account at the end of a number of years for both compound and simple interest. The table indicates that interest on interest, even though it may amount to only a few dollars in the beginning, can soon become a significant amount.

The year-by-year computation in order to determine the compound amount, as in the illustration at the beginning of this section, soon becomes very tedious. So one looks for an easier method. Consider the general case where P dollars are invested (or borrowed) at a rate of i per year. The computations are as follows:

TABLE 9.3 ACCOUNT BALANCE ON $1,000 INVESTED AT 8%

AT END OF YEAR	COMPOUND INTEREST	SIMPLE INTEREST
1	$1,080.00	$1,080.00
2	1,166.40	1,160.00
3	1,259.71	1,240.00
4	1,360.49	1,320.00
5	1,469.33	1,400.00
6	1,586.87	1,480.00
7	1,713.82	1,560.00
8	1,850.93	1,640.00
9	1,999.00	1,720.00
10	2,158.93	1,800.00
15	3,172.17	2,200.00
20	4,660.96	2,600.00

FIRST YEAR
Interest: $P \times i = Pi$ Since Time = 1, it is omitted.
Compound amount: $P + Pi = P(1 + i)$

SECOND YEAR
Interest: $P(1 + i) \times i = P(1 + i)i$
Compound amount: $P(1 + i) + P(1 + i)i = P(1 + i)(1 + i)$
$= P(1 + i)^2$

THIRD YEAR
Interest: $P(1 + i)^2 \times i = P(1 + i)^2 i$
Compound amount: $P(1 + i)^2 + P(1 + i)^2 i = P(1 + i)^2(1 + i)$
$= P(1 + i)^3$

In each case, the compound amount at the end of the nth year is $P(1 + i)^n$. Continued calculations would verify that the same expression gives the compound amount for any number of years. Consequently, if we denote the compound amount by S, then S is given by the equation

$$S = P(1 + i)^n \qquad (1)$$

EXAMPLE 1 Find the compound amount of $1,000 invested for 4 years at 8%.

Solution In this case $P = 1,000$, $i = 0.08$, and $n = 4$. From equation (1),

$$S = 1,000(1 + 0.08)^4 = 1,000(1.08)^4 \approx 1,000(1.36049) = 1,360.49$$

The compound amount is $1,360.49. (Compare this result with the illustration at the beginning of this section.) ■

Equation (1) was derived on the assumption that i is the annual rate and that the interest is compounded annually. Although i is normally stated on an annual basis, the interest is usually compounded for such periods as daily, monthly, or quarterly. Equation (1) is still valid if n is taken to be the number of conversion periods (a **conversion period** being the length of time between interest computations) and i is the interest *per conversion period*.

EXAMPLE 2 Find the compound amount of $1,000 invested for 4 years at 8% compounded semiannually.

Solution The conversion period is 6 months, or half a year. The interest rate for half a year is $i = \frac{8}{2}\%$, or 0.04, and the number of semiannual conversion periods is $n = 8$. The compound amount is

$$S = 1,000(1 + 0.04)^8 \approx 1,000(1.36857) = 1,368.57$$

or $1,368.57. Note the larger amount here in comparison to that of Example 1, where the interest was compounded annually. ■

The compound amounts at the end of the conversion periods determine a sequence of numbers,

$$P(1 + i), P(1 + i)^2, P(1 + i)^3, P(1 + i)^4, \ldots$$

Does this look familiar? It is a geometric progression with $r = 1 + i$. These compound amounts form a geometric progression with r greater than one and therefore *increase geometrically*. With good reason Baron Rothschild, who was more than a little familiar with banking, referred to compound interest as the eighth wonder of the world!

The value of $(1 + i)^n$ can be determined using your hand calculator. If it does not have a button for evaluating powers, the value of $(1 + i)^n$ for n fairly large can be calculated using the multiplication button and properties of exponents. For example, $(1 + 0.04)^{12} = (1.04)^{12}$ can be calculated using the multiplication button in the following steps:

$$(1.04)^2 = (1.04)(1.04) = 1.0816$$
$$(1.04)^4 = (1.04)^2(1.04)^2 \approx 1.1698586$$
$$(1.04)^8 = (1.04)^4(1.04)^4 \approx 1.3685691$$
$$(1.04)^{12} = (1.04)^4(1.04)^8 \approx 1.6010323$$

As a last resort, values of $(1 + i)^n$ for various values of i and n are given in Table VI of Appendix B.

EXAMPLE 3 Tom borrowed $2,500 for 3 years at 16% compounded quarterly. Find the amount he has to pay in 3 years and the interest that he pays for his loan.

Solution His total payment, including principal and interest, will be the compound amount when $P = 2,500$, $i = \frac{16}{4}\%$, or 4%, and $n = 12$. The value of $(1 + 0.04)^{12}$ is 1.60103, rounded to five decimal places. Therefore,

$$S = 2,500 \times 1.60103 = 4,002.58$$

Tom will have to repay $4,002.58. His interest is

$$\$4,002.58 - 2,500.00 = \$1,502.58$$

In order to determine how much money P should be invested now with compound interest in order to yield a desired amount S, we rewrite equation (1) as

$$\frac{S}{(1 + i)^n} = P$$

or

$$\boxed{P = S(1 + i)^{-n}} \qquad (2)$$

PRACTICE PROBLEM 1

Find the compound amount of $1,500 invested for 3 years at each rate.

(a) 8% compounded quarterly
(b) 12% compounded monthly

Answer

(a) $1,902.36   (b) $2,146.15

PRACTICE PROBLEM 2

If you borrow $500 for 2 years at 9% compounded monthly, find the amount you have to repay in 2 years and the interest that you pay for the loan.

Answer

$598.21; $98.21

As with simple interest, the quantity P is called the *present value* of S. The value of $(1 + i)^{-n}$ can be determined in several ways using a calculator. The quickest way is to use the button for evaluating powers, if there is one. Otherwise, the value of $(i + 1)^n$ can be determined as outlined earlier, and its reciprocal, $(1 + i)^{-n}$, can be determined using the $\boxed{1/x}$ button or else the division button, since

$$(1 + i)^{-n} = 1 \boxed{\div} (1 + i)^n$$

Again, as a last resort, values of $(1 + i)^{-n}$ for some values of i and n are given in Table VII of Appendix B.

EXAMPLE 4 How much money should be invested now at 12% compounded monthly to yield $1,800 in 2 years?

Solution $S = 1,800$, $i = \frac{12}{12}\%$, or 1%, $n = 24$, and $(1 + 0.01)^{-24} = 0.78757$, rounded to five decimal places. Therefore,

$$P = 1,800 \times 0.78757 = 1,417.63$$

That is, $1,417.63 invested now will yield $1,800 in 2 years.

EXAMPLE 5 What is the present value at 10% compounded semiannually of $2,400 due the bank in 2 years?

Solution $S = 2,400$, $i = \frac{10}{2}\%$, or 5%, $n = 4$; therefore,

$$P = 2,400 \times (1 + 0.05)^{-4} = 1,974.49$$

The present value is $1,974.49.

Note that the four-step procedure was involved in the application of mathematics in Example 4. The physical situation was that of an individual or institution investing a sum at 12% compounded monthly. The physical question was, How much should be invested now in order to yield $1,800 in 2 years? The mathematical model was the equation

$$P = S(1 + i)^{-n}$$

and the mathematical question was, What is the value of P when $i = 0.01$, $n = 24$, and $S = 1,800$? The answer to the mathematical question was 1,417.63. The interpretation was that $1,417.63 should be invested now to yield $1,800 in 2 years.

3 PRACTICE PROBLEM 3

What is the present value of $2,400 due the bank in 2 years at 12% compounded quarterly?

Answer

$1,894.58

EXERCISES

Determine the compound amount of $2,000 invested for 4 years compounded annually at 7% in each manner.

1. By calculating the interest for each of the 4 years, as in the illustration on page 351
2. By first calculating the value of $(1 + 0.07)^4$, as in Example 1
3. Using Table VI

Find the compound amount in each case.

4. $700 for 3 years at 12% compounded quarterly
5. $2,500 for 4 years at 10% compounded semiannually
6. $1,200 for $2\frac{1}{2}$ years at 8% compounded semiannually
7. $1,500 for 1 year at 18% compounded monthly

Find the total amount to be repaid and the total interest on each loan.

8. $800 for 3 years at 8% compounded quarterly
9. $500 for 2 years at 12% compounded monthly
10. $690 for 6 months at 15% compounded monthly
11. $475 for 15 months at 6% compounded quarterly
12. *Savings account* A father started a savings account for his daughter with a deposit of $200 the day she was born. The account pays 12% compounded quarterly. Determine the amount in the account on her sixth birthday. (Assume that her birthday occurs at the end of a conversion period.)
13. *Savings certificate* A savings certificate for $2,000 earns 9% compounded monthly. Determine the value of the certificate after 2 years.
14. Bob bought a piece of land for $2,750. He expects it to increase in value by about 7% a year. On this basis, what will the land be worth in 5 years?

Determine the present value of each.

15. $250 at 8% compounded quarterly due in 5 years
16. $1,500 at 12% compounded monthly due in 4 years
17. $1,750 at 10% compounded semiannually due in 3 years
18. $875 at 9% compounded monthly due in 3 years
19. In Example 4 of Section 9.2, if Eric has the $6,000 invested at $5\frac{1}{2}$% compounded monthly, should he pay for the land now or should he wait? (Having the $6,000 invested with compound interest is probably more realistic than having it invested with simple interest.)
20. From a financial point of view which is worth more to you, $600 now that you can invest at 10% compounded semiannually or $1,000 in 5 years?

Identify the steps of the four-step procedure involved in the application of mathematics (Section 1.3) in each case.

21. *Savings certificate* Exercise 13   22. Exercise 20
23. Explain the difference between simple interest and compound interest.

9.5 ORDINARY ANNUITIES

Table 9.4 gives the account activity for an individual who, for 5 consecutive years, deposits $1,000 into an account at the end of each year. The money earns 8% interest compounded annually. The last entry in each row gives the compound amount of the deposit in that row. For

TABLE 9.4

END OF YEAR	DEPOSIT	COMPOUND AMOUNT AT THE END OF 5 YEARS
1	$1,000	$1,000(1.08)^4 = \$1,360.49$
2	1,000	$1,000(1.08)^3 = 1,259.71$
3	1,000	$1,000(1.08)^2 = 1,166.40$
4	1,000	$1,000(1.08)^1 = 1,080.00$
5	1,000	$1,000 = 1,000.00$
		Total: $\$5,866.60$

example, the first deposit, which is made at the end of the first year, will draw compound interest for 4 years. The sum of all these compound amounts is the total amount in the account at the end of the fifth year.

Examine carefully the sum that gives this total. In reverse order, it is

$$1{,}000 + 1{,}000(1.08) + 1{,}000(1.08)^2 + 1{,}000(1.08)^3 + 1{,}000(1.08)^4$$

This is the sum of the first five terms of a geometric progression with $a_1 = 1,000$ and $r = 1.08$. Using equation (1) of Section 9.1, this total should be

$$\frac{1{,}000[1 - (1.08)^5]}{1 - 1.08}$$

which simplifies to 5,866.60.

Any series of equal payments or deposits made at regular intervals, as in this illustration, is called an **annuity**. This includes car payments, house payments, and insurance payments. The details of an annuity can be quite varied, so that there are different types of annuities. We shall consider only the basic type, called an **ordinary annuity**. In an ordinary annuity, payments are made at the end of the payment period, interest is compound interest, and the conversion period agrees with the payment period—that is, if the payments are monthly, the interest is compounded monthly; if the payments are semiannual, the interest is compounded semiannually, and so on.

The **amount**, or value, of an annuity at the end of any payment period is just the sum of the compound amounts of the individual payments at that time. The amount of the annuity in the last illustration at the end of 5 years is $5,866.60. We are interested in a general expression for the amount of an annuity. Suppose that the periodic payments are R dollars each and that the rate of interest per payment period is i. The compound amounts of these payments form a geometric progression with $a_1 = R$ and $r = (1 + i)$. In the last illustration, R is $1,000 and i is 0.08. The sum of the n terms of the progression is

$$S = \frac{R[1 - (1 + i)^n]}{1 - (1 + i)} = \frac{R[1 - (1 + i)^n]}{-i}$$

9.5 ORDINARY ANNUITIES

or

$$S = R\frac{[(1+i)^n - 1]}{i} \tag{1}$$

Equation (1) gives the amount of an ordinary annuity when the nth payment is made.

Using the standard notation

$$s_{\overline{n}|i} = \frac{(1+i)^n - 1}{i}$$

we can rewrite equation (1) in the form

$$S = R\left[\frac{(1+i)^n - 1}{i}\right] = Rs_{\overline{n}|i} \tag{2}$$

where:

R is the periodic payment.
i is the interest per payment period.
S is the amount of the annuity when the nth payment is made.

The quantity $s_{\overline{n}|i}$ (read, "s angle n at i") is the *amount of $1 per period for n periods at the rate i per period*. Values of $s_{\overline{n}|i}$ can be determined using your calculator with the value of $(1+i)^n$ determined by one of the methods indicated in the last section.

EXAMPLE 1 Find the amount of an annuity of $50 per quarter for 3 years at 8% compounded quarterly.

Solution The interest per quarter is $i = \frac{8}{4}\%$, or 2%, $R = 50$, and the number of payments is $n = 12$. Also,

$$s_{\overline{12}|0.02} = \frac{(1+0.02)^{12} - 1}{0.02} \approx \frac{1.26824 - 1}{0.02} = 13.412$$

so that

$$S = 50(13.412) = 670.60$$

The amount of the annuity is $670.60. ∎

EXAMPLE 2 Marilyn, who plans to go to Europe in 3 years, puts $100 per month into a special account, which pays 12% compounded monthly. How much will she have for her trip at the end of 3 years?

Solution The payments constitute an annuity with $i = \frac{12}{12} = 1\%$, $R = 100$, and $n = 36$. Since

$$s_{\overline{36}|0.01} = \frac{(1+0.01)^{36} - 1}{0.01} \approx \frac{1.43077 - 1}{0.01} = 43.077$$

1 PRACTICE PROBLEM 1

Find the amount of an annuity of $25 every 6 months for 4 years at 10% compounded semiannually.

Answer

$238.73

she will have

$$S = 100(43.077) = 4{,}307.70$$

or $4,307.70 for her trip.

Although we use equation (2) to compute the amount of an annuity, we should not lose sight of the fact that we are finding the sum of terms in a geometric progression. In fact, the mathematical model of the compound amounts at the end of the payment periods is the geometric progression. The sum of the terms answers the question, What is the amount after n periods? Does this process sound familiar?

A fund accumulated systematically to obtain a sum of money needed at some future date, such as that in Example 2, is called a **sinking fund**. Sinking funds are used to pay off debts, replace old equipment, expand manufacturing facilities, and so forth. A common type of sinking fund is an ordinary annuity; all the sinking funds discussed here are of this type.

In particular, we are interested in how much the periodic payments (R) should be in order that the amount (S) in the sinking fund will be a specified value on the required date. The size of the payments is the value of R in equation (1). Solving this equation for R gives

$$R = S\left[\frac{i}{(1+i)^n - 1}\right] \qquad (3)$$

EXAMPLE 3   A company plans to replace a piece of equipment at an estimated cost of $25,000 in 4 years. A sinking fund is established by making fixed monthly payments into an account paying 12% interest compounded monthly. How much should each payment be so that the amount in the account at the end of 4 years will be $25,000?

Solution   In equation (3), $i = 0.01$, $n = 48$, and $S = 25{,}000$. Therefore,

$$R = 25{,}000\left[\frac{0.01}{(1+0.01)^{48} - 1}\right] \approx 25{,}000\left[\frac{0.01}{1.61223 - 1}\right]$$

$$= \frac{250}{0.61223} \approx 408.34$$

The monthly payments should be $408.34.

2 PRACTICE PROBLEM 2

In Example 3, if the replacement is to be made in 3 years instead of 4, how much should each payment be?

Answer

$580.36

EXERCISES

Determine the amount of each of the following annuities:

1. $300 a year for 10 years at 8% compounded annually
2. $25 a month for 2 years at 12% compounded monthly
3. $60 per quarter for 5 years at 6% compounded quarterly

9.5 ORDINARY ANNUITIES

4. $75 every 4 months for 6 years at 9% compounded every 4 months
5. $500 every 6 months for 7 years at 14% compounded semiannually
6. $1,000 every month for 2 years at 6% compounded monthly
7. Betty wants to start a small business in 5 years, at which time she must have $10,000 in capital. At the end of every 3 months she puts $400 into an account, which pays 6% compounded quarterly. Will she have enough in her account at the end of 5 years?
8. Payments into a Christmas Club are $15 per month starting December 31 and ending November 31. On December 1, a check for $180 is sent to the members for their holiday expenses. This is the amount deposited and includes no interest. How much money would be available for the holidays if the payments were made into an account that pays 12% compounded monthly?
9. *Machine replacement* A manufacturing company expects to replace one of its machines in 8 years and estimates that the replacement will cost $250,000. At the end of every 6 months, the company puts $10,568 into a special fund to pay for the replacement. If the fund draws 10% interest compounded semiannually, verify that the amount in the fund will pay for the replacement at the end of 8 years.
10. Determine the amount in the college-fund account in the illustration at the beginning of Section 9.1.
11. At the end of each month, Jim deposits $35.00 into an account that pays 9% compounded monthly. After 4 years he discontinues the payments but leaves the total amount in the account to collect interest for 2 more years. Determine the balance in the account after the 6 years.
12. At the end of each 6 months, Pat deposits $150.00 into an account that pays 8% compounded semiannually. After 3 years she discontinues the payments and moves the total amount into an account that pays 12% compounded monthly. Determine her balance after the money has been in the second account for 18 months.
13. How much should be deposited at the end of every 3 months in an account that pays 8% compounded quarterly in order to have a balance of $1,200 after 4 years?
14. *Warehouse expansion* A company plans to expand its warehouse in 3 years at an estimated cost of $700,000. A sinking fund to pay for the expansion is established by making fixed monthly payments into an account paying 18% interest compounded monthly. How much should each payment be so that the amount in the account at the end of 3 years will be $700,000?
15. *Municipal bonds* An ordinary annuity paying 12% compounded quarterly is established by a city to retire $5,000,000 in bonds issued for street repairs. The bonds are to be retired in 10 years. How much should each quarterly payment be?
16. A couple establishes an ordinary annuity to accumulate a $15,000 down payment on a house in 5 years. The annuity pays 12% compounded monthly. How much should each of their monthly payments be?
17. In Exercise 7, how much should Betty put into the account every 3 months in order to accumulate $10,000?

Identify the steps of the four-step procedure involved in the application of mathematics (Section 1.3) in each case.

18. Municipal bonds   Exercise 15
19. Warehouse expansion   Exercise 14

9.6 PRESENT VALUE OF AN ANNUITY

In the previous section, we were concerned with the amount in an annuity after the annuity payments are made, including how much the payments should be to obtain a desired annuity amount. In this section the point of view is changed. We are now interested in establishing an account that pays an annuity of fixed payments for a specified period of time.

How much, for example, should you deposit in an account (that pays 10% compounded semiannually) in order to withdraw $1,000 every 6 months for the next 2 years? After the last withdrawal is made, the account is to have a zero balance.

The amount to be deposited is the present value of the four withdrawals. The present values are summarized in Table 9.5. The individual present values are calculated using equation (2) of Section 9.4.

TABLE 9.5   PAYMENT PERIOD = CONVERSION PERIOD: 6 MONTHS

WITHDRAWAL NUMBER	NUMBER OF PERIODS WITHDRAWAL EARNS INTEREST	PRESENT VALUE OF WITHDRAWAL
1	1	$1{,}000(1.05)^{-1}$
2	2	$1{,}000(1.05)^{-2}$
3	3	$1{,}000(1.05)^{-3}$
4	4	$1{,}000(1.05)^{-4}$

The amount to be deposited is equal to the sum of the present values. These values form a geometric progression with $a_1 = 1{,}000(1.05)^{-1}$ and $r = (1.05)^{-1}$. Using the formula of Section 9.1 for the sum of a geometric progression, we get a total of

$$\frac{a_1(1 - r^n)}{1 - r} = \frac{1{,}000(1.05)^{-1}[1 - (1.05)^{-4}]}{1 - (1.05)^{-1}}$$

$$= \frac{1{,}000[1 - (1.05)^{-4}]}{1.05 - 1} \approx \frac{1{,}000[0.177298]}{0.05}$$

$$= 3{,}545.96$$

The amount to be deposited is $3,545.96.

In general, if an ordinary annuity consists of n payments of R dollars each and earns interest at rate i per payment period, the present values of the payments form the geometric progression

$$R(1 + i)^{-1}, R(1 + i)^{-2}, R(1 + i)^{-3}, \ldots, R(1 + i)^{-n}$$

9.6 PRESENT VALUE OF AN ANNUITY

in which $a_1 = R(1 + i)^{-1}$ and $r = (1 + i)^{-1}$. The **present value (A) of the annuity** is the sum of the terms of this progression. Consequently,

$$A = \frac{R(1 + i)^{-1}[1 - (1 + i)^{-n}]}{1 - (1 + i)^{-1}}$$

$$= \frac{R(1 + i)^{-1}[1 - (1 + i)^{-n}]}{1 - (1 + i)^{-1}} \cdot \frac{(1 + i)}{(1 + i)}$$

$$= \frac{R[1 - (1 + i)^{-n}]}{(1 + i) - 1}$$

$$= R\left[\frac{1 - (1 + i)^{-n}}{i}\right]$$

$$= Ra_{\overline{n}|i}$$

where

$$a_{\overline{n}|i} = \frac{1 - (1 + i)^{-n}}{i}$$

The quantity $a_{\overline{n}|i}$ (read, "a angle n at i") is the *present value of $1 per period for n periods at the rate i per period*.

To summarize, the present value of an ordinary annuity is given by

$$A = R\left[\frac{1 - (1 + i)^{-n}}{i}\right] = Ra_{\overline{n}|i} \quad (1)$$

where:

R is the periodic payment.

i is the interest rate per payment period.

n is the number of payments.

As with $s_{\overline{n}|i}$, the values of $a_{\overline{n}|i}$ can be determined using your calculator. If necessary, the value of $(1 + i)^{-n}$, for some i and n, can be found in Table VII of Appendix B.

EXAMPLE 1 What is the present value of an annuity that pays $200 per month for 4 years if the money is worth 12% compounded monthly?

Solution From equation (1) with $R = 200$, $n = 48$, and $i = 0.01$, we get

$$A = 200\left[\frac{1 - (1 + 0.01)^{-48}}{0.01}\right]$$

$$\approx 200\left[\frac{1 - 0.62026}{0.01}\right]$$

$$= 200[37.974] = 7,594.80$$

The present value is $7,594.80. ■

EXAMPLE 2 A woman wants to provide a $700 scholarship every year for 10 years, starting 1 year from now. If the college can get 8% compounded annually on its investment, how much should the woman give now?

1 PRACTICE PROBLEM 1

A man's will specifies that his son is to be paid $5,000 a year for 20 years, starting 1 year from now. How much should the man's executor place into an account earning 9% compounded annually to make the payments?

Answer

$45,642.73

2 PRACTICE PROBLEM 2

An accounting firm buys some new teleprocessing equipment for $40,500, paying $10,000 down. The balance is to be paid by equal monthly payments over 2 years with 12% interest compounded monthly.

(a) How much will their monthly payments be?

(b) Find the present value of an annuity that pays the amount determined in part (a) each month for 2 years and earns 12% compounded monthly.

Answer

(a) $1,435.74   (b) $30,500

Solution   The gift is to provide an annuity of ten payments of $700 each. From equation (1) with $R = 700$, $n = 10$, and $i = 0.08$ we get

$$A = 700\left[\frac{1 - (1 + 0.08)^{-10}}{0.08}\right]$$

$$\approx 700\left[\frac{1 - 0.46319}{0.08}\right]$$

$$= 700[6.71013] = 4{,}697.09$$

The woman should make a gift of $4,697.09.

Suppose you take out a car loan of $2,000, which you are to repay at 18% interest in equal monthly payments over the next 3 years. How much should your monthly payments be? In this case your payments form an annuity whose present value is $2,000, and we need to determine the value of R in equation (1). Solving equation (1) for R gives

$$R = A\left[\frac{i}{1 - (1 + i)^{-n}}\right] \quad (2)$$

Your monthly payments would be

$$R = 2{,}000\left[\frac{0.015}{1 - (1 + 0.015)^{-36}}\right]$$

$$\approx 2{,}000\left[\frac{0.015}{1 - 0.58509}\right]$$

$$= 72.30$$

or $72.30.

The process of paying off an interest-bearing debt, such as your car loan, by a series of equal payments made at equal time intervals is called **amortization**. After the debt is fully paid, it is said to be *amortized*.

If you have difficulty in determining whether a particular annuity problem is an amount problem (Section 9.5) or a present-value problem (this section), it is helpful to keep in mind that in an amount problem, the payments precede the single sum. In a present-value problem, the payments follow the single sum.

EXERCISES

Determine the present value of an annuity that satisfies the given conditions.

1. Pays $2,000 annually for 8 years and earns 14% interest compounded annually

2. Pays $500 semiannually for 10 years and earns 10% interest compounded semiannually

3. Pays $150 monthly for 2 years and earns 12% interest compounded monthly

4. Pays $300 quarterly for 5 years and earns 14% compounded quarterly

5. *Endowment* An individual wants to set up an account (that earns 12% interest compounded seminannually) to provide a grant of $2,000 to the city's orchestra every 6 months, beginning 6 months from now, for 7 years. How much should the person donate?

6. The parents of a college student set up an annuity to pay for the student's college expenses. How much should they deposit now, at 10% compounded semiannually, for the annuity to pay $2,500 semiannually, beginning 6 months from now, for 4 years?

7. *Charitable trust* A trust is established to contribute $1,500 to a charitable foundation every 6 months for 5 years, beginning 6 months from now. If the trust earns 13% compounded semiannually, how much should be deposited in the trust?

8. What sum invested today at 9% compounded annually will yield the same amount as payments of $1,000 at the end of the year for 10 years, also earning 9% compounded annually?

9. What sum invested today at $11\frac{1}{2}$% compounded monthly will yield the same amount as payments of $50.00 at the end of each month for 2 years, also earning $11\frac{1}{2}$% compounded monthly?

Determine the periodic payments for the following loans, each of which is to be amortized over 2 years:

10. $750 with semiannual payments and interest at 12% compounded semiannually

11. $1,500 with annual payments and interest at 10% compounded annually

12. $275 with monthly payments and interest at $15\frac{3}{4}$% compounded monthly

13. $21,250 with quarterly payments and interest at 12% compounded quarterly

Determine the total amount of interest paid on the loan in the indicated exercise.

14. Exercise 10   15. Exercise 11   16. Exercise 12   17. Exercise 13

18. Joan buys some stereo equipment for $575, which she intends to pay for in 12 equal monthly payments. If interest is $1\frac{1}{2}$% per month compounded monthly, how much will her payments be? What is the total amount of interest that she will pay?

19. *Automobile loan* If you buy a car for $3,000, pay $500 down, and pay the balance in 36 equal monthly payments, how much will the payments be if interest is 12% compounded monthly? Determine the total amount of the 36 payments.

20. *Automobile loan* In Exercise 19, if you pay the balance in 48 equal monthly payments, how much will the payments be? Determine the amount of the 48 payments.

21. *Home mortgage* A couple buys a house for $80,000, paying 20% down. The balance is to be amortized in equal monthly payments over 30 years.

If money costs 14% compounded monthly, how much will each of the monthly payments be? What is the total amount of interest they will pay over the 30 years?

22. *Home mortgage* If the couple in Exercise 21 amortizes the balance in equal monthly payments over 25 years, how much will each monthly payment be? What is the total amount of interest they will pay over the 25 years?

23. Explain the relationship between present value and amortization.

9.7 CHAPTER SUMMARY

This chapter began with a discussion of geometric progressions, a basic concept in the study of compound interest. A **geometric progression** is a sequence of numbers in which each number, after the first, is obtained from the previous by multiplying by some constant r, called the **common ratio.** The two principal relationships for working with geometric progressions give an expression for the nth term and for the sum of the first n terms of the progression in terms of the common ratio and the first term a_1:

$$a_n = r^{n-1}a_1 \quad \text{and} \quad S_n = \frac{a_1(1-r^n)}{1-r}$$

Simple interest is interest calculated only on the principal. The main relationships for working with simple interest are the following:

$$\begin{aligned}
\text{Interest:} \quad & I = PRT \\
\text{Rate:} \quad & R = \frac{I}{PT} \\
\text{Present value:} \quad & P = \frac{S}{1+RT}
\end{aligned}$$

where

$$P = \text{Principal} \quad R = \text{Rate} \quad T = \text{Time}$$
$$I = \text{Interest} \quad S = \text{Amount} \quad \text{Principal} + \text{Interest}$$

The **present value** of an amount S of money is the principal P that should be invested now to obtain amount S in time T at rate R. When using any of the above expressions, the rate and time must both be expressed in terms of the same unit of time (years, months, quarters, and so forth).

We considered various situations involving simple interest, including a method for calculating the average daily balance. This is just a short method for calculating the average (or mean) of the balance in the account at the end of the days of a 1-month period.

9.7 CHAPTER SUMMARY

Compound interest involves interest on previously earned interest as well as on the principal. The main relationships for working with compound interest are the following:

$$\text{Compound amount:} \quad S = P(1 + i)^n$$
$$\text{Present value:} \quad P = S(1 + i)^{-n}$$

where

P = Principal
S = Compound amount   Principal + Interest
i = Interest per conversion period
n = Number of conversion periods

The **present value** in this case is the principal that should be invested now to obtain compound amount S after n conversion periods at interest i per conversion period.

An **annuity** is a series of equal payments or deposits made at regular intervals. **Ordinary annuities,** in which payments are made at the end of the payment period, interest is compound interest, and the conversion period agrees with the payment period, were the only type considered in this chapter. The main relationships for working with ordinary annuities are the following:

$$\text{Amount when } n\text{th payment is made:} \quad S = R\left[\frac{(1 + i)^n - 1}{i}\right]$$

$$\text{Periodic payment:} \quad R = S\left[\frac{i}{(1 + i)^n - 1}\right]$$

$$\text{Present value:} \quad A = R\left[\frac{1 - (1 + i)^{-n}}{i}\right]$$

$$\text{Periodic payment:} \quad R = A\left[\frac{i}{1 - (1 + i)^{-n}}\right]$$

where

S = Amount   Sum of all payments and their accumulated interest
R = Periodic payment A = Present value
i = Interest per payment (conversion) period
n = Number of payments

Here the present value indicates how much money should be invested now in order to make n future payments of R dollars each with the money earning i interest each payment period.

There are two expressions for the periodic payment R. The first, in terms of amount S, gives the size of the payments for obtaining amount S after n payments. The second, in terms of present value, A, gives the size of the payments for an annuity whose present value is A. The first is used for accumulating a fund of S dollars, the second for paying off a debt of A dollars.

REVIEW EXERCISES

Give the first four terms of each of the following geometric progressions:

1. $a_1 = -3$ and $r = 2$
2. $a_1 = 15$ and $r = -\frac{1}{5}$
3. $a_1 = 6$ and $a_2 = 3$
4. $a_1 = 5$ and $a_2 = -10$
5. $a_2 = 0.4$ and $a_3 = -0.16$
6. $a_2 = \frac{2}{5}$ and $a_3 = \frac{1}{5}$
7. $a_1 = 2$ and $a_3 = 18$
8. $a_1 = \frac{4}{3}$ and $a_3 = \frac{1}{3}$

Without calculating any other terms, find the indicated term in the given geometric progression.

9. a_4 if $a_1 = 5$ and $r = 3$
10. a_5 if $a_1 = 7$ and $r = -2$
11. a_3 if $a_1 = -1$ and $r = \frac{2}{3}$
12. a_7 if $a_1 = 3$ and $r = \sqrt{5}$
13. a_5 if $a_1 = 0.2$ and $a_2 = -0.1$
14. a_4 if $a_1 = 4$ and $a_2 = 3$

Which of the following could be the first three terms of a geometric progression?

15. $1, 2, 3$
16. $1, -2, 4$
17. $5, \frac{1}{5}, \frac{1}{125}$
18. $-\frac{2}{3}, 2, -6$
19. $0.1, 0.001, 0.0001$
20. $\frac{7}{8}, -\frac{7}{8}, -\frac{7}{8}$

Find the sum of the indicated number of terms using equation (1) of Section 9.1:

21. The first four terms; $a_1 = 4$, $r = 3$
22. The first six terms; $a_1 = -7$, $r = 2$
23. The first five terms; $a_1 = 8$, $r = -\frac{1}{3}$
24. The first six terms; $a_1 = 729$, $a_2 = 486$
25. The first four terms; $a_1 = 1$, $a_2 = 1/\sqrt{3}$
26. The first six terms; $a_1 = 5$, $r = 0.1$
27. The first 50 terms; $a_1 = 4$, $r = -1$
28. Suppose you start a Christmas fund with 5¢ in January, an additional 10¢ in February, and so forth, each month doubling the previous month's amount.
    (a) How much will you put into the fund in October?
    (b) How much will you have in the fund after you put in the December amount?
29. At each stroke, a pump removes $\frac{1}{3}$ of the liquid remaining in a tank. If the tank originally contains 100 liters, how much liquid remains in the tank after the fifth stroke?
30. *Profit increase* A company's profits have been increasing 10% a year. This year their profit is $250,000. If the 10% rate continues, what will their profit be 4 years from now?
31. *Spending cycles* Suppose that 85% of the money earned in the United States is spent on American products and services. Then, on the first

cycle, an American earning a salary spends 85% of it on American goods and services; on the second cycle, the American individuals and companies with whom he spends his money will, in turn, spend 85% that they receive on American individuals and companies. This spending of 85% continues through the third cycle, fourth cycle, and so forth. The United States government creates a job which pays $15,000 a year.
(a) How much money is put into the American economy on the fourth cycle as a result of the first year's salary?
(b) How much money is put into the American economy in the first four cycles as a result of the first year's salary?

Determine the amount of simple interest earned or paid on each amount.

32. $1,500 invested for 18 months at 10%
33. $350 invested for 7 months at 11%
34. $2,400 borrowed for 3 years at 13%
35. $1,825 borrowed for 30 months at 14%

Determine the rate of interest involved if you earn or pay interest as indicated.

36. $22.00 earned on $550 invested for 4 months
37. $85.00 earned on $425 invested for 15 months
38. $286.00 paid for $1,100 borrowed for 2 years
39. $122.47 paid for $835 borrowed for 16 months
40. *Certificate of deposit* An $8,000, 24-month C.D. pays 10% interest. The owner is paid simple interest every 3 months. Determine the quarterly interest.
41. Ann borrowed $600 discounted at 13.5% for 9 months.
    (a) How much money did she receive when the loan was made?
    (b) What rate of interest is she paying for the money actually received?
42. *Treasury bill* Ben bought a 15-month, $20,000 U.S. Treasury bill discounted at 14%. What rate of interest will he be making on the money that he paid for the bill?
43. What is the present value of each?
    (a) $1,300 due in 8 months at 9%   (b) $875 due in 15 months at 12.5%
44. How much money invested now at 8% will amount to $1,200 in 18 months?
45. From a financial point of view, which is worth more to you: $500 now, which you can invest at 11%, or $525 in 6 months?
46. *Charge-card account* Determine the average daily balance and the interest charged at 21% on the average daily balance for a charge-card account with the given activity during 1 month. Opening balance: $143.50.

DAY	ACTIVITY	
4	Charge:	$ 17.25
8	Charge:	39.18
15	Charge:	125.23
17	Payment:	143.50
22	Charge:	32.00
30	Customer billed	

47. *Charge-card account* Determine the average daily balance and the interest charged in Exercise 46 if the payment of $143.50 is made on day 3 instead of day 17.

48. *NOW account* Determine the average daily balance and the interest earned at the rate of 6% on the average daily balance for a NOW account with the given activity during 1 month. Opening balance: $637.19.

DAY	ACTIVITY	
2	Check cleared:	$ 42.38
5	Check cleared:	21.72
12	Checks cleared:	12.98
		27.75
14	Deposit:	376.95
16	Checks cleared:	142.50
		29.88
20	Check cleared:	67.76
28	Deposit	376.95
30	Interest paid on account at close of business day	

49. *NOW account* Determine the average daily balance and interest earned in the account of Exercise 48 if the check for $142.50 clears on day 27 instead of on day 16.

Find the compound amount for each investment.

50. $750 for 15 months at 18% compounded monthly
51. $1,230 for 18 months at 12% compounded quarterly

Find the total amount to be repaid and the total interest on each loan.

52. $975 for 2 years at 9% compounded monthly
53. $1,550 for 3 years at 10% compounded semiannually
54. *Savings certificate* A savings certificate for $1,500 earns 15% compounded every 4 months. Determine the value of the certificate after 3 years.
55. *Savings account* If you open a savings account with $400 which earns 10% compounded quarterly, how much will be in the account after 4 years assuming no other deposits and no withdrawals?

What is the present value in each case?

56. $750 at 18% compounded monthly due in 15 months
57. $1,230 at 12% compounded quarterly due in 18 months
58. From a financial point of view which is worth more to you: $600 now, which you can invest at 9% compounded every 4 months, or $750 in 3 years?

Determine the amount of the following annuities:

59. $150 a year for 9 years at 9% compounded annually
60. $35 a month for 2 years at 15% compounded monthly
61. $50 per quarter for 18 months at 14% compounded quarterly

62. $100 every 6 months for 18 months at 14% compounded semiannually

63. *Maintenance fund* The management of a shopping mall intends to repave the parking lot in 5 years at an estimated cost of $250,000. At the end of every 3 months, the management puts $9,305 into a sinking fund, which earns 9% compounded quarterly. Will the fund have the required amount after 5 years?

64. Paul wants to buy a new sports car, estimated to cost $15,000 in 4 years. He intends to establish a special fund to save the money for the car. If the fund pays 10% compounded semiannually, what should be his 6-month deposits into the fund so that he accumulates the required amount in 4 years?

65. *Company expansion* A company plans a $3,000,000 expansion of its research and development division in 3 years. It is establishing a sinking fund to pay for the expansion, making fixed monthly payments into the fund. If the fund earns 15% compounded monthly, how much should each payment be so that the amount in the fund at the end of 3 years will be $3,000,000?

Determine the present value of the following annuities, which pay the amounts indicated.

66. Pays $750 annually for 4 years and earns 11% compounded annually

67. Pays $400 every 6 months for $2\frac{1}{2}$ years and earns 10% compounded semiannually

Determine the periodic payments for each of the following:

68. A loan of $3,500 with monthly payments and interest at 18% compounded monthly, to be amortized over 2 years

69. A loan of $43,275 with quarterly payments and interest at 16% compounded quarterly, to be amortized over 5 years

70. *Patent rights payment* A company agrees to pay for patent rights by establishing an account that will pay the patent owner $1,500 every 6 months for 12 years, starting 6 months from now. If the account earns 11% interest compounded semiannually, how much must the company put into the account now in order to make the payments?

71. *Endowment* A foundation is establishing a fund that will be used to contribute $10,000 a year for cancer research, starting 1 year from now, for 6 years. If the fund earns 9% compounded annually, how much should the foundation put into the fund now in order to make the contributions?

72. *Municipal loan* A school district borrows $500,000 to be repaid in equal annual payments over a period of 10 years. If interest is 9% compounded annually, how much must they budget for the annual payments?

73. *Home mortgage* A couple buys a house for $75,000, paying 25% down, with the balance to be amortized in equal monthly payments over 25 years. If the interest rate is 15% compounded monthly, how much will each payment be? What is the total amount of interest the couple will pay over the 25 years?

CHAPTER 10

FUNCTIONS AND LIMITS

OUTLINE

10.1 Functions
10.2 Graphs of Functions
10.3 Limits
10.4 Continuity
*10.5 Rational Functions
10.6 Chapter Summary
   Review Exercises

Most of the computations in calculus involve operations on functions which, for the most part, are specified by a special type of equation. The purpose of this chapter is to examine the concept of a function and some related mathematical ideas.

* This section can be omitted without loss of continuity.

10.1 FUNCTIONS

The cost of shipping a piece of equipment by truck from one city to another depends upon the distance between the cities; to each distance there is associated a corresponding shipping cost. Similarly, the weekly wages of an hourly employee depend upon the time worked; to each amount of time there is associated a corresponding weekly wage. This association of one number with another, in mathematics, is called a *function*.

A ***function*** is a rule that associates a unique number to each member of some set D. For example, we might want to associate to each number in the set of numbers greater than zero its positive square root; or we might associate to each x in the set of real numbers the value of $2x$, and so forth. In the first case, the number associated with 9, for example, is $\sqrt{9} = 3$; in the second case, $2(9) = 18$.

Lowercase letters—usually f, g, and h—are most often used to designate functions. Then the number associated with a value x is designated by $f(x)$, $g(x)$, or $h(x)$, and the functions are usually described by an equation such as

$$f(x) = 3x + 1 \quad \text{for all real numbers } x$$

This equation indicates that f is the function (rule) that associates with each number x the number equal to three times x plus 1. Specifically, the number associated with 2 (called the value of f at 2) is

$$f(2) = 3(2) + 1 = 7$$

The number associated with 0 (called the value of f at 0) is

$$f(0) = 3(0) + 1 = 1$$

The number associated with -0.5 (called the value of f at -0.5) is

$$f(-0.5) = 3(-0.5) + 1 = -0.5$$

and so forth.
Similarly,

$$g(x) = x^2 + 4 \quad \text{for all } x \text{ greater than 0}$$

indicates that g is the function (rule) that associates with each positive number x the number equal to x squared plus four. Specifically, $g(1) = 1^2 + 4 = 5$, $g(7) = 7^2 + 4 = 53$, $g(10) = (10)^2 + 4 = 104$, and so forth.

As indicated earlier, the number that a function f associates with a value x, denoted by $f(x)$ (read, "f of x"), is called the ***value of f at x***. The same is true for $g(x)$, $h(x)$, and so on.

Functions are used to model physical situations when the value of one quantity depends uniquely on the value of another. The area of a circle depends on its radius; the area is a function of the radius. If f is

10.1 FUNCTIONS

the area function, then

$$f(r) = \pi r^2 \quad \text{for } r > 0$$

gives the area of a circle with radius r.

If speed is held constant at 55 miles per hour, the distance traveled is a function of the time driven. If h is the distance function, then

$$h(t) = 55t \quad \text{for } t \geq 0$$

gives the distance traveled for any time t.

The commission earned by a stockbroker for selling your stocks depends on the amount received for the stock sold and is usually specified by a rule that, although somewhat complicated, determines a unique commission for each amount of stock sold. The number of units of a product the public is willing to buy depends upon the price.

Some additional examples of functions are

$$f(x) = 2x + 1 \quad \text{for each real number } x$$
$$g(t) = 6t^2 - t \quad \text{for } -1 \leq t \leq 2$$
$$h(x) = \frac{x^3 - 5x}{x - 1} \quad \text{for each real number } x \neq 1$$

The set D referred to in the definition of a function is called the **domain** of the function. For the function f just given, the domain is

$$D = \{x : x \text{ is a real number}\}$$

For the function g, the domain is

$$D = \{t : -1 \leq t \leq 2\}$$

For the function h, the domain is

$$D = \{x : x \text{ is a real number}, x \neq 1\}$$

If the domain is not specified (and this is usually the case), it is understood to be all real numbers for which the associated value exists. With this convention, each of the three above functions is the same as its counterpart in the following:

$$f(x) = 2x + 1$$
$$g(t) = 6t^2 - t \quad -1 \leq t \leq 2$$
$$h(x) = \frac{x^3 - 5x}{x - 1}$$

EXAMPLE 1 For the function f defined by

$$f(x) = \frac{x + 4}{x + 2}$$

determine the value of $f(-3)$, $f(0)$, and $f(7)$; also determine the domain of f.

Solution
$$f(-3) = \frac{-3+4}{-3+2} = \frac{1}{-1} = -1$$
$$f(0) = \frac{0+4}{0+2} = 2$$
$$f(7) = \frac{7+4}{7+2} = \frac{11}{9}$$

Since $f(x)$ has a value for all x except $x = -2$ (which would make the denominator zero), the domain of f is the set of all real numbers except -2. ∎

EXAMPLE 2   For the function g defined by
$$g(x) = \log x + 3$$
determine the value of $g(1)$ and $g(10)$; also determine the domain of g.

Solution   In this case, $g(1) = \log 1 + 3 = 3$ and $g(10) = \log 10 + 3 = 4$. Since $\log x$ has a value only for numbers $x > 0$, the domain of g is the set of all positive numbers. 1

EXAMPLE 3   If $f(x) = 2x^2 + 1$, determine $f(x + 1)$ and $f(x + h)$.

Solution   The expressions for $f(x + 1)$ and $f(x + h)$ are obtained from $f(x)$ by direct substitution for x in the right-hand side of $f(x) = 2x^2 + 1$. The results are then simplified.
$$f(x + 1) = 2(x + 1)^2 + 1 = 2(x^2 + 2x + 1) + 1$$
$$= 2x^2 + 4x + 3$$
$$f(x + h) = 2(x + h)^2 + 1 = 2(x^2 + 2xh + h^2) + 1$$
$$= 2x^2 + 4xh + 2h^2 + 1$$
∎

EXAMPLE 4   If
$$g(t) = \frac{t^2}{t + 1}$$
determine $g(2a)$ and $g(t + h)$.

Solution
$$g(2a) = \frac{(2a)^2}{2a + 1} = \frac{4a^2}{2a + 1}$$

---

1 PRACTICE PROBLEM 1

(a) For $f(x) = \frac{x+5}{x^2-1}$ determine $f(3)$, $f(-4)$, and the domain of f.

(b) For $g(x) = \sqrt{x + 2}$, determine $g(4)$, $g(-2)$, and the domain of g.

Answer

(a) 1; $\frac{1}{15}$; the set of all real numbers except $x = \pm 1$
(b) $\sqrt{6}$; 0; the set of all real numbers greater than or equal to -2

and
$$g(t+h) = \frac{(t+h)^2}{(t+h)+1} = \frac{t^2+2h+h^2}{t+h+1}$$

EXAMPLE 5 If $g(t) = t^2 + 2$, determine $g(t+h) - g(t)$.

Solution
$$g(t+h) - g(t) = [(t+h)^2 + 2] - [t^2 + 2]$$
$$= [t^2 + 2ht + h^2 + 2] - [t^2 + 2]$$
$$= 2ht + h^2$$

EXAMPLE 6 If $f(x) = 3x^2$, determine $\frac{f(x+h) - f(x)}{h}$.

Solution
$$\frac{f(x+h) - f(x)}{h} = \frac{[3(x+h)^2] - [3x^2]}{h}$$
$$= \frac{3(x^2 + 2xh + h^2) - 3x^2}{h} = \frac{3x^2 + 6xh + 3h^2 - 3x^2}{h}$$
$$= \frac{6xh + 3h^2}{h} = \frac{h(6x + 3h)}{h} = 6x + 3h$$

In our next example, we construct a function to represent a physical situation.

EXAMPLE 7 In addition to a service charge of $10.00 per order, a textbook publisher charges bookstores $15.00 per copy for an algebra text on orders of up to 100 copies and $13.75 per copy on orders of over 100 copies. Construct a function f where $f(x)$ gives the billing amount in terms of the number x of copies ordered.

Solution
On orders up to 100 copies, the amount is $10 + 15x$, and on orders of over 100, the amount is $10 + 13.75x$. We can combine these expressions in the following manner:
$$f(x) = \begin{cases} 10 + 15x & 0 < x \leq 100 \\ 10 + 13.75x & 100 < x \end{cases}$$

PRACTICE PROBLEM 2

If $f(t) = t^2 - 2t$, determine $f(3a)$ and $f(t+h)$.

Answer

$9a^2 - 6a$; $t^2 + 2ht + h^2 - 2t - 2h$

PRACTICE PROBLEM 3

If $f(x) = x^2 - 2$, determine each of the following:

(a) $f(x+h) - f(x)$

(b) $\frac{f(x+h) - f(x)}{h}$

Answer

(a) $2xh + h^2$ (b) $2x + h$

EXERCISES

For the function f defined by $f(x) = 3x^2 + x$, determine each of the following:

1. $f(2)$ 2. $f(-3)$ 3. $f(0)$ 4. $f(a)$ 5. $f(x+h)$

For the function g defined by $g(t) = 2(3^t)$, determine each of the following:

6. $g(1)$ 7. $g(-4)$ 8. $g(0)$ 9. $g(b)$ 10. $g(t+2)$

For the function f defined by $f(t) = (t+2)^2$, determine each of the following:

11. $f(-5)$ 12. $f(-2)$ 13. $f(\frac{1}{2})$ 14. $f(3a)$ 15. $f(t+3)$

For the function g defined by $g(x) = (x+1)/(x^2 - 2)$, determine each of the following:

16. $g(-10)$ 17. $g(4)$ 18. $g(0.1)$ 19. $g(-a)$ 20. $g(1+a)$

For the function f defined by $f(x) = x^2 - 1$, determine each of the following:

21. $f(7)$ 22. $f(-5)$ 23. $f(x+h)$

24. $f(x+h) - f(x)$ 25. $\dfrac{f(x+h) - f(x)}{h}$

For the function g defined by $g(t) = 2t + 3$, determine each of the following:

26. $g(100)$ 27. $g(-0.4)$ 28. $g(3t)$

29. $g(3t - 1)$ 30. $g(t+h)$ 31. $\dfrac{g(t+h) - g(t)}{h}$

For the function f defined by $f(x) = x^2 - 2x$, determine each of the following:

32. $f(x+h)$ 33. $f(x+h) - f(x)$ 34. $\dfrac{f(x+h) - f(x)}{h}$

For the function g(x) defined by $g(x) = x^3$, determine each of the following:

35. $g(x+h)$ 36. $g(x+h) - g(x)$ 37. $\dfrac{g(x+h) - g(x)}{h}$

For the function f defined by

$$f(x) = \begin{cases} x+1 & x \geq 1 \\ x^2 & x < 1 \end{cases}$$

determine each of the following:

38. $f(7)$ 39. $f(0)$ 40. $f(1)$ 41. $f(-3)$ 42. $f(25)$

For the function f defined by

$$f(x) = \begin{cases} (2x-1)^2 & x \leq -1 \\ 2x+1 & -1 < x < 1 \\ (2x+1)^2 & x \geq 1 \end{cases}$$

determine each of the following:

43. $f(5)$ 44. $f(0.5)$ 45. $f(-3)$ 46. $f(1)$ 47. $f(-1)$

48. (a) Construct a function h such that $h(x)$ gives the surface area of a cube in terms of its length x in feet. Indicate the domain of h.
   (b) Find $h(2)$.   (c) Find $h(3.5)$.

49. *Sales commission* A salesperson's monthly salary is $500 plus 15% of his sales.
   (a) Construct a function f such that $f(x)$ gives his monthly salary in terms of the amount x of his sales. Indicate the domain of f.
   (b) Determine $f(7,500)$.   (c) Determine $f(10,000)$.

50. *Taxi fares* A taxi company charges $1.25 plus 50¢ a mile for the first 5 miles and 40¢ a mile for each mile over 5 miles.
   (a) Construct a function f similar to those given for Exercises 38–42 and 43–47, such that $f(x)$ gives the fare in terms of the number x of miles ridden. Indicate the domain of f.
   (b) Determine $f(4.5)$.   (c) Determine $f(7)$.

51. *Income taxes*
   (a) A state income tax rate table is shown here. Construct a function g similar to those given for Exercises 38–42 and 43–47, such that $g(x)$ gives the tax on a net taxable income of x dollars. Indicate the domain of g.
   (b) Determine $g(9,500)$.   (c) Determine $g(35,750)$.

TAX RATES

IF NET TAXABLE INCOME IS		
Over	But not over	The tax is
0	$2,000	3% of net taxable income
$2,000	$4,000	$60 + 4% of amount over $2,000
$4,000	$6,000	$140 + 5% of amount over $4,000
$6,000	$10,000	$240 + 6% of amount over $6,000
$10,000		$480 + 7% of amount over $10,000

Determine the domain of each of the following functions:

52. $f(x) = 2x^3 - x$ **53.** $g(x) = (x+3)^2$ **54.** $f(t) = \dfrac{t}{3-t}$

55. $h(t) = 3(2^t)$ **56.** $h(x) = \sqrt{x}$ **57.** $g(x) = \ln x$

58. $g(t) = \dfrac{2}{1-3^t}$ **59.** $f(x) = \dfrac{x}{(x+1)(x-2)}$

60. $g(x) = \dfrac{x+1}{x(x-4)},\ 1 \le x \le 3$ **61.** $f(x) = x^2 - 1,\ x > 0$

The **range** of a function f is the set of all values of $f(x)$; that is, the range of f is the set

$$R = \{f(x) : x \text{ is in the domain of } f\}$$

For the function f defined by $f(x) = \sqrt{x} + 1$, 5 is in the range of f since $f(16) = \sqrt{16} + 1 = 5$; $\sqrt{3} + 1$ is in the range of f, since $f(3) = \sqrt{3} + 1$. In fact, since $\sqrt{x}$ takes all values greater than or equal to zero, the range of f is the set of all numbers 1 and greater, $\{x:x \geq 1\}$. Determine the range of each of the following functions:

62. $f(x) = x^2$
63. $f(x) = x^2 + 1$
64. $f(x) = x + 1$
65. $h(t) = t^2 - 1$
66. $g(t) = e^t$
67. $g(x) = \ln x$

Explain the meaning of each term.

68. Function
69. Domain of a function

10.2 GRAPHS OF FUNCTIONS

Just as the graph of an equation gives a geometric picture of the equation, the graph of a function gives a geometric picture of the function. The procedure for graphing a function is exactly the same as that for graphing an equation because the **graph of a function f** is the same as the graph of the equation $y = f(x)$. For example, the graph of the function f defined by $f(x) = 6x - 4$ is the graph of the equation $y = 6x - 4$. The graph of this last equation is a line; the graph of f is the same line.

The graph of the function g defined by $g(x) = 2x^2 + x$ is the graph of the equation $y = 2x^2 + x$. This is a quadratic equation whose graph is a parabola; the graph of g is the same parabola.

EXAMPLE 1   Graph the function f defined by $f(x) = -3x + 2$.

Solution   The graph of f is the graph of the linear equation $y = -3x + 2$.

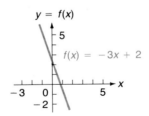

For any point (x, y) in the graph of a function, *the second coordinate is the number that the function associates with the first coordinate*. In Example 1, the graph of f passes through the point $(-1, 4)$, corresponding to the fact that $f(-1) = 4$. Similarly, the graph passes through the point $(0, 2)$ because $f(0) = 2$. Consequently, the coordinates of points in the graph can be labeled $(x, f(x))$. It is important to realize this relationship between the coordinates of the points in the graph of a function in order to appreciate how the graph represents the function geometrically. Also, because of this relationship, the vertical axis is sometimes labeled $f(x)$, $g(x)$, or $h(x)$ (whichever is appropriate), as in Example 2.

EXAMPLE 2 Graph the function g defined by $g(x) = 2x^2 + 1$.

Solution The graph of g is the graph of the quadratic equation $y = 2x^2 + 1$. By completing the square, we can rewrite this equation in the form

$$\tfrac{1}{2}(y - 1) = x^2$$

whose graph is a parabola with vertex at (0, 1) and opening upward.

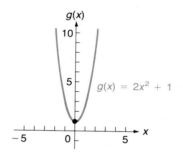

EXAMPLE 3 Graph the function f defined by $f(x) = 2^x$, $1 \le x \le 3$.

Solution The graph of f is the graph of the equation $y = 2^x$, $1 \le x \le 3$. That is, the graph of f is that part of the graph of $y = 2^x$ corresponding to values of x between 1 and 3. In the figure the entire graph of $y = 2^x$ is indicated, but only the solid portion is the graph of f.

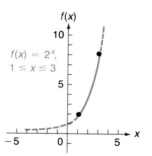

---

1 PRACTICE PROBLEM 1

Graph each function.

(a) $f(x) = 2x - 3$ (b) $f(x) = x^2 - 1$

Answer

(a)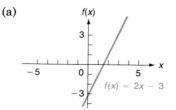

(b)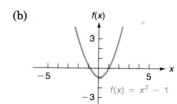

EXAMPLE 4 Graph the function h defined by
$$h(x) = \begin{cases} x+1 & x<1 \\ \ln x & x \geq 1 \end{cases}$$

Solution The graph of $y = h(x)$ uses parts of the graphs of $y = x + 1$ and $y = \ln x$. In the figure the entire graphs of these equations are indicated. However, only the solid portion is the graph of h.

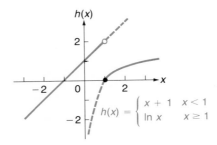

EXAMPLE 5 Graph the function f defined by $f(x) = \dfrac{x^2 - 1}{x - 1}$.

Solution Since
$$f(x) = \frac{x^2 - 1}{x - 1} = \frac{(x+1)(x-1)}{(x-1)} = x + 1, \quad x \neq 1$$

the graph of $y = f(x)$ is the graph of $y = x + 1$ with no point corresponding to $x = 1$.

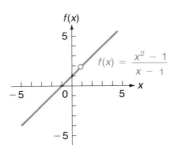

---

2 PRACTICE PROBLEM 2

Graph each function.

(a) $f(x) = \begin{cases} e^x & x<0 \\ x+1 & x \geq 0 \end{cases}$

(b) $h(x) = \dfrac{x^2 - 1}{x + 1}$

Answer

(a)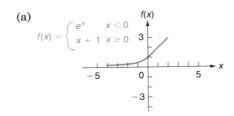

(b)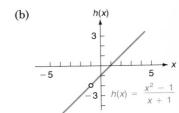

Hereafter, we shall follow the usual procedure of abbreviating the terminology for functions from, for example, "the function f defined by $f(x) = -2x^2 + 3$" to simply "the function $f(x) = -2x^2 + 3$." Furthermore, when the function notation is not required, we may simplify this last terminology to "the function $y = -2x^2 + 3$." Consequently, we shall encounter phrases such as:

The function $g(x) = 2(\ln x)$

The function $h(x) = x^2 + e^{3x}$

The function $y = 3x - 4$, $0 \le x \le 7$

The function $y = \log_2(x - 1)$

This is simply a change in notation or terminology and does not alter the concept of a function as a rule that associates a unique number to each number in some set D. It is just that the number that a given function associates with a value of x can be written in different ways and can be used to denote the function itself.

When notation such as $y = -2x^2 + 3$ is used to denote a function, y is called the **dependent variable** and x is called the **independent variable.** This terminology emphasizes the fact that the value of y depends upon the value of x. Letters other than x and y are also used in this context. For example, in the function $d = 55t$, t is the independent variable and d is the dependent variable.

Functions are classified by types in much the same way as equations are classified. The definitions of the various types, and some examples of each type, are given in Table 10.1. The graphs of each of the first four types of functions are familiar to us from earlier sections. We have not considered the last type before. In general, a **polynomial** in x is an expression of the form $a_n x^n + a_{n-1} x^{n-1} + \cdots + a_1 x + a_0$, where

TABLE 10.1

THE FUNCTION f IS	IF $f(x)$ CAN BE WRITTEN IN THE FORM	EXAMPLES
Linear	$f(x) = ax + b$, where a and b are constants	$f(x) = 2x - 3$ $f(x) = 3x + 7$
Quadratic	$f(x) = ax^2 + bx + c$, where a, b, and c, are constants, $a \ne 0$	$f(x) = 3x^2 + 2x - 1$ $y = 2x^2 - x$ $g(x) = x^2 - x + 2$
Exponential	$f(x) = Ab^{kx}$, where A, b, and k are constants, $b > 0$	$f(x) = 3(4^{2x})$ $y = 23^{-x}$ $g(t) = -3(5^t)$
Logarithmic	$f(x) = A \log_b(kx)$ where A, b, and k are constants, $b > 0$, $b \ne 1$	$f(x) = 3 \log_2(4x)$ $g(x) = \ln x$ $y = \log(\frac{1}{2}t)$
Polynomial	$f(x) = a_n x^n + a_{n-1} x^{n-1} + \cdots + a_1 x + a_0$, where the a_i's are constants	$f(x) = 2x^3 + 2x^2 + x + 7$ $g(x) = 2x^4 + x^3$ $y = x^3 + 2x^2 + 1$

the a_i's are constants. If $a_n \neq 0$, the **degree** of the polynomial is n. Examples of polynomials are $4x^3 + 2x^2 - 6x + 7$ (of degree 3), $-3x^2 + 2x + 4$ (of degree 2), $x^5 + 2x - 3$ (of degree 5), and so forth. Note that if the degree of a polynomial is 1 or 2, then the corresponding functions are, respectively, linear or quadratic.

EXERCISES

Graph each of the given functions.

1. $f(x) = 2x + 3$
2. $g(x) = -x + 4$
3. $y = 4 - 2x, \quad x \geq 0$
4. $y = x, \quad -1 \leq x \leq 1$
5. $f(x) = 3$
6. $y = -x^2 + 2x - 1$
7. $h(x) = x^2 + 2x - 1, \quad x \leq 0$
8. $y = x^2, \quad -1 \leq x \leq 1$
9. $y = 5 \ln x$
10. $y = \log(2x), \quad \frac{1}{2} \leq x \leq 5$
11. $g(x) = 3^{-x}$
12. $y = 2^x + 1$
13. $f(x) = \begin{cases} -x & x \leq 1 \\ x - 2 & x > 1 \end{cases}$
14. $y = \begin{cases} 4 & x < 2 \\ x^2 & x \geq 2 \end{cases}$
15. $g(x) = \begin{cases} x - 3 & x \leq 0 \\ x^2 + 2 & x > 0 \end{cases}$
16. $g(x) = \begin{cases} -2x & x \leq 0 \\ 2^x & x > 0 \end{cases}$
17. $y = \dfrac{x^2 - 4}{x - 2}$
18. $f(x) = \dfrac{x^2 - 9}{x + 3}$
19. $y = \sqrt{x}$
20. $g(x) = -\sqrt{x}$

Determine the type of function given in each exercise.

21. Exercise 1
22. Exercise 2
23. Exercise 3
24. Exercise 4
25. Exercise 5
26. Exercise 6
27. Exercise 7
28. Exercise 8
29. Exercise 9
30. Exercise 10
31. Exercise 11
32. *Taxi fares* Graph the taxi-fare function constructed in Exercise 50 of Section 10.1.
33. Graph the surface area function constructed in Exercise 48 of Section 10.1. (You will probably have to use different units on the two axes.)
34. *Sales commission* Graph the salary function constructed in Exercise 49 of Section 10.1. (You will probably have to use different units on the two axes.)
35. *Income tax* Graph the state income tax function constructed in Exercise 51 of Section 10.1. (You will probably have to use different units on the two axes.)
36. Explain why $y^2 = x$ is *not* a function with x the independent variable.

Explain the meaning of each term.

37. Graph of a function
38. Polynomial

39. Linear function
40. Quadratic function
41. Exponential function
42. Logarithmic function
43. Polynomial function

10.3 LIMITS

The dividing line between calculus and precalculus mathematics is the limit concept introduced in this section. As we shall see, the two major concepts of calculus (the derivative and the integral) are special types of limits.

The graph of the function

$$f(x) = \frac{x^2 - 1}{x - 1}$$

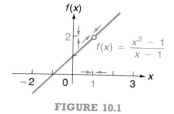

FIGURE 10.1

is given again in Figure 10.1. Consider what happens when values of x are chosen closer and closer to $x = 1$, indicated by the arrows along the x-axis. The corresponding points in the graph move closer and closer to the missing point, indicated by the arrows along the graph. The second coordinates [which are the values of $f(x)$] of these points move closer and closer to 2, indicated by the arrows along the vertical axis. Again, as the x-values move close to 1, the corresponding points in the graph move toward the missing point, and the second coordinates [which are the values of $f(x)$] move closer and closer to 2. In fact, we can get the values of $f(x)$ to be as close to 2 as we wish by taking the x-values sufficiently close to 1. Because of this behavior, we say that the *limit* of $f(x)$ as x approaches 1 is 2. To indicate that this limit is 2, we write

$$\lim_{x \to 1} \frac{x^2 - 1}{x - 1} = 2$$

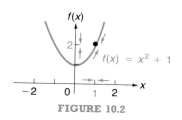

FIGURE 10.2

We consider the similar behavior of the function values of $f(x) = x^2 + 1$ in Figure 10.2. As the x-values move close to 1, the corresponding points in the graph move close to the point $(1, 2)$ and the function values (second coordinates) move close to 2 on the vertical axis. Again, we can obtain function values as close to 2 as we wish by choosing x-values sufficiently close to 1 so that the limit of $f(x)$ as x approaches 1 is again 2. That is,

$$\lim_{x \to 1} (x^2 + 1) = 2$$

Note that in this case our function has a value at $x = 1$; in the first case it did not. Whether the function has such a value or not does not affect the limit.

A similar analysis of the behavior of the values of

$$f(x) = \begin{cases} x + 2 & x \leq 1 \\ x^2 & x > 1 \end{cases}$$

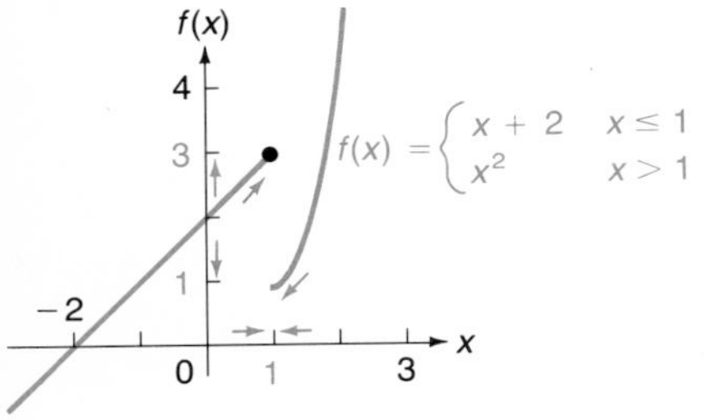

indicates that in this case, there is no limiting value as x approaches 1 (see Figure 10.3). As x gets close to 1 from the right, the function values (the second coordinates of the corresponding points on the graph) get close to 1; as x gets close to 1 from the left, the function values get close to 3. There is no single number to which the function values get close as x gets close to 1 as occurred in the two previous illustrations. In this case, there is *no limit* of $f(x)$ as x approaches 1.

In general, the **limit** of $f(x)$ as x approaches a is L provided that the values of $f(x)$ get closer and closer to L as x gets closer and closer to a [in the sense that the values of $f(x)$ can be made to be as close to L as desired by choosing values of x sufficiently close to a] without regard to the value of f at $x = a$. If such a number L exists, we write

$$\lim_{x \to a} f(x) = L$$

(for the left-hand side, read "the limit of f of x as x approaches a"). In the first two illustrations, a was 1 and L was 2.

We want to be able to determine the value of L, if it exists, without graphing the functions if this is possible. The procedure is illustrated in the next several examples.

FIGURE 10.3

EXAMPLE 1 For $f(x) = x^2 + 3$, determine each limit.

(a) $\lim_{x \to 2} f(x)$ (b) $\lim_{x \to -3} f(x)$

Solution   (a) As x gets close to 2, x^2 gets close to 4, so $x^2 + 3$ gets close to 7; that is, $\lim_{x \to 2} (x^2 + 3) = 7$.

(b) As x gets close to -3, x^2 gets close to 9, so $x^2 + 3$ gets close to 12; consequently, $\lim_{x \to -3} (x^2 + 3) = 12$. ■

EXAMPLE 2 For

$$g(x) = \begin{cases} 3x - 2 & x \leq 2 \\ 2^x & x > 2 \end{cases}$$

determine each limit.

(a) $\lim_{x \to 0} g(x)$ (b) $\lim_{x \to 2} g(x)$

Solution   (a) For values of x close to 0, all values of $g(x)$ are given by $3x - 2$. Thus, for x getting close to 0, $3x$ also gets close to 0 and $3x - 2$ gets close to -2; therefore, $\lim_{x \to 0} g(x) = -2$.

(b) As x gets close to 2 from the right, the function values, given by 2^x, get close to 4; as x gets close to 2 from the left, the function values, given by $3x - 2$, also get close to 4. Since the function values get close to 4 for values of x from the left and from the right, $\lim_{x \to 2} g(x) = 4$.

EXAMPLE 3 For

$$f(x) = \frac{x^2 - 9}{x + 3}$$

determine each limit.

(a) $\lim_{x \to 6} f(x)$ (b) $\lim_{x \to -3} f(x)$

Solution (a) In the numerator of $f(x)$, as x gets close to 6, x^2 gets close to 36 and $x^2 - 9$ gets close to 27. In the denominator, as x gets close to 6, $x + 3$ gets close to 9. The quotient, therefore, gets close to $\frac{27}{9}$. That is, $\lim_{x \to 6} f(x) = \frac{27}{9} = 3$.

(b) In the numerator, as x gets close to -3, x^2 gets close to 9 and $x^2 - 9$ gets close to 0. In the denominator, as x gets close to -3, $x + 3$ gets close to 0 also. Therefore, the limit takes the form $\frac{0}{0}$. The expression $\frac{0}{0}$ has no value, but this does not mean that there is no limit. In fact, for all x except $x = -3$,

$$\frac{x^2 - 9}{x + 3} = \frac{(x + 3)(x - 3)}{x + 3} = x - 3$$

In particular, for all x close to -3 but not equal to -3,

$$\frac{x^2 - 9}{x + 3} = x - 3$$

Consequently, since the limit at -3 does not depend on the value of the function at -3,

$$\lim_{x \to -3} \frac{x^2 - 9}{x + 3} = \lim_{x \to -3} (x - 3) = -6$$

Geometrically, the situation here is similar to that of

$$\lim_{x \to 1} \frac{x^2 - 1}{x - 1}$$

discussed in the beginning of this section; the graph of

$$y = \frac{x^2 - 9}{x + 3}$$

PRACTICE PROBLEM 1

For $f(x) = 4x - 3$ and

$$g(x) = \begin{cases} x^2 & x \leq 2 \\ x + 3 & x > 2 \end{cases}$$

determine each value.

(a) $\lim_{x \to 3} f(x)$ (b) $\lim_{x \to -2} f(x)$
(c) $\lim_{x \to 0} g(x)$ (d) $\lim_{x \to 2} g(x)$

Answer
(a) 9 (b) -11 (c) 0
(d) No limit

is the same as the graph of $y = x - 3$ except that there is no point corresponding to $x = -3$. ∎

EXAMPLE 4 Evaluate: $\lim\limits_{x \to 0} \dfrac{2}{x}$

Solution We graph the function as shown in the margin to observe the behavior of the values of the function for x close to 0. As x gets close to 0 from the right, the function values grow without bound. For example, if $x = 1/1,000,000$ (one one-millionth), $y = 2,000,000$; if $x = 1/1,000,000,000$ (one one-billionth), $y = 2,000,000,000$; and so forth. Similarly, as x gets close to 0 from the left, the function values decrease without bound through negative values. There is no one number to which the function values get close, so that

$$\lim_{x \to 0} \frac{2}{x}$$

does not exist. ∎

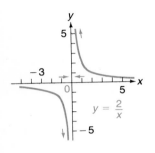

The forms of the three limits in the last two examples are fairly typical of those you can expect when evaluating the limit of a quotient. The three general forms are given in Table 10.2.

TABLE 10.2 GENERAL FORMS OF THE LIMIT OF A QUOTIENT

1. $\dfrac{a}{b}$, where a and b are constants, $b \neq 0$ (as in the first limit in Example 3, where a was 27 and b was 9): In this case, the limit is a/b.

2. $\dfrac{0}{0}$ (as in the second limit in Example 3): In this case, the value of the limit, if it has a value, cannot be determined until the function is rewritten and the limit is reevaluated.

3. $\dfrac{a}{0}$, where a is a constant, $a \neq 0$ (as in Example 4, where a was 2): In this case, the limit does not exist. In fact, the function values will increase or decrease without bound as x gets close to the value of interest.

These three forms are considered again in Example 5.

EXAMPLE 5 For

$$f(x) = \frac{x^2 - x - 2}{x^2 + 3x - 10}$$

determine each limit.

(a) $\lim\limits_{x \to 0} f(x)$ (b) $\lim\limits_{x \to 2} f(x)$ (c) $\lim\limits_{x \to -5} f(x)$

Solution (a) As x gets close to 0, the numerator in $f(x)$ gets close to -2 and the denominator gets close to -10. The quotient, therefore, gets close to $\frac{-2}{-10} = \frac{1}{5}$, and

$$\lim_{x \to 0} f(x) = \frac{1}{5}$$

This is the first case in Table 10.2 with $a = -2$ and $b = -10$.

(b) As x gets close to 2, the numerator in $f(x)$ gets close to 0 and the denominator also gets close to 0. This is the second case of Table 10.2; the function must be rewritten and the limit reevaluated:

$$\frac{x^2 - x - 2}{x^2 + 3x - 10} = \frac{(x+1)(x-2)}{(x+5)(x-2)} = \frac{x+1}{x+5} \quad \text{for } x \neq 2$$

Therefore, since the limit is not concerned with the value of f at 2,

$$\lim_{x \to 2} f(x) = \lim_{x \to 2} \frac{x+1}{x+5} = \frac{3}{7}$$

(c) As x gets close to -5, the numerator in $f(x)$ gets close to 28 and the denominator gets close to 0. This is the third case of Table 10.2 with $a = 28$. There is no limit—that is,

$$\lim_{x \to -5} f(x)$$

does not exist.

> **2 PRACTICE PROBLEM 2**
>
> If
>
> $$f(x) = \frac{x^2 + x - 12}{x^2 + 6x + 8} = \frac{(x-3)(x+4)}{(x+2)(x+4)}$$
>
> determine each value.
>
> (a) $\lim_{x \to 1} f(x)$ (b) $\lim_{x \to -2} f(x)$
> (c) $\lim_{x \to -4} f(x)$
>
> **Answer**
>
> (a) $-\frac{2}{3}$ (b) No limit (c) $\frac{7}{2}$

You may have noticed that limits of functions are defined and evaluated at individual points on the x-axis ($x = a$ in the definition). The existence and numerical values of limits vary from one point to another and must be evaluated at one point at a time.

Our final example is a type of limit that will be of particular interest in the next chapter. It is a limit of the form discussed in the second case in Table 10.2.

EXAMPLE 6 For $f(x) = 2x^2$, evaluate

$$\lim_{h \to 0} \frac{f(3+h) - f(3)}{h}$$

Solution

$$\frac{f(3+h) - f(3)}{h} = \frac{2(3+h)^2 - 2(3^2)}{h}$$

$$= \frac{2(9 + 6h + h^2) - 2(9)}{h} = \frac{12h + 2h^2}{h}$$

Therefore, the desired limit is equal to

$$\lim_{h \to 0} \frac{12h + 2h^2}{h}$$

which is of the form $\frac{0}{0}$ and must be rewritten:

$$\lim_{h \to 0} \frac{12h + 2h^2}{h} = \lim_{h \to 0} \frac{h(12 + 2h)}{h} = \lim_{h \to 0} (12 + h) = 12$$

That is,
$$\lim_{h \to 0} \frac{f(3+h) - f(3)}{h} = 12$$

EXERCISES

For $f(x) = x^2 + 3$, evaluate each of the following limits:

1. $\lim_{x \to 2} f(x)$
2. $\lim_{x \to -1} f(x)$
3. $\lim_{x \to 0} f(x)$

For $f(x) = 2x^3 - x$, evaluate each of the following limits:

4. $\lim_{x \to 2} f(x)$
5. $\lim_{x \to 1} f(x)$
6. $\lim_{x \to -4} f(x)$

For $h(t) = 3^t + t$, determine each of the following limits:

7. $\lim_{t \to 4} h(t)$
8. $\lim_{t \to 0} h(t)$
9. $\lim_{t \to -2} h(t)$

For $f(x) = 4^x + x^2$, evaluate each of the following limits:

10. $\lim_{x \to 1} f(x)$
11. $\lim_{x \to -1} f(x)$
12. $\lim_{x \to 1/2} f(x)$

For $f(x) = \ln x + 2x$, evaluate each of the following limits:

13. $\lim_{x \to 1} f(x)$
14. $\lim_{x \to e} f(x)$
15. $\lim_{x \to 0} f(x)$

For the function given below, evaluate each of the following limits:
$$f(t) = \begin{cases} t^2 & t \leq 5 \\ 4t + 5 & t > 5 \end{cases}$$

16. $\lim_{t \to 0} f(t)$
17. $\lim_{t \to 5} f(t)$
18. $\lim_{t \to 7} f(t)$

For the function given below, evaluate each of the following limits:
$$y = \begin{cases} 2t + 3 & t \leq 0 \\ t^3 + 3 & t > 0 \end{cases}$$

19. $\lim_{t \to -4} y$
20. $\lim_{t \to 0} y$
21. $\lim_{t \to 2} y$

For the function given below, evaluate each of the following limits:
$$g(x) = \begin{cases} 3^x & x \leq 0 \\ (4x + 1)^2 & x > 0 \end{cases}$$

22. $\lim_{x \to -2} g(x)$
23. $\lim_{x \to 0} g(x)$
24. $\lim_{x \to 2} g(x)$

For the function given below, evaluate each of the following limits:
$$f(x) = \begin{cases} \dfrac{x^2}{x - 1} & x < 1 \\ \dfrac{x - 1}{x^2} & x > 1 \end{cases}$$

25. $\lim_{x \to 0} f(x)$
26. $\lim_{x \to 1} f(x)$
27. $\lim_{x \to 3} f(x)$

For the function given below, determine each of the following limits:
$$h(x) = \frac{(x-1)(x-2)}{(x-2)(x-3)}$$

28. $\lim\limits_{x \to 0} h(x)$ 29. $\lim\limits_{x \to 2} h(x)$ 30. $\lim\limits_{x \to 3} h(x)$

For the function given below, determine each of the following limits:
$$f(x) = \frac{x-3}{x(x+3)}$$

31. $\lim\limits_{x \to 0} f(x)$ 32. $\lim\limits_{x \to 3} f(x)$ 33. $\lim\limits_{x \to 5} f(x)$

For the function given below, evaluate each of the following limits:
$$y = \frac{x^2 + 2x - 15}{x^2 - 5x + 6}$$

34. $\lim\limits_{x \to 2} y$ 35. $\lim\limits_{x \to 3} y$ 36. $\lim\limits_{x \to -5} y$

For the function given below, determine each of the following limits:
$$r(t) = \frac{t^2 + t - 2}{2t^2 + t - 6}$$

37. $\lim\limits_{t \to -2} r(t)$ 38. $\lim\limits_{t \to 1/2} r(t)$ 39. $\lim\limits_{t \to 3/2} r(t)$

Evaluate each of the following limits:

40. $\lim\limits_{x \to 1} \dfrac{x - 1}{\sqrt{x} - 1}$ 41. $\lim\limits_{x \to 1} \dfrac{\sqrt{x} - 1}{x - 1}$

42. For $f(x) = 2x + 1$, evaluate: $\lim\limits_{h \to 0} \dfrac{f(3+h) - f(3)}{h}$

43. For $g(x) = x^2$, evaluate: $\lim\limits_{h \to 0} \dfrac{g(2+h) - g(2)}{h}$

44. For $f(x) = x^2 + x$, evaluate: $\lim\limits_{h \to 0} \dfrac{f(4+h) - f(4)}{h}$

45. *Sales commission* Evaluate $\lim\limits_{x \to 5,000} f(x)$, where f is the salary function constructed in Exercise 49 of Section 10.1.

46. Evaluate $\lim\limits_{x \to 4} h(x)$, where h is the area function constructed in Exercise 48 of Section 10.1.

Taxi fares If f is the taxi-fare function in Exercise 50 of Section 10.1, determine each of the following limits:

47. $\lim\limits_{x \to 4} f(x)$ 48. $\lim\limits_{x \to 5} f(x)$ 49. $\lim\limits_{x \to 10} f(x)$

Income taxes If g is the income tax function in Exercise 51 of Section 10.1, determine each of the following limits:

50. $\lim\limits_{x \to 2,000} g(x)$ 51. $\lim\limits_{x \to 4,500} g(x)$

52. $\lim\limits_{x \to 6,000} g(x)$ 53. $\lim\limits_{x \to 12,000} g(x)$

54. Explain the meaning of the limit of a function f as x approaches a.

10.4 CONTINUITY

In our discussion of calculus, we shall at times be interested in functions that are called *continuous* functions. Intuitively, a continuous function is a function whose graph can be drawn by a continuous curve, without lifting the pencil from the paper. In order to make this idea mathematically precise, consider the graphs of the four functions shown in Figure 10.4. Only $f(x) = \frac{1}{2}x^3$ is continuous at $x = 1$.

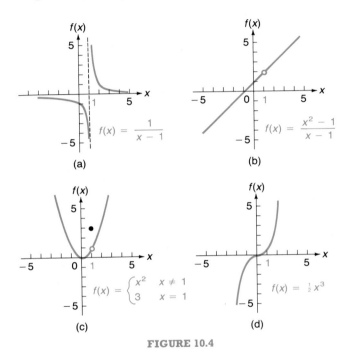

FIGURE 10.4

In terms of $f(1)$ and the $\lim_{x \to 1} f(x)$, what distinguishes the function $f(x) = \frac{1}{2}x^3$ from the other three functions?

For $f(x) = \dfrac{1}{x - 1}$, neither $f(1)$ nor the $\lim_{x \to 1} f(x)$ has a value.

For $f(x) = \dfrac{x^2 - 1}{x - 1}$, $f(1)$ has no value, but $\lim_{x \to 1} f(x) = 2$.

For $f(x) = \begin{cases} x^2 & x \neq 1 \\ 3 & x = 1 \end{cases}$, $f(1) = 3$, but $\lim_{x \to 1} f(x) = 1$.

For $f(x) = \frac{1}{2}x^3$, $f(1) = \frac{1}{2}$ and $\lim_{x \to 1} f(x) = \frac{1}{2}$.

It is because $f(1)$ and $\lim_{x \to 1} f(x)$ have the same value in the case of $f(x) = \frac{1}{2}x^3$ that this latter function is continuous at $x = 1$. Formally, a function f is **continuous at $x = a$** if $\lim_{x \to a} f(x) = f(a)$.

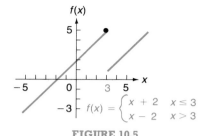

FIGURE 10.5

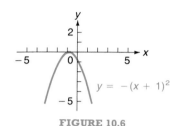

FIGURE 10.6

The graph of

$$f(x) = \begin{cases} x + 2 & x \leq 3 \\ x - 2 & x > 3 \end{cases}$$

shown in Figure 10.5 indicates that $f(x)$ is not continuous at $x = 3$. Although $f(3) = 5$, $\lim_{x \to 3} f(x)$ does not exist and therefore cannot possibly equal 5. However, f is continuous at every other value of x. For example, both $f(0) = 2$ and $\lim_{x \to 0} f(x) = 2$, so that f is continuous at 0; also, $f(6) = 4$ and $\lim_{x \to 6} f(x) = 4$, so that f is continuous at 6; and so forth.

The graph of $y = -(x + 1)^2$ shown in Figure 10.6 indicates that this function is continuous at every value of x.

In order for a function f to be continuous at $x = a$, the definition requires that three conditions be satisfied:

1. $f(a)$ exists (has a value).
2. $\lim_{x \to a} f(x)$ exists (has a value).
3. $\lim_{x \to a} f(x) = f(a)$

We want to be able to determine whether or not a given function is continuous at particular values of x without graphing the function. These three conditions give a checklist to use in determining whether a function is continuous at a given value of x. If even one of the conditions is not satisfied at $x = a$, the function is not continuous at a; and if all three are satisfied, the function is continuous at a.

EXAMPLE 1 Given the function f defined by

$$f(x) = \begin{cases} 2x^2 & x \leq -3 \\ 2x - 1 & x > -3 \end{cases}$$

(a) Verify that f is continuous at $x = 4$.
(b) Verify that f is not continuous at $x = -3$.

Solution (a) Following the checklist above with $a = 4$, we have:

1. $f(4) = 7$
2. $\lim_{x \to 4} f(x) = 7$
3. $f(4) = \lim_{x \to 4} f(x)$

Since the three conditions are satisfied, f is continuous at $x = 4$.

(b) Following the checklist with $a = -3$, we have:

1. $f(-3) = 18$
2. As x approaches -3 from the left, $f(x)$ approaches 18; as x approaches -3 from the right, $f(x)$ approaches -7. Therefore, $\lim_{x \to -3} f(x)$ has no value.

Because the second and the third of the three required conditions are not satisfied, f is not continuous at -3.

EXAMPLE 2 Given the function

$$g(x) = \frac{x}{(x+2)^2}$$

(a) Verify that g is continuous at $x = 0$.
(b) Verify that g is not continuous at $x = -2$.

Solution (a) Following the checklist with $a = 0$, we have:

1. $g(0) = 0$
2. $\lim_{x \to 0} g(x) = 0$
3. $g(0) = \lim_{x \to 0} g(x)$

Since the three conditions are satisfied, g is continuous at $x = 0$.

(b) Following the checklist with $a = -2$, we have:

1. $g(-2)$ does not exist. Therefore, g is not continuous at $x = -2$.

EXAMPLE 3 Determine whether

$$f(x) = \begin{cases} x^2 - x & x \leq 5 \\ 6x - 10 & x > 5 \end{cases}$$

is continuous at $x = 5$.

Solution Following the checklist for $a = 5$, we have:

1. $f(5) = 20$
2. As x approaches 5 from the right, $f(x)$ approaches 20; as x approaches 5 from the left, $f(x)$ approaches 20. Consequently, $\lim_{x \to 5} f(x) = 20$.

PRACTICE PROBLEM 1

If

$$f(x) = \begin{cases} \ln x & x \leq 1 \\ x^2 - 1 & x > 1 \end{cases}$$

verify each of the following:

(a) f is continuous at $x = 1$.
(b) f is not continuous at $x = 0$.

Answer

(a) $\lim_{x \to 1} f(x) = 0 = f(1)$
(b) $f(0)$ does not exist.

3. $f(5) = \lim_{x \to 5} f(x)$

Since the three conditions are satisfied, f is continuous at $x = 5$.

Just as with limits, continuity of a function is a pointwise property; that is, continuity is defined in terms of a point $x = a$ and must be checked for continuity at one value of x at a time.

Although we shall not prove it and shall not state the fact formally as a theorem, each function of the types listed in Table 10.1 (linear, quadratic, exponential, logarithmic, and polynomial) is continuous at each value in its domain. You may have observed that each graph of these types of functions that we have encountered was a continuous curve (line, in the case of a linear function).

Finally, the definition of continuity is extended to intervals on the x-axis as follows. A function is **continuous on an interval** if it is continuous at each point in the interval. The function f in Example 1 is continuous on any interval not containing -3; g in Example 2 is continuous on any interval not containing -2; and f in Example 3 is continuous on every interval.

PRACTICE PROBLEM 2

Determine whether

$$f(x) = \begin{cases} \dfrac{x+1}{x^2-1} & x \neq -1 \\ 2 & x = -1 \end{cases}$$

is continuous at $x = -1$.

Answer

Not continuous

PRACTICE PROBLEM 3

In what intervals is $g(x) = e^x$ continuous? (*Hint:* Consider the graph of g.)

Answer

Every interval

EXERCISES

Given the function below, verify each of the following statements:

$$f(x) = \begin{cases} 2x & x \leq 1 \\ x^2 & x > 1 \end{cases}$$

1. f is continuous at $x = 3$.
2. f is not continuous at $x = 1$.

Given the function below, verify each of the following statements:

$$g(x) = \begin{cases} 4 - x & x \leq 2 \\ x^2 - 2 & x > 2 \end{cases}$$

3. g is continuous at $x = -1$.
4. g is continuous at $x = 2$.

Given the function below, verify each of the following statements:

$$f(x) = \begin{cases} 3^x & x \neq 0 \\ 4 & x = 0 \end{cases}$$

5. f is continuous at $x = 4$.
6. f is not continuous at $x = 0$.

Given the function $f(t) = 3t^2 + 2t$, verify each of the following statements:

7. f is continuous at $t = 1$.
8. f is continuous at $t = -2$.

Given the function below, verify each of the following statements:

$$g(t) = \begin{cases} \dfrac{t^2 - 2t - 3}{t^2 - t - 6} & t \neq -2, 3 \\ \frac{4}{5} & t = 3 \\ -\frac{2}{3} & t = -2 \end{cases}$$

9. g is continuous at $t = 3$.
10. g is not continuous at $t = -2$.

Determine whether

$$y = \frac{3}{x+3}$$

is continuous at:

11. $x = -3$ 12. $x = 0.4$

Determine whether

$$g(t) = \frac{t^2 + 2t - 15}{t^2 - 5t + 6}$$

is continuous at:

13. $t = 2$ 14. $t = 3$

Determine whether

$$h(x) = \begin{cases} x^3 - x^2 & x \leq 5 \\ (2x)^2 & x > 5 \end{cases}$$

is continuous at:

15. $x = 3$ 16. $x = 5$

Determine whether

$$y = \frac{1}{\sqrt{x-4}}$$

is continuous at:

17. $x = 13$ 18. $x = 4$ 19. $x = -5$

Considering the graph of f, determine whether $f(x) = \ln x$ is continuous at:

20. $x = 1$ 21. $x = -1$ 22. $x = 0$

For which values of x is the function f whose graph is shown in the margin (domain: $-3 < x \leq 3$):

23. Continuous? 24. Discontinuous?

On which intervals are the following continuous?

25. The function f in Exercises 1–2
26. The function g in Exercises 3–4
27. The function g in Exercises 9–10
28. The function y in Exercises 11–12
29. The function g in Exercises 13–14
30. The function y in Exercises 17–19
31. The function f in Exercises 20–22
32. *Sales commission* If f is the salary function in Exercise 49 of Section 10.1, at what values in its domain is f not continuous?
33. If h is the area function in Exercise 48 of Section 10.1, at what values in its domain is h not continuous?
34. *Taxi fares* If f is the taxi-fare function in Exercise 50 of Section 10.1, at what values in its domain is f continuous?

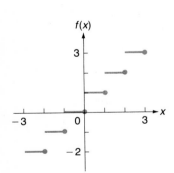

35. *Income taxes* If g is the income tax function in Exercise 51 of Section 10.1, at what values in its domain is g continuous?

Explain the meaning of each term.

36. A function f being continuous at $x = a$
37. A function f being continuous on an interval

*10.5 RATIONAL FUNCTIONS

We conclude this chapter by considering one more class of functions, the rational functions. A **rational function** is a function of the form

$$f(x) = \frac{p(x)}{q(x)}$$

where $p(x)$ and $q(x)$ are polynomials. Examples of rational functions are

$$f(x) = \frac{2x^2 - 3x + 7}{x^2 + 4x - 6} \qquad y = \frac{2x^3 + 3x + 4}{5x^2 - 3x}$$

$$f(x) = \frac{1}{x - 1} \qquad h(x) = \frac{2(x - 2)(x + 3)}{(x - 1)(x - 2)}$$

In the last example, $h(x)$, the numerator and denominator are written in factored form; if the factors were multiplied together, the result would be a polynomial in the numerator and in the denominator, as the definition of a rational function requires. The 1 in the numerator of $f(x) = 1/(x - 1)$ is a nonzero constant, which is also considered to be a polynomial of degree zero.

Rational functions are used as mathematical models in a variety of situations. For example, if the cost y (in millions of dollars) for a chemical company to remove x percent of the pollutants from the smoke emission at one of their plants is

$$y = \frac{x}{100 - x} \qquad 0 \leq x < 100$$

then y is a rational function.

Our major concern will be to graph rational functions. We shall assume throughout that the numerator and denominator do not have any common factor, as $h(x)$ does in the examples above. The next several graphs indicate some properties of the type of graphs to be expected.

Several things should be noted in the graph of f shown in Figure 10.7:

1. f is continuous except at $x = 1$, which corresponds to a zero in the denominator of $f(x)$.

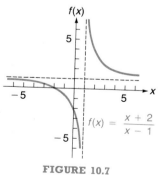

FIGURE 10.7

This section can be omitted without loss of continuity.

2. For values of x close to 1, the function values become unbounded through positive and negative values.
3. As x gets large through positive or negative values, the function values approach a limiting value of $y = 1$. The extreme right side and extreme left side of the graph approach the line $y = 1$.

The graph of another rational function is shown in Figure 10.8. Again, note the following:

1. g is continuous except at $x = 2$ and $x = -2$, when the denominator is zero.
2. For values of x close to 2 or to -2, the function values become unbounded through positive and negative values.
3. In this case, as x gets large through positive values, so do the function values; as x gets large through negative values, so do the function values. The left side extends downward indefinitely and the right side extends upward indefinitely.

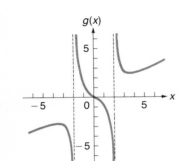

$$g(x) = \frac{x^3}{2(x^2 - 4)} = \frac{x^3}{2(x + 2)(x - 2)}$$

FIGURE 10.8

In general, a rational function is continuous except at those values of x for which the denominator is zero. At those values for which the denominator is zero, the function becomes unbounded through positive or negative values. As x gets large through positive or negative values, the function values may approach a limiting value ($y = 1$ in the case of f above) or may become unbounded (as in the case of g above). The vertical lines corresponding to values of x for which the denominator of a rational function is zero and at which the values of f become unbounded are called *vertical asymptotes*. The lines $x = 1$ for f and $x = 2$ and $x = -2$ for g are vertical asymptotes.

The horizontal lines that the function values approach as x gets large through positive or negative values are called *horizontal asymptotes*. For f, the line $y = 1$ is a horizontal asymptote; g does not have a horizontal asymptote because its values become unbounded and do not approach a limiting value at the "tails."

Our procedure for graphing rational functions is as follows:

1. Determine the vertical asymptotes.
2. Determine the horizontal asymptotes (if any).
3. Plot enough points to determine the shape of the graph.

We illustrate the procedure by graphing the function

$$f(x) = \frac{3}{(x - 1)(x + 3)} = \frac{3}{x^2 + 2x - 3}$$

10.5 RATIONAL FUNCTIONS

The vertical asymptotes are $x = 1$ and $x = -3$, corresponding to the fact that the denominator is zero for these values of x.

In order to determine the horizontal asymptotes, divide the numerator and the denominator by the highest power of x appearing in the denominator and then examine what happens to the values of the function for very large values of x. In this case, the highest power of x in the denominator is x^2. Dividing numerator and denominator by x^2 gives

$$\frac{\frac{3}{x^2}}{\frac{x^2 + 2x - 3}{x^2}} = \frac{\frac{3}{x^2}}{1 + \frac{2}{x} - \frac{3}{x^2}}$$

As x increases without bound, $3/x^2$ and $2/x$ both approach 0, so that the numerator approaches 0 and the denominator approaches 1. Therefore, the quotient approaches $\frac{0}{1} = 0$. We indicate this fact by writing

$$\lim_{x \to \infty} f(x) = 0.$$

The expression $\lim_{x \to \infty}$ is read, "the limit as x approaches infinity." It is not a limit in the previous sense, but indicates the behavior of the function values as x gets arbitrarily large. The symbol ∞ for infinity is not a number; $x \to \infty$ just indicates that x is getting arbitrarily large.

Because $\lim_{x \to \infty} f(x) = 0$, the line $y = 0$ is a horizontal asymptote for f; the graph of f approaches this line as x becomes larger and larger. For a rational function, if its graph approaches a horizontal asymptote as x increases without bound, then its graph approaches the same asymptote as x decreases without bound (becomes more and more negative). That is, for a rational function f, if

$$\lim_{x \to \infty} f(x) = a \quad \text{then} \quad \lim_{x \to -\infty} f(x) = a$$

where $\lim_{x \to -\infty}$ is read, "the limit as x approaches negative infinity."

For our illustration, this means that both tails of the graph of the function approach the line $y = 0$.

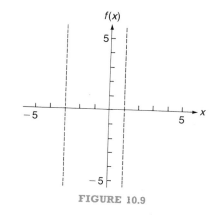

FIGURE 10.9

We first graph the vertical and horizontal asymptotes, as shown in Figure 10.9. Note that the line $y = 0$ is just the x-axis.

The next step is to pick some values of x and determine the corresponding values of $f(x)$. The results are given in the table.

x	−5	−4	−3.5	−2.5	−2	−1	0	0.5	1.5	2	3
$f(x)$	0.25	0.60	1.33	−1.71	−1	−0.75	−1	−1.71	1.33	0.60	0.25

Locating the corresponding points and connecting them with a curve yields the graph shown in Figure 10.10

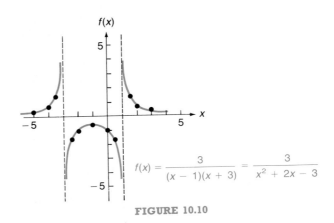

$$f(x) = \frac{3}{(x-1)(x+3)} = \frac{3}{x^2 + 2x - 3}$$

FIGURE 10.10

EXAMPLE 1 Graph the function $y = \dfrac{x^2}{1 - x^2}$.

Solution Factoring the denominator

$$y = \frac{x^2}{1 - x^2} = \frac{x^2}{(1 - x)(1 + x)}$$

we see that the graph has $x = 1$ and $x = -1$ as vertical asymptotes. Dividing numerator and denominator by x^2 (the highest power in the denominator), we get

$$y = \frac{\dfrac{x^2}{x^2}}{\dfrac{1}{x^2} - \dfrac{x^2}{x^2}} = \frac{1}{\dfrac{1}{x^2} - 1}$$

As x gets large, $1/x^2$ approaches 0; consequently $\lim\limits_{x \to \infty} y = -1$. The graph has $y = -1$ as a horizontal asymptote.

For locating points on the graph, we choose some values of x and determine the corresponding values of y. The results are given in the table.

x	-3	-2	-1.5	-0.5	0	0.5	1.5	2	3
y	-1.125	-1.33	-1.8	0.33	0	0.33	-1.8	-1.33	-1.125

To determine the graph, we locate the asymptotes in a coordinate system, plot the points with the coordinates given in the table, and then connect the points.

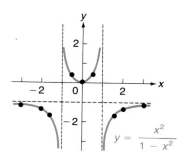

EXAMPLE 2 Graph the function $g(x) = \dfrac{x^2 - 9}{x}$.

Solution Since the denominator is zero when $x = 0$, the graph has the y-axis as a vertical asymptote. Dividing numerator and denominator by the highest power in the denominator gives

$$\frac{x^2 - 9}{x} = \frac{\dfrac{x^2}{x} - \dfrac{9}{x}}{\dfrac{x}{x}} = \frac{x - \dfrac{9}{x}}{1} = x - \frac{9}{x}$$

When x increases without bound, so does $x - (9/x)$; the right tail of the graph extends indefinitely upward. We indicate this by writing $\lim_{x \to \infty} g(x) = \infty$.

Similarly, when x decreases without bound through negative values, so does $x - (9/x)$. The left tail of the graph extends indefinitely downward, and we write $\lim_{x \to -\infty} g(x) = -\infty$. The graph does not have a horizontal asymptote.

A table of values is given below, and the graph is shown on the next page.

x	-6	-5	-3	-1	1	3	5	6
$g(x)$	-4.5	-3.2	0	8	-8	0	3.2	4.5

PRACTICE PROBLEM 1

Given $y = \dfrac{x^2}{x^2 - 4}$.

(a) Determine vertical asymptotes.
(b) Determine horizontal asymptotes.
(c) Graph the function.

Answer

(a) $x = 2$ and $x = -2$ (b) $y = 1$

(c)

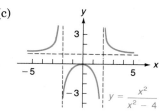

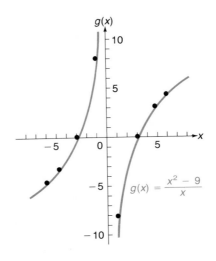

It could happen that as x gets large through positive values, the values of the function decrease without bound (become more and more negative). For example, if

$$f(x) = \frac{9-x^2}{x} = \frac{\frac{9-x^2}{x}}{\frac{x}{x}} = \frac{9}{x} - x$$

the values of $f(x)$ are large negatively for large positive values of x. We write $\lim_{x \to \infty} f(x) = -\infty$, and the right tail of the graph extends indefinitely downward.

Similarly, for the function f, as x decreases through negative values, $f(x)$ increases without bound. We write $\lim_{x \to -\infty} f(x) = \infty$; the left tail of the graph extends indefinitely upward.

 PRACTICE PROBLEM 2

Given $f(x) = \dfrac{4 - x^2}{x + 1}$.

(a) Determine the vertical asymptotes.

(b) Determine $\lim_{x \to \infty} f(x)$ and $\lim_{x \to -\infty} f(x)$.

(c) Graph the function.

Answer

(a) $x = -1$

(b) $\lim_{x \to \infty} f(x) = -\infty$;

$\lim_{x \to -\infty} f(x) = \infty$

(c)

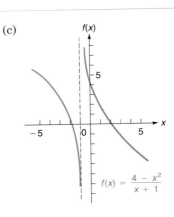

EXERCISES

Find the vertical asymptotes for each of the following:

1. $f(x) = \dfrac{(x+1)(x-2)}{(x-4)(x-1)}$
2. $f(x) = \dfrac{3(x-1)(x-2)}{(x-3)(x+4)}$
3. $g(x) = \dfrac{4}{x-1}$
4. $g(t) = \dfrac{(t+2)(t+3)}{(t-4)(t-3)^2}$
5. $f(x) = \dfrac{-4(5x+2)}{(3x-2)(2x+1)}$
6. $g(x) = \dfrac{x^2+3x+2}{x^2+5x}$

For each of the following:

(a) Determine the horizontal asymptote, if there is one.
(b) If there is no horizontal asymptote, determine whether

$$\lim_{x \to \infty} f(x) = \infty \quad \text{or} \quad \lim_{x \to \infty} f(x) = -\infty$$

(c) If there is no horizontal asymptote, determine whether

$$\lim_{x \to -\infty} f(x) = \infty \quad \text{or} \quad \lim_{x \to -\infty} f(x) = -\infty$$

7. $f(x) = \dfrac{3x^2+6x-1}{x^2-x+7}$
8. $f(x) = \dfrac{2x^3+6x^2-7}{x^2+3x}$
9. $f(x) = \dfrac{x^3+4x-5}{x+1}$
10. $f(x) = \dfrac{-x^2+4x-5}{x+1}$
11. $f(x) = \dfrac{4}{x-3}$
12. $f(x) = \dfrac{2x}{(x+1)(x-1)}$

For each of the following:

(a) Find the vertical asymptotes, if any.
(b) Determine the horizontal asymptote, if there is one.
(c) If there is no horizontal asymptote, determine whether

$$\lim_{x \to \infty} f(x) = \infty \quad \text{or} \quad \lim_{x \to \infty} f(x) = -\infty$$

(d) If there is no horizontal asymptote, determine whether

$$\lim_{x \to -\infty} f(x) = \infty \quad \text{or} \quad \lim_{x \to -\infty} f(x) = -\infty$$

13. $f(x) = \dfrac{3x+2}{x-1}$
14. $f(x) = \dfrac{(x+2)(x+1)}{(x+4)(x-1)}$
15. $f(x) = \dfrac{(x+3)(x-2)}{x+1}$
16. $f(x) = \dfrac{7}{x+2}$
17. $f(x) = \dfrac{-2x^2-4x+6}{x-5}$
18. $f(x) = \dfrac{3x^2}{x-2}$

Graph each of the following:

19. $f(x) = \dfrac{x+3}{x-1}$
20. $f(x) = \dfrac{2x+3}{x+3}$
21. $f(x) = \dfrac{-x}{x^2-1}$

22. $f(x) = \dfrac{x^2 - 1}{-x}$ 23. $g(x) = \dfrac{x^2 - 4}{x}$ 24. $g(x) = \dfrac{x^2 - 1}{x}$

25. $y = \dfrac{8}{x - 4}$ 26. $y = \dfrac{1}{x}$

27. $y = \dfrac{2 - x^2}{x}$ 28. $g(x) = \dfrac{x^2}{(x + 1)(x - 3)}$

29. $y = \dfrac{3x + 3}{x^2 + x - 2}$ 30. $g(x) = \dfrac{x^2 + 9x + 18}{x^2 - 6x}$

31. *Productivity* The number of items produced per hour by a new production worker on a particular job is given by

$$y = \dfrac{5x}{2 + x}$$

where x is the number of weeks the employee has been on the job.
(a) Graph the given function.
(b) Which points in the graph are relevant to the physical situation?
(c) About how many items per hour can be expected from the employee after he or she has been on the job for a long time?

32. *Cost function*
(a) Graph the cost function illustration given at the beginning of this section,

$$y = \dfrac{x}{100 - x} \quad 0 \le x < 100$$

where y is the cost (in millions of dollars) for removing x percent of the pollutants from the smoke emission.
(b) Which points in the graph are relevant to the physical situation?
(c) On the basis of your graph, discuss the cost as the company attempts to remove virtually all pollution.

33. *Drug concentration* The percent of concentration of a drug in the bloodstream x hours after it is administered is given by

$$y = \dfrac{3x}{4x^2 + 10}$$

Without graphing the function, determine the approximate percent of concentration a long time after the drug is administered.

Evaluate the following limits:

34. $\lim\limits_{x \to \infty} \dfrac{4x^3 - 6x + 3}{2x^2 + 4x + 1}$ 35. $\lim\limits_{x \to \infty} \dfrac{3x^2 - 5x + 1}{4x^2 + 2x - 7}$

36. $\lim\limits_{x \to -\infty} \dfrac{x^2 - 5x}{2x^3 - 3x^2 + 7}$ 37. $\lim\limits_{x \to -\infty} \dfrac{4x^3 - 6x + 3}{2x^2 + 4x + 1}$

38. $\lim\limits_{t \to \infty} \dfrac{12t^3 - 5t^2 + t}{4t^3 + 9t^2 + 5}$ 39. $\lim\limits_{t \to \infty} \dfrac{t^3 + 1}{t^4 + t}$

Explain the meaning of each term.

40. Rational function
41. A vertical asymptote for the graph of a rational function
42. A horizontal asymptote for the graph of a rational function

10.6 CHAPTER SUMMARY

Functions are a central concept in many areas of mathematics, one of which is calculus, as we shall see in the next several chapters. A *function* is a rule that associates a unique number to each member of some set D. The set D is called the *domain* of the function.

Lowercase letters are most often used to designate functions, usually f, g, and h. Then the number associated with a value x is designated $f(x)$, $g(x)$, $h(x)$, and so forth. In mathematics, functions are usually described by an equation such as

$$f(x) = x^2 + 1$$

which indicates that f is the function that associates with each number x the number equal to x^2 plus 1. If the domain of a function is not specified, it is understood to be all real numbers for which the associated value exists.

The *graph of a function* f is the same as the graph of the equation $y = f(x)$. For any point in the graph of a function, the second coordinate is the value of the function at the first coordinate. For example, the point (3, 10) in the graph of $f(x) = x^2 + 1$ corresponds to the fact that $f(3) = 10$.

When the function notation $f(x)$, $g(x)$, and so forth is not required, these can be replaced by a single letter, usually y. Functions are then presented by an equation such as

$$y = x^2 + 1$$

In this case, x is called the *independent variable* and y is called the *dependent variable;* the value of y "depends" on the value of x.

The *limit* of a function f as x approaches a is L, provided that the values of $f(x)$ get closer and closer to L as x gets closer and closer to a [in the sense that the values of $f(x)$ can be made to be as close to L as desired by choosing values of x sufficiently close to a] without regard to the value of f at $x = a$. If such a number L exists, we write

$$\lim_{x \to a} f(x) = L$$

The limit concept is another central concept in the development of calculus.

If the limit of f as x approaches a is equal to $f(a)$—that is, if

$$\lim_{x \to a} f(x) = f(a)$$

—then f is said to be *continuous at a.* If f is continuous at each number in an interval, then f is said to be *continuous on the interval.* Geometrically, a function is continuous on an interval if its graph can be drawn as a continuous curve without lifting the pencil from the paper.

In the last section we considered a special type of function—the rational function. A **rational function** is a function that can be written as the quotient of two polynomials. Two principal characteristics of the graphs of rational functions are vertical and horizontal asymptotes. As long as the numerator and denominator of a rational function have no common factors, the vertical asymptotes are the lines $x = a$ corresponding to values a which make the denominator zero. If $\lim_{x \to \infty} f(x) = k$ or $\lim_{x \to -\infty} f(x) = k$, then $y = k$ is a horizontal asymptote. If there is no such number k, then there is no horizontal asymptote, and the tails of the graph extend indefinitely upward or downward.

REVIEW EXERCISES

For the function f determined by $f(x) = x^2 - 2x + 4$, find each of the following:

1. $f(-3)$
2. $f(0)$
3. $f(2)$
4. $f(t + 1)$
5. Domain of f

For the function g determined by

$$g(x) = \frac{2x}{(x + 1)(x - 2)}$$

find each of the following:

6. $g(-2)$
7. $g(0)$
8. $g(0.3)$
9. $g(2.5)$
10. Domain of g

For the function f determined by $f(t) = \sqrt{t + 2}$, find each of the following:

11. $f(-1)$
12. $f(-1.99)$
13. $f(7)$
14. $f(125)$
15. Domain of f

For the function h determined by $h(x) = 3(4^x)$, find each of the following:

16. $h(-2)$
17. $h(-\frac{1}{2})$
18. $h(0)$
19. $h(3)$
20. Domain of h

For the function f determined by $f(x) = \log x$, find each of the following:

21. $f(1,000)$
22. $f(10)$
23. $f(1)$
24. $f(0.01)$
25. Domain of f

For the function f determined by

$$f(x) = \begin{cases} \dfrac{x}{x + 1} & x \geq 0 \\ \dfrac{x}{x - 1} & x < 0 \end{cases}$$

find each of the following:

26. $f(-3)$
27. $f(-1)$
28. $f(0)$

29. $f(7)$ **30.** Domain of f

For the function g determined by

$$g(x) = \begin{cases} 2x^2 + 1 & x \geq 10 \\ 2x^2 & -10 < x < 10 \\ 2x^2 - 1 & x \leq -10 \end{cases}$$

find each of the following:

31. $g(-2)$ **32.** $g(-10)$ **33.** $g(2)$
34. $g(12)$ **35.** Domain of g

Determine

$$\frac{f(x+h) - f(x)}{h}$$

for each of the following functions:

36. $f(x) = 3x - 2$ **37.** $f(x) = 2x^2 - x + 1$
38. $f(x) = -x^2 + 7$ **39.** $f(x) = -(x+2)^2$

Graph each of the following functions:

40. $y = 4 - 2x$ **41.** $f(x) = 2$
42. $y = -x, \quad x \leq 0$ **43.** $g(x) = x^2 - 6x + 7$
44. $f(x) = -x^2 + 4x - 3$ **45.** $y = x^2 - 4, \quad -2 \leq x \leq 2$
46. $g(x) = \log \dfrac{x}{2}$ **47.** $y = 3^x$

48. $y = \begin{cases} 2x & x \geq 2 \\ x - 2 & x < 2 \end{cases}$ **49.** $f(x) = \begin{cases} x^2 & x \geq 1 \\ e^{x-1} & x < 1 \end{cases}$

50. $y = \dfrac{x^2 - 25}{x + 5}$ **51.** $g(x) = \sqrt{x - 2}$

For $f(x) = 2x - x^2$, determine each of the following limits:

52. $\lim\limits_{x \to -3} f(x)$ **53.** $\lim\limits_{x \to 0} f(x)$ **54.** $\lim\limits_{x \to 4} f(x)$

For $g(x) = 2^x + 3$, determine each of the following limits:

55. $\lim\limits_{x \to -2} g(x)$ **56.** $\lim\limits_{x \to 1} g(x)$ **57.** $\lim\limits_{x \to 3} g(x)$

For the function given below, determine each of the following limits:

$$f(t) = \begin{cases} 2t^2 - 1 & t \leq 1 \\ \ln t & t > 1 \end{cases}$$

58. $\lim\limits_{t \to 0} f(t)$ **59.** $\lim\limits_{t \to 1} f(t)$ **60.** $\lim\limits_{t \to 2} f(t)$

For the function given below, determine each of the following limits:

$$f(x) = \begin{cases} -2 & x > 2 \\ x - x^2 & x \leq 2 \end{cases}$$

61. $\lim\limits_{x \to 1} f(x)$ **62.** $\lim\limits_{x \to 2} f(x)$ **63.** $\lim\limits_{x \to 4} f(x)$

For the function given below, determine each of the following limits:

$$f(x) = \frac{x^2 + x - 6}{x^2 + 2x - 8}$$

64. $\lim_{x \to 3} f(x)$ 65. $\lim_{x \to 2} f(x)$ 66. $\lim_{x \to -4} f(x)$

For the function given below, determine each of the following limits:

$$g(x) = \frac{2x^2 + 6x - 8}{x^2 + 2x - 3}$$

67. $\lim_{x \to -4} g(x)$ 68. $\lim_{x \to -3} g(x)$ 69. $\lim_{x \to 1} g(x)$

Evaluate the limit given below for each of the following functions:

$$\lim_{h \to 0} \frac{f(3+h) - f(3)}{h}$$

70. $f(x) = x^2 + 1$ 71. $f(x) = x - x^2$

Determine whether the function f in Exercises 52–54 is continuous at:

72. $x = -3$ 73. $x = 0$ 74. $x = 4$

Determine whether the function g in Exercises 55–57 is continuous at:

75. $x = -2$ 76. $x = 1$ 77. $x = 3$

Determine whether the function f in Exercises 58–60 is continuous at:

78. $t = 0$ 79. $t = 1$ 80. $t = 2$

Determine whether the function f in Exercises 61–63 is continuous at:

81. $x = 1$ 82. $x = 2$ 83. $x = 4$

Determine whether the function f in Exercises 64–66 is continuous at:

84. $x = 3$ 85. $x = 2$ 86. $x = -4$

Determine whether the function g in Exercises 67–69 is continuous at:

87. $x = -4$ 88. $x = -3$ 89. $x = 1$

90. Why is the function given below not continuous at $x = -3$?

$$y = \begin{cases} \dfrac{x^2 - 9}{x + 3} & x \neq -3 \\ 7 & x = -3 \end{cases}$$

On which intervals are the following continuous?

91. The function f in Exercises 52–54
92. The function g in Exercises 55–57
93. The function f in Exercises 58–60
94. The function f in Exercises 61–63
95. The function f in Exercises 64–66
96. The function g in Exercises 67–69
97. $f(x) = \dfrac{1}{\sqrt{x^2 - 9}}$

98. *Sales commission* A salesperson's weekly salary is $200 plus a commission of 5% on his first $1,000 in sales and 10% on the sales over $1,000.
   (a) Construct a function f similar to those given for Exercises 26–35, such that $f(x)$ gives the salesperson's weekly salary in terms of his sales x. Indicate the domain of f.
   (b) Determine the value of $f(575)$ and $f(1,500)$.
   (c) Graph the function f. (You may have to use different units on the two axes.)
   (d) Determine $\lim_{x \to 575} f(x)$ and $\lim_{x \to 1,000} f(x)$.
   (e) Determine whether f is continuous at 500 and at 1,000.
   (f) On which intervals is f continuous?

99. *Income tax* The table gives income tax rates for a particular state.

TAX TABLE

TAXABLE INCOME	TAX
$0–5,000	$\frac{1}{2}$% of taxable income
$5,000–10,000	$25 plus 1% of excess over $5,000
$10,000–15,000	$75 plus 2% of excess over $10,000
$15,000–20,000	$175 plus $2\frac{1}{2}$% of excess over $15,000
$20,000–40,000	$300 plus 3% of excess over $20,000
Over $40,000	$900 plus $3\frac{1}{2}$% of excess over $40,000

   (a) Construct a function g similar to those given for Exercises 26–35, such that $g(x)$ gives the tax on a person's taxable income x. Indicate the domain of g.
   (b) Determine $g(7,500)$, $g(12,750)$, and $g(48,500)$.
   (c) Determine $\lim_{x \to 5,000} g(x)$, $\lim_{x \to 17,000} g(x)$, and $\lim_{x \to 40,000} g(x)$.
   (d) Determine whether g is continuous at 5,000, at 17,000, and at 40,000.
   (e) On which intervals is g continuous?

For Exercises 100–105:
(a) Find the vertical asymptotes, if any.
(b) Determine the horizontal asymptote, if there is one.
(c) If there is no horizontal asymptote, determine whether

$$\lim_{x \to \infty} f(x) = \infty \quad \text{or} \quad \lim_{x \to \infty} f(x) = -\infty$$

(d) If there is no horizontal asymptote, determine whether

$$\lim_{x \to -\infty} f(x) = \infty \quad \text{or} \quad \lim_{x \to -\infty} f(x) = -\infty$$

100. $f(x) = \dfrac{2}{x+3}$

101. $f(x) = \dfrac{-3x^2}{x^2+5}$

102. $f(x) = \dfrac{4x+3}{(x-1)(x+5)}$

103. $f(x) = \dfrac{(x-1)(x+5)}{4x+3}$

104. $f(x) = \dfrac{2x^3+4}{x^2+5x+6}$

105. $f(x) = \dfrac{x^3-1}{2x+1}$

Graph each of the following functions:

106. $f(x) = \dfrac{2x}{x+3}$ **107.** $g(x) = \dfrac{x^2+9}{x^2-9}$ **108.** $y = \dfrac{4}{x-2}$

109. $y = \dfrac{x^2}{x+3}$ **110.** $f(x) = \dfrac{x^2+1}{x}$ **111.** $y = \dfrac{x^2}{2-x}$

112. *Employee errors* The average number of errors committed each day by a production worker in a particular factory is given by

$$y = \dfrac{x+5}{2x+1}$$

where x is the number of days the worker has been on the job. Without graphing the function, estimate the average number of errors committed by a worker after he or she has been on the job a long time.

CHAPTER 11

THE DERIVATIVE*

OUTLINE

11.1 Average Rate of Change
11.2 The Derivative
11.3 Interpretation of the Derivative
11.4 Rules for Derivatives
11.5 The Chain Rule
11.6 The Product and Quotient Rules
11.7 Derivatives of Exponential and Logarithmic Functions
11.8 Differentiable Functions
11.9 Chapter Summary

Review Exercises

Two of the problems considered by mathematicians that led to the development of calculus were the following:

1. To find the tangent to the graph of a function f at particular values of x (see Figure 11.1)
2. To find the area under the graph of a function f between two values of x, such as $x = a$ and $x = b$ (see Figure 11.2)

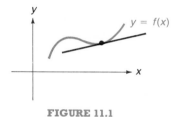

FIGURE 11.1

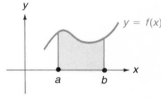

FIGURE 11.2

The mathematical concept introduced to solve the tangent problem is called the *derivative;* that for the area problem is called the *integral*. These are the two basic concepts of calculus. There is, as we shall see, an amazing relationship between the two, which is not apparent from the type of problems that led to their discovery. Chapters 11 and 12 are concerned with the derivative; Chapters 13 and 14 discuss the integral.

At this point it is probably not obvious how the solutions to the two problems above can lead to the solutions of many physical problems. However, it is because they do lead to solutions of many such problems that you are studying calculus today. For example, a recycling center sells shredded paper as packing to a pharmaceutical company. The center sells and ships the paper in units of 64 cubic feet. What size carton, with volume 64 cubic feet, would be the most economical (least surface area) to use—$2 \times 4 \times 8$ feet, $4 \times 4 \times 4$ feet, $2.5 \times 2.5 \times 10.24$ feet, or some other dimensions?

---

* It may be helpful to review Sections A.5 and A.6 of the algebra review in Appendix A at this time.

Or, if the unit cost y (in dollars) for producing a product when production lot size is x hundred units is given by

$$y = \frac{x^2 - x + 2}{x + 1} \qquad x > 0$$

what lot size results in the smallest unit cost?

Using calculus, we will be able to determine that the most economical carton for the recycling center has dimensions $4 \times 4 \times 4$ feet and that a lot size of 200 units gives the smallest unit cost in the second situation.

The development of calculus is the result of the work of many mathematicians. However, credit for its founding is generally given to the English physicist and mathematician Isaac Newton (1642–1727) and the German philosopher and mathematician Gottfried Leibniz (1646–1716).

11.1 AVERAGE RATE OF CHANGE

Suppose a test car is traveling in such a way that the total distance (in miles) traveled at any time x is given by $f(x) = 15x^2$ for $0 \leq x \leq 7$, where x denotes the number of hours the car has been moving. After 2 hours, the car has traveled $f(2) = 60$ miles; after 3 hours, it has traveled $f(3) = 135$ miles; after 4 hours, it has traveled $f(4) = 240$ miles, and so forth. We are interested first in calculating its average speed over various intervals of time.

Since the average speed of the car is the distance traveled divided by the time, the average speed over the time interval from 3 to 5 hours is

$$\frac{f(5) - f(3)}{5 - 3} = \frac{375 - 135}{5 - 3} = \frac{240}{2} = 120 \qquad (1)$$

or 120 miles per hour.

The average speed over the time interval from 3 to $4\tfrac{1}{2}$ hours is

$$\frac{f(4.5) - f(3)}{4.5 - 3} = \frac{303.75 - 135}{1.5} = 112.5 \qquad (2)$$

or 112.5 miles per hour. In general, the average speed over any time interval of length h beginning at 3 hours is

$$\frac{f(3 + h) - f(3)}{(3 + h) - 3} = \frac{f(3 + h) - f(3)}{h}$$

miles per hour. In equation (1), the length of the time interval is $h = 2$; in equation (2), the length of the interval is $h = 1.5$.

In the same way, the average speed of the car over any time interval of length h beginning at x hours is

$$\frac{f(x + h) - f(x)}{(x + h) - x} = \frac{f(x + h) - f(x)}{h} \qquad (3)$$

11.1 AVERAGE RATE OF CHANGE

Note that the average speed (the average rate of change of distance) is the change in distance divided by the corresponding change in time or, as given in (3), the change in the values of f divided by the corresponding change in the values of x:

$$\frac{f(x+h) - f(x)}{h}$$

We extend this concept to arbitrary functions by defining the **average rate of change** of a function f on any interval beginning at x and of length h—or of length $-h$ for h negative—to be

$$\frac{f(x+h) - f(x)}{h} \qquad (4)$$

Since

$$\frac{f(x+h) - f(x)}{h} = \frac{f(x+h) - f(x)}{(x+h) - x}$$

the average rate of change over an interval is just the change in the values of f divided by the corresponding change in the values of x. The expression in equation (4) is sometimes called the **difference quotient** for f.

EXAMPLE 1 Determine the average rate of change of the function $f(x) = x^2$ on the interval of length 0.5 beginning at $x = 1$.

Solution From equation (4) the average rate of change on the indicated interval is

$$\frac{(1.5)^2 - 1^2}{0.5} = \frac{2.25 - 1}{0.5} = \frac{1.25}{0.5} = 2.5$$

That is, values of f increase at an average rate of 2.5 units for each change of 1 unit in x.

1 PRACTICE PROBLEM 1

Determine the average rate of change of the function $f(x) = x^2$ on the interval of length 0.4 beginning at $x = 2$.

Answer

4.4

If h is positive, then for any fixed value of x, the value of $x + h$ is greater than x, and the interval beginning at x with its endpoint determined by $x + h$ lies to the right of x, as shown in Figure 11.3. But if h is negative, $x + h$ is less than x, and the interval beginning at x with its endpoint determined by $x + h$ lies to the left of x. By the expression "the interval of length h beginning at x," we shall mean the interval to the right of x if h is positive and the interval to the left of x if h is negative. This allows us to refer to arbitrary intervals of either type without having to distinguish between the two. However, the average rate of change of a function for either type is always given by the expression in (4).

FIGURE 11.3

EXAMPLE 2 Determine the average rate of change of the function $f(x) = x^2$ on an arbitrary interval of length h beginning at x.

2 PRACTICE PROBLEM 2

Use the result in Example 2 to calculate the average rate of change in Practice Problem 1.

Answer

4.4

Solution In this case, equation (4) gives

$$\frac{(x+h)^2 - x^2}{h} = \frac{(x^2 + 2xh + h^2) - x^2}{h}$$

$$= \frac{2xh + h^2}{h}$$

$$= \frac{h(2x+h)}{h} = 2x + h$$

Note that if we substitute $x = 1$ and $h = 0.5$ in $2x + h$, we get the result obtained in Example 1. ▧

EXAMPLE 3 Determine the average rate of change of the function $g(x) = x^2 + 2x$ on the interval from $x = 3$ to $x = 0$.

Solution The value of $h = 0 - 3 = -3$ is now negative since the interval is taken from right to left. The average rate of change is

$$\frac{[0^2 + 2(0)] - [3^2 + 2(3)]}{-3} = \frac{0 - 15}{-3} = 5$$ ▪

EXAMPLE 4 Determine the average rate of change of the function $g(x) = x^2 + 2x$ on an arbitrary interval of length h starting at x.

Solution The difference quotient in this case is

$$\frac{[(x+h)^2 + 2(x+h)] - [x^2 + 2x]}{h} = \frac{[x^2 + 2xh + h^2 + 2x + 2h] - [x^2 + 2x]}{h}$$

$$= \frac{2xh + h^2 + 2h}{h} = \frac{h(2x + h + 2)}{h} = 2x + h + 2$$

If $x = 3$ and $h = -3$ in $2x + h + 2$, the result is the same as that obtained in Example 3. ▪

EXAMPLE 5 Determine the average rate of change of the function $y = -2x + 5$ on an arbitrary interval of length h beginning at x.

Solution The difference quotient is

$$\frac{[-2(x+h) + 5] - [-2x + 5]}{h} = \frac{-2h}{h} = -2$$

Since the rate of change is independent of the values of x and of h, the rate is the same for every possible interval. Furthermore, the negative value indicates that the function decreases on every such interval. ▨

3 PRACTICE PROBLEM 3

Determine the average rate of change of the function $f(x) = x^2 - x$ in each case.

(a) On an arbitrary interval of length h starting at x

(b) On the interval from $x = 5$ to $x = 3$, using the result obtained in part (a).

Answer

(a) $2x + h - 1$ (b) 7

11.1 AVERAGE RATE OF CHANGE

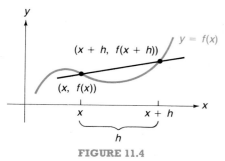

FIGURE 11.4

In order to determine the geometric significance of the average rate of change of a function f on an interval, consider the slope of the line passing through the points $((x + h, f(x + h))$ and $(x, f(x))$ in the graph of the function f shown in Figure 11.4.

The slope is

$$\frac{f(x + h) - f(x)}{(x + h) - x} = \frac{f(x + h) - f(x)}{h}$$

Geometrically, the average rate of change is the slope of the indicated line! In the graph shown in Figure 11.4, h was chosen to be positive; a similar geometric result is obtained if h is chosen to be negative.

4 PRACTICE PROBLEM 4

Determine the slope of the line passing through the points $(3, f(3))$ and $(5, f(5))$ if $f(x) = x^2 - x$. Compare the result with the result obtained in Practice Problem 3.

Answer

7

EXERCISES

Find the average rate of change of the given function on the indicated intervals.

1. $f(x) = x^2 + 1$
   (a) The interval of length 2 beginning at $x = 1$
   (b) An arbitrary interval of length h beginning at x

2. $f(x) = x^2 - 3$
   (a) The interval of length 0.5 beginning at $x = 0$
   (b) An arbitrary interval of length h beginning at x

3. $f(x) = x^2 + 3x$
   (a) The interval of length 1 beginning at $x = 10$
   (b) An arbitrary interval of length h beginning at $x = 10$

4. $g(x) = x^2 - 2x$
   (a) The interval from $x = 0$ to $x = 3$
   (b) An arbitrary interval of length h beginning at $x = 0$

5. $y = 3x + 7$
   (a) The interval from $x = 3$ to $x = 0$
   (b) An arbitrary interval of length h beginning at x

6. $y = -2x + 3$
   (a) The interval from $x = 12$ to $x = 10.5$
   (b) An arbitrary interval of length h beginning at $x = 12$

7. $f(x) = 2x^2$
   (a) The interval of length 2 beginning at $x = -2$
   (b) An arbitrary interval of length h beginning at x

8. $f(x) = -3x^2 + 2$
   (a) On an arbitrary interval of length h beginning at x
   (b) On the interval from $x = 0$ to $x = 4$, using the result obtained in part (a)
   (c) On the interval beginning at $x = 2$ with $h = -3$, using the result obtained in part (a)

9. $g(x) = 4x^2 - 3x$
   (a) On an arbitrary interval of length h beginning at x
   (b) On the interval from $x = 1$ to $x = -1$, using the result obtained in part (a)
   (c) On the interval from $x = -2$ to $x = 3$, using the result obtained in part (a)

10. $g(x) = x^3$
    (a) On an arbitrary interval
    (b) On the interval from $x = 2$ to $x = 0$, using the result obtained in part (a)

11. $y = \dfrac{1}{x}$
    (a) On an arbitrary interval
    (b) On the interval of length 3 starting at $x = 1$, using the result obtained in part (a)

12. (a) Determine the slope of the line passing through the points $(x, f(x))$ and $(x + h, f(x + h))$ in the graph of $f(x) = 2x^2 - 1$, where x and h are arbitrary.
    (b) Use the result obtained in part (a) to find the slope of the line passing through the points $(-2, f(-2))$ and $(1, f(1))$.

13. (a) Determine the slope of the line passing through the points $(x, f(x))$ and $(x + h, f(x + h))$ in the graph of $f(x) = -x^2 + x$, where x and h are arbitrary.
    (b) Use the result obtained in part (a) to obtain the slope of the line passing through the points $(0, f(0))$ and $(3, f(3))$.

14. (a) Determine the slope of the line passing through the points $(x, g(x))$ and $(x + h, g(x + h))$ in the graph of $g(x) = x^3$, where x and h are arbitrary.
    (b) Use the result obtained in part (a) to obtain the slope of the line passing through the points corresponding to $x = 0$ and $x = 2$.

Explain the meaning of each term.

15. Average rate of change of a function f on an interval of length h beginning at x

16. Difference quotient for a function f

11.2 THE DERIVATIVE

We return now to the test car considered at the beginning of Section 11.1. We want to determine how fast it is traveling at any instant when all we know is that it has traveled a total of $f(x) = 15x^2$ miles after x hours.

Table 11.1 gives its average speed on intervals beginning at $x = 3$ for various values of h. None of the average speeds in the table tell us

11.2 THE DERIVATIVE

TABLE 11.1

h	$3+h$	AVERAGE SPEED $\dfrac{f(3+h)-f(3)}{h}$ (miles per hour)	h	$3+h$	AVERAGE SPEED $\dfrac{f(3+h)-f(3)}{h}$ (miles per hour)
1.0	4.0	105	−1.0	2.0	75
0.5	3.5	97.5	−0.5	2.5	82.5
0.2	3.2	93	−0.2	2.8	87
0.1	3.1	91.5	−0.1	2.9	88.5
0.01	3.01	90.15	−0.01	2.99	89.85

how fast the car was going at exactly $x = 3$ hours. We might expect to get a reasonable approximation to the speed at $x = 3$ from the shorter time intervals—that is, using values of h close to 0. But even 0.01 hour is equal to 36 seconds, and the speed can vary considerably over a 36-second period.

In order to determine the instantaneous speed at $x = 3$, we take the limit of the average speed as the length of the time interval becomes closer and closer to 0. That is, the instantaneous speed is determined by

$$\lim_{h \to 0} \frac{f(3+h) - f(x)}{h} = \lim_{h \to 0} \frac{15(3+h)^2 - 15(3^2)}{h}$$

$$= \lim_{h \to 0} \frac{15(9 + 6h + h^2) - 15(9)}{h}$$

$$= \lim_{h \to 0} \frac{15(9) + 90h + 15h^2 - 15(9)}{h}$$

$$= \lim_{h \to 0} \frac{h(90 + 15h)}{h} = \lim_{h \to 0} (90 + 15h) = 90$$

The instantaneous speed at $x = 3$ is 90 miles per hour.

Similarly, its instantaneous speed at any time x is given by

$$\lim_{h \to 0} \frac{f(x+h) - f(x)}{h} = \lim_{h \to 0} \frac{15(x+h)^2 - 15x^2}{h}$$

$$= \lim_{h \to 0} \frac{15(x^2 + 2xh + h^2) - 15x^2}{h}$$

$$= \lim_{h \to 0} \frac{15x^2 + 30xh + 15h^2 - 15x^2}{h}$$

$$= \lim_{h \to 0} \frac{h(30x + 15h)}{h} = \lim_{h \to 0} (30x + 15h) = 30x$$

Note that if we set $x = 3$ in this last expression, we get the value 90 miles per hour, also obtained above.

Just as with the average speed, we extend this idea of the instantaneous speed or instantaneous rate of change of distance to arbitrary functions by defining the ***instantaneous rate of change*** of a function

f at a value of x to be

$$\lim_{h \to 0} \frac{f(x+h) - f(x)}{h} \quad (1)$$

provided this limit exists.

Note that the instantaneous rate of change of a function f at x is the limit of the average rate of change of f on an interval of length h beginning at x, the limit being taken as the length of the interval becomes closer and closer to 0. In order to evaluate the instantaneous rate of change, all that is required is to set up the difference quotient and then take the limit as h approaches zero.

EXAMPLE 1 Find the instantaneous rate of change of the function $f(x) = x^2$ at $x = 1$.

Solution   The instantaneous rate of change is given by

$$\lim_{h \to 0} \frac{(1+h)^2 - 1^2}{h} = \lim_{h \to 0} \frac{(1 + 2h + h^2) - 1}{h}$$

$$= \lim_{h \to 0} \frac{2h + h^2}{h}$$

$$= \lim_{h \to 0} \frac{h(2 + h)}{h} = 2 \quad \blacksquare$$

EXAMPLE 2 Find the instantaneous rate of change of the function $f(x) = x^2$ at an arbitrary value of x.

Solution   In this case, the limit of equation (1) becomes

$$\lim_{h \to 0} \frac{(x+h)^2 - x^2}{h} = \lim_{h \to 0} \frac{(x^2 + 2xh + h^2) - x^2}{h}$$

$$= \lim_{h \to 0} \frac{2xh + h^2}{h}$$

$$= \lim_{h \to 0} \frac{h(2x + h)}{h} = 2x$$

By setting $x = 1$ in this last expression, we get the result obtained in Example 1. $\blacksquare$

EXAMPLE 3 Determine the instantaneous rate of change of the function $f(x) = x^2 + 2x$ in each case.

(a) At an arbitrary value of x (b) At $x = 3$

Solution   (a) The limit in equation (1) becomes

$$\lim_{h \to 0} \frac{[(x+h)^2 + 2(x+h)] - (x^2 + 2x)}{h} = \lim_{h \to 0} \frac{(x^2 + 2xh + h^2 + 2x + 2h) - (x^2 + 2x)}{h}$$

$$= \lim_{h \to 0} \frac{h(2x + h + 2)}{h} = 2x + 2$$

1 PRACTICE PROBLEM 1

Determine the instantaneous rate of change of the function $f(x) = x^2 - x$ in each case.

(a) At an arbitrary value of x
(b) At $x = 3$, using the result obtained in part (a).
(c) At $x = -2$

Answer

(a) $2x - 1$ (b) 5 (c) -5

(b) Setting $x = 3$ in the result obtained in part (a) gives an instantaneous rate of change of 8. That is, at $x = 3$, the function values are changing at a rate of 8 units per 1 unit change in x.

Just as in Example 3, when the instantaneous rate of change of a function f is evaluated at an arbitrary value of x, a new function f' is determined. In Example 3, the function f was given by $f(x) = x^2 + 2x$; the function f' was given by $f'(x) = 2x + 2$. This new function f' is called the *derivative* of f.

The **derivative** of a function f is the function f', where

$$f'(x) = \lim_{h \to 0} \frac{f(x+h) - f(x)}{h} \qquad (2)$$

The domain of the derivative is all values of x for which the limit exists.

The procedure for evaluating derivatives is exactly the same as that for evaluating instantaneous rates of change—after all, they are the same thing. All that is required is to set up the difference quotient and then evaluate the limit as h approaches zero.

The usual procedure in evaluating derivatives is to determine $f'(x)$ for an arbitrary value of x and then to obtain values of the derivative for specific values of x by substitution.

EXAMPLE 4 Determine the value of the derivative of the function $f(x) = -3x^2 + x$ in each case.

(a) At an arbitrary value of x (b) At $x = -4$

Solution (a) From equation (2), for an arbitrary value of x,

$$f'(x) = \lim_{h \to 0} \frac{[-3(x+h)^2 + (x+h)] - (-3x^2 + x)}{h}$$

$$= \lim_{h \to 0} \frac{(-3x^2 - 6xh - 3h^2 + x + h) - (-3x^2 + x)}{h}$$

$$= \lim_{h \to 0} \frac{h(-6x - 3h + 1)}{h}$$

$$= \lim_{h \to 0} (-6x - 3h + 1) = -6x + 1$$

That is, $f'(x) = -6x + 1$.

(b) Using the result obtained in part (a), $f'(-4) = 25$.

PRACTICE PROBLEM 2

Determine the derivative of $f(x) = -2x^2 + 3$ in each case.

(a) At an arbitrary value of x
(b) At $x = 4$ (c) At $x = -2$

Answer

(a) $f'(x) = -4x$ (b) -16 (c) 8

In part (a) of Example 4, as well as in the previous examples, the quotient had to be rewritten until the h in the denominator was removed before the limit could be evaluated

If a function is denoted by another letter such as g, h, or r, its derivative is denoted correspondingly by g', h', or r'. Also, for historical reasons, if no other, derivatives are denoted by a variety of other symbols. One of the most common, when x is the independent variable, is to denote the derivative functions by

$$\frac{df}{dx}, \quad \frac{dg}{dx}, \quad \text{or} \quad \frac{dr}{dx}$$

and the corresponding values of the derivatives by

$$\frac{d}{dx}f(x), \quad \frac{d}{dx}g(x), \quad \text{or} \quad \frac{d}{dx}r(x)$$

When the form $y = f(x)$ is used to denote the function (for example, $y = 2x^3 - x^2$), the notation

$$\frac{dy}{dx} \quad \text{or} \quad y'$$

is used to denote the derivative function, and the corresponding values of the derivative dy/dx at a fixed value of x, such as $x = a$, are denoted by

$$\left.\frac{dy}{dx}\right|_{x=a}$$

In Example 4, if we had written the function in the form $y = -3x^2 + x$, then

$$\frac{dy}{dx} = -6x + 1 \quad \text{and} \quad \left.\frac{dy}{dx}\right|_{x=-4} = 25$$

EXERCISES

Determine the instantaneous rate of change of each of the following functions at the points indicated:

1. $f(x) = x^2 + 1$
   (a) An arbitrary value of x (b) $x = 1$ (c) $x = -2$
2. $f(x) = -x^2$
   (a) An arbitrary value of x (b) $x = 3$ (c) $x = \frac{1}{2}$
3. $g(x) = x^2 - 2x$
   (a) An arbitrary value of x (b) $x = 0$ (c) $x = -4$
4. $g(x) = 2x^2 + x - 3$
   (a) An arbitrary value of x (b) $x = 0.5$ (c) $x = \sqrt{2}$

5. $f(x) = x - 3x^2$
   (a) An arbitrary value of x (b) $x = 2$ (c) $x = 0$
6. $y = 3x + 7$
   (a) An arbitrary value of x (b) $x = 3$ (c) $x = 5$
7. $h(x) = 4 - 2x$
   (a) An arbitrary value of x (b) $x = 17$ (c) $x = -0.1$
8. $f(x) = x^3$
   (a) An arbitrary value of x (b) $x = 1.5$ (c) $x = 10$

Determine each of the following values for the given function.

9. $f(x) = 2x^2 - 1$
   (a) $f'(x)$ (b) $f'(-2)$ (c) $f'(1.3)$
10. $g(x) = -x^2 + x$
    (a) $g'(x)$ (b) $g'(0)$ (c) $g'(7)$
11. $y = x^2 - 3x + 2$
    (a) $\dfrac{dy}{dx}$ (b) $\left.\dfrac{dy}{dx}\right|_{x=-3}$ (c) $\left.\dfrac{dy}{dx}\right|_{x=0.7}$
12. $f(x) = 7$
    (a) $f'(x)$ (b) $f'(10)$ (c) $f'(-12)$
13. $y = -3$
    (a) $\dfrac{dy}{dx}$ (b) $\left.\dfrac{dy}{dx}\right|_{x=2}$ (c) $\left.\dfrac{dy}{dx}\right|_{x=-1}$
14. $h(x) = \dfrac{1}{x}$
    (a) $h'(x)$ (b) $h'(5)$ (c) $h'(-3)$
    (d) Why does $h'(0)$ have no value?
15. $g(x) = \dfrac{2}{x+1}$
    (a) $g'(x)$ (b) $g'(4)$ (c) $g'(-2)$
    (d) Why does $g'(-1)$ have no value?

Explain the meaning of each term.

16. Instantaneous rate of change of a function f at a value of x
17. Derivative of a function f at x

11.3 INTERPRETATION OF THE DERIVATIVE

Recall that the geometric interpretation of the average rate of change of a function f on an interval of length h starting at x, given by the difference quotient

$$\frac{f(x+h) - f(x)}{h}$$

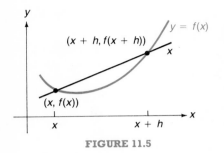

FIGURE 11.5

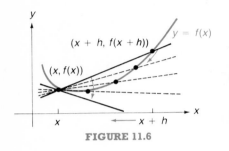

FIGURE 11.6

is the slope of the line passing through the points $(x, f(x))$ and $(x + h, f(x + h))$ in the graph of the function f (see Figure 11.5).

Then, as we let h get smaller and smaller, taking the limit of the difference quotient as h approaches zero to obtain the derivative, the point $x + h$ moves down the x-axis toward x. As it does so, the point $(x + h, f(x + h))$ moves along the graph of f toward the point $(x, f(x))$, and the lines through the corresponding points approach the position of the tangent to the graph at $(x, f(x))$, as shown in Figure 11.6.

Consequently, the slopes of the lines through the various points $(x, f(x))$ and $(x + h, f(x + h))$ approach the value of the slope of the tangent at $(x, f(x))$—that is,

$$\lim_{h \to 0} \frac{f(x+h) - f(x)}{h} = \text{Slope of the tangent}$$

Since the left side of this last equation is the derivative, $f'(x)$ *is the slope of the tangent to the graph of f at the point $(x, f(x))$*.

Using the derivative to find the slope, the equation of the tangent to the graph at a fixed value of $x = x_1$ can be found from the point-slope form

$$y - y_1 = m(x - x_1) \tag{1}$$

where the point (x_1, y_1) on the line is taken to be $(x_1, f(x_1))$.

EXAMPLE 1 Determine the equation of the tangent to the graph of

$$f(x) = x^2 - 3 \quad \text{at } x = 1$$

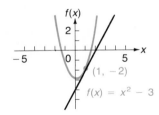

Solution Because $f(1) = -2$, $(1, -2)$ is a point on the tangent. For the slope,

$$f'(x) = \lim_{h \to 0} \frac{[(x+h)^2 - 3] - (x^2 - 3)}{h}$$

$$= \lim_{h \to 0} \frac{x^2 + 2xh + h^2 - 3 - x^2 + 3}{h}$$

$$= \lim_{h \to 0} \frac{h(2x + h)}{h} = 2x$$

so that $f'(1) = 2$. From (1), the equation of the tangent is

$$y + 2 = 2(x - 1)$$
$$y = 2x - 4$$

1 PRACTICE PROBLEM 1

Determine the equation of the tangent to the graph of the function f in Example 1 at $x = 2$.

Answer

$y = 4x - 7$

EXAMPLE 2 Find the equation of the tangent to the graph of $y = 3x^2 - x$ at $x = -2$.

Solution When $x = -2$, $y = 14$; therefore, $(-2, 14)$ is a point on the tangent. For the slope,

$$\frac{dy}{dx} = \lim_{h \to 0} \frac{[3(x+h)^2 - (x+h)] - (3x^2 - x)}{h}$$

$$= \lim_{h \to 0} \frac{3x^2 + 6xh + h^2 - x - h - 3x^2 + x}{h}$$

$$= \lim_{h \to 0} \frac{h(6x + h - 1)}{h} = 6x - 1$$

Therefore, the slope is

$$\left.\frac{dy}{dx}\right|_{x=-2} = -13$$

From (1), the equation of the tangent is

$$y - 14 = -13(x + 2)$$
$$y = -13x - 12$$

In economics, the (instantaneous) rate of change of a quantity is called a ***marginal change.*** We have already seen that the derivative gives this type of change. Consequently, when a quantity is represented by a function, the derivative of the function gives the marginal change of the quantity.

We shall encounter terms such as *marginal profit, marginal revenue,* and *marginal cost.* Marginal profit, for instance, gives the approximate change in profit resulting from the sale of an additional unit; marginal revenue gives the approximate change in revenue from the sale of one more unit, and so forth.

EXAMPLE 3 The total cost to a company for producing x items is given by

$$C(x) = 16x^2 - 3x + 25 \qquad x \geq 0$$

(a) Determine the marginal cost for any value of x.

(b) If the company has produced 25 items, approximately how much will it cost the company to produce one more item?

Solution (a) The marginal cost is given by

$$C'(x) = \lim_{h \to 0} \frac{[16(x+h)^2 - 3(x+h) + 25] - (16x^2 - 3x + 25)}{h}$$

$$= \lim_{h \to 0} \frac{16x^2 + 32xh + 16h^2 - 3x - 3h + 25 - 16x^2 + 3x - 25}{h}$$

$$= \lim_{h \to 0} \frac{h(32x + 16h - 3)}{h} = 32x - 3$$

---

2 PRACTICE PROBLEM 2

Determine the equation of the tangent to the graph of $g(x) = 2x^2$ at $x = -1$.

Answer

$y = -4x - 2$

(b) $C'(25) = 797$; that is, cost is increasing at the rate of $797 per unit change in x. It will cost about $797 to produce one more unit. ∎

Note the four-step process involved in the application of mathematics at work again in Example 3(b). The physical situation was a manufacturing company that wants to know the approximate cost of producing one more item when 25 items have already been produced. The mathematical model was the cost function $C(x) = 16x^2 - 3x + 25$; the mathematical question was: What is the value of $C'(25)$? The mathematical solution was found to be 797; the interpretation was that it would cost about $797 to produce the extra unit.

Suppose now that an object moves in a straight line in such a way that its distance s from some fixed reference point can be determined at any time t. Such an object might be a ball thrown straight up into the air with its distance measured from the ground, or it might be an arrow shot straight into a target with its distance measured from its starting position. The *velocity* of such an object is defined to be the (instantaneous) rate of change of distance with respect to time—or simply ds/dt. If the velocity is positive, it indicates the speed at which the object is moving away from the reference point (s is increasing). If the velocity is negative, it indicates the speed at which the object is moving toward the reference point (s is decreasing).

3 PRACTICE PROBLEM 3

The profit (in dollars) for a company from the sale of x items is given by

$$P(x) = 3x^2 + 2x - 10$$

(a) Determine the marginal profit for any value of x.
(b) Approximately how much profit will the company make from the sale of the eleventh item?

Answer

(a) $P'(x) = 6x + 2$
(b) $P'(10) = 62$; $62

EXAMPLE 4   An object is thrown into the air in such a way that its distance s in feet from the ground at any time t in seconds is given by

$$s = 192t - 16t^2$$

(a) How fast is it moving after 2 seconds?
(b) How high does it go?
(c) In what direction is it moving after 7 seconds?

Solution   (a) In this case,

$$\frac{ds}{dt} = 192 - 32t$$

Thus,

$$\left.\frac{ds}{dt}\right|_{t=2} = 192 - 64 = 128$$

The object is moving at the rate of 128 feet per second after 2 seconds.

(b) The object will continue rising until its velocity is zero, after which the velocity becomes negative and the object starts falling. So we determine the value of t for which the velocity is zero:

$$192 - 32t = 0$$

for $t = \frac{192}{32} = 6$. For this value of t,
$$s = 192(6) - 16(6)^2 = 576$$
so the height is 576 feet.

(c) $$\left.\frac{ds}{dt}\right|_{t=7} = 192 - 32(7) = -32$$

The negative value of the derivative indicates that s is decreasing; the object is therefore falling after 7 seconds.

We consider additional applications of the derivative in the following sections and especially in Chapter 12.

4 PRACTICE PROBLEM 4

In Example 4:

(a) How many feet above the ground is the object 3 seconds after it is thrown?

(b) How fast is the object moving after 3 seconds?

(c) In what direction is the object moving after 3 seconds?

Answer

(a) 432   (b) 96 feet/second

(c) Upward

EXERCISES

1. Determine the equation of the line tangent to the graph of $f(x) = 2x^2 - 1$ at $x = 1$. (Compare with Exercise 9 of Section 11.2.)

2. Determine the equation of the line tangent to the graph of $g(x) = -x^2 + x$ at each value.
   (a) $x = 3$ (b) $x = -3$
   (Compare with Exercise 10 of Section 11.2.)

3. Determine the equation of the line tangent to the graph of $y = x^2 - 3x + 2$ at each value.
   (a) $x = 0$ (b) $x = 4$
   (Compare with Exercise 11 of Section 11.2.)

4. Determine the equation of the line tangent to the graph $f(x) = 1/x$ at $x = -2$. (Compare with Exercise 14 of Section 11.2.)

5. Determine the equation of the line tangent to the graph of $g(x) = x^2 + 4x + 7$ at $x = -3$.

6. Determine the equation of the tangent to the graph of $y = 4 - x^2$ at $x = \frac{1}{2}$.

7. *Marginal cost*   The total cost, in dollars, to a company for producing x units of a product is given by
   $$C(x) = 2x^2 + 16x \qquad x \geq 0$$
   (a) Determine the marginal cost for any value of x.
   (b) If the company has produced 20 units, how much will it cost to produce one more unit?

8. *Marginal profit*   The total profit, in hundreds of dollars, to a company for selling x items is given by
   $$P(x) = -x^2 + 300x - 25$$
   (a) Determine the marginal profit for any value of x.
   (b) Determine the approximate additional profit to the company for selling one more item when it has sold 100 items.

9. *Drug experiment*   An experimental drug is being administered intravenously to a laboratory monkey. The weight of the monkey, in pounds,

is determined to be
$$f(x) = 70 - 0.1x^2 \qquad 0 \le x \le 5$$
where x is the time, in minutes, since the administration of the drug began. Determine the rate at which the monkey is losing weight 3 minutes after the administration of the drug is started.

10. *Projectile* An object is projected into the air in such a way that its distance s, in feet, from the ground at any time t, in seconds, is given by
$$s = 160t - 16t^2$$
   (a) How high does the object go?
   (b) At what time does it reach its greatest height?
   (c) At what time does it hit the ground?
   (d) How fast is it traveling when it hits the ground?

11. *Free-fall* If air resistance is ignored, a freely falling object travels approximately
$$s = 16t^2$$
feet in t seconds. A ball is dropped from the top of a building 144 feet tall.
   (a) How long does it take for the ball to reach the ground?
   (b) How fast is it moving when it hits the ground?
   (c) Why is the velocity positive when the ball hits the ground even though the ball is falling?

12. *Bacteria population* A culture of bacteria initially (at $t = 0$) contains 10,000 individuals. If the population of the culture after t hours is given by
$$f(t) = 10{,}000(4t - t^2) \qquad 0 \le t \le 4$$
determine the rate at which the population is changing after 3 hours.

13. *Learning experiment* In a learning experiment, a psychologist has determined that the number of new words, chosen from a select list of words, memorized by an average individual to be about
$$f(x) = 20(x - 0.2x^2) \qquad 0 \le x \le 4$$
after x hours of continuous study. Determine the rate at which an individual learns new words after 2 hours.

Identify the steps of the four-step procedure involved in the application of mathematics (Section 1.3) in each case.

14. *Marginal profit* Exercise 8(b)
15. *Drug experiment* Exercise 9
16. *Free-fall* Exercise 11(a)
17. Explain the meaning of a marginal change.

11.4 RULES FOR DERIVATIVES

In this and the next several sections we consider some theorems of calculus that provide rules for determining the derivatives of functions. These rules give methods for evaluating derivatives without resorting

11.4 RULES FOR DERIVATIVES

to the definition of a derivative each time. The proofs are not all given; however, each depends upon the definition of the derivative given in Section 11.2.

The first rule is concerned with the derivative of powers of x. Consider first

$$\frac{d}{dx}(x^2) = \lim_{h \to 0} \frac{(x+h)^2 - x^2}{h}$$

$$= \lim_{h \to 0} \frac{x^2 + 2xh + h^2 - x^2}{h}$$

$$= \lim_{h \to 0} \frac{h(2x+h)}{h} = 2x$$

and

$$\frac{d}{dx}(x^3) = \lim_{h \to 0} \frac{(x+h)^3 - x^3}{h}$$

$$= \lim_{h \to 0} \frac{x^3 + 3x^2h + 3xh^2 + h^3 - x^3}{h}$$

$$= \lim_{h \to 0} \frac{h(3x^2 + 3xh + h^2)}{h} = 3x^2$$

Similarly,

$$\frac{d}{dx}(x^4) = 4x^3 \quad \text{and} \quad \frac{d}{dx}(x^{-2}) = -2x^{-3}$$

In each case, the derivative of x raised to a power is equal to the power times x raised to a new power which is 1 less than the original. The first theorem states that the derivative of any power of x is in the same general form.

THEOREM 1   THE POWER RULE

For any constant n,

$$\frac{d}{dx}(x^n) = nx^{n-1}$$

EXAMPLE 1

(a) $\dfrac{d}{dx}(x^5) = 5x^4$

(b) $\dfrac{d}{dx}(x^{10}) = 10x^9$

(c) $\dfrac{d}{dx}(x^{-3}) = -3x^{-4} = \dfrac{-3}{x^4}$

(d) $\dfrac{d}{dx}(x^{-7}) = -7x^{-8} = \dfrac{-7}{x^8}$

(e) $\dfrac{d}{dx}(x^{2/3}) = \dfrac{2}{3}x^{(2/3)-1} = \dfrac{2}{3}x^{-1/3} = \dfrac{2}{3x^{1/3}}$

1 PRACTICE PROBLEM 1

Determine the value of each of the following derivatives:

(a) $\dfrac{d}{dx}(x^9)$ (b) $\dfrac{d}{dx}(x^{-5})$

(c) $\dfrac{d}{dx}(x^{3/5})$

Answer

(a) $9x^8$ (b) $-5x^{-6}$ (c) $\tfrac{3}{5}x^{-2/5}$

(f) $\dfrac{d}{dx}\sqrt{x} = \dfrac{d}{dx}(x^{1/2}) = \dfrac{1}{2}x^{(1/2)-1} = \dfrac{1}{2}x^{-1/2} = \dfrac{1}{2\sqrt{x}}$

The second rule gives a method for determining the derivative of a constant times a function.

THEOREM 2

If $f(x)$ has a derivative and c is any constant, then

$$\dfrac{d}{dx}[c \cdot f(x)] = c\left[\dfrac{d}{dx}f(x)\right]$$

That is, the derivative of a constant times a function is equal to the constant times the derivative of the function.

For the derivative of $5x^7$, c of Theorem 2 is 5 and $f(x)$ is x^7, so that

$$\dfrac{d}{dx}(5x^7) = 5\dfrac{d}{dx}(x^7) = 5(7x^6) = 35x^6$$

We shall have occasion to use Theorem 2 very often; its use is illustrated further in the next example.

EXAMPLE 2 (a) $\dfrac{d}{dx}(3x^2) = 3\dfrac{d}{dx}(x^2) = 3(2x) = 6x$

(b) $\dfrac{d}{dx}(-4x^6) = -4\dfrac{d}{dx}(x^6) = -4(6x^5) = -24x^5$

(c) $\dfrac{d}{dx}(12x^{-4}) = 12\dfrac{d}{dx}(x^{-4}) = 12(-4x^{-5}) = -48x^{-5} = \dfrac{-48}{x^5}$

(d) $\dfrac{d}{dx}\left(\dfrac{1}{2}x^{-3}\right) = \dfrac{1}{2}\dfrac{d}{dx}(x^{-3}) = \dfrac{1}{2}(-3x^{-4}) = -\dfrac{3}{2}x^{-4} = \dfrac{-3}{2x^4}$

(e) $\dfrac{d}{dx}(-9x^{2/3}) = -9\dfrac{d}{dx}(x^{2/3}) = -9\left(\dfrac{2}{3}x^{-1/3}\right) = -6x^{-1/3} = \dfrac{-6}{x^{1/3}}$

2 PRACTICE PROBLEM 2

Determine the value of each of the following derivatives:

(a) $\dfrac{d}{dx}(4x^2)$ (b) $\dfrac{d}{dx}(-2x^3)$

(c) $\dfrac{d}{dx}(5\sqrt{x})$

Answer

(a) $8x$ (b) $-6x^2$

(c) $\dfrac{5}{2}x^{-1/2} = \dfrac{5}{2\sqrt{x}}$

The proof of Theorem 2 is not difficult to establish. By the definition of the derivative, for a constant c and a function f,

$$\dfrac{d}{dx}(c \cdot f(x)) = \lim_{h \to 0}\dfrac{c \cdot f(x+h) - c \cdot f(x)}{h}$$

$$= \lim_{h \to 0} c \cdot \dfrac{f(x+h) - f(x)}{h}$$

Using the fact that the limit of a constant times a function is equal to the constant times the limit of the function, a property of limits that

11.4 RULES FOR DERIVATIVES

is proved in more advanced courses, this last expression can be written

$$c \cdot \lim_{h \to 0} \frac{f(x+h) - f(x)}{h} = c \left[\frac{d}{dx} f(x) \right]$$

Therefore,

$$\frac{d}{dx} [c \cdot f(x)] = c \left[\frac{d}{dx} f(x) \right]$$

which is the relationship given in Theorem 2.

Recall that a function f is a rule that assigns a unique number to each element in its domain. Theorem 3 is concerned with a special type of function that assigns the same number to each element in its domain. For example, $f(x) = 7$ associates the number 7 with each number x and $g(x) = -3$ assigns the number -3 to each number x. Such functions are called **constant functions**. The graph of a constant function is always a line parallel to the x-axis. The graphs of several such functions are given in the coordinate system in Figure 11.7.

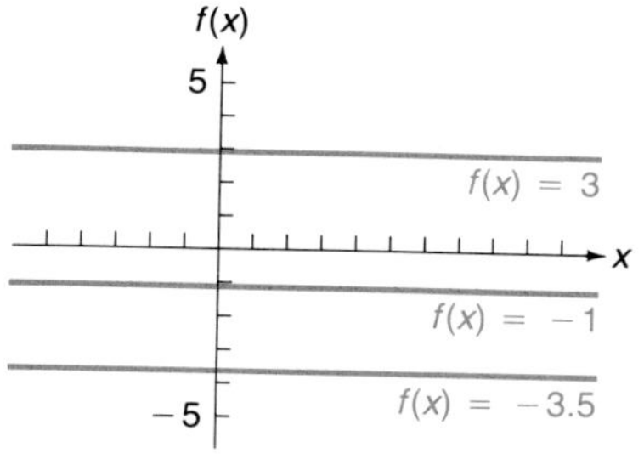

For any constant c, if $f(x) = c$,

$$f'(x) = \lim_{h \to 0} \frac{f(x+h) - f(x)}{h} = \lim_{h \to 0} \frac{c - c}{h} = \lim_{h \to 0} 0 = 0$$

Consequently, we have the following theorem:

FIGURE 11.7

THEOREM 3

The derivative of any constant function is zero.

EXAMPLE 3 (a) $\dfrac{d}{dx}(4) = 0$ (b) $\dfrac{d}{dx}\left(-\dfrac{2}{3}\right) = 0$ (c) $\dfrac{d}{dx}(2^4) = 0$ ▪ 3

Using Theorems 1, 2, and 3 to determine the derivatives of functions is considerably faster than resorting to the definition, as we did in Section 11.2. The additional theorems in this and in the next several sections are intended to serve the same purpose.

---

3 PRACTICE PROBLEM 3

Determine the value of each of the following derivatives:

(a) $\dfrac{d}{dx}(6)$ (b) $\dfrac{d}{dx}(-10)$

(c) $\dfrac{d}{dx}(\sqrt{2})$

Answer

(a) 0 (b) 0 (c) 0

THEOREM 4 THE ADDITION RULE

If $u(x)$ and $v(x)$ have derivatives, then

$$\frac{d}{dx}[u(x) \pm v(x)] = \frac{d}{dx} u(x) \pm \frac{d}{dx} v(x)$$

That is, the derivative of the sum (difference) of two functions is the sum (difference) of their derivatives.

EXAMPLE 4 Find the derivative of $f(x) = 4x^2 + 3x^5$.

Solution $f(x)$ is the sum of the two functions,

$$u(x) = 4x^2 \quad \text{and} \quad v(x) = 3x^5$$

According to the addition rule, then,

$$f'(x) = u'(x) + v'(x) = 8x + 15x^4$$

EXAMPLE 5 (a) $\dfrac{d}{dx}(2x^5 - 3x^3) = 10x^4 - 9x^2$

(b) $\dfrac{d}{dx}(6x^{-1} + 4) = -6x^{-2} + 0 = -6x^{-2} = \dfrac{-6}{x^2}$

(c) $\dfrac{d}{dx}(4x^2 - 3x + 7) = 8x - 3$

(d) $\dfrac{d}{dx}(5x^{1/3} - 2x^{-1/2} + x) = \dfrac{5}{3}x^{-2/3} + x^{-3/2} + 1$

---

4 PRACTICE PROBLEM 4

Determine the derivative of:

(a) $f(x) = 2x^3 + 4x$
(b) $f(x) = x^2 - 2$
(c) $g(x) = \frac{1}{2}x^2 + 3x^{1/3}$

Answer

(a) $6x^2 + 4$ (b) $2x$
(c) $x + x^{-2/3}$

We give the proof for the sum part of the addition rule. Observe how it depends on the definition of the derivative. According to the definition of a derivative,

$$\frac{d}{dx}[u(x) + v(x)] = \lim_{h \to 0} \frac{[u(x+h) + v(x+h)] - [u(x) + v(x)]}{h}$$

Regrouping the terms in the limit on the right gives

$$\lim_{h \to 0} \frac{[u(x+h) - u(x)] + [v(x+h) - v(x)]}{h}$$

Using the fact that the limit of the sum of two functions is the sum of the limits of the functions, a property of limits that is proved in more advanced courses, this last expression can be written

$$\lim_{h \to 0} \frac{u(x+h) - u(x)}{h} + \lim_{h \to 0} \frac{v(x+h) - v(x)}{h} = u'(x) + v'(x)$$

That is,

$$\frac{d}{dx}[u(x) + v(x)] = u'(x) + v'(x)$$

which completes the proof.

All the examples just given determine the value of the derivatives at arbitrary values of x. If we want to determine the derivative at a specific value of x—for example, $f'(6)$ when $f(x) = -4x^3$—we must first apply the rules to obtain $f'(x)$ for arbitrary x and then use the result to obtain $f'(6)$:

$$f'(x) = -12x^2 \quad \text{so that} \quad f'(6) = -12(36) = -432$$

11.4 RULES FOR DERIVATIVES

Similarly, we determine $g'(-2)$ when $g(x) = x^2 + x$:
$$g'(x) = 2x + 1 \quad \text{so that} \quad g'(-2) = -3 \quad \boxed{5}$$

Since the derivative gives the instantaneous rate of change and the slope of the tangent to the graph of a function as well as the marginal change of economic quantities, the rules that we have been considering apply to evaluate these values also.

EXAMPLE 6 Determine the instantaneous rate of change of the function
$$g(x) = x^3 - 2x^2$$
at $x = -4$.

Solution $g'(x) = 3x^2 - 4x$, so that $g'(-4) = 48 + 16 = 64$; g is increasing at the rate of 64 units per unit change in x at $x = -4$.

EXAMPLE 7 Find the equation of the tangent to the graph of the function $f(x) = x^{-1} + 3x$ at $x = 2$.

Solution $f'(x) = -x^{-2} + 3$, so that $f'(2) = -\frac{1}{4} + 3 = \frac{11}{4}$. Also, $f(2) = \frac{13}{2}$. The equation of the tangent is
$$y - \tfrac{13}{2} = \tfrac{11}{4}(x - 2)$$
$$4y = 11x + 4$$

In all the rules and examples thus far, the independent variable has always been x. To emphasize the role of x, the derivatives considered are said to be **derivatives with respect to x.** Other variables, such as t or u, are just as satisfactory. Then the derivatives are said to be *derivatives with respect to t* or *derivatives with respect to u*. These variables replace x in the notation as well; for example,
$$f'(t), \quad \frac{df}{dt}, \quad \frac{d}{dt}f(t)$$
represent derivatives with respect to t and
$$g'(u), \quad \frac{df}{du}, \quad \frac{d}{du}g(u)$$
represent derivatives with respect to u. For $f(t) = t^2 - 2t$, $f'(t) = 2t - 2$. For $g(u) = u^{-2} + u^3 - 3$, $g'(u) = -2u^{-3} + 3u^2$.

$\boxed{5}$ **PRACTICE PROBLEM 5**

Determine the value of:

(a) $f'(3)$ if $f(x) = 4x^2 - x$

(b) $g'(-1)$ if $g(x) = x^{-1} + 4$

Answer

(a) 23   (b) -1

EXERCISES

Find the indicated derivative.

1. $f'(x)$ for $f(x) = x^6$
2. $f'(x)$ for $f(x) = x^{11}$
3. $g'(x)$ for $g(x) = x^{3/4}$
4. $\dfrac{df}{dx}$ for $f(x) = x^{-4}$
5. $g'(t)$ for $g(t) = t^{-2/3}$
6. $g'(2)$ for $g(t) = t^3$
7. $f'(3)$ for $f(x) = \dfrac{1}{x^2}$
8. $f'(-1)$ for $f(x) = \dfrac{1}{x^3}$
9. $h'(x)$ for $h(x) = x$
10. $\dfrac{dg}{dt}$ for $g(t) = \dfrac{1}{\sqrt{t}}$
11. y' for $y = 2x^4$
12. $f'(x)$ for $f(x) = -3x^{-1}$
13. $\dfrac{df}{du}$ for $f(u) = \dfrac{2}{3}u^{3/2}$
14. $\dfrac{dh}{dt}$ for $h(t) = -t^{-3}$
15. $f'(-2)$ for $f(x) = \dfrac{-4}{x^3}$
16. $g'(8)$ for $g(x) = \dfrac{5}{x^{2/3}}$
17. $f'(x)$ for $f(x) = 0$
18. $f'(x)$ for $f(x) = -1.6$
19. $h'(t)$ for $h(t) = e^2$
20. $g'(4)$ for $g(x) = 5$
21. $\dfrac{df}{dx}$ for $f(x) = 5x^4 + 3x^5$
22. $f'(x)$ for $f(x) = 9x^2 - 3x$
23. $g'(-5)$ for $g(x) = x^2 - x + 7$
24. $\dfrac{dg}{dx}$ for $g(x) = 4x^3 - 6x + 7$
25. y' for $y = 3x^2 - 6\sqrt[3]{x}$
26. $f'(3)$ for $f(t) = t - \dfrac{1}{t} + 1$
27. $f'(u)$ for $f(u) = u^{-1} + u - 14$
28. $g'(x)$ for $g(x) = x^2 + 4\sqrt{x}$
29. $f'(9)$ for $f(x) = 2x^{1/2} + x^{3/2}$
30. $\dfrac{df}{dx}$ for $f(x) = \dfrac{1}{x^3} - x^3 + 6$

Find the instantaneous rate of change in each case.

31. $f(x) = 2x^3 - 3x + 1$ at an arbitrary value of x
32. $g(x) = x^{-2} - 4x$ at $x = 1$
33. $g(x) = 3x^{1/3} + 2x^{1/2} + x$ at $x = 64$
34. $f(x) \dfrac{2}{x^2} + \dfrac{3}{x^3}$ at $x = -2$

Find the equation of the tangent to the graph of each of the following functions at the given value:

35. $f(x) = x^2 - x$ at $x = -5$
36. $y = x^{3/2} + 4$ at $x = 1$
37. $g(x) = 2x^3 + 3x^2 + 6x$ at $x = -1$
38. $f(x) = x^{1/2} + x^{-1/2}$ at $x = 4$

39. *Velocity* A particle is moving in such a way that the distance (in feet) traveled at any time t (in seconds) is given by $f(t) = 3t^3 + t$. Determine its velocity at
    (a) $t = 5$ (b) $t = 1.5$

40. *Marginal revenue* The revenue, in dollars, from the sale of x units of a product is given by

$$R(x) = 3x^2 + 6x \qquad x \geq 0$$

   (a) Determine the marginal revenue for any value of x.
   (b) Determine the approximate increase in revenue the company receives from selling one more unit when it has sold 50 units.
   (c) Determine the approximate increase in revenue the company receives from selling the 61st unit.

41. *Marginal supply* The quantity, in hundreds of units, of a product that industry is willing to supply in terms of the price x, in dollars, is given by

$$S(x) = \frac{x^3}{6} - 3x^2 + 16x \qquad x \geq 0$$

   (a) Determine the marginal supply for any price x.
   (b) Determine the approximate change in supply if the price changes from $10 to $11.
   (c) Determine the approximate change in supply if the price changes from $11 to $12.

42. *Animal population* The population of an experimental group of laboratory animals after t months is given by

$$f(t) = 15(1 + 0.2t + 0.01t^2)$$

   How fast is the population increasing when the group contains 60 animals?

43. *Volume* The volume of a sphere with radius r is given by the function

$$V = \tfrac{4}{3}\pi r^3$$

   How fast is the volume of a balloon changing with respect to the radius when the volume is 972π cubic centimeters?

44. *Marginal supply* Identify the steps of the four-step procedure involved in the application of mathematics (Section 1.3) in Exercise 41(b).

45. Prove that if $f(x) = u(x) - v(x)$, then $f'(x) = u'(x) - v'(x)$, assuming that $u'(x)$ and $v'(x)$ exist.

46. Without referring to the proof given in the text, see if you can construct a proof for Theorem 3: "The derivative of any constant function $f(x) = c$ is zero."

Explain the meaning of each term. (State 48–50 verbally without using mathematical symbols.)

47. The power rule
48. The rule for the derivative of a constant times a function
49. The rule for the derivative of a constant function
50. The addition rule

11.5 THE CHAIN RULE

The power rule of Section 11.4 gives a method for calculating the derivative of power functions of the form $f(x) = x^n$. It does not apply to more complex power functions, such as

$$y = (3x^2 - 6x)^{1/2} \quad \text{or} \quad y = (x^4 + 1)^{-2}$$

The chain rule enables us to extend the power rule to these more complex power functions and similarly enables us to extend other rules considered later in this chapter.

The first function above,

$$y = (3x^2 - 6x)^{1/2}$$

can be written

$$y = u^{1/2} \quad \text{where} \quad u = 3x^2 - 6x$$

That is, y can be written as a function of u where u is a function of x. The original function, $y = (3x^2 - 6x)^{1/2}$, is said to be the **composition** of the two functions $y = u^{1/2}$ and $u = 3x^2 - 6x$.

Similarly, since

$$y = (x^4 + 1)^{-2}$$

can be written

$$y = u^{-2} \quad \text{where} \quad u = x^4 + 1$$

y is the composition of the two functions $y = u^{-2}$ and $u = x^4 + 1$. If u is replaced by its value $x^4 + 1$ in $y = u^{-2}$, the result is the original function

$$y = (x^4 + 1)^{-2}$$

Again, using function notation,

$$f(x) = \sqrt{x^3 - 3x}$$

can be written in terms of g where

$$g(u) = \sqrt{u} \quad \text{and} \quad u(x) = x^3 - 3x$$

The given function $f(x)$ is the composition of the two functions $g(u) = \sqrt{u}$ and $u(x) = x^3 - 3x$. Note that, in the usual function notation, we can write

$$g[u(x)] = \sqrt{x^3 - 3x} = f(x)$$

EXAMPLE 1 Write each of the following as a function of u, where u is a function of x:

(a) $y = \sqrt{2x - 3}$ (b) $y = e^{x^2 + x}$ (c) $y = \log(x^2 + 2x + 1)$
(d) $y = e^{(x+2)^3}$ (e) $y = \ln \sqrt{x^4 - 6x}$ (f) $y = 3^{(x^6 - 1)^{-2}}$

Solution (a) $y = u^{1/2}$, where $u = 2x - 3$
(b) $y = e^u$, where $u = x^2 + x$
(c) $y = \log u$, where $u = x^2 + 2x + 1$

11.5 THE CHAIN RULE

(d) $y = e^u$, where $u = (x + 2)^3$
(e) $y = \ln u$, where $u = \sqrt{x^4 - 6x}$
(f) $y = 3^u$, where $u = (x^6 - 1)^{-2}$

The chain rule provides a method for determining derivatives of functions y such as those in Example 1, which can be written as the composition of two simple functions.

> **THEOREM 5   THE CHAIN RULE**
>
> If y can be written as a function of u where u is a function of x, then
>
> $$\frac{dy}{dx} = \frac{dy}{du} \cdot \frac{du}{dx}$$
>
> That is, the derivative of y with respect to x is equal to the derivative of y with respect to u times the derivative of u with respect to x.

The chain rule presumes that the derivatives dy/du and du/dx exist.
To illustrate the use of the chain rule, suppose we want to determine the derivative of

$$y = (3x^2 - 6x)^{1/2}$$

We saw previously that

$$y = u^{1/2} \quad \text{where} \quad u = 3x^2 - 6x$$

According to the chain rule,

$$\frac{dy}{dx} = \frac{dy}{du} \cdot \frac{du}{dx} = \left(\frac{1}{2} u^{-1/2}\right) \cdot (6x - 6)$$

To obtain the derivative in terms of x alone (which is usually what is required), we substitute the expression for u in terms of x:

$$\frac{dy}{dx} = \frac{1}{2}(3x^2 - 6x)^{-1/2}(6x - 6) = \frac{3x - 3}{(3x^2 - 6x)^{1/2}}$$

EXAMPLE 2   Find the derivative of

$$y = (2x^3 + 4x)^6$$

Solution   The function y can be written

$$y = u^6 \quad \text{where} \quad u = 2x^3 + 4x$$

By the chain rule,

$$\frac{dy}{dx} = \frac{dy}{du} \cdot \frac{du}{dx} = (6u^5) \cdot (6x^2 + 4)$$

---

1 PRACTICE PROBLEM 1

Write each of the following as a function of u where u is a function of x:

(a) $y = (x^2 - 3x)^4$ (b) $y = e^{2x+1}$

Answer

(a) $y = u^4$, $u = x^2 - 3x$
(b) $y = e^u$, $u = 2x + 1$

2 PRACTICE PROBLEM 2

Determine the derivative of:

(a) $y = (x^2 - 3x)^4$ (b) $y = (1 - 4x)^3$

Answer

(a) $4(x^2 - 3x)^3(2x - 3)$
(b) $3(1 - 4x)^2(-4)$

Substituting for u in order to express the derivative in terms of x gives

$$\frac{dy}{dx} = 6(2x^3 + 4x)^5(6x^2 + 4) \qquad \boxed{2}$$

Suppose now that y is any function of the form $y = u^n$, where u is a function of x. (Note that the function in Example 2 and in the illustration just preceding Example 2 were in this form.) Then, by the chain rule,

$$\frac{dy}{dx} = \frac{dy}{du} \cdot \frac{du}{dx} = (nu^{n-1}) \cdot \frac{du}{dx}$$

This last equation extends the power rule of Theorem 1 to more-general types of functions, specifically to situations in which a function, rather than a single variable, is raised to a power. We express this general form as Theorem 1(a).

THEOREM 1(a) THE POWER RULE: GENERAL FORM

If u is a function of x having a derivative du/dx, then

$$\frac{d}{dx}(u^n) = (nu^{n-1}) \cdot \frac{du}{dx}$$

EXAMPLE 3 Find the derivative of each of the following functions:

(a) $y = (x^2 + 1)^3$ (b) $y = (6x + 2)^{2/3}$ (c) $f(x) = (x^2 - x)^{-2}$
(d) $g(t) = \sqrt{2t^2 + 4t}$ (e) $y = 7(x^4 - 3)^2$

Solution (a) The u of Theorem 1(a) is $(x^2 + 1)$ and the n is 3; the derivative is

$$\frac{dy}{dx} = 3(x^2 + 1)^2 \cdot \frac{du}{dx} = 3(x^2 + 1)^2(2x)$$

$$= 6x(x^2 + 1)^2$$

(b) In this case, $u = 6x + 2$ and n is $\frac{2}{3}$; the derivative is

$$\frac{dy}{dx} = \frac{2}{3}(6x + 2)^{-1/3} \cdot \frac{du}{dx}$$

$$= \frac{2}{3}(6x + 2)^{-1/3}(6) = \frac{4}{(6x + 2)^{1/3}}$$

(c) $u = x^2 - x$ and $n = -2$:

$$f'(x) = -2(x^2 - x)^{-3} \cdot \frac{du}{dx}$$

$$= -2(x^2 - x)^{-3}(2x - 1)$$

$$= \frac{2(1 - 2x)}{(x^2 - x)^3}$$

(d) $u = 2t^2 + 4t$ and $n = \frac{1}{2}$:

$$g'(t) = \frac{1}{2}(2t^2 + 4t)^{-1/2} \cdot \frac{du}{dt}$$

$$= \frac{1}{2}(2t^2 + 4t)^{-1/2}(4t + 4) = \frac{2(t+1)}{\sqrt{2t^2 + 4t}}$$

(e) In this case y is not exactly in the form $y = u^n$ because of the 7; that is, we have a constant times the function $(x^4 - 3)^2$, so we have to start with Theorem 2. Starting with Theorem 2 and then applying the general power rule gives

$$\frac{dy}{dx} = 7 \cdot \frac{d}{dx}(x^4 - 3)^2 = 7[2(x^4 - 3)(4x^3)]$$

$$= 56x^3(x^4 - 3)$$

3 PRACTICE PROBLEM 3

Determine the derivative of:

(a) $y = (2 - 3x)^4$ (b) $y = 5(2 - 3x)^4$

Answer

(a) $-12(2 - 3x)^3$ (b) $-60(2 - 3x)^3$

EXERCISES

Find each indicated derivative using the general form of the power rule.

1. $\dfrac{dy}{dx}$ for $y = (x^3 + 2)^2$
2. $\dfrac{dy}{dx}$ for $y = (-x^2 + 7)^3$
3. $\dfrac{dy}{dx}$ for $y = (2x + 1)^{-4}$
4. $f'(x)$ for $f(x) = (4x - x^3)^{-2}$
5. $g'(t)$ for $g(t) = (4t^2 + t^{-1})^3$
6. $\dfrac{dy}{dx}$ for $y = \sqrt{x^2 - x}$
7. $f'(x)$ for $f(x) = \sqrt{x + \dfrac{1}{x}}$
8. $f'(x)$ for $f(x) = (\sqrt{x} + 2x)^3$
9. $f'(x)$ for $f(x) = (2x^{-2} - 3)^4$
10. $\dfrac{dy}{dx}$ for $y = \left(\dfrac{3}{x^2} + \dfrac{2}{x^3}\right)^{-2}$
11. $g'(x)$ for $g(x) = (x^2 - x)^{-1/2}$
12. $f'(t)$ for $f(t) = 4(t^3 - 2t)^{3/4}$
13. $\dfrac{dy}{dx}$ for $y = 2(3x^4 - x^2)^3$
14. $f'(x)$ for $f(x) = -3(x + 6)^{-2}$
15. $g'(x)$ for $g(x) = \frac{1}{2}(x^{-1} + 3x)^2$
16. $g'(x)$ for $g(x) = 3(2x^2 - 7)^{1/3}$
17. $f'(3)$ for $f(x) = (4 - 2x)^3$
18. $f'(-2)$ for $f(t) = \sqrt{t^2 + 1}$
19. $g'(\frac{1}{2})$ for $g(x) = -3(x^2 - 1)^4$
20. $f'(1)$ for $f(x) = (x^2 + 2x + 1)^3$

Use the rules of Sections 11.4 and 11.5 to find each indicated derivative.

21. $f'(x)$ for $f(x) = 2x^3 - 6x + 7$
22. $f'(x)$ for $f(x) = 2x - 3$
23. $\dfrac{dy}{dx}$ for $y = \dfrac{5}{x}$
24. $g'(2)$ for $g(t) = 3t^{-2}$
25. y' for $y = 16$
26. $\dfrac{dy}{dt}$ for $y = 4(3t^5 + 6t^2)^3$
27. $f'(x)$ for $f(x) = -3(x + 6)^{-2}$
28. $f'(3)$ for $f(x) = 7$

29. $\dfrac{d}{dx}(3x - 6x^2)^{2/3}$

30. $f'(x)$ for $f(x) = (2x + 1)^2 - 5$

31. $g'(-2)$ for $g(x) = \dfrac{2}{(x + 3)^2}$

32. $\dfrac{d}{dx}(x^7 - x^6)$

33. $f'(x)$ for $f(x) = 9x^3 + (x + 2)^3$

34. $\dfrac{dy}{dx}$ for $y = 7(3 - 2x - x^2)^{-1}$

35. $f'(-1)$ for $f(x) = \dfrac{1}{x^2}$

36. $\dfrac{dy}{dx}$ for $y = \dfrac{-2}{(x^2 + 3)^2}$

37. Find the instantaneous rate of change of the function $y = (x^2 - 2x^3)^2$ at:
   (a) $x = 1$ (b) $x = -1$

38. Find the equation of the tangent to the graph of $f(x) = \sqrt{x^3 + 1}$ at:
   (a) $x = 2$ (b) $x = 0$

39. State the chain rule and the general power rule.

11.6 THE PRODUCT AND QUOTIENT RULES

The rules of the previous two sections help in determining the derivatives of products or quotients only in special situations. In this section we present and use general rules for determining derivatives of functions of this type.

THEOREM 6   THE PRODUCT RULE

If $u(x)$ and $v(x)$ are functions having derivatives $u'(x)$ and $v'(x)$ and if $y = u(x) \cdot v(x)$, then

$$\dfrac{dy}{dx} = u(x) \cdot v'(x) + v(x) \cdot u'(x)$$

That is, the derivative of the product of two functions is the first function times the derivative of the second plus the second function times the derivative of the first.

EXAMPLE 1   Determine the derivative of each of the following functions:

(a) $y = x^3(x + 1)^{1/2}$ (b) $y = (x^2 + 2)(3x - 1)$

Solution   (a) The functions $u(x)$ and $v(x)$ of the product rule are

$$u(x) = x^3 \quad \text{and} \quad v(x) = (x + 1)^{1/2}$$

Then

$$u'(x) = 3x^2 \quad \text{and} \quad v'(x) = \tfrac{1}{2}(x + 1)^{-1/2}$$

so that the product rule gives

$$\frac{dy}{dx} = u(x) \cdot v'(x) + v(x) \cdot u'(x)$$

$$= x^3 \cdot \left[\frac{1}{2}(x+1)^{-1/2}\right] + (x+1)^{1/2} \cdot 3x^2$$

$$= x^2(x+1)^{-1/2}\left[\frac{x}{2} + 3(x+1)\right]$$

$$= \frac{x^2(7x+6)}{2\sqrt{x+1}}$$

(b) $\quad u(x) = x^2 + 2 \quad$ and $\quad v(x) = (3x-1)$
$\quad u'(x) = 2x \quad$ and $\quad v'(x) = 3$

By the product rule,

$$\frac{dy}{dx} = (x^2+2)(3) + (3x-1)(2x)$$

which can be simplified by performing the indicated multiplication and collecting like terms to obtain

$$\frac{dy}{dx} = 9x^2 - 2x + 6 \qquad \boxed{1}$$

1 PRACTICE PROBLEM 1

Determine the derivative of:

(a) $y = (2x+1)(3x^2 - 7)$

(b) $y = \sqrt{x}(4 - 3x)$

Answer

(a) $y' = (2x+1)(6x) + (3x^2 - 7)(2)$
$= 18x^2 + 6x - 14$

(b) $y' = x^{1/2}(-3) + (4-3x)\left(\frac{1}{2}x^{-1/2}\right)$

$= \frac{4 - 9x}{2\sqrt{x}}$

THEOREM 7   THE QUOTIENT RULE

If $u(x)$ and $v(x)$ are functions having derivatives $u'(x)$ and $v'(x)$ and if $y = u(x)/v(x)$, then when $v(x) \neq 0$,

$$\frac{dy}{dx} = \frac{v(x) \cdot u'(x) - u(x) \cdot v'(x)}{v(x)^2}$$

That is, the derivative of a quotient is the denominator times the derivative of the numerator minus the numerator times the derivative of the denominator, all divided by the denominator squared.

EXAMPLE 2 Determine the derivative of each of the following functions:

(a) $y = \dfrac{x^2 - 3x}{x+1}$ $\qquad$ (b) $y = \dfrac{2x-1}{x^3 - 4x}$

Solution $\quad$ (a) The functions $u(x)$ and $v(x)$ of the quotient rule are

$$u(x) = x^2 - 3x \quad \text{and} \quad v(x) = x+1$$

Then

$$u'(x) = 2x - 3 \quad \text{and} \quad v'(x) = 1$$

so that
$$\frac{dy}{dx} = \frac{(x+1)(2x-3) - (x^2-3x)(1)}{(x+1)^2}$$

which simplifies to
$$\frac{2x^2 - x - 3 - x^2 + 3x}{(x+1)^2} = \frac{x^2 + 2x - 3}{(x+1)^2}$$

(b) $\quad u(x) = 2x - 1 \quad$ and $\quad v(x) = x^3 - 4x$
$\quad\quad u'(x) = 2 \quad$ and $\quad v'(x) = 3x^2 - 4$

By the quotient rule,
$$\frac{dy}{dx} = \frac{2(x^3 - 4x) - (2x-1)(3x^2 - 4)}{(x^3 - 4x)^2}$$

which simplifies to
$$\frac{2x^3 - 8x - (6x^3 - 3x^2 - 8x + 4)}{(x^3 - 4x)^2} = \frac{-4x^3 + 3x^2 - 4}{(x^3 - 4x)^2}$$

2 PRACTICE PROBLEM 2

Determine the derivative of:

(a) $y = \dfrac{2x-3}{1-4x}$ (b) $y = \dfrac{x^2}{(x+1)^2}$

Answer

(a) $y' = \dfrac{(1-4x)(2) - (2x-3)(-4)}{(1-4x)^2}$

$= \dfrac{-10}{(1-4x)^2}$

(b) $y' = \dfrac{(x+1)^2(2x) - x^2(2)(x+1)}{(x+1)^4}$

$= \dfrac{2x}{(x+1)^3}$

EXERCISES

Use the product and quotient rules to find the indicated derivative in each case.

1. $f'(x)$ for $f(x) = x^2(2x^3 - 4)$
2. $f'(x)$ for $f(x) = 2x(x^2 + 4)$
3. $f'(x)$ for $f(x) = (2x - 1)(4x + 6)$
4. y' for $y = \sqrt{x}\,(x + 1)$
5. $g'(1)$ for $g(x) = x^{-2}(x^2 + 9x)$
6. y' for $y = (x^3 - 1)(x - 3)^4$
7. $\dfrac{dy}{dt}$ for $y = (2t^2 + t)^2(-4t^3)$
8. $f'(1)$ for $f(x) = (x^{1/2} - x^{-1/2})(2x^3 + 5x^4)^2$
9. $f'(x)$ for $f(x) = \dfrac{x}{x^2 - 4}$
10. $g'(x)$ for $g(x) = \dfrac{6x}{4x^2 + 1}$
11. y' for $y = \dfrac{5x + 1}{x - 7x^2}$
12. $f'(2)$ for $f(x) = \dfrac{4}{(x-1)^3}$
13. $f'(t)$ for $f(t) = \dfrac{-4}{t^3 - 1}$
14. $g'(t)$ for $g(t) = \dfrac{(2t - t)^3}{(3t + 6)^4}$
15. $f'(x)$ for $f(x) = \left(\dfrac{x}{2x - 3}\right)^2$ (*Hint:* Start with the general power rule.)
16. y' for $y = \left(\dfrac{x - 2}{x^2}\right)^3$ (*Hint:* Start with the general power rule.)
17. $g'(x)$ for $g(x) = 6\left(\dfrac{5x}{x + 3}\right)^{-1}$ (*Hint:* Start with Theorem 2.)
18. $f'(0)$ for $f(x) = -3\left(\dfrac{x^2}{1 - 4x}\right)^2$ (*Hint:* Start with Theorem 2.)

11.6 THE PRODUCT AND QUOTIENT RULES

19. y' for $y = \sqrt{\dfrac{2x}{x^2 + 1}}$ (*Hint:* Start with the general power rule.)

20. $\dfrac{dy}{dt}$ for $y = -3\left(\dfrac{t^2}{4t - 9}\right)^{2/3}$ (*Hint:* Start with Theorem 2.)

21. $f'(x)$ for $f(x) = 4[(2x - 3)(1 - x^2)]$ (*Hint:* Start with Theorem 2.)

22. $g'(x)$ for $g(x) = -4[\sqrt{x}(3 + 2x)]$ (*Hint:* Start with Theorem 2.)

23. *Productivity* The number of items produced per hour by a new production worker on a particular job is given by

$$y = \dfrac{10x}{5 + 2x} \qquad x \geq 0$$

where x is the number of hours the employee has been on the job. Determine the rate at which the employee's productivity is increasing after the employee has been on the job for 5 hours.

24. *Marketing* Marketing surveys lead a company to estimate that the number of units of a new product that will be sold each day is given by

$$f(x) = \dfrac{500x}{10x + 17} \qquad x \geq 0$$

where x is the number of days the product is on the market. Determine the rate at which sales are increasing after the product has been on the market for 10 days.

25. *Drug concentration* The percent of concentration of a drug in the bloodstream after it has been administered x hours is given by

$$y = \dfrac{3x}{4x^2 + 10} \qquad x \geq 0$$

Determine the rate at which the percent of concentration is decreasing 2 hours after the drug is administered.

26. *Marginal revenue* The revenue, in dollars, from the sale of x units of a product is given by

$$R(x) = \dfrac{7x^2}{x + 1} \qquad x \geq 0$$

(a) Determine the marginal revenue for any value of x.
(b) Determine the revenue obtained from the sale of 100 units.
(c) Determine the approximate increase in revenue obtained from the sale of the 101st unit.

27. Determine the equation of the tangent to the graph of

$$4x^{-1}(2 - x^2)$$

at $x = 4$.

Identify the steps of the four-step procedure involved in the application of mathematics (Section 1.3) in each case.

28. *Productivity* Exercise 23
29. *Drug concentration* Exercise 25

State each of the following verbally, without using any mathematical symbols:

30. Product rule
31. Quotient rule

11.7 DERIVATIVES OF EXPONENTIAL AND LOGARITHMIC FUNCTIONS

This section presents our final group of rules for determining derivatives. The rules to be considered give the derivatives for exponential and logarithmic functions. The proofs for these rules will not be given; however, they are also based on the definition of a derivative given in Section 11.2.

THEOREM 8

For a constant $b > 0$,

$$\frac{d}{dx}(b^x) = b^x(\ln b)$$

and, therefore,

$$\frac{d}{dx}(e^x) = e^x$$

EXAMPLE 1 Find the derivative of each of the following functions:

(a) $y = 3^x$ (b) $y = e^x$ (c) $f(x) = 4(7^x)$

(d) $y = 6e^t + t^2$ (e) $g(x) = x^2(5^x)$ (f) $f(x) = \dfrac{x}{e^x}$

Solution For (a) and (b), we can apply Theorem 8 immediately:

(a) $y' = 3^x(\ln 3)$ (b) $y' = e^x$

(c) $f(x)$ is a constant times a function, so we start with Theorem 2:

$$f'(x) = 4\left[\frac{d}{dx}(7^x)\right] = 4(7^x)(\ln 7)$$

(d) y is the sum of two functions, so we start with the addition rule:

$$y' = \frac{d}{dt}(6e^t) + \frac{d}{dt}(t^2) = 6e^t + 2t$$

(e) $g(x)$ is a product, so we start with the product rule:

$$g'(x) = x^2 \cdot \frac{d}{dx}(5^x) + 5^x \cdot \frac{d}{dx}(x^2)$$
$$= x^2(5^x)(\ln 5) + 5^x(2x)$$
$$= (\ln 5)x^2 5^x + 2x 5^x$$

(f) $f(x)$ is a quotient, so we start with the quotient rule:

$$f'(x) = \frac{e^x \cdot \frac{d}{dx}(x) - x \cdot \frac{d}{dx}(e^x)}{(e^x)^2}$$

$$= \frac{e^x - xe^x}{(e^x)^2} = \frac{e^x(1-x)}{e^{2x}} = \frac{1-x}{e^x}$$

> **PRACTICE PROBLEM 1**
>
> Determine the derivative of:
>
> (a) $f(x) = 5^x$ (b) $y = -2e^x$
> (c) $y = x^3(e^x)$
>
> Answer
>
> (a) $5^x(\ln 5)$ (b) $-2e^x$
> (c) $x^3 e^x + e^x(3x^2) = e^x x^2(x+3)$

The rules given in Theorem 8 for determining the derivatives of exponential functions require that the exponent be the independent variable. By itself, Theorem 8 cannot handle more complex exponential functions such as

$$y = e^{x^2} \quad \text{or} \quad y = 4^{2x-3}$$

The chain rule enables us to extend the rules of Theorem 8 to handle more general exponential functions such as these. For example,

$$y = e^{x^2}$$

can be written as the composition

$$y = e^u \quad \text{where} \quad u = x^2$$

By the chain rule,

$$\frac{dy}{dx} = \frac{dy}{du} \cdot \frac{du}{dx} = e^u(2x) = e^{x^2}(2x)$$

Similarly, applying the chain rule to $y = 4^{2x-3}$, which can be written $y = 4^u$, where $u = 2x - 3$, gives

$$\frac{dy}{dx} = \frac{dy}{du} \cdot \frac{du}{dx} = 4^u(\ln 4)(2) = 2(\ln 4)4^{2x-3}$$

The general rules combining the chain rule with the derivatives in Theorem 8 are given in Theorem 8(a).

> **THEOREM 8(a)**
>
> If b is a constant, $b > 0$, then
>
> $$\frac{d}{dx} b^u = b^u(\ln b) \cdot \frac{du}{dx}$$
>
> and, therefore,
>
> $$\frac{d}{dx} e^u = e^u \cdot \frac{du}{dx}$$

EXAMPLE 2 Determine the derivative of each of the following functions:

(a) $y = 5^{x^3}$ (b) $y = e^{3x^2}$ (c) $f(x) = e^{2x-1} + 4x$

(d) $f(x) = 6^{x^2} - 3e^x$ (e) $g(x) = xe^{2x}$ (f) $y = \dfrac{e^{-t}}{t^2}$

Solution (a) $u = x^3$, so that
$$y' = 5^u(\ln 5)3x^2 = 5^{x^3}(\ln 5)3x^2 = 3(\ln 5)x^2 5^{x^3}$$

(b) $u = 3x^2$, so that by Theorem 8(a),
$$y' = e^u(6x) = e^{3x^2}(6x) = 6xe^{3x^2}$$

(c) $f(x)$ is the sum of two functions, so we start with the addition rule, applying Theorem 8(a) to the first function on the right:
$$f'(x) = e^{2x-1}(2) + 4$$

(d) This function is similar to that in part (c):
$$f'(x) = 6^{x^2}(\ln 6)(2x) - 3e^x = 2(\ln 6)x6^{x^2} - 3e^x$$

(e) The function g is a product, so we start with the product rule:
$$g'(x) = x \cdot \frac{d}{dx}(e^{2x}) + e^{2x} \cdot \frac{d}{dx}(x)$$
$$= xe^{2x}(2) + e^{2x}(1) = e^{2x}(2x + 1)$$

(f) The function is a quotient:
$$\frac{dy}{dt} = \frac{t^2 \cdot \dfrac{d}{dt}(e^{-t}) - e^{-t} \cdot \dfrac{d}{dt}(t^2)}{(t^2)^2}$$
$$= \frac{t^2(-e^{-t}) - e^{-t}(2t)}{t^4} = -\frac{t^2 + 2t}{t^4 e^t} = -\frac{t + 2}{t^3 e^t} \quad \boxed{2}$$

Rules for calculating the derivatives of logarithmic functions are given in Theorem 9.

THEOREM 9

For a constant $b > 0$, $b \neq 1$,
$$\frac{d}{dx}(\log_b x) = \frac{1}{(\ln b)x}$$
and, therefore,
$$\frac{d}{dx}(\ln x) = \frac{1}{x}$$

2 PRACTICE PROBLEM 2

Determine the derivative of:

(a) $y = 6^{2x-1}$ (b) $f(x) = e^{x^{-1}}$

(c) $f(t) = \dfrac{e^{t^2}}{t+1}$

Answer

(a) $6^{2x-1}(\ln 6)(2)$ (b) $e^{x^{-1}}(-x^{-2})$

(c) $\dfrac{(t+1)e^{t^2}(2t) - e^{t^2}}{(t+1)^2}$

$= \dfrac{e^{t^2}(2t^2 + 2t - 1)}{(t+1)^2}$

EXAMPLE 3 Determine the derivative of each of the following functions:

(a) $f(x) = \log_4 x$ (b) $y = \ln x$ (c) $y = (\ln x)^2$

(d) $g(x) = -2(\log_3 x) + 6x$ (e) $f(x) = \dfrac{2(\ln x)}{x+1}$ (f) $y = x(\log_2 x)$

Solution   In (a) and (b) we can apply Theorem 9 directly:

(a) $f'(x) = \dfrac{1}{(\ln 4)x}$ (b) $y' = \dfrac{1}{x}$

(c) This function is in the general power form u^n with $u = \ln x$ and $n = 2$. Applying the general power rule gives

$$y' = 2(\ln x) \cdot \dfrac{d}{dx}(\ln x) = 2(\ln x)\left(\dfrac{1}{x}\right) = \dfrac{2(\ln x)}{x}$$

(d) We start with the addition rule, obtaining

$$g'(x) = -2\left[\dfrac{1}{(\ln 3)x}\right] + 6 = \dfrac{-2}{(\ln 3)x} + 6$$

(e) We start with the quotient rule:

$$f(x) = \dfrac{(x+1)\cdot \dfrac{d}{dx}(2\ln x) - (2\ln x)\cdot \dfrac{d}{dx}(x+1)}{(x+1)^2}$$

$$= \dfrac{(x+1)\cdot \dfrac{2}{x} - (2\ln x)(1)}{(x+1)^2} = \dfrac{2(x+1) - 2x(\ln x)}{x(x+1)^2}$$

(f) The function is a product, so we start with the product rule:

$$y' = x \cdot \dfrac{d}{dx}(\log_2 x) + (\log_2 x) \cdot \dfrac{d}{dx}(x)$$

$$= x\left[\dfrac{1}{(\ln 2)x}\right] + (\log_2 x)(1) = \dfrac{1}{\ln 2} + \log_2 x$$

The use of Theorem 9 is somewhat restricted because it gives the derivatives of functions involving the logarithms of the independent variable only. By itself, Theorem 9 cannot be used to determine the derivative of functions such as

$$y = \log_3 x^2 \quad \text{or} \quad y = \ln(2x + 7)$$

As you might suspect by now, Theorem 9 can be extended to composite functions such as these by means of the chain rule.

The function $y = \log_3 x^2$ can be written $y = \log_3 u$, where $u = x^2$. By the chain rule,

$$\dfrac{dy}{dx} = \dfrac{dy}{du} \cdot \dfrac{du}{dx} = \dfrac{1}{(\ln 3)u} \cdot (2x) = \dfrac{1}{(\ln 3)x^2} \cdot (2x) = \dfrac{2}{(\ln 3)x}$$

3 PRACTICE PROBLEM 3

Determine the derivative of:

(a) $f(x) = \log_3 x$ (b) $f(x) = 6 \ln x$
(c) $f(x) = (5 \ln x)^2$

Answer

(a) $\dfrac{1}{(\ln 3)x}$ (b) $\dfrac{6}{x}$

(c) $2(5 \ln x)\dfrac{5}{x} = \dfrac{50 \ln x}{x}$

Also, $y = \ln(2x + 7)$ can be written as $y = \ln u$, where $u = 2x + 7$. By the chain rule,

$$y' = \frac{1}{u} \cdot 2 = \frac{1}{2x + 7} \cdot 2 = \frac{2}{2x + 7}$$

The general rules combining the chain rule with those of Theorem 9 are given in Theorem 9(a).

THEOREM 9(a)

For a constant $b > 0$, $b \neq 1$,

$$\frac{d}{dx}(\log_b u) = \frac{1}{(\ln b)u} \cdot \frac{du}{dx}$$

and, therefore,

$$\frac{d}{dx}(\ln u) = \frac{1}{u} \cdot \frac{du}{dx}$$

EXAMPLE 4 Determine the derivative of each of the following functions:

(a) $y = \log_5(3x + 6)$ (b) $y = \ln(-x^2 + x)$

(c) $f(x) = x \ln(x^3)$ (d) $g(t) = \dfrac{\log(t^2 + 3)}{t}$

Solution (a) $y' = \dfrac{1}{(\ln 5)(3x + 6)}(3) = \dfrac{1}{(\ln 5)(x + 2)}$

(b) $y' = \dfrac{1}{-x^2 + x}(-2x + 1) = \dfrac{-2x + 1}{-x^2 + x}$

(c) $f'(x) = x \cdot \dfrac{d}{dx}(\ln x^3) + (\ln x^3) \cdot \dfrac{d}{dx}(x)$

$= x \cdot \dfrac{3x^2}{x^3} + \ln x^3 = 3 + \ln x^3$

(d) $g'(t) = \dfrac{t \cdot \dfrac{d}{dt}[\log(t^2 + 3)] - \log(t^2 + 3) \cdot \dfrac{d}{dt}(t)}{t^2}$

$= \dfrac{\dfrac{2t^2}{(\ln 10)(t^2 + 3)} - \log(t^2 + 3)}{t^2}$

$= \dfrac{2t^2 - (\ln 10)(t^2 + 3) \log(t^2 + 3)}{(\ln 10)t^2(t^2 + 3)}$

4 PRACTICE PROBLEM 4

Determine the derivative of:

(a) $y = \ln(x^2 - 3x)$
(b) $y = \log(3x^4)$
(c) $y = (\ln x^2)^{-1}$

Answer

(a) $\dfrac{2x - 3}{x^2 - 3x}$ (b) $\dfrac{4}{(\ln 10)x}$

(c) $-(\ln x^2)^{-2}\left(\dfrac{1}{x^2}\right)(2x) = \dfrac{-2}{x(\ln x^2)^2}$

Why could we not start with Theorem 9(a) in parts (c) and (d) of Example 4? **4**

EXERCISES

1. List the three rules for determining derivatives that were used in Example 4(c).
2. List the five rules for determining derivatives that were used in Example 4(d).

Find the derivative of each of the following functions:

3. $f(x) = 3(7^x)$
4. $f(x) = -e^x$
5. $y = xe^x$
6. $g(x) = \dfrac{4^x}{x^2}$
7. $y = 3e^x + 2x$
8. $y = 2(\log_3 x)$
9. $g(t) = -\ln t$
10. $f(x) = \dfrac{\log x}{x + 1}$
11. $y = x(\ln x)$
12. $y = 2(\ln x) - \log x$
13. $y = e^{2x}$
14. $f(t) = e^{t^2}$
15. $g(x) = \dfrac{3^x}{x^2}$
16. $f(x) = x^2 e^x$
17. $f(x) = (x^2 + 1)e^{-x-1}$
18. $y = \left(\dfrac{e^x}{x}\right)^3$
19. $y = \ln t^3$
20. $y = x^2(\ln x^3)$
21. $y = (\log x)^2$
22. $y = x - \log_2 x^{-1}$
23. $f(x) = \ln(x^2 + x)$
24. $g(x) = \dfrac{2x}{\log x}$
25. $y = \left(\dfrac{\ln x}{x}\right)^4$
26. $f(x) = -7(\log x)^{-1}$
27. $y = -3$
28. $g(x) = x$

29. *Employee error* After an initial training period, the average number of errors committed per day by an employee on a particular job is given by
$$y = 3(1 + 2^{-x})$$
where x is the number of days completed on the job following the training period.
   (a) Determine the rate at which the average number of errors is decreasing when 3 days are completed on the job.
   (b) What happens to the values of y when x gets very large? Physically, what does this indicate?

30. *Continuous compounding* The balance in a savings account with a $1,000 deposit and drawing 6% interest compounded continuously is given by
$$y = 1,000(e^{0.06x})$$
where x is the number of years since the deposit was made. Determine how fast the balance is increasing 18 months after the deposit is made.

31. *Radioactive decay* If we now have 15 grams of a radioactive substance and the amount (in grams) that will remain t years from now is
$$y = 15(\tfrac{1}{2})^{t/5,000}$$
determine the rate at which the substance will be decaying 5,000 years from now.

32. *Product demand* The monthly demand q (in hundreds) for a product in terms of the price p (in dollars) is given by
$$q = \log_{1/2}\left(\frac{p}{50}\right) \qquad 0 < p \leq 50$$
Find dq/dp when $p = 20$, and interpret the result.

33. *Motion* A particle moves in a straight line in such a way that its distance d (in feet) from its starting point at any time t (in seconds) is given by
$$d = \ln(3t^2 + 1) \qquad 0 \leq t \leq 6$$
Determine how fast the particle is moving 2 seconds after it starts.

34. Determine the equation of the tangent to the graph of $y = e^{2x}$ at $x = 0$.

35. Determine the equation of the tangent to the graph of the equation $y = \log_3 x$ at $x = 9$.

36. Give two functions, each of which is the same as its own derivative.

11.8 DIFFERENTIABLE FUNCTIONS

Most of the functions discussed in this chapter have a derivative at each point in their domains. However, some exceptions do arise. In this section, we first examine situations in which a function does not have a derivative at one or more points.

Probably the easiest of these situations to recognize is that in which the expression giving the derivative has no value for particular values of x. For example, if $f(x) = \sqrt{x^2}$, then

$$f'(x) = \frac{1}{2}(x^2)^{-1/2}(2x) = \frac{x}{\sqrt{x^2}}$$

Since $f'(x)$ has no value for $x = 0$, the function f has no derivative at $x = 0$. However, f does have a derivative at every other value of x. Similarly, if $g(x) = \sqrt{x}$, then

$$g'(x) = \frac{1}{2}x^{-1/2} = \frac{1}{2\sqrt{x}}$$

which has no value for $x \leq 0$. Consequently, g has no derivative for $x \leq 0$.

1 PRACTICE PROBLEM 1

For what values of x does $f(x)$ not have a derivative if $f(x) = \sqrt{x - 1}$?

Answer

$x \leq 1$

11.8 DIFFERENTIABLE FUNCTIONS

Another situation in which a function will not have a derivative at a particular value of x is when the function is not continuous at that value of x. Recall that a function is continuous at $x = a$ if

$$\lim_{x \to a} f(x) = f(a)$$

For example,

$$f(x) = \frac{1}{x - 2}$$

is not continuous at $x = 2$, so that f cannot have a derivative at $x = 2$. Similarly,

$$g(x) = \begin{cases} x - 1 & x \leq -1 \\ x^2 & x > -1 \end{cases}$$

is not continuous at -1, so that g cannot have a derivative at -1.

The fact that a function cannot have a derivative at a value of x at which it is not continuous is a direct result of the following theorem.

THEOREM 10

If $f'(a)$ exists, then f is continuous at $x = a$.

To say that $f'(a)$ exists means

$$\lim_{h \to 0} \frac{f(a + h) - f(a)}{h} = f'(a)$$

That is, as h gets close to zero,

$$\frac{f(a + h) - f(a)}{h}$$

gets close to $f'(a)$ and, therefore, multiplying each expression by h,

$$h \left[\frac{f(a + h) - f(a)}{h} \right]$$

get close to $hf'(a)$. Since $hf'(a)$ gets close to zero with h, so does

$$h \left[\frac{f(a + h) - f(a)}{h} \right] = f(a + h) - f(a)$$

That is,

$$\lim_{h \to 0} [f(a + h) - f(a)] = 0$$

Now let $x = a + h$ and note that when h gets close to 0, x gets close to a. With this substitution, the last limit above becomes

$$\lim_{x \to a} [f(x) - f(a)] = 0$$

If the difference $f(x) - f(a)$ is getting arbitrarily small, then $f(x)$ must be getting closer and closer to $f(a)$. Therefore,

$$\lim_{x \to a} f(x) = f(a)$$

which means that f is continuous at $x = a$. This completes the proof.

The theorem states that a function must be continuous at $x = a$ if it has a derivative at $x = a$; therefore, if it is not continuous at $x = a$, it cannot have a derivative there.

Recall that three conditions must be satisfied in order for a function f to be continuous at $x = a$:

1. $f(a)$ must have a value.
2. $\lim_{x \to a} f(x)$ must have a value.
3. These two values must be the same: $\lim_{x \to a} f(x) = f(a)$

If any one of these three conditions are not satisfied, f is not continuous at a and, therefore, $f'(a)$ cannot exist or cannot have a value.

EXAMPLE 1 (a) The function $f(x) = 1/(x - 3)$ does not satisfy condition 1 at $x = 3$; therefore, f is not continuous at $x = 3$ and $f'(3)$ does not exist.

(b) The function

$$g(x) = \begin{cases} x^3 & x \geq 0 \\ 4x + 2 & x < 0 \end{cases}$$

satisfies condition 1 but not condition 2, so g is not continuous at $x = 0$; therefore, $g'(0)$ does not exist.

(c) The function

$$h(x) = \begin{cases} x^2 & x \neq 2 \\ 5 & x = 2 \end{cases}$$

satisfies both of the first two conditions but not the third. Therefore, f is not continuous at $x = 2$ and $f'(2)$ does not exist.

Finally, a function f is said to be **differentiable** at a value of x if f has a derivative at that value of x, and is **differentiable on the interval from a to b** if it has a derivative at each value of x between a and b.

PRACTICE PROBLEM 2

Verify that

$$f(x) = \begin{cases} \dfrac{x^2 - 1}{x - 1} & x \neq 1 \\ 3 & x = 1 \end{cases}$$

does not have a derivative at $x = 1$.

Answer

f does not satisfy condition 3 at $x = 1$ and is not continuous at $x = 1$; therefore $f'(1)$ does not exist.

The function g in Example 1 is differentiable at every value of x except $x = 0$; it is differentiable on every interval that does not contain 0. Moreover, its derivative is

$$g'(x) = \begin{cases} 3x^2 & x > 0 \\ 4 & x < 0 \end{cases}$$

PRACTICE PROBLEM 3

On what intervals is $f(x)$ differentiable if $f(x)$ is the function in:

(a) Example 1(a)?
(b) Practice Problem 1?

Answer

(a) Any interval not containing 3
(b) Any interval not containing numbers less than or equal to 1

EXERCISES

The given functions are not differentiable (do not have a derivative) at the indicated values of x. Determine which of the following reasons is correct:
(a) The function is not continuous at that value of x.
(b) The expression that gives the derivative does not have a value at the value of x even though the function is continuous there.

1. $f(x) = x^{2/3}$ at $x = 0$
2. $g(x) = \begin{cases} 2x + 4 & x \leq 3 \\ x - 7 & x > 3 \end{cases}$ at $x = 3$
3. $f(x) = \dfrac{x^2}{x + 4}$ at $x = -4$
4. $f(x) = x^{1/4}$ at $x = 0$
5. $h(x) = \sqrt{(x + 2)^2}$ at $x = -2$
6. $f(x) = \begin{cases} 1 & x \leq 5 \\ -1 & x > 5 \end{cases}$ at $x = 5$
7. $h(x) = \ln x$ at $x = 0$
8. $g(x) = \dfrac{x^2 - 1}{x + 1}$ at $x = -1$

Determine the intervals in which the function in the indicated exercise is differentiable.

9. Exercise 1
10. Exercise 2
11. Exercise 3
12. Exercise 4
13. Exercise 5
14. Exercise 6
15. Exercise 7
16. Exercise 8

17. For the function f with domain $\{x: -1 < x < 2\}$, whose graph is given in the margin, determine each of the following:
    (a) The values of x at which the function is differentiable
    (b) The value of $f'(x)$ for values of x at which the function is differentiable

18. For the function f, where

$$f(x) = \begin{cases} x^2 & x \leq 2 \\ 3 - x & x > 2 \end{cases}$$

whose graph is given in the margin, determine each of the following:
(a) The values of x at which f is differentiable
(b) The value of $f'(x)$ for the values of x at which f is differentiable

Explain what it means for a function to be:

19. Differentiable at a value of x
20. Differentiable on an interval

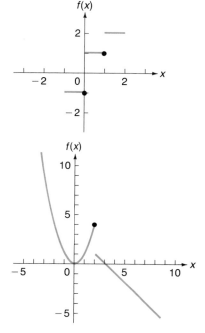

11.9 CHAPTER SUMMARY

The derivative is concerned with the instantaneous rate of change of a function f. In order to obtain this instantaneous rate, we first considered the average rate of change of f. The expression

$$\frac{f(x+h)-f(x)}{h}$$

called the **difference quotient** of f, gives the average rate of change of f on an interval of length h beginning at x.

The instantaneous rate of change of f at x is then the limit of the average change as the length of the interval, given by h, becomes arbitrarily small:

$$\lim_{h \to 0} \frac{f(x+h)-f(x)}{h}$$

This limit, denoted by $f'(x)$, is called the **derivative** of f at x.

Geometrically, the derivative $f'(x)$ gives the slope of the tangent to the graph of f at x. As an instantaneous rate of change, the derivative $f'(x)$ is given different names in different situations. We considered two in this chapter. If $f(x)$ denotes an economic quantity, $f'(x)$ is called the **marginal change** in f at x. If $f(x)$ denotes the position (at time x) of an object moving in a straight line, $f'(x)$ is called the **velocity** of the moving object.

If a function f has a derivative at a value of x, f is said to be **differentiable** at x. If f has a derivative at each value of x in some interval I, f is said to be **differentiable on I**.

The calculation of derivatives using only the definition is extremely cumbersome, if not impossible, for all except the most elementary functions. The rules for determining derivatives that we considered in this chapter make derivatives a viable mathematical tool when working with more complex functions. For easy reference, these rules are presented in concise form in Table 11.2.

A variety of functions, which are similar to those of the previous sections and whose derivatives require the use of these rules, are given in the following exercises. One of the major requirements in determining the derivative of a function is to start with the correct rule from among those in the table. In doing Exercises 13–52, note the rule that you use to start the process of determining the derivative for each of the given functions.

TABLE 11.2 TABLE OF DERIVATIVES

b, c, and n are constants, $b > 0$, $b \neq 1$; u is a function of x

1. $\dfrac{d}{dx} x^n = nx^{n-1}$

1(a). $\dfrac{d}{dx} u^n = nu^{n-1} \cdot \dfrac{du}{dx}$ General power rule

2. $\dfrac{d}{dx}[cf(x)] = c \cdot f'(x)$

3. $\dfrac{d}{dx} c = 0$

4. $\dfrac{d}{dx}[u(x) \pm v(x)] = u'(x) \pm v'(x)$ Addition rule

5. $\dfrac{d}{dx}[f(u)] = \dfrac{d}{du} f(u) \cdot \dfrac{du}{dx}$ Chain rule

6. $\dfrac{d}{dx}[u(x) \cdot v(x)] = u(x) \cdot v'(x) + v(x) \cdot u'(x)$ Product rule

7. $\dfrac{d}{dx}\left[\dfrac{u(x)}{v(x)}\right] = \dfrac{v(x) \cdot u'(x) - u(x) \cdot v'(x)}{v(x)^2}$ Quotient rule

8. $\dfrac{d}{dx} b^x = b^x (\ln b)$ and $\dfrac{d}{dx} e^x = e^x$

8(a). $\dfrac{d}{dx} b^u = b^u (\ln b) \cdot \dfrac{du}{dx}$ and $\dfrac{d}{dx} e^u = e^u \cdot \dfrac{du}{dx}$

9. $\dfrac{d}{dx} (\log_b x) = \dfrac{1}{(\ln b)x}$ and $\dfrac{d}{dx} (\ln x) = \dfrac{1}{x}$

9(a). $\dfrac{d}{dx} (\log_b u) = \dfrac{1}{(\ln b)u} \cdot \dfrac{du}{dx}$ and $\dfrac{d}{dx} (\ln u) = \dfrac{1}{u} \cdot \dfrac{du}{dx}$

REVIEW EXERCISES

1. Determine the average rate of change of $f(x) = x^2 - 1$ in each case.
   (a) On an arbitrary interval of length h starting at x
   (b) On the interval of length 2 starting at 3

2. Determine the average rate of change of $g(x) = 2x - x^2$ in each case.
   (a) On an arbitrary interval of length h starting at x
   (b) On the interval of length 2 starting at -1

3. Determine the average rate of change of $f(t) = 3t + 4$ in each case.
   (a) On an arbitrary interval of length h starting at t
   (b) On the interval from $t = 4$ to $t = 0$

4. Determine the average rate of change of $f(x) = 1/(x - 1)$ in each case.
   (a) On an arbitrary interval of length h starting at x
   (b) On the interval from $x = 0.5$ to $x = -1$

5. Using the definition, determine the instantaneous rate of change of $f(x) = x^2 - 1$ in each case.
   (a) At $x = 3$ (b) At an arbitrary value of x

6. Using the definition, determine the instantaneous rate of change of $g(x) = 2 - x^2$ in each case.
   (a) At an arbitrary value of x (b) At $x = -1$

7. Using the definition, determine the derivative of $f(x) = 2x + x^2$ in each case.
   (a) At $x = 6$ (b) At an arbitrary value of x

8. Using the definition, determine the derivative of $g(x) = 1 - 2x^2$ in each case.
   (a) At $x = 0$ (b) At an arbitrary value of x

9. Using the definition, determine the derivative of $f(t) = 3t + 4$ in each case.
   (a) At an arbitrary value of t (b) At $t = 4$

10. Using the definition, determine the derivative of $f(x) = 1/(x - 1)$ in each case.
    (a) At an arbitrary value of x (b) At $x = 0.5$

11. Using the definition, determine the derivative of $g(x) = x^2 + x + 7$ in each case.
    (a) At an arbitrary value of x (b) At $x = -2$

12. Using the definitions, determine both the instantaneous rate of change and the derivative of $f(x) = (x^2 + x)/2$ at an arbitrary value of x.

Determine the derivative of the given function. In each case, give the rule (by number) with which you start in determining the derivative.

13. $f(x) = 3x^4$
14. $f(x) = 2x^2 - 6x$
15. $g(x) = (4x)^{-1} + 4x^{-2}$
16. $y = \sqrt{x} + \sqrt[3]{x^2}$
17. $f(t) = t^2 + 2t - 3$
18. $y = -x^5$
19. $y = (-x)^5$
20. $g(x) = (2x + 3)^4$
21. $g(x) = (x^2 + 1)^{-3}$
22. $f(t) = (\ln t + 2)^4$
23. $y = -4(3x + x^2)^2$
24. $g(x) = 10$
25. $h(t) = t$
26. $f(x) = \sqrt{x + \sqrt{x}}$
27. $f(x) = e^{\sqrt[3]{x}}$
28. $g(x) = e^{\sqrt{x}} + e^{-\sqrt{x}}$
29. $y = 7e^{\sqrt{x}}$
30. $y = 2(\ln x^3)$
31. $y = (\ln x)^3$
32. $f(x) = \dfrac{x + 1}{2x - 3}$
33. $y = \ln\left(\dfrac{x + 1}{x - 1}\right)$
34. $f(x) = x^3(\ln x)$
35. $g(x) = x^3 e^{-x}$
36. $g(x) = -4xe^x$
37. $h(x) = e^{\sqrt{x+3}}$
38. $g(t) = \sqrt{e^{t+3}}$
39. $h(t) = \dfrac{2(\ln t)}{t}$
40. $g(r) = \left(\dfrac{r}{r^2 + 7}\right)^4$

41. $f(x) = \dfrac{2}{x-4}$

42. $g(x) = \dfrac{\log x}{x^2}$

43. $y = x(\log x)^3$

44. $h(x) = 4^x + 4^{-x}$

45. $f(x) = \dfrac{1}{1+e^x}$

46. $y = (1-4x)(x^2+3)$

47. $f(x) = (x+x^2)(4-x^{-2})$

48. $y = (3x^4+1)^2(2x-7)^3$

49. $g(t) = \dfrac{-4t^2}{3t+5}$

50. $f(x) = \dfrac{x+1}{(x-1)^2}$

51. $f(x) = \left(\dfrac{x^2}{x-1}\right)^4$

52. $f(t) = 2\left(\dfrac{t}{3t-4}\right)^3$

Determine the equation of the tangent to the graph of each of the following functions at the indicated value of x:

53. $f(x) = x^3 - 2x$ at $x = 2$

54. $f(x) = e^{x^2-9}$ at $x = -3$

55. $f(x) = \ln x^2$ at $x = -1$

56. $g(x) = \dfrac{2x}{x+2}$ at $x = 0.5$

Determine the instantaneous rate of change of each of the following functions at the indicated value of x:

57. $f(x) = (x^2-3)^2$ at $x = 1$

58. $g(x) = \dfrac{2}{x+4}$ at $x = -1$

59. $f(x) = 2^x$ at $x = 0$

60. $g(x) = x\ln(x^2-8)$ at $x = 3$

61. *Human population* The population of a rural county was found to be 2,400 at the end of 1980. Its future population (y), which is expected to grow rapidly, is estimated to be

$$y = 2{,}400(3^{0.1x})$$

where x denotes number of years after 1980. How fast will the population be increasing at the end of 1990?

62. A 20-gallon gasoline tank springs a leak; the amount (in gallons) of gasoline remaining in the tank t seconds later is given by

$$V = 20\left(1 - \dfrac{t^2}{400}\right)$$

How fast is the gasoline leaking out 10 seconds after the leak occurs?

63. *Production cost* The average cost (in thousands of dollars) for producing x hundred units of a product each month is given by

$$g(x) = \dfrac{x^3+30}{x} \qquad x > 0$$

Determine how the average cost is changing when the monthly production level is 200 units.

64. *Bacteria population* The population of a bacteria colony is currently estimated to be 50 million. Its population y (in millions) t seconds from now is expected to be

$$y = 0.3t^3 - 6t^2 + 30t + 50 \qquad t \geq 0$$

How fast is the population changing 11 seconds from now?

65. *Marginal cost* The monthly cost (in thousands of dollars) to manufacture and market x hundred units of a product is

$$C(x) = x^2 + 70x + 310 \qquad x \geq 0$$

Determine the marginal cost when the monthly production level is 300 units.

66. *Marginal revenue* The revenue, in dollars, that a record manufacturer receives from producing x records per week is approximately

$$R(x) = 7.5x - 0.0001x^2$$

Determine the marginal revenue when 10,000 records are manufactured a week.

67. *Production* The profit, in dollars, that a manufacturer of electric toasters makes from producing x toasters per day is

$$P(x) = 18x - 0.01x^2 - 4{,}000$$

Determine the approximate profit from producing the 51st toaster.

68. *Velocity* An object dropped from the top of a tall building falls about $16t^2$ feet in t seconds. Determine its velocity $1\tfrac{1}{2}$ seconds after it is dropped.

69. *Projectile* An object is fired straight up into the air in such a way that its distance (in feet) from the earth t seconds later is

$$s(t) = -16t^2 + 256t$$

Determine its velocity and the direction in which it is moving after 10 seconds.

Identify the steps of the four-step procedure involved in the application of mathematics (Section 1.3) in each case.

70. Exercise 62
71. *Production* Exercise 67

Explain the meaning of each term.

72. Function
73. Derivative of a function at $x = a$

CHAPTER 12

ADDITIONAL TOPICS ON THE DERIVATIVE

OUTLINE

12.1 Derivatives of Higher Order
12.2 Increasing and Decreasing Functions
12.3 Relative Maxima and Minima
12.4 Absolute Maxima and Minima
12.5 Maxima–Minima Applications
*12.6 Inflection Points
12.7 Chapter Summary

Review Exercises

In Chapter 11, we discussed the meaning of the derivative, procedures for finding the derivatives of different types of functions, and some applications of derivatives in solving physical problems. Applications of the derivative occur when the mathematical model representing the physical situation is a function and the derivative is the principle device for obtaining the mathematical solution. Most of these applications depend on the fact that the derivative gives the rate of change or the slope of the tangent to the graph of a function. In this chapter we are primarily concerned with the use of the derivative in graphing functions, in determining the maximum or minimum value of a function, and in solving related physical problems.

Consider, for example, the following problem for a truck-rental company. The company wishes to determine the most cost-efficient time to replace a truck. There are two types of costs to be taken into consideration: capital cost (original price less salvage value) and operating cost (insurance, maintenance, and so forth).

If a truck costs $25,000 new and its salvage value is $23,200 - 0.10x$ (dollars), where x is the number of miles the truck has been driven, then the capital cost after x miles is the difference

$$25,000 - (23,200 - 0.10x) = 1,800 + 0.10x$$

dollars. The *average* capital cost per mile then is

$$\frac{1,800 + 0.10x}{x} = \frac{1,800}{x} + 0.10$$

Also, if it costs a total of

$$0.0000005x^2 + 0.10x$$

dollars to operate the truck for x miles, then the *average* operating cost per mile is

$$\frac{0.0000005x^2 + 0.10x}{x} = 0.0000005x + 0.10$$

dollars. (Note that the average capital cost decreases but the average operating cost increases as the truck gets older; that is, as x increases.)

* This section can be omitted without loss of continuity.

Adding these two averages gives a total average cost per mile of

$$f(x) = \frac{1,800}{x} + 0.0000005x + 0.20$$

dollars if the truck is driven x miles. The value of x for which this average is minimal is the number of miles at which the truck should be traded in order to obtain the minimal average cost per mile.

Later in this chapter, we shall use derivatives to determine that $f(x)$ is minimal when $x = 60,000$; that is, the company should trade in the truck after it has been driven 60,000 miles.

12.1 DERIVATIVES OF HIGHER ORDER

This section introduces derivatives of higher order. We shall need derivatives of this type in several of the following sections.

The derivative of a function is always another function. For example, the derivative of

$$f(x) = 4x^3 + 2x$$

is the function

$$f'(x) = 12x^2 + 2$$

Consequently, we can determine the derivative of this new function; the result is the **second derivative** of the original function. If f denotes the original function, then f'' denotes its second derivative. For $f(x) = 4x^3 + 2x$, $f'(x) = 12x^2 + 2$, and

$$f''(x) = 24x$$

Or, if $f(x) = x^4 + 6x^2$, then $f'(x) = 4x^3 + 12x$ and

$$f''(x) = 12x^2 + 12$$

Second derivatives are also denoted by

$$y'', \qquad \frac{d^2}{dx^2} f(x) \qquad \text{or} \qquad \frac{d^2}{dx^2} g(x)$$

EXAMPLE 1 (a) For $y = 6x^5 + 2x^3$,

$$y' = 30x^4 + 6x^2 \qquad \text{and} \qquad y'' = 120x^3 + 12x$$

(b) For $g(x) = -3x$,

$$\frac{d}{dx} g(x) = -3 \qquad \text{and} \qquad \frac{d^2}{dx^2} g(x) = 0 \qquad \boxed{1}$$

1 PRACTICE PROBLEM 1

(a) Determine $f''(x)$ if
$$f(x) = 3x^4 - 2x + 7.$$
(b) Determine y'' if $y = \sqrt{x+1}$.

Answer

(a) $36x^2$ (b) $-\frac{1}{4}(x+1)^{-3/2}$

12.1 DERIVATIVES OF HIGHER ORDER

Similarly, the derivative of the second derivative is the **third derivative** of the original function. If f denotes the original function, then f''' denotes its third derivative. For $f(x) = 12x^4 - 3x^2 + 2$,

$$f'(x) = 48x^3 - 6x, \qquad f''(x) = 144x^2 - 6, \qquad f'''(x) = 288x$$

Other notation for third derivatives includes

$$y''', \qquad \frac{d^3}{dx^3} f(x), \qquad \text{or} \qquad \frac{d^3}{dx^3} g(x)$$

EXAMPLE 2 (a) For $y = x^4 + 6x^3$,

$$y' = 4x^3 + 18x^2, \qquad y'' = 12x^2 + 36x, \qquad \text{and} \qquad y''' = 24x + 36$$

(b) For $g(x) = \ln x$,

$$\frac{d}{dx} g(x) = \frac{1}{x} = x^{-1}, \qquad \frac{d^2}{dx^2} g(x) = -x^{-2}, \qquad \text{and} \qquad \frac{d^3}{dx^3} g(x) = 2x^{-3}$$

2 PRACTICE PROBLEM 2

(a) Determine $f'''(x)$ if
$$f(x) = 5x^3 - 4x^2 + 2$$

(b) Determine y''' if $y = e^{2x}$.

Answer

(a) 30 (b) $8e^{2x}$

EXERCISES

Find the second and third derivatives of each given function.

1. $f(x) = x^4 - 2x^3 + x$
2. $f(x) = 6x^3 - 2x^2 + 7$
3. $f(x) = 3x^2 - 2x^5$
4. $g(x) = \frac{1}{4}x^4 - \frac{1}{3}x^3 + \frac{1}{2}x^2$
5. $y = x^4 + x^3 + x^2 + x + 1$
6. $y = \frac{1}{6}x^3 + \frac{2}{3}x^2 - \frac{4}{5}$
7. $g(x) = 7x - 4$
8. $f(x) = x^2 - 2x - 8$
9. $y = \sqrt{x}$
10. $f(x) = \sqrt{3x}$
11. $f(x) = x^{5/2} + x^{3/2}$
12. $g(x) = 9x^{4/3} - 27x^{1/3}$
13. $f(t) = 2t^{-1}$
14. $g(t) = t^2 + t^{-2}$
15. $y = (x - 3)^3$
16. $f(x) = (2x + 1)^4$
17. $f(x) = \sqrt{3 - 5x}$
18. $f(t) = (1 - 2t)^{4/3}$
19. $f(t) = -e^t$
20. $f(t) = e^{-t}$
21. $g(x) = xe^x$
22. $f(x) = x^2 e^x$
23. $f(x) = \ln x$
24. $g(x) = x \ln x$
25. $f(x) = \ln x^2$
26. $y = \ln(2x + 1)$
27. $y = \dfrac{x + 1}{x}$
28. $f(t) = \dfrac{t}{t + 1}$

Explain the meaning of each term.

29. Second derivative of a function f
30. Third derivative of a function f
31. What do you think the fourth derivative of a function f would be?

12.2 INCREASING AND DECREASING FUNCTIONS

We have already seen that the derivative of a function at a value of x gives the rate of change of the function at that value of x. The values of the function are increasing on intervals where the derivative is positive and decreasing on intervals where the derivative is negative. In this section, we want to use this property of the derivative to determine the intervals in which a function is increasing and the intervals in which it is decreasing; we then want to use this information as an aid in graphing additional types of functions.

First, there are several things to be noted in the graph (See Figure 12.1) of the function

$$y = x^3 - 6x^2 + 9x^2 + 1$$

1. At the values of x for which the derivative

$$y' = 3x^2 - 12x + 9 = 3(x-1)(x-3)$$

is zero—namely, $x = 1$ and $x = 3$—the slope of the tangent is also zero; the tangents are therefore parallel to the x-axis. It is at these points that the graph has either a peak or a valley.

2. For all values of x less than 1, the derivative is positive and the values of the function increase as x increases. For values of x between 1 and 3, the derivative is negative and the values of the function decrease as x increases. For values of x greater than 3, the derivative is again positive and the values of the function increase as x increases.

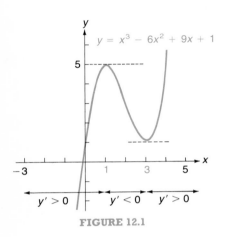

FIGURE 12.1

In general, as x increases the graph of a function will rise on any interval in which its derivative is positive and fall on any interval in which the derivative is negative. Also, as long as the derivative is continuous, the peaks and valleys of the graph occur when the derivative is zero. With this information, along with the coordinates of a few points in the graph, we can obtain fairly accurate graphs of some additional functions.

The procedure is illustrated in obtaining the graph of the function

$$y = x^3 + 2x^2 - 4x - 2$$

The derivative is

$$y' = 3x^2 + 4x - 4 = (3x - 2)(x + 2)$$

We first determine the values of x for which the derivative is zero; from the factored form of the derivative, we see that these are $x = \frac{2}{3}$ and $x = -2$. In each interval on the x-axis determined by these values we pick an additional value of x—for example, $x = -3$, $x = 0$, and $x = 2$ (see Figure 12.2).

FIGURE 12.2

12.2 INCREASING AND DECREASING FUNCTIONS

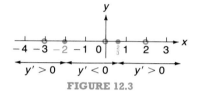

x	y	y'
−3	1	11
−2	6	0
0	−2	−4
$\frac{2}{3}$	$\frac{-94}{27}$	0
2	6	16

For these five values of x, we determine the values of y and y', as shown in the table.

We now have a considerable amount of information about the graph:

The coordinates of five points are known.

The points at which the graph has a peak or a valley are known.

The function is increasing (left to right) for values of x less than -2, as indicated by the positive derivative at $x = -3$.

The function is decreasing for values of x between -2 and $\frac{2}{3}$, as indicated by the negative derivative at $x = 0$.

The function is increasing for all values of x greater than $\frac{2}{3}$, as indicated by the derivative at $x = 2$.

This information is presented in summary form along the x-axis in Figure 12.3.

All that remains is to locate the five points (x, y) from the table in a coordinate system and connect them with a curve, which increases or decreases as indicated (Figure 12.4).

The procedure used to graph the function in this illustration is given in summary form:

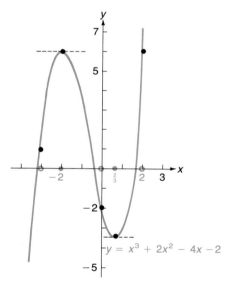

FIGURE 12.3

FIGURE 12.4

1. Determine the derivative of the function.
2. Determine the values of x at which the derivative is zero.
3. In each interval on the x-axis determined by the values found in Step 2, choose an additional value of x.
4. For each value of x found in Steps 2 and 3, determine the value of the function and the value of the derivative. (The derivative will always be zero at the values of x found in Step 2.)
5. On the basis of the derivatives found in Step 4, determine the intervals on which the function is increasing and the intervals on which it is decreasing.
6. Locate the points whose coordinates were found in Step 4 in a coordinate system and connect them with a curve which increases or decreases on the appropriate intervals.

The procedure is illustrated again in the next several examples. The steps are numbered as indicated in the summary.

EXAMPLE 1 Graph the function $y = -x^3 + 3x^2 + 2$.

Solution **Step 1.** The derivative is

$$y' = -3x^2 + 6x = -3x(x - 2)$$

Step 2. The derivative is zero for $x = 0$ and $x = 2$.

Step 3. In the intervals determined by 0 and 2 we choose $x = -1, 1,$ and 3.

Step 4. In table form:

x	y	y'
-1	6	-9
0	2	0
1	4	3
2	6	0
3	2	-9

Step 5.

Step 6.

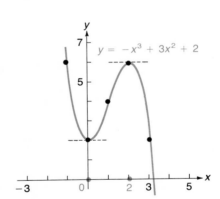

PRACTICE PROBLEM 1

Given $y = \frac{1}{3}x^3 - x^2 + 1$:

(a) For which values of x is the derivative zero?

(b) In which intervals is y increasing and in which is y decreasing?

(c) Graph the function.

Answer

(a) $x = 0$ and $x = 2$

(b)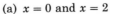

(c)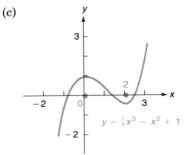

EXAMPLE 2 Graph the function $f(x) = x^4 + \tfrac{8}{3}x^3$.

Solution **Step 1.** The derivative is
$$f'(x) = 4x^3 - 8x^2 = 4x^2(x-2)$$

Step 2. The derivative is zero for $x = 0$ and $x = 2$.

Step 3. In the intervals determined by 0 and 2, we choose $x = -1, 1,$ and 2.5.

Step 4.

x	$f(x)$	$f'(x)$
-1	$\tfrac{11}{3}$	-12
0	0	0
1	$-\tfrac{5}{3}$	-4
2	$-\tfrac{16}{3}$	0
2.5	-2.6	12.5

Step 5.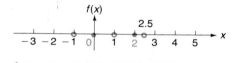

Step 6.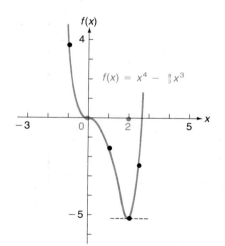

An important thing to note in Example 2 is that even though the derivative is zero at $x = 0$, the graph has neither a peak nor a valley there. This is because the derivative is negative for values of x on either side of $x = 0$, and therefore the function is decreasing on each side. Consequently, although the peaks and valleys for functions with continuous derivatives occur at values of x for which the derivatives are zero, the converse need not be true—the derivative can be zero for values of x at which the graph has neither a peak nor a valley. The deciding factor is whether or not the derivative changes sign (positive to negative or negative to positive) at such values of x.

Finally, if the derivative is not continuous (which includes the possibility that the derivative does not exist) at a single point in the domain of a function, the graph of the function may or may not have a peak or valley at that point. Each function has to be examined individually at such points. The only change in procedure is that the isolated values of x at which the derivative is not continuous must be treated in the same way as points at which the derivative is zero. The function will or will not have a peak or valley depending on whether or not the derivative changes sign at that value of x. The procedure is illustrated in Example 3.

EXAMPLE 3 Graph the function $y = x^{2/3}$.

Solution **Step 1.** The derivative is

$$y' = \frac{2}{3} x^{-1/3} = \frac{2}{3x^{1/3}}$$

Step 2. The derivative is never zero and fails to exist at $x = 0$.

Step 3. In the intervals determined by $x = 0$, we choose $x = -1$ and $x = 1$.

Step 4.

x	y	y'
-1	1	$-\frac{2}{3}$
0	0	
1	1	$\frac{2}{3}$

Step 5.

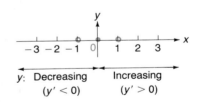

y: Decreasing ($y' < 0$)   Increasing ($y' > 0$)

PRACTICE PROBLEM 2

Given $f(x) = \frac{2}{3}x^3 - \frac{1}{4}x^4$:

(a) For which values of x is the derivative zero?

(b) In which intervals is f increasing? Decreasing?

(c) Graph the function.

Answer

(a) $x = 0$ and $x = 2$

(b)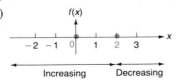

(c)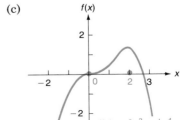

$f(x) = \frac{2}{3}x^3 - \frac{1}{4}x^4$

Step 6. Because the derivative changes sign, from negative to positive, the graph of the function will have a valley at $x = 0$.

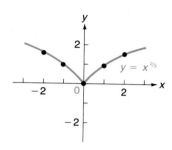

Two additional points $(2, 1.6)$ and $(-2, 1.6)$ were plotted to help determine the shape of the graph. ∎

Note that $x = 0$, where the derivative did not exist in Example 3, was treated the same as values of x at which the derivative is zero. Otherwise, the procedure is the same as that in the preceding examples. Additional points can be determined for the graph of any function when more information is desired about the shape of the graph. ③

EXAMPLE 4 The total monthly profit from the sales of x units of a product is given by

$$P(x) = 500x - \tfrac{1}{2}x^2 \qquad x \geq 0$$

Without graphing the function, determine the sales levels at which profit is increasing and the levels at which profit is decreasing.

Solution   The marginal profit is given by

$$P'(x) = 500 - x$$

which is equal to zero for $x = 500$. Following the procedure discussed

---

③ **PRACTICE PROBLEM 3**

Given $y = x^{1/3}$:

(a) For which values of x is the derivative zero or does the derivative fail to exist?

(b) In which intervals is y increasing? Decreasing? (*Hint:* The graph in Example 3 indicates that $x^{2/3}$ is positive if $x \neq 0$.)

(c) Graph the function.

Answer

(a) No derivative at $x = 0$

(b)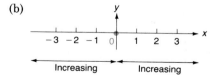

(c)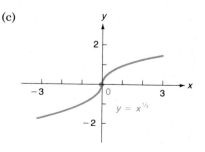

above, we have (in summary form):

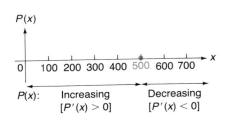

Profit will increase as sales increase from 0 to 500 units, after which it will start decreasing. ∎

How can profits decrease as the number of units sold increases, as happens in this case for $x > 500$? It may be that the sales force, manufacturing facilities, or other similar factors would have to be increased when sales reach the 500 mark, thereby making sales beyond that level less profitable.

From the point of view of the four-step procedure in the application of mathematics (Section 1.3) in Example 4, the physical question was: At what sales levels are profits increasing and at what sales levels are profits decreasing? The mathematical model was the given profit function $P(x)$ and the mathematical question was: For what values of x is $P'(x)$ positive and for what values is $P'(x)$ negative? The mathematical solution was that $P'(x)$ is positive for x between 0 and 500 and negative for x greater than 500. The interpretation was that profits will increase as sales increase from 0 to 500 units and will decrease after 500.

EXERCISES

Graph each function using the procedure discussed in this section.

1. $y = \frac{1}{3}x^3 - x^2 + 2$
2. $f(x) = \frac{1}{3}x^3 + \frac{1}{2}x^2 - 6x$
3. $y = x^3 - 3x + 1$
4. $y = \frac{1}{3}x^3 - x^2 + x + 1$
5. $f(x) = -2x^3 + 6x^2 - 4$
6. $y = -2x^3 + 3x^2 + 12x - 10$
7. $y = x^2 + 4x + 6$
8. $f(x) = 2x^2 - 12x + 18$
9. $y = -x^2 + x - 3$
10. $y = -3x^2 + 4x + 2$
11. $f(x) = -x^4$
12. $y = x^4 - 4x + 3$
13. $y = 1 + x^{1/3}$
14. $f(x) = 1 - x^{2/3}$
15. $y = (1 + x)^{1/3}$
16. $y = x - e^x$
17. $f(x) = \frac{1}{2}x^2 - \ln x$ (Note that the domain of f is $x > 0$ only.)
18. *Profit* For each of the following functions giving the monthly profit (in hundreds of dollars) from the sale of x units, determine the sales levels

at which profits are increasing and the sales levels at which profits are decreasing:

(a) $P(x) = 100x - x^2$, $x \geq 0$ (b) $P(x) = x^3 - 6x^2 - 96x$, $x \geq 0$

19. *Supply* Suppose that the quantity, in the number of units indicated, of a product that industry is willing to supply each month is given in terms of the price x in dollars by the following supply functions. For each, determine the price range at which industry will increase the supply and the range at which industry will decrease the supply.

(a) $S(x) = 3x^2 + 60x$, $x \geq 0$ (in single units)

(b) $S(x) = \dfrac{x^2}{x+2}$, $x \geq 0$ (in hundreds of units)

20. *Cost-revenue-profit* The weekly cost (in thousands of dollars) to manufacture and market x units (in hundreds of units) is given by

$$C(x) = 2x + 4 \qquad x \geq 0$$

The weekly revenue (in thousands of dollars) from the sale of x hundred units is

$$R(x) = 5x^2 + 10x \qquad x \geq 0$$

(a) Determine marginal cost.
(b) Determine marginal revenue.
(c) Use the relationship

$$\text{Profit} = \text{Revenue} - \text{Cost}$$

to construct the profit function.
(d) Determine marginal profit, the sales levels for which profit is decreasing, and the sales levels for which profit is increasing.
(e) At what sales level is weekly profit minimal, and what is the corresponding minimal profit?

21. *Cost-revenue-profit* Repeat Exercise 20 if the cost and revenue (in hundreds of dollars) for x units are given by

$$C(x) = 0.02x^3 - 0.15x^2 + 2x + 10 \qquad x \geq 0$$
$$R(x) = 5x \qquad x \geq 0$$

22. *Cost* A corporation has a chain of stores across the country. Weekly administrative costs at corporate headquarters (in thousands of dollars) are given by the cost function

$$C(x) = \dfrac{3x^2}{x+7} \qquad x \geq 0$$

where x is the number of stores in operation. Determine the marginal cost, the number of stores for which administrative costs are decreasing, and the number of stores for which administrative costs are increasing.

Identify the steps of the four-step procedure involved in the application of mathematics (Section 1.3) in each case.

23. *Profit* Exercise 18(a)   24. *Supply* Exercise 19(a)

12.3 RELATIVE MAXIMA AND MINIMA

What we have been calling peaks and valleys in the graphs of functions are more commonly called *relative maximum points* and *relative minimum points*. Formally, a function f has a **relative maximum** at $x = a$ if $f(a) > f(x)$ for all x in some interval containing a in its interior, and f has a **relative minimum** at $x = a$ if $f(a) < f(x)$ for all x in some interval containing a in its interior. The graph of the function f shown in Figure 12.5 indicates that f has a relative minimum at $x = c$; $f(c) < f(x)$ for all values of x close to c. Similarly, f has a relative maximum at $x = b$ and $x = d$; $f(b) > f(x)$ for all values of x close to b and $f(d) > f(x)$ for all values of x close to d.

Our objective is now to determine the points at which a function has a relative maximum or minimum without graphing the function. From Section 12.2, we know that relative maxima and minima occur at values of x at which the derivative is zero (corresponding to a horizontal tangent) or at isolated points in the domain of the function at which the derivative does not exist. Collectively, these values of x—having a zero derivative or no derivative—are called **critical values** of the function. Consequently, the relative maxima and relative minima of a function always occur at the critical values of the function. However, there need not be a maximum or minimum at *every* critical value.

There are two ways of checking whether a function has a relative maximum, relative minimum, or neither at a critical value. The first of these, which should be fairly obvious from our work in the previous section, amounts to observing the sign of the first derivative. We state it here formally as follows:

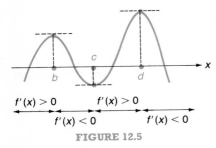

FIGURE 12.5

THE FIRST DERIVATIVE TEST

Let $x = a$ be a critical value of the function f:

1. If $f'(x) > 0$ for all values of x just to the left of a and $f'(x) < 0$ for all values of x just to the right of a, then f has a relative maximum at $x = a$.
2. If $f'(x) < 0$ for all values of x just to the left of a and $f'(x) > 0$ for all values of x just to the right of a, then f has a relative minimum at $x = a$.

Less formally, the first derivative test says that f has a relative maximum at $x = a$ if $f'(x)$ changes from positive to negative at a and a relative minimum if $f'(x)$ changes from negative to positive at a, as indicated in Figure 12.5.

12.3 RELATIVE MAXIMA AND MINIMA

The procedure for using the first derivative test is not new to us; first determine the critical values and then determine if and how the derivative changes sign at each critical value. As long as the derivative is continuous, the question of whether and how the derivative changes sign can be answered by checking the value of the derivative at a value of x in each of the intervals determined by the critical values.

EXAMPLE 1 Determine the relative maxima and relative minima of

$$f(x) = x^3 + 3x^2 - 24x - 20$$

Solution The derivative

$$f'(x) = 3x^2 + 6x - 24 = 3(x-2)(x+4)$$

x	$f'(x)$	$f(x)$
-5	21	
-4	0	60
0	-24	
2	0	-48
3	21	

is zero for $x = 2$ and $x = -4$. In the intervals determined by 2 and -4, we choose the values $x = -5, 0,$ and 3, as shown in the margin. Then we construct a table similar to those constructed before. Now, however, we need the values of the function only at the critical values. The table indicates that $f'(x)$ changes from positive to negative at $x = -4$, so that f has a relative maximum of 60 at -4. Also, $f'(x)$ changes from negative to positive at $x = 2$, so that f has a relative minimum of -48 at 2. ∎

EXAMPLE 2 Determine the relative maxima and relative minima of

$$y = 3x^2 - 9x + 1$$

Solution The derivative

$$y' = 6x - 9$$

x	y'	y
0	-9	—
$\frac{3}{2}$	0	$-\frac{23}{4}$
2	3	—

is zero for $x = \frac{3}{2}$. In the intervals determined by $\frac{3}{2}$, we choose $x = 0$ and $x = 2$. The table shows that y has a minimum of $-\frac{23}{4}$ at $x = \frac{3}{2}$. ∎

EXAMPLE 3 Determine the relative maxima and minima of

$$f(x) = x^3$$

Solution The derivative

$$f'(x) = 3x^2$$

---

1 PRACTICE PROBLEM 1

Determine the relative maxima and relative minima of

$$f(x) = \tfrac{1}{3}x^3 - 2x^2 + 3x - 1$$

Answer

Relative maximum of $\frac{1}{3}$ at $x = 1$; relative minimum of -1 at $x = 3$

x	y'	y
-1	3	
0	0	0
1	3	

is zero only for $x = 0$. In the intervals determined by $x = 0$, we choose $x = -1$ and 1. Since the derivative does not change sign, there is neither a relative maximum nor a relative minimum at $x = 0$. Nor can there be any at any other value of x, as there are no more critical values.

PRACTICE PROBLEM 2

Determine the relative maxima and relative minima of

$$f(x) = x^3 + 6x^2 + 12x - 8$$

Answer

No relative maximum nor minimum

The second test referred to earlier for determining relative maxima and minima is based on the fact that the second derivative gives the rate of change of the first derivative—just as the first derivative gives the rate of change of the function. In the graph in Figure 12.6(a), f has a relative maximum at $x = a$ and the first derivative decreases—positive to zero to negative—as x increases through a. Therefore, the second derivative is negative at $x = a$, corresponding to a decreasing first derivative. In the graph in Figure 12.6(b), the first derivative increases—negative to zero to positive—as x passes through a. Therefore, the second derivative is positive at $x = a$.

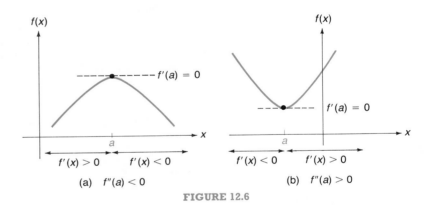

FIGURE 12.6

If both $f'(a) = 0$ and $f''(a) = 0$, we cannot be sure about how the first derivative is changing without additional information. These relationships form the basis for the second test.

THE SECOND DERIVATIVE TEST

Suppose that f is differentiable on an interval containing a in its interior. Also, suppose $f'(a) = 0$:

1. If $f''(a) < 0$, f has a relative maximum at $x = a$.
2. If $f''(a) > 0$, f has a relative minimum at $x = a$.
3. If $f''(a) = 0$, the test fails and the first derivative test must be used.

While the second derivative test is easier to use at times, it is not always applicable because it requires the function to have a second derivative at the critical value in question; furthermore, if the second derivative is zero at the critical point, the test fails and one must resort to the previous test. The first derivative test, on the other hand, is applicable at every critical value and becomes the crucial test when the second derivative test fails.

EXAMPLE 4 Determine the relative maxima and minima of

$$f(x) = \tfrac{1}{6}x^3 + \tfrac{1}{2}x^2 - 4x - \tfrac{10}{3}$$

Solution   The derivative

$$f'(x) = \tfrac{1}{2}x^2 + x - 4 = \tfrac{1}{2}(x - 2)(x + 4)$$

is zero for $x = 2$ and $x = -4$. The second derivative test requires that we evaluate

$$f''(x) = x + 1$$

x	$f''(x)$	$f(x)$
-4	-3	10
2	3	-8

at these two values of x. The results are shown in the table. The sign of the second derivative indicates that f attains a relative maximum value of 10 at $x = -4$ and a relative minimum value of -8 at $x = 2$.

3 PRACTICE PROBLEM 3

Use the second derivative test to determine the relative maxima and relative minima of the function in Practice Problem 1.

The graphs of the function f of Example 4 and of its first and second derivative (Figure 12.7, page 470) indicate the geometric relationship among the three at the relative maximum and minimum of f. At $x = -4$, the first derivative is zero and the second derivative is negative, corresponding to the fact that f has a relative maximum at $x = -4$. At $x = 2$, the first derivative is zero and the second derivative is positive, corresponding to the relative minimum of f.

In Figure 12.7, note how $f(x)$ decreases on intervals on which $f'(x)$ is negative and increases on intervals on which $f'(x)$ is positive. Similarly, $f'(x)$ decreases on intervals on which $f''(x)$ is negative and increases on intervals on which $f''(x)$ is positive.

EXAMPLE 5 Determine the relative maxima and minima of

$$y = \frac{x}{x^2 + 4}$$

Solution   The derivative

$$y' = \frac{(x^2 + 4) - x(2x)}{(x^2 + 4)^2} = \frac{4 - x^2}{(x^2 + 4)^2} = \frac{(2 + x)(2 - x)}{(x^2 + 4)^2}$$

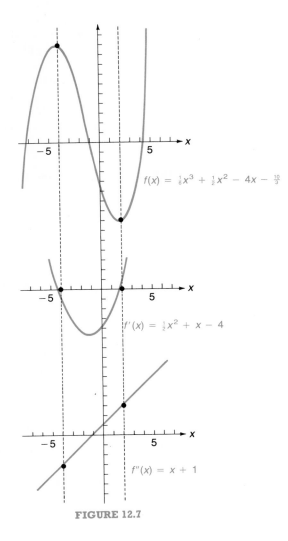

FIGURE 12.7

is zero for $x = 2$ and $x = -2$. We want to evaluate

$$y'' = \frac{2x(x^2 - 12)}{(x^2 + 4)^3}$$

at $x = 2$ and -2. The results are shown in the table. The sign of the second derivative indicates that y has a relative minimum of $-\frac{1}{4}$ at $x = -2$ and a relative maximum of $\frac{1}{4}$ at $x = 2$.

x	y''	y
-2	$\frac{1}{16}$	$-\frac{1}{4}$
2	$-\frac{1}{16}$	$\frac{1}{4}$

---

4 PRACTICE PROBLEM 4

Use the second derivative test to determine the relative maxima and minima of $y = xe^x$.

Answer

Relative minimum of $-e^{-1}$ at $x = -1$

12.3 RELATIVE MAXIMA AND MINIMA

EXAMPLE 6 Determine the relative maxima and minima of
$$g(x) = 3x^4 - 7$$

Solution The derivatives are
$$g'(x) = 12x^3 \quad \text{and} \quad g''(x) = 36x^2$$

The first derivative is zero for $x = 0$. At $x = 0$, the second derivative is also 0, so the second derivative test gives no information. The first derivative test must be used.

In the intervals determined by $x = 0$, we choose $x = -2$ and $x = 2$. Then we construct a table to present the required information. The first derivative changes sign from negative to positive at $x = 0$, indicating that g has a relative minimum there. The corresponding minimal value is $g(0) = -7$.

x	$g'(x)$	$g(x)$
-2	-96	
0	0	-7
2	96	

5 PRACTICE PROBLEM 5

Determine the relative maxima and minima of $f(x) = 3 - 2x^4$.

Answer

Relative maximum of 3 at $x = 0$

EXERCISES

Determine all the critical values and then determine the relative maxima and minima using the first derivative test.

1. $y = x^2 - 6x + 4$
2. $y = -6x^2 - 12x$
3. $f(x) = -3x^2 + 9x + 2$
4. $f(x) = 6x^2 - 6x$
5. $f(x) = \frac{1}{3}x^3 - \frac{1}{2}x^2 - 2x + 1$
6. $y = -x^3 + 6x^2$
7. $y = x^3 + 5x^2 - 8x - 2$
8. $y = x^3 - 6x^2 + 12x$
9. $g(x) = 2x^3 + 6$
10. $f(x) = 2x^3 - 4x^2 - 8x + 1$
11. $f(x) = x^4 - 2x^2 + 1$
12. $y = x^4 + 2x^3 - 2$
13. $y = 3x^{2/3}$
14. $y = x^{2/3} + \frac{2}{3}x$
15. $f(x) = e^{2x} - 2x$
16. $y = x^2 - 18(\ln x)$

Determine all the critical values and then determine the relative maxima and minima using the second derivative test when applicable.

17. $f(x) = x^2 + 4x - 3$
18. $y = -5x^2 + 20x$
19. $f(x) = -2x^2 + 8x + 1$
20. $y = x^2 + 5x - 2$
21. $y = \frac{1}{3}x^3 - \frac{1}{2}x^2 - 6x + 1$
22. $g(x) = 9x^2 - x^3$
23. $y = x^3 + 3x^2 + 3x - 2$
24. $f(x) = (x + 1)^3$
25. $f(x) = \frac{2}{3}x^3 - \frac{x^2}{2}$
26. $y = x^3 + x^2 + x + 1$
27. $f(x) = x^4 + 32x$
28. $f(x) = 3x^4 - 4$
29. $y = 2(x - 3)^4$
30. $y = (2x - 3)^4$
31. $f(x) = \frac{x^2 + 25}{x}$
32. $g(x) = \frac{x}{x^2 + 9}$
33. $y = \sqrt{(x + 2)^2}$
34. $y = 3x^{1/3}$
35. $y = 3^x - (\ln 3)x$
36. $y = 7x - 3$

Construct a graph for the function given in the indicated exercise, along with the graphs of its first and second derivatives, as in Figure 12.7.

37. Exercise 19
38. Exercise 21
39. Verify that the exponential function
$$f(x) = Ab^{kx}$$
(where A, b, and k are constants with $A \neq 0$, $k \neq 0$, $b > 0$, and $b \neq 1$) has no relative maxima or minima.

40. Verify that the logarithmic function
$$f(x) = A \log_b(kx)$$
(where A, b, and k are constants with $A \neq 0$, $k \neq 0$, $b > 0$, and $b \neq 1$) has no relative maxima or minima.

41. State the first derivative test.
42. State the second derivative test.
43. Explain the meaning of relative maximum and relative minimum.

12.4 ABSOLUTE MAXIMA AND MINIMA

In Section 12.3, we considered methods for finding values of the independent variable at which a function took on the largest or smallest value *in some interval in the domain* of the function. This section is concerned with determining the values of the independent variable at which a function takes on the largest or smallest value *in its whole domain*. These largest and smallest values are called the *absolute maximum* and *absolute minimum* of the function. Formally, a function f has an **absolute maximum** at $x = a$ if $f(x) \leq f(a)$ for all values of x in the domain of f, and f has an **absolute minimum** at $x = a$ if $f(x) \geq f(a)$ for all values of x in the domain of f.

The function whose graph is shown in Figure 12.8 has an absolute maximum at $x = c$ and an absolute minimum at $x = b$. (It also has relative maxima at $x = a$ and $x = c$ and a relative minimum at $x = b$.)

We want to determine the values of x at which a function has an absolute maximum and an absolute minimum. The problem is that many functions might not have one or the other or either. The graph of

$$y = \frac{2}{x - 3}$$

in Figure 12.9 shows that this function has neither.

There is no general procedure to determine whether or not an arbitrary function has an absolute maximum or an absolute minimum. Each function has to be examined individually; graphing the function is particularly helpful here, as in the illustration of Figure 12.9.

Consequently, we shall consider more restricted situations, as suggested by the following theorem from advanced calculus.

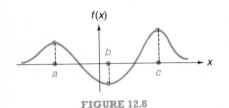

FIGURE 12.8

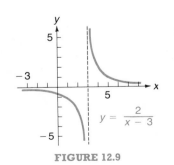

FIGURE 12.9

12.4 ABSOLUTE MAXIMA AND MINIMA

THEOREM

A function continuous on a finite closed interval attains an absolute maximum and absolute minimum on that interval.

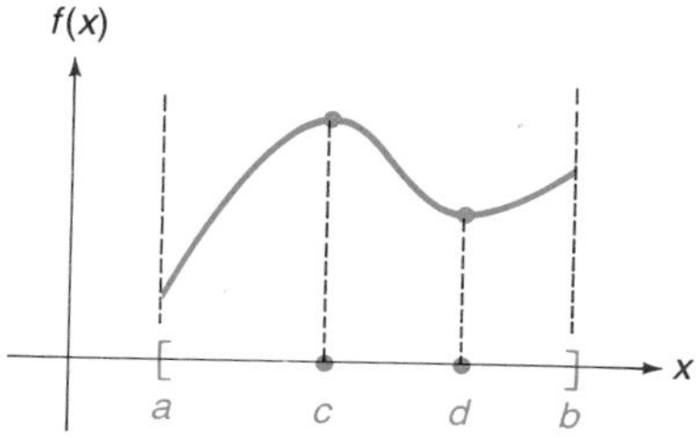

Recall that a closed interval is a set of real numbers of the form $\{x : a \leq x \leq b\}$, where a and b are constants. The theorem guarantees that for a function f continuous on such a domain, there are values of x in the interval at which f attains its absolute maximum and absolute minimum on that interval. Figure 12.10 shows the graph of a function continuous on the closed interval from a to b. It attains its absolute maximum on that interval at $x = c$ and its absolute minimum at $x = a$.

As the graph suggests, the absolute maximum and absolute minimum will occur at values of x at which the function has a relative maximum or minimum ($x = c$ and $x = d$ in Figure 12.10) or at the endpoints ($x = a$ and $x = b$ in Figure 12.10). Consequently, in order to determine the location of the absolute maximum and absolute minimum of a function f continuous on a closed interval, it is sufficient to determine the critical values of f and then to examine the values of the function at the critical values and at the endpoints of the interval. The greatest of these function values will be the absolute maximum and the smallest will be the absolute minimum. There is no need to apply any of the tests for relative maxima and minima at the critical values.

This procedure is illustrated in the next several examples.

FIGURE 12.10

EXAMPLE 1 Determine the absolute maximum and minimum of the function

$$f(x) = x^3 - 6x^2 + 9x - 5 \qquad 2 \leq x \leq 5$$

Solution The derivative

$$f'(x) = 3x^2 - 12x + 9 = 3(x - 1)(x - 3)$$

is zero for $x = 3$; $x = 1$ is outside the domain of f. The only critical value of f is $x = 3$. We determine the values of the function at 3 and at the endpoints of the domain, $x = 2$ and $x = 5$, as given in the table. From the table, f has an absolute maximum of 15 at $x = 5$ and an absolute minimum of -5 at $x = 3$.

x	$f(x)$
2	-3
3	-5
5	15

---

1 PRACTICE PROBLEM 1

Determine the absolute maximum and minimum of the function in Example 1 if the domain is changed to the interval [0, 2].

Answer

Maximum of -1 at $x = 1$; minimum of -5 at $x = 0$

PRACTICE PROBLEM 2

Determine the absolute maximum and minimum of $y = x^2 - 6x + 10$ for each domain.

(a) [0, 5]   (b) [4, 10]

Answer

(a) Maximum of 10 at $x = 0$; minimum of 1 at $x = 3$
(b) Maximum of 50 at $x = 10$; minimum of 2 at $x = 4$

x	$g(x)$
−1	1
0	0
10	10

PRACTICE PROBLEM 3

Determine the absolute maximum and minimum of

$$f(x) = x^{2/3} \quad -1 \leq x \leq 8$$

Answer

Maximum of 4 at $x = 8$; minimum of 0 at $x = 0$

EXAMPLE 2 Determine the absolute maximum and minimum of
$$y = -x^2 - 8x - 11 \quad -6 \leq x \leq 0$$

Solution The derivative
$$y' = -2x - 8 = -2(x + 4)$$
is zero for $x = -4$, which is the only critical value. We examine the value of the function at -4 and at the endpoints of the domain, $x = -6$ and $x = 0$. From the table, y has an absolute maximum of 5 at $x = -4$ and an absolute minimum of -11 at $x = 0$.

x	y
−6	1
−4	5
0	−11

EXAMPLE 3 Determine the absolute maximum and minimum of
$$g(x) = \sqrt{x^2} \quad -1 \leq x \leq 10$$

Solution The derivative
$$g'(x) = \frac{x}{\sqrt{x^2}}$$
is never zero, but $x = 0$ is a critical value of g (because it is an isolated point in the domain of g at which there is no derivative). We examine the values of g at $x = 0$ and at the endpoints of the domain of g. From the table, g has an absolute maximum of 10 at $x = 10$ and an absolute minimum of 0 at $x = 0$.

In the next section, we shall see how to apply these maxima–minima methods to solving physical problems.

EXERCISES

Determine the absolute maximum and absolute minimum for each of the following functions:

1. $f(x) = x^2 - 4x + 7, \quad 1 \leq x \leq 4$
2. $f(x) = x^2 + 5x + 13, \quad -4 \leq x \leq 0$
3. $y = -x^2 + 4x + 3, \quad 0 \leq x \leq 3$
4. $y = -x^2 + 4x + 3, \quad 4 \leq x \leq 7$
5. $f(x) = x^3 + 3x^2 - 24x + 1, \quad 0 \leq x \leq 4$
6. $f(x) = x^3 + 3x^2 - 24x + 1, \quad -6 \leq x \leq 0$

7. $f(x) = x^3 + 3x^2 - 24x + 1$, $-6 \le x \le 4$
8. $g(x) = -x^3 + x^2 + 3x + 1$, $-2 \le x \le 0$
9. $g(x) = -x^3 - x^2 + 8x + 2$, $0 \le x \le 2$
10. $y = x^4 + 2x^3 - 2$, $-1 \le x \le 1$
11. $y = x^4 - 2x^2 + 1$, $-2 \le x \le 2$
12. $y = 3x^{1/3}$, $-1 \le x \le 8$
13. $f(x) = \sqrt{(x-3)^2}$, $0 \le x \le 4$
14. $f(x) = 3e^x$, $0 \le x \le 1$
15. $g(x) = x^2 - 8(\ln x)$, $1 \le x \le e$
16. Explain the procedure for determining the absolute maximum and absolute minimum of a function continuous on a closed interval.
17. Explain the meaning of absolute maximum and absolute minimum.

12.5 MAXIMA–MINIMA APPLICATIONS

If the mathematical model of a physical situation is a function, we can apply the procedures of the last section to determine the maximum or minimum of the quantity represented by the function. This is one of the more important uses of the derivative in solving physical problems. The procedure is illustrated in the following examples.

EXAMPLE 1 A company's facilities are limited to producing 600 units of a product each month. Its profit in thousands of dollars from the sales of x units (in hundreds of units) is given by

$$P(x) = x^3 - \tfrac{15}{2}x^2 + 12x \qquad 0 \le x \le 6$$

(The upper limit on the domain of P is determined by production facilities.) Determine the sales level at which profit is maximum and the corresponding maximum profit if the company can sell all the units that it manufactures each month.

Solution The derivative

$$P'(x) = 3x^2 - 15x + 12 = 3(x-1)(x-4)$$

is zero for $x = 1$ and $x = 4$. Since we want to determine the absolute maximum of P, it is sufficient to examine the value of P at these critical values and at the endpoints of the domain. From the table, P attains a maximum value of 18 at $x = 6$. The company will have a maximum monthly profit of $18,000 from the sale of this item if they sell 600 units.

x	$P(x)$
0	0
1	5.5
4	8
6	18

EXAMPLE 2 A box is to be made from a piece of metal 12 inches by 12 inches by cutting a square from each corner and then folding up the edges. What is the maximum volume that can be obtained by constructing a box in this manner?

Solution In this case, we first have to construct the function representing the volume, the quantity to be maximized. In Example 1 the function P was given; we did not have to construct it.

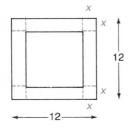

x	$V(x)$
0	0
2	128
6	0

1 PRACTICE PROBLEM 1

What maximum volume could be obtained in Example 2 if the piece of metal were 6 inches by 6 inches?

Answer

16 cubic inches

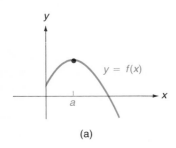

(a)

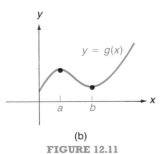

(b)

FIGURE 12.11

Let x denote the length of each side of the square to be cut from the corners, as shown in the margin. The volume (Length × Width × Height) as a function of x is

$$V(x) = x(12 - 2x)(12 - 2x) = 4x^3 - 48x^2 + 144x$$

Since a square is to be cut from each corner, the maximum length of the side of each square is 6 inches; that is, the maximum possible value of x is 6. Also, the length of a square cannot be negative, so the minimum possible value of x is 0. The domain of V is therefore $0 \leq x \leq 6$. We need to determine the (absolute) maximum of the function

$$V(x) = 4x^3 - 48x^2 + 144x \qquad 0 \leq x \leq 6$$

The derivative

$$V'(x) = 12x^2 - 96x + 144 = 12(x - 6)(x - 2)$$

is zero for $x = 6$ and $x = 2$. The table gives the value of V at these critical values and at the endpoints of the domain. The maximum volume that can be obtained is $V(2) = 128$ cubic inches when a 2-inch square is cut from each corner. ■

Note the four-step procedure involved in the application of mathematics in Example 2. The physical situation was as described in the beginning of the example and the question was: What is the maximum volume that can be obtained? The mathematical model was the function

$$V(x) = 4x^3 - 48x^2 + 144x \qquad 0 \leq x \leq 6$$

The mathematical question then was: For what value of x does $V(x)$ have a maximum value and what is the maximum value?

The mathematical solution was that a maximum of 128 occurs at $x = 2$, and the interpretation of this solution was that a maximum volume of 128 cubic inches could be obtained by cutting 2-inch squares from each corner.

Suppose now that a company's annual profit (in hundreds of dollars) from the sale of x hundred units of one of its products is

$$P(x) = 96x + 6x^2 - x^3, \quad x > 0$$

and we want to determine the (absolute) maximum annual profit.

Here we have a problem! The procedure of the last section for determining the absolute maximum or minimum applies only when the function is continuous on a finite closed interval (as was the case in Examples 1 and 2 above). In this case, the domain (all $x > 0$) is not such an interval, so that the previous procedure is no longer applicable.

Fortunately, an alternate procedure, which is applicable in many similar situations, is available. Compare the graphs shown in Figure 12.11 of two functions f and g whose domains are each all $x > 0$ and which are differentiable at each x in their domain. Both functions have a relative maximum at $x = a$. The function f whose graph is given in Figure 12.11(a) also has an absolute maximum at a; This is not so for the function g in Figure 12.11(b). The reason for this difference is that

12.5 MAXIMA—MINIMA APPLICATIONS

g has another critical value at $x = b$, where a relative minimum occurs. This allows the values of $g(x)$ to increase for $x > b$.

However, f has no other critical value; f cannot have a relative minimum, which would allow the values of $f(x)$ to increase to values greater than $f(a)$.

A similar situation occurs with respect to minimum values. Our discussion leads to the following alternate test for absolute maxima and minima:

For a function f differentiable on an interval:

1. if f has a relative maximum (minimum) at $x = a$ in the interval, and
2. if f has no other critical value in the interval,

then f has an absolute maximum (minimum) in the interval at $x = a$.

We can apply the alternate test to the profit function

$$P(x) = 96x + 6x^2 - x^3 \quad x > 0$$

considered above. Its domain is the interval consisting of all numbers greater than zero, written $(0, \infty)$; also, P has a derivative at each x in this interval.

Setting the derivative equal to zero gives

$$96 + 12x - 3x^2 = 0$$

or, dividing by -3,

$$x^2 - 4x - 32 = 0$$

Factoring this gives

$$(x - 8)(x + 4) = 0$$

The only critical value for P is $x = 8$; since its domain is $x > 0$, we exclude $x = -4$.

Also,

$$P''(x) = 12 - 6x$$

and $P''(8) = -36$, so that P has a relative maximum at $x = 8$ by the second derivative test. Since P has no other critical values in its domain, the interval $(0, \infty)$, P also has an absolute maximum at $x = 8$ by the alternate test. Since $P(8) = 640$, the company will make a maximum annual profit of $64,000 from the sale of 800 units.

EXAMPLE 3 If the total cost of producing a batch or lot of x units is given by the cost function $C(x)$, the ***average cost*** per unit when lot size is x units is given by the quotient $C(x)/x$ for $x > 0$. If the cost in dollars is given by

$$C(x) = x^3 - 14x^2 + 105x$$

determine the lot size (value of x) for which the average unit cost is minimal and the corresponding minimal average cost.

Solution We apply the alternate test to the average cost function

$$A(x) = \frac{C(x)}{x} = x^2 - 14x + 105$$

with domain $(0, \infty)$. The derivative

$$A'(x) = 2x - 14$$

is zero only for $x = 7$. The second derivative is

$$A''(x) = 2$$

In particular, $A''(7) = 2$, so that A has a relative minimum at $x = 7$. There are no other critical values in the domain $(0, \infty)$; by the alternate test, A also has an absolute minimum at $x = 7$.

The average unit cost will be minimal with lot size 7. Furthermore, with this lot size, the average unit cost will be $A(7) = 56$, or $56. (What would the total cost per lot with size 7 be?)

2 PRACTICE PROBLEM 2

The monthly profit (in thousands of dollars) for a company from the sale of x hundred units of one of its products is

$$P(x) = -x^3 + 4.5x^2 + 12x - 14.5$$

How many units should the company sell each month for maximum monthly profit?

Answer

400 units

EXERCISES

1. *Volume* A box is to be made from a piece of metal 16 inches square by cutting a square from each corner and folding up the edges. What is the maximum volume that can be attained by constructing a box in this manner?

2. *Cost-revenue-profit* A company's facilities are limited to producing 100 units of a product each month. Its monthly revenue in dollars from the sale of x units is given by

$$R(x) = 150x - x^2$$

Its monthly cost in dollars to produce x units is given by

$$C(x) = x^2 + 18x + 225$$

   (a) Determine the monthly sales level at which revenue is maximum.
   (b) Determine the monthly production level at which cost is minimum.
   (c) Determine the monthly sales level at which profit (revenue minus cost) is maximum.
   (d) What is the maximum monthly profit?

3. *Supply* A company has decided that after the price for one of its products reaches $80.00, the demand is so low that it will no longer be profitable to manufacture the product. Otherwise, the number of units it will supply consumers each week is given, in terms of price x in dollars, by

$$S(x) = 100x - x^2$$

Determine the price at which the company will supply the maximum number of units.

12.5 MAXIMA—MINIMA APPLICATIONS

4. *Surface area* Determine the least amount of material that can be used to construct a box with a square base, open top, and volume of 108 cubic inches. (*Hint:* Height $= 108/x^2$, where x is side of base; the material required consists of the four sides and the base.)

5. *Apartment rental* A real estate management company manages an apartment complex with 20 units. The manager estimates that all 20 units can be rented if the rent is $150 per unit per month and that for each increase in rent of $25, one apartment will be vacated. Considering each empty apartment to correspond to a $25 increase in rent, determine the maximum monthly rent the company can obtain from the complex. (*Hint:* See Exercise 17, Section 2.1.)

6. *Airline revenue* An airline company has scheduled a 100-seat airplane for a charter flight to the Super Bowl. The price per person is $140 plus $5.00 additional for each empty seat. If there are more than 15 empty seats, the flight will be canceled.
   (a) Construct a function that will give the revenue to the company from the flight in terms of the number of empty seats. (*Hint:* See Exercise 14, Section 2.1.)
   (b) Determine the maximum revenue the company can receive from the flight.

7. *Depot dimensions* The owners of a company intend to convert a rectangular part of its property into a small storage depot and have 60 feet of fencing with which to surround the depot. At first they plan to allow 15 feet of fencing per side, but they suspect that if they vary the length of the sides, they might possibly obtain a larger area. Determine the maximum area that can be surrounded with the 60 feet of fencing. (*Hint:* See Exercise 16, Section 2.1.)

8. *Area* A rectangular field along a river is to be fenced in with 80 feet of fencing, with no fence along the river. Determine the dimensions of the largest field that can be enclosed.

Profit Each of the following functions gives the profit (in hundreds of dollars) from the sale of x units. Determine the sales level at which profit is maximum and the corresponding maximum profit.

9. $P(x) = 500x - x^2$, $x \geq 0$
10. $P(x) = -x^3 + 6x^2 + 96x$, $x \geq 0$

Average cost The monthly total cost (in thousands of dollars) of producing x units, where x denotes hundreds of units, is given by each of the following cost functions. Determine the monthly production level at which $C(x)/x$, the average cost per unit, is minimal.

11. $C(x) = x^3 - 8x^2 + 27x$, $x > 0$
12. $C(x) = x^3 + 250$, $x > 0$

13. *Cost-revenue-profit* The monthly cost (in thousands of dollars) to manufacture and market x units (in hundreds of units) is given by

$$C(x) = x^2 + 70x - 310 \quad x \geq 0$$

The monthly revenue (in thousands of dollars) from the sale of x hundred units is given by

$$R(x) = 4x^2 + 10x \quad x \geq 0$$

Using the relationship

$$\text{Profit} = \text{Revenue} - \text{Cost}$$

construct the profit fucntion and determine the monthly sales level at which profit is minimal. What is the monthly profit at this minimal level?

14. *Cost-revenue-profit* The monthly cost (in thousands of dollars) to manufacture and market x units (in hundreds of units) is given by
$$C(x) = 2x^2 + 5x + 5 \qquad x \geq 0$$
The monthly revenue (in thousands of dollars) is given by
$$R(x) = x^2 + 15x \qquad x \geq 0$$
Using the relationship
$$\text{Profit} = \text{Revenue} - \text{Cost}$$
construct the profit function and determine the monthly sales level at which profit is maximum. What is the monthly profit at this maximum level?

15. It is sometimes said that maximum profit is attained when marginal revenue equals marginal cost. Answer the following in view of Exercises 13 and 14.
    (a) Can you justify this statement?
    (b) Is it always true that when marginal revenue equals marginal cost the maximum profit is attained?

16. *Agribusiness* A farmer has an apple orchard with 90 trees. He estimates that his annual profit per tree is $100.00 and that for each additional tree planted in the orchard, his profit per tree will decrease by $1.00.
    (a) Contruct a function that will give the profit in terms of the number of additional trees planted. (*Hint:* See Exercise 15, Section 2.1.)
    (b) Determine the number of trees to be planted to obtain the maximum profit. What is the maximum profit?

17. *Physiology* When an individual coughs, his or her trachea expands and contracts to increase or decrease the velocity of the flow of air. For a section of trachea 1 centimeter in radius, the velocity in centimeters per second is given in terms of the radius r of the trachea by
$$f(r) = cr^2(r - 1) \qquad \tfrac{1}{2} \leq r \leq 1$$
where c is a negative constant. Verify that the maximum velocity occurs when the radius is contracted by one-third of its original size.

18. *Population growth* A culture of bacteria initially (at $t = 0$) contains 10,000 individuals. If the population of the culture after t hours is given by
$$f(t) = 10{,}000(1 + 4t - t^2) \qquad 0 \leq t \leq 4$$
determine the time at which the population is maximum and the time at which it is minimum. What are the maximum and minimum populations?

19. *Learning experiment* In a learning experiment, a psychologist has determined that the number of new words chosen from a select list of words and memorized by an average individual is about
$$f(x) = 20(x - 0.2x^2) \qquad 0 \leq x \leq 4$$
after x hours of continuous study. After how many hours of study will an individual have memorized the most number of words? What is this maximum number of words?

20. *Drug concentration* The percent of concentration of a drug in the bloodstream x hours after it is administered is given by
$$y = \frac{3x}{4x^2 + 16}$$
Determine the time at which the percent of concentration is the greatest. What is the percent at that time?

21. *Population growth* The population (in millions) of a bacteria colony at any time t (in seconds) is estimated to be
$$y = 3t^3 - 60t^2 + 300t + 50 \qquad t \geq 0$$
Determine the time at which the population has a relative maximum. What is the population at that time? Is there an absolute maximum?

22. *Operating cost* Verify that the minimal cost per mile is obtained if the truck in the illustration at the beginning of this chapter is driven 60,000 miles.

23. *Operating cost* If in the illustration at the beginning of this chapter, the truck had cost $25,378 when new, determine the number of miles at which the average cost per mile is minimal.

Identify the steps of the four-step procedure involved in the application of mathematics (Section 1.3) in each case.

24. *Airline revenue* Exercise 6(b)
25. *Learning experiment* Exercise 19

*12.6 INFLECTION POINTS

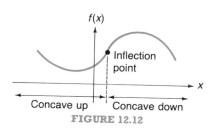

FIGURE 12.12

In Figure 12.12, one portion of the graph is curved upward and one portion is curved downward. That portion of the curve which is curved upward is said to be **concave up,** and that portion which is curved downward is said to be **concave down.** The point at which the concavity changes is called an **inflection point.**

The method of Section 12.2 for graphing functions will generally result in the proper concavity; however, the accuracy of the graph can be improved by determining just where the concavity changes—that is, by determining the location of the inflection points. Our purpose is to consider a method for finding the inflection points.

Consider the slopes of tangents to the portion of the graph shown in Figure 12.13 that is concave up. As x increases from left to right, the slopes of the tangents increase—that is, the derivative of the function is increasing. Since the second derivative gives the rate of change of the first derivative, this means that the second derivative is positive in intervals in which the graph is concave up.

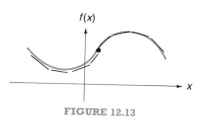

FIGURE 12.13

Similarly, for the portion of the curve that is concave down, as x increases, the slopes of the tangents decrease. Consequently, the first derivative is decreasing, so the second derivative (which gives the rate of change of the first derivative) is negative.

* This section can be omitted without loss of continuity.

A point of inflection thus occurs *when the second derivative changes sign*—positive to negative or negative to positive. This will happen at values of x at which the second derivative is zero or at values of x at which there is no second derivative. The process of determining these values of x is similar to that used to find the values of x at which a function had a relative maximum or relative minimum (Section 12.3).

EXAMPLE 1   Determine the coordinates of the inflection points in the graph of

$$f(x) = \tfrac{1}{6}x^3 - \tfrac{1}{2}x^2 - \tfrac{3}{2}x + 1$$

Solution   The derivatives are

$$f'(x) = \tfrac{1}{2}x^2 - x - \tfrac{3}{2}$$
$$f''(x) = x - 1$$

The second derivative is zero for $x = 1$. We choose a point in each of the intervals determined by $x = 1$—for example, $x = 0$ and $x = 2$—and determine the value of the second derivative at these values of x. The results are shown in the table. Since the second derivative changes sign, the graph has an inflection point at $x = 1$. The coordinates of the point are $(1, -\tfrac{5}{6})$.

The graph of the function is shown below.

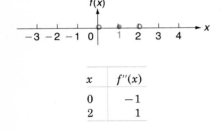

x	$f''(x)$
0	-1
2	1

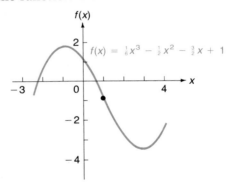

PRACTICE PROBLEM 1

Graph the function

$$y = \tfrac{1}{3}x^3 - x^2 - 3x + 4$$

indicating the coordinates of the relative maximum and minimum points and the point of inflection.

Answer

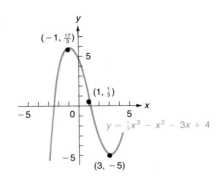

12.6 INFLECTION POINTS

EXAMPLE 2 Determine the coordinates of the inflection points in the graph of
$$y = x^4 - 4$$

Solution The derivatives are
$$y' = 4x^3$$
$$y'' = 12x^2$$

The second derivative is zero for $x = 0$. In the intervals determined by $x = 0$, we choose $x = 2$ and $x = -1$ and then determine the value of the second derivative at these values of x. The results are shown in the table. Since the second derivative does not change sign at $x = 0$, there is no inflection point at $x = 0$. Nor are there any other values of x at which the graph could have an inflection point. The graph of y is given below. Compare the concavity of the graph with the values of y'', which are always nonnegative.

x	y''
-1	12
2	48

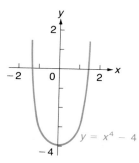

PRACTICE PROBLEM 2

Determine the coordinates of the inflection points in the graph of
$$y = 4x^3 + x + 4$$

Answer

(0, 4)

EXAMPLE 3 Determine the coordinates of the inflection points in the graph of
$$g(x) = 3(x + 2)^{5/3}$$

Solution The derivatives are
$$g'(x) = 5(x + 2)^{2/3}$$
$$g''(x) = \frac{10}{3}(x + 2)^{-1/3} = \frac{10}{3(x + 2)^{1/3}}$$

The second derivative is not zero for any value of x; however, $x = -2$ is an isolated point in the domain of g at which there is no second derivative.

In the intervals determined by $x = -2$, we choose $x = -3$ and $x = -1$ and find the values given in the table. Since the second derivative changes sign, the graph of g has an inflection point at $x = -2$. The coordinates of the point are $(-2, 0)$.

PRACTICE PROBLEM 3

Determine the coordinates of the inflection points in the graph of
$$f(x) = x^{2/3}$$

Answer

No inflection point

EXERCISES

Determine the coordinates of the inflection points, if any, in the graph of each of the following functions:

1. $f(x) = x^3 - 3x^2 - 9x + 6$
2. $y = 2x^3 + 3x^2$
3. $y = 4x^2 + 2x - 3$
4. $f(x) = (2x - 3)^2$
5. $f(x) = \frac{1}{2}x^4 - x^3 - 18x^2$
6. $g(x) = (x - 2)^4$
7. $f(x) = (2x + 4)^{2/3}$
8. $y = 3x^{1/3} + 2x$
9. $y = xe^x$
10. $y = \ln x$

Determine the coordinates of the inflection points, if any, in each of the graphs that you constructed in the following exercises of Section 12.2 Locate the inflection points in the graphs you constructed.

11. Exercise 1
12. Exercise 2
13. Exercise 3
14. Exercise 4
15. Exercise 5
16. Exercise 6
17. Exercise 7
18. Exercise 8
19. Exercise 9
20. Exercise 10
21. Exercise 11
22. Exercise 12
23. Exercise 13
24. Exercise 14
25. Exercise 15
26. Exercise 16
27. Exercise 17

12.7 CHAPTER SUMMARY

The derivative of a function f gives some important information about the graph of f. On intervals in which the derivative is positive, f is increasing and the graph of f rises left to right. Similarly, on intervals in which the derivative is negative, f is decreasing and the graph of f falls left to right. In addition, if the derivative is continuous, the peaks and valleys occur at values of x for which the derivative is zero. The following procedure, based on this information obtained from the derivative, was used for obtaining the graphs of functions.

1. Determine the derivative of the function.
2. Determine the values of x at which the derivative is zero.
3. In each interval on the x-axis determined by the values found in Step 2, choose an additional value of x.
4. For each value of x found in Steps 2 and 3, determine the value of the function and the value of the derivative.
5. On the basis of the derivatives found in Step 4, determine the intervals on which the function is increasing and on which it is decreasing.
6. Locate the points whose coordinates were found in Step 4 in a coordinate system and connect them with a curve, which increases or decreases on the appropriate intervals.

The **second derivative**, f'', of a function f is the derivative of f'; the **third derivative**, f''', of f is the derivative of f''. We used the second derivative for determining the location of the maxima and minima as well as the concavity of functions.

A function f has a **relative maximum** at $x = a$ if $f(x) < f(a)$ for all values of x close to a and a **relative minimum** at $x = a$ if $f(x) > f(a)$ for all values of x close to a. These values a will be values of x for which the derivative of f is zero or isolated points in the domain of f at which there is no derivative. These two types of values are called the **critical values** of f.

We considered two types of tests for determining whether f actually has a relative maximum or minimum at a critical value a.

THE FIRST DERIVATIVE TEST

Let $x = a$ be a critical value of the function f:

1. If $f'(x) > 0$ for all values of x just to the left of a and $f'(x) < 0$ for all values of x just to the right of a, then f has a relative maximum at $x = a$.
2. If $f'(x) < 0$ for all values of x just to the left of a and $f'(x) > 0$ for all values of x just to the right of a, then f has a relative minimum at $x = a$.

While the second derivative test is easier to use at times, it is not always applicable, and if the second derivative is zero at a critical value, the test fails. The first derivative test, on the other hand, is applicable at every critical value and is used when the second derivative test fails.

THE SECOND DERIVATIVE TEST

Suppose that f is differentiable on an interval containing a in its interior. Also, suppose $f'(a) = 0$:

1. If $f''(a) < 0$, f has a relative maximum at $x = a$.
2. If $f''(a) > 0$, f has a relative minimum at $x = a$.
3. If $f''(a) = 0$, the test fails and the first derivative test must be used.

The first and second derivative tests enable us to find the largest and smallest values of a function on *some interval in the domain* of the function. The largest and smallest values *on the whole domain* of a function are called the *absolute maximum and minimum*. Formally, a function f has an **absolute maximum** at $x = a$ if $f(x) \leq f(a)$ for all values of x in the domain of f. Similarly, f has an **absolute minimum** at $x = a$ if $f(x) \geq f(a)$ for all values of x in the domain of f.

Not every function has an absolute maximum and minimum. However, a function continuous on a closed interval $[a, b]$ will have both. To find the absolute maximum and minimum in such a case, one need only examine the value of the function at the critical values in the interval and at the endpoints of the interval. The largest and smallest of these values give the required maximum and minimum.

If a function is not continuous on a finite closed interval, there is no guarantee that the function has an absolute maximum or minimum on the interval. The following alternate test gives a second method for determining the absolute maximum and minimum on an interval, when applicable.

For a function f differentiable on an interval:

1. if f has a relative maximum (minimum) at $x = a$ in the interval, and
2. if f has no other critical value in the interval,

then f has an absolute maximum (minimum) in the interval at $x = a$.

Finally, the sign of the second derivative of a function indicates whether the graph curves upward or downward. On intervals in which the second derivative is positive, the graph is **concave upward;** on intervals in which the second derivative is negative, the graph is **concave downward.** Points at which the concavity changes (that is, the second derivative changes sign) are called **points of inflection.** These occur at values of x at which the second derivative is zero or at values of x at which there is no second derivative. The process of determining these values is similar to that used to find the values of x at which a function has a relative maximum or minimum.

REVIEW EXERCISES

Determine the second and third derivatives of each of the following functions:

1. $f(x) = 2x^4 - 5x^2 + 7$
2. $f(x) = \frac{1}{5}x^5 + x^3 - 5x + 3$
3. $y = x^{1/2} - x^{-1/2}$
4. $g(x) = 8x^{3/4} + (3x)^{-1/3}$
5. $g(t) = \sqrt{2 - 3t}$
6. $y = (2x - 1)^3$
7. $f(x) = e^{x^2}$
8. $g(x) = x^2(\ln x)$
9. $g(x) = \ln(2 - x)$
10. $f(x) = \dfrac{x}{x - 1}$

For each of the functions given below:

(a) *Determine the values of x at which the derivative is zero and the values of x in the domain of the function at which there is no derivative.*
(b) *Determine the intervals on which the function is increasing and the intervals on which it is decreasing.*
(c) *Determine the coordinates of the points of inflection.*
(d) *Graph the function, indicating the location and the coordinates of the relative maximum and minimum points and of the points of inflection.*

11. $y = x^3 - 3x^2 + 2$
12. $f(x) = -\frac{1}{3}x^3 + x^2 + 3x - 3$
13. $f(x) = -x^2 - 3x + 4$
14. $g(x) = 4x^2 - 4x + 1$
15. $y = x^4 - x^3$
16. $y = 2 - (x+1)^4$
17. $f(x) = (2-x)^{2/3}$
18. $y = xe^{-x}$
19. $f(x) = x^{1/2} - x^{-1/2}$ (*Hint:* First determine the domain of f.)
20. $f(x) = x - \ln x$ (See the hint in Exercise 19.)

21. **Profit** For each of the following functions giving the weekly profit (in thousands of dollars) from the sale of x hundred units, determine the sales levels at which profits are increasing and the sales levels at which profits are decreasing:

(a) $P(x) = -\frac{1}{3}x^3 + x^2 + 3x - 4.5, \quad x \geq 0$
(b) $P(x) = x^2 - 5x + 5.25, \quad x \geq 0$

22. **Demand** Suppose that the quantity of a product that consumers are willing to buy each month is given in terms of the price x (in dollars) by the following demand functions. For each, determine the price range at which the demand will increase and the range at which the demand will decrease.

(a) $D(x) = -x^2 + 4x + 396, \quad 0 < x < 22$ (in single units)
(b) $D(x) = 5e^{-(x-1)^2}, \quad 0 < x$ (in hundreds of units)

23. **Inventory** The total cost (ordering cost, purchasing price, and inventory costs) for a retailer to stock an item he sells in terms of the number x of cases ordered at a time is

$$C(x) = \frac{720}{x} + 5x \qquad x > 0$$

Determine the number of cases per order for which this cost is increasing and the number for which this cost is decreasing.

For each of the following functions, determine all the critical values and then determine the relative maxima and minima using the first derivative test.

24. $y = x^2 + 6x - 8$
25. $y = -4x^2 + 10x$
26. $y = 3x^2 - 2x^3$
27. $f(x) = \frac{1}{3}x^3 - x^2 - x + 2$
28. $f(x) = x^3 - 3x^2 + 3x - 1$
29. $g(x) = x^4 - 2x^2 + 3$
30. $y = (x+5)^{1/3}$
31. $y = e^{-x^2}$
32. $y = 2x + \dfrac{8}{x}, \quad x > 0$
33. $f(x) = \dfrac{x^2}{2x+3}$

For each of the following functions, determine all the critical values and then determine the relative maxima and minima using the second derivative test when applicable.

34. $f(x) = x^2 - 3x + 1$
35. $y = 4 - 6x - x^2$
36. $y = -2x^3 + 3x^2 - 12x - 10$
37. $f(x) = x^4 - 8x^2 + 1$
38. $y = x^4 + 3$
39. $y = x^3 + 4$
40. $f(x) = x^3 + 6x^2 + 12x - 5$
41. $g(x) = 3x^4 + 12x$
42. $y = 3x^4 + 12x^3$
43. $f(x) = \dfrac{x^2 + 2}{x}$
44. $f(x) = 4^{x-1} - (\ln 4)x$
45. $y = (x + 5)^{2/3}$

Determine the absolute maximum and minimum of each function.

46. $y = 3x^2 - 4x + 6, \quad 0 \le x \le 4$
47. $y = 3x^2 - 4x + 6, \quad -4 \le x \le 0$
48. $f(x) = x^3 - 12x - 5, \quad -1 \le x \le 3$
49. $f(x) = x^3 - 12x - 5, \quad -3 \le x \le 1$
50. $f(x) = x^3 - 12x - 5, \quad -3 \le x \le 3$
51. $y = \dfrac{x^3}{3} - 1.25x^2 - 1.5x + 1, \quad 0 \le x \le 4$
52. $y = \dfrac{x^3}{3} - 1.25x^2 - 1.5x + 1, \quad -1 \le x \le 4$
53. $y = e^x, \quad 0 \le x \le \ln 2$
54. $f(x) = xe^{-x}, \quad 0 \le x \le \ln 3$
55. $g(x) = x^4 - 4x, \quad 0 < x < \infty$ (absolute minimum only)
56. $y = 36x + 3x^2 - 2x^3, \quad 0 < x < \infty$ (absolute maximum only)

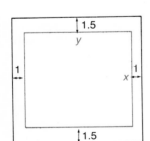

57. *Page dimensions* Each page of a book is to contain 54 square inches of print. In addition, the left and right margins are to be 1 inch each; the top and bottom margins are to be 1.5 inches each (see the figure in the margin). Determine the dimensions of the page with minimum area that can be used. [*Hint:* Area is $(x + 3)(y + 2)$; $y = 54/x$.]

58. *Page dimensions* In Exercise 57, suppose instead that the area of the whole page is to be 54 square inches, with the same margins as given in Exercise 57. Determine the dimensions of the page which gives the maximum printing space area.

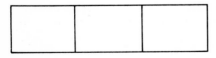

59. *Lot dimensions* A company has 4,800 feet of fencing to enclose a storage lot with three sections of equal area, as indicated in the margin. Determine the maximum area that can be enclosed.

60. *Container cost* A company is designing a rectangular box with a square top and bottom and with a capacity of 50 cubic feet to be used as a shipping container. The material for the top and bottom costs $0.25 per square foot; the more durable material for the sides costs $0.32 per square foot. What are the dimensions of the box with minimal cost?

61. *Profit* The monthly profit (in thousands of dollars) from the sale of x hundred units of a product is

$$P(x) = -x^3 + 9x^2 + 18x - 90$$

Determine the sales level at which profit is maximum.

62. *Cost* The weekly total cost (in thousands of dollars) for producing x hundred units of a product is

$$C(x) = x^3 + 16$$

Determine the weekly production level at which the average cost per unit is minimal.

63. *Cost-revenue-profit* If a company's daily revenue and cost (in dollars) from the production of x hand calculators are

$$C(x) = 0.0002x^3 - 0.04x^2 + 10.90x + 150$$
$$R(x) = 14.50x - 0.007x^2$$

what should the daily production be for maximum profit? What is the maximum daily profit?

64. *Chemical concentration* A chemical is distributed over the positive x-axis in such a way that its concentration is given by

$$C(x) = \frac{x}{x^2 + 25}$$

(A unit on the x-axis is 1 centimeter and the concentration is measured in milligrams per centimeter.) Where is the concentration the greatest?

Identify the steps of the four-step procedure involved in the application of mathematics (Section 1.3) in each case.

65. *Container cost* Exercise 60
66. *Profit* Exercise 61

CHAPTER 13
INTEGRATION

OUTLINE

13.1 Indefinite Integrals
13.2 Some Applications of Indefinite Integrals
13.3 Definite Integrals
13.4 Evaluating Definite Integrals
13.5 Properties of Definite Integrals
13.6 Some Applications of Definite Integrals
13.7 Chapter Summary

Review Exercises

In this chapter we begin consideration of the second basic concept of calculus, the integral. Actually, there are two different types of integrals, *indefinite integrals* and *definite integrals*. Both types will be considered in this chapter. Integrals are used to solve some interesting physical problems.

Suppose, for example, that you want to make an ornament out of a thin sheet of gold in the shape of the area between the graphs of $y = 8 - x^2$ and $y = x^2$ in the coordinate system shown in Figure 13.1. A unit on each axis represents 1 inch. You want to determine the exact area of the part in order to determine its cost.

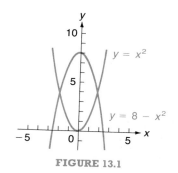

FIGURE 13.1

For areas of such figures as rectangles, triangles, or circles, there are standard formulas that quickly give the answer. But there are no standard formulas for figures such as the one in Figure 13.1. However, we shall see that we can quickly determine the area of the figure to be $21\frac{1}{3}$ square units by using integration.

13.1 INDEFINITE INTEGRALS

The process of determining indefinite integrals is the reverse of that for finding derivatives. That is, it involves finding a function F whose derivative is equal to a given function f. The function

$$F(x) = x^2$$

is an indefinite integral of

$$f(x) = 2x$$

because $F'(x) = 2x = f(x)$. Similarly,

$$G(x) = \frac{x^3}{3}$$

is an indefinite integral of

$$g(x) = x^2$$

because $G'(x) = x^2 = g(x)$. One problem that might suggest itself is that x^2 is not the only function whose derivative is $2x$. Each of the following also has $2x$ as its derivative:

$$x^2 + 3$$
$$x^2 - 7$$
$$x^2 + 4.7$$
$$x^2 - \tfrac{1}{2}$$

Thus each of these is also an indefinite integral of $2x$. In fact, the same can be said of any function of the form $x^2 + C$, where C is an arbitrary constant. (In the illustrations just given, C had the values 3, -7, 4.7, and $-\tfrac{1}{2}$.) The most general function having $2x$ as its derivative, $x^2 + C$, is called *the* indefinite integral of $2x$.

Formally, **the indefinite integral** of a function $f(x)$ is the function $F(x) + C$, where $F'(x) = f(x)$ and C denotes an arbitrary constant. The indefinite integral of x^2 is $(x^3/3) + C$ because the derivative of $x^3/3$ is x^2.

The usual notation for the indefinite integral is

$$\int f(x)\,dx$$

[read, "the integral of $f(x)$"]. With this notation, the definition can be rewritten

$$\int f(x)\,dx = F(x) + C$$

where $F'(x) = f(x)$ and C denotes an arbitrary constant.

Then

$$\int x^2\,dx = \frac{x^3}{3} + C$$

because the derivative of $x^3/3$ is x^2;

$$\int 8x^{-3}\,dx = -4x^{-2} + C$$

13.1 INDEFINITE INTEGRALS

because the derivative of $-4x^{-2}$ is $8x^{-3}$, and

$$\int 2xe^{x^2}\,dx = e^{x^2} + C$$

because the derivative of e^{x^2} is $2xe^{x^2}$.

The process of evaluating integrals, of recapturing a function when its derivative is known, consists for very simple functions of being familiar with and using a few formulas and properties of integrals to be discussed below. Each formula given in Table 13.1 follows from the differentiation formula to its right.

TABLE 13.1

INTEGRATION	DIFFERENTIATION
1. $\int x^n\,dx = \dfrac{x^{n+1}}{n+1} + C,\quad n \neq -1$	$\dfrac{d}{dx}\left(\dfrac{x^{n+1}}{n+1}\right) = x^n$
2. $\int e^x\,dx = e^x + C$	$\dfrac{d}{dx}e^x = e^x$
3. $\int b^x\,dx = \dfrac{b^x}{\ln b} + C,\quad b > 0$	$\dfrac{d}{dx}\left(\dfrac{b^x}{\ln b}\right) = b^x$
4. $\int \dfrac{1}{x}\,dx = \ln x + C,\quad x > 0$	$\dfrac{d}{dx}(\ln x) = \dfrac{1}{x}$

EXAMPLE 1 (a) $\int x^4\,dx$ is in the form of the first type in Table 13.1 with $n = 4$, so that

$$\int x^4\,dx = \frac{x^5}{5} + C$$

(b) $\int x^{-1/3}\,dx$ is also in the form of the first type in Table 13.1 with $n = -\frac{1}{3}$ so that

$$\int x^{-1/3}\,dx = \frac{x^{2/3}}{\frac{2}{3}} + C = \frac{3}{2}x^{2/3} + C$$

(c) $\int 3^x\,dx$ is of the third type in Table 13.1 with $b = 3$ so that

$$\int 3^x\,dx = \frac{3^x}{\ln 3} + C$$

(d) $\int \dfrac{1}{x}\,dx$ is in the exact form of the fourth type in Table 13.1, so that

$$\int \frac{1}{x}\,dx = \ln x + C \qquad x > 0 \qquad \boxed{1}$$

PRACTICE PROBLEM 1

Evaluate the following integrals:

(a) $\int x^7\,dx$ (b) $\int 5^x\,dx$

Answer

(a) $\dfrac{x^8}{8} + C$ (b) $\dfrac{5^x}{\ln 5} + C$

The properties of integrals given in the following two theorems enable us to extend the above four integration formulas to many more different functions. Their proofs follow from the corresponding properties of derivatives.

> **THEOREM 1**
>
> For a constant k
> $$\int kf(x)\,dx = k\int f(x)\,dx$$
>
> That is, the integral of a constant times a function is equal to the constant times the integral of the function.

EXAMPLE 2 Evaluate each of the following:

(a) $\int 4x^2\,dx$ (b) $\int -3e^x\,dx$

Solution (a) The k of Theorem 1 is 4, and by Theorem 1,

$$\int 4x^2\,dx = 4\int x^2\,dx = 4\left(\frac{x^3}{3} + C_1\right) = \frac{4}{3}x^3 + C$$

Note that if C_1 is an arbitrary constant, then so is $4C_1$; the usual procedure is to represent this type of combination simply as the arbitrary constant C.

(b) The k is -3, and by Theorem 1,

$$\int -3e^x\,dx = -3\int e^x\,dx = -3(e^x + C_1) = -3e^x + C$$

▣ **PRACTICE PROBLEM 2**

Evaluate the following integrals:

(a) $\int -5x^7\,dx$ (b) $\int 2e^x\,dx$

Answer

(a) $-\frac{5}{8}x^8 + C$ (b) $2e^x + C$

> **THEOREM 2**
>
> $$\int [f(x) \pm g(x)]\,dx = \int f(x)\,dx \pm \int g(x)\,dx$$
>
> That is, the integral of the sum (difference) of two functions is the sum (difference) of their integrals.

For example, in

$$\int (x^3 - x^{-2})\,dx$$

$f(x)$ is x^3 and $g(x)$ is x^{-2}. By Theorem 2,

$$\int (x^3 - x^{-2})\,dx = \int x^3\,dx - \int x^{-2}\,dx$$
$$= \left(\frac{x^4}{4} + C_1\right) - \left(\frac{x^{-1}}{-1} + C_2\right)$$
$$= \frac{x^4}{4} - \frac{x^{-1}}{-1} + C = \frac{x^4}{4} + \frac{1}{x} + C$$

13.1 INDEFINITE INTEGRALS

Note that if C_1 and C_2 are arbitrary constants, then so are their sum and difference, $C_1 - C_2$ in this last example. Again, the usual procedure is to represent this type of combination simply by the arbitrary constant C.

In

$$\int \left(\sqrt{x} + \frac{1}{x} \right) dx$$

$f(x)$ is $\sqrt{x}$ or $x^{1/2}$ and $g(x)$ is $1/x$. By Theorem 2,

$$\int \left(\sqrt{x} + \frac{1}{x} \right) dx = \int x^{1/2} \, dx + \int \frac{1}{x} \, dx$$

$$= \left(\frac{x^{3/2}}{\frac{3}{2}} + C_1 \right) + (\ln x + C_2)$$

$$= \frac{2}{3} x^{3/2} + \ln x + C \qquad x > 0$$

EXAMPLE 3 Evaluate

$$\int \left(3e^x - \frac{x^3}{2} \right) dx$$

Solution By Theorem 2,

$$\int \left(3e^x - \frac{x^3}{2} \right) dx = \int 3e^x \, dx - \int \frac{x^3}{2} \, dx$$

Applying Theorem 1 to each of the integrals on the right gives

$$\int \left(3e^x - \frac{x^3}{2} \right) dx = 3 \int e^x \, dx - \frac{1}{2} \int x^3 \, dx$$

$$= 3e^x - \frac{1}{2} \left(\frac{x^4}{4} \right) + C = 3e^x - \frac{1}{8} x^4 + C \qquad \boxed{3}$$

EXAMPLE 4 (a) $\int (2t^{1/3} + 4^t) \, dt = \int 2t^{1/3} \, dt + \int 4^t \, dt$

$$= 2 \int t^{1/3} \, dt + \int 4^t \, dt = \frac{2t^{4/3}}{\frac{4}{3}} + \frac{4^t}{\ln 4} + C$$

$$= \frac{3}{2} t^{4/3} + \frac{4^t}{\ln 4} + C$$

---

3 PRACTICE PROBLEM 3

Evaluate the following integrals:

(a) $\int \left(\frac{1}{x} + x^3 \right) dx$

(b) $\int (4x^3 - 3e^x) \, dx$

Answer

(a) $\ln x + \frac{x^4}{4} + C, \quad x > 0$

(b) $x^4 - 3e^x + C$

(b) $\int x^2(x - 2)\,dx = \int (x^3 - 2x^2)\,dx$
$= \int x^3\,dx - 2\int x^2\,dx = \tfrac{1}{4}x^4 - \tfrac{2}{3}x^3 + C$

(c) $\int \dfrac{x^3 + 3x^2}{x^2}\,dx = \int (x + 3)\,dx = \int x^1\,dx + \int 3x^0\,dx$
$= \dfrac{x^2}{2} + 3x + C$

(d) $\int \left(x^2 - 2x + \dfrac{3}{x}\right)dx = \int x^2\,dx - 2\int x\,dx + 3\int \dfrac{1}{x}\,dx$
$= \dfrac{x^3}{3} - x^2 + 3(\ln x) + C \qquad x > 0$ ∎

In part (c) of Example 4, since $x^0 = 1$,
$$3 = 3(1) = 3x^0$$
Therefore, by the first formula in Table 13.1,
$$\int 3\,dx = 3\int x^0\,dx = 3\left(\dfrac{x^1}{1}\right) + C = 3x + C$$
Similarly, for any constant k,
$$\int k\,dx = k\int x^0\,dx = kx + C$$

Part (d) in Example 4 illustrates the fact that the property of integrals in Theorem 2 extends to the integral of the sum or difference of three or more functions.

The symbol $\int$ in
$$\int f(x)\,dx$$
is called the **integral sign,** the function $f(x)$ is called the **integrand,** and the process of evaluating an indefinite integral is called **integrating.** The dx in the notation is called the **differential of x** (we shall discuss differentials further in a later section) and x is called the **variable of integration;** the integral is said to be evaluated with respect to x. In Example 4(a), the integration was performed with respect to t.

Because of the relationship between a function and its indefinite integral, the term **antiderivative** is sometimes used in place of indefinite integral.

The last integration formula given in Table 13.1 is

> 4. $\int \dfrac{1}{x}\,dx = \ln x + C \qquad x > 0$

4 PRACTICE PROBLEM 4

Evaluate the following integrals:

(a) $\int \left(2t^{1/2} - \dfrac{3}{t}\right)dt$

(b) $\int x(x + 1)\,dx$

Answer

(a) $\tfrac{4}{3}t^{3/2} - 3(\ln t) + C, \quad t > 0$

(b) $\dfrac{x^3}{3} + \dfrac{x^2}{2} + C$

13.1 INDEFINITE INTEGRALS

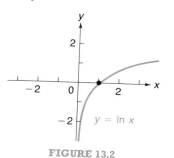

FIGURE 13.2

The restriction $x > 0$ is required because $\ln x$ has no value for $x \leq 0$, as indicated in the graph of the natural log function (Figure 13.2).

However, if $x < 0$, then $-x > 0$ and $\ln(-x)$ does have a value. For example, if $x = -3$, then $\ln x = \ln(-3)$ has no value, but $\ln(-x) = \ln[-(-3)] = \ln 3$ does have a value. Furthermore,

$$\frac{d}{dx}\ln(-x) = \frac{1}{-x}(-1) = \frac{1}{x}$$

so that $\ln(-x)$ is also an indefinite integral of $1/x$ for negative values of x.

To allow for negative values of x, we replace the fourth formula of Table 13.1 by the more general form

$$\textbf{4.} \int \frac{1}{x} dx = \ln|x| + C$$

where

$$|x| = \begin{cases} x & \text{if } x > 0 \\ -x & \text{if } x < 0 \end{cases}$$

This general form, which we shall use hereafter, applies whether x is positive or negative, so the previous restriction on x is no longer necessary.

EXAMPLE 5 (a) $\int \frac{3}{x} dx = 3 \int \frac{1}{x} dx = 3\ln|x| + C$

(b) $\int \left(4z^2 - \frac{6}{z}\right) dz = 4 \int z^2\, dz - 6 \int \frac{1}{z} dz = \frac{4}{3} z^3 - 6\ln|z| + C$

In general, for any expression a, the symbol $|a|$ denotes the **absolute value** of a, which is defined by

$$|a| = \begin{cases} a & a \geq 0 \\ -a & a < 0 \end{cases}$$

For example,

$$|17| = 17, \quad |-17| = -(-17) = 17,$$
$$|-5| = -(-5) = 5, \quad |10| = 10, \quad |0| = 0$$

If $a < 0$, then $-a$ is greater than zero, so that $|a|$ is always greater than zero or equal to zero when a is equal to 0.

EXAMPLE 6 Determine the value of $|9 - x^2|$ for $x = -4, -3, 0,$ and 5.

---

5 PRACTICE PROBLEM 5

Evaluate the following integrals:

(a) $\int \frac{4}{x} dx$ (b) $\int \left(\frac{1}{t} - 6t^2\right) dt$

Answer

(a) $4(\ln|x|) + C$ (b) $\ln|t| - 2t^3 + C$

Solution

For $x = -4$, $|9 - x^2| = |9 - 16| = |-7| = -(-7) = 7$.
For $x = -3$, $|9 - x^2| = |9 - 9| = |0| = 0$.
For $x = 0$, $|9 - x^2| = |9 - 0| = 9$.
For $x = 5$, $|9 - x^2| = |9 - 25| = |-16| = 16$.

6 PRACTICE PROBLEM 6

Evaluate $|x(x - 5)|$ for each value of x.

(a) $x = 4$ (b) $x = -7$ (c) $x = 5$

Answer

(a) 4 (b) 84 (c) 0

EXERCISES

Evaluate each of the following integrals.

1. $\int x^2 \, dx$
2. $\int x^{1/2} \, dx$
3. $\int t^{-4} \, dt$
4. $\int 5 \, dx$
5. $\int -3 \, dt$
6. $\int t \, dt$
7. $\int dx$
8. $\int 5^x \, dx$
9. $\int \sqrt{t} \, dt$
10. $\int \frac{dt}{t}$
11. $\int 3x^7 \, dx$
12. $\int -26 \, dt$
13. $\int 2(3^x) \, dx$
14. $\int \frac{-2}{t} \, dt$
15. $\int \frac{1}{4} e^x \, dx$
16. $\int (x^4 + x^{-4}) \, dx$
17. $\int (x^{-1/2} - e^x) \, dx$
18. $\int (6x - 6^x) \, dx$
19. $\int (2t + 5e^t) \, dt$
20. $\int (4x^3 - 3x^2 + 1) \, dx$
21. $\int (x^{1/2} - x^{1/3} + 2x^{1/4}) \, dx$
22. $\int x^2(x^3 + 6x) \, dx$
23. $\int t\left(4 + \frac{1}{t^2}\right) dt$
24. $\int \frac{x^4 + 6x}{x^2} \, dx$
25. $\int \frac{x^3 - x^2 - x}{x} \, dx$
26. $\int \frac{t^2 - 1}{t + 1} \, dt$
27. $\int \frac{x^2 + x - 6}{x - 2} \, dx$

Evaluate the given expression for the indicated values of x.

28. $|x^2 - 4|$; $x = -1, 0, 2,$ and 5
29. $\left|\frac{x + 1}{x - 1}\right|$; $x = -3, -1, 0,$ and 4
30. $|(x + 1)(x - 2)|$; $x = -2, 0, 2,$ and 10
31. Explain the meaning of the indefinite integral of a function $f(x)$.

13.2 SOME APPLICATIONS OF INDEFINITE INTEGRALS

Just as the derivative gives the rate of change of a function, the slope of the tangent to the graph of a function, the marginal cost for a cost function, and so forth, the definite integral reverses these processes

13.2 SOME APPLICATIONS OF INDEFINITE INTEGRALS

when the rate of change or the slope of the tangent are known. However, in each case some additional information must be available in order to determine the value of the arbitrary constant. The process is illustrated in the next several examples.

EXAMPLE 1 Determine the function whose graph has a tangent with slope $f'(x) = 3x^2 - 1$ at each value of x and whose graph passes through the point $(2, 9)$.

Solution The integral

$$\int (3x^2 - 1)\, dx = x^3 - x + C$$

gives a whole family of functions $f(x) = x^3 - x + C$, each of which has slope given by $3x^2 - 1$. To find the one whose value is 9 when $x = 2$, we substitute these values and solve the resulting equation,

$$9 = 2^3 - 2 + C$$

for C, obtaining $C = 3$. The desired function is $f(x) = x^3 - x + 3$.

PRACTICE PROBLEM 1

Determine the function whose graph has a tangent with slope $f'(x) = 2x^2 + x$ for each value of x and whose graph passes through the point $(0, -3)$.

Answer

$f(x) = \tfrac{2}{3}x^3 + \tfrac{1}{2}x^2 - 3$

EXAMPLE 2 Determine the cost function if the marginal cost for a company is given by $C'(x) = \sqrt{x}$, where x is the number of units produced and fixed costs are $900.

Solution The integral

$$\int \sqrt{x}\, dx = \tfrac{2}{3}x^{3/2} + C$$

gives a whole family of cost functions, each of which has a marginal cost of $\sqrt{x}$. Fixed costs are the costs incurred regardless of the number of items produced, whether it is 0, 10, 200, 1,000, or any other positive number. In particular, if 0 items are produced, the fixed costs are the total costs. We use this fact to determine the value of C by setting $x = 0$ and solving the equation

$$900 = \tfrac{2}{3}0^{3/2} + C$$

for C, obtaining $C = 900$. The cost function then is

$$C(x) = \tfrac{2}{3}x^{3/2} + 900$$

PRACTICE PROBLEM 2

Determine the cost function if the marginal cost is $C'(x) = 3$, where x is the number of units produced, and fixed costs are $2,560.

Answer

$C(x) = 3x + 2,560$

EXAMPLE 3 There are currently 10,000 citizens of voting age in a small city. Demographics indicate that over the next 5 years, the voting population will change at the rate of

$$f'(t) = 1.8t - 0.6t^2$$

where $f(t)$ denotes citizens of voting age, in thousands, and t denotes time in years. How many citizens of voting age will there be 4 years from now?

Solution The integral

$$\int (1.8t - 0.6t^2)\, dt = 0.9t^2 - 0.2t^3 + C$$

except for the constant C, gives the number of citizens of voting age for any number of years t, $0 \le t \le 5$. Using the fact that currently ($t = 0$) there are 10,000 citizens of voting age and that the citizen count is given in thousands, we see that C must equal 10. Consequently,

$$f(t) = 0.9t^2 - 0.2t^3 + 10 \qquad 0 \le t \le 5$$

and $f(4) = 11.6$. There will be 11,600 citizens of voting age 4 years from now.

3 PRACTICE PROBLEM 3

In Example 3, what would the voting population be in 4 years if there were currently 11,550 citizens of voting age?

Answer

13,150

EXERCISES

Evaluate the given integrals.

1. $\int \dfrac{3}{x}\, dx$

2. $\int \left(x - \dfrac{1}{x}\right) dx$

3. $\int \left(4^t + \dfrac{2}{t}\right) dt$

4. $\int (1 + x + x^2)\, dx$

5. $\int (2x^{-3} + 5)\, dx$

6. $\int (x^{-2} - x^{-1})\, dx$

7. $\int \left(\sqrt{x} + \dfrac{4}{\sqrt{x}}\right) dx$

8. $\int (2x^{1/3} + 3x^{-2/3})\, dx$

9. $\int \dfrac{1}{3}\left(\dfrac{1}{x} - 1\right) dx$

10. $\int -2(e^x + 4)\, dx$

11. $\int t\left(\dfrac{1}{t} + \dfrac{1}{t^4}\right) dt$

12. $\int (x + 1)(x - 1)\, dx$

13. $\int \dfrac{3z^3 + 3}{z}\, dz$

14. $\int \dfrac{x^3 - 2x^2 + 3x^{1/3}}{x^{1/3}}\, dx$

15. $\int dx$

16. $\int 0\, dx$

Determine the function whose graph has the given slope.

17. $f'(x) = 2x - 1$; passes through the point $(0, 3)$

18. $f'(x) = \sqrt{x}$; passes through the point $(1, -\tfrac{1}{3})$

19. $y' = 3e^x$; passes through the point $(0, 3)$

20. $y' = \dfrac{1}{x} + 1$; passes through the point $(1, -4)$

21. $g'(t) = 2(t^3 - 3t)$; passes through the point $(-1, -\tfrac{3}{2})$

22. $g'(x) = \dfrac{x + 2}{4x^3}$; passes through the point $(\tfrac{1}{2}, -\tfrac{3}{2})$

Cost Each of the following gives the marginal cost for producing x units of a product. For each, determine the cost function $C(x)$.

23. $C'(x) = 5$ (in dollars; fixed costs equal $15)
24. $C'(x) = 4x + 16$ (in dollars; zero fixed costs)
25. $C'(x) = 4$ (in hundreds of dollars; fixed costs equal $300)
26. $C'(x) = \dfrac{x^3}{3} - 18x + 27$ (in thousands of dollars; fixed costs equal $4,500)

Demand Each of the following gives the marginal demand in terms of the price x (in dollars). For each, determine the demand function $D(x)$.

27. $D'(x) = -2x$ (in hundreds of units; demand is for 5,100 units when the price is $7.00)
28. $D'(x) = \dfrac{-4}{x}$ (in thousands of units; demand disappears when the price reaches $1.00)

Revenue Each of the following gives the marginal revenue (in dollars) from the sale of x items. For each determine the revenue function $R(x)$. (Note: Revenue is zero when zero units are sold.)

29. $R'(x) = 100$
30. $R'(x) = 6x + 6$

31. *Cost-revenue-profit* The marginal cost (in thousands of dollars) for manufacturing and marketing x hundred units is

$$C'(x) = 2x + 70$$

with fixed costs equal to $1,500.
The marginal revenue (in thousands of dollars) from the sale of x hundred units is given by

$$R'(x) = 8x + 10$$

(a) Determine the cost function $C(x)$.
(b) Determine the revenue function $R(x)$.
(c) Determine the profit function $P(x)$.

32. *Resource depletion* The water currently used from a lake is estimated to amount to 100 million gallons a month. Water usage (in millions of gallons) for the next 2 years is expected to increase at the rate of

$$f'(x) = 2 + 0.01x$$

after x months. Determine the monthly water usage a year from now.

33. *Free-fall* An object dropped from an airplane 1,600 feet above the ground falls at the rate of

$$v(t) = 32t \quad \text{(feet per second)}$$

where t is given in seconds. If the object has fallen 1,024 feet after 8 seconds, how long after it is dropped will the object hit the ground?

34. *Chemical reaction* The end product of a chemical reaction is produced at the rate of

$$\dfrac{\sqrt{t} - 1}{t} \quad \text{(milligrams per minute)}$$

for $1 \leq t \leq 10$. If the reaction is started at time $t = 1$, determine the amount produced during the first 3 minutes of the reaction.

35. *Psychology* A psychologist working with severely handicapped individuals has determined that the time required for a handicapped person to perform a task originally taking 12 minutes decreased at the rate of $(0.9)^t$ minutes after t minutes of continuous external support and motivation. How long would it take such a person to complete the task after 10 minutes of external support and motivation?

13.3 DEFINITE INTEGRALS

In this section we consider the second type of integrals, definite integrals, discussing their definition and their geometric interpretation as the area under a curve.

Suppose we want to determine the area under the graph of

$$f(x) = \frac{x^2}{4} + 1$$

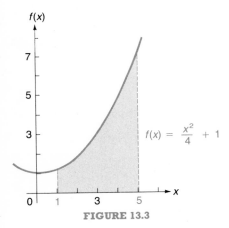

FIGURE 13.3

and above the x-axis from $x = 1$ to $x = 5$ (Figure 13.3). Since this is not a standard geometric figure, the usual formulas for calculating area do not apply. A different method for calculating the area must be developed.

We start by approximating the area with that of a series of rectangles with bases on the interval [1, 5] and heights determined by values of the function for particular values of x.

Suppose we start with two such rectangles, as shown in Figure 13.4. The width of the first rectangle is

$$\Delta x_1 = 3 - 1 = 2$$

and that of the second is

$$\Delta x_2 = 5 - 3 = 2$$

(The symbol Δ is the capital Greek letter delta; it is the standard notation used for indicating the width of such rectangles.)

The height of the first rectangle is $f(2) = 2$ and that of the second is $f(4) = 5$. The sum of the areas of the two rectangles is then

$$f(2)\Delta x_1 + f(4)\Delta x_2 = 2(3 - 1) + 5(5 - 3) = 4 + 10 = 14$$

or 14 square units. This sum gives a first approximation to the desired area. It differs from the true area by an error equal to the net effect of the areas of the shaded regions in Figure 13.5. The true area, as we shall see later, is $\frac{43}{3}$ square units. The error in our approximation then is $\frac{43}{3} - 14 = \frac{1}{3}$ square unit.

If we increase the number of rectangles, we might expect the approximations to be better, since the corresponding shaded areas would shrink

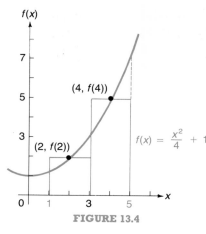

FIGURE 13.4

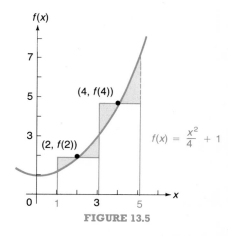

FIGURE 13.5

in size. In Figure 13.6, four rectangles are used. The widths of the bases of the four rectangles are

$$\Delta x_1 = 2 - 1 = 1 \quad \Delta x_2 = 3 - 2 = 1$$
$$\Delta x_3 = 4 - 3 = 1 \quad \Delta x_4 = 5 - 4 = 1$$

and the respective heights are $f(1.5)$, $f(2.5)$, $f(3.5)$, and $f(4.5)$. The sum of the areas of the four rectangles is

$$f(1.5)\Delta x_1 + f(2.5)\Delta x_2 + f(3.5)\Delta x_3 + f(4.5)\Delta x_4$$
$$= (1.5625)(1) + (2.5625)(1) + (4.0625)(1) + (6.0625)(1)$$
$$= 14.25$$

square units. The error now is $\frac{43}{3} - 14.25$, or about 0.08 square unit.

If we use six rectangles, as indicated in Figure 13.7, the sum of the areas of the six rectangles is

$$f(\tfrac{4}{3})\Delta x_1 + f(2)\Delta x_2 + f(\tfrac{8}{3})\Delta x_3 + f(\tfrac{10}{3})\Delta x_4 + f(4)\Delta x_4 + f(\tfrac{14}{3})\Delta x_6$$
$$= (1.444)(\tfrac{2}{3}) + (2)(\tfrac{2}{3}) + (2.777)(\tfrac{2}{3}) + (3.777)(\tfrac{2}{3}) + 5(\tfrac{2}{3}) + (6.444)(\tfrac{2}{3})$$
$$\approx 14.296$$

that is, 14.296 square units. The error is $\frac{43}{3} - 14.296 \approx 0.037$ square unit.

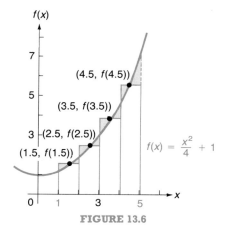

FIGURE 13.6

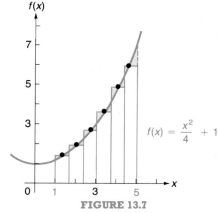

FIGURE 13.7

As the number of rectangles gets larger and larger, the shaded regions seem to disappear, the error approaches zero, and the sum of the areas of the rectangles approach the area under the curve. For an arbitrary positive integer n, we can write the sum of the areas of n rectangles as

$$f(x_1^*)\Delta x_1 + f(x_2^*)\Delta x_2 + f(x_3^*)\Delta x_3 + \cdots + f(x_{n-1}^*)\Delta x_{n-1} + f(x_n^*)\Delta x_n$$

where x_1^* is the value of x used to determine the height of the first rectangle, x_2^* is the value used to determine the height of the second rectangle, and so forth.

This general sum can be written more briefly as

$$\sum_{i=1}^{n} f(x_i^*)\Delta x_i$$

This expression means that i (the index of summation) is to have each of the values $1, 2, 3, \ldots, n$ in the expression

$$f(x_i^*)\Delta x_i$$

and the results are to be added—that is, the sum represented by this notation is exactly the same as that considered in the last paragraph.

The area under the curve, which is obtained by letting n become larger and larger, is the limit of this sum as n becomes unbounded; that is,

$$\text{Area} = \lim_{n \to \infty} \sum_{i=1}^{n} f(x_i^*)\Delta x_i$$

where $n \to \infty$ indicates that the number of rectangles becomes unbounded. At the same time, this limit is the *definite integral* of the function $f(x) = (x^2/4) + 1$ on the interval from $x = 1$ to $x = 5$.

Before considering some further examples of calculating the area under a curve, we give the formal definition of a definite integral on an interval $[a, b]$. The definition is essentially a description of the process just used to calculate the area under the curve except that it is not given an area interpretation. The formal definition is as follows.

Suppose that the function f is continuous on the interval $[a, b]$. For each positive integer n, partition the interval from $x = a$ to $x = b$ into n subintervals by choosing points $x_0, x_1, x_2, \ldots, x_n$ such that $x_0 = a < x_1 < x_2 < \cdots < x_{n-1} < x_n = b$ (Figure 13.8):

FIGURE 13.8

Pick a point in each subinterval by choosing points $x_1^*, x_2^*, \ldots, x_n^*$ such that $x_0 \le x_1^* \le x_1, x_1 \le x_2^* \le x_2, \ldots, x_{n-1} \le x_n^* \le x_n$ (Figure 13.9):

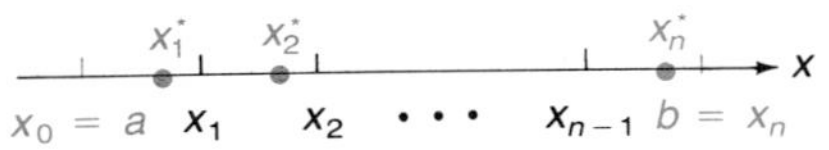

FIGURE 13.9

Then form the sum

$$\sum_{i=1}^{n} f(x_i^*)\Delta x_i$$

where $\Delta x_1 = x_1 - x_0, \Delta x_2 = x_2 - x_1, \ldots, \Delta x_n = x_n - x_{n-1}$. The **definite integral** of the function f on the interval $[a, b]$, written

$$\int_a^b f(x)\,dx$$

is the limit of this sum as n becomes unbounded in such a way that each Δx_i approaches zero; that is:

$$\int_a^b f(x)\,dx = \lim_{n \to \infty} \sum_{i=1}^{n} f(x_i^*)\Delta x_i$$

If $\int_a^b f(x)\,dx$ exists (has a value), f is said to be **integrable** on the interval $[a, b]$. A theorem of calculus, which we shall not attempt to prove here, states that any function continuous on an interval is integrable on that interval. Since this is the only type of function we shall consider, the question of whether or not a given integral exists will never be a problem for us. Our concern will be the evaluation of integrals, which is the subject of the next section (and which will be much easier than the definition might suggest!).

As with the illustration in the beginning of this section, for a function f positive on the interval $[a, b]$, geometrically the integral $\int_a^b f(x)\,dx$ is the area between the graph of f and the x-axis over the interval $[a, b]$. If f is everywhere negative on $[a, b]$, the integral will be the negative of this area, since each $f(x_i^*)$ in its calculations will be negative.

EXAMPLE 1 Approximate the area between the graph of the function $f(x) = x^3 + 1$ and the x-axis over the interval $[-1, 1]$ using two rectangles of equal width whose heights are determined by the value of the function at the right endpoints of the bases. Determine the error in the approximation by comparing the result with the true area which, as we shall see later, is 2 square units.

Solution   The length of the interval $[-1, 1]$ is $1 - (-1) = 2$ units long. If the rectangles are to have equal widths, each must be $2 \div 2 = 1$ unit wide. The points of subdivision are $x = -1, 0$ and 1.

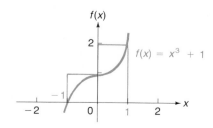

If the heights of the rectangles are equal to the values of f at the right endpoints of the bases, the sum of their areas is

$$f(0)\Delta x_1 + f(1)\Delta x_2 = 1(1) + 2(1) = 3 \text{ square units}$$

The error in this approximation is $3 - 2 = 1$ square unit.

EXAMPLE 2 Repeat the process in Example 1 using four rectangles.

Solution In this case, the width of each base is $2 \div 4 = \frac{1}{2}$ unit. The points of subdivision are $x = -1, -\frac{1}{2}, 0, \frac{1}{2},$ and 1.

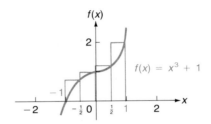

The sum of the areas of the four rectangles is

$$f(-\tfrac{1}{2})\Delta x_1 + f(0)\Delta x_2 + f(\tfrac{1}{2})\Delta x_3 + f(1)\Delta x_4$$
$$= \tfrac{7}{8}(\tfrac{1}{2}) + 1(\tfrac{1}{2}) + \tfrac{9}{8}(\tfrac{1}{2}) + 2(\tfrac{1}{2}) = 2.5 \text{ square units}$$

The error now is $2.5 - 2 = 0.5$ square unit.

Eventually we shall use definite integrals to evaluate the exact values of areas such as the one considered in Examples 1 and 2. However, we could also obtain the exact value by letting the number of rectangles get arbitrarily large, as the result obtained in this manner is precisely the definite integral of the function f over the interval involved. This latter method becomes so cumbersome even in the most simple cases that we gladly postpone the determination of the true area until we have learned to evaluate definite integrals in the next section.

PRACTICE PROBLEM 1

Repeat the process of Example 1 to approximate the area between the graph of $f(x) = x^2$ and the x-axis over the interval [1, 3] using the given number of rectangles. In each case, determine the error in the approximation by comparing the result with the true area, which is $8\frac{2}{3}$ square units.

(a) 2 rectangles   (b) 4 rectangles

Answer

(a) 13 square units; error is $4\frac{1}{3}$ square units

(b) 10.75 square units; error is $2\frac{1}{12}$ square units

EXERCISES

For each of the following, approximate the area between the graph of the function and the x-axis over the given interval.

(a) *First, use two rectangles of equal width whose heights are determined by the value of the function at the right endpoint of the bases. Determine the error in the approximation by comparing the result with the exact area, which is given.*

(b) Repeat the process in part (a) with four rectangles.

1. $f(x) = \dfrac{x^2}{4} + 1$; [2, 4]; exact area: $6\frac{2}{3}$ square units
2. $f(x) = x^2 + 2$; [−1, 1]; exact area: $4\frac{2}{3}$ square units
3. $f(x) = 4 - x^2$; [0, 2]; exact area: $5\frac{1}{3}$ square units
4. $f(x) = 4 - x^2$; [−2, 2]; exact area: $10\frac{2}{3}$ square units
5. $g(x) = x + 2$; [1, 5]; exact area: 20 square units
6. $f(x) = 6 - 2x$; [−1, 2]; exact area: 15 square units
7. $y = 3$; [0, 3]; exact area: 9 square units
8. $f(x) = e^x$; [0.6, 1]; exact area: 0.9 square unit
9. $g(x) = \dfrac{1}{x}$; [1, 5]; exact area: 1.6 square units

By decomposing the area of interest into a triangle and a rectangle, verify that the exact area given is correct in each of the indicated exercises.

10. Exercise 5
11. Exercise 6
12. Explain the meaning of the definite integral of a function f on an interval $[a, b]$.

13.4 EVALUATING DEFINITE INTEGRALS

The two major concepts of calculus, the derivative and the definite integral, are both limits, but they are limits of completely different types. On the surface it would seem that there is no relationship between these two concepts. The theorem considered in this section indicates that this is not the case; in fact, it shows that there is a fundamental relationship between the two. At the same time, it gives a method for evaluating definite integrals that is much easier than applying the definition directly. The proof of the theorem, one of the most important in mathematics, is given in more advanced courses in calculus.

THEOREM 3   THE FUNDAMENTAL THEOREM OF CALCULUS

If the function $f(x)$ is continuous on the interval $[a, b]$, then

$$\int_a^b f(x)\,dx = F(b) - F(a)$$

where $F(x)$ is an indefinite integral of $f(x)$.

The meaning of $F(x)$ being an indefinite integral of $f(x)$ is that $F'(x) = f(x)$; consequently, the fundamental theorem states that, under the continuity condition,

$$\int_a^b f(x)\,dx = F(b) - F(a)$$

where $F'(x) = f(x)$. This establishes the relationship between derivatives and definite integrals to which we referred.

We illustrate the use of the fundamental theorem for evaluating integrals by evaluating

$$\int_3^5 (2x - 1)\,dx$$

We first find the value of the indefinite integral

$$\int (2x - 1)\,dx = x^2 - x + C$$

The fundamental theorem indicates that any indefinite integral can be used, so we take $C = 0$; then

$$F(x) = x^2 - x$$

$F(5) = 20$, and $F(3) = 6$. Consequently,

$$\int_3^5 (2x - 1)\,dx = 20 - 6 = 14$$

A short way of writing $F(b) - F(a)$, which simplifies notation in evaluating definite integrals, is to use the notation

$$F(x)\Big|_a^b = F(b) - F(a)$$

The evaluation of the integral just given can then be written

$$\int_3^5 (2x - 1)\,dx = (x^2 - x)\Big|_3^5 = 20 - 6 = 14$$

The numbers a and b in

$$\int_a^b f(x)\,dx$$

are called the **limits of integration**, with a the **lower limit** of integration and b the **upper limit** of integration.

EXAMPLE 1 Evaluate $\int_0^2 (x^2 + 3)\,dx$.

Solution We first need the value of the corresponding indefinite integral

$$\int (x^2 + 3)\,dx = \frac{x^3}{3} + 3x + C$$

Then

$$\int_0^2 (x^2 + 3)\,dx = \left(\frac{x^3}{3} + 3x\right)\Big|_0^2 = \frac{26}{3} - 0$$

$$= \frac{26}{3}$$

1 PRACTICE PROBLEM 1

Evaluate $\int_1^2 (x - 1)\,dx$.

Answer

$0 - (-\tfrac{1}{2}) = \tfrac{1}{2}$

13.4 EVALUATING DEFINITE INTEGRALS

EXAMPLE 2 Evaluate $\int_1^4 (x + \sqrt{x})\,dx$.

Solution The corresponding indefinite integral is

$$\int (x + \sqrt{x})\,dx = \frac{x^2}{2} + \frac{2}{3}x^{3/2} + C$$

Therefore,

$$\int_1^4 (x + \sqrt{x})\,dx = \left(\frac{x^2}{2} + \frac{2}{3}x^{3/2}\right)\bigg|_1^4 = \frac{40}{3} - \frac{7}{6} = \frac{73}{6}$$

2 PRACTICE PROBLEM 2

Evaluate $\int_0^1 \sqrt{x}\,dx$.

Answer

$\frac{2}{3}$

EXAMPLE 3 (a) $\int_{-1}^{1} 3e^t\,dt = 3e^t\bigg|_{-1}^{1} = 3e^1 - 3e^{-1}$, or approximately 7.05

(b) $\int_1^e -\frac{2}{x}\,dx = -2(\ln|x|)\bigg|_1^e = -2 - 0 = -2$

3 PRACTICE PROBLEM 3

Evaluate each of the following integrals:

(a) $\int_{-e}^{-1} -\frac{2}{x}\,dx$ (b) $\int_{-1}^{1} 2(3x+1)\,dx$

Answer

(a) 2 (b) 4

EXERCISES

Evaluate each of the following integrals:

1. $\int_1^3 2x\,dx$

2. $\int_{-1}^{2} -6x^2\,dx$

3. $\int_0^3 \frac{1}{3}x^{1/3}\,dx$

4. $\int_4^9 3\sqrt{x}\,dx$

5. $\int_1^5 x^{-2}\,dx$

6. $\int_1^4 -x^{-1/2}\,dx$

7. $\int_2^3 \frac{4}{t^3}\,dt$

8. $\int_1^5 \frac{-5}{t^2}\,dt$

9. $\int_1^e \frac{dt}{t}$

10. $\int_{-2}^{0} -(3^x)\,dx$

11. $\int_1^3 2^x\,dx$

12. $\int_{-3}^{-1} 4\,dt$

13. $\int_{\ln 1}^{\ln 3} e^y\,dy$

14. $\int_{-1}^{2} (x-2)\,dx$

15. $\int_0^{1.5} (5x+1)\,dx$

16. $\int_2^4 (3x^2 + 2x)\,dx$

17. $\int_{-2}^{2} (3x - 6x^2)\,dx$

18. $\int_1^2 \left(t + \frac{1}{t}\right)\,dt$

19. $\int_0^{\ln 2} (2e^x - 3x)\,dx$

20. $\int_1^9 (x^{-1/2} + 4x)\,dx$

21. $\int_0^4 2(\sqrt{x} - \frac{1}{2})\,dx$

22. $\int_0^1 4(3^x + x)\,dx$

23. $\int_3^5 -2(x^{-2} + 1)\,dx$

24. $\int_{-1}^{2} x(3x-2)\,dx$

25. $\int_1^5 2x(3x^{-3} - 4)\,dx$

26. $\int_{-4}^{1} (3t^2 + t + 1)\,dt$

27. $\int_{-1}^{3} (-x^2 - 2x + 4)\,dx$

28. $\int_{-2}^{2} \frac{y^3 - 5}{2}\,dy$

29. $\int_2^6 \frac{4+t}{t}\,dt$

30. $\int_{0.5}^{1.5} 2\left(\frac{x^2 - 1}{x+1}\right)\,dx$

31. $\int_{25}^{100} dt$

32. $\int_1^{100} 0\,dt$

Find $F'(x)$ for each of the following functions:

33. $F(x) = \int_0^x (t^2 - 2t)\, dt$
34. $F(x) = \int_1^x (e^t + 3)\, dt$
35. State the fundamental theorem of calculus.

13.5 PROPERTIES OF DEFINITE INTEGRALS

As indicated in Section 13.3, if the function f is positive on the interval $[a, b]$, then the definite integral

$$\int_a^b f(x)\, dx$$

gives the area between the graph of f and the x-axis over the interval $[a, b]$. All the exact areas given in that section were determined by means of definite integrals.

EXAMPLE 1 Determine the area between the graph of the function $f(x) = \frac{1}{4}(e^x - 1)$ and the x-axis over the interval $[0, 3]$.

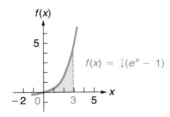

Solution The area is

$$\int_0^3 \frac{1}{4}(e^x - 1)\, dx = \frac{1}{4}(e^x - x)\Big|_0^3 = 4.27 - 0.25 = 4.02 \text{ square units} \quad \boxed{1}$$

If the function is negative for all values of x in the interval $[a, b]$, the graph of the function lies below the x-axis. In this case, the area between the graph of the function and the x-axis *below* the interval

---

1 PRACTICE PROBLEM 1

Determine the area between the graph of the function $f(x) = x + 2$ and the x-axis over the interval $[1, 3]$.

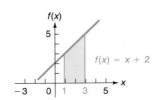

Answer

8 square units

13.5 PROPERTIES OF DEFINITE INTEGRALS

$[a, b]$ is given by the negative of the corresponding integral; that is, the area is given by

$$-\int_a^b f(x)\,dx$$

EXAMPLE 2 Determine the area between the graph of the function $f(x) = x^2 - 4$ below the interval $[-1, 2]$.

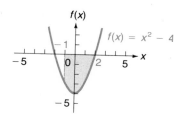

Solution Because the graph lies below the x-axis, the area is

$$-\int_{-1}^{2}(x^2-4)\,dx = -\left(\frac{x^3}{3}-4x\right)\Big|_{-1}^{2} = -\left(-\frac{16}{3}-\frac{11}{3}\right) = 9 \text{ square units}$$

PRACTICE PROBLEM 2

Determine the total area bounded by the graph of the function $f(x) = x^2 - 4$ in Example 2 and the x-axis.

Answer

$-\int_{-2}^{2}(x^2-4)\,dx = \frac{32}{3}$ square units

The definition of the definite integral

$$\int_a^b f(x)\,dx$$

requires that $a < b$. Consequently, if $a > b$, the integrals

$$\int_a^b f(x)\,dx \quad \text{and} \quad \int_a^a f(x)\,dx$$

in which the upper limit of integration is greater than or equal to the lower limit, do not have any meaning according to the definition. In order to give meaning to these integrals, their values are defined as follows:

$$\int_a^b f(x)\,dx = -\int_b^a f(x)\,dx \quad \text{and} \quad \int_a^a f(x)\,dx = 0$$

The use of these definitions is illustrated in the next example.

EXAMPLE 3 Evaluate the integrals

$$\int_1^{-2}(x+2)\,dx \quad \text{and} \quad \int_3^3 (x^6 - 4x^{-7})\,dx$$

3 PRACTICE PROBLEM 3

Evaluate each of the following integrals:

(a) $\int_2^1 (x - 3x^2)\,dx$

(b) $\int_4^4 (x - 3x^2)\,dx$

Answer

(a) $-\frac{11}{2}$ (b) 0

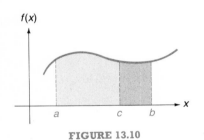

FIGURE 13.10

Solution In the first of these integrals, the upper limit of integration is less than the lower limit. By the above definition,

$$\int_1^{-2} (x+2)\,dx = -\int_{-2}^1 (x+2)\,dx = -[\tfrac{5}{2} - (-2)] = -\tfrac{9}{2}$$

In the second integral, both limits of integration are the same. By the above definition,

$$\int_3^3 (x^6 - 4x^{-7})\,dx = 0$$

We conclude this section with some basic properties of definite integrals, given in Theorem 4.

THEOREM 4

If k is a constant and c is a number in the domain of f, then the following properties hold:

(i) $\int_a^b k \cdot f(x)\,dx = k \cdot \int_a^b f(x)\,dx$

(ii) $\int_a^b [f(x) \pm g(x)]\,dx = \int_a^b f(x)\,dx \pm \int_a^b g(x)\,dx$

(iii) $\int_a^b f(x)\,dx = \int_a^c f(x)\,dx + \int_c^b f(x)\,dx$

Properties (i) and (ii) follow from the corresponding properties of indefinite integrals. Property (iii) states that the integral of a function over the interval $[a, b]$ is the sum of the integral of the function over the interval $[a, c]$ and the integral over the interval $[c, b]$. This latter property might seem obvious from geometric considerations (Figure 13.10). The area between the graph of the function and the x-axis over the interval $[a, b]$ is the sum of the corresponding areas over the intervals $[a, c]$ and $[c, b]$. The area over the whole interval $[a, b]$ is given by the integral on the left in property (iii); the areas over the subintervals are given by the integrals on the right.

We illustrate each of these three properties by means of an example. For property (i), consider the integral

$$\int_1^2 3(2x + 1)\,dx$$

The k of property (i) is 3 and $f(x)$ is $2x + 1$. If we evaluate this integral in the form on the left in property (i), we get

$$\int_1^2 3(2x + 1)\,dx = (3x^2 + 3x)\Big|_1^2 = 18 - 6 = 12$$

If we evaluate it in the form on the right, we get

$$3\int_1^2 (2x+1)\,dx = 3\left[(x^2+x)\Big|_1^2\right] = 3(6-2) = 3(4) = 12$$

In each case we get the same result, as property (i) guarantees.

For property (ii), consider the integral

$$\int_{-1}^3 (2x+3x^2)\,dx$$

The $f(x)$ of property (ii) is $2x$ and $g(x)$ is $3x^2$. If we evaluate this integral in the form on the left of property (ii), we get

$$\int_{-1}^3 (2x+3x^2)\,dx = (x^2+x^3)\Big|_{-1}^3 = 36 - 0 = 36$$

If we evaluate it in the form on the right, we get

$$\int_{-1}^3 2x\,dx + \int_{-1}^3 3x^2\,dx = x^2\Big|_{-1}^3 + x^3\Big|_{-1}^3$$
$$= (9-1) + [27-(-1)] = 8 + 28 = 36$$

In each case we get the same result, as property (ii) guarantees.

Finally, for property (iii), consider the integral

$$\int_1^5 (2x-2)\,dx$$

with $c = 3$. If we evaluate this integral in the form on the left in property (iii), we get

$$\int_1^5 (2x-2)\,dx = (x^2-2x)\Big|_1^5 = 15 - (-1) = 16$$

If we evaluate it in the form on the right, we get

$$\int_1^3 (2x-2)\,dx + \int_3^5 (2x-2)\,dx = (x^2-2x)\Big|_1^3 + (x^2-2x)\Big|_3^5$$
$$= [3-(-1)] + (15-3) = 4 + 12 = 16$$

In each case we get the same result, as property (iii) guarantees.

---

4 PRACTICE PROBLEM 4

Evaluate each side of the equation

$$\int_1^3 2(x-4)\,dx = 2\int_1^3 (x-4)\,dx$$

to verify that they are in fact equal.

Answer

Each side equals -8.

5 PRACTICE PROBLEM 5

Evaluate each side of the equation

$$\int_0^2 (x^2-3x)\,dx = \int_0^2 x^2\,dx - \int_0^2 3x\,dx$$

to verify that they are in fact equal.

Answer

Each side equals $-\frac{10}{3}$.

6 PRACTICE PROBLEM 6

Evaluate each side of the equation

$$\int_{-1}^3 (x+4)\,dx = \int_{-1}^1 (x+4)\,dx + \int_1^3 (x+4)\,dx$$

to verify that they are in fact equal.

Answer

Each side equals 20.

EXERCISES

Determine the area between the graph of each of the following functions and the x-axis over the indicated interval:

1. $f(x) = 6x + 2$; $[2, 5]$
2. $f(x) = 6x^2 + 4x$; $[0, 1]$
3. $f(x) = x^3 - 1$; $[1, 4]$
4. $g(x) = \dfrac{1}{x^2}$; $[-3, -0.5]$

5. $y = 2x^{1/3} + 1$; [1, 8] 6. $f(x) = 3^x$; [−1, 3]

7. $g(x) = x + \dfrac{1}{\sqrt{x}}$; [4, 9] 8. $y = x + \dfrac{1}{x}$; [2, 6]

Verify that the areas given in each of the indicated exercises from Section 13.3 are correct.

9. Exercise 1 10. Exercise 2 11. Exercise 3
12. Exercise 4 13. Exercise 5 14. Exercise 6
15. Exericse 7 16. Exercise 8 17. Exericse 9

Determine the area between the graph of each of the following functions and the x-axis below the indicated interval:

18. $f(x) = -x^2 - 1$; [−1, 1] 19. $f(x) = 4x - 8$; [0, 2]

20. $y = (-3)2^x$; [2, 4] 21. $g(x) = \dfrac{2}{x}$; [−5, −3]

Graph each of the following functions. Then determine the area bounded by the graph and the x-axis.

22. $y = -x^2 + 4$ 23. $y = -x^2 + 6x$
24. $y = x^2 - 9$ 25. $y = x^3 - 3x$

Evaluate each of the following integrals:

26. $\displaystyle\int_4^2 (6x^2 - 4x)\,dx$ 27. $\displaystyle\int_1^{-1} x^{-1/3}\,dx$

28. $\displaystyle\int_{-5}^{-5} x\,dx$ 29. $\displaystyle\int_{20}^{20} \ln x\,dx$

Rewrite each of the following integrals using property (i) of Theorem 4; evaluate the rewritten integrals:

30. $\displaystyle\int_0^2 -4(x+3)\,dx$ 31. $\displaystyle\int_{-2}^2 (6x^2 - 3x)\,dx$

Rewrite each of the following integrals using property (ii) of Theorem 4; evaluate the rewritten integrals:

32. $\displaystyle\int_0^1 (2x^5 - 7x^{4/3})\,dx$ 33. $\displaystyle\int_5^7 4(2-3x)\,dx$

Rewrite each of the following integrals using property (iii) of Theorem 4 and the indicated value of c; evaluate the rewritten integrals:

34. $\displaystyle\int_{-4}^3 (3-x^2)\,dx$; $c = 0$ 35. $\displaystyle\int_1^5 (x^{-2} + x^{-3})\,dx$; $c = 3$

State the following in words without using any mathematical symbols:

36. Property (i) of Theorem 4 37. Property (ii) of Theorem 4

13.6 SOME APPLICATIONS OF DEFINITE INTEGRALS

We first extend the use of definite integrals for evaluating the area under a curve to evaluating the area between two curves. In Figure 13.11, suppose we want to determine the area between the graphs of

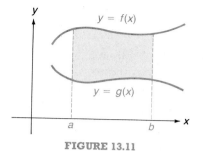

FIGURE 13.11

$y = f(x)$ and $y = g(x)$ over the interval $[a, b]$. The indicated area is the same as the total area under the graph of $y = f(x)$ over the interval $[a, b]$ minus the corresponding area under the graph of $y = g(x)$. That is, the indicated area is given by the difference

$$\int_a^b f(x)\,dx - \int_a^b g(x)\,dx$$

Combining these two integrals into a single integral gives

$$\text{Area} = \int_a^b [f(x) - g(x)]\,dx$$

This last equation indicates that the area is given by the integral of the difference obtained by subtracting the values of the function corresponding to the lower curve from those corresponding to the higher curve.

EXAMPLE 1   Determine the area between the graphs of $y = 2x + 1$ and $y = x^2$ over the interval from $x = 1$ to $x = 2$.

Solution   The graphs of the given equations indicate that the higher curve over the interval $[1, 2]$ corresponds to $y = 2x + 1$ and the lower to $y = x^2$.

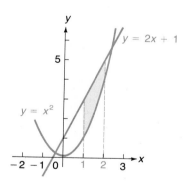

Consequently, the area is given by

$$\int_1^2 [(2x + 1) - x^2]\,dx = \left(x^2 + x - \frac{x^3}{3}\right)\bigg|_1^2 = \frac{5}{3} \text{ square units}$$

PRACTICE PROBLEM 1

Determine the area between the graphs of $y = 4 - x^2$ and $y = x + 2$ over the interval $[-2, 1]$.

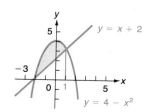

Answer

$\frac{9}{2}$ square units

Suppose the price for a product when the demand is for x units is given by $y = P(x)$. Suppose also that the price for the product when the supply is x units is given by $y = Q(x)$. The point (x_0, y_0) where the graphs of the equations $y = P(x)$ and $y = Q(x)$ intersect is called the **equilibrium point** (see Figure 13.12). The coordinate y_0, which indicates the price at which supply is equal to demand, is called the **equilibrium price**.

If the product sells at the equilibrium price, then those consumers who would be willing to buy at a higher price save money. The total amount they save is called the **consumer surplus**. Consumer surplus is represented geometrically by the area between the graph of $y = P(x)$ and $y = y_0$, as indicated in Figure 13.12. Consequently,

$$\text{Consumer surplus} = \int_0^{x_0} [P(x) - y_0] \, dx \tag{1}$$

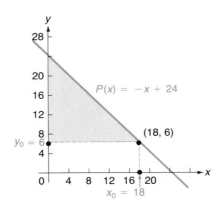

FIGURE 13.12

Similarly, those producers who would be willing to sell at a lower price will make money when the product sells at the equilibrium price. The total amount they make is called the **producer surplus**. Producer surplus is represented geometrically by the area between the graphs of $y = y_0$ and $y = Q(x)$, as indicated in Figure 13.12. Consequently,

$$\text{Producer surplus} = \int_0^{x_0} [y_0 - Q(x)] \, dx \tag{2}$$

EXAMPLE 2 Find the consumer surplus if the unit price (in dollars) for an item is given in terms of demand x by

$$P(x) = -x + 24$$

and the equilibrium price is $y_0 = \$6$.

Solution

To find the equilibrium quantity at which supply equals demand, set $P(x) = 6$ and solve for x_0:

$$6 = -x_0 + 24$$

2 PRACTICE PROBLEM 2

Find the consumer surplus if the equilibrium price in Example 2 is $y_0 = 8$.

Answer

$128

gives $x_0 = 18$. From equation (1),

$$\text{Consumer surplus} = \int_0^{18} [(-x + 24) - 6] \, dx = \int_0^{18} (-x + 18) \, dx$$

$$= \left(-\frac{x^2}{2} + 18x\right)\Big|_0^{18} = 162$$

so the consumer surplus is $162.

EXAMPLE 3 In Example 2, suppose that instead of an equilibrium price of $6.00, the unit price of the item is given in terms of the supply x by

$$Q(x) = \tfrac{1}{6}(x^2 + x)$$

Determine the consumer surplus and producer surplus.

Solution

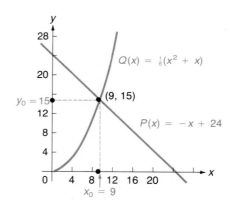

In this case, the coordinates of the equilibrium point are determined by first setting $P(x) = Q(x)$ and solving the resulting equation for x_0:

$$-x + 24 = \tfrac{1}{6}(x^2 + x)$$
$$x^2 + 7x - 144 = 0$$
$$(x + 16)(x - 9) = 0$$

Since -16 would not have any meaning in this situation, $x = 9$. Evaluating either $P(9)$ or $Q(9)$ gives $y_0 = 15$. Then,

$$\text{Consumer surplus} = \int_0^9 [(-x + 24) - 15] \, dx = \int_0^9 (-x + 9) \, dx$$

$$= \left(-\frac{x^2}{2} + 9x\right)\Big|_0^9 = 40.50$$

$$\text{Producer surplus} = \int_0^9 \left[15 - \frac{x^2 + x}{6}\right] dx$$

$$= \left(15x - \frac{x^3}{18} - \frac{x^2}{12}\right)\Big|_0^9 = 87.75$$

3 PRACTICE PROBLEM 3

If the unit price (in dollars) for an item in terms of demand x is

$$P(x) = 27 - x$$

and in terms of supply x is

$$Q(x) = \frac{x}{2}$$

determine the consumer and producer surpluses.

Answer

Consumer surplus: $162; producer surplus: $81

Therefore, the consumer surplus is $40.50 and the producer surplus is $87.75.

For a third type of application of definite integrals, consider the following relationship between a function and its derivative:

$$\int_a^b \underbrace{f'(x)}_{\substack{\uparrow \\ \text{Instantaneous rate} \\ \text{of change of } f}} dx = \underbrace{f(b) - f(a)}_{\substack{\uparrow \\ \text{Total change of} \\ f \text{ on the interval } [a, b]}} \qquad (3)$$

This equation follows from the fundamental theorem of calculus with a slight change in notation and is valid on any interval $[a, b]$ on which $f'(x)$ is continuous. It gives a method of determining the total change of a function on an interval when the rate at which it changes (its derivative) is known.

EXAMPLE 4 The marginal cost (in dollars) of producing x units is given by $C'(x) = 4x + 16$. Determine the total change in cost if production is increased from 30 to 40 units.

Solution According to the relationship expressed in equation (3), the total change in the cost function $C(x)$ is obtained from the integral of its derivative $C'(x)$. The total change is

$$\int_{30}^{40} (4x + 16) \, dx = (2x^2 + 16x)\Big|_{30}^{40} = 1{,}560$$

The total change is $1,560.

4 PRACTICE PROBLEM 4

A company is conducting an intensive advertising campaign to promote a new product. If the rate of sales of the product during week t of the campaign is $200t - (2/t^2)$ units per week, determine the total number of units sold during the tenth through the fifteenth weeks of the advertising campaign.

Answer

About 12,500 units

EXERCISES

Find each area between the graphs of the given functions over the indicated interval.

1. $f(x) = x + 4$, $g(x) = \dfrac{x}{2} + 1$; $[0, 3]$
2. $y = x^2$, $y = -x + 6$; $[-2, 2]$
3. $f(x) = -x^2 + 5$, $g(x) = x$; $[0, 1]$
4. $y = 5$, $y = x + 2$; $[-1, 3]$
5. $y = 2x + 5$, $2y = -x^2 + 4x + 2$; $[1, 4]$
6. $f(x) = 10$, $g(x) = 3^x$; $[0, 2]$

Determine the area bounded by the graphs of each pair of functions.

7. $f(x) = 4$ and $g(x) = x^2$
8. $y = x^2$ and $y = \sqrt{x}$
9. $y = 4 - x^2$ and the x-axis
10. $f(x) = x^3$ and $f(x) = x$, above the x-axis

11. Verify that the metal part illustrated in Figure 13.1 at the beginning of this chapter will use $21\frac{1}{3}$ square inches of metal.

12. If the graph of $f(x)$ lies above the x-axis and that of $g(x)$ lies below the x-axis for x in the interval $[a, b]$, the area between the two curves bounded by $x = a$ and $x = b$ (see the figure in the margin) is also given by the integral

$$\int_a^b [f(x) - g(x)]\, dx$$

Explain why.

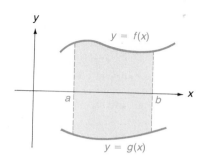

Using the integral given in Exercise 12, determine the area bounded by the graphs of:

13. $f(x) = 9 - x^2$, $g(x) = x^2 - 9$, $x = -1$, $x = 2$

14. $y = \dfrac{-3x}{2} - 3$, $y = x + 2$, $x = 4$

15. Find the area bounded by the graphs of $y = x^3$, $y = 0$, $x = -1$, and $x = 1$.

16. *Consumer surplus* If the unit price of an item is given in terms of demand x by

$$P(x) = -\tfrac{1}{4}x + 20$$

and the equilibrium point is (4, 16), find the consumer surplus.

17. *Producer surplus* If the unit price of an item is given in terms of supply x by

$$Q(x) = \tfrac{3}{2}x$$

and the equilibrium point is (10, 15), find the producer surplus.

18. *Producer surplus* If the unit price of an item is given in terms of supply x by

$$Q(x) = x^2 + x$$

and the equilibrium price is \$12, find the producer surplus.

19. *Consumer surplus* If the unit price of an item is given in terms of demand x by

$$P(x) = 12 - \tfrac{1}{2}x^2$$

and the equilibrium price is \$4, find the consumer surplus.

20. *Consumer-producer surplus* Find the consumer surplus and the producer surplus if the unit price of an item is given in terms of demand x by

$$P(x) = -2x + 98$$

and in terms of supply x by

$$Q(x) = x^2 + 5x$$

21. *Producer surplus* Find the producer surplus if

$$P(x) = 15 \quad \text{and} \quad Q(x) = 2^x - 1$$

22. *Cost* The marginal cost (in dollars) of producing x units is given by $C'(x) = 3x^2 + 18$. Determine the total change in cost if production is changed from 80 to 100 units.

23. *Cost* The marginal cost (in hundreds of dollars) of producing x units is given by $C'(x) = 6x + 2$. Determine the total change in cost if production is changed from 12 to 15 units.

24. *Profit* The marginal profit (in hundreds of dollars) from the sale of x units is given by $P'(x) = 3x^2 - 12x$. Determine the increase in profit if the sales level increases from 20 to 30 units.

25. *Profit* The marginal profit (in dollars) from the sale of x units is given by $P'(x) = e^x - 1$. Determine the change in profit if the sales level changes from 10 to 12 units.

26. *Population* The population (in hundreds) of a bacteria colony is changing at the rate of $9t^2 + 120t + 300$ per second. Determine the change in population from $t = 15$ to $t = 20$ seconds after an initial population estimate ($t = 0$).

27. *Motion* An object is moving along a straight path in such a way that its velocity (speed) is $160 - 32t$ feet per second. How far does it travel from $t = 2$ to $t = 3$ seconds after its movement begins?

28. *Demographics* Demographics indicate that over the next 5 years the voting-age population in a small city will change at the rate of

$$f'(t) = 1.8t - 0.6t^2$$

where $f(t)$ denotes citizens of voting age (in thousands) and t denotes time (in years). Determine the change in the voting-age population over the next 3 years.

29. *Learning experiment* In a learning experiment, a psychologist has determined that the number of new words chosen from a select list of words and memorized by an average individual changes at the rate of $20(1 - 0.4x)$ words per hour after x hours of continuous study, $0 \leq x \leq 4$. How many words will an average individual memorize during the second and third hours of continuous study?

30. *Chemical reaction* The end product of a chemical reaction is produced at the rate of

$$\frac{\sqrt{t} - 1}{t} \quad \text{(milligrams per minute)}$$

for $1 \leq t \leq 10$. If the reaction is started at time $t = 1$, determine the amount produced during the fifth through the eighth minutes of the reaction.

Identify the steps of the four-step procedure involved in the application of mathematics (Section 1.3) in each case.

31. *Consumer surplus* Exercise 16

32. *Chemical reaction* Exercise 30

33. What relationship between the instantaneous rate of change of a function and the total change in the values of the function over an interval is given by the fundamental theorem of calculus?

13.7 CHAPTER SUMMARY

The two basic concepts of calculus are derivatives and integrals. Derivatives were discussed in Chapters 11 and 12; in this chapter we began the discussion of integrals. There are two types—indefinite and definite integrals.

13.7 CHAPTER SUMMARY

The *indefinite integral* of a function f is defined by the equation

$$\int f(x)\,dx = F(x) + C$$

where $F'(x) = f(x)$ and C is an arbitrary constant. Because of the relationship $F'(x) = f(x)$, indefinite integrals are also called **antiderivatives**. Also, the formulas for evaluating indefinite integrals come from "reversing" derivative formulas, as indicated in the table:

INTEGRATION	DIFFERENTIATION
1. $\int x^n\,dx = \dfrac{x^{n+1}}{n+1} + C,\ n \neq -1$	$\dfrac{d}{dx}\left(\dfrac{x^{n+1}}{n+1}\right) = x^n$
2. $\int e^x\,dx = e^x + C$	$\dfrac{d}{dx}(e^x) = e^x$
3. $\int b^x\,dx = \dfrac{b^x}{\ln b} + C,\ b > 0$	$\dfrac{d}{dx}\left(\dfrac{b}{\ln b}\right) = b^x$
4. $\int \dfrac{1}{x}\,dx = \ln |x| + C$	$\dfrac{d}{dx}(\ln |x|) = \dfrac{1}{x}$

The following two properties of integrals enable us to evaluate the integrals of many more functions, which are combinations of the four basic types given in the table:

$$\int k f(x)\,dx = k \int f(x)\,dx \qquad k \text{ a constant}$$
$$\int [f(x) \pm g(x)]\,dx = \int f(x)\,dx \pm \int g(x)\,dx$$

Indefinite integrals enable us to determine which functions have a given rate of change—for example, to find the function when we are given the slope of its tangents and the value of the function for one value of x, or to determine the cost function when we know the marginal and fixed costs.

The **definite integral** of a function f continuous on an interval $[a, b]$ is defined as the limit of a sum:

$$\int_b^a f(x)\,dx = \lim_{n \to \infty} \sum_{i=1}^n f(x_i^*)\Delta x_i$$

where n becomes unbounded in such a way that each Δx_i approaches zero. This limit has a value for any interval $[a, b]$ on which f is continuous; that is, f is **integrable** on any interval $[a, b]$ on which f is a continuous function.

The two definitions

$$\int_a^b f(x)\,dx = -\int_b^a f(x)\,dx \qquad \int_a^a f(x)\,dx = 0$$

give a value to indefinite integrals in which the upper limit of integration is less than or equal to the lower limit of integration.

Definite integrals are evaluated using the **fundamental theorem of calculus**. This theorem states that if $f(x)$ is continuous on the interval $[a, b]$, then

$$\int_a^b f(x)\,dx = F(b) - F(a)$$

where $F(x)$ is an indefinite integral of $f(x)$. The techniques for evaluating indefinite integrals become techniques for evaluating definite integrals because of the fundamental theorem.

Definite integrals enable us to calculate various areas associated with the graphs of functions; this is also the basis for the use of integrals in calculating **consumer surplus** and **producer surplus**. They also give a method for determining the total change of a function on an interval when the rate at which it changes (its derivative) is known.

REVIEW EXERCISES

Evaluate each integral.

1. $\int 3\sqrt{x}\,dx$
2. $\int -5e^x\,dx$
3. $\int 5^x\,dx$
4. $\int \dfrac{dy}{y}$
5. $\int (3x^2 - 2)\,dx$
6. $\int (4x - 8x^3)\,dx$
7. $\int (x^{1/2} + 2x^{1/3})\,dx$
8. $\int (x^{1/4} + x^{-1/4})\,dx$
9. $\int (2x^{-2} + x^{-1})\,dx$
10. $\int (5x + 5^x)\,dx$
11. $\int \left(t^5 + \dfrac{5}{t}\right)dt$
12. $\int (3e^x - 4)\,dx$
13. $\int (3x^2 - 2x + 1)\,dx$
14. $\int (x + 2e^x - 3)\,dx$
15. $\int y\left(3y + \dfrac{1}{y^2}\right)dy$
16. $\int t(\sqrt{t} + t^{-1})\,dt$
17. $\int \dfrac{x^2 - 9}{x - 3}\,dx$
18. $\int \dfrac{x^2 + 3x - 10}{x + 5}\,dx$
19. $\int_{-1}^{3} (1 + 3x^2)\,dx$
20. $\int_0^1 (x + \sqrt{x})\,dx$
21. $\int_{-2}^{2} (2t + 2^t)\,dt$
22. $\int_{-1}^{1} (y^3 - 4y)\,dy$
23. $\int_2^4 \left(\dfrac{2}{x^2} + \dfrac{1}{3}\right)dx$
24. $\int_{-3}^{-1} (3x^2 - 2x + 2)\,dx$
25. $\int_{0.5}^{0.5} 3(1 - x)\,dx$
26. $\int_1^{-2} -4(1 + t + t^2)\,dt$
27. $\int_1^2 x\left(\dfrac{2}{x} + \dfrac{1}{x^2}\right)dx$
28. $\int_{-1}^{2} \dfrac{t - 6}{3}\,dt$
29. $\int_0^1 \dfrac{2x^2 + 6x + 4}{x + 1}\,dx$
30. $\int_{30}^{70} dx$

Determine the function whose graph passes through the given point and has tangents with the given slope.

31. $g'(x) = 12x^2 - 1$; passes through the point $(1, 5)$

32. $f'(x) = \dfrac{2 + x^2}{x^2}$; passes through the point $(-4, -3.5)$

33. *Cost* If the marginal cost for producing x units of a product is $C'(x) = x^3 - 2x^2$ dollars and fixed costs equal \$760, determine the cost function $C(x)$.

34. *Demand* If the marginal demand (in hundreds of units) is $D'(x) = -1/2x^2$ in terms of the price x (in dollars) and if the demand is for 260 when the price is \$5.00, determine the demand function $D(x)$.

35. *Revenue* If the marginal revenue (in dollars) from the sale of x units is $R'(x) = 480 - 0.04x$, determine the revenue function $R(x)$.

36. *Cost-revenue-profit* The marginal revenue (in dollars) from the sale of x units is $R'(x) = 24$. The marginal cost (also in dollars) for manufacturing and marketing x units is $C'(x) = 144 - 0.024x$. Fixed costs are \$1,800.
    (a) Determine the cost function $C(x)$.
    (b) Determine the revenue function $R(x)$.
    (c) Determine the profit function $P(x)$.

37. *Projectile* An object, fired straight up from the ground, moves with a velocity of

    $$v(t) = -32t + 160 \quad \text{(feet per second)}$$

    (a) If distance is measured in feet from the ground upward, determine its position at time t. (*Hint:* Its position is 0 when $t = 0$.)
    (b) At what time will the object return to the ground?

Determine the area between the graph of each of the following functions and the x-axis over the indicated intervals:

38. $y = x^2 + 1$; $[-3, 1]$
39. $f(x) = e^x - x$; $[0, 6]$

Determine the area between the graph of each of the following functions and the x-axis below the indicated interval:

40. $f(x) = 4 - 6x$; $[2, 5]$
41. $y = -(x + 2)^2$; $[-2, 0]$

Determine the area bounded by the x-axis and the graph of each of the following functions:

42. $y = 1 - x^2$
43. $y = -x^2 + 4x - 1$

Determine the area between the graphs of the given functions over the indicated interval.

44. $y = 2x + 3$, $y = -x + 4$; $[2, 4]$
45. $f(x) = 5 - x^2$, $g(x) = 4^x$; $[-1, 1]$

Determine the area bounded by the graphs of each of the following pairs of functions:

46. $y = 1 + x^2$ and $y = 5$
47. $f(x) = x + 4$ and $f(x) = x^2 + 2x - 2$

48. *Consumer surplus*   If the unit price of an item is given in terms of demand x by
$$P(x) = 25 - 0.2x$$
and the equilibrium point is (10, 23), find the consumer surplus.

49. *Producer surplus*   If the unit price of an item is given in terms of supply x by
$$Q(x) = 5x$$
and the equilibrium point is (7, 35), find the producer surplus.

50. *Consumer-producer surplus*   Find the consumer and producer surplus if the unit price of an item is given in terms of demand x by
$$P(x) = 50 - \tfrac{1}{4}x^2$$
and in terms of supply x by
$$Q(x) = \tfrac{5}{2}x$$

51. *Cost*   The marginal cost (in hundreds of dollars) of producing x units is given by $C'(x) = 0.03x^2 - 0.4x + 3$. Determine the total change in cost if production is changed from 10 to 30 units.

52. *Profit*   The marginal profit (in dollars) from the sale of x units is given by $P'(x) = 4x + 8$. Determine the increase in profit if the sales level increases from 50 to 60 units.

53. *Motion*   If an object is moving in a straight line with a velocity of
$$v(t) = -32t + 160$$
feet per second, how many feet does the object travel in the first 3 seconds?

54. *Population*   After a killing agent is introduced into a bacteria colony, the bacteria population decreases at a rate of $2e^{-0.2t}$ thousand per second after t seconds. Determine the change in the population from the second through the fifth seconds after the killing agent is introduced.

Identify the steps in the four-step procedure involved in the application of mathematics (Section 1.3) in each case.

55.   *Cost*   Exercise 51   **56.**   *Population*   Exercise 54

CHAPTER 14

ADDITIONAL TOPICS IN INTEGRATION

OUTLINE

14.1 Integration by Substitution
14.2 Use of Integration Tables
14.3 Integration by Parts
14.4 Differential Equations
14.5 Improper Integrals
*14.6 Probability Density Functions
14.7 Chapter Summary

Review Exercises

In Chapter 13 we discussed the basic concepts of integration. In this chapter we consider some additional methods for evaluating integrals and some additional areas in which integrals are used for solving physical problems.

* This section can be omitted without loss of continuity.

14.1 INTEGRATION BY SUBSTITUTION

The derivative of the function $y = (x^2 + 1)^3$ is

$$y' = 3(x^2 + 1)^2 \, 2x$$

Therefore, when integrating

$$\int 3(x^2 + 1)^2 (2x) \, dx = (x^2 + 1)^3 + C$$

However, the previous integration formulas would not allow us to evaluate the integral in this form. The method of substitution considered in this section allows us to extend the previous formulas to more complex functions such as y. In this sense it is much like the chain rule for derivatives, which allowed us to extend the differentiation formulas for elementary functions to more complex functions.

The method of integration by substitution requires the use of **differentials**. For independent variables x and t, differentials are denoted by dx and dt, respectively. For other independent variables, the letter d also precedes the variable. For functions or dependent variables, their differential is their derivative multiplied by the differential of the independent variable. These too are denoted by the letter d—df, dy, and so forth. Several examples are given in Table 14.1.

TABLE 14.1

FOR THE FUNCTION:	THE DERIVATIVE IS:	THE DIFFERENTIAL IS:
$f(x) = e^{2x}$	$2e^{2x}$	$df = (2e^{2x}) \, dx$
$g(t) = t^2 - 3t$	$2t - 3$	$dg = (2t - 3) \, dt$
$y = \ln x^2$	$\dfrac{1}{x^2}(2x) = \dfrac{2}{x}$	$dy = \dfrac{1}{x^2}(2x) \, dx = \dfrac{2}{x} \, dx$
$u = x^3 - 2x^2 + 1$	$3x^2 - 4x$	$du = (3x^2 - 4x) \, dx$

The rules for finding differentials are exactly the same as the rules for finding derivatives. One simply multiplies the result by the differential of the independent variable.

Previously, symbols such as dy/dx and df/dt for derivatives could not be considered as quotients because dy, dx, df, and dt alone had no meaning. Now that they do, these symbols can be treated as quotients when convenient. In fact, differentials can be considered to be obtained from the equation giving the derivative by multiplying each side of the equation by the differential of the independent variable. For example, if

$$y = (4x^2 + 3)^{-3}$$

---

1 PRACTICE PROBLEM 1

Determine the differential of each of the following functions:

(a) $y = 2x^3 + 4x$ (b) $f(x) = 4e^x$
(c) $g(t) = (2t - 3)^4$

Answer

(a) $dy = (6x^2 + 4) \, dx$
(b) $df = 4e^x \, dx$
(c) $dg = 8(2t - 3)^3 \, dt$

then

$$\frac{dy}{dx} = -3(4x^2 + 3)^{-4}(8x)$$

and the differential of y is obtained by multiplying each side of this last equation by dx:

$$dy = -3(4x^2 + 3)^{-4}(8x)\,dx$$

Before examining the method of integration by substitution, we restate the four basic integration formulas of Section 13.1 with u in place of x:

1. $\int u^n\,du = \dfrac{u^{n+1}}{n+1} + C, \quad n \neq -1$
2. $\int e^u\,du = e^u + C$
3. $\int b^u\,du = \dfrac{b^u}{\ln b} + C, \quad b > 0$
4. $\int \dfrac{1}{u}\,du = \ln|u| + C$

The method of integration by substitution in this section is still based on these four basic formulas. Now, however, u denotes *a function* and not necessarily a single variable.

To illustrate the procedure, suppose we want to evaluate

$$\int (x^2 + 6)^2 (2x)\,dx$$

If we write $u = x^2 + 6$, then $du = 2x\,dx$ and the given integral can be rewritten

$$\int u^2\,du$$

which is in form (1) in the table with $n = 2$. Consequently,

$$\int (x^2 + 6)^2 (2x)\,dx = \int u^2\,du = \frac{u^3}{3} + C$$

In order to obtain the value of the integral in terms of the original variable x, we use the relationship $u = x^2 + 6$ to substitute $x^2 + 6$ for u. Then

$$\int (x^2 + 6)^2 (2x)\,dx = \frac{(x^2 + 6)^3}{3} + C$$

EXAMPLE 1 Evaluate

$$\int e^{x^2}(2x)\,dx$$

Solution If $u = x^2$, then $du = 2x\,dx$ and the integral is in form (2) with $u = x^2$. Substituting $u = x^2$ and $du = 2x\,dx$, the given integral can be rewritten as

$$\int e^{x^2}(2x)\,dx = \int e^u\,du = e^u + C = e^{x^2} + C$$

EXAMPLE 2 Evaluate

$$\int \frac{1}{-3x+5}(-3)\,dx$$

Solution This integral is in form (4) with $u = -3x + 5$, in which case $du = -3\,dx$. With this substitution,

$$\int \frac{1}{-3x+5}(-3)\,dx = \int \frac{1}{u}\,du = \ln|u| + C = \ln|-3x+5| + C \quad \boxed{2}$$

The integral

$$\int (x^3 + 4)^2 (x^2)\,dx$$

looks as though it might be in form (1) in the table with $u = x^3 + 4$ and $n = 2$. Then $du = 3x^2\,dx$, but the 3 required for the differential is not in the integral! However, we can rewrite the integral, multiplying by 1 in the form $\frac{1}{3} \cdot 3$:

$$\int (x^3 + 4)^2 (x^2)\,dx = \tfrac{1}{3} \cdot 3 \int (x^3 + 4)^2 (x^2)\,dx$$

Since a constant factor can be moved across the integral sign, this last integral is equal to

$$\tfrac{1}{3} \int (x^3 + 4)^2 (3x^2)\,dx$$

Substituting $u = x^3 + 4$ and $du = 3x^2\,dx$ gives

$$\int (x^3 + 4)^2 x^2\,dx = \frac{1}{3}\int (x^3 + 4)^2 (3x^2)\,dx = \frac{1}{3}\int u^2\,du$$

$$= \frac{1}{3}\left(\frac{u^3}{3}\right) + C = \frac{1}{9}(x^3 + 4)^3 + C$$

In general, if a given integral is in one of the four basic forms in the table except for a constant factor in the differential, the integral can be rewritten in the desired form by multiplying by 1 in a form similar to that just illustrated and then moving the required constant factor across the integral sign. The procedure is illustrated again in the next several examples. $\boxed{3}$

EXAMPLE 3 Evaluate

$$\int \frac{1}{3x^2 - 1} x\,dx$$

---

2 PRACTICE PROBLEM 2

(a) Use form (1) in the table to evaluate

$$\int (4x^2 - 5)^3 (8x)\,dx$$

(b) Use form (4) in the table to evaluate

$$\int \frac{1}{x^2}(2x)\,dx$$

(c) Use form (2) in the table to evaluate

$$\int e^{3x}(3)\,dx$$

Answer

(a) $\dfrac{(4x^2 - 5)^4}{4} + C$ (b) $\ln x^2 + C$

(c) $e^{3x} + C$

---

3 PRACTICE PROBLEM 3

Evaluate $\int (x^2 + 4)^3 x\,dx$.

Answer

$\dfrac{(x^2 + 4)^4}{8} + C$

14.1 INTEGRATION BY SUBSTITUTION

Solution If one were to attempt to put this integral into form (4) with $u = 3x^2 - 1$, then $du = 6x\,dx$. But there is no 6 available for the differential. The 6 can be obtained by multiplying the integral by 1 as follows:

$$\int \frac{1}{3x^2 - 1} x\,dx = \frac{1}{6} \cdot 6 \int \frac{1}{3x^2 - 1} x\,dx = \frac{1}{6} \int \frac{1}{3x^2 - 1} (6x)\,dx$$

With the above substitutions, this last integral can be rewritten

$$\frac{1}{6} \int \frac{1}{u}\,du = \frac{1}{6} \ln|u| + C = \frac{1}{6} \ln|3x^2 - 1| + C$$

▸ **4 PRACTICE PROBLEM 4**

Evaluate $\int \frac{1}{x^2 - 1} x\,dx$.

Answer

$\frac{1}{2}(\ln|x^2 - 1|) + C$

EXAMPLE 4 Evaluate

$$\int e^{t/2}\,dt$$

Solution This integral is in form (2) with $u = t/2$ and $du = \frac{1}{2}dt$ except for the $\frac{1}{2}$. The $\frac{1}{2}$ can be obtained by multiplying the integral by 1 as follows:

$$2 \cdot \frac{1}{2} \int e^{t/2}\,dt = 2 \int e^{t/2}(\tfrac{1}{2})\,dt = 2 \int e^u\,du$$
$$= 2e^u + C = 2e^{t/2} + C$$

▸ **5 PRACTICE PROBLEM 5**

Evaluate $\int e^{-t/3}\,dt$. (*Hint:* Let $u = -\frac{1}{3}t$.)

Answer

$-3e^{-t/3} + C$

The next two examples combine the properties of integrals stated in Theorems 1 and 2 of Section 13.1 with the method of substitution.

EXAMPLE 5 Evaluate

$$\int 4^{x^2}(6x)\,dx$$

Solution With $u = x^2$ and $du = 2x\,dx$, this integral is in form (3) except for an extra factor of 3. According to Theorem 1 of Section 13.1, the 3 can be moved across the integral sign:

$$\int 4^{x^2}(3 \cdot 2x)\,dx = 3 \int 4^{x^2}(2x)\,dx = 3 \int 4^u\,du$$
$$= 3\left(\frac{4^u}{\ln 4}\right) + C = 3\left(\frac{4^{x^2}}{\ln 4}\right) + C$$

▸ **6 PRACTICE PROBLEM 6**

Evaluate $\int (x^2 + 3)^4 (4x)\,dx$.

Answer

$2\dfrac{(x^2 + 3)^5}{5} + C$

EXAMPLE 6 Evaluate

$$\int [8x + (x^2 - 1)^3 x]\,dx$$

Solution The given integral can be written as the sum of two integrals:

$$\int 8x\,dx + \int (x^2 - 1)^3 x\,dx$$

The first of these integrals has the value $4x^2 + C$. The second is in form (1) with $u = x^2 - 1$, $n = 3$, and $du = 2x\,dx$. By the substitution method,

$$\int (x^2 - 1)^3 x \, dx = \frac{(x^2 - 1)^4}{8} + C$$

By adding these two values, the original integral

$$\int [8x + (x^2 - 1)^3 x] \, dx = 4x^2 + \tfrac{1}{8}(x^2 - 1)^4 + C$$

After enough practice, some of the intermediate details, including the actual substitution, can be omitted. For example,

$$\int \frac{x}{3x^2 - 2} \, dx = \frac{1}{6} \int \frac{6x}{3x^2 - 2} \, dx = \frac{1}{6} \ln|3x^2 - 2| + C$$

Without stating the fact, we mentally used the substitution $u = 3x^2 - 2$ and $du = 6x \, dx$ in form (4).

EXAMPLE 7 Evaluate

$$\int_0^1 (t^2 + 1)^2 t \, dt$$

Solution   We first use the substitution method as usual to find an indefinite integral:

$$\int (t^2 + 1)^2 t \, dt = \frac{1}{2} \int (t^2 + 1)^2 (2t) \, dt = \frac{1}{2} \int u^2 \, du$$

$$= \frac{1}{2} \cdot \frac{u^3}{3} + C = \frac{1}{6}(t^2 + 1)^3 + C$$

Therefore,

$$\int_0^1 (t^2 + 1)^2 t \, dt = \frac{1}{6}(t^2 + 1)^3 \bigg|_0^1 = \frac{8}{6} - \frac{1}{6} = \frac{7}{6}$$

7 PRACTICE PROBLEM 7

Evaluate $\int [e^{2t} + (3t + 2)^2] \, dt$.

Answer

$\tfrac{1}{2}e^{2t} + \tfrac{1}{9}(3t + 2)^3 + C$

8 PRACTICE PROBLEM 8

Evaluate $\int_0^2 e^{3t} \, dt$.

Answer

$\tfrac{1}{3}(e^6 - 1)$

EXERCISES

Determine the differential of each given function.

1. $f(x) = x^2 + 3$
2. $f(x) = 2x^4 + 6x$
3. $f(x) = 4(x^3 + x)$
4. $y = e^x$
5. $g(t) = \dfrac{2}{t}$
6. $g(t) = \ln t^2$
7. $y = (2x + 3)^4$
8. $y = \sqrt{5x - 1}$
9. $f(x) = (4 - x)^{-1}$
10. $f(t) = (t^2 + t^{-1})^3$
11. $y = 3x^2 + 2x + 1$
12. $y = 3x^3 - x^2 + 4x$
13. $f(x) = x^2(3x - 4)$
14. $y = \dfrac{x^2}{3x - 4}$

Evaluate each given integral.

15. $\int (x^2 + 5)^3 x \, dx$

16. $\int (x^4 - 3)^2 x^3 \, dx$

17. $\int (2x^3 + 1)^2 x^2 \, dx$

18. $\int (4t^2 + 7)^{-2} t \, dt$

19. $\int (3 - 6x^2)^{-3}(-2x) \, dx$

20. $\int (2 - x^4)^{-3}(-2x^3) \, dx$

21. $\int \sqrt{t^2 + 1}\,(t) \, dt$

22. $\int \sqrt{4 - x^2}\,(x) \, dx$

23. $\int (3x + 1)^{-2} \, dx$

24. $\int (7 - 5x)^4 \, dx$

25. $\int e^{x^3} x^2 \, dx$

26. $\int e^{2t} 2 \, dt$

27. $\int e^{-3t} \, dt$

28. $\int e^{t/4} \, dt$

29. $\int e^{3x^2}(2x) \, dx$

30. $\int e^{2y-1} \, dy$

31. $\int e^{x^2 - 2} x \, dx$

32. $\int e^{2-t^2} t \, dt$

33. $\int 4^{x^3}(3x^2) \, dx$

34. $\int 5^{-3y} \, dy$

35. $\int 6^{1-4x^2}(4x) \, dx$

36. $\int 3^{-\sqrt{x}}\left(\dfrac{1}{\sqrt{x}}\right) dx$

37. $\int \dfrac{1}{2x + 1}(2) \, dx$

38. $\int \dfrac{1}{x^2 + 1}(2x) \, dx$

39. $\int \dfrac{1}{4 - 3x} \, dx$

40. $\int \dfrac{1}{4 - t^2}(t) \, dt$

41. $\int \dfrac{x}{5 + x^2} \, dx$

42. $\int \dfrac{3x^2}{1 + 2x^3} \, dx$

43. $\int (x^2 + 2x)^3 (x + 1) \, dx$

44. $\int (2x^3 - 3x)(2x^2 - 1) \, dx$

45. $\int e^{t^2 - 2t}(t - 1) \, dt$

46. $\int e^{x^4 + 2x}(2x^3 + 1) \, dx$

47. $\int \dfrac{2x + 3}{3x^2 + 9x} \, dx$

48. $\int \dfrac{2t - 1}{4t - 4t^2} \, dt$

49. $\int (t^2 + 3)^3 (8t) \, dt$

50. $\int (4 - 2t^3)^2 (-12t^2) \, dt$

51. $\int \sqrt{2x^2 - 1}\,(8x) \, dx$

52. $\int \sqrt{2x - x^2}\,(4x - 4) \, dx$

53. $\int e^{5t^2}(20t) \, dt$

54. $\int \tfrac{1}{6} e^{x/3} \, dx$

55. $\int 7^{2x^3 + 2}(24x^2) \, dx$

56. $\int 5^{-4x^3}(2x^2) \, dx$

57. $\int (\ln 3) 3^{x^2} x \, dx$

58. $\int \dfrac{4t}{t^2 + 4} \, dt$

59. $\int \dfrac{10}{2t - 1} \, dt$

60. $\int \dfrac{6}{5 - 3t} \, dt$

61. $\int \dfrac{5}{2x + 1} \, dx$

62. $\int 3(x^2 + 1)^2 x \, dx$

63. $\int (5t)e^{-3t^2}\, dt$

64. $\int \dfrac{4}{(5x+1)^2}\, dx$

65. $\int \dfrac{4(x^2+1)}{2x^3+6x}\, dx$

66. $\int (x^3 + xe^{x^2})\, dx$

67. $\int \left(\dfrac{1}{2x} - \dfrac{1}{3x+1}\right) dx$

68. $\int (2e^t + 3e^{-t})\, dt$

69. $\int (\sqrt{2x} + 5x)\, dx$

70. $\int [4^{3x} - (3x)^4]\, dx$

71. $\int \left(2 + \dfrac{x}{x^2+2}\right) dx$

72. $\int \dfrac{1}{x}(1 + \ln^2 x)\, dx$

Evaluate the following definite integrals:

73. $\int_1^2 (2x-1)^3 2\, dx$

74. $\int_0^1 (6x-2)^2 3\, dx$

75. $\int_0^1 3x(2x^2-1)^2\, dx$

76. $\int_{-2}^{-1} 4(3x-2)^{-2}\, dx$

77. $\int_1^2 4^{2x}\, dx$

78. $\int_{-1}^0 3^{-x}\, dx$

79. $\int_1^3 (\ln 5) 5^x\, dx$

80. $\int_0^1 xe^{x^2}\, dx$

81. $\int_0^e \dfrac{x}{x^2+1}\, dx$

82. $\int_4^{12} \sqrt{2x+1}\, dx$

83. $\int_3^{12} (\sqrt{3x} - 2)\, dx$

84. $\int_0^4 \left(\dfrac{1}{2x+1} + 3\right) dx$

Determine the area between the graph of each of the following functions and the x-axis over the given interval:

85. $f(x) = x(x^2-1)^2;\quad [0, 1]$

86. $f(x) = 3^{2x};\quad [-1, 2]$

Determine the area between the graph of each of the following functions and the x-axis below the given interval:

87. $f(x) = x(x^2-1)^2;\quad [-1, 0]$

88. $y = -xe^{3x^2};\quad [0.5, 1.2]$

14.2 USE OF INTEGRATION TABLES

A table of integrals is given in Table 14.2. The value given for each of the integrals can be verified by determining that the derivative of the function on the right is equal to the integrand in the corresponding integral on the left. This is a relatively short table of integrals. The table of integrals appearing in some of the standard handbooks of mathematical tables might contain several hundred integrals of different types.

The use of a table in integration is the same as the method of substitution based on the four integral forms considered in the last section. In fact, the first four forms in the table are the four forms used in the previous section. The only additional difficulty might be in

TABLE 14.2 TABLE OF INTEGRALS

Note: a, b, and *r* denote constants in the formulas in which they appear.

1. $\int u^n \, du = \dfrac{u^{n+1}}{n+1} + C, \quad n \neq -1$

2. $\int e^u \, du = e^u + C$

3. $\int b^u \, du = \dfrac{b^u}{\ln b} + C, \quad b > 0$

4. $\int \dfrac{1}{u} \, du = \ln|u| + C$

5. $\int \dfrac{du}{u(a+bu)} = \dfrac{1}{a}\left(\ln\left|\dfrac{u}{a+bu}\right|\right) + C$

6. $\int \dfrac{u \, du}{a+bu} = \dfrac{1}{b^2}[a + bu - a(\ln|a+bu|)] + C$

7. $\int \dfrac{u \, du}{(a+bu)^2} = \dfrac{1}{b^2}\left(\ln|a+bu| + \dfrac{a}{a+bu}\right) + C$

8. $\int \dfrac{du}{u(a+bu)^2} = \dfrac{1}{a(a+bu)} + \dfrac{1}{a^2}\left(\ln\left|\dfrac{u}{a+bu}\right|\right) + C$

9. $\int \dfrac{du}{u^2 - a^2} = \dfrac{1}{2a}\left(\ln\left|\dfrac{u-a}{u+a}\right|\right) + C, \quad u^2 > a^2$

10. $\int \dfrac{du}{a^2 - u^2} = \dfrac{1}{2a}\left(\ln\left|\dfrac{a+u}{a-u}\right|\right) + C, \quad u^2 < a^2$

11. $\int \dfrac{du}{\sqrt{u^2 \pm a^2}} = \ln|u + \sqrt{u^2 \pm a^2}| + C$

12. $\int \dfrac{du}{\sqrt{2au + u^2}} = \ln|u + a + \sqrt{2au + u^2}| + C$

13. $\int \ln u \, du = u(\ln u) - u + C$

14. $\int u^r (\ln u) \, du = \dfrac{u^{r+1}}{r+1}\left(\ln u - \dfrac{1}{r+1}\right) + C, \quad r \neq -1$

15. $\int u e^u \, du = u e^u - e^u + C$

16. $\int u^2 e^u \, du = e^u(u^2 - 2u + 2) + C$

17. $\int \dfrac{\ln u}{u} \, du = \dfrac{(\ln u)^2}{2} + C$

---

fitting a given integral into one of the forms in the table because so many additional forms are now available. Practice should alleviate this difficulty. The procedure for using the table is illustrated in the following examples.

EXAMPLE 1 Evaluate

$$\int \dfrac{dx}{\sqrt{4x^2 + 9}}$$

Solution This looks like form (11) in Table 14.2 with $u = 2x$ and $a = 3$, in which case $du = 2 \, dx$. The 2 for du is missing, but we can obtain this 2 using the method of the previous section:

$$\int \dfrac{dx}{\sqrt{4x^2 + 9}} = \dfrac{1}{2}\int \dfrac{2 \, dx}{\sqrt{4x^2 + 9}}$$

$$= \dfrac{1}{2}\int \dfrac{du}{\sqrt{u^2 + 9}}$$

$$= \dfrac{1}{2}(\ln|u + \sqrt{u^2 + 9}|) + C$$

$$= \dfrac{1}{2}(\ln|2x + \sqrt{4x^2 + 9}|) + C$$

1 PRACTICE PROBLEM 1

Evaluate $\int \dfrac{dx}{\sqrt{9x^2 + 16}}$.

Answer

$\frac{1}{3}(\ln|3x + \sqrt{9x^2 + 16}|) + C$

EXAMPLE 2 Evaluate

$$\int 3xe^{3x}\,dx$$

Solution This looks like form (15) in Table 14.2 with $u = 3x$. Then $du = 3\,dx$; we need a factor of 3 for du in order to make the substitution.

$$\begin{aligned}\int 3xe^{3x}\,dx &= \tfrac{1}{3}\int (3x)e^{3x}(3\,dx)\\ &= \tfrac{1}{3}\int ue^u\,du\\ &= \tfrac{1}{3}(ue^u - e^u) + C\\ &= \tfrac{1}{3}(3xe^{3x} - e^{3x}) + C\\ &= xe^{3x} - \tfrac{1}{3}e^{3x} + C\end{aligned}$$

2 PRACTICE PROBLEM 2

Evaluate $\int -5xe^{-5x}\,dx$.

Answer

$xe^{-5x} + \tfrac{1}{5}e^{-5x} + C$

EXAMPLE 3 Evaluate

$$\int \frac{dx}{x^2 - 16}$$

Solution This looks like form (9) in Table 14.2 with $u = x$ and $a = 4$, in which case $du = dx$.

$$\begin{aligned}\int \frac{dx}{x^2 - 16} &= \int \frac{du}{u^2 - 16}\\ &= \frac{1}{8}\left(\ln\left|\frac{u - 4}{u + 4}\right|\right) + C\\ &= \frac{1}{8}\left(\ln\left|\frac{x - 4}{x + 4}\right|\right) + C\end{aligned}$$

3 PRACTICE PROBLEM 3

Letting $u = 3x$ and $a = 1$, evaluate

$$\int \frac{dx}{9x^2 - 1}$$

Answer

$\dfrac{1}{6}\left(\ln\left|\dfrac{3x - 1}{3x + 1}\right|\right) + C$

EXAMPLE 4 Evaluate

$$\int \frac{2x^3\,dx}{3 + 5x^2}$$

Solution It might not seem so at first, but the given integral is in form (6) with $u = x^2$, $a = 3$, and $b = 5$. Then $du = 2x\,dx$; we need one of the powers of x in the numerator for du. Then

$$\begin{aligned}\int \frac{2x^3\,dx}{3 + 5x^2} &= \int \frac{x^2(2x\,dx)}{3 + 5x^2}\\ &= \int \frac{u\,du}{3 + 5u}\\ &= \frac{1}{25}[3 + 5u - 3(\ln|3 + 5u|)] + C\\ &= \frac{1}{25}[3 + 5x^2 - 3(\ln|3 + 5x^2|)] + C\end{aligned}$$

4 PRACTICE PROBLEM 4

Letting $u = x^2$, $a = 1$, and $b = -5$, evaluate

$$\int \frac{x^3\,dx}{1 - 5x^2}$$

Answer

$\tfrac{1}{50}(1 - 5x^2 - \ln|1 - 5x^2|) + C$

EXAMPLE 5 Evaluate

$$\int_1^e \frac{\ln x^2}{x}\,dx$$

Solution First we must evaluate the indefinite integral

$$\int \frac{\ln x^2}{x}\,dx$$

We use form (17) with $u = x^2$ so that $du = 2x\,dx$. In this case, the integral is rewritten twice, once to get the x for du and then to get the 2 for du:

$$\int \frac{\ln x^2}{x}\,dx = \int \frac{\ln x^2}{x^2}\, x\,dx = \frac{1}{2}\int \frac{\ln x^2}{x^2}(2x)\,dx$$

$$= \frac{1}{2}\int \frac{\ln u}{u}\,du = \frac{1}{2}\frac{(\ln u)^2}{2} + C$$

$$= \frac{1}{4}(\ln x^2)^2 + C$$

Returning to the original integral gives

$$\int_1^e \frac{\ln x^2}{x}\,dx = \frac{1}{4}(\ln x^2)^2 \Big|_1^e = \frac{1}{4}(4) - 0 = 1 \qquad \blacksquare$$

Note the different procedures used to obtain the 2 and the x for du in Example 5. Since a constant can be moved across the integral sign, it makes no difference whether the $\frac{1}{2}$ is written inside the integral sign or outside.

The same is not true for $1/x$. In other words,

$$\int \frac{\ln x^2}{x}\,dx = \int \frac{x}{x}\left(\frac{\ln x^2}{x}\right)dx = \int \frac{x \ln x^2}{x^2}\,dx$$

is *not* equal to

$$x \int \frac{\ln x^2}{x^2}\,dx$$

In general, *no expression involving the variable of integration can be moved from inside the integral sign to outside the integral sign*. Doing so is a mistake that is often made in the substitution process when rewriting a given integral to get it into one of the forms in the table!

---

5 PRACTICE PROBLEM 5

Evaluate $\int_1^2 \frac{\ln x^2}{x} 2\,dx$.

Answer

$\frac{1}{2}(\ln 4)^2$

EXERCISES

Evaluate each of the following integrals:

1. $\int \frac{dx}{9 - 4x^2}$ Use form (10) with $u = 2x$.

2. $\int \dfrac{dx}{\sqrt{9 + 4x^2}}$ Use form (11) with $u = 2x$.

3. $\int (5x)^2 \ln(5x)\,dx$ Use form (14) with $u = 5x$, $r = 2$.

4. $\int x^2 e^{3x}\,dx$ Use form (16) with $u = 3x$.

5. $\int \dfrac{dx}{\sqrt{9x^2 - 16}}$ Use form (11) with $u = 3x$.

6. $\int \dfrac{dx}{x(3 + 4x^2)}$ Use form (5) with $u = x^2$.

7. $\int e^{-3x}\,dx$

8. $\int (3 - x^2)^4 x\,dx$

9. $\int \dfrac{dx}{x(1 + 3x)}$ Use form (5) with $u = x$.

10. $\int \dfrac{dx}{7x(2 + 6x)^2}$ Use form (8) with $u = x$.

11. $\int \dfrac{2t\,dt}{9 + 3t}$ Use form (6) with $u = t$.

12. $\int \dfrac{-t\,dt}{(7 + 2t)^2}$ Use form (7) with $u = t$.

13. $\int \dfrac{dx}{\sqrt{6x + x^2}}$ Use form (12) with $u = x$.

14. $\int x^4 (\ln x)\,dx$ Use form (14) with $u = x$.

15. $\int 3t e^{4t}\,dt$ Use form (15) with $u = 4t$.

16. $\int -2(\ln 5t)\,dt$ Use form (13) with $u = 5t$.

17. $\int \dfrac{e^x\,dx}{\sqrt{e^{2x} + 4}}$ Use form (11) with $u = e^x$.

18. $\int \dfrac{3e^x\,dx}{e^{2x} - 4}$ Use form (9) with $u = e^x$.

19. $\int \dfrac{dx}{4x^2 - 9}$

20. $\int \dfrac{5\,dx}{\sqrt{x^2 + 25}}$

21. $\int \dfrac{x\,dx}{2 + 6x^2}$

22. $\int \dfrac{-3x\,dx}{(2 + 6x)^2}$

23. $\int x^2 e^x\,dx$

24. $\int x e^{x^2}\,dx$

25. $\int_0^2 \dfrac{dx}{\sqrt{9 + 4x^2}}$

26. $\int_1^2 \dfrac{3\,dx}{x(1 + 2x)}$

27. $\int_2^6 \dfrac{dx}{\sqrt{8x + x^2}}$

28. $\int_0^1 \dfrac{x\,dx}{16 - x^4}$

29. Look up the number of integration formulas in one of the handbooks of mathematical formulas, such as *Mathematical Handbook of Formulas and Tables* by Murray R. Spiegel, Schaum's Outline Series, McGraw-Hill Book Company, or *Mathematical Tables from Handbook of Chemistry and Physics,* Chemical Rubber Publishing Co.

14.3 INTEGRATION BY PARTS

In this section we consider one more method for evaluating integrals when a table of integral formulas is not available. It is based on the product rule for derivatives:

$$\frac{d}{dx}(uv) = u\frac{dv}{dx} + v\frac{du}{dx}$$

where u and v are functions of x. First, multiplying each side of the equation by dx gives

$$d(uv) = u\,dv + v\,du$$

Then, integrating each side gives

$$\int d(uv) = \int (u\,dv + v\,du) = \int u\,dv + \int v\,du$$

or

$$uv = \int u\,dv + \int v\,du$$

We can rewrite this last equation in the form

$$\int u\,dv = uv - \int v\,du \tag{1}$$

Equation (1), called the **integration by parts formula,** enables us to evaluate some types of integrals that cannot be evaluated using only the four basic forms of Section 14.1. Its use is illustrated in the next several examples; the method of integration is called *integration by parts.*

EXAMPLE 1 Evaluate $\int xe^x\,dx$.

Solution The given integral is in the form of the left side of equation (1) with

$$u = x \qquad dv = e^x\,dx$$

Then

$$du = dx \qquad v = \int e^x\,dx = e^x$$

(In determining v we have taken the constant of integration C to be zero.) With these substitutions, equation (1) becomes

$$\int xe^x \, dx = \int u \, dv = uv - \int v \, du = xe^x - \int e^x \, dx$$
$$= xe^x - e^x + C$$

That is,

$$\int xe^x \, dx = xe^x - e^x + C$$

In Example 1, the direct integration of $\int xe^x \, dx$ was replaced by the integration of the simpler integral $\int e^x \, dx$. This is typical of the integration by parts method—the integration of a given integral is replaced by the integration of a more simple integral. **1**

PRACTICE PROBLEM 1

Let $u = x$ and $dv = 2e^{2x} \, dx$.

(a) Determine du and v.
(b) Evaluate $\int 2xe^{2x} \, dx$.

Answer

(a) $du = dx$, $v = e^{2x}$
(b) $xe^{2x} - \frac{1}{2}e^{2x} + C$

EXAMPLE 2 Evaluate $\int x\sqrt{x+4} \, dx$.

Solution This integral is in the form of the left side of equation (1) with

$$u = x \qquad dv = \sqrt{x+4} \, dx$$

Then

$$du = dx \qquad v = \tfrac{2}{3}(x+4)^{3/2}$$

With these substitutions, equation (1) becomes

$$\int x\sqrt{x+4} \, dx = \int u \, dv = uv - \int v \, du$$
$$= \tfrac{2}{3}x(x+4)^{3/2} - \int \tfrac{2}{3}(x+4)^{3/2} \, dx$$
$$= \tfrac{2}{3}x(x+4)^{3/2} - \tfrac{4}{15}(x+4)^{5/2} + C \qquad \boxed{2}$$

PRACTICE PROBLEM 2

Let $u = x$ and $dv = \sqrt{x-3} \, dx$.

(a) Determine du and v.
(b) Evaluate $\int x\sqrt{x-3} \, dx$.

Answer

(a) $du = dx$, $v = \tfrac{2}{3}(x-3)^{3/2}$
(b) $\tfrac{2}{3}x(x-3)^{3/2} - \tfrac{4}{15}(x-3)^{5/2} + C$

Several things that are true of the integration by parts method in general should be noted in Examples 1 and 2.

1. The integrand in each case was the product of two functions—x and e^x in Example 1 and x and $\sqrt{x+4}$ in Example 2.
2. The functions u and dv were chosen so that du and v were easily determined.
3. The resulting integral $\int v \, du$ on the right side of equation (1) could be evaluated. This was $\int e^x \, dx$ in Example 1 and $\int \tfrac{2}{3}(x+4)^{3/2} \, dx$ in Example 2.

EXAMPLE 3 Evaluate $\int x(\ln x) \, dx$.

Solution If we let

$$u = x \qquad dv = \ln x \, dx$$

then

$$du = dx \qquad v = \int \ln x \, dx$$

14.3 INTEGRATION BY PARTS

This would be a poor choice for u and dv because we do not yet know how to evaluate $\int \ln x \, dx$.

However, if we let

$$u = \ln x \qquad dv = x \, dx$$

then

$$du = \frac{1}{x} dx \qquad v = \frac{x^2}{2}$$

Substitution into equation (1) gives

$$\int x(\ln x) \, dx = \int u \, dv = uv - \int v \, du$$

$$= \frac{x^2}{2} \ln x - \int \frac{x^2}{2} \cdot \frac{1}{x} dx$$

$$= \frac{x^2}{2} \ln x - \int \frac{x}{2} dx$$

$$= \frac{x^2}{2} \ln x - \frac{x^2}{4} + C$$

3 PRACTICE PROBLEM 3

(a) Determine eight possible choices of u and dv for evaluating $\int x^3 e^{x^2} dx$.

(b) Evaluate the integral in part (a).

Answer

(a)

u	dv	u	dv
x^3	$e^{x^2} dx$	e^{x^2}	$x^3 dx$
x^2	$xe^{x^2} dx$	xe^{x^2}	$x^2 dx$
x	$x^2 e^{x^2} dx$	$x^2 e^{x^2}$	$x \, dx$
1	$x^3 e^{x^2} dx$	$x^3 e^{x^2}$	dx

(b) $\frac{1}{2} x^2 e^{x^2} - \frac{1}{2} e^{x^2} + C$

EXERCISES

Evaluate each of the following integrals using integration by parts:

1. $\int x\sqrt{x + 3} \, dx$
2. $\int x\sqrt{x - 1} \, dx$
3. $\int x\sqrt{2x - 1} \, dx$
4. $\int x\sqrt{3x + 4} \, dx$
5. $\int x(x + 4)^2 \, dx$
6. $\int x(x - 7)^2 \, dx$
7. $\int 2x(x + 4)^2 \, dx$
8. $\int -3x(x - 4)^2 \, dx$
9. $\int 2x(-3x + 4)^2 \, dx$
10. $\int x(5x + 2)^2 \, dx$
11. $\int 3xe^{3x} \, dx$
12. $\int 5xe^{5x} \, dx$
13. $\int xe^{3x} \, dx$
14. $\int xe^{5x} \, dx$
15. $\int xe^{-x} \, dx$
16. $\int xe^{-4x} \, dx$
17. $\int 3xe^{-x} \, dx$
18. $\int 5xe^{x} \, dx$
19. $\int_{2}^{6} x\sqrt{2x - 3} \, dx$
20. $\int_{1}^{2} x(3x + 2)^2 \, dx$
21. $\int \ln x \, dx$ (*Hint:* Let $u = \ln x$ and $dv = dx$.)
22. Why can we *not* evaluate

$$\int x(x + 8)^2 \, dx$$

using the substitution method of Section 14.1 with $u = x + 8$ and $n = 2$ in form (1)?

23. Why can we *not* evaluate

$$\int 2xe^{2x}\,dx$$

using the substitution method of Section 14.1 with $u = 2x$ in form (2)?

24. State the integration by parts formula.

14.4 DIFFERENTIAL EQUATIONS

It seems reasonable to assume that the rate of sales of a reasonably priced fad item is proportional to the number of individuals who do not own the item. For example, the rate of sales of a new record by a popular rock group would be proportional to the number of the group's fans who do not own the record. In order to express this relationship mathematically, let y denote the number of fans in a particular city who do own the record; then the rate of change in sales can be approximated by the rate of change of y. In addition, if the group has 10,000 fans in the city, the number of fans not owning the record is $10{,}000 - y$. The rate of sales of the record in that city would be approximately

$$\frac{dy}{dt} = K(10{,}000 - y) \tag{1}$$

where t denotes time and K is the constant of proportionality. If we solve this equation for y, we would know (approximately) how many fans own the record at any time t.

Equation (1) is an example of a ***differential equation;*** that is, it is an equation containing the derivative of an unknown function. Some additional examples of differential equations are

$$\frac{dy}{dx} = \frac{y}{x}, \qquad 3y\frac{dy}{dx} = x^2, \qquad \frac{1}{y}\,dy = 2x\,dx$$

The last of these three equations may not look as though it contains a derivative. However, by dividing each side by dx, it can be rewritten in the form

$$\frac{1}{y}\frac{dy}{dx} = 2x$$

in which the appearance of the derivative is more obvious.

A ***solution*** of a differential equation is a function y that satisfies the equation in the sense that when y and its derivative (or differential) are substituted into the equation, equality holds. For example, $y = 3x$ is a solution for the equation $dy/dx = y/x$ because when $3x$ is substituted for y and its derivative 3 is substituted for dy/dx,

$$\frac{dy}{dx} = \frac{y}{x} \quad \text{becomes} \quad 3 = \frac{3x}{x}$$

The two sides of this last equation are obviously equal as long as $x \neq 0$. Similarly, $y = e^{x^2}$ is a solution for the third equation

$$\frac{1}{y} dy = 2x \, dx$$

above because when e^{x^2} is substituted for y and its differential $2xe^{x^2} dx$ is substituted for dy, equality does hold:

$$\frac{1}{e^{x^2}} (2x)e^{x^2} dx = 2x \, dx$$

Note that the left side of this equation simplifies to equal the right side.

To solve a differential equation means to determine a function y that is a solution of the equation. In general, the method of solving a differential equation depends on the form of the equation. If the equation can be written in the form

$$\boxed{\frac{dy}{dx} = f(x)} \qquad (2)$$

in which the derivative is expressed in terms of x alone, then the solution can be determined by integrating $f(x)$. For example, if

$$\frac{dy}{dx} = 3x^2 + 5$$

then

$$y = \int (3x^2 + 5) \, dx = x^3 + 5x + C$$

A solution such as that obtained in the last paragraph, which contains an arbitrary constant, is called the **general solution** of the given equation. If we add the requirement that $y = 2$ when $x = 1$, by substitution into the general solution we get

$$2 = 1^3 + 5(1) + C$$

This last equation requires that $C = -4$. With this value of C,

$$y = x^3 + 5x - 4$$

This solution is called the **particular solution** satisfying the *initial condition* $y = 2$ when $x = 1$.

EXAMPLE 1 Find the general solution and the particular solution satisfying the initial condition $y = 0$ when $x = -2$ of

$$dy = (x^2 - 4x) \, dx$$

Solution This equation can be written in the form of equation (2) by dividing both sides by dx. Then

$$y = \int (x^2 - 4x)\,dx = \frac{x^3}{3} - 2x^2 + C$$

is the general solution. For the particular solution, setting $y = 0$ and $x = -2$ in the general solution gives

$$0 = \frac{(-2)^3}{3} - 2(-2)^2 + C$$

Solving for C gives $C = \frac{32}{3}$. The required particular solution is

$$y = \frac{x^3}{3} - 2x^2 + \frac{32}{3}$$ ■

Note that the procedure is first to determine the general solution and then to use the initial condition to determine the value of the arbitrary constant for a particular solution.

If a given differential equation contains both y and its derivative, then it is impossible to write the equation in the form of equation (2). For example, the equation

$$y\,dy = x\,dx$$

cannot be written in that form. When both y and its derivative appear, the problem of solving the equation is much more complicated and at times is impossible.

We consider only one method for solving such equations. The method to be considered is for solving equations that can be written in the form

$$\boxed{f(y)\,dy = g(x)\,dx} \qquad (3)$$

That is, y is the only variable in the coefficient of dy and x is the only variable in the coefficient of dx.

When a differential equation can be written in the form of equation (3), it is called a *separable* equation; the solution is obtained by integrating both sides of the equation.

EXAMPLE 2 Find the general solution of

$$\frac{dy}{dx} = \frac{2x + 1}{2y}$$

Solution This equation can be written in the form of equation (3) by multiplying each side by $2y\,dx$ to obtain

$$2y\,dy = (2x + 1)\,dx$$

---

1 PRACTICE PROBLEM 1

Find (a) the general solution and (b) the particular solution of $dy = (2x + 1)\,dx$ satisfying the initial condition $y = 5$ when $x = 1$.

Answer

(a) $y = x^2 + x + C$
(b) $y = x^2 + x + 3$

14.4 DIFFERENTIAL EQUATIONS

Then integrating each side gives

$$\int 2y\,dy = \int (2x+1)\,dx$$

or

$$y^2 + C_1 = x^2 + x + C_2$$

If we denote the difference of the arbitrary constants $C_2 - C_1$ by C, this last equation can be written

$$y^2 = x^2 + x + C$$

The general solution is, therefore,

$$y = \pm\sqrt{x^2 + x + C}$$

PRACTICE PROBLEM 2

Find the general solution of

$$2\frac{dy}{dx} = \frac{2x-3}{y}$$

Answer

$$y = \pm\sqrt{x^2 - 3x + C}$$

EXAMPLE 3 Determine the particular solution of the differential equation in Example 2 satisfying the given conditions.

(a) $y = -2$ when $x = 0$ (b) $y = 3$ when $x = 1$

Solution (a) In order for the value of y to be negative, we must choose the negative square root in the general solution found in Example 2. Setting $y = -2$ and $x = 0$ gives

$$-2 = -\sqrt{C}$$

so that $C = 4$; the required particular solution is

$$y = -\sqrt{x^2 + x + 4}$$

(b) We use the positive square root. Setting $y = 3$ and $x = 1$ gives

$$3 = \sqrt{2 + C}$$

Therefore, $2 + C = 9$ and $C = 7$. The required particular solution is

$$y = \sqrt{x^2 + x + 7}$$

EXAMPLE 4 Determine the general solution of

$$\frac{dQ}{dt} = kQ \qquad (4)$$

where $Q > 0$ and k is a constant.

PRACTICE PROBLEM 3

Find the particular solution of the differential equation in Practice Problem 2 satisfying each of the following:

(a) $y = 4$ when $x = 0$
(b) $y = -3$ when $x = 2$

Answer

(a) $y = \sqrt{x^2 - 3x + 16}$
(b) $y = -\sqrt{x^2 - 3x + 11}$

Solution Separating variables to obtain the form of equation (3) gives

$$\frac{dQ}{Q} = k\,dt$$

Solving this last equation by integrating both sides results in the general solution

$$\ln|Q| = kt + C$$

or, since $Q > 0$,

$$\ln Q = kt + C$$

Then, from this last equation,

$$e^{\ln Q} = e^{kt+C} = e^{kt} \cdot e^C$$

Since $e^{\ln Q} = Q$, the general solution can finally be written

$$Q = Ae^{kt}$$

where A is written in place of the constant e^C.

Equation (4) is one of the more important differential equations in the application of mathematics. It states that Q is changing at a rate proportional to itself. Consequently, it serves to describe physical quantities which, at any time, are changing in proportion to the amount present.

Quantities with this property—that the rate of change is proportional to the amount present—are said to *increase* (or *decrease,* as the case may be) *exponentially.* If the quantity is increasing, the constant k is positive and is called the **growth factor.** If the quantity is decreasing, then k is negative and is called the **decay factor.**

The general solution

$$Q(t) = Ae^{kt} \qquad (5)$$

obtained in Example 4 gives the amount present at any time t. Note that

$$Q(0) = Ae^0 = A$$

so that A in equation (5) denotes the quantity present when the observation begins—that is, at $t = 0$.

EXAMPLE 5 Money invested at 6% interest compounded continuously grows exponentially with a growth factor of 0.06. If $1,000 is invested in this manner, what will be the amount (the original $1,000 plus the interest) 1 year later?

Solution The rate of change in the amount of money is given by

$$\frac{dQ}{dt} = 0.06Q$$

The solution given by (5) can be applied directly;

$$Q(t) = 1{,}000 e^{0.06t}$$

gives the amount at any time t. At $t = 1$, we have

$$Q(1) = 1{,}000 e^{0.06(1)} = 1{,}061.84$$

or $1,061.84.

Observe the four-step process involved in the application of mathematics in Example 5. The physical situation was the investment of $1,000 as described and the physical question was: What will be the amount after 1 year? The mathematical model was the differential equation

$$\frac{dQ}{dt} = 0.06 Q$$

The mathematical question was: What is the value of $Q(1)$? The mathematical solution was $Q(1) = 1{,}061.84$, and the interpretation was that the amount will be $1,061.84.

4 PRACTICE PROBLEM 4

If $1,000 is invested at 7% compounded continuously, determine the amount 2 years later.

Answer

$1,150.27

EXERCISES

For each of the following differential equations, verify that the given function is in fact a solution:

1. $y \dfrac{dy}{dx} = x$;   solution: $y = \sqrt{x^2}$

2. $y \dfrac{dy}{dx} = 2x^3 + 2x$;   solution: $y = x^2 + 1$

3. $x^3 \, dy = -\dfrac{1}{y} \, dx$;   solution: $y = \dfrac{1}{x}$

4. $dy = 2y \, dt$;   solution: $y = e^{2t}$

5. $dt = t \, dy$;   solution: $y = \ln t + C$, where C is an arbitrary constant

6. $\dfrac{1}{2} \dfrac{dy}{dx} = x + 1$;   solution: $y = (x+1)^2 + C$, where C is an arbitrary constant

Find the general solution and the particular solution if initial conditions are given.

7. $\dfrac{dy}{dx} = 2x + 4$

8. $\dfrac{dy}{dx} = x^3 - x$

9. $dy = (2 - x^{-2}) \, dx$

10. $dy = \dfrac{dx}{x}$

11. $dy = (3x^2 - 2x) \, dx$; $y = 2$ when $x = 1$

12. $dy = \sqrt{x}\, dx$; $y = -5$ when $x = 0$
13. $t\, dy = dt$; $y = -10$ when $t = 1$
14. $dy = e^t\, dt$; $y = 1$ when $t = 0$
15. $y\, dy = x\, dx$
16. $y\, dy = (e^x + 3x^2)\, dx$
17. $dy = 2yx\, dx$
18. $2y\, dy = (x^2 + 1)\, dx$
19. $\dfrac{dy}{dx} = \dfrac{3x^2}{2y}$; $y = -3$ when $x = 2$
20. $\dfrac{dy}{dx} = y^2(x + 1)$; $y = -1$ when $x = 2$
21. $\dfrac{3y^2\, dy}{2t + 1} = dt$; $y = 5$ when $t = 10$
22. $x\, dx - \dfrac{dy}{\sqrt{y}} = 0$; $y = 4$ when $x = 4$

23. *Continuous compounding*   If $5,000 is invested at 9% compounded continuously, what will be the amount 16 months ($\tfrac{4}{3}$ years) later?

24. *Population growth*   The population in a developing country was 10,000,000 in 1980. The population increases exponentially with a growth factor of 0.08 per year.
    (a) Determine a function that will give the population at any future time.
    (b) What will the population be in 1990?

25. *Population decrease*   After a killing agent is introduced into a bacteria colony, the bacteria population decreases exponentially at a decay rate of -0.20 per second. If there are 10,000 bacteria present when the agent is introduced, what will the population be 10 seconds later?

26. *Radioactive decay*   A radioactive chemical has a decay rate of -0.012 per year. There are 1,000 grams present initially.
    (a) Determine a function that will give the number of grams at any future time.
    (b) How much of the chemical will be present 6 months later?

27. *Present value*   If a sum of money P is invested now at some fixed rate of interest, it will yield a certain amount A at any time in the future. The sum P is the **present value** of the future amount A. For example, $100 invested at 8% simple interest will yield $116 in 2 years; the $100 is the present value of the $116 in 2 years. The present value of money invested at 9% interest compounded continuously decreases exponentially with a decay factor of -0.09.
    (a) Determine a function that will give the present value of an amount of $5,000 resulting from an investment compounded continuously at 9% for any time t.
    (b) Determine the present value if the time t in part (a) is 2 years.

28. *Temperature*   Newton's law of cooling states that the rate at which a body changes temperature is proportional to the difference between its temperature and the temperature of the surrounding medium. If a body is in air at 40°F and cools from 120°F to 60°F in 30 minutes, determine its temperature at any time t.

29. *Animal population* The rate of change of the number N of fish in a particular lake is approximately

$$\frac{dN}{dt} = -KN$$

where $-K$ is due to the natural mortality of the fish. If the number of fish is initially $(t = 0)$ estimated to be P, determine an expression that gives the number of fish at any future time.

30. *Continuous compounding* If $5,000 invested now at interest compounded continuously will amount to $6,549.80 at the end of 3 years, determine the amount at the end of 5 years.

31. Verify that the time it takes a quantity that increases exponentially with a growth factor k to double is given by $t = (\ln 2)/k$. (*Hint:* Solve the equation $2A = Ae^{kt}$ for t.)

Using the result given in Exercise 31, determine each of the following:

32. *Continuous compounding* How long will it take in Exercise 23 for the amount to be $10,000?

33. *Population growth* How long will it take the population in Exercise 24 to double?

34. Explain why the following statement is true: Any sum of money invested at $k\%$ compounded continuously will double in approximately $70/k$ years.

The time it takes a quantity that decreases exponentially with a decay factor k to halve is given by $t = (\ln 2)/-k$. Using this result, determine each of the following:

35. *Population decrease* How long will it take the population in Exercise 25 to halve?

36. *Radioactive decay* How long will it take until the original amount of chemical in Exercise 26 reduces to 500 grams?

37. *Present value* How long will it take the $5,000 in Exercise 27 to halve? (Compare the result here with that of Exercise 32.)

38. In the record illustration at the beginning of this section, determine how many fans own the record at any time t if 500 fans own the record initially (at $t = 0$) and 2,400 own the record 2 weeks later.

Identify the steps of the four-step procedure involved in the application of mathematics (Section 1.3) in each case.

39. *Continuous compounding* Exercise 23

40. *Population decrease* Exercise 25

Explain the meaning of each term.

41. Differential equation
42. Solution of a differential equation
43. General solution of a differential equation
44. Particular solution of a differential equation

45. *Bonus question* What is the difference between a quantity increasing geometrically (Chapter 9) and increasing exponentially?

14.5 IMPROPER INTEGRALS

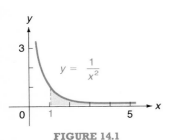

FIGURE 14.1

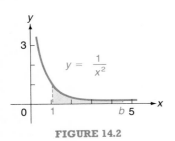

FIGURE 14.2

All the integrals we have considered to this point have been integrals on finite intervals. The integral $\int_2^5 (x^2 + 1)\, dx$, for example, is the integral of $x^2 + 1$ on the finite interval [2, 5]. The integral $\int_{-1}^3 e^x\, dx$ is the integral of e^x on the interval [−1, 3]. In some applications, it is necessary to evaluate integrals on infinite intervals. These integrals, called *improper integrals,* are considered in this section.

Suppose we want to determine the area between the graph of $y = 1/x^2$ and the x-axis to the right of $x = 1$ (see Figure 14.1). The area in which we are interested is the area over an infinite interval. We cannot simply set up and evaluate an integral as we have done previously for finite intervals. Rather, we determine the area over an arbitrary interval [1, b] (see Figure 14.2):

$$\int_1^b \frac{1}{x^2}\, dx = -\frac{1}{x}\Big|_1^b = 1 - \frac{1}{b} \text{ square units}$$

The total area to the right of 1 is found by examining the value of the area $1 - (1/b)$ as b gets arbitrarily large:

$$\text{Area} = \lim_{b \to \infty}\left(1 - \frac{1}{b}\right) = 1 \text{ square unit}$$

or

$$\text{Area} = \lim_{b \to \infty} \int_1^b \frac{1}{x^2}\, dx$$

If we denote the area by the integral

$$\int_1^\infty \frac{1}{x^2}\, dx$$

we have

$$\int_1^\infty \frac{1}{x^2}\, dx = \lim_{b \to \infty} \int_1^b \frac{1}{x^2}\, dx$$

We extend this relationship to integrals of arbitrary functions by the following definition. If $f(x)$ is continuous for $x \geq a$, then

$$\boxed{\int_a^\infty f(x)\, dx = \lim_{b \to \infty} \int_a^b f(x)\, dx}$$

The definition indicates that first the integration on the right is to be performed over the finite interval [a, b] and then the limit of the result is to be evaluated as b becomes arbitrarily large. If the limit does not have a value, the improper integral has no value.

EXAMPLE 1 Evaluate $\int_0^\infty e^{-2x}\, dx$.

14.5 IMPROPER INTEGRALS

Solution According to the definition,

$$\int_0^\infty e^{-2x}\,dx = \lim_{b\to\infty}\int_0^b e^{-2x}\,dx = \lim_{b\to\infty}\left(-\frac{1}{2}e^{-2x}\bigg|_0^b\right)$$

$$= \lim_{b\to\infty}\left(\frac{-1}{2e^{2b}} + \frac{1}{2}\right) = \frac{1}{2}$$

PRACTICE PROBLEM 1

Evaluate $\int_3^\infty \frac{1}{x^3}\,dx$.

Answer

$\frac{1}{18}$

Note that the evaluation of the limit is the last step in the solution in Example 1.

EXAMPLE 2 Evaluate $\int_3^\infty x\,dx$.

Solution

$$\int_3^\infty x\,dx = \lim_{b\to\infty}\int_3^b x\,dx = \lim_{b\to\infty}\left(\frac{x^2}{2}\bigg|_3^b\right)$$

$$= \lim_{b\to\infty}\left(\frac{b^2}{2} - \frac{9}{2}\right)$$

Since

$$\frac{b^2}{2} - \frac{9}{2}$$

becomes arbitrarily large as b does, this limit does not exist. Consequently,

$$\int_3^\infty x\,dx$$

has no value.

PRACTICE PROBLEM 2

Evaluate $\int_1^\infty \sqrt{x}\,dx$.

Answer

The integral has no value.

A second type of improper integral is one of the form

$$\int_{-\infty}^b f(x)\,dx$$

in which the lower limit of integration is negative infinity. If $f(x)$ is continuous for $x \leq b$, the value of such an integral is defined by the equation

$$\int_{-\infty}^b f(x)\,dx = \lim_{a\to-\infty}\int_a^b f(x)\,dx$$

Again, if this limit does not have a value, the improper integral has no value.

EXAMPLE 3 Evaluate $\int_{-\infty}^0 e^x\,dx$.

Solution According to the definition just given,

$$\int_{-\infty}^0 e^x\,dx = \lim_{a\to-\infty}\int_a^0 e^x\,dx = \lim_{a\to-\infty}\left(e^x\bigg|_a^0\right)$$

$$= \lim_{a\to-\infty}(1 - e^a) = 1 - 0 = 1$$

PRACTICE PROBLEM 3

Evaluate $\int_{-\infty}^4 4x\,dx$.

Answer

The integral has no value.

A final type of improper integral is one of the form

$$\int_{-\infty}^{\infty} f(x)\,dx$$

in which both the upper and lower limits of integration are infinite. If $f(x)$ is continuous for every x, the value of such an integral is given by the equation

$$\boxed{\int_{-\infty}^{\infty} f(x)\,dx = \int_{-\infty}^{0} f(x)\,dx + \int_{0}^{\infty} f(x)\,dx}$$

That is,

$$\int_{-\infty}^{\infty} f(x)\,dx = \lim_{a \to -\infty} \int_{a}^{0} f(x)\,dx + \lim_{b \to \infty} \int_{0}^{b} f(x)\,dx$$

In this case, if even one of the two limits does not exist, the improper integral has no value.

EXAMPLE 4 Evaluate $\int_{-\infty}^{\infty} 2x\,dx$.

Solution

$$\int_{-\infty}^{\infty} 2x\,dx = \lim_{a \to -\infty} \int_{a}^{0} 2x\,dx + \lim_{b \to \infty} \int_{0}^{b} 2x\,dx$$

$$= \lim_{a \to -\infty} (0 - a^2) + \lim_{b \to \infty} (b^2 - 0)$$

Since neither of these last two limits exists, the improper integral has no value.

4 PRACTICE PROBLEM 4

Evaluate $\int_{-\infty}^{\infty} \dfrac{2x}{(x^2 + 3)^2}\,dx$.

Answer

0

An improper integral of the three types we have considered is said to be **convergent** if it has a value according to the definitions given. If it has no value, the improper integral is said to be **divergent.**

EXERCISES

Determine whether each of the following improper integrals is convergent or divergent; if it is convergent, give the value of the integral:

1. $\int_{2}^{\infty} \dfrac{1}{x^2}\,dx$ 2. $\int_{1}^{\infty} \dfrac{-2}{x^3}\,dx$ 3. $\int_{0}^{\infty} 2x^2\,dx$

4. $\int_{1}^{\infty} \dfrac{1}{x}\,dx$ 5. $\int_{0}^{\infty} 3e^{-x}\,dx$ 6. $\int_{4}^{\infty} \dfrac{dx}{\sqrt{x}}$

7. $\int_{-\infty}^{-1} \dfrac{5}{x^3}\,dx$ 8. $\int_{-\infty}^{0} 3^x\,dx$ 9. $\int_{-\infty}^{1} \sqrt[3]{x}\,dx$

10. $\int_{-\infty}^{5} 2\,dx$ 11. $\int_{1}^{\infty} \dfrac{dx}{(x+2)^2}$ 12. $\int_{0}^{\infty} \dfrac{2x\,dx}{(x^2+3)^3}$

13. $\int_{-\infty}^{0} \dfrac{2x\,dx}{x^2+3}$ 14. $\int_{-\infty}^{-2} x3^{-x^2}\,dx$ 15. $\int_{-\infty}^{\infty} 3x^2\,dx$

16. $\int_{-\infty}^{\infty} 2^x \, dx$ 17. $\int_{-\infty}^{\infty} 2xe^{-x^2} \, dx$ 18. $\int_{-\infty}^{\infty} 2xe^{x^2} \, dx$

19. Give a geometric interpretation of the value of the improper integral in Example 3.

Explain the meaning of each term.

20. Convergent improper integral   21. Divergent improper integral

*14.6 PROBABILITY DENSITY FUNCTIONS

FIGURE 14.3

If the arrow on the spinner in Figure 14.3 is spun, what is the probability that it lands on a number between 1 and 3?

The previous methods for calculating probabilities do not apply here because the arrow can land on any number between 0 and 5, and—consequently—there are an infinite number of possible outcomes when the arrow is spun. The previous methods for calculating probabilities always presumed that the sample space consisted of a finite number of elements, which is not the case here.

Intuitively, the probability of obtaining a number between 1 and 3 should be $\frac{2}{5}$ because the interval from 1 to 3 makes up $\frac{2}{5}$ of the total distance around the dial. Note that we get the same result by evaluating the integral

$$\int_1^3 \frac{1}{5} \, dx = \frac{1}{5} x \Big|_1^3 = \frac{3}{5} - \frac{1}{5} = \frac{2}{5}$$

Similarly, the probability of the arrow landing on a number in the interval $[c, d]$, where $0 \le c \le d \le 5$, is given by the integral

$$\int_c^d \frac{1}{5} \, dx$$

If X is the random variable describing the outcome when the arrow is spun, then the function $f(x) = \frac{1}{5}$ is called the **density function** for X. In general, a density function for a random variable X with possible values in the interval $[a, b]$ is a function f with the following two properties:

> 1. $\int_a^b f(x) \, dx = 1$
>
> 2. $f(x) \ge 0$ for each x in the interval $[a, b]$

Then the probability that X takes a value in the interval $[c, d] \subset [a, b]$ is given by

$$P(c \le X \le d) = \int_c^d f(x) \, dx$$

* This section can be omitted without loss of continuity.

The first property above of a density function guarantees that the probability of the random variable taking a value in the sample space $S = [a, b]$ is 1 since

$$P(a \leq X \leq b) = \int_a^b f(x)\,dx = 1$$

The second property guarantees that the probability of the random variable taking a value in any interval is greater than or equal to zero—that is, probabilities will never be negative.

If $f(x)$ is the density function for a random variable X, then X is said to have the **distribution** specified by $f(x)$.

EXAMPLE 1 Verify that $f(x) = x/36$ is a density function for the random variable X that takes values in the sample space $S = [3, 9]$.

Solution Since

$$\int_3^9 \frac{x}{36}\,dx = \left.\frac{x^2}{72}\right|_3^9 = 1$$

the first requirement of a density function is satisfied. For the second,

$$\frac{x}{36} \geq 0 \quad \text{for each} \quad x \in [3, 9]$$

as required.

EXAMPLE 2 Determine the probability that the random variable X of Example 1 takes a value in the interval $[3, 5]$.

Solution
$$P(3 \leq X \leq 5) = \int_3^5 \frac{x}{36}\,dx = \left.\frac{x^2}{72}\right|_3^5 = \frac{2}{9}$$

Since a density function $f(x)$ is never negative, its integral over any interval $[c, d]$ is equal to the area under the graph of the density function over the interval $[c, d]$ (see Figure 14.4). Consequently, the probability that the random variable takes a value in the interval $[c, d]$ is the same as this area:

$$\text{Area} = \int_c^d f(x)\,dx = P(c \leq X \leq d)$$

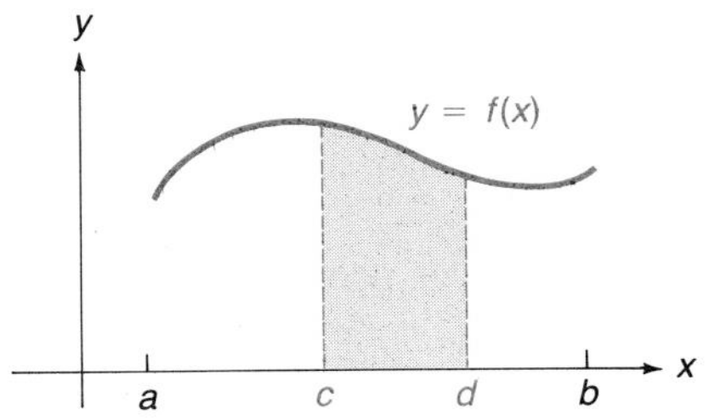

FIGURE 14.4

Up to this point we have considered random variables on a sample space consisting of a finite interval $S = [a, b]$. The discussion extends

---

1 PRACTICE PROBLEM 1

(a) Verify that $f(x) = 2x/25$ is a density function for a random variable X that takes values in the sample space $S = [0, 5]$.

(b) Determine the probability that the random variable X in part (a) takes a value between 0 and 2.

Answer

(a) $\int_0^5 \frac{2x}{25}\,dx = 1$ and $f(x) \geq 0$ for each x in $[0, 5]$

(b) $\frac{4}{25}$

14.6 PROBABILITY DENSITY FUNCTIONS

EXAMPLE 3 Verify that

$$f(x) = \tfrac{1}{5} e^{-x/5}$$

is a density function for the random variable X that takes values in the sample space $S = \{x : x \geq 0\}$.

Solution Since

$$\int_0^\infty \tfrac{1}{5} e^{-x/5}\, dx = \lim_{c \to \infty} \int_0^c \tfrac{1}{5} e^{-x/5}\, dx = \lim_{c \to \infty} (1 - e^{-c/5}) = 1$$

the first requirement of a density function is satisfied. Also, since the value of e^t is positive for every value of t,

$$\tfrac{1}{5} e^{-x/5}$$

is also positive for every value of x, in particular, for all $x \geq 0$. The given function satisfies the requirements for a density function. ■

2 PRACTICE PROBLEM 2

Verify that $f(x) = \tfrac{2}{3} e^{-2x/3}$ is a density function for the random variable X that takes values in the sample space $S = \{x : x \geq 0\}$.

Answer

$\int_0^\infty \tfrac{2}{3} e^{-2x/3}\, dx = 1$ and $f(x) \geq 0$ for each $x \geq 0$

EXAMPLE 4 If the random variable in Example 3 denotes the time spent by a customer in the checkout line at a grocery store, determine the probability that a randomly chosen customer will spend between 3 and 5 minutes in the line.

Solution

$$P(3 \leq X \leq 5) = \int_3^5 \tfrac{1}{5} e^{-x/5}\, dx = -e^{-x/5} \Big|_3^5 = 0.18093 \quad ■$$

EXAMPLE 5 Determine the probability that the customer in Example 4 will spend more than 6 minutes in line.

Solution

$$P(X \geq 6) = \int_6^\infty \tfrac{1}{5} e^{-x/5}\, dx = \lim_{c \to \infty} \int_6^c \tfrac{1}{5} e^{-x/5}\, dx$$
$$= \lim_{c \to \infty} (e^{-6/5} - e^{-c/5}) = 0.30119$$

Geometrically, this probability is the area under the graph of the density function and over the interval $x \geq 6$.

3 PRACTICE PROBLEM 3

If the random variable in Practice Problem 2 denotes the length in hours of the downtime of a particular computer, determine the probability that the downtime will be:

(a) Between 2 and 3 hours
(b) More than $1\tfrac{1}{2}$ hours

Answer

(a) 0.12827   (b) 0.36788

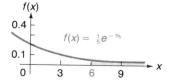

The most important distribution is undoubtedly the normal distribution (Section 7.5). The density function is

$$f(x) = \frac{1}{\sigma\sqrt{2\pi}} e^{-(x-u)^2/2\sigma^2}$$

To integrate $f(x)$ requires more advanced techniques than we have considered here. However, the normal probabilities given in Table II in Appendix B are determined in the manner discussed above with $\sigma = 1$ and $u = 0$.

EXERCISES

1. (a) Verify that $f(x) = \frac{1}{4}$ is a density function for the random variable X that takes values in the sample space $S = [2, 6]$.
   (b) Determine the probability that the random variable in part (a) takes a value in the interval $[3, 5]$.
   (c) Describe the geometric meaning of the probability found in part (b).

2. (a) Verify that $f(x) = \frac{1}{2}$ is a density function for the random variable X that takes values in the sample space $[-1, 1]$.
   (b) Determine the probability that the random variable in part (a) takes a value in the interval $[0, 0.5]$.
   (c) Describe the geometric meaning of the probability found in part (b).

3. (a) Verify that $f(x) = \frac{3}{64}x^2$ is a density function for the random variable X that takes values in the sample space $X = [0, 4]$.
   (b) Determine the probability that the random variable in part (a) takes a value in the interval $[0, 2]$.
   (c) Describe the geometric meaning of the probability found in part (b).

4. (a) Verify that $f(x) = 1 - (1/\sqrt{x})$ is a density function for the random variable X that takes values in the sample space $S = [1, 4]$.
   (b) Determine the probability that the random variable in part (a) takes a value in the interval $[2, 3]$.
   (c) Describe the probability determined in part (b) geometrically.

5. (a) Verify that $f(x) = e^{-x}$ is a density function for the random variable X that takes values in the sample space $S = \{x : x \geq 0\}$.
   (b) Determine $P(0 \leq X \leq 5)$, $P(X \geq 4)$, and $P(X \leq 2)$.
   (c) Describe each of the probabilities found in part (b) geometrically.

6. (a) Verify that $f(x) = 0.1e^{-0.1x}$ is a density function for the random variable X that takes values in the sample space $S = \{x : x \geq 0\}$.
   (b) Determine $P(X \leq 20)$, $P(10 \leq X \leq 20)$, and $P(X \geq 35)$.
   (c) Describe each of the probabilities found in part (b) geometrically.

7. Determine the value of k for which $f(x) = kx$ is a density function for the random variable X that takes a value in the sample space $S = [6, 10]$.

8. Why is $f(x) = x^3/60$ not a density function for the random variable X that takes values in the sample space $S = [-2, 4]$?

14.6 PROBABILITY DENSITY FUNCTIONS

9. *Service time*   The time required (in minutes) for a bank teller in a particular bank to wait on a customer is a random variable X with density function

$$f(x) = \tfrac{1}{2} e^{-x/2} \qquad 0 \leq x$$

What is the probability that the time required to wait on an arbitrarily chosen customer will take:
(a) From 2 to 4 minutes?   (b) More than 4 minutes?

10. *Performance time*   The time required (in hours) for a crew to unload a truck at the service dock is a random variable X with density function

$$f(x) = \tfrac{1}{3} e^{-x/3} \qquad 0 \leq x$$

What is the probability that the crew will unload the next truck to arrive in less than 3 hours?

11. *Employee learning*   The time required (in days) for a new employee to learn a particular job is a random variable with density function

$$f(x) = e^{-x} \qquad 0 \leq x$$

Determine the probability that a randomly chosen future employee will learn the job in less than 2 days.

12. *Product life*   The life (in days) of a particular type of transistor is a random variable with density function

$$f(x) = \frac{400}{x^2} \qquad x \geq 400$$

Determine the probability that one of the transistors, arbitrarily chosen, will have a life of more than 500 days.

13. *Selling price*   The selling price (in dollars) of a particular item is a random variable with density function

$$f(x) = \tfrac{2}{3} \qquad 3 \leq x \leq 4.5$$

If a store selling the item is arbitrarily chosen, determine the probability that the store charges between \$3.75 and \$4.00 for the item.

14. *Plant height*   The height (in inches) at maturity of a particular plant is a random variable with density function

$$f(x) = \tfrac{1}{6} \qquad 18 \leq x \leq 24$$

Determine the probability that one of the plants, arbitrarily chosen, will grow to a height of less than 20 inches.

15. *Patient stay*   If the length of a patient's stay in a hospital (in days) is a random variable with density function

$$f(x) = \frac{2}{(2x+1)^2} \qquad x \geq 0$$

determine the probability that the next patient to be admitted will stay less than 1 week.

Identify the steps of the four-step procedure involved in the application of mathematics (Section 1.3) in each case.

16. *Service time*   Exercise 9(b)
17. *Plant height*   Exercise 14
18. Explain the meaning of a density function for a random variable X with possible values in the interval $[a, b]$.

14.7 CHAPTER SUMMARY

Three different methods of evaluating integrals were considered in this chapter:

1. Integration by substitution
2. Use of integral tables
3. Integration by parts

Integration by substitution was based on the following four basic integral formulas:

$$1. \int u^n \, du = \frac{u^{n+1}}{n+1} + C, \quad n \neq -1$$

$$2. \int e^u \, du = e^u + C$$

$$3. \int b^u \, du = \frac{b^u}{\ln b} + C, \quad b > 0$$

$$4. \int \frac{1}{u} \, du = \ln|u| + C$$

In these formulas, u represents a function and du represents its differential. The **differential** of an independent variable x is simply denoted by dx; that of a function $f(x)$ is $df = f'(x) \, dx$. The key idea in integration by substitution is to recognize when a given integral is in one of the four forms above for a particular function u.

Integration using an integral table is essentially the same as integration by substitution. Instead of using just the four forms above, we used the expanded table given in Section 14.2.

Integration by parts is based on the relationship

$$\int u \, dv = uv - \int v \, du$$

In this equation u and v represent functions. The advantage of the integration by parts method is that evaluation of a given integral $\int u \, dv$ is replaced by evaluation of an integral $\int v \, du$ that is, we hope, more simple.

Techniques for evaluating integrals are also used to solve differential equations. A **differential equation** is an equation containing the derivative of an unknown function. A **solution** of a differential equation is a function y that satisfies the equation in the sense that when y and its derivative (or its differential) are substituted into the equation, the equation is true. The technique for solving a differential equation depends on the form of the equation. We considered two general forms:

$$\frac{dy}{dx} = f(x) \quad \text{or} \quad dy = f(x)\,dx$$

and

$$f(y)\,dy = g(x)\,dx$$

(Is the first form a special case of the second?) The solution of the first form requires an integration of the right side; that of the second requires that each side be integrated.

There are also two types of solutions for a differential equation, **general** and **particular**. The general solution of a differential equation contains an arbitrary constant C. A particular solution is obtained from the general solution when initial conditions are given enabling one to determine a specific value for C.

Integrals with infinite limits of integration are called **improper integrals**. The three types together with the value defined for them are given in the following table:

1. $\int_a^\infty f(x)\,dx = \lim_{b \to \infty} \int_a^b f(x)\,dx$ if $f(x)$ is continuous for all $x \geq a$

2. $\int_{-\infty}^b f(x)\,dx = \lim_{a \to -\infty} \int_a^b f(x)\,dx$ if $f(x)$ is continuous for all $x \leq b$

3. $\int_{-\infty}^\infty f(x)\,dx = \int_{-\infty}^0 f(x)\,dx + \int_0^\infty f(x)\,dx$ if $f(x)$ is continuous for all x

For each of the first two types, an ordinary integral with finite limits of integration is evaluated; then, as the last step, the indicated limits are evaluated. An integral of the third type is first written as the sum of two integrals of the first types; if either of these first two types does not have a value, then the given integral does not have a value.

Improper integrals are used in conjunction with density functions of a particular type. In general, a **density function** for a random variable X with possible values in the interval $[a, b]$ is a function f with the following two properties:

1. $\int_a^b f(x)\,dx = 1$

2. $f(x) \geq 0$ for each x in the interval $[a, b]$

Then the probability that X takes a value in the interval $[c, d] \subset [a, b]$ is given by

$$P(c \leq X \leq d) = \int_c^d f(x)\, dx$$

If the random variable takes values on an infinite interval, the integrals in the first requirement of a density function and in the calculation of probabilities are replaced by appropriate improper integrals.

REVIEW EXERCISES

Evaluate each of the following integrals using the substitution method:

1. $\int (x^3 - 4)^2 x^2\, dx$

2. $\int \dfrac{5x^2}{(x^3 - 4)^2}\, dx$

3. $\int \dfrac{x}{3x^2 - 4}\, dx$

4. $\int \dfrac{7}{5 + 4y}\, dy$

5. $\int t e^{2t^2}\, dt$

6. $\int e^{x^3 - 3x}(x^2 - 1)\, dx$

7. $\int_1^2 (\ln 3) 3^{2x}\, dx$

8. $\int_0^2 (x + 1)(2x^2 + 4x)^2\, dx$

Use Table 14.2 (in Section 14.2) to evaluate each of the following integrals:

9. $\int \dfrac{x\, dx}{(1 + 2x)^2}$

10. $\int t^3 (\ln t)\, dt$

11. $\int \dfrac{3\, dx}{9 - 4x^2}$

12. $\int \dfrac{dx}{\sqrt{6x + x^2}}$

13. $\int (2x)^2 e^x\, dx$

14. $\int \dfrac{3y}{\sqrt{1 + y^2}}\, dy$

15. $\int_0^1 \dfrac{x\, dx}{1 + 3x}$

16. $\int_{\ln 3}^{\ln 5} \dfrac{e^x\, dx}{4 - e^{2x}}$

Evaluate the following using integration by parts:

17. $\int x(x - 7)^2\, dx$

18. $\int 2x\sqrt{3x - 1}\, dx$

19. $\int 6x e^{-x}\, dx$

20. $\int x(\ln x)\, dx$

21. $\int_4^{12} x\sqrt{2x + 1}\, dx$

22. $\int_0^{\ln 5} x e^{2x}\, dx$

Evaluate the following integrals:

23. $\int x\sqrt{2x^2 + 3}\, dx$

24. $\int -3x(x + 5)^2\, dx$

25. $\int \dfrac{dx}{\sqrt{4x + x^2}}$

26. $\int (8x + 6) e^{2x^2 + 3x}\, dx$

27. $\int (e^x - xe^x)\, dx$

28. $\int \dfrac{x\, dx}{2 + 5x}$

29. $\int_1^5 (2x - 4)^2\, dx$

30. $\int_{\ln 2}^{\ln 4} 3x e^{-x}\, dx$

Determine whether each of the following integrals are convergent or divergent; if it is convergent, give its value:

31. $\int_3^\infty \dfrac{5}{x^2} \, dx$

32. $\int_1^\infty -6x \, dx$

33. $\int_{-\infty}^{-1} -2e^{-x} \, dx$

34. $\int_{-\infty}^0 \dfrac{x \, dx}{(x^2+5)^2}$

35. $\int_{-\infty}^\infty 2x3^{-x^2} \, dx$

36. $\int_{-\infty}^\infty 7x^{1/3} \, dx$

Determine the area between the graph of each of the following functions and the x-axis:

37. $y = x\sqrt{2x-1}$ from $x = 1$ to $x = 5$

38. $y = x2^{x^2}$ from $x = -2$ to $x = -1$

Determine the area between the graph of $y = xe^x$ and the x-axis in each case.

39. From $x = 0$ to $x = 2$

40. From $x = -1$ to $x = 0$

A possible solution is given with each of the following differential equations. Determine whether the given function is, in fact, a solution of the corresponding equation.

41. $dy = (3x^2 + 2x) \, dx;\quad y = x^3 + x^2 - 1$

42. $x \, dy = 2x;\quad y = \ln x^2 + C$

43. $y^2 \, dy = x \, dx;\quad y = e^x$

44. $dy = \dfrac{x}{y} \, dy;\quad y = x^2 + C$

For each of the following differential equations find the general solution and the particular solution if initial conditions are given:

45. $\dfrac{dy}{dx} = x - \dfrac{1}{x^2}$

46. $\dfrac{dy}{dx} = 3e^x + 1$

47. $dy = 2(x+1) \, dx;\quad y = -2$ when $x = 0$

48. $dy = 4(t+1)^2 \, dt;\quad y = 5$ when $t = 0$

49. $3y^2 \, dy = 2x \, dx$

50. $dy = xy \, dx$

51. $2yt^2 \dfrac{dy}{dt} = 2t^2 - 1;\quad y = -\sqrt{3}$ when $t = 1$

52. $y' = y^2\sqrt{x};\quad y = 1$ when $x = 0$

53. *Continuous compounding* Suppose $6,000 is invested at 8% compounded continuously.
(a) Determine a function which will give the amount at any future time.
(b) What will be the amount 18 months later?

54. *Present value*
(a) Determine a function that will give the present value of $6,000 resulting from an investment compounded continuously at 8% for any time t.
(b) Determine the present value if the investment in part (a) is for 2 years.

55. *Continuous compounding* If $1,000 invested now with interest compounded continuously will amount to $1,221.40 at the end of 2 years, determine the amount after 4 years.

56. *Continuous compounding and present value*
(a) How long will it take the $6,000 in Exercise 53 to double?
(b) How long will it take the $6,000 in Exercise 54 to halve?

57. *Population growth* The population in a certain country was 150 million in 1980 ($t = 0$) and was 160 million in 1985. Assume that at any time the population increases at a rate proportional to the current population.
    (a) Determine the population at any time t.
    (b) Determine the population in 1995.

58. *Radioactive decay* A radioactive substance decays at a rate proportional to the amount present at any time. That is, if y denotes the amount present at time t, then the rate of change of the amount is
$$\frac{dy}{dt} = -Ky$$
    where K is a positive constant.
    (a) Solve this differential equation to verify that the amount present at any time t is $y = Ae^{-Kt}$, where A is the amount present initially (at $t = 0$).
    (b) How long will it take for the substance to reduce to one-half of the initial amount?
    (c) If $K = 0.00017$ (when time is measured in years) for a particular substance and there are 200 grams present initially, how much of the substance will remain after 1,000 years?
    (d) How long will it take for the amount remaining after 1,000 years in part (c) to halve?

59. *Animal population* Suppose that the fish population (y) in a particular lake is changing at the rate
$$\frac{dy}{dt} = \sqrt{y}$$
    where t denotes time measured in months.
    (a) Determine the population at any time t if the lake is initially (at $t = 0$) stocked with 400 fish.
    (b) Determine the population 8 months after the lake is stocked.

60. (a) Verify that $f(x) = \frac{1}{5}$ is a density function for the random variable X which takes values in the sample space $S = [0, 5]$.
    (b) Determine the probability that the random variable in part (a) takes a value in the interval $[1, 3]$.

61. (a) Verify that $f(x) = \frac{1}{2}e^{-x/2}$ is a density function for the random variable X which takes values in the sample space $S = \{x : x \geq 0\}$.
    (b) Determine the probability that the random variable in part (a) takes a value in the interval $[0, 1]$.
    (c) Determine the probability that the random variable in part (a) takes a value greater than or equal to 1.

62. (a) Determine the value of k for which $f(x) = k(1 + x)$ is a density function for the random variable X that takes values in the interval $[0, 4]$.
    (b) If k has the value determined in part (a), determine the probability that X takes a value in the interval $[2, 4]$.

63. Determine whether $f(x) = 4/x$ can be a density function for the random variable that takes values in $S = \{x : x \geq 0\}$.

64. *Immunization experiment* The length of time (in years) that an experimental treatment for immunizing rabbits is effective has been found to be

a random variable with density function

$$f(t) = \tfrac{3}{4}(2t - t^2) \qquad t \in [0, 2]$$

Determine the probability that the treatment on one of the rabbits will be effective less than 6 months.

65. *Customer arrivals* The time (in minutes) between customer arrivals at a particular gasoline station is a random variable with density function

$$f(x) = \tfrac{1}{3}e^{-x/3} \qquad x \geq 0$$

Determine the probability that the time between customer arrivals will be:

(a) Less than 2 minutes (b) Greater than 3 minutes

Identify the steps in the four-step procedure involved in the application of mathematics (Section 1.3) in each case.

66. *Continuous compounding* Exercise 53(b)
67. *Immunization experiment* Exercise 64.

CHAPTER 15

MULTIVARIABLE CALCULUS

OUTLINE

15.1 Functions of Several Variables
15.2 Graphs in Three Dimensions
15.3 Partial Derivatives
15.4 Interpretation of Partial Derivatives
15.5 Second-Order Partial Derivatives
15.6 Maxima and Minima
15.7 Lagrange Multipliers
15.8 Chapter Summary

Review Exercises

All the functions we have considered so far have been functions of one variable—that is, the value of each function depended only on the value of a single independent variable. For example, the value of y in

$$y = 3x^2 + 1$$

depends only on the value of x; the value of $f(t)$ in

$$f(t) = 2t - 6$$

depends only on the value of t.

However, some physical situations cannot be described by functions of this type. For example, if \$1,000 is invested, the interest earned, given by

$$1{,}000 \times R \times T$$

depends on both the value of R (rate) and T (time). Similarly, the volume of a gas depends on both the temperature of the gas and the pressure under which it is stored.

The purpose of this chapter is to investigate some properties of functions whose values depend on more than one independent variable.

15.1 FUNCTIONS OF SEVERAL VARIABLES

The value of the expression
$$5x - 4xy^2$$
depends on the values of both x and y. For example, if $x = 2$ and $y = 3$, the value of the expression is
$$5(2) - 4(2)(3)^2 = -62$$
if $x = -1$ and $y = 2$, the value of the expression is 11. An expression such as this, which associates a unique value to pairs of values of x and y, is said to be a *function of the two variables x and y*. Formally, a **function of the two variables x and y** is a rule that associates a unique value to each pair of values of x and y in some set D.

The notation used is similar to that for functions of a single variable. If f is the function (rule) that associates the value of
$$5x - 4xy^2$$
to each pair of values of x and y, we write
$$f(x, y) = 5x - 4xy^2$$

Then $f(2, 3) = -62$, $f(-1, 2) = 11$, $f(0, 16) = 0$, and so forth. In general, $f(x, y)$ denotes the value that the function f associates with a pair of values of x and y.

EXAMPLE 1 If $f(x, y) = x^2 + y^2$, then
$$f(2, 4) = 2^2 + 4^2 = 20 \quad \text{and} \quad f(-3, 0) = (-3)^2 + 0 = 9$$
If $g(x, t) = 2(x - t)^3$, then
$$g(1, 2) = 2(1 - 2)^3 = -2 \quad \text{and} \quad g(10, 6) = 2(10 - 6)^3 = 128 \quad \blacksquare$$

Sometimes a single variable is used in place of $f(x, y)$ to indicate the value of the function. For example,
$$z = x^2 + y^2$$
might be used in place of
$$f(x, y) = x^2 + y^2$$
Then $z = 20$ when $x = 2$ and $y = 4$. Similarly, $z = 9$ when $x = -3$ and $y = 0$, and so forth. In such a situation, z is called the **dependent variable**; x and y are called the **independent variables**. The value of z depends on the values of the independent variables. To indicate that the value of z depends upon the values of x and y, we shall sometimes write $z = f(x, y)$.

PRACTICE PROBLEM 1

If $f(x, y) = 3x^2 - xy$, determine the value of:

(a) $f(2, 4)$ (b) $f(-1, 3)$
(c) $f(0, -5)$

Answer

(a) 4 (b) 6 (c) 0

15.1 FUNCTIONS OF SEVERAL VARIABLES

EXAMPLE 2 Given the function

$$z = \frac{x^2 - y^2}{\sqrt{2xy}}$$

determine the value of z when:

(a) $x = 3$ and $y = 6$ (b) $x = -4$ and $y = -1$

Solution (a) $$z = \frac{3^2 - 6^2}{\sqrt{2(3)(6)}} = \frac{-27}{6}$$

(b) $$z = \frac{(-4)^2 - (-1)^2}{\sqrt{2(-4)(-1)}} = \frac{15}{2\sqrt{2}}$$

The set D of pairs of values of x and y referred to in the definition of a function of two variables above is called the **domain** of the function. Unless otherwise indicated, the domain of a function will be all pairs of values of the independent variables for which the function has a value. Consequently, the domain of the function

$$z = \frac{x^2 - y^2}{\sqrt{2xy}}$$

is all pairs of values of x and y for which $2xy > 0$. The domain of the function f defined by

$$f(x, y) = 3xy + y^2$$

is all pairs of values of x and y.

The idea of a function of two variables extends to functions of three or more variables. For example, if

$$f(x, y, z) = 3xy + 2yz^2$$

then

$$f(1, 2, -3) = 3(1)(2) + 2(2)(-3)^2 = 42$$

Similarly, if

$$g(x, y, z) = \sqrt{xyz}$$

then

$$g(-2, -3, 6) = \sqrt{(-2)(-3)(6)} = \sqrt{36} = 6$$

2 PRACTICE PROBLEM 2

Given the function $z = \sqrt{\dfrac{2x + y}{2}}$, determine the value of z if:

(a) $x = 5$ and $y = 8$
(b) $x = 7$ and $y = -4$

Answer

(a) 3   (b) $\sqrt{5}$

3 PRACTICE PROBLEM 3

If $w = 2(xy - xz)$, determine the value of w if:

(a) $x = 2$, $y = -3$, and $z = 5$
(b) $x = -4$, $y = 0.5$, and $z = 1$

Answer

(a) -32 (b) 4

EXERCISES

Determine each of the following values for the function $f(x, y) = 2x - 3y$:

1. $f(4, 5)$ 2. $f(0, 2)$ 3. $f(-1, -1)$

Determine each of the following values for the function $f(x, y) = \dfrac{x^2}{y - 2}$:

4. $f(-3, 4)$ 5. $f(2, 2)$ 6. $f(0, 0)$

Determine the value of z if $z = \sqrt{x+y}$ for the given values of x and y.

7. $x = 1, \quad y = 3$ **8.** $x = 7, \quad y = -7$ **9.** $x = \frac{1}{2}, \quad y = \frac{1}{3}$

Determine the value of z if $z = x(2x - y)$ for the given values of x and y.

10. $x = 1, \quad y = 0$ **11.** $x = 3, \quad y = 6$ **12.** $x = \frac{1}{2}, \quad y = -5$

Determine the value of $f(x, y, z) = x + y - z$ for the given values of x, y, and z.

13. $x = 2, \quad y = 3, \quad z = -4$ **14.** $x = 1, \quad y = 0, \quad z = 3$

15. $x = 12, \quad y = 17, \quad z = 9$

Determine the value of w if $w = xy + xz$ for the given values of x, y, and z.

16. $x = 2, \quad y = 3, \quad z = 5$ **17.** $x = 6, \quad y = -3, \quad z = \frac{1}{2}$

18. $x = 0, \quad y = 1, \quad z = 1$

19. *Area* The area of a rectangle is the product of its length and width. Construct a function that will give the area of a rectangle (z) in terms of its width (x) and length (y).

20. *Volume* The volume of a right circular cylinder (a tin can) is the area of the base times the height, and the area of the base is its radius squared times π (see the figure). Construct a function that gives the volume (z) of a right circular cylinder in terms of its radius (r) and height (h).

21. *Interest* Simple interest is given by the product principal times rate times time. Construct a function that gives the interest (I) in terms of the principal (P), rate (R), and time (T).

Evaluate the limit below for each of the following functions:

$$\lim_{h \to 0} \frac{f(x + h, y) - f(x, y)}{h}$$

22. $f(x, y) = 2x - 3y^2$ **23.** $f(x, y) = 4x^2 y$

Evaluate the limit below for each of the following functions:

$$\lim_{h \to 0} \frac{f(x, y + h) - f(x, y)}{h}$$

24. $f(x, y) = y^2 - x^2$ **25.** $f(x, y) = x(x + 2y)$

15.2 GRAPHS IN THREE DIMENSIONS

The graphs of functions of one independent variable were given in two dimensions, one dimension for the independent variable and one for the dependent variable—an x-axis and a y-axis. The graph of a function such as

$$z = x^2 + y^2$$

15.2 GRAPHS IN THREE DIMENSIONS

requires three dimensions—an extra dimension for the additional variable.

To construct a coordinate system for graphs in three dimensions, we start with three number lines intersecting at right angles to each other at their origins, as shown in Figure 15.1.

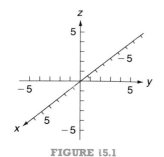

FIGURE 15.1

The result is a rectangular coordinate system in three-dimensional space. Every point in space can be associated with a triple of numbers (x, y, z) and conversely, every triple can be associated with a point (see Figure 15.2).

The numbers in the triple (x, y, z) associated with a point are called the **coordinates** of the point, the first number being the **x-coordinate**, the second, the **y-coordinate**, and the third, the **z-coordinate**. We shall describe a three-step method for locating a point with a given set of coordinates and illustrate the method by locating the point with coordinates $(3, 4, -2)$.

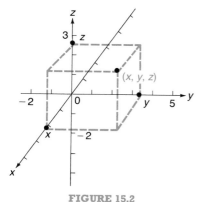

FIGURE 15.2

1. Draw the line through the x-axis at the x-coordinate and parallel to the y-axis. For $(3, 4, -2)$, the x-coordinate is 3.

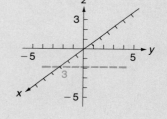

2. Draw the line through the y-axis at the y-coordinate and parallel to the x-axis. For $(3, 4, -2)$, the y-coordinate is 4.

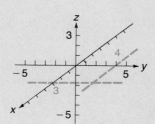

3. At the point where these two lines intersect, draw a third line parallel to the z-axis; then move up or down this line the number of units indicated by the z-coordinate to locate the point. For $(3, 4, -2)$ the z-coordinate is -2.

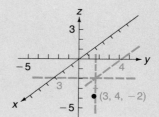

EXAMPLE 1 The following figure shows the location of the point with coordinates $(-3, 4, 5)$ obtained following the procedure just outlined:

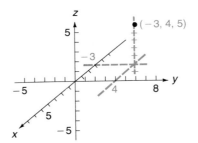

1 PRACTICE PROBLEM 1

Locate the points $(2, -3, -4)$ and $(2, 3, 0)$ in a coordinate system.

Answer

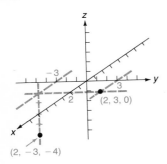

Figure 15.3 gives the coordinates for some additional points in three dimensions. As in one and two dimensions, the expression "the point with coordinates (x, y, z)" is shortened to "the point (x, y, z)."

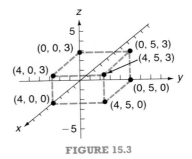

FIGURE 15.3

The graph of a function of two variables forms a surface in three dimensions (Figure 15.4). For any point on the surface, the z-coordinate is the value of the function when x has the value of the first coordinate and y has the value of the second coordinate. For example, the point $(1, -2, 6)$ in the graph of $z = 2x + y^2$ corresponds to the fact that z has the value 6 when $x = 1$ and $y = -2$.

The graphs in Figures 15.5 and 15.6 indicate the type of surfaces obtained when functions of two variables are graphed. Obtaining graphs

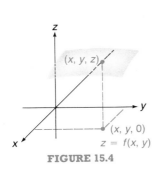

FIGURE 15.4

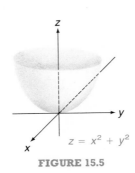

FIGURE 15.5

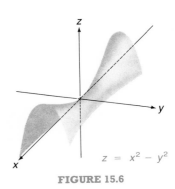

FIGURE 15.6

of functions such as these is quite difficult and will not be attempted here. However, in Section 15.4 it will be important to appreciate the form of the graph of the simple equations $x = k$, $y = k$, and $z = k$ where k is a constant.

The graph of the equation $y = 3$ (Figure 15.7), for example, is the set of all points with y-coordinate equal to 3. This set contains all points in the plane through the point (0, 3, 0) and perpendicular to the y-axis.

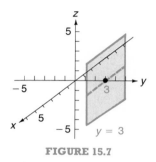

FIGURE 15.7

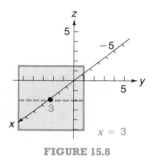

FIGURE 15.8

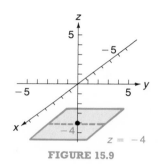

FIGURE 15.9

Similarly, the graph of $x = 3$ (Figure 15.8) is the set of all points with x-coordinate equal to 3; these are the points in the plane through the point (3, 0, 0) and perpendicular to the x-axis.

Finally, the graph of $z = -4$ (Figure 15.9), being the set of all points with z-coordinate -4, is the plane through the point (0, 0, -4) and perpendicular to the z-axis.

The graphs of functions of three variables would require four dimensions. There is no satisfactory method of drawing four dimensions on a sheet of paper. Consequently, graphs of such functions cannot be obtained.

2 PRACTICE PROBLEM 2

Graph the equation $y = -5$ in a three-dimensional coordinate system.

Answer

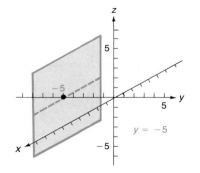

EXERCISES

Locate each of the following points in a coordinate system:

1. (2, 5, 4)
2. (3, 5, -4)
3. (-3, 5, 4)
4. (-3, 5, -4)
5. (3, -5, 4)
6. (0, 2, 3)
7. (-2, 0, 3)
8. (-3, 2, 0)
9. (0, 0, 4)
10. (0, 4, 0)
11. (4, 0, 0)
12. (0, 0, 0)

13.–22. Give the coordinates of each point indicated in the following coordinate system:

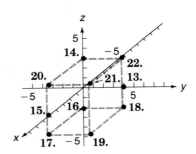

Graph each given equation in a three-dimensional coordinate system.

23. $x = 5$ 24. $x = -3$ 25. $y = 2$ 26. $y = -6$
27. $z = 5$ 28. $z = 0$ 29. $x = 0$ 30. $y = 0$
31. $x = 4$ 32. $y = 4$ 33. $z = 4$

15.3 PARTIAL DERIVATIVES

Consider again the function that gives the area of a rectangle (z) in terms of its width (x) and length (y),

$$z = xy$$

If the length y is held at a fixed value, such as 3 or 4 or 10, and x is allowed to vary, the area z becomes a function of x alone. We can then determine the derivative of z as a function of the single variable x just as we did in Chapter 11. The resulting derivative is called the **partial derivative of z with respect to x** and is denoted by

$$z_x \quad \text{or} \quad \frac{\partial z}{\partial x}$$

Similarly, if x is held constant at some value and y is allowed to vary, z becomes a function of y alone. We can then determine the derivative of z as a function of the single variable y. In this case, the resulting derivative is called the **partial derivative of z with respect to y,** denoted by

$$z_y \quad \text{or} \quad \frac{\partial z}{\partial y}$$

The mechanics for determining partial derivatives are exactly the same as those used previously for determining ordinary derivatives.

15.3 PARTIAL DERIVATIVES

We just have to remember that the variable not entering into the differentiation is to be treated as a constant.

EXAMPLE 1 Determine z_x and z_y for

$$z = x^2y + y^2 - 3$$

Solution   With y constant,

$$z_x = 2xy$$

With x constant,

$$z_y = x^2 + 2y$$

EXAMPLE 2 Determine z_x and z_y for

$$z = (xy - y^3)^4$$

Solution   Applying the power rule with y constant gives

$$z_x = 4(xy - y^3)^3 y$$

With x constant,

$$z_y = 4(xy - y^3)^3(x - 3y^2)$$

If the notation $f(x, y)$ is used to denote the value of the function, then the partial derivatives are denoted by

$$f_x, \quad f_y \quad \text{or} \quad \frac{\partial f}{\partial x}, \quad \frac{\partial f}{\partial y}$$

For example, if

$$f(x, y) = 3x^3y^2 - 7x^2 + 2y - 4$$

then

$$f_x(x, y) = 9x^2y^2 - 14x$$
$$f_y(x, y) = 6x^3y + 2$$

Also,

$$f_x(2, 3) = 9(2^2)(3^2) - 14(2) = 296$$
$$f_y(2, 3) = 6(2^3)(3) + 2 = 146$$

EXAMPLE 3 Determine $f_x(-2, 4)$ if

$$f(x, y) = \frac{x^2 + y^2}{2x + 3}$$

Solution   Applying the quotient rule with y held constant gives

$$f_x(x, y) = \frac{(2x + 3)(2x) - (x^2 + y^2)2}{(2x + 3)^2} = \frac{2x^2 + 6x - 2y^2}{(2x + 3)^2}$$

---

1 PRACTICE PROBLEM 1

Determine z_x and z_y for:

(a) $z = xy^2 - x^2$ (b) $z = (xy + 3y)^2$

Answer

(a) $z_x = y^2 - 2x$; $z_y = 2xy$

(b) $z_x = 2(xy + 3y)y = 2xy^2 + 6y^2$;
$z_y = 2(xy + 3y)(x + 3)$
$ = 2x^2y + 12xy + 18y$

2 PRACTICE PROBLEM 2

Determine $f_y(x, y)$ and $f_y(9, -2)$ if $f(x, y) = 5\sqrt{2x + y}$.

Answer

$f_y(x, y) = \dfrac{5}{2\sqrt{2x+y}}$; $f_y(9, -2) = \tfrac{5}{8}$

Then, for $x = -2$ and $y = 4$,

$$f_x(-2, 4) = \frac{2(-2)^2 + 6(-2) - 2(4^2)}{[2(-2) + 3]^2} = \frac{-36}{1} = -36$$

The idea and notation for partial derivatives of functions of two variables extend to functions of three or more variables. The mechanics for determining these derivatives are also similar—the variables not entering into the differentiation are treated as constants. The procedure is illustrated in Example 4.

EXAMPLE 4 Determine f_x, f_y, and f_z if

$$f(x, y, z) = x^2 y - 2xz + yz^2$$

Solution With y and z constant,

$$f_x(x, y, z) = 2xy - 2z$$

With x and z constant,

$$f_y(x, y, z) = x^2 + z^2$$

With x and y constant,

$$f_z(x, y, z) = -2x + 2yz$$

3 PRACTICE PROBLEM 3

For $f(x, y, z) = x^2 z - yz^2 + 3xy$, determine:

(a) $f_y(x, y, z)$ (b) $f_z(x, y, z)$

Answer

(a) $-z^2 + 3x$ (b) $x^2 - 2yz$

EXERCISES

Find each indicated partial derivative.

1. $z = x^2 - y^2 + 3xy$
   (a) z_x (b) z_y

2. $z = xy^2 + 3xy - 7$
   (a) z_x (b) z_y

3. $f(x, y) = 2x^3 y^2 + x^{-1} y^4$
   (a) $f_x(x, y)$ (b) $f_y(x, y)$

4. $f(x, y) = 2x + 2y + 2$
   (a) $f_x(x, y)$ (b) $f_y(x, y)$

5. $z = \sqrt{xy}$
   (a) z_x (b) z_y

6. $f(x, y) = (4x^5 - y)^3$
   (a) $\dfrac{\partial f}{\partial x}$ (b) $\dfrac{\partial f}{\partial y}$

7. $f(x, y) = (x^2 y - y^3)^2$
   (a) $f_x(3, 1)$ (b) $f_y(3, 1)$

8. $f(x, y) = (xy + yx^2)^{-2}$
   (a) $f_x(1, 2)$ (b) $f_y(1, -2)$

9. $z = 2y(3x^2 - y)^2$
   (a) $\dfrac{\partial z}{\partial x}$ (b) $\dfrac{\partial z}{\partial y}$

10. $f(x, y) = (x + y)(x - y)^{1/2}$
    (a) $f_x(5, 1)$ (b) $f_y(1, 0)$

11. $f(x, y) = x^2 e^{x+y}$
    (a) $f_x(x, y)$ (b) $f_y(x, y)$

12. $f(x, y) = (2xy)\ln(x^2 + y^2)$
    (a) $\dfrac{\partial f}{\partial x}$ (b) $\dfrac{\partial f}{\partial y}$

13. $z = e^{xy}$

    (a) z_x (b) z_y

14. $z = \ln(2xy)$

    (a) $\dfrac{\partial z}{\partial x}$ (b) $\dfrac{\partial z}{\partial y}$

15. $f(x, y) = \dfrac{x + y}{x - y}$

    (a) $f_x(3, -4)$ (b) $f_y(0, 2)$

16. $f(x, y) = \dfrac{x^2 y}{x^2 + y^2}$

    (a) $f_x(1, 1)$ (b) $f_y(0, -1)$

17. $z = \left(\dfrac{x}{2x + y^2}\right)^2$

    (a) z_x (b) z_y

18. $z = \sqrt{\dfrac{2x}{x - 3y}}$

    (a) z_x (b) z_y

19. $f(x, y, z) = xy + xz + yz$

    (a) $f_x(x, y, z)$ (b) $f_y(x, y, z)$
    (c) $f_z(x, y, z)$

20. $f(x, y, z) = x^2 y - 3y^2 z$

    (a) $f_x(x, y, z)$ (b) $f_y(x, y, z)$
    (c) $f_z(x, y, z)$

21. $f(x, y, z) = \dfrac{xyz}{x + y + z}$

    (a) $f_x(1, 2, 3)$ (b) $f_y(-1, 0, 2)$
    (c) $f_z(0, 1, -2)$

22. $f(x, y, z) = (x^2 - yz)(2x + 3z)$

    (a) $f_x(1, 1, 1)$ (b) $f_z(2, 4, -5)$
    (c) $f_y(0.1, 0, 0.3)$

If f is a function of the two variables x and y, explain what is meant by each expression.

23. The partial derivative of f with respect to x
24. The partial derivative of f with respect to y

15.4 INTERPRETATION OF PARTIAL DERIVATIVES

What do we obtain geometrically when we evaluate partial derivatives? As with functions of one variable, we obtain the slope of a tangent line with a partial derivative of a function of two variables. Recall that the graph of such a function is a surface in three dimensions. When y is held constant—for instance, $y = k$—we restrict our attention to that part of the graph lying in the plane $y = k$. We illustrate this situation with the graph of the function $f(x, y) = 16 - x^2 - y^2$ in Figure 15.10 (page 574). The plane $y = 2$ cuts out the curve ABC in the graph of the function. Then, just as with a function of one variable, $f_x(x, 2)$ gives the slope of the line in the plane $y = 2$ tangent to the graph at the point $(x, 2, z)$, where $z = f(x, 2)$.

For example, if $x = 1$, then $z = 16 - 1^2 - 2^2 = 11$; the corresponding point $(1, 2, 11)$ is indicated in Figure 15.11. The slope of the line in the plane $y = 2$ tangent to the graph at $(1, 2, 11)$ is given by $f_x(1, 2) = -2$.

In addition, $f_x(x, 2)$ gives the (instantaneous) rate of change of the function along the curve ABC in the plane $y = 2$. For example, when $x = -3$, the function is increasing at the rate of $f_x(-3, 2) = 6$ units per unit change in x along the curve ABC.

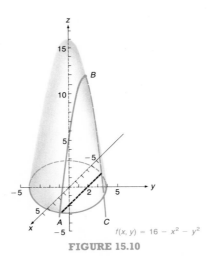

FIGURE 15.10

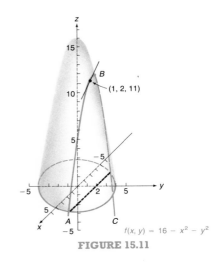

FIGURE 15.11

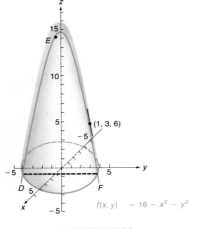

FIGURE 15.12

Similarly, if x is held constant at $x = 1$, for example, we restrict our attention to the curve DEF cut out by the plane $x = 1$ in the graph shown in Figure 15.12. The slope of any line in the plane tangent to the graph is then given by $f_y(1, y) = -2y$ for the appropriate value of y. For example, if $y = 3$, $f(1, 3) = 6$ and the slope of the tangent in the plane $x = 1$ through the point $(1, 3, 6)$ is given by $f_y(1, 3) = -6$.

In addition, $f_y(1, y)$ gives the (instantaneous) rate of change of the function along the curve DEF in the plane $x = 1$. If, for instance, $y = -4$, the function is increasing at the rate of $f_y(1, -4) = 8$ units per unit change of y along the curve DEF.

EXAMPLE 1 Determine the slope of the tangent to the curve cut out by the plane $y = 3$ in the surface $f(x, y) = x^2y - 3xy$ at $x = 4$.

Solution For any value of x, the slope is $f_x(x, 3) = 6x - 9$. In particular, at $x = 4$ the slope is $f_x(4, 3) = 15$.

EXAMPLE 2 Determine the slope of the tangent to the curve cut out by the plane $x = -2$ in the surface $f(x, y) = x^2y^2 - 2y$ at $y = 1$.

Solution For any value of y, the slope is $f_y(-2, y) = 8y - 2$. For $y = 1$, the slope is $f_y(-2, 1) = 6$.

---

1 PRACTICE PROBLEM 1

Determine the slope of the tangent to the curve cut out by each plane.

(a) $y = 1$ in the surface $z = 2x^2y - 3y^2$ at $x = -1$

(b) $x = 0$ in the surface $z = 2x^2y - 3y^2$ at $y = 2$

Answer

(a) -4 (b) -12

EXAMPLE 3
Determine how fast the function $f(x, y) = 3xy^{-1}$ is changing along the curve cut out by the plane $x = 4$ at $y = 3$.

Solution For any value of y the rate of change is given by $f_y(4, y) = -3(4)y^{-2} = -12y^{-2}$. At $y = 3$, the rate of change is $f_y(4, 3) = -\frac{12}{9} = -\frac{4}{3}$. The function is decreasing at the rate of $\frac{4}{3}$ units per unit change in y. ▨2

PRACTICE PROBLEM 2

The amount (principal plus interest) resulting from investing $5,000 at an annual interest rate r compounded continuously for t years is given by $A = 5{,}000e^{rt}$.

(a) If r is constant, determine how fast the amount is changing after 2 years.

(b) Determine the rate of change in part (a) if $r = 8\%$.

Answer

(a) $5{,}000re^{2r}$ (b) $469.40 per year

EXERCISES

1. Determine the slope of the tangent to the curve cut out by the plane $y = 3$ in the surface $f(x, y) = x^2 + y^2$ at $x = 1$.

2. Determine the slope of the tangent to the curve cut out by the plane $x = 0$ in the surface $f(x, y) = x^2 + xy + y^2$ at $y = -4$.

3. Determine the slope of the tangent to the curve cut out by the plane $x = 0$ in the surface $z = e^{x^2 + y^2}$ at the point $(0, 1, e)$.

4. Determine the slope of the tangent to the curve cut out by the plane $y = 3$ in the surface $z = (x^2 - 2y)/xy$ at the point $(-2, 3, \frac{1}{3})$.

5. At what point is the slope of the tangent to the curve cut out by the plane $y = -4$ in the surface $f(x, y) = x^2 + xy + y^2$ equal to zero?

6. At what point is the slope of the tangent to the curve cut out by the plane $x = 3$ in the surface $z = -y^3 + 4y^2 x - 40x$ the greatest?

7. Determine how fast the function $z = \ln(xy)$ is changing along the curve cut out by the plane $x = 2$ at $y = 3$.

8. Determine how fast the function $f(x, y) = (2x + 3y)^2$ is changing along the curve cut out by the plane $y = -5$ at the point $(8, -5, 1)$.

9. *Marginal profit* A company makes two types of computers, a personal model and a commercial model. Its weekly profit from the sale of x personal computers and y commercial computers is given by
$$P(x, y) = 3x^2 + 5y^2 - 2xy$$
(a) Determine the marginal profit from the sale of y commercial computers if the sale of personal computers is held constant at 200.
(b) Determine the marginal profit from the sale of commercial computers at the 250 level if the sale of personal computers is held constant at 200.

10. *Marginal profit*
(a) Determine the marginal profit in Exercise 9 from the sale of x personal computers if the sale of commercial computers is held constant at 100.
(b) Determine the marginal profit from the sale of personal computers at the 20 level if the sale of commercial computers is held constant at 100.

11. *Marginal cost* The daily manfacturing costs of a small company are given by
$$C(x, y) = xy - \tfrac{1}{2}x^2 - y^2 + 10$$
where x gives the labor costs and y gives the cost of materials. All units are in thousands of dollars.
(a) Evaluate $C_x(x, 6)$. Interpret the result.
(b) Evaluate $C_x(4, 6)$. Interpret the result.

12. *Marginal cost*
    (a) In Exercise 11, evaluate $C_y(x, y)$. Interpret the result.
    (b) In Exercise 11, evaluate $C_y(8.5, 4)$. Interpret the result.
13. *Gas law* The ideal-gas law of physics states that the pressure (P) exerted by a confined gas is given in terms of its volume (V) and its temperature (T) by the equation

    $$P = k\frac{T}{V}$$

    where k is a constant whose value depends on the amount of gas and the units of measurement. Suppose that for a certain gas, $k = 9$.
    (a) Find the instantaneous rate of change of pressure (in pounds per square inch) with respect to temperature when the temperature is 90°C if the volume remains constant at 60 cubic inches.
    (b) Find the instantaneous rate of change of pressure with respect to volume when the volume is 60 cubic inches if the temperature remains constant at 90°C.
    (c) Find the instantaneous rate of change of the volume with respect to pressure when the pressure is 14 pounds per square inch if the temperature remains constant at 90°C.
14. *Marginal profit* Identify the steps of the four-step procedure involved in the application of mathematics (Section 1.3) in Exercise 9(b).

15.5 SECOND-ORDER PARTIAL DERIVATIVES

If we take the partial derivative of the function

$$f(x, y) = 3x^2 - 4y^2 + x^2y$$

with respect to x, we obtain

$$\frac{\partial f}{\partial x} = 6x + 2xy$$

The result is another function, and we can determine its partial derivatives just as we would with any other function. We obtain

$$\frac{\partial}{\partial x}\left(\frac{\partial f}{\partial x}\right) = 6 + 2y \qquad \text{Denoted by } \frac{\partial^2 f}{\partial x^2}$$

and

$$\frac{\partial}{\partial y}\left(\frac{\partial f}{\partial x}\right) = 2x \qquad \text{Denoted by } \frac{\partial^2 f}{\partial y\, \partial x}$$

These latter are called **second-order partial derivatives** of the original function f.

15.5 SECOND-ORDER PARTIAL DERIVATIVES

There are four second-order partial derivatives for a function of two variables. Their definitions, along with alternate notation, are given in Table 15.1.

TABLE 15.1 SECOND PARTIAL DERIVATIVES OF $f(x, y)$

DEFINITION	ALTERNATE NOTATION
$\dfrac{\partial^2 f}{\partial x^2} = \dfrac{\partial}{\partial x}\left(\dfrac{\partial f}{\partial x}\right)$	$f_{xx}, f_{xx}(x, y),$ or $\dfrac{\partial^2 z}{\partial x^2}$ if $z = f(x, y)$
$\dfrac{\partial^2 f}{\partial y^2} = \dfrac{\partial}{\partial y}\left(\dfrac{\partial f}{\partial y}\right)$	$f_{yy}, f_{yy}(x, y),$ or $\dfrac{\partial^2 z}{\partial y^2}$
$\dfrac{\partial^2 f}{\partial y\, \partial x} = \dfrac{\partial}{\partial y}\left(\dfrac{\partial f}{\partial x}\right)$	$f_{xy}, f_{xy}(x, y),$ or $\dfrac{\partial^2 z}{\partial y\, \partial x}$
$\dfrac{\partial^2 f}{\partial x\, \partial y} = \dfrac{\partial}{\partial x}\left(\dfrac{\partial f}{\partial y}\right)$	$f_{yx}, f_{yx}(x, y),$ or $\dfrac{\partial^2 z}{\partial x\, \partial y}$

EXAMPLE 1 Find the two partial derivatives and four second partial derivatives of

$$f(x, y) = x^4 y + 3xy^2 + 7$$

Solution

$$\frac{\partial f}{\partial x} = 4x^3 y + 3y^2 \qquad \frac{\partial f}{\partial y} = x^4 + 6xy$$

$$\frac{\partial^2 f}{\partial x^2} = 12x^2 y \qquad \frac{\partial^2 f}{\partial y^2} = 6x$$

$$\frac{\partial^2 f}{\partial y\, \partial x} = 4x^3 + 6y \qquad \frac{\partial^2 f}{\partial x\, \partial y} = 4x^3 + 6y \qquad \boxed{1}$$

EXAMPLE 2 Find $f_{xx}(2, 0)$ and $f_{xy}(-1, 3)$ for

$$f(x, y) = 4x^3 y^2 - 3x^{-1}$$

Solution For any x and y in the domain of f,

$$f_x(x, y) = 12x^2 y^2 + 3x^{-2}$$
$$f_{xx}(x, y) = 24xy^2 - 6x^{-3}$$
$$f_{xy}(x, y) = 24x^2 y$$

Then

$$f_{xx}(2, 0) = 24(2)(0^2) - 6(2^{-3}) = 0 - \tfrac{6}{8} = -\tfrac{3}{4}$$

---

1 PRACTICE PROBLEM 1

Find the two partial derivatives and four second partial derivatives of $f(x, y) = x^2 y - 4x + 3y^2$.

Answer

$f_x(x, y) = 2xy - 4; \quad f_y(x, y) = x^2 + 6y;$
$f_{xx}(x, y) = 2y; \quad f_{xy}(x, y) = 2x; \quad f_{yy}(x, y) = 6; \quad f_{yx}(x, y) = 2x$

and
$$f_{xy}(-1, 3) = 24(-1)^2(3) = 72$$

EXAMPLE 3 Find the two partial derivatives and four second partial derivatives of
$$z = \frac{x}{x+y}$$

Solution
$$\frac{\partial z}{\partial x} = \frac{(x+y) - x}{(x+y)^2} = \frac{y}{(x+y)^2}$$

$$\frac{\partial z}{\partial y} = \frac{(x+y)0 - x}{(x+y)^2} = \frac{-x}{(x+y)^2}$$

$$\frac{\partial^2 z}{\partial x^2} = \frac{(x+y)^2 0 - 2y(x+y)}{(x+y)^4} = \frac{-2y}{(x+y)^3}$$

$$\frac{\partial^2 z}{\partial y^2} = \frac{(x+y)^2 0 + 2x(x+y)}{(x+y)^4} = \frac{2x}{(x+y)^3}$$

$$\frac{\partial^2 z}{\partial y \, \partial x} = \frac{(x+y)^2 - 2y(x+y)}{(x+y)^4} = \frac{(x+y)(x+y-2y)}{(x+y)^4} = \frac{x-y}{(x+y)^3}$$

$$\frac{\partial^2 z}{\partial x \, \partial y} = \frac{(x+y)^2(-1) + 2x(x+y)}{(x+y)^4} = \frac{(x+y)(-x-y+2x)}{(x+y)^4} = \frac{x-y}{(x+y)^3}$$

In Examples 1 and 3, the mixed partials
$$\frac{\partial^2 f}{\partial x \, \partial y} \quad \text{and} \quad \frac{\partial^2 f}{\partial y \, \partial x}$$
were equal. This does not always have to be true. However, it is true of the functions ordinarily encountered in calculus, such as those encountered in this text.

The notation and definition for second-order partial derivatives extend to functions of three or more variables. For example, if
$$f(x, y, z) = 2x^2 y^2 - 3x^2 z^2$$
then
$$\frac{\partial f}{\partial x} = 4xy^2 - 6xz^2, \quad \frac{\partial f}{\partial z} = -6x^2 z$$

PRACTICE PROBLEM 2

Find $f_{xy}(1, 3)$ and $f_{yy}(0, -3)$ for $f(x, y) = (2x - y)^3$.

Answer

$f_{xy}(1, 3) = 12; \quad f_{yy}(0, -3) = 18$

and
$$\frac{\partial^2 f}{\partial x^2} = 4y^2 - 6z^2, \qquad \frac{\partial^2 f}{\partial y\, \partial x} = 8xy, \qquad \frac{\partial^2 f}{\partial z\, \partial x} = -12xz$$

How many second-order partial derivatives of a function of three variables are there?

EXERCISES

Find the two partial derivatives and four second partial derivatives of each of the following:

1. $f(x, y) = x^3 + y^3 + 7xy$
2. $f(x, y) = x^2 y^3 - 3x + 7y$
3. $z = 3x^2 y - 4xy^3 + 2$
4. $z = 2x^{-1} + 3y^{-2}$
5. $f(x, y) = 2x^4 y^{-1} + 3xy$
6. $f(x, y) = x^2 + y^2 - 1$
7. $z = (x^2 y - 1)^2$
8. $z = (x + y)^3$
9. $z = \sqrt{xy}$
10. $z = \sqrt{x - y}$
11. $f(x, y) = e^{xy}$
12. $f(x, y) = \ln(xy)$

Find each indicated partial derivative.

13. $z = 3x^3 - 2y^4$
    (a) $\dfrac{\partial^2 z}{\partial x^2}$ (b) $\dfrac{\partial^2 z}{\partial y\, \partial x}$ (c) $\dfrac{\partial^2 z}{\partial y^2}$
14. $f(x, y) = x^2 y^2 - x^3$
    (a) f_{xx} (b) f_{yy} (c) f_{yx}
15. $f(x, y) = x^3 + 3xy^2 + y^3$
    (a) $f_{xx}(x, y)$ (b) $f_{xx}(1, -2)$ (c) $f_{xy}(2, 3)$
16. $f(x, y) = \dfrac{x}{x + y}$
    (a) $f_{yy}(x, y)$ (b) $f_{yy}(1, 1)$ (c) $f_{xy}(-1, 2)$
17. $f(x, y) = 2\sqrt{x - y}$
    (a) $f_{xx}(1, 0)$ (b) $f_{xy}(5, 1)$ (c) $f_{yx}(5, 1)$
18. $f(x, y) = (x + 2y)^2$
    (a) $f_{xy}(2, 3)$ (b) $f_{yx}(2, 3)$ (c) $f_{yx}(3, -1)$

Find the indicated partial derivatives of each given function of three variables.

19. $f(x, y, z) = x^3 y + 2z^3 y - 4$
    (a) $f_{xy}(x, y, z)$ (b) $f_{xx}(x, y, z)$ (c) $f_{xx}(1, 2, -1)$
20. $f(x, y, z) = x^2 y^2 + x^2 z^2 + y^2 z^2$
    (a) $f_{yz}(x, y, z)$ (b) $f_{yx}(x, y, z)$ (c) $f_{zz}(3, -2, 1)$

21. $f(x, y, z) = x^{-1}z - yz^{-1}$
   (a) $f_{zx}(x, y, z)$ (b) $f_{yy}(1, 2, -1)$ (c) $f_{zy}(-2, 4, 2)$
22. $f(x, y, z) = \sqrt{xyz}$
   (a) $f_{xx}(1, 1, 1)$ (b) $f_{yy}(2, 1, 2)$ (c) $f_{zz}(2, 2, 2)$

15.6 MAXIMA AND MINIMA

In Chapter 12 we saw how derivatives can be used to determine the maximum and minimum of a function of one variable. In this section we shall see how partial derivatives are used to determine points at which a function of two variables has a maximum or a minimum. The function whose graph is given in the coordinate system in Figure 15.13 has a maximum at point P. The z-coordinate of P is larger than the z-coordinate of any nearby point. The function also has a minimum at point Q. The z-coordinate of Q is smaller than the z-coordinate of any nearby point.

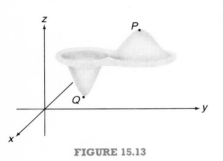

FIGURE 15.13

Unless we are familiar with the whole graph of the function, we can conclude only that the function has a maximum at P relative to all nearby points and a minimum at Q relative to all nearby points. There may be points elsewhere in the graph with a larger z-coordinate than that of P, and there may be other points with a smaller z-coordinate than that of Q. We shall be interested in locating the relative maximum and relative minimum points, such as P and Q, for a given function.

Formally, a function $z = f(x, y)$ has a **relative maximum** at (x_0, y_0) if

$$f(x_0, y_0) > f(x, y)$$

for all points (x, y) near (x_0, y_0). Similarly, $z = f(x, y)$ has a **relative minimum** at (x_0, y_0) if

$$f(x_0, y_0) < f(x, y)$$

for all points (x, y) near (x_0, y_0).

The function whose graph is given in Figure 15.14 has a relative maximum of z_0 at (x_0, y_0). [Recall that the third coordinate of a point in the graph of a function $z = f(x, y)$ is the value of the function at the first two coordinates.] In Figure 15.14(a), the tangent to the curve P_0PP_1 cut out by the plane $y = y_0$ has slope 0; that is, $f_x(x_0, y_0) = 0$. Similarly, in Figure 15.14(b), the tangent to the curve P_2PP_3 cut out by the plane $x = x_0$ has slope 0; that is, $f_y(x_0, y_0) = 0$. This suggests that both partial derivatives $f_x(x_0, y_0)$ and $f_y(x_0, y_0)$ must be zero in order for a function to have a relative maximum or minimum at (x_0, y_0). However, this alone is not sufficient (just as in the case of a function of one variable, the first derivative being zero does not by itself guarantee a relative maximum or minimum).

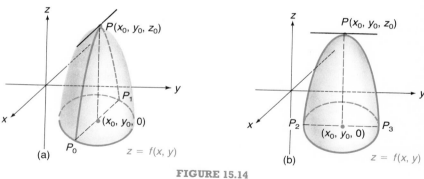

FIGURE 15.14

Sufficient conditions for a function of two variables to have a relative maximum or minimum at a point are a bit more complicated than for a function of one variable. These conditions are stated in the following theorem, which we shall not attempt to prove here:

THEOREM 1

Suppose that f has continuous first-order and second-order partial derivatives. If at (x_0, y_0)

$$f_x = 0 \quad \text{and} \quad f_y = 0$$

and if:

(i) $(f_{xy})^2 - f_{xx}f_{yy} < 0$ and $f_{xx} < 0$ at (x_0, y_0),
then $f(x, y)$ has a relative maximum at (x_0, y_0);

(ii) $(f_{xy})^2 - f_{xx}f_{yy} < 0$ and $f_{xx} > 0$ at (x_0, y_0),
then $f(x, y)$ has a relative minimum at (x_0, y_0);

(iii) $(f_{xy})^2 - f_{xx}f_{yy} > 0$ at (x_0, y_0),
then $f(x, y)$ has neither a relative maximum nor a relative minimum at (x_0, y_0);

(iv) $(f_{xy})^2 - f_{xx}f_{yy} = 0$ at (x_0, y_0),
the test fails.

Use of the theorem to determine the maximum and minimum points of functions of two variables is illustrated in the following examples.

EXAMPLE 1 Determine the points at which the function

$$f(x, y) = 2x^2 + y^2 - xy + 7x$$

has a relative maximum or a relative minimum.

Solution   We first need both partial derivatives:
$$f_x = 4x - y + 7 \quad \text{and} \quad f_y = 2y - x$$

In order to determine the value of x and of y for which both partials are zero, we must solve the system of equations

$$4x - y + 7 = 0$$
$$2y - x = 0$$

Solving the second equation for y in terms of x gives $y = x/2$; substituting for y in the first equation gives:

$$4x - \frac{x}{2} + 7 = 0$$
$$8x - x + 14 = 0$$
$$7x = -14$$
$$x = -2$$

Then $y = x/2$ gives $y = -1$; that is, both partial derivatives are zero for $x = -2$ and $y = -1$.

The required second-order partial derivatives in this case are constant:

$$f_{xy} = -1, \quad f_{xx} = 4, \quad f_{yy} = 2$$

for all values of x and y. In particular, for $x = -2$ and $y = -1$,

$$(f_{xy})^2 - f_{xx}f_{yy} = 1 - (4)(2) < 0$$

In addition, $f_{xx} = 4 > 0$. Consequently, f has a relative minimum of $f(-2, -1) = -7$ at $x = -2$ and $y = -1$.

> **1 PRACTICE PROBLEM 1**
>
> Determine the points at which the function
> $$f(x, y) = 3x + xy - x^2 - y^2$$
> has a relative maximum or relative minimum.
>
> Answer
> f has a relative maximum of 3 at (2, 1)

EXAMPLE 2   Determine the points at which the function

$$f(x, y) = \tfrac{3}{2}y - \tfrac{1}{2}y^3 + x^2 + 1$$

has a relative maximum or a relative minimum.

Solution   We first determine both partial derivatives,

$$f_x = 2x \quad \text{and} \quad f_y = \tfrac{3}{2} - \tfrac{3}{2}y^2$$

and set them equal to zero:

$$2x = 0$$
$$\tfrac{3}{2} - \tfrac{3}{2}y^2 = 0$$

This is not a system of linear equations; however, the first equation requires that $x = 0$. The second equation can be rewritten

$$\tfrac{3}{2}(1 - y^2) = 0$$

15.6 MAXIMA AND MINIMA

which requires that $y = 1$ or $y = -1$. We have two points at which a relative maximum or minimum could exist, $(0, 1)$ and $(0, -1)$.

For each of these points, we must determine which conditions of Theorem 1 hold. The required information is summarized in the table. Note that

$$f_{xx} = 2$$
$$f_{yy} = -3y$$
$$f_{xy} = 0$$

POINT	f_{xy}	f_{xx}	f_{yy}	$(f_{xy})^2 - f_{xx}f_{yy}$
$x = 0, y = 1$	0	2	-3	6
$x = 0, y = -1$	0	2	3	-6

At $(0, -1)$, $(f_{xy})^2 - f_{xx}f_{yy} < 0$ and $f_{xx} > 0$; therefore, f has relative minimum of $f(0, -1) = 0$ there. At $(0, 1)$, $(f_{xy})^2 - f_{xx}f_{yy} > 0$, so that f has neither a relative maximum or minimum there.

2 PRACTICE PROBLEM 2

Determine the points at which $f(x, y) = 2y^2 - 4y + x^3 - 12x$ has a relative maximum or relative minimum.

Answer

Minimum of -18 at $(2, 1)$; neither at $(-2, 1)$

EXAMPLE 3 The annual profit (in hundred thousands of dollars) to a small company from the manufacture of x thousand units of item A and y thousand units of item B is given by

$$z = 16xy - 17x^2 - 4y^2 + 3x$$

Determine the number of units of each item the company should produce each year to maximize its profit.

Solution

$$\frac{\partial z}{\partial x} = 16y - 34x + 3$$

$$\frac{\partial z}{\partial y} = 16x - 8y$$

The location of a possible relative maximum or minimum is then given by the solution of the system of linear equations:

$$16y - 34x + 3 = 0$$
$$16x - 8y = 0$$

The second equation gives $y = 2x$; substituting into the first equation gives

$$16(2x) - 34x = -3 \quad \text{or} \quad x = 1.5$$

then $y = 2x = 3$.

The partial derivatives required by Theorem 1 are

$$\frac{\partial^2 z}{\partial y\, \partial x} = 16, \qquad \frac{\partial^2 z}{\partial x^2} = -34, \qquad \frac{\partial^2 z}{\partial y^2} = -8$$

These partials are constant for all values of x and y. In particular, for $x = 1.5$ and $y = 3$,

$$\left(\frac{\partial^2 z}{\partial y\, \partial x}\right)^2 - \left(\frac{\partial^2 z}{\partial x^2}\right)\left(\frac{\partial^2 z}{\partial y^2}\right) = 16^2 - (-34)(-8) = -16 < 0$$

and

$$\frac{\partial^2 z}{\partial x^2} = -34 < 0$$

so that z has a relative maximum at $x = 1.5$ and $y = 3$.

Consequently, the company will obtain maximum profit if it manufactures 1.5 thousand (1,500) units of item A and 3 thousand (3,000) units of item B annually.

Its corresponding profit is given by

$$z = 16(1.5)(3) - 17(1.5^2) - 4(3^2) + 3(1.5) = 2.25$$

Hence, the profit is $225,000.

3 PRACTICE PROBLEM 3

If the annual profit function in Example 3 is $z = 4x + xy - x^2 - \frac{1}{2}y^2$, determine the number of units to be produced in order to maximize its annual profit and the corresponding maximum profit.

Answer

4,000 units of each item for a maximum profit of $800,000

EXERCISES

Use Theorem 1 to determine the points at which each of the following functions has a relative maximum or minimum. Determine the maximum or minimum value of the function at each such point.

1. $f(x, y) = x^2 + y^2 - 6x$
2. $f(x, y) = 7y - x^2 - y^2$
3. $f(x, y) = xy - x^2 - y^2 - 3y + 1$
4. $f(x, y) = x^2 + y^2 - xy + 6x + 2$
5. $z = 2xy - 3x^2 - 2y^2 + 4y - 2x + 5$
6. $z = 3x^2 - 2xy + 2y^2 - 11x - 3y + 4$
7. $f(x, y) = y^2 - x^2 - 2xy - 4y$
8. $f(x, y) = 2x^2 - 3xy - 4y^2 + 7x + 5y - 3$
9. $z = e^{xy}$
10. $f(x, y) = (x - 2)^2 + (y - 3)^2 - x + 7y$
11. $f(x, y) = 2x^3 + 2y^3 - 6xy + 5$
12. $f(x, y) = xy^2 - 4x^2 - 8y^2$

13. *Research-labor-profit* A company estimates that its annual profit for next year in terms of the amount x that it spends on research and the amount y that it spends on labor will be

$$z = 10x + 30y - x^2 - y^2 - 160$$

where all units are in millions of dollars. Determine the amount the company should spend on research and on labor to maximize its profit. Determine the maximum profit.

14. *Bid preparation* A computer software firm is preparing a bid to provide the programming for a government project. It estimates that its labor costs (in thousands of dollars) to provide the programming will be

$$f(x, y) = 1{,}200 + x^2 + \tfrac{1}{3}y^3 - 9y - 4xy$$

where x is the number of programmers and y is the number of systems analysts assigned to the task. Determine the number of employees of each type that should be assigned to the task to minimize labor costs. What is the minimal cost?

15. *Campaign strategy* The campaign staff of a mayoral candidate in a medium-sized city estimates the number of additional votes their candidate will pick up during the final week of the campaign in terms of the amount x spent on television advertising and the amount y spent on radio advertising that week to be

$$z = 3xy - y^3 - x^2$$

where x and y are in units of \$10,000 and z is in units of 10,000 votes. Determine the amount to be spent on each type of advertising to maximize the number of additional votes. What is the maximum number of additional votes?

16. *Space-inventory-profit* A department-store manager believes that the monthly profit from one of the departments (in dollars) in terms of the amount of floor space x allotted to the department and the amount y of inventory maintained by the department is

$$P(x, y) = 3{,}000 - (10 - x)^2 - (12 - y)^2$$

where x is in hundreds of square feet and y is in thousands of dollars. Determine the amount of floor space and inventory that would maximize the department's monthly profit. What is the monthly profit?

17. *Inventory* A company sells a computer component to the electronics industry on a steady basis and orders the component from the manufacturer on a regular basis. The inventory cost (ordering cost, storage cost, and component cost) for the company is given by

$$I(S, Q) = \frac{250{,}000}{Q} + \frac{S^2}{2Q} + \frac{(Q - S)^2}{2Q} + 2{,}000$$

where S is the quantity on hand when an order is placed and Q is the quantity ordered. Determine the values of S and of Q that minimize inventory cost. What is the minimum inventory cost?

18. *Surface area* A scientist is to design a box with an open top and a volume of 32 cubic inches to hold the apparatus for a scientific experiment aboard a space satellite. Determine the dimensions of the lightest (minimal surface area) box which can be made from a given material.

Identify the steps of the four-step procedure involved in the application of mathematics (Section 1.3) in each case.

19. *Profit* Example 3
20. *Bid preparation* Exercise 14
21. *Campaign strategy* Exercise 15

15.7 LAGRANGE MULTIPLIERS

In Example 3 of the last section we were told that the annual profit (in hundred thousands of dollars) for a company was

$$f(x, y) = 16xy - 17x^2 - 4y^2 + 3x$$

from manufacturing x thousand units of item A and y thousand units of item B. We were to determine how many units of each item the company should produce in order to maximize profit. An implicit assumption in the example was that the company can manufacture (and sell) any number of units of each item. This might not be realistic. Perhaps the company's facilities are limited to producing a total of 4,000 units a year; that is,

$$x + y = 4$$

Now the problem is changed considerably. We now want to determine the maximum value of the profit function subject to the constraint $x + y = 4$.

In this final section we consider a method for solving a maximum or minimum problem of this type. The method is called the **method of Lagrange multipliers,** named after the French mathematician Joseph Louis Lagrange (1736–1813) who discovered the method.

In general, we want to determine the relative maximum or relative minimum of a function $f(x, y)$ subject to a constraint $g(x, y) = 0$. In the profit function discussed above, the functions f and g would be

$$f(x, y) = 16xy - 17x^2 - 4y^2 + 3x$$

and

$$g(x, y) = x + y - 4$$

[Note that $g(x, y) = 0$ is equivalent to $x + y = 4$.]

The method of Lagrange multipliers is based on the following theorem, which is proved in more-advanced calculus courses:

15.7 LAGRANGE MULTIPLIERS

> **THEOREM 2**
>
> Let f and g be functions with continuous partial derivatives and let
> $$F(x, y, \lambda) = f(x, y) + \lambda g(x, y)$$
> If, subject to the constraint $g(x, y) = 0$, f has a maximum or minimum at (x_0, y_0), then
> $$F_x(x_0, y_0, \lambda_0) = 0, \quad F_y(x_0, y_0, \lambda_0) = 0, \quad F_\lambda(x_0, y_0, \lambda_0) = 0 \quad (1)$$
> for some number λ_0.

The Greek letter λ (lambda) is called the **Lagrange multiplier**. It is the standard notation used in this context.

We illustrate the use of Theorem 2 in the next several examples.

EXAMPLE 1 Determine the minimum value of $f(x, y) = x^2 + y^2$ subject to the constraint $x + 2y = 10$.

Solution The function $g(x, y)$ in this case is $x + 2y - 10$ and
$$F(x, y, \lambda) = x^2 + y^2 + \lambda(x + 2y - 10)$$
The partial derivatives of F are
$$F_x(x, y, \lambda) = 2x + \lambda, \quad F_y(x, y, \lambda) = 2y + 2\lambda, \quad F_\lambda(x, y, \lambda) = x + 2y - 10$$
In view of equations (1), we set each of these derivatives equal to zero and solve the resulting system:
$$2x + \lambda = 0$$
$$2y + 2\lambda = 0$$
$$x + 2y - 10 = 0$$
From the first of these equations, $\lambda = -2x$; from the second, $\lambda = -y$. Equating these two values of λ gives
$$\lambda = -2x = -y$$
or $y = 2x$. Substituting into the third of the above equations gives $x + 4x - 10 = 0$, or $x = 2$. Then $y = 4$ and $\lambda = -4$ (although the actual value of λ is never required).

The minimum value of $f(x, y)$, subject to the constraint $x + 2y = 10$, is 20; it is obtained when $x = 2$ and $y = 4$. ∎

There is a six-step general procedure that can be followed in solving a Lagrange multiplier problem. We list the six steps and illustrate the procedure in solving for the maximum value of $f(x, y) = 4 - x^2 - y^2$ subject to the constraint $2x - y = 6$.

1. Write the constraint in the form $g(x, y) = 0$ and construct the function $$F(x, y, \lambda) = f(x, y) + \lambda g(x, y)$$	1. The constraint becomes $2x - y - 6 = 0$ and $$F(x, y, \lambda) = 4 - x^2 - y^2 \\ + \lambda(2x - y - 6)$$
2. Construct the three equations $$F_x(x, y, \lambda) = 0 \\ F_y(x, y, \lambda) = 0 \\ F_\lambda(x, y, \lambda) = 0$$	2. $$-2x + 2\lambda = 0 \\ -2y - \lambda = 0 \\ 2x - y - 6 = 0$$
3. As a first step in solving the system of equations in Step 2, solve each of the first two equations for λ.	3. From the first equation, $$\lambda = x$$ From the second, $$\lambda = -2y$$
4. Equate the values of λ obtained in Step 3 and obtain an expression for x in terms of y or for y in terms of x.	4. $\lambda = x = -2y$, or $$x = -2y$$
5. Substitute the expression obtained in Step 4 into the third equation of Step 2 and solve for the remaining variable.	5. Substituting $x = -2y$ into $2x - y - 6 = 0$ gives $$-4y - y - 6 = 0 \\ y = -\tfrac{6}{5}$$
6. Determine the corresponding value of the other variable using the last expression obtained in Step 4. The constrained maximum or minimum value of f occurs at the values x_0 and y_0 obtained in these last two steps.	6. Setting $y = -\tfrac{6}{5}$ in the expression $x = -2y$ gives $x = \tfrac{12}{5}$. The maximum value of f, subject to the given constraint, is $$f(\tfrac{12}{5}, -\tfrac{6}{5}) = -\tfrac{16}{5}$$

EXAMPLE 1 Solve for the constrained maximum in the profit function illustrated at the beginning of this section.

15.7 LAGRANGE MULTIPLIERS

Solution We number the steps in accordance with those in the general procedure.

Step 1. The constraint becomes $x + y - 4 = 0$, and
$$F(x, y, \lambda) = 16xy - 17x^2 - 4y^2 + 3x + \lambda(x + y - 4)$$

Step 2. Setting the partial derivatives equal to 0 gives
$$16y - 34x + 3 + \lambda = 0$$
$$16x - 8y + \lambda = 0$$
$$x + y - 4 = 0$$

Step 3. From the first of these equations
$$\lambda = -16y + 34x - 3$$
From the second,
$$\lambda = -16x + 8y$$

Step 4. Equating these two values of λ and solving for y gives
$$-16y + 34x - 3 = -16x + 8y$$
$$-24y = -50x + 3$$
$$y = \frac{50x - 3}{24}$$

Step 5. Substituting this expression for y into the third equation of Step 2 gives
$$x + \frac{50x - 3}{24} - 4 = 0$$
$$24x + 50x - 3 = 96$$
$$x = \frac{99}{74}$$
or, rounded to three decimal places, $x = 1.338$.

Step 6. Thus $y = \frac{1}{24}[50(\frac{99}{74}) - 3] = 2.662$. The maximum value of f is then
$$f(1.338, 2.662) = 2.223$$

The company should manufacture 1,338 units of item A and 2,662 units of item B for a maximum profit of $222,300.

The equations obtained by setting the partial derivatives of F equal to zero were linear equations in each of the examples we saw. They need not always be linear, as illustrated in the following example in which there is both a constrained maximum and minimum.

EXAMPLE 2 Determine the maximum and minimum values of $f(x, y) = 2x^3 + y^3$ subject to the constraint $x + 2y = 15$.

---

1 PRACTICE PROBLEM 1

Determine the minimum value of $f(x, y) = 2x^2 + y^2 + 2xy$ subject to the constraint $x - y = 1$.

Answer

Minimum of $\frac{1}{5}$ at $x = \frac{2}{5}$ and $y = -\frac{3}{5}$.

Solution In this example,
$$F(x, y, \lambda) = 2x^3 + y^3 + \lambda(x + 2y - 15)$$

The equations obtained by setting the partial derivatives equal to zero are
$$6x^2 + \lambda = 0$$
$$3y^2 + 2\lambda = 0$$
$$x + 2y - 15 = 0 \qquad (2)$$

From the first two of these equations,
$$\lambda = -6x^2 \quad \text{and} \quad \lambda = -\tfrac{3}{2}y^2$$

Equating these two values of λ gives
$$-6x^2 = -\tfrac{3}{2}y^2$$
$$4x^2 = y^2$$
$$y = \pm 2x$$

For $y = 2x$, from equation (2), $x = 3$; then $y = 6$ and $f(3, 6) = 270$. Similarly, for $y = -2x$, $x = -5$; then $y = 10$ and $f(-5, 10) = 750$. Comparing these two values of f, we see that a minimum of 270 is obtained at $x = 3$ and $y = 6$; a maximum of 750 is obtained at $x = -5$ and $y = 10$. ◼

The Lagrange multiplier method for locating constrained maximum and minimum points extends to functions of three variables x, y, and z. Setting the partial derivatives equal to zero in this case results in a system of four equations. The corresponding changes in the procedure are illustrated in Example 3.

2 PRACTICE PROBLEM 2

Determine the maximum and minimum value of $f(x, y) = x^3 - 3y^2$ subject to the constraint $2y - x = 15$.

Answer

Minimum of -216 at $x = 3$ and $y = 9$; maximum of $-\tfrac{2{,}125}{16}$ at $x = -\tfrac{5}{2}$ and $y = \tfrac{25}{4}$.

EXAMPLE 3 A closed box with a volume of 8 cubic meters is to be constructed. Determine the dimensions of such a box, which uses the least amount of material (has the least surface area).

Solution If the dimensions of the box are x meters by y meters by z meters high, the surface area of the six sides will be $2xy + 2xz + 2yz$ square meters (see the figure). The volume will be xyz cubic meters. The function to be minimized is, therefore,
$$f(x, y, z) = 2xy + 2xz + 2yz$$

subject to
$$xyz = 8$$

In this case,
$$F(x, y, z, \lambda) = 2xy + 2xz + 2yz + \lambda(xyz - 8)$$

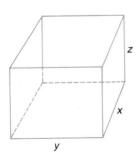

Setting F_x, F_y, F_z, and F_λ equal to zero and solving for λ gives the equations

$$2y + 2z + \lambda yz = 0 \quad \text{or} \quad \lambda = -\frac{2y + 2z}{yz}$$

$$2x + 2z + \lambda xz = 0 \quad \text{or} \quad \lambda = -\frac{2x + 2z}{xz}$$

$$2x + 2y + \lambda xy = 0 \quad \text{or} \quad \lambda = -\frac{2x + 2y}{xy}$$

$$xyz - 8 = 0$$

Equating the first two expressions for λ gives

$$-\frac{2y + 2z}{yz} = -\frac{2x + 2z}{xz}$$

$$xz(2y + 2z) = yz(2x + 2z)$$

$$2xyz + 2xz^2 = 2xyz + 2yz^2$$

$$2xz^2 = 2yz^2$$

and, finally,

$$x = y$$

Similarly, equating the second and third expressions for λ,

$$-\frac{2x + 2z}{xz} = -\frac{2x + 2y}{xy}$$

gives

$$z = y$$

Substituting $x = y$ and $z = y$ into $xyz - 8 = 0$ (the equation obtained above by setting F_λ equal to zero) gives

$$y^3 - 8 = 0$$

so that $y = 2$. Then $x = 2$ and $z = 2$.

The minimum amount of material will be used if the box is 2 meters long, 2 meters wide, and 2 meters high. For these dimensions, the amount of material will be $f(2, 2, 2) = 24$ square meters.

3 PRACTICE PROBLEM 3

If the box in Example 3 is to be an open box (no top) with a volume of 4 cubic meters, determine the dimensions that use the least amount of material.

Answer

2 meters long, 2 meters wide, and 1 meter high

EXERCISES

1. Determine the minimum of $f(x, y) = x^2 + y^2$ subject to the constraint $2x + y = 5$.

2. Determine the maximum of $f(x, y) = 8 - 3x^2 - y^2$ subject to the constraint $3x - y = 8$.

3. Determine the maximum of $f(x, y) = 2xy - x^2 - 4y^2$ subject to the constraint $2x + 5y = 12$.

4. Determine the minimum of $f(x, y) = 2x^2 + 2xy + y^2$ subject to the constraint $x = y$.

5. Determine the minimum of $z = x^2 + 3y^2 - 2xy$ subject to the constraint $3x - 4y = 1$.

6. Determine the maximum of $z = 3xy - 3x^2 - 4y^2$ subject to the constraint $2x + y = 5$.

7. Determine the maximum and minimum of $z = x^3 + 2y^2$ subject to the constraint $x - y = 1$.

8. Determine the maximum and minimum of $f(x, y) = 2x^3 - \frac{3}{2}y^2$ subject to the constraint $y - 2x = 15$.

9. Determine the minimum of $f(x, y, z) = x^2 + y^2 + z^2$ subject to the constraint $x + 2y - 2z = 2$.

10. Determine the maximum and minimum of $f(x, y, z) = x + 2y + 4z$ subject to the constraint $x^2 + y^2 + z^2 = 21$.

11. *Production-profit* A company's monthly profit (in thousands of dollars) from manufacturing and selling x hundred units of item A and y hundred units of item B is
$$f(x, y) = 12xy - 3x^2 - 5y^2$$
The company's facilities limit production to 1,000 units each month ($x + y = 10$). Also, the company can sell all of the units it manufactures. Determine how many units of each item the company should manufacture to obtain the maximum monthly profit. What is the maximum monthly profit?

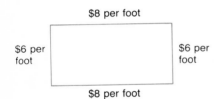

12. *Fencing cost* Two sides of a fence can be built for $6.00 a foot; the other two sides, for $8.00 a foot. Determine the dimensions of the field of maximum area that can be enclosed if $4,800 is available for the project.

13. *Production function* A **Cobb-Douglas production function** is a function of the form
$$f(x, y) = x^\alpha y^\beta \qquad (\alpha + \beta = 1)$$
where x is the number of units of labor and y is the number of units of capital used. The following is an example of this type of function: Suppose that x units of labor and y units of capital can produce
$$f(x, y) = 40x^{3/4}y^{1/4}$$
units of a particular product. If each unit of labor costs $250, each unit of capital costs $100, and $52,000 is available for production, how many units of labor and how many units of capital will maximize production? What is the maximum number of units that can be produced?

14. A piece of wire 144 inches long is to be cut into pieces and each piece formed into a square. Use Lagrange multipliers to determine what the size of each piece should be in order to enclose the minimum total area in each case.
(a) The wire is cut into two pieces. (b) The wire is cut into three pieces. What is the minimum total area in each case?

15. The material for the top and bottom of a rectangular closed box costs $3.00 per square foot; the material for the sides costs $1.50 per square foot. If the box is to have a volume of 54 cubic feet, determine the dimensions of the box with the minimal cost. What is the minimal cost?

Identify the steps of the four-step procedure involved in the application of mathematics (Section 1.3) in each case.

16. *Fencing cost*   Exercise 12
17. *Production function*   Exercise 13

15.8 CHAPTER SUMMARY

If a sum of money is invested at simple interest, the amount of interest earned depends upon both the rate of interest and the time the money is invested. Furthermore, the rate of interest and the time are independent of each other. Quantities such as the amount of interest earned, whose value depends on the value of two or more other quantities, are represented mathematically by functions of two or more variables.

The graphs of functions of two variables are surfaces in a three-dimensional coordinate system. For any point (x_0, y_0, z_0) in the surface, z_0 is the value of the function when $x = x_0$ and $y = y_0$. The only graphs that we constructed were planes corresponding to the simple equations $x = k$, $y = k$, and $z = k$, where k is a constant. The graphs of functions of three or more variables require coordinate systems that cannot be represented satisfactorily on paper. Consequently, we cannot construct graphs of such functions.

If f is a function of two variables x and y and if y is held constant, then f becomes a function of x alone, and we can determine the derivative of f as we did previously for functions of a single variable. The resulting derivative is the **partial derivative of f with respect to x.** Similarly, if x is held constant, we can determine the derivative of f with respect to y. The resulting derivative is the **partial derivative of f with respect to y.** This partial derivative concept extends to functions of several variables.

The mechanics for determining partial derivatives are exactly the same as those used for determining ordinary derivatives. We just have to remember that the variables not entering into the differentiation are to be treated as constants.

Recall that the graph of a function $f(x, y)$ is a surface in three dimensions. If y is held constant—for example, $y = k$—we restrict our attention to that part of the graph consisting of a curve in the plane $y = k$. The partial derivative of f with respect to x is then the slope of the tangent to this curve in the plane $y = k$. If f represents an economic quantity, the partial derivative of f with respect to x gives the marginal change in f as x varies and y is held fixed.

Similarly, if x is held constant—for example, $x = k$—the partial derivative of f with respect to y gives the slope of the tangent to the corresponding curve in the plane $x = k$. And if f represents an economic quantity, the partial derivative of f with respect to y gives the marginal change in f as y varies and x is held fixed.

In general, if f is a function of several variables, the partial derivative of f with respect to one of the variables gives the instantaneous rate of change of f as that variable varies and the other variables are held fixed.

A function $f(x, y)$ has a **relative maximum** at (x_0, y_0) if

$$f(x_0, y_0) > f(x, y)$$

for all points (x, y) near (x_0, y_0). Similarly, $f(x, y)$ has a **relative minimum** at (x_0, y_0) if

$$f(x_0, y_0) < f(x, y)$$

for all points (x, y) near (x_0, y_0). The following test, sometimes called the **second derivative test for functions of two variables**, gives a method for determining relative maxima and minima:

Suppose that f has continuous first-order and second-order partial derivatives. If at (x_0, y_0),

$$f_x = 0 \quad \text{and} \quad f_y = 0$$

and if:

(i) $(f_{xy})^2 - f_{xx}f_{yy} < 0$ and $f_{xx} < 0$ at (x_0, y_0),
then $f(x, y)$ has a relative maximum at (x_0, y_0);

(ii) $(f_{xy})^2 - f_{xx}f_{yy} < 0$ and $f_{xx} > 0$ at (x_0, y_0),
then $f(x, y)$ has a relative minimum at (x_0, y_0);

(iii) $(f_{xy})^2 - f_{xx}f_{yy} > 0$ at (x_0, y_0),
then $f(x, y)$ has neither a relative maximum nor a relative minimum at (x_0, y_0);

(iv) $(f_{xy})^2 - f_{xx}f_{yy} = 0$ at (x_0, y_0),
the test fails.

Finally, the method of **Lagrange multipliers** gives a method for locating the maxima and minima of functions of two variables when the two variables are required to satisfy an additional condition (called a *constraint*). The method is based on the following theorem:

Let f and g be functions with continuous partial derivatives and let

$$F(x, y, \lambda) = f(x, y) + \lambda g(x, y)$$

If, subject to the constraint $g(x, y) = 0$, f has a maximum or minimum at (x_0, y_0), then

$$F_x(x_0, y_0, \lambda_0) = F_y(x_0, y_0, \lambda_0) = F_\lambda(x_0, y_0, \lambda_0) = 0$$

for some number λ_0.

Our ability to use the Lagrange method depends to a large extent on our ability to solve the system of equations obtained by setting each of the three partial derivatives equal to zero.

REVIEW EXERCISES

Determine each of the following values for the function $f(x, y) = \dfrac{-x^2}{(x+y)^2}$:

1. $f(2, 1)$
2. $f(0, -1)$
3. $f(-0.5, 1.5)$

Determine the value of z if $z = \sqrt{\dfrac{x-y}{2y}}$ for the given values of x and y.

4. $x = 2, \; y = 1$
5. $x = -9, \; y = -1$
6. $x = 5, \; y = 5$

Determine the value of $f(x, y, z) = z(x + y)^2$ for the given values of x, y, and z.

7. $x = 4, \; y = -1, \; z = 7$
8. $x = 0.3, \; y = -0.5, \; z = 0.1$

Determine the value of w if $w = 4^{x+y}z^{-2}$ for the given values of x, y, and z.

9. $x = 3, \; y = 0, \; z = -1$
10. $x = -2, \; y = 2.5, \; z = \frac{1}{4}$

11. The area of a triangle is the product of one-half its base and height (see the figure). Construct a function that will give the area of a triangle (a) in terms of its base (x) and height (y).

Locate each of the following points in a coordinate system:

12. $(2, -1, -2)$
13. $(-3, 2, 2)$
14. $(-2, -4, 3)$
15. $(-1, 4, 0)$
16. $(0, 3, -3)$
17. $(0, 0, 0)$

Graph each of the following in a three-dimensional coordinate system:

18. $y = 6$
19. $x = 3.5$
20. $z = -5$

Determine each of the following partial derivatives for $f(x, y) = 3x^2y - 2x + y^2 + 7$:

21. $f_x(x, y)$
22. $f_y(x, y)$
23. $f_{xx}(x, y)$
24. $f_{xy}(x, y)$
25. $f_{yy}(x, y)$
26. $f_y(2, -3)$
27. $f_{xx}(2, -3)$
28. $f_x(-4, -5)$
29. $f_{xy}(0, 7)$

Determine each of the following partial derivatives for $f(x, y) = (x^2 - y^2)^3$:

30. $f_x(x, y)$
31. $f_y(x, y)$
32. $f_{xx}(x, y)$
33. $f_{yx}(x, y)$
34. $f_{yy}(x, y)$
35. $f_x(2, 1)$
36. $f_{yy}(0, -2)$
37. $f_y(1, \frac{1}{2})$
38. $f_{yx}(1, -1)$

Determine each of the following partial derivatives for $z = \dfrac{x}{x-y}$:

39. $\dfrac{\partial z}{\partial x}$
40. $\dfrac{\partial z}{\partial y}$
41. $\dfrac{\partial^2 z}{\partial x \, \partial x}$
42. $\dfrac{\partial^2 z}{\partial x \, \partial y}$
43. $\dfrac{\partial^2 z}{\partial y \, \partial y}$
44. $\dfrac{\partial^2 z}{\partial y \, \partial x}$

Determine each of the following partial derivatives for $z = (x + y)e^{x^2}$:

45. z_x
46. z_y
47. z_{xx}
48. z_{yy}
49. z_{xy}
50. z_{yx}

Determine each of the following partial derivatives for

$$f(x, y, z) = x^2z + z^2y - 3y^3$$

51. $f_x(x, y, z)$
52. $f_y(x, y, z)$
53. $f_z(x, y, z)$
54. $f_{xz}(x, y, z)$
55. $f_{zy}(x, y, z)$
56. $f_{zz}(x, y, z)$
57. $f_y(3, -2, -2)$
58. $f_{xx}(4, 7, 9)$
59. $f_{yy}(1, \frac{1}{2}, \frac{1}{3})$

Determine each of the following partial derivatives for

$$f(x, y, z) = (x + z^2)(y^2 - z)$$

60. $f_x(x, y, z)$
61. $f_y(x, y, z)$
62. $f_z(x, y, z)$
63. $f_{xz}(x, y, z)$
64. $f_{zy}(x, y, z)$
65. $f_{yy}(x, y, z)$
66. $f_{xz}(-1, 2, 3)$
67. $f_{xx}(-6, -3, -2)$
68. $f_z(0, 2, 1)$

69. Determine the slope of the tangent to the curve cut out by the plane $x = 1$ in the surface $z = \ln(2xy)$ at $y = \frac{1}{2}$.

70. At what point is the slope of the tangent to the curve cut out by the plane $y = 2$ in the surface $z = x^2y - 2xy^2$ equal to zero?

71. At what point is the slope of the tangent to the curve cut out by the plane $y = 3$ in the surface $z = yx^2 - x^3 - 2$ the greatest?

72. *Marginal profit* A company's monthly profit (in dollars) from the sale of x units of item A and y units of item B is given by

$$P(x, y) = xy^2 + 2x^2 - 6xy$$

(a) Determine the marginal profit from the sale of y units of item B if sales of item A are held constant at 20 units.
(b) Determine the marginal profit from the sale of item B at the 30-unit level if sales of item A are held constant at 20 units.

73. *Marginal profit*
(a) Determine the marginal profit in Exercise 72 from the sale of x units of item A if sales of item B are held constant at 30 units.
(b) Determine the marginal profit from the sale of item A at the 20-unit level if sales of item B are held constant at 30 units.

Determine the points at which each of the following functions has a relative maximum or minimum. Determine the maximum or minimum value of the function at each such point.

74. $f(x, y) = 3x^2 + y^2 - 8y + 1$
75. $f(x, y) = 8x + 4y - 2x^2 - y^2 - 3$
76. $z = 3x^2 - 2xy + 4y^2 + 11y$
77. $z = x^2 - 4xy - y^2 + 6y - 2x$
78. $f(x, y) = x(\ln y)$
79. $f(x, y) = 2x^3 - 6xy + 3y^2 + 5$

80. *Labor-advertising-profit* If a company's profit next year in terms of the amount x that it spends on labor and the amount y that it spends on

advertising is given by

$$P(x, y) = 10x + 16y - x^2 - 2y^2 + xy + 90$$

where all units are in millions of dollars, determine the amount the company should spend on labor and on advertising to obtain the maximum profit. What is the maximum profit?

81. *Daily cost* If the daily cost (in hundreds of dollars) for a company to manufacture x hundred units of item A and y hundred units of item B is

$$C(x, y) = 4x + 9y + 36\left(\frac{x+y}{xy}\right) - 10$$

determine how many units of each item should be produced to minimize the daily cost. What is the minimum daily cost?

82. Determine the maximum of $f(x, y) = 4 - x^2 - 3y^2$ subject to the constraint $x - 6y = 7$.

83. Determine the maximum and minimum of $f(x, y) = 8x^3 + 6y^2 - 25$ subject to the constraint $3x + 6y = 2$.

84. Determine the maximum of $w = 10 - x^2 - 2y^2 - 3z^2$ subject to the constraint $3x + 2y + z = 11$.

85. *Labor-capital-production* Suppose that x units of labor and y units of capital can produce

$$f(x, y) = 60x^{2/3}y^{1/3}$$

units of a product. If each unit of labor costs $500, each unit of capital costs $350, and $270,000 is available for production, how many units of labor and of capital will maximize production? What is the maximum number of units that can be produced?

86. The material for the bottom of a rectangular box costs $4.00 a square foot; that for the sides costs $2.50 a square foot. If the box is to have no top and a volume of 100 cubic feet, determine the dimensions of the box with the minimal cost. What is the minimal cost?

Identify the steps of the four-step procedure involved in the application of mathematics (Section 1.3) in each case.

87. *Daily cost* Exercise 81   88. Exercise 86

APPENDIX A

ALGEBRA REVIEW

OUTLINE

A.1 Evaluating Expressions
A.2 Algebraic Expressions
A.3 Factoring
A.4 Solving Linear and Quadratic Equations
A.5 Algebraic Fractions
A.6 Solving Additional Types of Equations
A.7 Exponents
A.8 Sets

A.1 EVALUATING EXPRESSIONS

Suppose that you were to evaluate the expression

$$2 - 4 + 6 \cdot 5$$

In what order should the operations be performed? The answer to this question is crucial because the value of the expression varies with the order in which the operations are chosen. Consider evaluating this expression in each of the following orders:

$$\left. \begin{array}{l} \text{Subtraction} \\ \text{Addition} \\ \text{Multiplication} \end{array} \right\} \text{gives} \quad \begin{array}{l} -2 + 6 \cdot 5 \\ 4 \cdot 5 \\ 20 \end{array}$$

$$\left. \begin{array}{l} \text{Multiplication} \\ \text{Addition} \\ \text{Subtraction} \end{array} \right\} \text{gives} \quad \begin{array}{l} 2 - 4 + 30 \\ 2 - 34 \\ -32 \end{array}$$

$$\left. \begin{array}{l} \text{Multiplication} \\ \text{Subtraction} \\ \text{Addition} \end{array} \right\} \text{gives} \quad \begin{array}{l} 2 - 4 + 30 \\ -2 + 30 \\ 28 \end{array}$$

How does one determine which is the correct value?

Mathematicians have adopted a convention that specifies the **order** in which operations must be performed in evaluating expressions such as the one just given. Operations are to be performed in the following order:

> 1. Evaluation of exponents
> 2. Multiplication and division, proceeding left to right
> 3. Addition and subtraction, proceeding left to right

Accordingly, the final order of operations, giving a result of 28, is the correct order for evaluating the expression in the last paragraph.

EXAMPLE 1   Evaluate $6 + 4 \cdot 5 - 9 \div 3$.

Solution   There are no exponents, so we can start with the second step.

$$\begin{aligned} 6 + 4 \cdot 5 - 9 \div 3 &= 6 + 20 - 9 \div 3 & &\text{Multiplication and division,} \\ &= 6 + 20 - 3 & &\text{left to right} \\ &= 26 - 3 = 23 & &\text{Addition and subtraction, left to right} \end{aligned}$$

EXAMPLE 2   Evaluate $-3 + 7^2 - 4 \cdot 3$.

A.1 EVALUATING EXPRESSIONS

Solution

$$-3 + 7^2 - 4 \cdot 3 = -3 + 49 - 4 \cdot 3 \quad \text{Evaluation of exponents}$$
$$= -3 + 49 - 12 \quad \text{Multiplication and division, left to right}$$
$$= 46 - 12 = 34 \quad \text{Addition and subtraction, left to right}$$

PRACTICE PROBLEM 1

Evaluate $4 \cdot 3 - 2^3 + 1$.

Answer

5

The given order for performing operations can be changed by use of parentheses or other grouping symbols. When parentheses occur, the expressions within parentheses are evaluated first. In doing so, the operations are again performed in the given order.

EXAMPLE 3 Evaluate $-2^3 + 3 \cdot (2^2 + 4 \cdot 5) + 16 \div 4$.

Solution First, within parentheses:

$$-2^3 + 3 \cdot (2^2 + 4 \cdot 5) + 16 \div 4$$
$$= -2^3 + 3 \cdot (4 + 4 \cdot 5) + 16 \div 4 \quad \text{Evaluation of exponents}$$
$$= -2^3 + 3 \cdot (4 + 20) + 16 \div 4 \quad \text{Multiplication and division}$$
$$= -2^3 + 3 \cdot 24 + 16 \div 4 \quad \text{Addition and subtraction}$$

Then, in the overall expression:

PRACTICE PROBLEM 2

Evaluate $3 - (10 - 3^2) \cdot 5$.

Answer

-2

$$= -8 + 3 \cdot 24 + 16 \div 4 \quad \text{Evaluation of exponents}$$
$$= -8 + 72 + 16 \div 4 = -8 + 72 + 4 \quad \text{Multiplication and division}$$
$$= 64 + 4 = 68 \quad \text{Addition and subtraction}$$

EXAMPLE 4 Evaluate $2 - (7 + 6 \cdot 5) + \dfrac{12 \cdot 6}{5 - 2}$.

Solution The bar appearing in the above expression serves a twofold purpose: It is a symbol for grouping and for the operation of division. The given expression could also be written

$$2 - (7 + 6 \cdot 5) + (12 \cdot 6) \div (5 - 2)$$

Note that the bar is replaced by parentheses and a division symbol. The numerator and denominator are treated as expressions in parentheses, unless they happen to be single numbers. It is essential to appreciate the double use of the bar when evaluating expressions.

To evaluate, we start within the parentheses:

$$2 - (7 + 6 \cdot 5) + \dfrac{12 \cdot 6}{5 - 2}$$
$$= 2 - (7 + 30) + \dfrac{72}{5 - 2} \quad \text{Multiplication and division}$$
$$= 2 - 37 + \dfrac{72}{3} \quad \text{Addition and subtraction}$$

3 PRACTICE PROBLEM 3

Evaluate $14 \cdot \dfrac{9-6}{2+1} \div 2$.

Answer

7

In the overall expression, we then have:

$$= 2 - 37 + 24 \quad \text{Multiplication and division}$$
$$= -35 + 24 = -11 \quad \text{Addition and subtraction} \quad \boxed{3}$$

Finally, if an expression contains nested grouping symbols—that is, grouping symbols within grouping symbols—the innermost parentheses must be evaluated first. Within each pair of grouping symbols, the operations are always performed in the order given above. The procedure is illustrated in Example 5.

EXAMPLE 5 Evaluate $7 - 2 \cdot [5^2 + 6 \cdot (3 \cdot 4 - 8)] + 9$.

Solution In the innermost parentheses:

$$7 - 2 \cdot [5^2 + 6 \cdot (3 \cdot 4 - 8)] + 9$$
$$= 7 - 2 \cdot [5^2 + 6 \cdot (12 - 8)] + 9 \quad \text{Multiplication and division}$$
$$= 7 - 2 \cdot [5^2 + 6 \cdot 4] + 9 \quad \text{Addition and subtraction}$$

In the outer brackets:

$$= 7 - 2 \cdot [25 + 6 \cdot 4] + 9 \quad \text{Evaluation of exponents}$$
$$= 7 - 2 \cdot [25 + 24] + 9 \quad \text{Multiplication and division}$$
$$= 7 - 2 \cdot 49 + 9 \quad \text{Addition and subtraction}$$

In the overall expression:

$$= 7 - 98 + 9 \quad \text{Multiplication and division}$$
$$= -91 + 9 = -82 \quad \text{Addition and subtraction} \quad \boxed{4}$$

4 PRACTICE PROBLEM 4

Evaluate $4 - [3^2 - (12 - 2 \cdot 5) + 3]$.

Answer

-6

If you have done computer programming, you may have noticed that the order (1) evaluation of exponents, (2) multiplication and division, left to right, and (3) addition and subtraction, left to right, is generally the order in which a computer evaluates a mathematical expression. Furthermore, expressions involving parentheses are evaluated just as we have done here.

EXERCISES

Evaluate the given expressions.

1. $4 + 3 \cdot 6 - 8 \div 2$
2. $(4 + 3) \cdot 6 - 8 \div 2$
3. $4 + 3 \cdot (6 - 8) \div 2$
4. $4 + 3 \cdot (6 - 8 \div 2)$
5. $5 - 6^2 \div 9 \cdot 2$
6. $5 - 6^2 \div (9 \cdot 2)$
7. $(5 - 6^2 \div 9) \cdot 2$
8. $(5 - 6)^2 \div 9 \cdot 2$
9. $5 + (7 - 3 \cdot 2) - 4$
10. $(20 - 4^2 + 2) + 2(8 - 5)$
11. $(3 \cdot 2)^3 - 3 \cdot 2^3$
12. $(3 \cdot 2^2 - 1) \cdot 4 - 2$

13. $5 + \dfrac{7-4}{3}$

14. $-1 - \dfrac{10+2}{9-3}$

15. $10 + 2 \cdot \dfrac{4-9}{5 \cdot 2} - 3$

16. $(17 - 3^2) - \dfrac{5+3}{7-2}$

17. $20 + \left(4 - 2 \cdot \dfrac{3+7}{5-2}\right)$

18. $2 \cdot (8-1) - (6 \cdot 5 - 8) \div 2$

19. $3 \cdot \left(\dfrac{4^2}{5+3} - \dfrac{7+5}{4-6}\right)^2$

20. $[2^2 \cdot (4+1) - (3^2 + 6)] \div 3$

21. $2 - [6 - 4 \cdot (3 + 2^2)]$

22. $3^2 \cdot 6 \div [2 + (6-1)^2]$

23. $\dfrac{(7+6) - 5 \cdot 3}{4^2 - 5 \cdot (2^2 - 1)}$

24. $\dfrac{5 \cdot (3 \cdot 10)^2}{2 \cdot 5^2 + (2 \cdot 5)^2}$

Give the order of performing operations in evaluating expressions in each case.

25. Without parentheses

26. With parentheses (unnested)

A.2 ALGEBRAIC EXPRESSIONS

In the previous section, all the expressions contained only numbers and all the operations were operations with numbers only. Therefore, the section should have properly been called *arithmetic* rather than *algebra*. The major difference between the two is that algebra involves the use of letters to denote numbers and arithmetic does not. The letters in algebra may stand for any one in a group of numbers (in which case they are called **variables**) or they may stand for a single number (in which case they are called **constants**). One example of a constant that occurs frequently is the use of the Greek letter pi (π) to denote the ratio of the circumference of a circle to its diameter. (The circumference C of a circle with diameter d is given by the equation $C = \pi d$; therefore, $\pi = C/d$.)

Hereafter, when there is no danger of confusion, we shall follow the usual convention of omitting the symbol for multiplication and shall indicate that two quantities are to be multiplied simply by writing them next to each other. For example:

$2x$ means $2 \cdot x$

$x(a+b)$ means $x \cdot (a+b)$

$-4(a+b)(x+y)$ means $-4 \cdot (a+b) \cdot (x+y)$

The result of multiplying two or more expressions together is called a ***product***. The expressions forming the product are called ***factors***. For example:

EXPRESSION	FACTORS
$2x$	2 and x
$x(a+b)$	x and $(a+b)$
$-4(a+b)(x+y)$	-4, $(a+b)$, and $(x+y)$

In other words, factors are quantities combined by multiplication.

On the other hand, quantities combined by addition are called ***terms.*** For example:

EXPRESSION	TERMS
$2x^2 + 4x + 6$	$2x^2$, $4x$, and 6
$a - 2b$ or $a + (-2b)$	a and $-2b$
$-5ax + 6by - 3$	$-5ax$, $6by$, and -3

1 PRACTICE PROBLEM 1

(a) Give the factors of $3(x - y)$.
(b) Give the terms of $3x - 3y$.

Answer

(a) 3 and $(x - y)$ (b) $3x$ and $-3y$

Terms in which the variable factors are exactly the same are called ***like terms.*** For example, $2xy$ and $-7xy$ are like terms; x^2 and $5x^2$ are like terms; and $6a^3y^2$ and $4a^3y^2$ are like terms. However, $2x^2$ and $5x^3$ are unlike terms.

Algebraic expressions containing variables have no numerical value unless a value is assigned to the variables. The expression

$$2x - 4(x^2 - 1)$$

has no numerical value. However, if x is given a value—say $x = 3$—the value of the expression is

$$2(3) - 4(3^2 - 1) = -26$$

When the variables in an algebraic expression are given a value, the expression is evaluated in the manner described in the previous section.

EXAMPLE 1   Evaluate $3x + 4(x^2 - 5x \div 4)$ if $x = 8$.

Solution   If $x = 8$, the given expression becomes

$$3(8) + 4[8^2 - 5(8) \div 4]$$

which, when evaluated according to the procedure of the previous section, is equal to 240.

2 PRACTICE PROBLEM 2

Evaluate $6(2x^2 - 3) \div 4x$ if $x = 3$.

Answer

7.5

When simplifying algebraic expressions, the operations are performed in exactly the same order as in numerical expressions:

> 1. Grouping symbols, innermost first
> 2. Evaluation of exponents
> 3. Multiplication and division, left to right
> 4. Addition and subtraction, left to right

The manner in which the operations are performed are different, of course. These operations are illustrated in the following examples.

EXAMPLE 2   Simplify $2(x + y) + 3(x - y)$.

Solution The expressions in parentheses, $x + y$ and $x - y$, cannot be simplified any further, since only like terms can be added or subtracted. Also, there are no exponents, so we can start with multiplication and division.

$$2(x + y) + 3(x - y) = 2x + 2y + 3(x - y) \quad \text{Multiplication/division}$$
$$= 2x + 2y + 3x - 3y$$
$$= 5x - y \quad \text{Addition/subtraction} \quad \blacksquare$$

EXAMPLE 3 Simplify $3[x - 2(x - 1)] + 4$.

Solution Again, the expression in the parentheses, $x - 1$, cannot be simplified any further. Nor are there any exponents. In the square brackets:

$$3[x - 2(x - 1)] + 4 = 3[x - 2x + 2] + 4 \quad \text{Multiplication/division}$$
$$= 3[-x + 2] + 4 \quad \text{Addition/subtraction}$$

In the overall expression:

$$= -3x + 6 + 4 \quad \text{Multiplication/division}$$
$$= -3x + 10 \quad \text{Addition/subtraction} \quad \blacksquare$$

EXAMPLE 4 Simplify $7ab + (2a - b)(a - b)$.

Solution
$$7ab + (2a - b)(a - b)$$
$$= 7ab + 2a(a - b) - b(a - b) \quad \text{Multiplication/division}$$
$$= 7ab + 2a^2 - 2ab - ba + b^2$$
$$= 4ab + 2a^2 + b^2 \quad \text{Addition/subtraction} \quad \blacksquare$$

EXAMPLE 5 Simplify $4 + (x - 1) - (3 - 2x)$.

Solution To add a quantity in parentheses, each term inside the parentheses keeps the same sign and the parentheses are dropped; to subtract, the sign of each term inside the parentheses is changed. We have, therefore,

$$4 + (x - 1) - (3 - 2x) = 4 + x - 1 - (3 - 2x) \quad \text{Addition/subtraction}$$
$$= 4 + x - 1 - 3 + 2x = 3x$$

3 PRACTICE PROBLEM 3

Simplify $3(1 + 2x) - (4x - 7)$.

Answer

$2x + 10$

EXAMPLE 6 Simplify $7ab - 4(2a - b)(a - b)$.

Solution The minus sign before the 4 makes the signs for the terms in the product a little tricky. Consequently, to remind ourselves that the whole product is to be subtracted, we first enclose the whole product in square brackets.

$$7ab - 4(2a - b)(a - b)$$
$$= 7ab - [4(2a - b)(a - b)] \quad \text{Rewriting}$$

Within grouping symbols:
$$= 7ab - [(8a - 4b)(a - b)] \quad \text{Multiplication/division}$$
$$= 7ab - [8a(a - b) - 4b(a - b)]$$
$$= 7ab - [8a^2 - 8ab - 4ab + 4b^2]$$
$$= 7ab - [8a^2 - 12ab + 4b^2] \quad \text{Addition/subtraction}$$

In the overall expression:
$$= 7ab - 8a^2 + 12ab - 4b^2 \quad \text{Addition/subtraction}$$
$$= -8a^2 + 19ab - 4b^2 \quad\blacksquare$$

In general, if a complicated product follows a minus sign, it is a good idea to enclose the whole product within grouping symbols, as we did in Example 6. It is not absolutely essential to do this, but it makes the proper signs in the product more obvious. Ending up with the wrong sign in situations such as this is one of the most common errors in algebra.

EXAMPLE 7 Simplify $-5[(2t + 1) - (t - 4)] + 3t$.

Solution We first write the whole product following the minus sign in braces.
$$-5[(2t + 1) - (t - 4)] + 3t$$
$$= -\{5[(2t + 1) - (t - 4)]\} + 3t \quad \text{Rewriting}$$
$$= -\{5[t + 5]\} + 3t \quad \text{Addition/subtraction}$$
$$= -\{5t + 25\} + 3t \quad \text{Multiplication/division}$$
$$= -5t - 25 + 3t = -2t - 25 \quad \text{Addition/subtraction} \quad \boxed{4}$$

The word *simplify* in algebra can mean different things in different situations. In each of the above examples, to simplify meant to perform the indicated operations until all parentheses were removed and then to combine like terms.

Our final example involves both of the procedures discussed in this section—simplifying expressions and evaluating expressions for given values of the variables.

EXAMPLE 8 Given $6 - 8x - \frac{2}{3}(9x - 30)$.

(a) Evaluate the given expression for $x = 2$.
(b) Simplify (in the sense just indicated) the given expression.
(c) Evaluate the result obtained in part (b) for $x = 2$.

---

4 PRACTICE PROBLEM 4

Simplify $b(6a - 3b) - 3(a + b)(3a - b)$.

Answer

$-9a^2$

Solution (a) If $x = 2$, the given expression becomes

$$6 - 8(2) - \tfrac{2}{3}[9(2) - 30] = 6 - 8(2) - \tfrac{2}{3}[18 - 30]$$
$$= 6 - 8(2) - \tfrac{2}{3}[-12]$$
$$= 6 - 16 + 8$$
$$= -2$$

(b)
$$6 - 8x - \tfrac{2}{3}(9x - 30) = 6 - 8x - [\tfrac{2}{3}(9x - 30)]$$
$$= 6 - 8x - [6x - 20]$$
$$= 6 - 8x - 6x + 20$$
$$= 26 - 14x$$

(c) If $x = 2$, the simplified expression in part (b) becomes

$$26 - 14(2) = 26 - 28 = -2$$

Note that the result in part (c) is the same as that obtained in part (a). The results in parts (a) and (c) would be the same for any value of x. ■

Throughout all the examples of this section, we have performed only one operation at a time, although at times it may have seemed obvious that several could be done at the same time to shorten the problem. One operation at a time is the recommended procedure until you are confident that combining several operations at the same step will not lead to errors.

You may have noticed that we have avoided division by algebraic expressions in all the examples. This is because such division has enough problems of its own—so we shall treat it by itself in a later section.

EXERCISES

Determine the factors in each of the following expressions:

1. $2x$
2. $-3xy$
3. $(a + b)(a - b)$
4. $(x - 2y)(3x + y)$
5. $(a - 2)(a^2 + 3)$
6. $4(x^2 + y^2)$
7. $(x + y)^2$
8. $-3(2x - 6y + 4)(x + y - 1)$
9. $(2x + 5) \div 3$
10. $(a - 3b) \div (7a + 2)$

Determine the terms in each of the following expressions:

11. $2x + 3$
12. $2x^2 - 3y$
13. $x^2y + 4y^2x - 9$
14. $2a - 3b + 4c - d$
15. $a + 2b - 6a^2b^2$
16. $x^2 + 2xy + y^2$
17. Determine the terms in each of the factors of $(x - 3y + 2)(z + y - 2x)$.
18. Determine the factors in each of the terms of $2xy - 6z$.

Evaluate the following expressions for the indicated values of the variables:

19. $2(3 - x^2); \quad x = 1$
20. $2(3 - x)^2; \quad x = -1$
21. $-4x^2 + 10; \quad x = 3$
22. $-4x^2 + 10; \quad x = -3$
23. $2 - [3x - 4(x + 2)]; \quad x = 4$
24. $[(y + 1) - (y^2 + 6)] \div 5; \quad y = 7$
25. $\dfrac{6}{5}\left(2t + \dfrac{t + 5}{t - 1}\right); \quad t = -2$
26. $3x^2 - (3x)^2; \quad x = 5$
27. $x(2x + y); \quad x = 2, y = -3$
28. $4x - 3(x + y)(2x + 3); \quad x = -5, y = 1.5$

Simplify (perform the indicated operations until all grouping symbols have been removed; then combine like terms) each of the following:

29. $2x - 3x[5 + 3(2 - x)]$
30. $y + 2y[5 - (3 - y)]$
31. $(2 - x)[3 - (2 + 2x)]$
32. $(2x - 3)(-6x + 1)$
33. $[6 + 2(x - 3)][-(x + 4) - 2]$
34. $2t - 3[(4t + 2) - (t + 6)]$
35. $2 + 8a - 3(2a - 5)$
36. $6 - (x + 4) - 3(2x - 1)$
37. $(x - 7) - 2(x + 1)(3x - 2)$
38. $2 - 3[2 - 3(2 - y)]$
39. $7\{x - 4[2 + x(2 - x)] + 8(x + 1)\}$
40. $(x - 1)\{-3 + 3[x - 4(3 + x)] + 10(x + 4)\}$

For Exercises 41–44:
(a) *Evaluate the given expression for the given value of the variable.*
(b) *Simplify the given expression.*
(c) *Evaluate the simplified expression for the given value of the variable.*

41. $4 - 2[x - (3 - x)]; \quad x = 3$
42. $[1 + 2(x - 1)][2 - 3(4 + x)]; \quad x = 0$
43. $\tfrac{1}{3}(3t - 6) + \tfrac{3}{2}(4t + 6); \quad t = -1$
44. $x - \{x - [x - 3(x - 3)] - 3\}; \quad x = 1$

45. Explain the difference between the letters in an algebraic expression that are variables and those that are constants.
46. Explain the difference between factors and terms.

A.3 FACTORING

In the last section we formed the product of given factors in a number of situations. In this section we reverse that procedure—that is, given a product, we want to reconstruct the factors making up the product. For example, given the expression $2x^2 - 6x + 4$, we shall want to determine the factors of this expression. This process of determining the factors of an expression is called **factoring**. Factoring will be essential in the next several sections on solving equations and on algebraic fractions.

We start with a few basic factorization formulas:

A.3 FACTORING

> Factoring expressions with two terms:
> 1. $x^2 - a^2 = (x + a)(x - a)$
> 2. $x^3 - a^3 = (x - a)(x^2 + ax + a^2)$
> 3. $x^3 + a^3 = (x + a)(x^2 - ax + a^2)$
>
> Factoring expressions with three terms:
> 4. $x^2 + 2ax + a^2 = (x + a)^2$
> 5. $x^2 - 2ax + a^2 = (x - a)^2$
> 6. $x^2 + (a + b)x + ab = (x + a)(x + b)$

Each of these formulas can be verified by forming the product indicated on the right and comparing the result with the expression on the left. The use of these forms is illustrated in the next several examples.

EXAMPLE 1 Factor $x^3 + 8$.

Solution This expression is in form 3 with $a = 2$. Therefore,
$$x^3 + 8 = (x + 2)(x^2 - 2x + 4)$$

EXAMPLE 2 Factor $x^2 - 8x + 16$.

Solution This expression is in form 5 with $a = 4$. Therefore,
$$x^2 - 8x + 16 = (x - 4)^2$$

EXAMPLE 3 Factor $x^2 + 5x + 6$.

Solution This expression contains three terms, and its factorization should, therefore, be based on one of the final three forms in the table. It is not a perfect square (forms 4 or 5), so we try form 6. For form 6 we need two numbers a and b such that $a + b = 5$ and $ab = 6$. We see that 2 and 3 satisfy these requirements. Therefore,
$$(x^2 + 5x + 6) = (x + 2)(x + 3)$$

EXAMPLE 4 Factor $x^2 + 5x - 14$.

PRACTICE PROBLEM 1

Factor $x^3 - 1$.

Answer

$(x - 1)(x^2 + x + 1)$

PRACTICE PROBLEM 2

Factor $x^2 + 8x + 16$.

Answer

$(x + 4)^2$

Solution   Again we try form 6. In this case we need two numbers a and b whose product is -14 (and therefore have opposite signs!) and whose sum is 5. By trial and error, we find that 7 and -2 satisfy these requirements. Therefore,
$$x^2 + 5x - 14 = (x + 7)(x - 2)$$
[3]

EXAMPLE 5   Factor $9x^4 - 4y^2$.

Solution   The letters x, a, and b in the forms in the table, particularly in forms 1, 2, and 3, can represent expressions rather than just single letters or numbers. The given expression in this case is the difference of two squares, so that it is in form 1 with the x of form 1 in this case being $3x^2$ and the a being $2y$. Therefore,
$$9x^4 - 4y^2 = (3x^2 - 2y)(3x^2 + 2y)$$
[4]

The expressions on the left in forms 1, 4, 5, and 6 are called **quadratic** in x because the highest power of x appearing in each is the second power. It is important to realize that not all quadratics can be factored. For example,
$$x^2 - 2x + 6$$
cannot be factored. In general, any quadratic
$$ax^2 + bx + c$$
for which $b^2 - 4ac$ is negative cannot be factored (unless one wants to get involved in complex numbers, and we do not). In our illustration, $a = 1$, $b = -2$, and $c = 6$, so that
$$b^2 = (-2)^2 = 4 \quad \text{and} \quad 4ac = 4(6) = 24$$
Since $b^2 - 4ac$ is negative, factorization cannot be performed. This is also the reason that the quadratic in Example 1 could not be factored further.

A major technique in factoring, in addition to using the given standard forms, is to factor out any factor that is common to every term in an expression. In fact, a basic rule of factoring is the following:

> First examine whether the terms in the given expression have any factors in common.

[3] **PRACTICE PROBLEM 3**

Factor $x^2 + 2x - 15$.

Answer

$(x + 5)(x - 3)$

[4] **PRACTICE PROBLEM 4**

Factor $27t^3 + x^3$.

Answer

$(3t + x)(9t^2 - 3tx + x^2)$

A.3 FACTORING

The procedure of factoring out common factors is illustrated in the next several examples.

EXAMPLE 6 Factor $6x^2 + 18x + 12$.

Solution Each term of the given expression has 6 as a factor. We first remove this common factor

$$6x^2 + 18x + 12 = 6(x^2 + 3x + 2)$$

The quadratic on the right can be factored further using form 6, so that we get

$$6x^2 + 18x + 12 = 6(x + 1)(x + 2)$$

EXAMPLE 7 Factor $2x^3 + 8x^2 + 8x$.

Solution In this case, each term has a factor of $2x$. Factoring out the $2x$ gives

$$2x^3 + 8x^2 + 8x = 2x(x^2 + 4x + 4)$$

Then, using form 4,

$$2x^3 + 8x^2 + 8x = 2x(x + 2)^2$$ [5]

EXAMPLE 8 Factor $2x^2(x - 1) - (8x - 6)(x - 1)$.

Solution In this case, each term has a factor of $2(x - 1)$. Factoring gives

$$2x^2(x - 1) - (8x - 6)(x - 1) = 2(x - 1)[x^2 - (4x - 3)]$$
$$= 2(x - 1)[x^2 - 4x + 3]$$

Then by form 6,

$$2x^2(x - 1) - (8x - 6)(x - 1) = 2(x - 1)[(x - 1)(x - 3)]$$
$$= 2(x - 1)^2(x - 3)$$

EXAMPLE 9 Factor $(x^2 - 1) + 3(x - 1)$.

Solution In this case we first factor the quadratic term:

$$(x^2 - 1) + 3(x - 1) = (x + 1)(x - 1) + 3(x - 1)$$
$$= (x - 1)[(x + 1) + 3] = (x - 1)[x + 4]$$ [6]

EXAMPLE 10 Simplify and then factor $(2x - 3)(x + 1) - (3x - 5)$.

---

[5] **PRACTICE PROBLEM 5**

Factor $2x^3 - 18x$.

Answer

$2x(x + 3)(x - 3)$

---

[6] **PRACTICE PROBLEM 6**

Factor $x^2(x - 1) + (5x + 6)(x - 1)$.

Answer

$(x - 1)(x + 2)(x + 3)$

Solution Simplifying gives
$$(2x - 3)(x + 1) - (3x - 5) = (2x^2 - x - 3) - (3x - 5)$$
$$= 2x^2 - 4x + 2$$

Factoring the result gives
$$= 2(x^2 - 2x + 1) = 2(x - 1)^2$$

Consequently,
$$(2x - 3)(x + 1) - (3x - 5) = 2(x - 1)^2 \qquad \blacksquare$$

EXERCISES

Factor each of the following expressions, if possible:

1. $x^2 - 9$
2. $4x^2 - 25$
3. $x^3 - 8$
4. $8y^3 - 1$
5. $t^3 + 27$
6. $8t^6 + 1$
7. $x^2 + 6x + 9$
8. $x^2 + 10x + 25$
9. $x^2 - 6x + 9$
10. $y^2 - 10y + 25$
11. $y^2 + 9y + 20$
12. $y^2 - y - 20$
13. $y^2 + y - 20$
14. $x^2 + 8x + 12$
15. $x^2 - 4x - 12$
16. $x^2 + 4x - 12$
17. $(x + y)^2 - 4$
18. $t^2 + 12 - 7t$
19. $y - 6 + y^2$
20. $(y - x)^3 + 8$
21. $x^2 + 3x + 9$
22. $x^2 + 1$
23. $9 - 6x + x^2$
24. $(x + 2)^3 - 1$
25. $9 + y^2$
26. $x^4 + x^2 - 2$
27. $x^4 - 1$
28. $16x^4 - 25$

Factor each of the following expressions, first factoring out any common factors:

29. $3x^2 - 12$
30. $4y^2 - 100$
31. $4y^3 - 32$
32. $4x^3 + 32$
33. $2x^2 - 12x + 18$
34. $2x^3 - 12x^2 + 18x$
35. $-3x^2 + 18x - 27$
36. $-2y^2 - 10y - 12$
37. $-x^3 - 8x^2 - 16x$
38. $4x + 8$
39. $x^3 + 2x^2$
40. $t^2 + 18t$
41. $100 - 25x$
42. $(x + 2)(x - 1) + (x + 2)(2 + 3x)$
43. $(3 - x)x + (2 + x)(x - 3)$
44. $x\left(2x + \dfrac{1}{2}\right) - \left(x^2 - \dfrac{x}{3}\right)$
45. $3x^2(x - 2) - 6(x + 4)(x - 2)$
46. $y^2(3 - y) + 9(y - 3)$
47. $3(x + 4) - (x^2 - x - 20)$
48. $2x^2(x - 3) + 8(x^2 - 2x - 3)$

Simplify each of the following expressions; then factor the result.

49. $(x + 2)(x - 3) - 2(x - 1)$
50. $(2x - 1)(x + 6) + 3(x + 10)$
51. $(x + 2)(x - 2) - 3x$
52. $(3x + 2)(x - 1) - (x - 2)(3x + 3)$
53. $x(x + 1) - 2(x - 5)$
54. $x^2(2x + 1) + x(6 - 2x^2)$
55. $x^3 + 5(x + 2) - (5x + 2)$
56. $x(x + 1)(x + 2) + 2x(2x + 5)$

57. $(x + 3)(x - 3) + 3(x + 4) - 1$

58. $(x + 4)(2x - 1) - x(2x + 3) + 4(1 - x)$

59. The expression in Exercise 43

60. The expression in Exercise 44

61. The expression in Exercise 47

A.4 SOLVING LINEAR AND QUADRATIC EQUATIONS

It is difficult to find any branch of mathematics that does not use equations. A major concern in any branch of mathematics is the equality of mathematical expressions, and an **equation** is a statement that two (or more) mathematical expressions are equal. In applications, equations are one of the most frequently used methods to express the relationships between physical quantities mathematically.

In algebra, one of the major topics is the solution of equations. For example, the equation $2x + 4 = 14$ is not true for every value of x. If x should be equal to 3, for instance, the equation becomes $2(3) + 4 = 10 \neq 14$. But if $x = 5$, we get $2(5) + 4 = 14$. For this reason, $x = 5$ is a solution of the given equation. In general, a **solution** of an equation is a value of the variable for which the equation is true or, as is sometimes said, a value of the variable that satisfies the equation. **To solve** an equation means to determine the value(s) of the variable for which the equation is true.

The procedure to be used for solving an equation depends on the type of equation. The first type we shall consider are called *linear equations*. Essentially, these are equations in which the variable appears only to the first power. The following are all linear equations:

$$5x + 6 = 0 \qquad x - \frac{1}{2} = 3x \qquad 3x + 6 = 2x - 8$$

$$-3(x + 1) = 6x + 13 \qquad \frac{x + 1}{2} = 4$$

Formally, a **linear equation** (in x) is an equation that can be written in the form

$$ax + b = 0$$

where a and b are constants, $a \neq 0$. None of the following is a linear equation:

$$3x^2 - 4x = 7 \qquad x + \frac{1}{x} = 4 \qquad (x + 2)(x + 3) = 0$$

The general idea in solving a linear equation is to add, subtract, multiply, or divide on each side of the equation by the same quantity to obtain an equation of the form

$$x = c \quad (c \text{ a constant})$$

which gives the solution.

EXAMPLE 1 Solve the equation $3x - 6 = x + 2$.

Solution We must convert this equation into the form indicated above.

$2x - 6 = 2$ Subtract x from each side.
$2x = 8$ Add 6 to each side.
$x = 4$ Divide each side by 2.

The final equation gives the solution. We can verify $x = 4$ is the solution by substituting into the original equation

$$3(4) - 6 = 4 + 2$$
$$6 = 6$$

which is obviously true.

---

1 PRACTICE PROBLEM 1

Solve the equation $5x + 7 = 2x - 5$.

Answer

$x = -4$

EXAMPLE 2 Solve the equation

$$\frac{x-2}{3} = \frac{x}{4} + 1$$

Solution If an equation contains fractions in which the denominators are constants, such as this one does, it usually simplifies the operation if we multiply each side of the equation by the least common denominator (L.C.D.) of the fractions, which in this case is 12.

$4(x - 2) = 3x + 12$ Multiply each side by 12.
$4x - 8 = 3x + 12$ Simplify left side.
$x - 8 = 12$ Subtract $3x$ from each side.
$x = 20$ Add 8 to each side.

The solution is $x = 20$.

EXAMPLE 3 Solve the equation

$$2 - \left(\frac{y+1}{3}\right) = y + 2(y - 1)$$

Solution
$6 - (y + 1) = 3y + 6(y - 1)$ Multiply each side by 3.
$5 - y = 9y - 6$ Simplify each side.
$5 - 10y = -6$ Subtract $9y$ from each side.
$-10y = -11$ Subtract 5 from each side.
$y = \frac{11}{10}$ Divide each side by -10.

---

2 PRACTICE PROBLEM 2

Solve the equation $\frac{x}{3} + 1 = \frac{x+2}{4}$.

Answer

$x = -6$

The solution is $y = \frac{11}{10}$.

The second type of equation to be considered is the quadratic equation, in which the variable or unknown appears to the second power but to no higher power. Each of the following are quadratic equations:

$$2x^2 - 3x + 4 = 0 \qquad x^2 + 5 = 3x - 7$$

$$\frac{x^2 - 2}{6} = x + 1 \qquad x^2 - 1 = 0$$

Formally, a **quadratic equation** (in x) is an equation that can be written in the form

$$ax^2 + bx + c = 0$$

where a, b, and c are constants, $a \neq 0$.

We shall consider two methods of solving quadratic equations, factoring and using the quadratic formula. The general idea in factoring is that if an equation can be written in the form

$$P \cdot Q = 0$$

in which the left hand side is the product of two factors, then at least one of the factors must be zero. That is, either $P = 0$ or $Q = 0$ (or perhaps both).

For example, in order to determine the values of x for which

$$x^2 + 2x - 15 = 0$$

we first factor the left-hand side to obtain

$$(x + 5)(x - 3) = 0$$

In order for the product on the left to be equal to zero, either

$$x + 5 = 0 \qquad \text{or} \qquad x - 3 = 0$$

The first of these two equations gives $x = -5$; the second, $x = 3$. The solutions of the quadratic equation are, therefore, $x = -5$ and $x = 3$.

3 PRACTICE PROBLEM 3

Solve the equation $x^2 + x - 2 = 0$.

Answer

$x = 1, -2$

EXAMPLE 4 Solve the equation $3x^2 + 8x = x^2 - 6$.

Solution   Before a quadratic equation is solved, the equation should be written so that one side (usually the right-hand side) is zero. In this case, we do the following:

$2x^2 + 8x + 6 = 0$ Subtract $x^2 - 6$ from each side.
$x^2 + 4x + 3 = 0$ Since each term on the left has a factor of 2, divide each side by 2.
$(x + 3)(x + 1) = 0$ Factor the left-hand side.
$x + 3 = 0$ or $x + 1 = 0$ Set each factor equal to zero.

From these last two equations the solutions are $x = -3$ and $x = -1$. (We can verify that these solutions are correct by substituting each into the original equation.)

EXAMPLE 5 Solve the equation

$$\frac{t(t+2)}{2} = t^2 + \frac{1}{2}$$

Solution

$t(t+2) = 2t^2 + 1$	Multiply each side by 2.
$t^2 + 2t = 2t^2 + 1$	Simplify the left side.
$-t^2 + 2t - 1 = 0$	Subtract $2t^2 + 1$ from each side.
$t^2 - 2t + 1 = 0$	Divide each side by -1.
$(t-1)(t-1) = 0$	Factor the left-hand side.
$t - 1 = 0$ or $t - 1 = 0$	Set each factor equal to 0.

The only solution is $t = 1$. ◢

4 PRACTICE PROBLEM 4

Solve the equation $\dfrac{x^2 + 3x}{-2} = x + 3$.

Answer

$x = -2, -3$

If the factoring involved in solving a quadratic equation seems difficult or impossible, an alternate approach is to use the quadratic formula, which avoids factoring altogether. The **quadratic formula** states that the solutions of the equation

$$ax^2 + bx + c = 0 \qquad a \neq 0 \qquad (1)$$

are

$$\boxed{x = \frac{-b \pm \sqrt{b^2 - 4ac}}{2a}} \qquad (2)$$

The quadratic formula (2) can be derived from equation (1) by a procedure called completing the square, which is described in any standard algebra text.

EXAMPLE 6 Solve the equation $2x^2 - 12 = x + 3$.

Solution In order to obtain a zero on the right-hand side, we first subtract $x + 3$ from each side to obtain

$$2x^2 - x - 15 = 0$$

The result is in the form of equation (1) with $a = 2$, $b = -1$, and $c = -15$. The quadratic formula gives the solutions

$$x = \frac{1 \pm \sqrt{(-1)^2 - 4(2)(-15)}}{2(2)}$$

$$= \frac{1 \pm \sqrt{1 + 120}}{4}$$

$$= \frac{1 \pm 11}{4}$$

The choice of $+11$ gives $x = \frac{12}{4} = 3$. The choice of -11 gives $x = -\frac{10}{4} = -\frac{5}{2}$. ◢

5 PRACTICE PROBLEM 5

Solve the equation $4x^2 + 5x = 6$ using the quadratic formula.

Answer

$x = -2, \frac{3}{4}$

EXAMPLE 7
Solve the equation in Example 5 using the quadratic formula.

Solution As in Example 5, we first rewrite the equation to get a zero on the right-hand side:
$$-t^2 + 2t - 1 = 0$$
Then $a = -1$, $b = 2$, and $c = -1$, so that the solutions are
$$t = \frac{-2 \pm \sqrt{2^2 - 4(-1)(-1)}}{2(-1)}$$
$$= \frac{-2 \pm \sqrt{4 - 4}}{-2}$$
$$= \frac{-2 \pm 0}{-2}$$

The only solution is $t = 1$.

If the quantity $b^2 - 4ac$ appearing under the radical sign in the quadratic formula is negative, there are no real solutions. This corresponds to the fact that the factorization of $ax^2 + bx + c$ cannot then be performed, as indicated in the last section. The quantity $b^2 - 4ac$ is called the **discriminant** of the quadratic equation.

6 PRACTICE PROBLEM 6
Solve the equation in Practice Problem 4 using the quadratic formula.

Answer

$x = -2, -3$

EXERCISES

Solve each of the following equations:

1. $2x - 3 = x + 1$
2. $x + 6 = 4x + 12$
3. $3x - 7 = 5x + 2$
4. $3 - 4x = 9$
5. $10 - 2x = 3x - 5$
6. $4 = 6 + 3y$
7. $0 = y - 7$
8. $2x + 12 = 3x + 12$
9. $5x - 6 = 0$
10. $4x = 0$
11. $\dfrac{x}{2} + 2 = \dfrac{x}{3}$
12. $3 - \dfrac{x}{4} = 2 + \dfrac{x}{2}$
13. $\dfrac{4 - x}{9} = 1 + \dfrac{x}{2}$
14. $\dfrac{2x - 3}{5} = 1$
15. $3 - \dfrac{t + 2}{2} = -\dfrac{3}{2} + \dfrac{t}{5}$
16. $\dfrac{4(y + 3)}{5} = \dfrac{1 + 3y}{2}$
17. $3 - 2(y + 7) = y + 4(1 - y)$
18. $6(x - 2) + 3 = 3(x - 1) - x$
19. $7(x + 1) - 4(x + 1) = 0$
20. $7(x + 1) - 4(x + 1) = 3$

Solve each of the following equations by factoring:

21. $x^2 + x - 12 = 0$
22. $x^2 + 3x = 10$
23. $x^2 - x = x + 3$
24. $2x^2 + 12x + 16 = 0$

25. $2x^2 + 2x = x^2 + 15$
26. $x^2 + 2x = 2x + 1$
27. $3x(x + 1) = 2x^2 - 2x - 6$
28. $4y(y + 1) = 3(y^2 + 7)$
29. $y^2 = -4(y + 1)$
30. $x^2 + 7(2 - x) = 2$
31. $\dfrac{2x^2}{3} = \dfrac{x^2 - 4x + 5}{3}$
32. $3(x^2 - x) = \dfrac{3x(3x + 1)}{4}$
33. $x^2 + 7x = 0$
34. $y^2 - y = 5 - y$

Solve each of the following equations using the quadratic formula. (Some equations have no real solution.)

35. $2x^2 + 3x - 5 = 0$
36. $x^2 - x - 20 = 0$
37. $2x^2 + x - 3 = 0$
38. $3x^2 + x - 10 = 0$
39. $6(x^2 - x) = 4 - x$
40. $x\left(\dfrac{6x - 1}{3}\right) = 5$
41. $3t^2 + 5 = t(t + 1)$
42. $(x + 4)(x - 4) = 9$
43. $y(y - 1) = -\tfrac{1}{4}$
44. $5x^2 + 1 = -2(x + 1)$
45. Exercise 27
46. Exercise 28
47. Exercise 31
48. Exercise 32
49. Exercise 33
50. Exercise 34

Explain the meaning of each term.

51. Equation
52. Solution of an equation

53. Name the two methods considered for solving quadratic equations.

A.5 ALGEBRAIC FRACTIONS

In Section A.2 we avoided algebraic expressions involving fractions. We did this because working with algebraic fractions generally involves factoring, and factoring had not been considered at that time. In this section we concentrate on operations with algebraic fractions—first multiplication and division and then addition and subtraction.

There is a standard three-step procedure involved in multiplying two fractions. The three steps are given below and are illustrated by performing the operations in calculating the product

$$\dfrac{2x - 4}{x^2 - 1} \cdot \dfrac{x^2 + 2x + 1}{3x - 6}$$

Step 1. Factor each numerator and denominator, if possible.

$$\dfrac{2x - 4}{x^2 - 1} \cdot \dfrac{x^2 + 2x + 1}{3x - 6}$$
$$= \dfrac{2(x - 2)}{(x + 1)(x - 1)} \cdot \dfrac{(x + 1)(x + 1)}{3(x - 2)}$$

A.5 ALGEBRAIC FRACTIONS

Step 2. Cancel any factors appearing in both numerator and denominator.
$$= \frac{2(x-2)}{(x+1)(x-1)} \cdot \frac{(x+1)(x+1)}{3(x-2)}$$

Step 3. Multiply the remaining factors in the numerator and in the denominator.
$$= \frac{2(x+1)}{3(x-1)} = \frac{2x+2}{3x-3}$$

EXAMPLE 1 Multiply:
$$\frac{x^2 + 4x}{125 - x^3} \cdot \frac{x^2 - 25}{x^2 + 5x}$$

Solution The steps are numbered according to the procedure outlined above.

Step 1. $\dfrac{x^2 + 4x}{125 - x^3} \cdot \dfrac{x^2 - 25}{x^2 + 5x} = \dfrac{x(x+4)}{(5-x)(25+5x+x^2)} \cdot \dfrac{(x+5)(x-5)}{x(x+5)}$

Step 2. $= \dfrac{x(x+4)}{-(x-5)(25+5x+x^2)} \cdot \dfrac{(x+5)(x-5)}{x(x+5)}$

Step 3. $= \dfrac{x+4}{(-1)(25+5x+x^2)} = \dfrac{x+4}{-25-5x-x^2}$

Note that in the second step of Example 1, we did an additional factorization
$$(5 - x) = -1(x - 5)$$
in order to cancel the factor $x - 5$ with the corresponding factor in the numerator.

If three or more fractions are to be multiplied, the three-step procedure is still used, as illustrated in Example 2. Note that the numerators and denominators in the first two fractions are considered to be already factored.

EXAMPLE 2 Multiply:
$$\frac{6x^2}{5} \cdot \frac{1}{3x} \cdot \frac{5x - 5}{2x^2 - 2x}$$

Solution
$$\frac{6x^2}{5} \cdot \frac{1}{3x} \cdot \frac{5x-5}{2x^2-2x} = \frac{6x^2}{5} \cdot \frac{1}{3x} \cdot \frac{5(x-1)}{2x(x-1)}$$
$$= \frac{6x^2}{5} \cdot \frac{1}{3x} \cdot \frac{5(x-1)}{2x(x-1)} = \frac{1}{1} = 1$$

EXAMPLE 3 Multiply:
$$4 \cdot \frac{t^2 - 1}{t^2 - 2t + 1} \cdot \frac{3t - 3}{12}$$

1 PRACTICE PROBLEM 1

Multiply: $\dfrac{x^2 - 4x}{x^2 + 7x} \cdot \dfrac{2x + 8}{x^2 - 16}$

Answer

$\dfrac{2}{x + 7}$

Solution

$$4 \cdot \frac{t^2 - 1}{t^2 - 2t + 1} \cdot \frac{3t - 3}{12} = \frac{\cancel{4}}{1} \cdot \frac{(t + 1)\cancel{(t - 1)}}{\cancel{(t - 1)}\cancel{(t - 1)}} \cdot \frac{3\cancel{(t - 1)}}{\cancel{12}}$$

$$= \frac{t + 1}{1} = t + 1$$ 2

To divide two fractions, we first invert the divisor and then proceed as in multiplication. Consequently, to divide, we first invert the divisor and then follow the three steps given for multiplication. The procedure is illustrated in the next several examples.

EXAMPLE 4 Divide:

$$\frac{2x + 2}{x^2 + x - 2} \div \frac{x + 1}{x^2 + 2x}$$

Solution

$$\frac{2x + 2}{x^2 + x - 2} \div \frac{x + 1}{x^2 + 2x} = \frac{2x + 2}{x^2 + x - 2} \cdot \frac{x^2 + 2x}{x + 1}$$

$$= \frac{2(x + 1)}{(x + 2)(x - 1)} \cdot \frac{x(x + 2)}{x + 1}$$

$$= \frac{2\cancel{(x + 1)}}{\cancel{(x + 2)}(x - 1)} \cdot \frac{x\cancel{(x + 2)}}{\cancel{x + 1}} = \frac{2x}{x - 1}$$

EXAMPLE 5 Divide:

$$\frac{9 - x^2}{2x} \div (2x - 6)$$

Solution

$$\frac{9 - x^2}{2x} \div (2x - 6) = \frac{9 - x^2}{2x} \div \frac{2x - 6}{1}$$

$$= \frac{9 - x^2}{2x} \cdot \frac{1}{2x - 6}$$

$$= \frac{\cancel{(3 - x)}(3 + x)}{2x} \cdot \frac{1}{\underset{-1}{2\cancel{(x - 3)}}} = \frac{3 + x}{-4x}$$ 3

---

2 **PRACTICE PROBLEM 2**

Multiply: $(1 - x) \cdot \dfrac{x + 3}{x^2 + 3x} \cdot \dfrac{4x^2}{x^2 - 1}$

Answer

$\dfrac{-4x}{x + 1}$

---

3 **PRACTICE PROBLEM 3**

Divide: $\dfrac{-3}{x^2 + 2x} \div \dfrac{12x}{x^2 - x - 6}$

Answer

$\dfrac{3 - x}{4x^2}$

A.5 ALGEBRAIC FRACTIONS

EXAMPLE 6 Divide:

$$\frac{\dfrac{x^2}{x+1}}{\dfrac{x^2+2x}{x^2-1}}$$

Solution A fraction such as this, in which the numerator or denominator also contains at least one fraction, is called a **complex fraction**. The given expression is the same as

$$\frac{x^2}{x+1} \div \frac{x^2+2x}{x^2-1}$$

Consequently, to perform the indicated division, we *invert the denominator and multiply*.

$$\frac{\dfrac{x^2}{x+1}}{\dfrac{x^2+2x}{x^2-1}} = \frac{x^2}{x+1} \cdot \frac{x^2-1}{x^2+2x} = \frac{x^2}{\cancel{x+1}} \cdot \frac{\cancel{(x+1)}(x-1)}{\cancel{x}(x+2)}$$

$$= \frac{x(x-1)}{x+2} \quad \text{or} \quad \frac{x^2-x}{x+2} \quad \blacksquare$$

We can perform the addition

$$\frac{2}{x^2+4} + \frac{x}{x^2+4} + \frac{2x-5}{x^2+4} = \frac{3x-3}{x^2+4}$$

by adding the numerators because the fractions on the left all have a common denominator. However, we cannot perform the addition

$$\frac{2}{9x^2} + \frac{1}{2x^2-2x} + \frac{x+3}{x^2-2x+1}$$

until all the fractions are rewritten with a common denominator.

In general, fractions can be added or subtracted only when they have the same denominator. If this is not the case to begin with, the fractions must first be rewritten with a common denominator, usually their least common denominator (L.C.D.). We first review the procedure for rewriting fractions in this way. This, again, is a three-step procedure. We list the three steps and illustrate the steps by rewriting the three fractions

$$\frac{2}{9x^2}, \quad \frac{1}{2x^2-2x}, \quad \frac{x+3}{x^2-2x+1}$$

with a common denominator.

Step 1. Factor each denominator. (Constant factors should be written as products of prime numbers.)

$$9x^2 = 3^2 x^2$$
$$2x^2 - 2x = 2x(x-1)$$
$$(x^2 - 2x + 1) = (x-1)^2$$

Step 2. Form the L.C.D. by multiplying together the factors occurring in the different denominators. The highest power to which a factor appears in any denominator is used.

$$\text{L.C.D.} = 3^2 \cdot 2 \cdot x^2(x-1)^2$$
$$= 18x^2(x-1)^2$$

Step 3. Rewrite each fraction with the L.C.D. as its denominator: Multiply the numerator and denominator of each fraction by the factors of the L.C.D. not in the original denominator.

$$\frac{2}{9x^2} \cdot \frac{2(x-1)^2}{2(x-1)^2}$$
$$= \frac{4(x-1)^2}{18x^2(x-1)^2}$$
$$\frac{1}{2x^2-2x} \cdot \frac{3^2 x(x-1)}{3^2 x(x-1)}$$
$$= \frac{9x(x-1)}{18x^2(x-1)^2}$$
$$\frac{x+3}{x^2-2x+1} \cdot \frac{3^2 \cdot 2 \cdot x^2}{3^2 \cdot 2 \cdot x^2}$$
$$= \frac{18x^2(x+3)}{18x^2(x-1)^2}$$

EXAMPLE 7 Rewrite with a common denominator:

$$\frac{3}{10x}, \quad \frac{x+2}{4(x+1)}, \quad \frac{x^2}{5x^2+10x+5}$$

Solution We number the steps according to the procedure just outlined.

Step 1. Factor the denominators.
$$10x = 2 \cdot 5 \cdot x$$
$$4(x+1) = 2^2(x+1)$$
$$5x^2 + 10x + 5 = 5(x+1)^2$$

Step 2. Form the L.C.D.
$$\text{L.C.D.} = 2^2 \cdot 5 \cdot x(x+1)^2 = 20x(x+1)^2$$

Step 3. Rewrite the fractions
$$\frac{3}{10x} \cdot \frac{2(x+1)^2}{2(x+1)^2} = \frac{6(x+1)^2}{20x(x+1)^2}$$
$$\frac{x+2}{4(x+1)} \cdot \frac{5x(x+1)}{5x(x+1)} = \frac{5x(x+2)(x+1)}{20x(x+1)^2}$$
$$\frac{x^2}{5x^2+10x+5} \cdot \frac{4x}{4x} = \frac{4x^3}{20x(x+1)^2}$$

4

---

4 PRACTICE PROBLEM 4

Rewrite with a common denominator

$$\frac{3}{x^2-x}, \quad \frac{x}{2(x-1)}, \quad \frac{-5}{4x^2}$$

Answer

$$\frac{12x}{4x^2(x-1)}, \quad \frac{2x^3}{4x^2(x-1)}, \quad \frac{-5x+5}{4x^2(x-1)}$$

A.5 ALGEBRAIC FRACTIONS

EXAMPLE 8 Rewrite with a common denominator:

$$3, \quad \frac{4}{25x^2}, \quad \frac{2}{3x}$$

Solution In this context, expressions without a denominator are considered to have a denominator of 1. The denominators in this case then are 1, $5^2 \cdot x^2$ and $3x$. The L.C.D. is $75x^2$, and the fractions can be rewritten

$$3 \cdot \frac{75x^2}{75x^2} = \frac{225x^2}{75x^2}$$

$$\frac{4}{25x^2} \cdot \frac{3}{3} = \frac{12}{75x^2}$$

$$\frac{2}{3x} \cdot \frac{25x}{25x} = \frac{50x}{75x^2}$$

Finally, there is one more three-step procedure involved in adding and subtracting fractions. We list the three steps and illustrate the procedure by evaluating

$$\frac{2}{x+1} + \frac{1}{x-1} - \frac{2}{x^2-1}$$

Step 1. Rewrite the fractions with a common denominator. The L.C.D. in this case is $x^2 - 1$.

$$\frac{2}{x+1} + \frac{1}{x-1} - \frac{2}{x^2-1}$$

$$= \frac{2(x-1)}{x^2-1} + \frac{x+1}{x^2-1} - \frac{2}{x^2-1}$$

Step 2. Add or subtract the numerators and write the result over the common denominator.

$$= \frac{(2x-2) + (x+1) - 2}{x^2-1}$$

$$= \frac{3x-3}{x^2-1}$$

Step 3. Simplify the result by canceling any factors common to both the numerator and denominator.

$$= \frac{3(x-1)}{(x+1)(x-1)} = \frac{3}{x+1}$$

EXAMPLE 9 Perform the indicated addition:

$$\frac{3}{2x+2} + \frac{5-x}{x^2-2x-3}$$

Step 1. Rewrite the fractions. The L.C.D. is $2(x-3)(x+1)$.

$$\frac{3}{2x+2} + \frac{5-x}{x^2-2x-3}$$

$$= \frac{3(x-3)}{2(x-3)(x+1)} + \frac{2(5-x)}{2(x-3)(x+1)}$$

Step 2. Add or subtract numerators.

$$= \frac{3(x-3) + 2(5-x)}{2(x-3)(x+1)}$$

$$= \frac{x+1}{2(x-3)(x+1)}$$

Step 3. Simplify.

$$= \frac{1}{2(x-3)}$$ [5]

EXAMPLE 10 Perform the indicated subtraction:

$$4 - \frac{4x+8}{x-2}$$

Solution The L.C.D. is $x - 2$. Then

$$4 - \frac{4x+8}{x-2} = \frac{4(x-2)}{(x-2)} - \frac{4x+8}{x-2} = \frac{(4x-8) - (4x+8)}{x-2}$$

$$= \frac{-16}{x-2} \quad \text{or} \quad \frac{16}{2-x} \qquad [6]$$

We are in a position now of being able to simplify expressions involving algebraic fractions. The operations are again performed in the standard order:

1. Grouping symbols, innermost first
2. Evaluation of exponents
3. Multiplication and division, left to right
4. Addition and subtraction, left to right

Recall that quantities in the numerator and denominator of a fraction, complex or otherwise, are to be considered as expressions within parentheses.

---

[5] **PRACTICE PROBLEM 5**

Add the three fractions given in Practice Problem 4.

Answer

$$\frac{2x^3 + 7x + 5}{4x^2(x-1)}$$

---

[6] **PRACTICE PROBLEM 6**

Subtract: $\dfrac{3x}{x-2} - x$

Answer

$$\frac{5x - x^2}{x - 2}$$

A.5 ALGEBRAIC FRACTIONS

EXAMPLE 11 Simplify:

$$\frac{3 + \dfrac{4x}{1-x}}{\dfrac{1}{1+x} + \dfrac{1}{1-x}}$$

Solution Start within grouping symbols.

$$\frac{3 + \dfrac{4x}{1-x}}{\dfrac{1}{1+x} + \dfrac{1}{1-x}} = \frac{\dfrac{3-3x}{1-x} + \dfrac{4x}{1-x}}{\dfrac{1-x}{1-x^2} + \dfrac{1+x}{1-x^2}}$$

$$= \frac{\dfrac{3+x}{1-x}}{\dfrac{2}{1-x^2}}$$

$$= \frac{3+x}{1-x} \cdot \frac{1-x^2}{2}$$

$$= \frac{3+x}{1-x} \cdot \frac{(1-x)(1+x)}{2} = \frac{3 + 4x + x^2}{2}$$

EXERCISES

Perform the indicated operations.

1. $\dfrac{x^2 + x}{x^2} \cdot \dfrac{2x}{x^2 - 1}$

2. $\dfrac{3x - 6}{8} \cdot \dfrac{4}{x^2 - 4x + 4}$

3. $\dfrac{5}{2 - x} \cdot \dfrac{x^2 + x - 6}{5x + 15}$

4. $\dfrac{-7}{2y + 10} \cdot \dfrac{2y^2 - 50}{7y - 35}$

5. $-3 \cdot \dfrac{t^3 - 1}{6t + 12} \cdot \dfrac{2t + 4}{t^2 + t + 1}$

6. $2 \cdot \dfrac{x^3}{x + 1} \cdot \dfrac{x^2 + 5x + 4}{x^3 + 4x^2}$

7. $\dfrac{x + 3}{2x^2 + 10x + 12} \cdot (x + 2)$

8. $\dfrac{t + 3}{27 - 3t^2} \cdot (t - 3)$

9. $\dfrac{5}{3x^2 - x} \div \dfrac{10x^2}{3x - 1}$

10. $\dfrac{2x^2 - 4x}{x + 1} \div \dfrac{x - 2}{x^2 - 1}$

11. $\dfrac{x^2 - 4}{x^2 + 5x + 6} \div \dfrac{x - 2}{x + 3}$

12. $\dfrac{3}{y^2 + 7y} \div \dfrac{1}{y^2}$

13. $\dfrac{2x^2 - 6x - 8}{3} \div (2x - 8)$

14. $\dfrac{x^2 - 7x + 10}{x + 12} \div (5 - x)$

15. $\dfrac{\dfrac{x + 2}{2x^2 - 2x}}{\dfrac{x^2 - 4}{x^3 - x}}$

16. $\dfrac{\dfrac{x^2 + 2x}{x^2 - 9}}{\dfrac{2x + 4}{3x - 9}}$

17. $\dfrac{\dfrac{x^2 + 4x - 5}{3x + 15}}{x + 1}$

18. $\dfrac{\dfrac{y^2 - 2y}{2y + 5}}{3y - 6}$

19. List the three-step procedure for multiplying algebraic fractions.
20. List the four-step procedure for dividing algebraic fractions.

Perform the indicated operations.

21. $\dfrac{4}{x} + \dfrac{1}{x^2}$

22. $\dfrac{2}{x + 1} + \dfrac{x - 1}{x + 1}$

23. $\dfrac{x}{x - 1} - \dfrac{3}{x}$

24. $5 - \dfrac{2x}{3x - 2}$

25. $\dfrac{3}{4} + \dfrac{1}{2x}$

26. $\dfrac{x}{3x^2 + 6} - \dfrac{5}{3}$

27. $\dfrac{1}{x + 1} + \dfrac{1}{x - 1} + \dfrac{2}{x^2 - 1}$

28. $\dfrac{4}{x + 2} - \dfrac{1}{x - 2} + \dfrac{4}{x^2 - 4}$

29. $\dfrac{2x - 9}{5x - 10} + \dfrac{1}{x - 2}$

30. $\dfrac{5y + 23}{4y + 12} - \dfrac{2}{y + 3}$

31. $\dfrac{x - 1}{x^3 - x} + \dfrac{3}{x + 1} - \dfrac{2}{x}$

32. $\dfrac{2}{x} - \dfrac{1}{x - 3} + \dfrac{9}{x^3 - 3x^2}$

33. $\dfrac{3}{3x^2 + 9x + 6} + \dfrac{1}{x^2 + 5x + 6}$

34. $\dfrac{-2}{x^2 + 4x + 3} + \dfrac{1}{x^2 + 3x + 2}$

35. $\dfrac{6}{5t} + \dfrac{4}{3t^2} - 2$

36. $\dfrac{2}{x - 2} - \dfrac{1}{x} + (x + 1)$

Simplify the following expressions:

37. $\dfrac{\dfrac{2}{y}}{1 - \dfrac{1}{y}}$

38. $\dfrac{1 - \dfrac{1}{x^2}}{\dfrac{1}{x} - 1}$

39. $\dfrac{\dfrac{2}{x + 1} - \dfrac{4}{x}}{2}$

40. $\dfrac{\dfrac{5}{t} + \dfrac{1}{t - 3}}{-3}$

41. $\dfrac{2 + \dfrac{x}{x - 3}}{1 - \dfrac{4}{x - 3}}$

42. $\dfrac{3 - \dfrac{x}{2 + x}}{x + 3}$

43. $\dfrac{\dfrac{1}{y} - \dfrac{3}{y - 2}}{\dfrac{y^2 + y}{4 - y^2}}$

44. $\dfrac{\dfrac{x}{x - 6} - \dfrac{1}{x}}{x - 2}$

45. $4 + \dfrac{2x - 6}{x + 4} \cdot \dfrac{x^2 - 16}{x - 3}$

46. $\left(4 + \dfrac{2x - 6}{x + 4}\right) \dfrac{x^2 - 16}{3x + 5}$

47. $\dfrac{(x + 1)(x - 1) - 8}{x(4x + 3) - (4x^2 + x + 6)}$

48. $\dfrac{(2x + 3)(x - 1) - x^2 + x - 5}{3x + (x + 4)(2x - 5)}$

49. $\dfrac{10 - (x + 2)(x - 1)}{(x - 3)^2}$

50. $\dfrac{(x + 3)(x + 7) - 4(x - 1) - 11x}{x^3 + 125}$

51. List the three-step procedure for rewriting fractions with a common denominator.
52. List the three-step procedure for adding and subtracting fractions.

A.6 SOLVING ADDITIONAL TYPES OF EQUATIONS

In Section A.4 we indicated that to solve an equation means to determine the value (or values) of the variable for which the equation is true. The procedure for solving an equation depends on the form of the equation. The important thing to notice about the form of linear equations is that they can be rewritten in such a way that the variable appears only to the first power. Similarly, in quadratic equations the variable appears to the second power and to no higher power. Consequently, the equation

$$2x^3 - 3x^2 - 5x + 6 = 0$$

is not of either type considered previously. Polynomial equations such as this, in which the variable appears to a higher power, can sometimes be solved by factoring. The general idea is that if an equation can be written in the form

$$P \cdot Q \cdot R = 0$$

in which the left-hand side is the product of three (or more) factors, then at least one of the factors must be zero; that is, either $P = 0$, $Q = 0$, or $R = 0$.

The left-hand side of the equation given above,

$$2x^3 - 3x^2 - 5x + 6 = 0$$

can be factored to give

$$(2x + 3)(x - 1)(x - 2) = 0$$

Setting the individual factors equal to zero, we get

$$2x + 3 = 0 \quad \text{or} \quad x = -\tfrac{3}{2}$$
$$x - 1 = 0 \quad \text{or} \quad x = 1$$
$$x - 2 = 0 \quad \text{or} \quad x = 2$$

That is, the solutions are $-\tfrac{3}{2}$, 1, and 2.

EXAMPLE 1 Solve the equation $(2x - 1) = (2x - 1)(2x - x^2 + 9)$.

Solution Again, the equation must be rewritten so that one side of the equation is zero. We subtract the expression on the right from each side of the equation to obtain

$$(2x - 1) - (2x - 1)(2x - x^2 + 9) = 0$$

Then, factor the left-hand side:

$$(2x - 1)[1 - (2x - x^2 + 9)] = 0$$
$$(2x - 1)[x^2 - 2x - 8] = 0$$
$$(2x - 1)(x - 4)(x + 2) = 0$$

Setting each of the factors from this last equation equal to zero gives the solutions $x = \frac{1}{2}$, 4, and -2.

EXAMPLE 2 Solve the equation
$$\frac{x^2(x-3)}{2} = -x$$

Solution In order to clear the equation of fractions, multiply each side by 2:
$$x^2(x-3) = -2x$$
Add $2x$ to each side:
$$x^2(x-3) + 2x = 0$$
Factor the left side; note that each term on the left has a factor of x.
$$x[x(x-3) + 2] = 0$$
$$x[x^2 - 3x + 2] = 0$$
$$x(x-1)(x-2) = 0$$
The solutions are $x = 0$, 1, and 2.

EXAMPLE 3 Solve the equation
$$\frac{2x+3}{x-2} = 4 - \frac{3x-13}{x-2}$$

Solution We multiply each side of the equation by $x - 2$, the L.C.D. of the fractions appearing in the equation, to obtain
$$2x + 3 = 4(x-2) - (3x-13)$$
This is a linear equation, which we can solve by the method discussed earlier:
$$2x + 3 = 4(x-2) - (3x-13)$$
$$2x + 3 = 4x - 8 - 3x + 13$$
$$2x + 3 = x + 5$$
$$x = 2$$

Substituting $x = 2$ into the original equation results in a zero in a denominator and, therefore, in a meaningless expression. The given equation has no solution.

PRACTICE PROBLEM 1

Solve
$2(3x - 2) + (3x - 2)(x^2 + 7x + 8) = 0$

Answer
$-2, -5, \frac{2}{3}$

PRACTICE PROBLEM 2

Solve $\dfrac{x(x^2 - 1)}{4} = x^2 - x$.

Answer
0, 1, 3

A.6 SOLVING ADDITIONAL TYPES OF EQUATIONS

Until now we have avoided solving equations involving algebraic fractions—that is, fractions containing the variable in the denominator—because of the problem encountered in Example 3. Any time each side of an equation is multiplied by an expression containing the variable, the resulting equation may have a solution that is not a solution of the original equation. Whenever this operation is used (multiplying each side of an equation by an expression containing the variable), the final solution *must* be checked in the original equation. In particular, this check must be performed when each side of an equation is multiplied by the L.C.D. of the algebraic fractions involved.

With all the previous equations, we had the *option* of verifying the solution in the original equation. With equations involving algebraic fractions, such a check is no longer optional; it is mandatory. Obtaining solutions that are not solutions of the original equation does not happen very often. However, it can happen, and so a check must be made each time.

EXAMPLE 4   Solve the equation

$$\frac{x^2 - 3x - 18}{x - 6} = 2$$

Solution   Multiplying each side by $x - 6$ gives

$$x^2 - 3x - 18 = 2(x - 6)$$

or

$$x^2 - 5x - 6 = 0$$

Factoring the left side of this last equation gives

$$(x - 6)(x + 1) = 0$$

This equation is true for $x = 6$ and $x = -1$. Both values of x must be checked in the original equation. For $x = -1$, the left side of the original equation is

$$\frac{1 + 3 - 18}{-1 - 6} = \frac{-14}{-7} = 2$$

so that $x = -1$ is a solution. However, $x = 6$ causes the denominator on the left side of the original equation to be zero. The only solution is therefore $x = -1$.

3 PRACTICE PROBLEM 3

Solve $\dfrac{2x + 5}{x + 1} = \dfrac{3}{x + 1}$.

Answer

No solution

EXERCISES

Solve the following equations:

1. $x^3 + 2x^2 - 15x = 0$
2. $-x^3 = 7x^2 + 12x$
3. $\dfrac{x^2(3x + 2)}{5} = x$
4. $\dfrac{x(14 - x)}{4} = x^3$
5. $(x - 1)(x^2 + 6) = 5x(x - 1)$
6. $(x + 1)(2x^2 - 3x - 12) = 2(x + 1)$

7. $\left(\dfrac{4x}{7} + 1\right)(6x^2 - 5) = x(4x + 7)$

8. $2x^4 + 2x^3 - 24x^2 = 0$

9. $\dfrac{1}{4x} - \dfrac{1}{x} = \dfrac{3}{2}$

10. $\dfrac{3}{5y} - 1 = \dfrac{1}{y}$

11. $\dfrac{1}{2y - 1} = \dfrac{3}{y + 7}$

12. $\dfrac{2}{x} - \dfrac{1}{x + 1} = \dfrac{2}{x^2 + x}$

13. $\dfrac{\dfrac{x + 1}{6}}{\dfrac{x - 1}{10}} = 2$

14. $\dfrac{\dfrac{2x + 2}{3}}{\dfrac{x - 1}{2}} = 1$

15. $\dfrac{2x^2 - x - 3}{(x + 1)^2} = 0$

16. $\dfrac{1}{x + 1} + \dfrac{2}{x + 2} - \dfrac{3}{x + 3} = 0$

17. $\dfrac{2t + 2}{t^2 - 1} = \dfrac{2}{t - 1} - 5$

18. $\dfrac{5 - 3x}{x + 4} + 3x = 5$

19. $\dfrac{x}{x^2 + x - 6} + \dfrac{1}{x^2 + 4x + 3} = 0$

20. $\dfrac{x + 1}{x + 4} + \dfrac{1}{x - 3} = \dfrac{7}{x^2 + x - 12}$

21. $\dfrac{3x^2(x + 1)^2 - 2x^3(x + 1)}{(x + 1)^4} = 0$

22. $\dfrac{2(x + 5)(x - 3) - (x - 3)^2}{(x + 5)^2} = 0$

(*Hint:* Factor and simplify the numerator first.)

23. $\dfrac{4(x - 1)^2(2x + 3) - 2(2x + 3)^2(x - 1)}{(x - 1)^3} = 0$

24. $\dfrac{(x - 3)(2x) - (x^2 - 1)}{(x - 3)^2} = 0$

25. $\dfrac{(x - 2)(2x + 1) - (x^2 + x)}{3x^2 - 2x + 5} = 0$

26. $\dfrac{(x + 1)(4x + 7) - (2x - 1)(x - 1)}{5x^2 + 3x + 1} = 0$

A.7 EXPONENTS

In the previous sections of this review we have had to use only elementary properties of exponents. This section considers operations with exponents in greater depth.

We start with four basic definitions and then give some additional rules of exponents, which follow from the definitions.

DEFINITION 1

For a positive integer n,
$$a^n = \underbrace{a \cdot a \cdot a \cdots a}_{n \text{ factors}}$$

A.7 EXPONENTS

EXAMPLE 1
(a) $2^3 = 2 \cdot 2 \cdot 2 = 8$
(b) $(\frac{1}{4})^2 = \frac{1}{4} \times \frac{1}{4} = \frac{1}{16}$
(c) $(-3)^4 = (-3)(-3)(-3)(-3) = 81$
(d) $(\frac{2}{5})^2 = \frac{2}{5} \times \frac{2}{5} = \frac{4}{25}$

1 PRACTICE PROBLEM 1

Evaluate each of the following:

(a) 3^4 (b) $(-\frac{2}{3})^3$

Answer

(a) 81   (b) $-\frac{8}{27}$

DEFINITION 2

For $a \neq 0$,
$$a^0 = 1$$

EXAMPLE 2
(a) $3^0 = 1$
(b) $(\frac{1}{4})^0 = 1$
(c) $(-4)^0 = 1$
(d) $-4^0 = 1$

2 PRACTICE PROBLEM 2

Evaluate each of the following:

(a) $(-2)^0$ (b) $(\frac{27}{35})^0$

Answer

(a) 1   (b) 1

DEFINITION 3

For $a \neq 0$,
$$a^{-n} = \frac{1}{a^n}$$

EXAMPLE 3
(a) $3^{-2} = \frac{1}{3^2} = \frac{1}{9}$
(b) $(-2)^{-3} = \frac{1}{(-2)^3} = \frac{1}{-8}$
(c) $\left(\frac{2}{3}\right)^{-2} = \frac{1}{(\frac{2}{3})^2} = \frac{1}{\frac{4}{9}} = \frac{9}{4}$

3 PRACTICE PROBLEM 3

Evaluate each of the following:

(a) 5^{-3} (b) $(-\frac{3}{4})^{-2}$

Answer

(a) $\frac{1}{125}$ (b) $\frac{16}{9}$

DEFINITION 4

For m and n integers,
$$a^{m/n} = (\sqrt[n]{a})^m = \sqrt[n]{a^m}$$

EXAMPLE 4
(a) $4^{3/2} = (\sqrt[2]{4})^3 = 2^3 = 8$
(b) $(\frac{1}{8})^{4/3} = (\sqrt[3]{\frac{1}{8}})^4 = (\frac{1}{2})^4 = \frac{1}{16}$
(c) $(-32)^{1/5} = \sqrt[5]{-32} = -2$

4 PRACTICE PROBLEM 4

Evaluate each of the following:

(a) $8^{4/3}$ (b) $(\frac{4}{9})^{3/2}$

Answer

(a) 16   (b) $\frac{8}{27}$

There are several properties that follow from the above definitions.

PROPERTY 1
$$a^m \cdot a^n = a^{m+n}$$

APPENDIX A ALGEBRA REVIEW

EXAMPLE 5 (a) $2^5 \cdot 2^3 = 2^8 = 256$ (b) $3^{-2} \cdot 3^4 = 3^2 = 9$
(c) $x^3 \cdot x^{-4} = x^{-1} = \dfrac{1}{x}$ (d) $(\tfrac{1}{4})^2 \cdot (\tfrac{1}{4})^{1/2} = (\tfrac{1}{4})^{5/2} = (\sqrt{\tfrac{1}{4}})^5 = (\tfrac{1}{2})^5 = \tfrac{1}{32}$

5 PRACTICE PROBLEM 5
Evaluate each of the following:
(a) $(-3)^3 \cdot (-3)^2$
(b) $\left(\dfrac{1}{x}\right)^{-5} \cdot \left(\dfrac{1}{x}\right)^3$

Answer
(a) $(-3)^5 = -243$ (b) $\left(\dfrac{1}{x}\right)^{-2} = x^2$

PROPERTY 2
$$\dfrac{a^m}{a^n} = a^{m-n}$$

EXAMPLE 6 (a) $\dfrac{2^5}{2^2} = 2^3 = 8$ (b) $\dfrac{5^4}{5^6} = 5^{-2} = \dfrac{1}{5^2} = \dfrac{1}{25}$
(c) $\dfrac{y^{10}}{y^8} = y^2$ (d) $\dfrac{(\tfrac{3}{2})^6}{(\tfrac{3}{2})^7} = \left(\dfrac{3}{2}\right)^{-1} = \dfrac{1}{\tfrac{3}{2}} = \dfrac{2}{3}$
(e) $\dfrac{4x^5}{x^3} = 4 \cdot \dfrac{x^5}{x^3} = 4x^2$

6 PRACTICE PROBLEM 6
Evaluate each of the following:
(a) $\dfrac{(-3)^2}{(-3)^4}$ (b) $\dfrac{x^5}{-3x^3}$

Answer
(a) $(-3)^{2-4} = \dfrac{1}{9}$
(b) $-\dfrac{1}{3}(x^{5-3}) = \dfrac{x^2}{-3}$

PROPERTY 3
$$(a^m)^n = a^{mn}$$

EXAMPLE 7 (a) $(2^3)^2 = 2^6 = 64$ (b) $(5^{-1})^2 = 5^{-2} = \dfrac{1}{5^2} = \dfrac{1}{25}$
(c) $(x^6)^3 = x^{18}$ (d) $(x^4)^{1/4} = x^1 = x$
(e) $(\sqrt{4})^6 = (4^{1/2})^6 = 4^3 = 64$

7 PRACTICE PROBLEM 7
Evaluate each of the following:
(a) $(2^3)^{-2}$ (b) $(\sqrt{x})^4$

Answer
(a) $2^{-6} = \tfrac{1}{64}$ (b) x^2

PROPERTY 4
$$\left(\dfrac{a}{b}\right)^m = \dfrac{a^m}{b^m}$$

EXAMPLE 8 (a) $\left(\dfrac{4}{5}\right)^2 = \dfrac{4^2}{5^2} = \dfrac{16}{25}$ (b) $\left(\dfrac{x}{y}\right)^{-3} = \dfrac{x^{-3}}{y^{-3}} = \dfrac{1/x^3}{1/y^3} = \dfrac{y^3}{x^3}$

A.7 EXPONENTS

8 PRACTICE PROBLEM 8

Evaluate each of the following:

(a) $\left(\dfrac{2}{5}\right)^{-2}$ (b) $\sqrt[3]{\dfrac{x}{8}}$

Answer

(a) $\dfrac{2^{-2}}{5^{-2}} = \dfrac{25}{4}$ (b) $\dfrac{\sqrt[3]{x}}{\sqrt[3]{8}} = \dfrac{\sqrt[3]{x}}{2}$

EXAMPLE 9

(a) $(2 \cdot 5)^3 = 2^3 \cdot 5^3 = 8 \cdot 125 = 1{,}000$
(b) $(3x)^4 = 3^4 \cdot x^4 = 81x^4$
(c) $\sqrt[3]{48} = \sqrt[3]{8 \cdot 6} = \sqrt[3]{8} \cdot \sqrt[3]{6} = 2\sqrt[3]{6}$
(d) $(2x^{-1})^2 = 2^2(x^{-1})^2 = 4x^{-2} = \dfrac{4}{x^2}$
(e) $\sqrt{ab} = \sqrt{a}\sqrt{b}$

(c) $\left(\dfrac{1}{2}\right)^4 = \dfrac{1^4}{2^4} = \dfrac{1}{16}$ (d) $\left(\dfrac{9}{25}\right)^{1/2} = \dfrac{9^{1/2}}{25^{1/2}} = \dfrac{3}{5}$

PROPERTY 5

$$(ab)^m = a^m b^m$$

9 PRACTICE PROBLEM 9

Evaluate each of the following:

(a) $(-2x)^3$ (b) $(4x^{-2})^3$

Answer

(a) $(-2)^3 x^3 = -8x^3$

(b) $4^3(x^{-2})^3 = \dfrac{64}{x^6}$

EXERCISES

Evaluate the given expression using the four definitions given in this section.

1. 3^2
2. $(-4)^3$
3. 5^3
4. $\left(-\tfrac{1}{2}\right)^4$
5. $\left(\tfrac{2}{3}\right)^2$
6. $(-2)^2$
7. -2^2
8. $(-3)^0$
9. $\left(\tfrac{1}{3}\right)^0$
10. 2^{-4}
11. $(-5)^{-3}$
12. $\left(\tfrac{3}{4}\right)^{-1}$
13. $9^{1/2}$
14. $\left(\tfrac{4}{25}\right)^{3/2}$
15. $(-8)^{1/3}$
16. $\left(\tfrac{1}{27}\right)^{2/3}$

Rewrite the given expression using the five properties of exponents given in this section. Simplify the result, expressing the answer without negative exponents.

17. $3^2 \cdot 3^3$
18. $4^{-5} \cdot 4^3$
19. $\left(\tfrac{2}{3}\right)^5 \left(\tfrac{2}{3}\right)^{-4}$
20. $\dfrac{5^5}{5^3}$
21. $\dfrac{6^{-4}}{6^{-2}}$
22. $\dfrac{x^2}{3x^4}$
23. $\dfrac{(-7)^2}{(-7)^2}$
24. $(x^2)^{10}$
25. $(4^{-1})^3$
26. $(\sqrt[3]{-8})^{-4}$
27. $(7^{1/4})^8$
28. $\left(\tfrac{16}{25}\right)^{1/2}$
29. $\left(\dfrac{-3}{5}\right)^3$
30. $\left(\dfrac{u}{v}\right)^{-5}$
31. $\left(\dfrac{\sqrt[3]{2}}{\sqrt{3}}\right)^6$
32. $(3 \cdot x)^2$
33. $(4^{-2} \cdot 25)^{1/2}$
34. $(2^{1/3} \cdot 9^{1/6})^3$
35. $(x^2 \cdot y^{-4})^3$
36. $\sqrt{75}$
37. $\sqrt{a^4 b^{-1}}$
38. $\sqrt[3]{-8x^3}$

Determine the value of x that satisfies the given equation.

39. $4^x = 16$
40. $3^x = 1$
41. $2^x = \tfrac{1}{4}$
42. $\left(\tfrac{2}{3}\right)^x = \tfrac{8}{27}$
43. $\left(\tfrac{8}{27}\right)^x = \tfrac{2}{3}$
44. $\left(\tfrac{2}{5}\right)^x = \tfrac{5}{2}$
45. $16^x = 4$
46. $16^x = 2$

47. $75^x = 1$
48. $10^x = 100$
49. $10^x = 1{,}000$
50. $10^x = 0.10$
51. $2^3 \cdot 2^x = 16$
52. $3^2 \cdot 3^x = 9$
53. $x^2 \cdot x^3 = 32$
54. $\dfrac{2^3}{2^x} = 2$
55. $\dfrac{4^x}{4^3} = 16$
56. $\dfrac{4^x}{4^3} = \dfrac{1}{16}$
57. $(3^{-2})^x = \tfrac{1}{81}$
58. $(3^{-x})^2 = 81$
59. $\left(\dfrac{4}{x}\right)^{-1} = 2$
60. $\left(\tfrac{27}{125}\right)^x = \tfrac{5}{3}$

Indicate whether the given equation is true or false; if false, change the right side to form a valid equation.

61. $a^m a^n = a^{m \cdot n}$
62. $\left(\dfrac{x}{y}\right)^m = x^{m-y}$
63. $a^n \cdot b^n = (a \cdot b)^n$
64. $(x^2 y)^{-1} = \dfrac{x^2}{y}$
65. $(x^2)^m = x^{2+m}$
66. $\dfrac{x^m}{x^n} = x^{m/n}$
67. $2^x \cdot 2^y = 2^{xy}$
68. $b^{2-x} = \dfrac{b^2}{b^x}$

A.8 SETS

Sets are used in so many diverse areas of mathematics that they have become part of the common mathematical language, along with concepts such as number, angle, or circle.

A **set** is a collection of objects. The collection of words on this page, the collection of books in a library, the collection of teams in the American League, the collection of people in Toledo, and the collection of numbers on a roulette wheel are all examples of sets. The members of a set are called the **elements** of the set; the Detroit Tigers Team is an element of the set of teams in the American League; 7 is an element of the set of odd numbers.

A bit of notation facilitates working with sets. Uppercase letters, A, B, C, ..., will be used to denote sets, and lowercase letters, a, b, c, ..., will be used to denote elements of sets.

Membership in a set is denoted by the symbol $\in$. To denote, for example, that a is an element of set B, write $a \in B$ (read, "a is an element of B"). Similarly, to denote that a is not an element of set C, write $a \notin C$ (read, "a is not an element of C"). If A is the collection of all odd numbers, then $7 \in A$ but $38 \notin A$.

One way to denote the elements of a set is to list all the elements of the set. This is usually done using braces, as the following examples illustrate. The set of all integers between 0 and 10 is written

$$\{1, 2, 3, 4, 5, 6, 7, 8, 9\}$$

1 PRACTICE PROBLEM 1

If A is the set of all integers divisible by 4, are the following true or false?

(a) $15 \in A$ (b) $-8 \in A$
(c) $27 \notin A$ (d) $20 \notin A$

Answer

(a) F   (b) T   (c) T   (d) F

A.8 SETS

The set of all vowels is written

$$\{a, e, i, o, u, y\}$$

Another way to describe the elements of a set is by use of **set-builder** notation. Using this notation the two sets just given are written

$$\{x : x \text{ is an integer}, \quad 0 < x < 10\}$$

(read, "the set of all x, where x is an integer, x is greater than 0, and x is less than 10"), and

$$\{x : x \text{ is a vowel}\}$$

(read, "the set of all x, where x is a vowel"), respectively.

EXAMPLE 1 Write the collection of all even integers between 0 and 10 (inclusive) as a set by each of the following methods:

(a) List all the elements in the set.

(b) Use set-builder notation.

Solution (a) $\{0, 2, 4, 6, 8, 10\}$

(b) $\{x : x \text{ is an even integer}, \quad 0 \le x \le 10\}$

Consider the two sets

$$A = \{0, 2, 4, 6, 8, 10\}$$

and

$$B = \{0, 1, 2, 3, 4, 5, 6, 7, 8, 9, 10\}$$

Every element of A is also an element of B; to describe this relationship between sets A and B, A is said to be a *subset* of B. Formally, set A is a **subset** of set B if and only if every element of A is also an element of B. To indicate that A is a subset of B, the following notation is used:

$$A \subset B$$

(read, "A is contained in B" or "A is a subset of B").

If B is the set of letters of the alphabet and D is the set of vowels, then $D \subset B$. To denote that a set C is *not* a subset of a set E, the notation

$$C \not\subset E$$

(read, "C is not contained in E" or "C is not a subset of E") is used.

Two sets A and B are **equal** if they contain the same elements, without regard to order. If

$$A = \{0, 2, 4, 6, 8, 10\}$$

and

PRACTICE PROBLEM 2

Write the collection of all odd integers between 1 and 13 (exclusive) as a set in the indicated manner.

(a) List all the elements in the set.

(b) Use set-builder notation.

Answer

(a) $\{3, 5, 7, 9, 11\}$

(b) $\{x : x \text{ is an odd integer}, 1 < x < 13\}$

PRACTICE PROBLEM 3

If $A = \{a, b, c, d, e, f\}$, $B = \{a, c, e\}$, and $C = \{a, c, e, g\}$, are the following true or false?

(a) $A \subset B$ (b) $B \subset C$ (c) $B = C$
(d) $B \subset A$ (e) $C \subset B$

Answer

(a) F (b) T (c) F (d) T (e) F

$$B = \{x:x \text{ is an even integer}, \ 0 \leq x \leq 10\}$$

then it follows that $A = B$. If $A = B$, then $A \subset B$ and $B \subset A$.

Two special sets are the *empty set* and the *universal set*. The **empty set** is the set that contains no elements. It is denoted by the symbol $\emptyset$, and is considered to be a subset of *every* set; that is, for every set A, $\emptyset \subset A$.

The set containing all the elements that enter into a particular discussion is called the **universal set,** denoted by U. If the universal set is not obvious in any particular discussion, then it must be specified.

For every set $A \subset U$, there is associated another set, called the **complement of A** (relative to U), which consists of all those elements of U that are *not* elements of A. The complement of A is denoted by A'.

If

$$U = \{x:x \text{ is an integer}, \ 0 < x < 20\}$$

and

$$A = \{1, 2, 3, 4, 5, 6, 7, 8, 19\}$$

then

$$A' = \{9, 10, 11, 12, 13, 14, 15, 16, 17, 18\}$$

Note that $U' = \emptyset$ and $\emptyset' = U$.

EXAMPLE 2 If $U = \{a, b, c, d, e, f, g\}$ and $A = \{b, f, g\}$, determine the elements in A'.

Solution Since A' consists of those elements in U that are not in A,

$$A' = \{a, c, d, e\}$$

PRACTICE PROBLEM 4

If U is the universal set in Example 2 and $B = \{a, b, e, g\}$, determine the elements in B'.

Answer

$$B' = \{c, d, f\}$$

Sets are usually represented pictorially by Venn diagrams, as shown in Figure A.1. In these diagrams the universal set is represented by a rectangle; the points interior to the rectangle represent the elements of the universal set. Subsets of the universal set are then represented by circles or other figures interior to the rectangle; the points inside these figures represent the elements of these subsets.

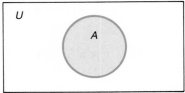

(a) Points in the shaded area represent the elements of set A

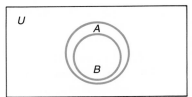

(b) Representation of B as a subset of A

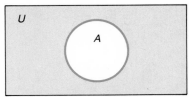

(c) Points in the shaded area represent the elements of A'

FIGURE A.1

A.8 SETS

The two main operations on pairs of sets are *union* and *intersection*. The **union** of two sets A and B is the set consisting of those elements that are either in A or in B or in both A and B. The union of two sets A and B is denoted by

$$A \cup B$$

(read, "A union B"); hence,

$$A \cup B = \{x : x \in A, \text{ or } x \in B, \text{ or } x \in A \text{ and } x \in B\}$$

See Figure A.2. If

$$A = \{0, 2, 4, 6, 8, 10\}$$
$$B = \{1, 2, 3, 4, 5\}$$

then

$$A \cup B = \{0, 1, 2, 3, 4, 5, 6, 8, 10\}$$

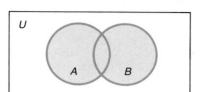

Points in the shaded portion represent the elements in $A \cup B$
FIGURE A.2

The **intersection** of two sets A and B is the set consisting of all those elements that are in both A and B. The intersection of A and B is denoted by

$$A \cap B$$

(read, "A intersect B"); hence,

$$A \cap B = \{x : x \in A \text{ and } x \in B\}$$

See Figure A.3. If A and B are the sets just given, then

$$A \cap B = \{2, 4\}$$

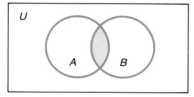

Representation of $A \cap B$
FIGURE A.3

EXAMPLE 3 For $A = \{1, 5, 9, 13, 17\}$ and $B = \{1, 9, 17, 25\}$, determine each of the following:

(a) $A \cup B$ (b) $A \cap B$

Solution   (a) $A \cup B = \{1, 5, 9, 13, 17, 25\}$ (b) $A \cap B = \{1, 9, 17\}$ 5

EXAMPLE 4 Indicate which portion of the Venn diagram below represents $A \cup B'$.

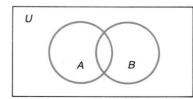

---

5 PRACTICE PROBLEM 5

If $C = \{a, e, i, o, u\}$ and $D = \{a, b, c, d, e\}$, determine:

(a) $C \cup D$ (b) $C \cap D$

Answer

(a) $C \cup D = \{a, b, c, d, e, i, o, u\}$
(b) $C \cap D = \{a, e\}$

Solution The points representing B' are all the points outside circle B. These points, together with all the points in A, are shaded in the following diagram:

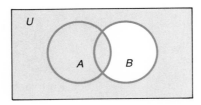

6 PRACTICE PROBLEM 6

For Example 4, indicate which portion of the Venn diagram represents $A' \cap B$.

Answer

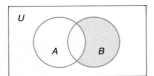

The following hold for every set $A \subset U$:

$$A \cap U = A \qquad A \cup \emptyset = A$$
$$A \cup U = U \qquad A \cup A' = U$$
$$A \cap \emptyset = \emptyset \qquad A \cap A' = \emptyset$$

EXERCISES

Let

$$U = \{-6, -5, -4, -3, -2, -1, 0, 1, 2, 3, 4, 5, 6\}$$
$$A = \{-6, -4, -2, 0, 2, 4, 6\}$$
$$B = \{x : x \text{ is an integer}, \ 0 < x < 7\}$$
$$C = \{x : x \text{ is an integer}, \ -7 < x < 7, \ x \text{ is a multiple of 3}\}$$

Determine each of the followimg:

1. $A \cup B$
2. $A \cap B$
3. $(A \cap B) \cup C$
4. A'
5. C'
6. $A' \cup C'$
7. $(A \cup C)'$
8. $(A \cup C)' \cap B'$

Let

$$U = \{-4, -3, -1, 0, 1, 2, 3, 4\}$$
$$A = \{-4, -1, 1, 4\}$$
$$B = \{x : x \in U, \ x \text{ is even}\}$$
$$C = \{x : x \in U, \ x \text{ is positive}\}$$

Determine each of the following:

9. $A \cup B$
10. $B \cap C'$
11. $(B \cap C)'$
12. $A' \cup B'$
13. $A' \cap (B \cup C)$
14. $(C \cup A)' \cup B$
15. $B \cap (A \cap B')$
16. $(A' \cap B') \cap C'$

Let

$$U = \{a, d, i, j, l, m, n, o, p, t, u\}$$
$$J = \{j, i, m\}$$
$$P = \{p, a, u, l\}$$
$$T = \{t, o, m\}$$
$$D = \{d, a, n\}$$
$$A = \{p, a, t\}$$

Determine each of the following:

17. $P \cap A$
18. $(P \cap A) \cap D$
19. $J \cup T$
20. $(J \cup T)'$
21. $J' \cap A'$
22. $P' \cap U$
23. $\emptyset \cup D$

Let U, A, B, and C be as in Exercises 1–8 and let $D = \{2, 4, 6\}$. Insert a proper symbol from the set $\{\in, \notin, \subset, \not\subset, =\}$ between each pair.

24. 2   A
25. D   A
26. A   B
27. {3}   C
28. ∅   C
29. U   B
30. 3   D

Let U, J, P, T, D, and A be as in Exercises 17–23. Insert a proper symbol as in Exercises 24–30.

31. m $(J \cap T)$
32. {m} $(J \cap T)$
33. D   J'
34. j   A
35. j   A'
36. ∅ $(J \cup P)$
37. P   D

Let U be the set of all students at a college (or university), F be the set of all female students, M be the set of all students taking a mathematics course, and B be the set of all students taking a business course. Describe in words the elements in each of the following sets:

38. $F \cap M$
39. F'
40. $(F \cap B)'$
41. $B \cup M$
42. $F \cup (M \cap B)'$
43. $F' \cap (B \cup M)$
44. $F \cup F'$
45. $F \cap F'$

Let U be the set of the months, R be the set of all months containing the letter r, J be the set of all months beginning with the letter J, and S be the set consisting of the first 6 months of the year. Give the elements in each of the following sets:

46. $R \cap J$
47. R'
48. $(S \cup J)'$
49. $R \cap J'$
50. $J \cup (S \cap R)'$
51. $S \cap (J \cap R)$

52. If U, A, B, and C are the sets given for Exercises 1–8, determine the number of elements in each of the eight sections of the Venn diagram in the margin.

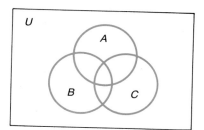

53. If U, J, R, and S are the sets given for Exercises 46–51, determine the number of elements in each of the eight sections of the Venn diagram:

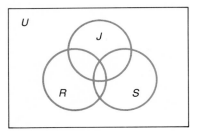

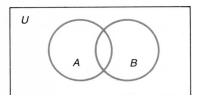

Use shading to indicate which portions of the Venn diagram at the left represent each of the following sets:

54. $(A \cup B)'$
55. B'
56. $A' \cup B$
57. $(A \cap B)'$
58. $A \cap B'$

Use shading to indicate which portions of the Venn diagram at the left represent each of the following sets:

59. $(A \cup B) \cup C$
60. $(A \cap B) \cap C$
61. $(A \cup B)' \cup C$
62. $(A \cup B)' \cap C$
63. $A \cap (B \cup C)$
64. $A' \cap (B' \cap C')$

State what is meant by each of the following:

65. Set
66. Empty set
67. Universal set
68. Subset
69. Complement of a set
70. Union of two sets
71. Intersection of two sets

APPENDIX B

TABLES

Table I   Binomial Probabilities
Table II Areas Under the Standard Normal Curve
Table III Values of e^x and e^{-x}
Table IV Natural Logarithms
Table V   Common Logarithms
Table VI Values of $(1+i)^n$
Table VII Values of $(1+i)^{-n}$

TABLE I BINOMIAL PROBABILITIES*

n	r	0.05	0.1	0.2	0.3	0.4	p 0.5	0.6	0.7	0.8	0.9	0.95
5	0	0.774	0.590	0.328	0.168	0.078	0.031	0.010	0.002			
	1	0.204	0.328	0.410	0.360	0.259	0.156	0.077	0.028	0.006		
	2	0.021	0.073	0.205	0.309	0.346	0.312	0.230	0.132	0.051	0.008	0.001
	3	0.001	0.008	0.051	0.132	0.230	0.312	0.346	0.309	0.205	0.073	0.021
	4			0.006	0.028	0.077	0.156	0.259	0.360	0.410	0.328	0.204
	5				0.002	0.010	0.031	0.078	0.168	0.328	0.590	0.774
6	0	0.735	0.531	0.262	0.118	0.047	0.016	0.004	0.001			
	1	0.232	0.354	0.393	0.303	0.187	0.094	0.037	0.010	0.002		
	2	0.031	0.098	0.246	0.324	0.311	0.234	0.138	0.060	0.015	0.001	
	3	0.002	0.015	0.082	0.185	0.276	0.312	0.276	0.185	0.082	0.015	0.002
	4		0.001	0.015	0.060	0.138	0.234	0.311	0.324	0.246	0.098	0.031
	5			0.002	0.010	0.037	0.094	0.187	0.303	0.393	0.354	0.232
	6				0.001	0.004	0.016	0.047	0.118	0.262	0.531	0.735
7	0	0.698	0.478	0.210	0.082	0.028	0.008	0.002				
	1	0.257	0.372	0.367	0.247	0.131	0.055	0.017	0.004			
	2	0.041	0.124	0.275	0.318	0.261	0.164	0.077	0.025	0.004		
	3	0.004	0.023	0.115	0.227	0.290	0.273	0.194	0.097	0.029	0.003	
	4		0.003	0.029	0.097	0.194	0.273	0.290	0.227	0.115	0.023	0.004
	5			0.004	0.025	0.077	0.164	0.261	0.318	0.275	0.124	0.041
	6				0.004	0.017	0.055	0.131	0.247	0.367	0.372	0.257
	7					0.002	0.008	0.028	0.082	0.210	0.478	0.698
8	0	0.663	0.430	0.168	0.058	0.017	0.004	0.001				
	1	0.279	0.383	0.336	0.198	0.090	0.031	0.008	0.001			
	2	0.051	0.149	0.294	0.296	0.209	0.109	0.041	0.010	0.001		
	3	0.005	0.033	0.147	0.254	0.279	0.219	0.124	0.047	0.009		
	4		0.005	0.046	0.136	0.232	0.273	0.232	0.136	0.046	0.005	
	5			0.009	0.047	0.124	0.219	0.279	0.254	0.147	0.033	0.005
	6			0.001	0.010	0.041	0.109	0.209	0.296	0.294	0.149	0.051
	7				0.001	0.008	0.031	0.090	0.198	0.336	0.383	0.279
	8					0.001	0.004	0.017	0.058	0.168	0.430	0.663
9	0	0.630	0.387	0.134	0.040	0.010	0.002					
	1	0.299	0.387	0.302	0.156	0.060	0.018	0.004				
	2	0.063	0.172	0.302	0.267	0.161	0.070	0.021	0.004			
	3	0.008	0.045	0.176	0.267	0.251	0.164	0.074	0.021	0.003		
	4	0.001	0.007	0.066	0.172	0.251	0.246	0.167	0.074	0.017	0.001	
	5		0.001	0.017	0.074	0.167	0.246	0.251	0.172	0.066	0.007	0.001
	6			0.003	0.021	0.074	0.164	0.251	0.267	0.176	0.045	0.008
	7				0.004	0.021	0.070	0.161	0.267	0.302	0.172	0.063
	8					0.004	0.018	0.060	0.156	0.302	0.387	0.299
	9						0.002	0.010	0.040	0.134	0.387	0.630
10	0	0.599	0.349	0.107	0.028	0.006	0.001					
	1	0.315	0.387	0.268	0.121	0.040	0.010	0.002				
	2	0.075	0.194	0.302	0.233	0.121	0.044	0.011	0.001			
	3	0.010	0.057	0.201	0.267	0.215	0.117	0.042	0.009	0.001		
	4	0.001	0.011	0.088	0.200	0.251	0.205	0.111	0.037	0.006		
	5		0.001	0.026	0.103	0.201	0.246	0.201	0.103	0.026	0.001	
	6			0.006	0.037	0.111	0.205	0.251	0.200	0.088	0.011	0.001
	7			0.001	0.009	0.042	0.117	0.215	0.267	0.201	0.057	0.010
	8				0.001	0.011	0.044	0.121	0.233	0.302	0.194	0.075
	9					0.002	0.010	0.040	0.121	0.268	0.387	0.315
	10						0.001	0.006	0.028	0.107	0.349	0.599
11	0	0.569	0.314	0.086	0.020	0.004						
	1	0.329	0.384	0.236	0.093	0.027	0.005	0.001				
	2	0.087	0.213	0.295	0.200	0.089	0.027	0.005	0.001			
	3	0.014	0.071	0.221	0.257	0.177	0.081	0.023	0.004			
	4	0.001	0.016	0.111	0.220	0.236	0.161	0.070	0.017	0.002		
	5		0.002	0.039	0.132	0.221	0.226	0.147	0.057	0.010		
	6			0.010	0.057	0.147	0.226	0.221	0.132	0.039	0.002	
	7			0.002	0.017	0.070	0.161	0.236	0.220	0.111	0.016	0.001
	8				0.004	0.023	0.081	0.177	0.257	0.221	0.071	0.014
	9				0.001	0.005	0.027	0.089	0.200	0.295	0.213	0.087
	10					0.001	0.005	0.027	0.093	0.236	0.384	0.329
	11							0.004	0.020	0.086	0.314	0.569

* All values omitted in this table for $5 \leq n \leq 15$ are 0.0005 or less.

APPENDIX B TABLES

n	r	0.05	0.1	0.2	0.3	0.4	p 0.5	0.6	0.7	0.8	0.9	0.95
12	0	0.540	0.282	0.069	0.014	0.002						
	1	0.341	0.377	0.206	0.071	0.017	0.003					
	2	0.099	0.230	0.283	0.168	0.064	0.016	0.002				
	3	0.017	0.085	0.236	0.240	0.142	0.054	0.012	0.001			
	4	0.002	0.021	0.133	0.231	0.213	0.121	0.042	0.008	0.001		
	5		0.004	0.053	0.158	0.227	0.193	0.101	0.029	0.003		
	6			0.016	0.079	0.177	0.226	0.177	0.079	0.016		
	7			0.003	0.029	0.101	0.193	0.227	0.158	0.053	0.004	
	8			0.001	0.008	0.042	0.121	0.213	0.231	0.133	0.021	0.002
	9				0.001	0.012	0.054	0.142	0.240	0.236	0.085	0.017
	10					0.002	0.016	0.064	0.168	0.283	0.230	0.099
	11						0.003	0.017	0.071	0.206	0.377	0.341
	12							0.002	0.014	0.069	0.282	0.540
13	0	0.513	0.254	0.055	0.010	0.001						
	1	0.351	0.367	0.179	0.054	0.011	0.002					
	2	0.111	0.245	0.268	0.139	0.045	0.010	0.001				
	3	0.021	0.100	0.246	0.218	0.111	0.035	0.006	0.001			
	4	0.003	0.028	0.154	0.234	0.184	0.087	0.024	0.003			
	5		0.006	0.069	0.180	0.221	0.157	0.066	0.014	0.001		
	6		0.001	0.023	0.103	0.197	0.209	0.131	0.044	0.006		
	7			0.006	0.044	0.131	0.209	0.197	0.103	0.023	0.001	
	8			0.001	0.014	0.066	0.157	0.221	0.180	0.069	0.006	
	9				0.003	0.024	0.087	0.184	0.234	0.154	0.028	0.003
	10				0.001	0.006	0.035	0.111	0.218	0.246	0.100	0.021
	11					0.001	0.010	0.045	0.139	0.268	0.245	0.111
	12						0.002	0.011	0.054	0.179	0.367	0.351
	13							0.001	0.010	0.055	0.254	0.513
14	0	0.488	0.229	0.044	0.007	0.001						
	1	0.359	0.356	0.154	0.041	0.007	0.001					
	2	0.123	0.257	0.250	0.113	0.032	0.006	0.001				
	3	0.026	0.114	0.250	0.194	0.085	0.022	0.003				
	4	0.004	0.035	0.172	0.229	0.155	0.061	0.014	0.001			
	5		0.008	0.086	0.196	0.207	0.122	0.041	0.007			
	6		0.001	0.032	0.126	0.207	0.183	0.092	0.023	0.002		
	7			0.009	0.062	0.157	0.209	0.157	0.062	0.009		
	8			0.002	0.023	0.092	0.183	0.207	0.126	0.032	0.001	
	9				0.007	0.041	0.122	0.207	0.196	0.086	0.008	
	10				0.001	0.014	0.061	0.155	0.229	0.172	0.035	0.004
	11					0.003	0.022	0.085	0.194	0.250	0.114	0.026
	12					0.001	0.006	0.032	0.113	0.250	0.257	0.123
	13						0.001	0.007	0.041	0.154	0.356	0.359
	14							0.001	0.007	0.044	0.229	0.488
15	0	0.463	0.206	0.035	0.005							
	1	0.366	0.343	0.132	0.031	0.005						
	2	0.135	0.267	0.231	0.092	0.022	0.003					
	3	0.031	0.129	0.250	0.170	0.063	0.014	0.002				
	4	0.005	0.043	0.188	0.219	0.127	0.042	0.007	0.001			
	5	0.001	0.010	0.103	0.206	0.186	0.092	0.024	0.003			
	6		0.002	0.043	0.147	0.207	0.153	0.061	0.012	0.001		
	7			0.014	0.081	0.177	0.196	0.118	0.035	0.003		
	8			0.003	0.035	0.118	0.196	0.177	0.081	0.014		
	9			0.001	0.012	0.061	0.153	0.207	0.147	0.043	0.002	
	10				0.003	0.024	0.092	0.186	0.206	0.103	0.010	0.001
	11				0.001	0.007	0.042	0.127	0.219	0.188	0.043	0.005
	12					0.002	0.014	0.063	0.170	0.250	0.129	0.031
	13						0.003	0.022	0.092	0.231	0.267	0.135
	14							0.005	0.031	0.132	0.343	0.366
	15								0.005	0.035	0.206	0.463

TABLE II AREAS UNDER THE STANDARD NORMAL CURVE

z	0.00	0.01	0.02	0.03	0.04	0.05	0.06	0.07	0.08	0.09
0.0	0.0000	0.0040	0.0080	0.0120	0.0160	0.0199	0.0239	0.0279	0.0319	0.0359
0.1	0.0398	0.0438	0.0478	0.0517	0.0557	0.0596	0.0636	0.0675	0.0714	0.0753
0.2	0.0793	0.0832	0.0871	0.0910	0.0948	0.0987	0.1026	0.1064	0.1103	0.1141
0.3	0.1179	0.1217	0.1255	0.1293	0.1331	0.1368	0.1406	0.1443	0.1480	0.1517
0.4	0.1554	0.1591	0.1628	0.1664	0.1700	0.1736	0.1772	0.1808	0.1844	0.1879
0.5	0.1915	0.1950	0.1985	0.2019	0.2054	0.2088	0.2123	0.2157	0.2190	0.2224
0.6	0.2257	0.2291	0.2324	0.2357	0.2389	0.2422	0.2454	0.2486	0.2517	0.2549
0.7	0.2580	0.2611	0.2642	0.2673	0.2704	0.2734	0.2764	0.2794	0.2823	0.2852
0.8	0.2881	0.2910	0.2939	0.2967	0.2995	0.3023	0.3051	0.3078	0.3106	0.3133
0.9	0.3159	0.3186	0.3212	0.3238	0.3264	0.3289	0.3315	0.3340	0.3365	0.3389
1.0	0.3413	0.3438	0.3461	0.3485	0.3508	0.3531	0.3554	0.3577	0.3599	0.3621
1.1	0.3643	0.3665	0.3686	0.3708	0.3729	0.3749	0.3770	0.3790	0.3810	0.3830
1.2	0.3849	0.3869	0.3888	0.3907	0.3925	0.3944	0.3962	0.3980	0.3997	0.4015
1.3	0.4032	0.4049	0.4066	0.4082	0.4099	0.4115	0.4131	0.4147	0.4162	0.4177
1.4	0.4192	0.4207	0.4222	0.4236	0.4251	0.4265	0.4279	0.4292	0.4306	0.4319
1.5	0.4332	0.4345	0.4357	0.4370	0.4382	0.4394	0.4406	0.4418	0.4429	0.4441
1.6	0.4452	0.4463	0.4474	0.4484	0.4495	0.4505	0.4515	0.4525	0.4535	0.4545
1.7	0.4554	0.4564	0.4573	0.4582	0.4591	0.4599	0.4608	0.4616	0.4625	0.4633
1.8	0.4641	0.4649	0.4656	0.4664	0.4671	0.4678	0.4686	0.4693	0.4699	0.4706
1.9	0.4713	0.4719	0.4726	0.4732	0.4738	0.4744	0.4750	0.4756	0.4761	0.4767
2.0	0.4772	0.4778	0.4783	0.4788	0.4793	0.4798	0.4803	0.4808	0.4812	0.4817
2.1	0.4821	0.4826	0.4830	0.4834	0.4838	0.4842	0.4846	0.4850	0.4854	0.4857
2.2	0.4861	0.4864	0.4868	0.4871	0.4875	0.4878	0.4881	0.4884	0.4887	0.4890
2.3	0.4893	0.4896	0.4898	0.4901	0.4904	0.4906	0.4909	0.4911	0.4913	0.4916
2.4	0.4918	0.4920	0.4922	0.4925	0.4927	0.4929	0.4931	0.4932	0.4934	0.4936
2.5	0.4938	0.4940	0.4941	0.4943	0.4945	0.4946	0.4948	0.4949	0.4951	0.4952
2.6	0.4953	0.4955	0.4956	0.4957	0.4959	0.4960	0.4961	0.4962	0.4963	0.4964
2.7	0.4965	0.4966	0.4967	0.4968	0.4969	0.4970	0.4971	0.4972	0.4973	0.4974
2.8	0.4974	0.4975	0.4976	0.4977	0.4977	0.4978	0.4979	0.4979	0.4980	0.4981
2.9	0.4981	0.4982	0.4982	0.4983	0.4984	0.4984	0.4985	0.4985	0.4986	0.4986
3.0	0.4987	0.4987	0.4987	0.4988	0.4988	0.4989	0.4989	0.4989	0.4990	0.4990

TABLE III VALUES OF e^x AND e^{-x}

x	e^x	e^{-x}	x	e^x	e^{-x}
.00	1.00000	1.00000	**.60**	1.82212	.54881
.01	1.01005	.99005	.61	1.84043	.54335
.02	1.02020	.98020	.62	1.85893	.53794
.03	1.03045	.97045	.63	1.87761	.53259
.04	1.04081	.96079	.64	1.89648	.52729
.05	1.05127	.95123	.65	1.91554	.52205
.06	1.06184	.94176	.66	1.93479	.51685
.07	1.07251	.93239	.67	1.95424	.51171
.08	1.08329	.92312	.68	1.97388	.50662
.09	1.09417	.91393	.69	1.99372	.50158
.10	1.10517	.90484	**.70**	2.01375	.49659
.11	1.11628	.89583	.71	2.03399	.49164
.12	1.12750	.88692	.72	2.05443	.48675
.13	1.13883	.87810	.73	2.07508	.48191
.14	1.15027	.86936	.74	2.09594	.47711
.15	1.16183	.86071	.75	2.11700	.47237
.16	1.17351	.85214	.76	2.13828	.46767
.17	1.18530	.84366	.77	2.15977	.46301
.18	1.19722	.83527	.78	2.18147	.45841
.19	1.20925	.82696	.79	2.20340	.45384
.20	1.22140	.81873	**.80**	2.22554	.44933
.21	1.23368	.81058	.81	2.24791	.44486
.22	1.24608	.80252	.82	2.27050	.44043
.23	1.25860	.79453	.83	2.29332	.43605
.24	1.27125	.78663	.84	2.31637	.43171
.25	1.28403	.77880	.85	2.33965	.42741
.26	1.29693	.77105	.86	2.36316	.42316
.27	1.30996	.76338	.87	2.38691	.41895
.28	1.32313	.75578	.88	2.41090	.41478
.29	1.33643	.74826	.89	2.43513	.41066
.30	1.34986	.74082	**.90**	2.45960	.40657
.31	1.36343	.73345	.91	2.48432	.40252
.32	1.37713	.72615	.92	2.50929	.39852
.33	1.39097	.71892	.93	2.53451	.39455
.34	1.40495	.71177	.94	2.55998	.39063
.35	1.41907	.70469	.95	2.58571	.38674
.36	1.43333	.69768	.96	2.61170	.38289
.37	1.44773	.69073	.97	2.63794	.37908
.38	1.46228	.68386	.98	2.66446	.37531
.39	1.47698	.67706	.99	2.69123	.37158
.40	1.49182	.67032	**1.00**	2.71828	.36788
.41	1.50682	.66365	1.01	2.74560	.36422
.42	1.52196	.65705	1.02	2.77319	.36059
.43	1.53726	.65051	1.03	2.80107	.35701
.44	1.55271	.64404	1.04	2.82922	.35345
.45	1.56831	.63763	1.05	2.85765	.34994
.46	1.58407	.63128	1.06	2.88637	.34646
.47	1.59999	.62500	1.07	2.91538	.34301
.48	1.61607	.61878	1.08	2.94468	.33960
.49	1.63232	.61263	1.09	2.97427	.33622
.50	1.64872	.60653	**1.10**	3.00417	.33287
.51	1.66529	.60050	1.11	3.03436	.32956
.52	1.68203	.59452	1.12	3.06485	.32628
.53	1.69893	.58860	1.13	3.09566	.32303
.54	1.71601	.58275	1.14	3.12677	.31982
.55	1.73325	.57695	1.15	3.15819	.31664
.56	1.75067	.57121	1.16	3.18993	.31349
.57	1.76827	.56553	1.17	3.22199	.31037
.58	1.78604	.55990	1.18	3.25437	.30728
.59	1.80399	.55433	1.19	3.28708	.30422

TABLE III VALUES OF e^x AND e^{-x} (CONTINUED)

x	e^x	e^{-x}	x	e^x	e^{-x}
1.20	3.32012	.30119	**1.80**	6.04965	.16530
1.21	3.35348	.29820	1.81	6.11045	.16365
1.22	3.38719	.29523	1.82	6.17186	.16203
1.23	3.42123	.29229	1.83	6.23389	.16041
1.24	3.45561	.28938	1.84	6.29654	.15882
1.25	3.49034	.28650	1.85	6.35982	.15724
1.26	3.52542	.28365	1.86	6.42374	.15567
1.27	3.56085	.28083	1.87	6.48830	.15412
1.28	3.59664	.27804	1.88	6.55350	.15259
1.29	3.63279	.27527	1.89	6.61937	.15107
1.30	3.66930	.27253	**1.90**	6.68589	.14957
1.31	3.70617	.26982	1.91	6.75309	.14808
1.32	3.74342	.26714	1.92	6.82096	.14661
1.33	3.78104	.26448	1.93	6.88951	.14515
1.34	3.81904	.26185	1.94	6.95875	.14370
1.35	3.85743	.25924	1.95	7.02869	.14227
1.36	3.89619	.25666	1.96	7.09933	.14086
1.37	3.93535	.25411	1.97	7.17068	.13946
1.38	3.97490	.25158	1.98	7.24274	.13807
1.39	4.01485	.24908	1.99	7.31553	.13670
1.40	4.05520	.24660	**2.00**	7.38906	.13534
1.41	4.09596	.24414	2.01	7.46332	.13399
1.42	4.13712	.24171	2.02	7.53832	.13266
1.43	4.17870	.23931	2.03	7.61409	.13134
1.44	4.22070	.23693	2.04	7.69061	.13003
1.45	4.26311	.23457	2.05	7.76790	.12873
1.46	4.30596	.23224	2.06	7.84597	.12745
1.47	4.34924	.22993	2.07	7.92482	.12619
1.48	4.39295	.22764	2.08	8.00447	.12493
1.49	4.43710	.22537	2.09	8.08491	.12369
1.50	4.48169	.22313	**2.10**	8.16617	.12246
1.51	4.52673	.22091	2.11	8.24824	.12124
1.52	4.57223	.21871	2.12	8.33114	.12003
1.53	4.61818	.21654	2.13	8.41487	.11884
1.54	4.66459	.21438	2.14	8.49944	.11765
1.55	4.71147	.21225	2.15	8.58486	.11648
1.56	4.75882	.21014	2.16	8.67114	.11533
1.57	4.80665	.20805	2.17	8.75828	.11418
1.58	4.85496	.20598	2.18	8.84631	.11304
1.59	4.90375	.20393	2.19	8.93521	.11192
1.60	4.95303	.20190	**2.20**	9.02501	.11080
1.61	5.00281	.19989	2.21	9.11572	.10970
1.62	5.05309	.19790	2.22	9.20733	.10861
1.63	5.10387	.19593	2.23	9.29987	.10753
1.64	5.15517	.19398	2.24	9.39333	.10646
1.65	5.20698	.19205	2.25	9.48774	.10540
1.66	5.25931	.19014	2.26	9.58309	.10435
1.67	5.31217	.18825	2.27	9.67940	.10331
1.68	5.36556	.18637	2.28	9.77668	.10228
1.69	5.41948	.18452	2.29	9.87494	.10127
1.70	5.47395	.18268	**2.30**	9.97418	.10026
1.71	5.52896	.18087	2.31	10.07442	.09926
1.72	5.58453	.17907	2.32	10.17567	.09827
1.73	5.64065	.17728	2.33	10.27794	.09730
1.74	5.69734	.17552	2.34	10.38124	.09633
1.75	5.75460	.17377	2.35	10.48557	.09537
1.76	5.81244	.17204	2.36	10.59095	.09442
1.77	5.87085	.17033	2.37	10.69739	.09348
1.78	5.92986	.16864	2.38	10.80490	.09255
1.79	5.98945	.16696	2.39	10.91349	.09163

x	e^x	e^{-x}	x	e^x	e^{-x}
2.40	11.02318	.09072	**3.00**	20.08554	.04979
2.41	11.13396	.08982	3.01	20.28740	.04929
2.42	11.24586	.08892	3.02	20.49129	.04880
2.43	11.35888	.08804	3.03	20.69723	.04832
2.44	11.47304	.08716	3.04	20.90524	.04783
2.45	11.58835	.08629	3.05	21.11534	.04736
2.46	11.70481	.08543	3.06	21.32756	.04689
2.47	11.82245	.08458	3.07	21.54190	.04642
2.48	11.94126	.08374	3.08	21.75840	.04596
2.49	12.06128	.08291	3.09	21.97708	.04550
2.50	12.18249	.08208	**3.10**	22.19795	.04505
2.51	12.30493	.08127	3.11	22.42104	.04460
2.52	12.42860	.08046	3.12	22.64638	.04416
2.53	12.55351	.07966	3.13	22.87398	.04372
2.54	12.67967	.07887	3.14	23.10387	.04328
2.55	12.80710	.07808	3.15	23.33606	.04285
2.56	12.93582	.07730	3.16	23.57060	.04243
2.57	13.06582	.07654	3.17	23.80748	.04200
2.58	13.19714	.07577	3.18	24.04675	.04159
2.59	13.32977	.07502	3.19	24.28843	.04117
2.60	13.46374	.07427	**3.20**	24.53253	.04076
2.61	13.59905	.07353	3.21	24.77909	.04036
2.62	13.73572	.07280	3.22	25.02812	.03996
2.63	13.87377	.07208	3.23	25.27966	.03956
2.64	14.01320	.07136	3.24	25.53372	.03916
2.65	14.15404	.07065	3.25	25.79034	.03877
2.66	14.29629	.06995	3.26	26.04954	.03839
2.67	14.43997	.06925	3.27	26.31134	.03801
2.68	14.58509	.06856	3.28	26.57577	.03763
2.69	14.73168	.06788	3.29	26.84286	.03725
2.70	14.87973	.06721	**3.30**	27.11264	.03688
2.71	15.02928	.06654	3.31	27.38512	.03652
2.72	15.18032	.06587	3.32	27.66035	.03615
2.73	15.33289	.06522	3.33	27.93834	.03579
2.74	15.48698	.06457	3.34	28.21913	.03544
2.75	15.64263	.06393	3.35	28.50273	.03508
2.76	15.79984	.06329	3.36	28.78919	.03474
2.77	15.95863	.06266	3.37	29.07853	.03439
2.78	16.11902	.06204	3.38	29.37077	.03405
2.79	16.28102	.06142	3.39	29.66595	.03371
2.80	16.44465	.06081	**3.40**	29.96410	.03337
2.81	16.60992	.06020	3.41	30.26524	.03304
2.82	16.77685	.05961	3.42	30.56941	.03271
2.83	16.94546	.05901	3.43	30.87664	.03239
2.84	17.11577	.05843	3.44	31.18696	.03206
2.85	17.28778	.05784	3.45	31.50039	.03175
2.86	17.46153	.05727	3.46	31.81698	.03143
2.87	17.63702	.05670	3.47	32.13674	.03112
2.88	17.81427	.05613	3.48	32.45972	.03081
2.89	17.99331	.05558	3.49	32.78595	.03050
2.90	18.17414	.05502	**3.50**	33.11545	.03020
2.91	18.35680	.05448	3.51	33.44827	.02990
2.92	18.54129	.05393	3.52	33.78443	.02960
2.93	18.72763	.05340	3.53	34.12397	.02930
2.94	18.91585	.05287	3.54	34.46692	.02901
2.95	19.10595	.05234	3.55	34.81332	.02872
2.96	19.29797	.05182	3.56	35.16320	.02844
2.97	19.49192	.05130	3.57	35.51659	.02816
2.98	19.68782	.05079	3.58	35.87354	.02788
2.99	19.88568	.05029	3.59	36.23408	.02760

TABLE III VALUES OF e^x AND e^{-x} (CONTINUED)

x	e^x	e^{-x}	x	e^x	e^{-x}
3.60	36.59823	.02732	**4.20**	66.68633	.01500
3.61	36.96605	.02705	4.21	67.35654	.01485
3.62	37.33757	.02678	4.22	68.03348	.01470
3.63	37.71282	.02652	4.23	68.71723	.01455
3.64	38.09184	.02625	4.24	69.40785	.01441
3.65	38.47467	.02599	4.25	70.10541	.01426
3.66	38.86134	.02573	4.26	70.80998	.01412
3.67	39.25191	.02548	4.27	71.52163	.01398
3.68	39.64639	.02522	4.28	72.24044	.01384
3.69	40.04485	.02497	4.29	72.96647	.01370
3.70	40.44730	.02472	**4.30**	73.69979	.01357
3.71	40.85381	.02448	4.31	74.44049	.01343
3.72	41.26439	.02423	4.32	75.18863	.01330
3.73	41.67911	.02399	4.33	75.94429	.01317
3.74	42.09799	.02375	4.34	76.70754	.01304
3.75	42.52108	.02352	4.35	77.47846	.01291
3.76	42.94843	.02328	4.36	78.25713	.01278
3.77	43.38006	.02305	4.37	79.04363	.01265
3.78	43.81604	.02282	4.38	79.83803	.01253
3.79	44.25640	.02260	4.39	80.64042	.01240
3.80	44.70118	.02237	**4.40**	81.45087	.01228
3.81	45.15044	.02215	4.41	82.26946	.01216
3.82	45.60421	.02193	4.42	83.09628	.01203
3.83	46.06254	.02171	4.43	83.93141	.01191
3.84	46.52547	.02149	4.44	84.77494	.01180
3.85	46.99306	.02128	4.45	85.62694	.01168
3.86	47.46535	.02107	4.46	86.48751	.01156
3.87	47.94238	.02086	4.47	87.35672	.01145
3.88	48.42421	.02065	4.48	88.23467	.01133
3.89	48.91089	.02045	4.49	89.12144	.01122
3.90	49.40245	.02024	**4.50**	90.01713	.01111
3.91	49.89895	.02004	4.51	90.92182	.01100
3.92	50.40044	.01984	4.52	91.83560	.01089
3.93	50.90698	.01964	4.53	92.75856	.01078
3.94	51.41860	.01945	4.54	93.69080	.01067
3.95	51.93537	.01925	4.55	94.63240	.01057
3.96	52.45732	.01906	4.56	95.58347	.01046
3.97	52.98453	.01887	4.57	96.54411	.01036
3.98	53.51703	.01869	4.58	97.51439	.01025
3.99	54.05489	.01850	4.59	98.49443	.01015
4.00	54.59815	.01832	**4.60**	99.48431	.01005
4.01	55.14687	.01813	4.61	100.48415	.00995
4.02	55.70110	.01795	4.62	101.49403	.00985
4.03	56.26091	.01777	4.63	102.51406	.00975
4.04	56.82634	.01760	4.64	103.54435	.00966
4.05	57.39745	.01742	4.65	104.58498	.00956
4.06	57.97431	.01725	4.66	105.63608	.00947
4.07	58.55696	.01708	4.67	106.69774	.00937
4.08	59.14547	.01691	4.68	107.77007	.00928
4.09	59.73989	.01674	4.69	108.85318	.00919
4.10	60.34029	.01657	**4.70**	109.94717	.00910
4.11	60.94671	.01641	4.71	111.05216	.00900
4.12	61.55924	.01624	4.72	112.16825	.00892
4.13	62.17792	.01608	4.73	113.29556	.00883
4.14	62.80282	.01592	4.74	114.43420	.00874
4.15	63.43400	.01576	4.75	115.58428	.00865
4.16	64.07152	.01561	4.76	116.74592	.00857
4.17	64.71545	.01545	4.77	117.91924	.00848
4.18	65.36585	.01530	4.78	119.10435	.00840
4.19	66.02279	.01515	4.79	120.30136	.00831

x	e^x	e^{-x}	x	e^x	e^{-x}
4.80	121.51041	.00823	**8.00**	2980.95779	.00034
4.81	122.73161	.00815	8.10	3294.46777	.00030
4.82	123.96509	.00807	8.20	3640.95004	.00027
4.83	125.21096	.00799	8.30	4023.87219	.00025
4.84	126.46935	.00791	8.40	4447.06665	.00022
4.85	127.74039	.00783	8.50	4914.76886	.00020
4.86	129.02420	.00775	8.60	5431.65906	.00018
4.87	130.32091	.00767	8.70	6002.91180	.00017
4.88	131.63066	.00760	8.80	6634.24371	.00015
4.89	132.95357	.00752	8.90	7331.97339	.00014
4.90	134.28978	.00745	**9.00**	8103.08295	.00012
4.91	135.63941	.00737	9.10	8955.29187	.00011
4.92	137.00261	.00730	9.20	9897.12830	.00010
4.93	138.37951	.00723	9.30	10938.01868	.00009
4.94	139.77024	.00715	9.40	12088.38049	.00008
4.95	141.17496	.00708	9.50	13359.72522	.00007
4.96	142.59379	.00701	9.60	14764.78015	.00007
4.97	144.02688	.00694	9.70	16317.60608	.00006
4.98	145.47438	.00687	9.80	18033.74414	.00006
4.99	146.93642	.00681	9.90	19930.36987	.00005
5.00	148.41316	.00674	**10.00**	22026.46313	.00005
5.10	164.02190	.00610	10.10	24343.00708	.00004
5.20	181.27224	.00552	10.20	26903.18408	.00004
5.30	200.33680	.00499	10.30	29732.61743	.00003
5.40	221.40641	.00452	10.40	32859.62500	.00003
5.50	244.69192	.00409	10.50	36315.49854	.00003
5.60	270.42640	.00370	10.60	40134.83350	.00002
5.70	298.86740	.00335	10.70	44355.85205	.00002
5.80	330.29955	.00303	10.80	49020.79883	.00002
5.90	365.03746	.00274	10.90	54176.36230	.00002
6.00	403.42877	.00248	**11.00**	59874.13477	.00002
6.10	445.85775	.00224	11.10	66171.15430	.00002
6.20	492.74903	.00203	11.20	73130.43652	.00001
6.30	544.57188	.00184	11.30	80821.63379	.00001
6.40	601.84502	.00166	11.40	89321.72168	.00001
6.50	665.14159	.00150	11.50	98715.75879	.00001
6.60	735.09516	.00136	11.60	109097.78906	.00001
6.70	812.40582	.00123	11.70	120571.70605	.00001
6.80	897.84725	.00111	11.80	133252.34570	.00001
6.90	992.27469	.00101	11.90	147266.62109	.00001
7.00	1096.63309	.00091			
7.10	1211.96703	.00083			
7.20	1339.43076	.00075			
7.30	1480.29985	.00068			
7.40	1635.98439	.00061			
7.50	1808.04231	.00055			
7.60	1998.19582	.00050			
7.70	2208.34796	.00045			
7.80	2440.60187	.00041			
7.90	2697.28226	.00037			

TABLE IV NATURAL LOGARITHMS

x		0	1	2	3	4	5	6	7	8	9
1.0	0.0	0000	0995	1980	2956	3922	4879	5827	6766	7696	8618
1.1		9531	*0436	*1333	*2222	*3103	*3976	*4842	*5700	*6551	*7395
1.2	0.1	8232	9062	9885	*0701	*1511	*2314	*3111	*3902	*4686	*5464
1.3	0.2	6236	7003	7763	8518	9267	*0010	*0748	*1481	*2208	*2930
1.4	0.3	3647	4359	5066	5767	6464	7156	7844	8526	9204	9878
1.5	0.4	0547	1211	1871	2527	3178	3825	4469	5108	5742	6373
1.6		7000	7623	8243	8858	9470	*0078	*0682	*1282	*1879	*2473
1.7	0.5	3063	3649	4232	4812	5389	5962	6531	7098	7661	8222
1.8		8779	9333	9884	*0432	*0977	*1519	*2058	*2594	*3127	*3658
1.9	0.6	4185	4710	5233	5752	6269	6783	7294	7803	8310	8813
2.0		9315	9813	*0310	*0804	*1295	*1784	*2271	*2755	*3237	*3716
2.1	0.7	4194	4669	5142	5612	6081	6547	7011	7473	7932	8390
2.2		8846	9299	9751	*0200	*0648	*1093	*1536	*1978	*2418	*2855
2.3	0.8	3291	3725	4157	4587	5015	5442	5866	6289	6710	7129
2.4		7547	7963	8377	8789	9200	9609	*0016	*0422	*0826	*1228
2.5	0.9	1629	2028	2426	2822	3216	3609	4001	4391	4779	5166
2.6		5551	5935	6317	6698	7078	7456	7833	8208	8582	8954
2.7		9325	9695	*0063	*0430	*0796	*1160	*1523	*1885	*2245	*2604
2.8	1.0	2962	3318	3674	4028	4380	4732	5082	5431	5779	6126
2.9		6471	6815	7158	7500	7841	8181	8519	8856	9192	9527
3.0		9861	*0194	*0526	*0856	*1186	*1514	*1841	*2168	*2493	*2817
3.1	1.1	3140	3462	3783	4103	4422	4740	5057	5373	5688	6002
3.2		6315	6627	6938	7248	7557	7865	8173	8479	8784	9089
3.3		9392	9695	9996	*0297	*0597	*0896	*1194	*1491	*1788	*2083
3.4	1.2	2378	2671	2964	3256	3547	3837	4127	4415	4703	4990
3.5		5276	5562	5846	6130	6413	6695	6976	7257	7536	7815
3.6		8093	8371	8647	8923	9198	9473	9746	*0019	*0291	*0563
3.7	1.3	0833	1103	1372	1641	1909	2176	2442	2708	2972	3237
3.8		3500	3763	4025	4286	4547	4807	5067	5325	5584	5841
3.9		6098	6354	6609	6864	7118	7372	7624	7877	8128	8379
4.0		8629	8879	9128	9377	9624	9872	*0118	*0364	*0610	*0854
4.1	1.4	1099	1342	1585	1828	2070	2311	2552	2792	3031	3270
4.2		3508	3746	3984	4220	4456	4692	4927	5161	5395	5629
4.3		5862	6094	6326	6557	6787	7018	7247	7476	7705	7933
4.4		8160	8367	8614	8840	9065	9290	9515	9739	9962	*0185
4.5	1.5	0408	0630	0851	1072	1293	1513	1732	1951	2170	2388
4.6		2606	2823	3039	3256	3471	3687	3902	4116	4330	4543
4.7		4756	4969	5181	5393	5604	5814	6025	6235	6444	6653
4.8		6862	7070	7277	7485	7691	7898	8104	8309	8515	8719
4.9		8924	9127	9331	9534	9737	9939	*0141	*0342	*0543	*0744
5.0	1.6	0944	1144	1343	1542	1741	1939	2137	2334	2531	2728
5.1		2924	3120	3315	3511	3705	3900	4094	4287	4481	4673
5.2		4866	5058	5250	5441	5632	5823	6013	6203	6393	6582
5.3		6771	6959	7147	7335	7523	7710	7896	8083	8269	8455
5.4		8640	8825	9010	9194	9378	9562	9745	9928	*0111	*0293

* The first two digits are those at the beginning of the next row.

x		0	1	2	3	4	5	6	7	8	9
5.5	1.7	0475	0656	0838	1019	1199	1380	1560	1740	1919	2098
5.6		2277	2455	2633	2811	2988	3166	3342	3519	3695	3871
5.7		4047	4222	4397	4572	4746	4920	5094	5267	5440	5613
5.8		5786	5958	6130	6302	6473	6644	6815	6985	7156	7326
5.9		7495	7665	7834	8002	8171	8339	8507	8675	8842	9009
6.0	1.7	9176	9342	9509	9675	9840	*0006	*0171	*0336	*0500	*0665
6.1	1.8	0829	0993	1156	1319	1482	1645	1808	1970	2132	2294
6.2		2455	2616	2777	2938	3098	3258	3418	3578	3737	3896
6.3		4055	4214	4372	4530	4688	4845	5003	5160	5317	5473
6.4		5630	5786	5942	6097	6253	6408	6563	6718	6872	7026
6.5		7180	7334	7487	7641	7794	7947	8099	8251	8403	8555
6.6		8707	8858	9010	9160	9311	9462	9612	9762	9912	*0061
6.7	1.9	0211	0360	0509	0658	0806	0954	1102	1250	1398	1545
6.8		1692	1839	1986	2132	2279	2425	2571	2716	2862	3007
6.9		3152	3297	3442	3586	3730	3874	4018	4162	4305	4448
7.0		4591	4734	4876	5019	5161	5303	5445	5586	5727	5869
7.1		6009	6150	6291	6431	6571	6711	6851	6991	7130	7269
7.2		7408	7547	7685	7824	7962	8100	8238	8376	8513	8650
7.3		8787	8924	9061	9198	9334	9470	9606	9742	9877	*0013
7.4	2.0	0148	0283	0418	0553	0687	0821	0956	1089	1223	1357
7.5		1490	1624	1757	1890	2022	2155	2287	2419	2551	2683
7.6		2815	2946	3078	3209	3340	3471	3601	3732	3862	3992
7.7		4122	4252	4381	4511	4640	4769	4898	5027	5156	5284
7.8		5412	5540	5668	5796	5924	6051	6179	6306	6433	6560
7.9		6686	6813	6939	7065	7191	7317	7443	7568	7694	7819
8.0		7944	8069	8194	8318	8443	8567	8691	8815	8939	9063
8.1		9186	9310	9433	9556	9679	9802	9924	*0047	*0169	*0291
8.2	2.1	0413	0535	0657	0779	0900	1021	1142	1263	1384	1505
8.3		1626	1746	1866	1986	2106	2226	2346	2465	2585	2704
8.4		2823	2942	3061	3180	3298	3417	3535	3653	3771	3889
8.5		4007	4124	4242	4359	4476	4593	4710	4827	4943	5060
8.6		5176	5292	5409	5524	5640	5756	5871	5987	6102	6217
8.7		6332	6447	6562	6677	6791	6905	7020	7134	7248	7361
8.8		7475	7589	7702	7816	7929	8042	8155	8267	8380	8493
8.9		8605	8717	8830	8942	9054	9165	9277	9389	9500	9611
9.0		9722	9834	9944	*0055	*0166	*0276	*0387	*0497	*0607	*0717
9.1	2.2	0827	0937	1047	1157	1266	1375	1485	1594	1703	1812
9.2		1920	2029	2138	2246	2354	2462	2570	2678	2786	2894
9.3		3001	3109	3216	3324	3431	3538	3645	3751	3858	3965
9.4		4071	4177	4284	4390	4496	4601	4707	4813	4918	5024
9.5		5129	5234	5339	5444	5549	5654	5759	5863	5968	6072
9.6		6176	6280	6384	6488	6592	6696	6799	6903	7006	7109
9.7		7213	7316	7419	7521	7624	7727	7829	7932	8034	8136
9.8		8238	8340	8442	8544	8646	8747	8849	8950	9051	9152
9.9		9253	9354	9455	9556	9657	9757	9858	9958	*0058	*0158
10.0	2.3	0259	0358	0458	0558	0658	0757	0857	0956	1055	1154

TABLE V COMMON LOGARITHMS*

x	0	1	2	3	4	5	6	7	8	9
1.0	0000	0043	0086	0128	0170	0212	0253	0294	0334	0374
1.1	0414	0453	0492	0531	0569	0607	0645	0682	0719	0755
1.2	0792	0828	0864	0899	0934	0969	1004	1038	1072	1106
1.3	1139	1173	1206	1239	1271	1303	1335	1367	1399	1430
1.4	1461	1492	1523	1553	1584	1614	1644	1673	1703	1732
1.5	1761	1790	1818	1847	1875	1903	1931	1959	1987	2014
1.6	2041	2068	2095	2122	2148	2175	2201	2227	2253	2279
1.7	2304	2330	2355	2380	2405	2430	2455	2480	2504	2529
1.8	2553	2577	2601	2625	2648	2672	2695	2718	2742	2765
1.9	2788	2810	2833	2856	2878	2900	2923	2945	2967	2989
2.0	3010	3032	3054	3075	3096	3118	3139	3160	3181	3201
2.1	3222	3243	3263	3284	3304	3324	3345	3365	3385	3404
2.2	3424	3444	3464	3483	3502	3522	3541	3560	3579	3598
2.3	3617	3636	3655	3674	3692	3711	3729	3747	3766	3784
2.4	3802	3820	3838	3856	3874	3892	3909	3927	3945	3962
2.5	3979	3997	4014	4031	4048	4065	4082	4099	4116	4133
2.6	4150	4166	4183	4200	4216	4232	4249	4265	4281	4298
2.7	4314	4330	4346	4362	4378	4393	4409	4425	4440	4456
2.8	4472	4487	4502	4518	4533	4548	4564	4579	4594	4609
2.9	4624	4639	4654	4669	4683	4698	4713	4728	4742	4757
3.0	4771	4786	4800	4814	4829	4843	4857	4871	4886	4900
3.1	4914	4928	4942	4955	4969	4983	4997	5011	5024	5038
3.2	5051	5065	5079	5092	5105	5119	5132	5145	5159	5172
3.3	5185	5198	5211	5224	5237	5250	5263	5276	5289	5302
3.4	5315	5328	5340	5353	5366	5378	5391	5403	5416	5428
3.5	5441	5453	5465	5478	5490	5502	5514	5527	5539	5551
3.6	5563	5575	5587	5599	5611	5623	5635	5647	5658	5670
3.7	5682	5694	5705	5717	5729	5740	5752	5763	5775	5786
3.8	5798	5809	5821	5832	5843	5855	5866	5877	5888	5899
3.9	5911	5922	5933	5944	5955	5966	5977	5988	5999	6010
4.0	6021	6031	6042	6053	6064	6075	6085	6096	6107	6117
4.1	6128	6138	6149	6160	6170	6180	6191	6201	6212	6222
4.2	6232	6243	6253	6263	6274	6284	6294	6304	6314	6325
4.3	6335	6345	6355	6365	6375	6385	6395	6405	6415	6425
4.4	6435	6444	6454	6464	6474	6484	6493	6503	6513	6522
4.5	6532	6542	6551	6561	6571	6580	6590	6599	6609	6618
4.6	6628	6637	6646	6656	6665	6675	6684	6693	6702	6712
4.7	6721	6730	6739	6749	6758	6767	6776	6785	6794	6803
4.8	6812	6821	6830	6839	6848	6857	6866	6875	6884	6893
4.9	6902	6911	6920	6928	6937	6946	6955	6964	6972	6981
5.0	6990	6998	7007	7016	7024	7033	7042	7050	7059	7067
5.1	7076	7084	7093	7101	7110	7118	7126	7135	7143	7152
5.2	7160	7168	7177	7185	7193	7202	7210	7218	7226	7235
5.3	7243	7251	7259	7267	7275	7284	7292	7300	7308	7316
5.4	7324	7332	7340	7348	7356	7364	7372	7380	7388	7396

* Leading decimal points are omitted in this table: e.g., log 1.01 = .0043.

x	0	1	2	3	4	5	6	7	8	9
5.5	7404	7412	7419	7427	7435	7443	7451	7459	7466	7474
5.6	7482	7490	7497	7505	7513	7520	7528	7536	7543	7551
5.7	7559	7566	7574	7582	7589	7597	7604	7612	7619	7627
5.8	7634	7642	7649	7657	7664	7672	7679	7686	7694	7701
5.9	7709	7716	7723	7731	7738	7745	7752	7760	7767	7774
6.0	7782	7789	7796	7803	7810	7818	7825	7832	7839	7846
6.1	7853	7860	7868	7875	7882	7889	7896	7903	7910	7917
6.2	7924	7931	7938	7945	7952	7959	7966	7973	7980	7987
6.3	7993	8000	8007	8014	8021	8028	8035	8041	8048	8055
6.4	8062	8069	8075	8082	8089	8096	8102	8109	8116	8122
6.5	8129	8136	8142	8149	8156	8162	8169	8176	8182	8189
6.6	8195	8202	8209	8215	8222	8228	8235	8241	8248	8254
6.7	8261	8267	8274	8280	8287	8293	8299	8306	8312	8319
6.8	8325	8331	8338	8344	8351	8357	8363	8370	8376	8382
6.9	8388	8395	8401	8407	8414	8420	8426	8432	8439	8445
7.0	8451	8457	8463	8470	8476	8482	8488	8494	8500	8506
7.1	8513	8519	8525	8531	8537	8543	8549	8555	8561	8567
7.2	8573	8579	8585	8591	8597	8603	8609	8615	8621	8627
7.3	8633	8639	8645	8651	8657	8663	8669	8675	8681	8686
7.4	8692	8698	8704	8710	8716	8722	8727	8733	8739	8745
7.5	8751	8756	8762	8768	8774	8779	8785	8791	8797	8802
7.6	8808	8814	8820	8825	8831	8837	8842	8848	8854	8859
7.7	8865	8871	8876	8882	8887	8893	8899	8904	8910	8915
7.8	8921	8927	8932	8938	8943	8949	8954	8960	8965	8971
7.9	8976	8982	8987	8993	8998	9004	9009	9015	9020	9025
8.0	9031	9036	9042	9047	9053	9058	9063	9069	9074	9079
8.1	9085	9090	9096	9101	9106	9112	9117	9122	9128	9133
8.2	9138	9143	9149	9154	9159	9165	9170	9175	9180	9186
8.3	9191	9196	9201	9206	9212	9217	9222	9227	9232	9238
8.4	9243	9248	9253	9258	9263	9269	9274	9279	9284	9289
8.5	9294	9299	9304	9309	9315	9320	9325	9330	9335	9340
8.6	9345	9350	9355	9360	9365	9370	9375	9380	9385	9390
8.7	9395	9400	9405	9410	9415	9420	9425	9430	9435	9440
8.8	9445	9450	9455	9460	9465	9469	9474	9479	9484	9489
8.9	9494	9499	9504	9509	9513	9518	9523	9528	9533	9538
9.0	9542	9547	9552	9557	9562	9566	9571	9576	9581	9586
9.1	9590	9595	9600	9605	9609	9614	9619	9624	9628	9633
9.2	9638	9643	9647	9652	9657	9661	9666	9671	9675	9680
9.3	9685	9689	9694	9699	9703	9708	9713	9717	9722	9727
9.4	9731	9736	9741	9745	9750	9754	9759	9763	9768	9773
9.5	9777	9782	9786	9791	9795	9800	9805	9809	9814	9818
9.6	9823	9827	9832	9836	9841	9845	9850	9854	9859	9863
9.7	9868	9872	9877	9881	9886	9890	9894	9899	9903	9908
9.8	9912	9917	9921	9926	9930	9934	9939	9943	9948	9952
9.9	9956	9961	9965	9969	9974	9978	9983	9987	9991	9996

TABLE VI VALUES OF $(1+i)^n$

n	0.005	0.01	0.015	0.02	0.025	0.03	0.035	0.04	0.045	0.05	0.055	0.06	0.07	0.08
1	1.00500	1.01000	1.01500	1.02000	1.02500	1.03000	1.03500	1.04000	1.04500	1.05000	1.05500	1.06000	1.07000	1.08000
2	1.01003	1.02010	1.03023	1.04040	1.05063	1.06090	1.07123	1.08160	1.09203	1.10250	1.11303	1.12360	1.14490	1.16640
3	1.01508	1.03030	1.04568	1.06121	1.07689	1.09273	1.10872	1.12486	1.14117	1.15763	1.17424	1.19102	1.22504	1.25971
4	1.02015	1.04060	1.06136	1.08243	1.10381	1.12551	1.14752	1.16986	1.19252	1.21551	1.23882	1.26248	1.31080	1.36049
5	1.02525	1.05101	1.07728	1.10408	1.13141	1.15927	1.18769	1.21665	1.24618	1.27629	1.30696	1.33822	1.40255	1.46933
6	1.03038	1.06152	1.09344	1.12616	1.15969	1.19405	1.22926	1.26532	1.30226	1.34010	1.37884	1.41852	1.50073	1.58687
7	1.03553	1.07214	1.10984	1.14869	1.18869	1.22987	1.27228	1.31593	1.36086	1.40710	1.45468	1.50363	1.60578	1.71382
8	1.04071	1.08286	1.12649	1.17166	1.21840	1.26678	1.31681	1.36857	1.42210	1.47746	1.53469	1.59385	1.71819	1.85093
9	1.04591	1.09368	1.14339	1.19509	1.24886	1.30477	1.36290	1.42331	1.48610	1.55133	1.61909	1.68948	1.83846	1.99900
10	1.05114	1.10462	1.16054	1.21899	1.28008	1.34392	1.41060	1.48024	1.55297	1.62889	1.70814	1.79085	1.96715	2.15893
11	1.05640	1.11567	1.17795	1.24337	1.31209	1.38423	1.45997	1.53945	1.62285	1.71034	1.80209	1.89830	2.10485	2.33164
12	1.06168	1.12683	1.19562	1.26824	1.34489	1.42576	1.51107	1.60103	1.69588	1.79586	1.90121	2.01220	2.25219	2.51817
13	1.06699	1.13809	1.21355	1.29361	1.37851	1.46853	1.56396	1.66507	1.77220	1.88565	2.00577	2.13293	2.40985	2.71962
14	1.07232	1.14947	1.23176	1.31948	1.41297	1.51259	1.61869	1.73168	1.85194	1.97993	2.11609	2.26090	2.57853	2.93719
15	1.07768	1.16097	1.25023	1.34587	1.44830	1.55797	1.67534	1.80094	1.93528	2.07893	2.23248	2.39656	2.75903	3.17217
16	1.08307	1.17258	1.26899	1.37279	1.48451	1.60471	1.73399	1.87298	2.02237	2.18287	2.35526	2.54035	2.95216	3.42594
17	1.08849	1.18430	1.28802	1.40024	1.52162	1.65285	1.79468	1.94790	2.11338	2.29202	2.48480	2.69277	3.15881	3.70002
18	1.09393	1.19615	1.30734	1.42825	1.55966	1.70243	1.85749	2.02582	2.20848	2.40662	2.62147	2.85434	3.37993	3.99602
19	1.09940	1.20811	1.32695	1.45681	1.59865	1.75351	1.92250	2.10685	2.30786	2.52695	2.76565	3.02560	3.61653	4.31570
20	1.10489	1.22019	1.34686	1.48595	1.63862	1.80611	1.98979	2.19112	2.41171	2.65330	2.91776	3.20714	3.86968	4.66096
24	1.12716	1.26973	1.42950	1.60844	1.80873	2.03279	2.28332	2.56330	2.87601	3.22510	3.61459	4.04893	5.07237	6.34118
30	1.16140	1.34785	1.56308	1.81136	2.09757	2.42726	2.80679	3.24340	3.74532	4.32194	4.98395	5.74349	7.61226	10.06266
36	1.19668	1.43077	1.70914	2.03989	2.43254	2.89828	3.45027	4.10393	4.87738	5.79182	6.87209	8.14725	11.42394	15.96817
40	1.22079	1.48886	1.81402	2.20804	2.68506	3.26204	3.95926	4.80102	5.81636	7.03999	8.51331	10.28572	14.97446	21.72452
48	1.27049	1.61223	2.04348	2.58707	3.27149	4.13225	5.21359	6.57053	8.27146	10.40127	13.06526	16.39387	25.72891	40.21057
50	1.28323	1.64463	2.10524	2.69159	3.43711	4.38391	5.58493	7.10668	9.03264	11.46740	14.54196	18.42015	29.45703	46.90161
60	1.34885	1.81670	2.44322	3.28103	4.39979	5.89160	6.87809	10.51963	14.02741	18.67919	24.83977	32.98773	57.94640	
70	1.41783	2.00676	2.83546	3.99956	5.63210	7.91782	11.11283	15.57162	21.78414	30.42643	42.42992	59.07607		
80	1.49034	2.21672	3.29066	4.87544	7.20957	10.64089	15.67574	23.04980	33.83010	49.56144	72.47643			
90	1.56655	2.44863	3.81895	5.94313	9.22886	14.30047	22.11218	34.11933	52.53711	80.73037				
100	1.64667	2.70481	4.43205	7.24465	11.81372	19.21863	31.19141	50.50495	81.58852					

TABLE VII VALUES OF $(1+i)^{-n}$

n	0.005	0.01	0.015	0.02	0.025	0.03	0.035	0.04	0.045	0.05	0.055	0.06	0.07	0.08
1	.99502	.99010	.98522	.98039	.97561	.97087	.96618	.96154	.95694	.95238	.94787	.94340	.93458	.92593
2	.99007	.98030	.97066	.96117	.95181	.94260	.93351	.92456	.91573	.90703	.89845	.89000	.87344	.85734
3	.98515	.97059	.95632	.94232	.92860	.91514	.90194	.88900	.87630	.86384	.85161	.83962	.81630	.79383
4	.98025	.96098	.94218	.92385	.90595	.88849	.87144	.85480	.83856	.82270	.80722	.79209	.76290	.73503
5	.97537	.95147	.92826	.90573	.88385	.86261	.84197	.82193	.80245	.78353	.76513	.74726	.71299	.68058
6	.97052	.94205	.91454	.88797	.86230	.83749	.81350	.79031	.76790	.74622	.72525	.70496	.66634	.63017
7	.96569	.93272	.90103	.87056	.84127	.81309	.78599	.75992	.73483	.71068	.68744	.66506	.62275	.58349
8	.96089	.92348	.88771	.85349	.82075	.78941	.75941	.73069	.70319	.67684	.65160	.62741	.58201	.54027
9	.95610	.91434	.87459	.83676	.80073	.76642	.73373	.70259	.67290	.64461	.61763	.59190	.54393	.50025
10	.95135	.90529	.86167	.82035	.78120	.74409	.70892	.67556	.64393	.61391	.58543	.55839	.50835	.46319
11	.94661	.89632	.84893	.80426	.76214	.72242	.68495	.64958	.61620	.58468	.55491	.52679	.47509	.42888
12	.94191	.88745	.83639	.78849	.74356	.70138	.66178	.62460	.58966	.55684	.52598	.49697	.44401	.39711
13	.93722	.87866	.82403	.77303	.72542	.68095	.63940	.60057	.56427	.53032	.49856	.46884	.41496	.36770
14	.93256	.86996	.81185	.75788	.70773	.66112	.61778	.57748	.53997	.50507	.47257	.44230	.38782	.34046
15	.92792	.86135	.79985	.74301	.69047	.64186	.59689	.55526	.51672	.48102	.44793	.41727	.36245	.31524
16	.92330	.85282	.78803	.72845	.67362	.62317	.57671	.53391	.49447	.45811	.42458	.39365	.33873	.29189
17	.91871	.84438	.77639	.71416	.65720	.60502	.55720	.51337	.47318	.43630	.40245	.37136	.31657	.27027
18	.91414	.83602	.76491	.70016	.64117	.58739	.53836	.49363	.45280	.41552	.38147	.35034	.29586	.25025
19	.90959	.82774	.75361	.68643	.62553	.57029	.52016	.47464	.43330	.39573	.36158	.33051	.27651	.23171
20	.90506	.81954	.74247	.67297	.61027	.55368	.50257	.45639	.41464	.37689	.34273	.31180	.25842	.21455
24	.88719	.78757	.69954	.62172	.55288	.49193	.43796	.39012	.34770	.31007	.27666	.24698	.19715	.15770
30	.86103	.74192	.63976	.55207	.47674	.41199	.35628	.30832	.26700	.23138	.20064	.17411	.13137	.09938
36	.83564	.69892	.58509	.49022	.41109	.34503	.28983	.24367	.20503	.17266	.14552	.12274	.08754	.06262
40	.81914	.67165	.55126	.45289	.37243	.30656	.25257	.20829	.17193	.14205	.11746	.09722	.06678	.04603
48	.78710	.62026	.48936	.38654	.30567	.24200	.19181	.15219	.12090	.09614	.07654	.06100	.03887	.02487
50	.77929	.60804	.47500	.37153	.29094	.22811	.17905	.14071	.11071	.08720	.06877	.05429	.03395	.02132
60	.74137	.55045	.40930	.30478	.22728	.16973	.12693	.09506	.07129	.05354	.04026	.03031	.01726	
70	.70530	.49831	.35268	.25003	.17755	.12630	.08999	.06422	.04590	.03287	.02357	.01693		
80	.67099	.45112	.30389	.20511	.13870	.09398	.06379	.04338	.02956	.02018	.01380			
90	.63834	.40839	.26185	.16826	.10836	.06993	.04522	.02931	.01903	.01239				
100	.60729	.36971	.22563	.13803	.08465	.05203	.03206	.01980	.01226					

APPENDIX C

ANSWERS TO ODD-NUMBERED EXERCISES AND TO REVIEW EXERCISES

CHAPTER 1 SECTION 1.1

1. | Vertex | Degree |
    |--------|--------|
    | A | 1 |
    | B | 3 |
    | C | 3 |
    | D | 1 |

3. | Vertex | Degree |
    |--------|--------|
    | A | 3 |
    | B | 2 |
    | C | 3 |
    | D | 2 |
    | E | 4 |
    | F | 4 |

5. | Vertex | Degree |
    |--------|--------|
    | A | 4 |
    | B | 2 |
    | C | 4 |
    | D | 4 |
    | E | 3 |
    | F | 3 |

7. | Vertex | Degree |
    |--------|--------|
    | A | 2 |
    | B | 2 |
    | C | 2 |
    | D | 2 |
    | E | 4 |

9. Even: None
    Odd: A, B, C, D
    Sum: 8
    Number of edges: 4

11. Even: B, D, E, F
    Odd: A, C
    Sum: 18
    Number of edges: 9

13. Even: A, B, C, D
    Odd: E, F
    Sum: 20
    Number of edges: 10

15. Even: A, B, C, D, E
    Odd: None
    Sum: 12
    Number of edges: 6

17. One of many possible graphs is:

19. One of many possible graphs is:

21. One of many possible graphs is:

23. This is not possible, since the sum of the degrees of all of the vertices must be an even number.
25. Cannot be drawn in the manner indicated.

27. Can be drawn in the manner indicated; follow the vertices in the order $A, B, C, D, A, E, F, E, F, C$ for one possible method.
29. Can be drawn in the manner indicated; follow the vertices in the order $E, A, B, C, A, D, C, F, E, D, F$ for one possible method.
31. Can be drawn in the manner indicated; follow the vertices in the order E, A, B, E, C, D, E for one possible method.
33.   35. A graph is a set of points and of lines connecting the points.

37. A vertex is odd if its degree is an odd number.
39. A graph is connected if, for any two vertices A and B in the graph, there is a path that begins at A and ends at B.

SECTION 1.2

1. By Theorem 2, there is an Euler path. One of many possible paths is $ABCDABDCA$.
3. By Theorem 3, there is an Euler path. One possible path is $ABABCBC$.
5. By Theorem 1, no Euler path is possible.
7. By Theorem 2, there is an Euler path. One possible path is $GBFDGAECG$.
9. Theorem 3 guarantees that such a route exists. One possible route is the route corresponding to the Euler path $ABGKJIHDACEBCDEIJFEFG$.
11. Since the desired route corresponds to an Euler path, Theorem 3 guarantees such a route exists. One possibility is the route corresponding to the Euler path $HBADEFCBEJIGHJ$.
13. (a) [graph]

    (b) No; the desired route would correspond to a path that visits each vertex (or room) except A, exactly once rather than traverse each edge (pass through each room) exactly once.
    (c) No such route exists; G has to be visited at least twice.
15. An Euler path for a graph is a path that contains each edge of the graph exactly once.
17. A graph has an Euler path if it has just two odd vertices or if all of its vertices are even.
19. A graph has an Euler path that begins and ends at the same vertex if all its vertices are even.

SECTION 1.3

1. The physical situation was the street inspector planning to inspect the streets in the neighborhood. The physical question was whether there was a route that traveled each street exactly once. The mathematical model was the given graph, and the mathematical question was whether the graph had an Euler path. The mathematical solution was that there was such a path on the basis of Theorem 3. The interpretation was that the desired route was possible; furthermore, the desired route corresponded to the given Euler path.
3. The physical situation was the floor plan of the exhibition hall, and the question was whether visitors could be routed through the exhibition in the manner indicated. The mathematical model was the given graph, and the mathematical question was whether there was an Euler path beginning at vertex H and ending at vertex J. The mathematical solution was that there was such a path on the basis of Theorem 3. The interpretation was that the desired routing was possible, and one such route corresponded to the given Euler path.
5. The physical situation was the gross national product, and the question was, What effect does government spending have on the gross national product? The mathematical model consisted of the two equations $Y = C + I + G + X_n$ and $C = a + 0.9Y$; the mathematical question was, How does Y change as G changes? The mathematical solution, based on the equation $Y = 10a + 10I + 10G + 10X_n$, was that the change in Y is ten times the change in G. The interpretation was that a \$1 change in government spending resulted in a corresponding \$10 change in the gross national product.
7. The physical situation was the age relationship between Betty and Paul, and the question was, How old is Betty? The mathematical model consisted of the two equations $Y = 2X$ and $X + 3 = 21$; the mathematical question was, What is the value of Y? The mathematical solution was found to be $Y = 36$, and the interpretation was that Betty is 36 years old.
9. A mathematical model is a mathematical description of a physical situation.

APPENDIX C ANSWERS TO ODD-NUMBERED AND REVIEW EXERCISES

CHAPTER 2 SECTION 2.1

1. (a) $y = 2.5x + 575$ (b) $5,575 3. (a) $y = 17x$ (b) $51,000 5. (a) $y = 0.1x$ (b) 50 books
7. (a) $y = 0.11x$ (b) 1,100 defective bulbs 9. (a) $y = 0.08x$ (b) 3,125 bills 11. (a) $y = x/0.8$ (b) $50
13. $y = 7x - 2,300$ 15. (a) $w = 90 + x$ (b) $z = 100 - x$ (c) $y = 9,000 + 10x - x^2$ (d) $8,800
17. (a) $r = 150 + 25x$ (b) $z = 20 - x$ (c) $y = 3,000 + 350x - 25x^2$ (d) $250
19. (a) 200 (b) 300 (c) 5 (d) 60 minutes
21. Equations are appropriate as mathematical models for stating equality between two or more physical quantities.

SECTION 2.2

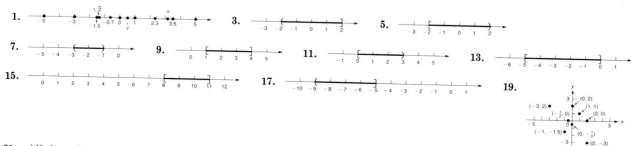

21. All the points on the line drawn through the point (5, 0) and perpendicular to the x-axis
23. All the points on the line drawn through the point (0, −3) and perpendicular to the y-axis
25. All the points on the y-axis
27. The points whose x-coordinates are equal to their y-coordinates are on the line l, which makes a 45° angle with the x-axis.

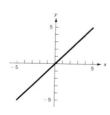

29. A real number line is a geometric representation of the real numbers whereby the points on a line are put in a one-to-one correspondence with the real numbers.
31. The unit distance on a number line is the distance between the two points 0 and 1.
33. Coordinate axes are two real number lines at right angles to each other that intersect at their origins.
35. A plane with coordinates assigned to each point by using perpendicular distances from a set of axes is a Cartesian plane.

SECTION 2.3

1. Yes 3. No 5. No 7. Yes 9. Yes 11. No 13. No 15. No 17. No 19. Yes 21. No
23. No
25. $y = x^2$ 27. $y = -x^2 + 1$ 29. $x - y = 3$ 31. $3x + y = 0$ 33. $y = x^3$

x	y
−2	4
−1	1
0	0
1	1
2	4

x	y
−2	−3
−1	0
0	1
1	0
2	−3

x	y
−2	−5
−1	−4
0	−3
1	−2
2	−1

x	y
−2	6
−1	3
0	0
1	−3
2	−6

x	y
−2	−8
−1	−1
0	0
1	1
2	8

660 APPENDIX C ANSWERS TO ODD-NUMBERED AND REVIEW EXERCISES

35. $y = x^2$ **37.** $y = -x^2 + 1$ **39.** $x - y = 3$ **41.** $3x + y = 0$ **43.** $y = x^3$

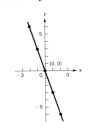
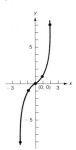

45. (a) 4 (b) 4 errors per hour (c) 1 (d) one
47. The graph of an equation is the set of all points (x, y) whose coordinates satisfy the equation.

SECTION 2.4

1. $m = \frac{2}{3}$ **3.** $m = -4$ **5.** No slope **7.** $m = 0$ **9.** $m = 1$

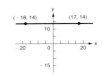
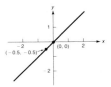

11. **13.** **15.** **17.**

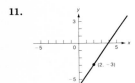
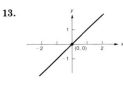

19. **23.** $y = 2x - 1$ **25.** $y = -4x + 2$ **27.** $x = -3$ **29.** $y = 14$ **31.** $y = -\frac{2}{3}x + 1$

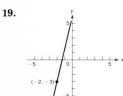

33. $y = 3x - 2$ **35.** $y = -4x + 1$ **37.** Point-slope form: $y - y_1 = m(x - x_1)$; slope-intercept form: $y = mx + b$
39. The slope of a line is the ratio $\dfrac{y_2 - y_1}{x_2 - x_1}$, where (x_1, y_1) and (x_2, y_2) are two points on the line.

SECTION 2.5

1. $x - 3y = -2$ **3.** $y = 2x + 3$ **5.** $x + y = 0$ **7.** $x = 1.5$ **9.** $0.5x - 0.2y = 0$

Note: Points emphasized in these, and many of the following graphs, are suggested check points to test the accuracy of your solution.

11. $m = \frac{1}{3}$ **13.** $m = 2$ **15.** $m = -1$ **17.** No slope **19.** $m = \frac{5}{2}$
21. (a) 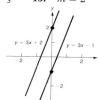 (b) Parallel lines have equal slopes.

23. A linear equation in x and y is any equation that can be written in the form $Ax + By + C = 0$, where A, B, and C are real numbers with A and B not both zero.

SECTION 2.6

1. (a) and (b) 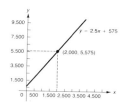   **3.** (a) and (b)   **5.** 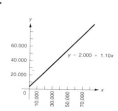   **7.** (a) and (b)

(c) x and y integers, $x \geq 0$, $y \geq 575$

9. (a) $y = 1{,}250x$ (b) $y = 15{,}000 - 1{,}250x$ (c) 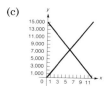   (d) $0 \leq x \leq 12$; $0 \leq y \leq 15{,}000$

11. (a) $d = 125t$ (b) $v = 1{,}200 - 125t$ (c)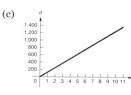
(d) Points for which $0 \leq t \leq 8$

13. (a) $I = PRT$ (b) $A = P(1 + RT)$ (c) $I = 55T$; $A = 500 + 55T$ (d)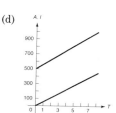

15. (a) $q = -12p + 185$ (b) 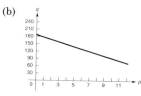   (c) $p = \$15.42$ (d) $p = -\frac{1}{12}q + \frac{185}{12}$ (e)

17. $°F = \frac{9}{5}°C + 32$

SECTION 2.7

1. $6y = 3x^2 - 6x + 9$ 3. $y = -\dfrac{x^2}{8} + x - 3$ 5. $y = 6x + x^2$ 7. $y = -x^2 - 5x - 4.75$

 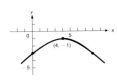 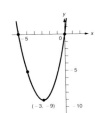 

9. $4y = -x^2 + 2x - 9$ 11. $y = x^2$ 13. $y = x^2 + 3$ 15. $y + 2 = x^2$

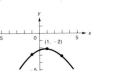 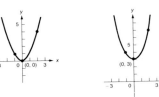 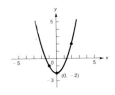

17. $(-1, 2)$, opens upward   19. $(5, -4)$, opens downward   21. $(0, 0)$, opens upward   23. $(4, 0)$, opens upward
25. The point (h, k) at which the parabola reaches its highest (or lowest) point is called the vertex.

SECTION 2.8

1. (a) $-(y - 36) = (x - 6)^2$

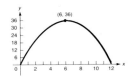

   (b) Day 6, 36 errors
   (c) Number of errors declines as x approaches 12. At $x = 12$, the number of errors is zero. For x greater than 12, the graph has no physical meaning since the number of errors must be equal to or greater than zero.
   (d) When the employee begins work, he or she makes an increasing number of errors. After a number of days pass, the employee begins to learn his or her job and the number of errors begins to decrease.

3. 10 hours   5. (a) $20,480   (b) 36 empty seats   (c) $320   7. 15 feet by 15 feet
9. (1) Physical situation: blood flow in a section of capillary; physical question: At what points will the velocity of the blood flow be maximum?
   (2) Mathematical model: $V = 1.2 - 19,000r^2$; mathematical question: What are the coordinates of the vertex of the parabola in the model?
   (3) Mathematical solution: $(0, 1.2)$
   (4) The maximum velocity of 1.2 centimeters per second occurs when $r = 0$ (at the center).
11. (1) Physical situation: renting of apartments; physical question: How many apartments can be rented and for what price in order to maximize income?
   (2) Mathematical model: $y = 3,000 + 350x - 25x^2$; mathematical question: What are the coordinates of the vertex of the parabola in the model?
   (3) Mathematical solution: $(7, 4,225)$
   (4) By increasing the rent seven times ($7 \times \$25$) to $325, seven apartments will be vacant and a maximum income of $4,225 will be received.

APPENDIX C ANSWERS TO ODD-NUMBERED AND REVIEW EXERCISES   663

REVIEW EXERCISES

1. Yes **2.** No **3.** No **4.** Yes **5.** No **6.** No **7.** Yes
8. $2x - 3y = 7$ **9.** $y = 4x^2 - 24x + 35$ **10.** $x = -2$ **11.** $2y = 9$

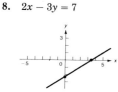 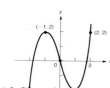 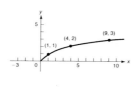 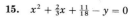

12. $y = x^3 - 3x$ **13.** $y = \sqrt{x}$ **14.** $0.2y - 0.3x = 1$ **15.** $x^2 + \frac{2}{3}x + \frac{11}{18} - y = 0$

16. $y - 2x^2 + 8x - 4 = 0$ **17.** $y = \dfrac{1}{x-1}$ **18.** Linear, $m = \frac{2}{3}$ **19.** Quadratic **20.** Linear, no slope

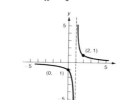

21. Linear, $m = 0$ **22.** Neither **23.** Neither **24.** Linear, $m = \frac{3}{2}$ **25.** Quadratic **26.** Quadratic
27. Neither **28.** **29.** **30.** **31.**

  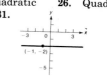

32. $y = \frac{4}{3}x - 3$ **33.** $y = -4x - 8$ **34.** $x = 1$ **35.** $y = 2$ **36.** $y = \frac{3}{5}x + \frac{18}{5}$ **37.** $y = -\frac{8}{3}x + \frac{1}{3}$ **38.** $y = 0$
39. $x = 1$ **40.** $y = -2x - \frac{2}{3}$
41. (a) $y = 9.25x + 450$
   (b)

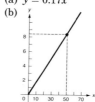

   (c) $x \geq 0, y \geq 450$
   (d) $2,300
   (e) $3,687.50

42. (a) $y = 0.17x$
   (b)

   (c) $x \geq 0, y \geq 0$
   (d) 8.5, or approximately 9
   (e) 12.75, or approximately 13

43. Total depreciation: $y = 2{,}500t$; value: $y = 35{,}000 - 2{,}500t$

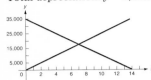

44. (a) $y = -20x + 600$
(b)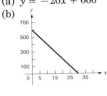
(c) $30

45. (a) $y = 4x - 0.01x^2$

(b) 200 mowers
(c) $40,000

46. Price: $250 each; maximum revenue: $12,500

47. (1) Physical situation: buying rats for experimental purposes; physical question: In a shipment of 50(75) rats how many are unsatisfactory for experimental use?
(2) Mathematical model: $y = 0.17x$; mathematical question: When $x = 50$, what is y? When $x = 75$, what is y?
(3) Mathematical solution: When $x = 50$, y is approximately 9. When $x = 75$, y is approximately 13.
(4) In a shipment of 50 rats, approximately 9 are unsatisfactory for experimental use. In a shipment of 75, approximately 13 are unsatisfactory.

48. (1) Physical situation: selling sofas; physical question: At what price should the sofas be sold to maximize revenue? What is the maximum revenue?
(2) Mathematical model: $y = (450 - 10x)(10 + 2x)$; mathematical question: What are the coordinates of the vertex of the parabola?
(3) Mathematical solution: (20, 12,500)
(4) The sofas should be sold for $250 each to receive a maximum revenue of $12,500.

CHAPTER 3   SECTION 3.1

1. $x = 2, y = -1$ **3.** $x = -1, y = 1$ **5.** No solution   **7.** Infinitely many solutions, which satisfy $y - x = 4$
9. $x = 0.1, y = -0.2$
11. (a) $x + y = 50$ (b) $60x + 70y = 3{,}200$ (c) $x = 30, y = 20$; he should buy 30 3-speed and 20 10-speed bicycles
13. 35 grams of mix 1 and 30 grams of mix 2   **15.** Option 1: When $x > 325$; option 2: $x < 325$
17. (a)

(b) $x < 150, C > R$
$x > 150, R > C$
(c) (150, 30,000)
(d) 150
(e) $30,000

19. (a) $y = 15x + 18{,}000$ (b) $y = 30x$ (c) (1,200, 36,000)   (d) 1,200 units, $36,000
21. (a)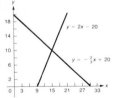
(b) $x = 15, y = 10$
(c) $15
(d) 1,000

23. (a) $x = 10, y = 20$ (b) $10   (c) 20
25. (1) Physical situation: Ordering bicycles; physical question: How many of the 50 bicycles should be three-speed and how many should be ten-speed?
(2) Mathematical model: The given system of equations

$x + y = 50$
$60x + 70y = 3{,}200$

and their graphs; mathematical question: What is the solution of the system? or What are the coordinates of the point of intersection of the graphs?
(3) Mathematical solution: $x = 30, y = 20$
(4) Eric should buy 30 three-speed bicycles and 20 ten-speed bicycles.

27. (1) Physical situation: police force of a city; physical question: How many male and female police officers are on the force?
(2) Mathematical model: the system
$$x + y = 104$$
$$9x - y = -4$$
mathematical question: What is the solution to this system of equations?
(3) Mathematical solution: $x = 10$, $y = 94$
(4) There are 10 female and 94 male police officers on the force.

29. A solution for a system of equations is a set of values for the variables that satisfy each of the equations.
31. The equilibrium price is the price at which supply and demand are equal.
33. The break-even point is the point at which revenue equals cost.
35. (1) Both sides of any equation can be multiplied by a constant other than zero.
(2) Both sides of any equation can be multiplied by a constant and the result added to one of the other equations.
(3) The order in which the equations are written can be changed.

SECTION 3.2

1. Echelon form   **3.** Conditions 3 and 4   **5.** Condition 3   **7.** Echelon form   **9.** Condition 2

11. $\begin{bmatrix} 1 & 0 & -4 \\ 0 & 1 & 1 \end{bmatrix}$ **13.** $\begin{bmatrix} 1 & -\frac{4}{3} \\ 0 & 0 \end{bmatrix}$ **15.** $\begin{bmatrix} 1 & 0 & 0 \\ 0 & 1 & 0 \\ 0 & 0 & 1 \end{bmatrix}$ **17.** $\begin{bmatrix} 1 & 0 & 0 \\ 0 & 1 & 0 \\ 0 & 0 & 1 \\ 0 & 0 & 0 \end{bmatrix}$ **19.** $\begin{bmatrix} 1 & 0 & 0 & \frac{4}{5} \\ 0 & 1 & 0 & -\frac{1}{5} \\ 0 & 0 & 1 & \frac{6}{5} \end{bmatrix}$

21. A linear equation in the variables $x_1, x_2, \ldots, x_n$ is an equation that can be written in the form $a_1 x_1 + a_2 x_2 + \cdots + a_n x_n = b$, where $a_1, a_2, \ldots, a_n$ and b are constants.
23. A solution for a system of equations is a set of values for the variables that satisfy each of the equations.
25. The augmented coefficient matrix is the matrix whose rows are the coefficients and constants of a corresponding system of equations.

SECTION 3.3

1. $w = -\frac{1}{2}, x = 2, y = 3, z = -1$ **3.** Infinitely many solutions, $z = 2, x = 8 + 2y, w = -1 - 2y$
5. $x_1 = 0.5, x_2 = -\frac{1}{3}, x_3 = 0$ **7.** $x_1 = \frac{1}{8}, x_2 = 0, x_3 = -\frac{1}{3}$ **9.** $x = \frac{4}{3}, y = \frac{5}{3}, z = 1$ **11.** $x = 1, y = \frac{1}{2}$
13. $x = y = z = 0$ **15.** $x = 3 + 5y$ **17.** No solution   **19.** $x_1 = \frac{1}{3}, x_2 = 0, x_3 = -1, x_4 = 2$
21. 30 3-speed and 20 10-speed bicycles   **23.** 10 female and 94 male police officers   **25.** 1,200 units; $36,000
27. 20 items of A, 24 of B, and 16 of C   **29.** $y = 2x^2 - x - 2$ **31.** $y = 1 + 3z, x = -2 - 7z$ **33.** No solution
35. $y = 5 + 3z, x = 3 + z$
37. When the number of equations is less than the number of variables, the solution, if it exists, will not be unique.
39. (1) Physical situation: Manufacturing three products, A, B, and C; physical question: How many of each of the products A, B, and C can be produced if all available worker-hours are to be used?
(2) Mathematical model: Given system of equations
$$2A + B + C = 80$$
$$A + B + 0.5C = 52$$
$$0.5A + 0.5B + 0.5C = 30$$
Mathematical question: What is the solution of the system?
(3) Mathematical solution: $A = 20, B = 24, C = 16$
(4) If all available worker-hours are to be used, then each day 20 units of A, 24 units of B, and 16 units of C must be produced.

SECTION 3.4

1.
	Poor	Satisfactory	Very Good
Northeast	46	30	24
South	21	27	52
Midwest	23	56	21
West	39	41	20

3. $\begin{bmatrix} 2 & 0 \\ 1 & 7 \end{bmatrix}$ **5.** $\begin{bmatrix} -2 & -2 \\ 3 & 1 \end{bmatrix}$ **7.** $\begin{bmatrix} 6 & 2 \\ -1 & 13 \end{bmatrix}$ **9.** $\begin{bmatrix} 4 & 3 \\ -4 & 2 \end{bmatrix}$

11. $\begin{bmatrix} 0 & 3 \\ 7 & 1 \\ -4 & 3 \end{bmatrix}$ 13. $\begin{bmatrix} 2 & 1 \\ 3 & 5 \\ -4 & -3 \end{bmatrix}$ 15. $\begin{bmatrix} -1 & 4 \\ 9 & -1 \\ -4 & 6 \end{bmatrix}$ 17. $\begin{bmatrix} -2 & -10 \\ -24 & -8 \\ 16 & -6 \end{bmatrix}$ 19. $\begin{bmatrix} -1 & 10 \\ 23 & 1 \\ -12 & 12 \end{bmatrix}$ 21. $\begin{bmatrix} 2 & 7 \\ 0 & -2.5 \end{bmatrix}$

23. $\begin{bmatrix} -2 & -5 \\ -2 & 3.5 \end{bmatrix}$ 25. $\begin{bmatrix} 6 & 5 \\ 9 & -2.5 \end{bmatrix}$ 27. $\begin{bmatrix} \frac{7}{3} & 4 \\ 3 & -\frac{17}{6} \end{bmatrix}$ 29. $\begin{bmatrix} -12 & -24 \\ 12 & 24 \\ 0 & -12 \end{bmatrix}$ 31. $\begin{bmatrix} -4 & -8 \\ -20 & -28 \\ 6 & 2 \end{bmatrix}$ 33. $\begin{bmatrix} 3 & 0 & 0 \\ 0 & 3 & 0 \\ 0 & 0 & 3 \end{bmatrix}$

35. $x = -3, y = 5, z = 3$ 39. Only S2 holds.
41. The numbers that form the entries of the matrices are called scalars.
43. When adding three matrices, the first may be added to the sum of the second and third, or the sum of the first two may be added to the third, and the result will be the same.

SECTION 3.5

1. $\begin{bmatrix} 3 & 6 \\ -3 & 6 \end{bmatrix}$ 3. $\begin{bmatrix} 42 \\ -3 \end{bmatrix}$ 5. $\begin{bmatrix} 7 & 24 \\ -1 & 17 \\ -8 & 14 \end{bmatrix}$ 7. $\begin{bmatrix} 11 \\ 53 \\ 35 \end{bmatrix}$ 9. $\begin{bmatrix} -5 & 4 & 9 \\ -1 & 12 & 5 \end{bmatrix}$ 11. $\begin{bmatrix} -15 & 12 & 27 \\ -3 & 36 & 15 \end{bmatrix}$ 13. $\begin{bmatrix} -3 & 8 & 7 \\ -5 & 18 & 13 \end{bmatrix}$

15. $\begin{bmatrix} 7 & 9 & 0 \\ 7 & 8 & -2 \end{bmatrix}$ 17. $\begin{bmatrix} 7 & 2 \\ 3 & 10 \end{bmatrix}$ 19. $\begin{bmatrix} -7 & 7 \\ 3 & 1 \\ -12 & 10 \end{bmatrix}$ 21. Not defined 23. $\begin{bmatrix} 18 & -24 \\ 15 & -6 \end{bmatrix}$ 25. $\begin{bmatrix} -5 & 6 \\ 3 & 2 \\ -9 & 8 \end{bmatrix}$

27. $\begin{bmatrix} 3 & 1 & 0 \\ 1 & 1 & 1 \\ 1 & -1 & 2 \end{bmatrix}$ 29. $\begin{bmatrix} 1 & 0 & 0 \\ 0 & 1 & 0 \\ 0 & 0 & 1 \end{bmatrix}$ 31. $x = 1, y = -2$

33. (a) $AB = \begin{bmatrix} 1 & 1 \\ 1 & 3 \end{bmatrix}, BA = \begin{bmatrix} 2 & 2 \\ 1 & 2 \end{bmatrix}$ (b) There is no commutative law for matrix multiplication.

37. $\begin{bmatrix} 3 & -2 & 1 & -1 \\ 2 & 1 & -2 & 0 \\ -2 & 0 & 0 & 1 \\ 0 & 1 & 1 & -3 \end{bmatrix} \begin{bmatrix} w \\ x \\ y \\ z \end{bmatrix} = \begin{bmatrix} 4 \\ -1 \\ 0 \\ 2 \end{bmatrix}$ 39. $\begin{aligned} 2x + 6y - z &= 1 \\ 4x + 3z &= 2 \\ -2x + y &= -2 \end{aligned}$

41. (a) $\begin{bmatrix} 18,840 \\ 495 \\ 1,620 \\ 100 \end{bmatrix}$; the amount of materials needed to produce the number of cabinets that the company intends to make next month

(b) $\begin{bmatrix} 8,640 \\ 145 \\ 620 \\ 0 \end{bmatrix}$; the amount of each material that must be ordered for the next month

43. The product of a first matrix times the sum of a second and third is equal to the sum of the products of the first times the second and of the first times the third.
45. A vector is a matrix consisting of a single row or a single column.

SECTION 3.6

1. $\begin{bmatrix} 1 & 2 & -1 \\ 2 & 1 & 0 \\ -1 & 0 & 3 \end{bmatrix} \begin{bmatrix} -\frac{3}{10} & \frac{6}{10} & -\frac{1}{10} \\ \frac{6}{10} & -\frac{2}{10} & \frac{2}{10} \\ -\frac{1}{10} & \frac{2}{10} & \frac{3}{10} \end{bmatrix} = \begin{bmatrix} 1 & 0 & 0 \\ 0 & 1 & 0 \\ 0 & 0 & 1 \end{bmatrix}$ 3. $\begin{bmatrix} \frac{3}{5} & \frac{2}{5} \\ \frac{1}{5} & -\frac{1}{5} \end{bmatrix}$ 5. $\begin{bmatrix} 7 & -6 & 1 \\ -1 & 1 & 0 \\ 4 & -4 & 1 \end{bmatrix}$ 7. $\begin{bmatrix} 1 & 2 & -7 & -16 \\ 0 & 1 & -2 & -7 \\ 0 & 0 & 1 & 2 \\ 0 & 0 & 0 & -1 \end{bmatrix}$

9. No inverse   11. No inverse   13. No inverse   15. $\begin{bmatrix} \frac{1}{14} & -\frac{4}{14} \\ \frac{3}{14} & \frac{2}{14} \end{bmatrix}$ 17. $\begin{bmatrix} \frac{7}{4.1} & \frac{-3}{4.1} \\ \frac{2}{4.1} & \frac{5}{4.1} \end{bmatrix}$

19. A square matrix is a matrix in which the number of rows is the same as the number of columns.
21. An identity matrix is an $n \times n$ matrix with 1's on its main diagonal and 0's everywhere else.

SECTION 3.7

1. R and S should each produce at the indicated level (in millions of dollars); R: 85.4; S: 145.4
3. R, S, and T should each produce at the indicated level (in millions of dollars); R: 94.3; S: 140.5; T: 205.3
5. R and S should each produce at the indicated level (in millions of dollars); R: 78.5; S: 134.5
7. R, S, and T should each produce at the indicated level (in millions of dollars); R: 36.4; S: 58.4; T: 82.01

REVIEW EXERCISES

1. $x = -3, y = 1$ **2.** $x = 4, y = -2$ **3.** No solution **4.** $x = 4 + 2y$ **5.** $x = 0, y = -1, z = 2$
6. $x = 0, y = 0.1, z = 0.2$ **7.** $x = 3, y = -2, z = 0$ **8.** No solution **9.** $x = 4, y = 5$ **10.** No solution
11. $x = 1 + 5z, y = 8z - 1$ **12.** $w = \frac{37}{47}z - \frac{65}{47}, x = \frac{39}{47}z - \frac{52}{47}, y = \frac{215}{47} - \frac{79}{47}z$ **13.** $\begin{bmatrix} 0 & 0 & -1 \\ 4 & 8 & 3 \\ 1 & 2 & 2 \end{bmatrix}$ **14.** Undefined
15. No inverse **16.** $\begin{bmatrix} 1 & 4 & -2 \\ 8 & 14 & 5 \end{bmatrix}$ **17.** $\begin{bmatrix} 0 & \frac{1}{2} \\ -1 & \frac{1}{2} \end{bmatrix}$ **18.** $\begin{bmatrix} 1 & \frac{1}{2} \\ -3 & \frac{1}{2} \\ -2 & -\frac{1}{2} \end{bmatrix}$ **19.** $\begin{bmatrix} -1 & 4 \\ -5 & 12 \end{bmatrix}$ **20.** $\begin{bmatrix} 2 & 0 & 2 \\ 8 & 12 & 4 \end{bmatrix}$
21. $\begin{bmatrix} 1 & -1 & 3 \\ 0 & -1 & 3 \end{bmatrix}$ **22.** No inverse exists **23.** $\begin{bmatrix} 5 & 4 \\ 6 & -1 \end{bmatrix}$ **24.** $\begin{bmatrix} 1 & 2 \\ 3 & -2 \end{bmatrix}$ **25.** $\begin{bmatrix} 0 & 8 \\ 12 & -12 \end{bmatrix}$ **26.** Undefined
27. $\begin{bmatrix} -1 & 1 & -1 \\ 4 & -2 & 3 \\ 2 & -1 & 1 \end{bmatrix}$ **28.** $\begin{bmatrix} \frac{1}{4} & \frac{1}{4} \\ \frac{3}{8} & -\frac{1}{8} \end{bmatrix}$ **29.** $\begin{bmatrix} 55{,}000 \\ 50{,}000 \\ 55{,}000 \end{bmatrix}$; the weekly income for each factory
30. [1,980 2,180 2,050]; income from each orchard   **31.** Control: 1.365 kilograms; experimental: 0.965 kilogram
32. (a) Model I: 150; model II: 100   (b) Model I: 40; model II: 60
33. When the quantity to be purchased is less than 150, use option 2; when the quantity is greater than 150, use option 1.
34. (a) $y = 90x + 15{,}015$ (b) $y = 145x$ (c) 273 bicycles   (d) $39,585   (e) (273, 39,585)
35. (a) 103   (b) $18,540   (c) (103, 18,540)   **36.** Equilibrium price = $2.45; equilibrium quantity is 98
37. Additive 1: 20; additive 2: 24; additive 3: 30   **38.** 33, 21, 12
39. R and S should each produce at the indicated level (in millions of dollars); R: 79.3; S: 115.3
40. Same as 39 except R: 77.9; S: 117.9
41. (1) Physical situation: The production of wheat and corn considering the cost of labor and fertilizer; physical question: From the given information, what does the farmer pay for fertilizer and labor?
   (2) Mathematical model:
   $5x + 13y = 46.30$
   $4x + 15y = 51.30$
   mathematical question: What is the solution of the system of equations?
   (3) Mathematical solution: $x = 1.2, y = 3.1$
   (4) The farmer pays $1.20 for a pound of fertilizer and $3.10 for an hour of labor.

CHAPTER 4   SECTION 4.1

1. $-x + y \leq 2$ **3.** $2x + 4y \geq 6$ **5.** $-x - 4y \leq 8$ **7.** $y \leq -2$

9. $y \leq 0$ **11.** $\frac{1}{2}x - \frac{1}{4}y \leq 1$ **13.**   **15.**

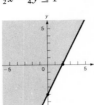
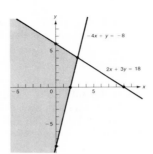
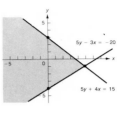

17.   **19.**   **21.**

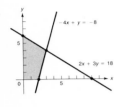
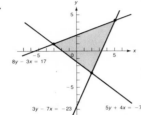
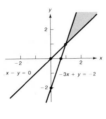

23. Unbounded   **25.** Unbounded   **27.** Bounded   **29.** Bounded   **31.** Unbounded

33. A linear inequality in two variables x and y is an expression that can be written in one of the forms

$$ax + by < c \qquad ax + by > c$$
$$ax + by \leq c \qquad ax + by \geq c$$

where a, b, and c are constants with a and b not both 0.

35. A solution of a linear inequality in two variables x and y is a set of pairs of values for x and y that satisfy the inequality.

37. The graph of a system of linear inequalities is the set of all points whose coordinates satisfy all the inequalities in the system.

39. The graph of a system of linear inequalities is said to be unbounded if it cannot be completely encircled by a circle with its center at the origin.

SECTION 4.2

1. Maximum value of z is 6, occurring at $x = 0$, $y = 6$.   **3.** Maximum value of z is 19, occurring at $x = 5$, $y = 4$.
5. Maximum value of z is 22, occurring at $x = 4$, $y = 1$.   **7.** Minimum value of z is -1, occurring at $x = 0$, $y = -2$.
9. No solution exists since we can make z as large as we wish by picking appropriate points in the region—for example, points on the line $4y - 3x = -12$, $x > 4$.
11. No solution exists since we can make z as small as we wish by selecting appropriate points in the region—for example, points on the x-axis, $x > 3$.
13. (a) $z = x + 0.75y$ (b) $\frac{3}{2}x + y \leq 336$ (c) $\frac{1}{4}x + \frac{2}{3}y \leq 116$ (d) $x \geq 0$, $y \geq 0$ (e) 144 units of A and 120 units of B
15. (a) $z = x + y$ (b) $x + 3y \leq 36$ (c) $3x + 5y \leq 84$ (d) $x + y \leq 20$ (e) $x \geq 0$, $y \geq 0$
   (f) 20 quarts, with the number of quarts of type I between 12 and 20 inclusive
17. 20 one-person rooms; 35 two-person rooms; maximum number of people is 90
19. Produce 1,833 units of A.
21. A total of 27 children with the number in the mathematics group between 12 and 15 inclusive.
23. (1) Physical situation: manufacture and sale of two items, A and B; physical question: How many units of the items, A and B, should the company produce each day in order to maximize its profit?
   (2) Mathematical model:
   Maximize
   $$z = x + 0.75y$$
   subject to
   $$\frac{3}{2}x + y \leq 336$$
   $$\frac{1}{4}x + \frac{2}{3}y \leq 116$$
   $$x \geq 0, \quad y \geq 0$$
   mathematical question: What is the solution to the given linear programming problem?

APPENDIX C   ANSWERS TO ODD-NUMBERED AND REVIEW EXERCISES   669

  (3) Mathematical solution: $z = 234$ is maximum for $x = 144$ and $y = 120$.
  (4) Interpretation: The company must produce and sell 144 units of item A and 120 units of item B each day to maximize its profit at $234.

SECTION 4.3

1. Basic variables: x, t; nonbasic variables: y, s; basic solution: $x = 7, t = -6, y = s = 0$
3. Basic variables: u, s; nonbasic variables: x, y, t; basic solution: $u = 3, s = 12, x = y = t = 0$
5. Basic variables: x, y, t; nonbasic variables: w, s; basic solution: $x = 4, y = -3, t = 1, w = s = 0$
7. Basic variables: x, y, w; nonbasic variables: s, t, u; basic solution: $x = -\frac{1}{3}, y = \frac{4}{9}, w = \frac{3}{5}, s = t = u = 0$
9. $x = \frac{26}{7}, y = -\frac{8}{7}, s = 0, t = 0$ 11. $x = 46, t = -18, y = 0, s = 0$ 13. $x = 3, y = \frac{11}{3}, t = -\frac{1}{3}, w = 0, s = 0$
15. $s = 4, t = 2, u = -1, x = y = w = 0$
17. Four of the following: $x = 6, y = -2, u = v = 0$; $u = -1, x = 5, y = v = 0$; $v = -1, x = 7, u = y = 0$; $y = 10, u = -6, x = v = 0$; $y = -14, v = 6, x = u = 0$; $u = -\frac{7}{2}, v = \frac{5}{2}, x = y = 0$

SECTION 4.4

1.

	x	y	s	t	
s	1	-2	1	0	10
t	2	4	0	1	18
z	-2	-3	0	0	0

3.

	x	y	w	s	t	
s	2	1	-1	1	0	10
t	1	3	1	0	1	14
z	-1	-2	-2	0	0	0

5.

	x	y	s	t	
x	1	0	$-\frac{1}{2}$	$\frac{9}{2}$	$\frac{5}{2}$
y	0	1	$\frac{1}{4}$	$-\frac{1}{4}$	$\frac{15}{4}$
z	0	0	$\frac{3}{4}$	$\frac{13}{4}$	$\frac{53}{4}$

7.

	x	y	w	s	t	
y	2	1	$\frac{1}{4}$	$\frac{1}{4}$	0	$\frac{5}{2}$
t	-11	0	$\frac{7}{4}$	$-\frac{5}{4}$	1	$\frac{15}{2}$
z	6	0	0	1	0	10

9. Maximum value of 51 at $x = 7, y = 5, s = t = 0$ 11. Maximum value of 287 at $w = 143, x = 67, y = s = t = 0$
13. $x = 5, y = 6, z = 17$ 15. $y = 9, w = 10, z = 37, x = 0$ 17. $w = 4, y = 6, z = 8, x = 0$
19. 224 units of item A, for a maximum profit of $224
21. (a) $z = 12x + 10y + 6w$ (b) $4x + 3y + 2w \leq 320$ (c) $x + y + w \leq 100$ (d) $x \geq 0, y \geq 0, w \geq 0$
    (e) 20 full crews and 80 partial crews, for a maximum of 1,040 tons
23. 60 items of product I, 2 of product II, and 16 of product III (round off)   25. This is a minimization problem.
27. One of the variables is restricted to being greater than or equal to 9.
29. (1) Physical situation: Using garbage crews of various sizes; physical question: How many crews of each type should be formed to maximize the amount of garbage to be hauled?
    (2) Mathematical model:
       Maximize
       $$z = 12x + 10y + 6w$$
       subject to
       $$4x + 3y + 2w \leq 320$$
       $$x + y + w \leq 100$$
       $$x \geq 0, \quad y \geq 0, \quad w \geq 0$$
       mathematical question: What is the solution to the given linear programming problem?
    (3) Mathematical solution: $z = 1{,}040$ is maximum for $x = 20, y = 80$, and $w = 0$.
    (4) Interpretation: The city should use 20 full crews and 80 partial crews to haul a maximum of 1,040 tons of garbage.

SECTION 4.5

1. $\begin{bmatrix} 1 & -4 \\ 2 & 4 \\ 3 & 9 \end{bmatrix}$ 3. $\begin{bmatrix} 4 & 10 & 19 \\ -7 & 12 & 0 \end{bmatrix}$

5. Minimize
$$w = 3y_1 + 10y_2 + 12y_3$$
subject to
$$y_1 + 2y_2 + 4y_3 \geq 2$$
$$-y_1 + 3y_2 + 6y_3 \geq 3$$
$$y_1 \geq 0, \quad y_2 \geq 0, \quad y_3 \geq 0$$

7. Maximize
$$z = 12x_1 + 18x_2$$
subject to
$$2x_1 + x_2 \leq 2$$
$$-x_1 + 3x_2 \leq 5$$
$$x_1 \geq 0, \quad x_2 \geq 0$$

9. $w = 34, y_1 = \frac{32}{5}, y_2 = \frac{6}{5}$ 11. $w = 37, y_1 = \frac{7}{3}, y_2 = \frac{5}{3}$ 13. $w = \frac{43}{2}, y_1 = \frac{5}{2}, y_2 = \frac{1}{4}, y_3 = 0$

15. (a) $w = 6y_1 + 8y_2$ (b) $y_1 + \frac{3}{2}y_2 \geq 8$ (c) $2y_1 + 2y_2 \geq 12$ (d) $y_1 \geq 0, y_2 \geq 0$
    (e) 44¢: 2 grams of supplement I and 4 of supplement II

17. Food I: 5 units; food II: 12 units; minimum calorie content: 920

REVIEW EXERCISES

1. $2x - y \leq 1$ 2. $-3x - 2y \geq 6$ 3. $-4x + 2y \geq 3$ 4. $7x + 9y \leq 12$

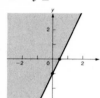 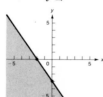 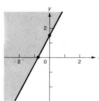 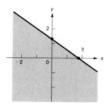

5. $5x - 10y \leq 10$ 6. $2x \geq 5$ 7. $y \leq -4$ 8. $y \geq 0$ 9. $x \leq 0$

10. $0.2x - 0.3y \leq 0$ 11. $7 \geq x - y$ 12. $4x + 3 \leq 2y$ 13.

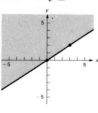 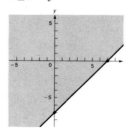 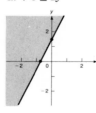 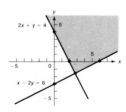

14.   15.   16.   17.

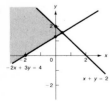 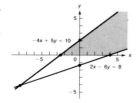 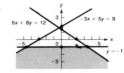 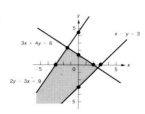

18. 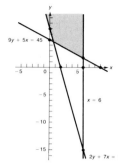 **19.** 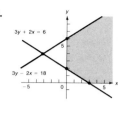 **20.**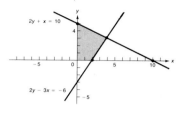

21. Maximum value of 19 at $x = -3, y = 5$ 22. Maximum value of 18 at $x = 6, y = 2$
23. Minimum value of -20 at $x = -6, y = -4$ 24. Maximum of 32 occurs anywhere on the line $4y - 5x = 16$ for $-4 \leq x \leq 4$
25. Minimum value of -15 occurs when $x = 4, y = 9$ 26. Minimum value of -27 occurs at $x = -3, y = 3$
27. Maximum value of 30 occurs anywhere on the line $3y - 2x = 15$ for $-3 \leq x \leq 0$ 28. Minimum value of -2 at $x = 4, y = 2$
29. No solution; we can make z as large as we wish by selecting appropriate points in the region.
30. Basic variables: s, t; nonbasic variables: x, y; basic solution: $s = 5, t = 7, x = y = 0$
31. Basic variables: u, y; nonbasic variables: x, t, v; basic solution: $u = -3, y = 7; x = t = v = 0$
32. Basic variables: x, y, s; nonbasic variables: w, t; basic solution: $x = 10, y = -19, s = 14, w = t = 0$
33. Basic variables: w, s, t; nonbasic variables: x, y, u; basic solution: $w = -3, s = -2, t = -1, x = y = u = 0$
34. $x = 37, y = -10, s = t = 0$ 35. $x = 5, t = 2, y = s = 0$ 36. $y = -\frac{3}{5}, u = \frac{16}{5}, x = s = t = 0$
37. $s = -\frac{1}{2}, t = 3, x = y = u = 0$ 38. $x = 6, y = 2, w = 3, s = t = u = 0$ 39. $x = 9, y = -\frac{5}{2}, t = -\frac{3}{2}, w = s = u = 0$

40.

	x	y	s	t	
y	$\frac{4}{3}$	1	$\frac{1}{3}$	0	4
t	$\frac{16}{3}$	0	$-\frac{2}{3}$	1	1
z	$\frac{11}{3}$	0	$\frac{5}{3}$	0	20

$y = 4, t = 1, z = 20, x = s = 0$

41.

	x	y	s	t	
y	0	1	$-\frac{1}{2}$	$\frac{9}{2}$	6
x	1	0	$\frac{1}{6}$	$-\frac{3}{2}$	3
z	0	0	$\frac{4}{3}$	-8	46

$z = 46, x = 3, y = 6, s = t = 0$

42.

	x	y	w	s	t	
x	1	$\frac{5}{2}$	0	$\frac{7}{4}$	$\frac{1}{4}$	$\frac{25}{4}$
w	0	$-\frac{37}{4}$	1	$-\frac{20}{3}$	-1	9
z	0	$\frac{41}{2}$	0	$\frac{93}{4}$	$\frac{11}{4}$	$\frac{459}{4}$

$x = \frac{25}{4}, w = 9, z = \frac{459}{4}, y = s = t = 0$

43.

	x	y	w	s	t	u	
x	1	$-\frac{3}{14}$	0	$\frac{1}{7}$	0	$\frac{13}{28}$	$\frac{68}{7}$
t	0	$\frac{110}{7}$	0	$-\frac{8}{7}$	1	$\frac{2}{7}$	$\frac{44}{7}$
w	0	$-\frac{11}{21}$	1	$\frac{4}{7}$	0	$\frac{101}{28}$	$\frac{776}{7}$
z	0	$-\frac{29}{14}$	0	$\frac{5}{7}$	0	$\frac{177}{28}$	$\frac{1,341}{7}$

$x = \frac{68}{7}, t = \frac{44}{7}, w = \frac{776}{7}, y = s = u = 0, z = \frac{1,341}{7}$

44. $x = 4, y = 6, z = 18$ 45. $x = 7, y = 8, z = 108$ 46. $y = 2, w = 4, z = 32, x = 0$ 47. $w = \frac{240}{7}, y = \frac{40}{7}, x = \frac{360}{7}, z = \frac{10,960}{7}$
48. (a) Let $x =$ the number of industrial models and $y =$ the number of personal models.
    Maximize
    $$z = 1,000x + 750y$$
    subject to
    $$10x + 8y \leq 440$$
    $$3x + 2y \leq 120$$
    $$x \geq 0, \quad y \geq 0$$
    (b) and (c) Maximum profit of $42,500 from 20 industrial and 30 personal models
49. (a) Let $x =$ the amount in stocks and $y =$ the amount in bonds.
    Maximize
    $$z = 0.11y + 0.09x$$
    subject to
    $$x + y \leq 24,000$$
    $$-\tfrac{1}{2}x + y \leq 0$$
    $$x \geq 0, \quad y \geq 0$$
    (b) and (c) Maximum return of $2,320 with $16,000 in stocks and $8,000 in bonds
50. Maximum profit of $4,693.33 from 5,333.$\bar{3}$ pounds of deluxe, 1,333.$\bar{3}$ pounds of superior, and 5,333.$\bar{3}$ pounds of special

51. Maximize
$$z = x_1 + 2x_2$$
subject to
$$3x_1 + x_2 \leq 21$$
$$2x_1 + 5x_2 \leq 40$$
$$x_1 \geq 0, \quad x_2 \geq 0$$
$w = 17, y_1 = \frac{1}{13}, y_2 = \frac{5}{13}$

52. Maximize
$$z = 3x_1 + x_2$$
subject to
$$x_1 + 2x_2 \leq 16$$
$$6x_1 - x_2 \leq 18$$
$$x_1 \geq 0, \quad x_2 \geq 0$$
$w = 18, y_1 = \frac{9}{13}, y_2 = \frac{5}{13}$

53. Maximize
$$z = 2x_1 + 3x_2 + x_3$$
subject to
$$x_1 + 2x_2 - x_3 \leq 8$$
$$4x_1 - x_2 + 2x_3 \leq 11$$
$$x_1 \geq 0, \quad x_2 \geq 0, \quad x_3 \geq 0$$
$w = 37, y_1 = \frac{7}{3}, y_2 = \frac{5}{3}$

54. Maximize
$$z = 15x_1 + 19x_2 + 20x_3$$
subject to
$$x_1 + 2x_2 + 4x_3 \leq 200$$
$$5x_1 + 6x_2 + 2x_3 \leq 360$$
$$3x_1 + 3x_2 + 2x_3 \leq 240$$
$$x_1 \geq 0, \quad x_2 \geq 0, \quad x_3 \geq 0$$
$w = \frac{10{,}960}{7}, y_1 = \frac{23}{7}, y_2 = \frac{5}{7}, y_3 = \frac{19}{7}$

55. (1) Physical situation: the manufacture and sale of two microcomputers, an industrial model and a personal model; physical question: How many of each model should the company manufacture in order to maximize its profit?
(2) Mathematical model:
Maximize
$$z = 1{,}000x + 750y$$
subject to
$$10x + 8y \leq 440$$
$$3x + 2y \leq 120$$
$$x \geq 0, \quad y \geq 0$$
mathematical question: What is the solution of the given linear programming problem?
(3) Mathematical solution: $z = 42{,}500$ is maximum when $x = 20$ and $y = 30$
(4) Interpretation: The computer company should produce and sell 20 industrial models and 30 personal models in order to maximize its profits at $42,500.

56. (1) Physical situation: the investing of money in stocks and bonds; physical question: What amount of the $24,000 should be invested in stocks and what amount in bonds?
(2) Mathematical model:
Maximize
$$z = 0.09x + 0.11y$$
subject to
$$(-\tfrac{1}{2})x + y \leq 0$$
$$x + y \leq 24{,}000$$
$$x \geq 0, \quad y \geq 0$$
mathematical question: What is the solution of the given linear programming problem?
(3) Mathematical solution: $z = 2{,}320$ is maximum when $x = 16{,}000$ and $y = 8{,}000$
(4) Interpretation: In order for the individual to maximize the return during the first year, $16,000 must be invested in stocks and $8,000 in bonds for a return of $2,320.

CHAPTER 5   SECTION 5.1

1. $\{1, 3, 5\}$ 3. $\{1, 2, 3, 5, 6\}$ 5. $\{HHH, HHT, HTH, THH\}$ 7. $\{(123), (124), (125), (134), (135), (145)\}$
9. $\{(234), (235), (245), (345)\}$ 11. $\{CJ, CQ, CK, DJ, DQ, DK, HJ, HQ, HK, SJ, SQ, SK\}$
13. $\{(1, 1), (1, 2), (1, 3), (1, 4), (1, 5), (1, 6), (2, 1), (2, 2) (2, 3), (2, 4), (2, 5), (2, 6), (3, 1), (3, 2), (3, 3), (3, 4), (3, 5), (3, 6), (4, 1), (4, 2), (4, 3),$
    $(4, 4), (4, 5), (4, 6), (5, 1), (5, 2), (5, 3), (5, 4), (5, 5), (5, 6), (6, 1), (6, 2), (6, 3), (6, 4), (6, 5), (6, 6)\}$
15. $S = \{(1, 2), (1, 3), (1, 4), (1, 5), (1, 6), (2, 1), (2, 3), (2, 4), (2, 5), (2, 6), (3, 1), (3, 2), (3, 4), (3, 5), (3, 6), (4, 1), (4, 2), (4, 3), (4, 5), (4, 6),$
    $(5, 1), (5, 2), (5, 3), (5, 4), (5, 6), (6, 1), (6, 2), (6, 3), (6, 4), (6, 5)\}$
17. $\{WDC, RDC, WSC, RSC, WDN, RDN, WSN, RSN\}$
19. $S = \{GGGG, GGGB, GGBG, GGBB, GBGG, GBGB, GBBG, GBBB, BGGG, BGGB, BGBG, BGBB, BBGG, BBGB, BBBG, BBBB\}$
21. $\{KPV, PKV, PVK, VPK, KVP, VKP\}$
23. $S = \{BBB, BBW, BWB, BWW, WBB, WBW, WWB, WWW\}$

25. (a) $E_1 \cap E_2 = \{(135), (235), (345)\}$ (b) $E_1 \cup E_2 = \{(123), (125), (134), (135), (145), (234), (235), (245), (345)\}$
    (c) $E_2' = \{(123), (124), (134), (234)\}$
27. (a) $E_1 \cap E_3 = \{(3, 5)\}$ (b) $E_1 \cup E_2 = \{(3, 1), (3, 2), (3, 4), (3, 5), (3, 6), (4, 1), (4, 2), (4, 3), (4, 5), (4, 6)\}$
    (c) $E_2' = \{(1, 2), (1, 3), (1, 4), (1, 5), (1, 6), (2, 1), (2, 3), (2, 4), (2, 5), (2, 6), (3, 1), (3, 2), (3, 4), (3, 5), (3, 6), (5, 1), (5, 2), (5, 3), (5, 4),$
        $(5, 6), (6, 1), (6, 2), (6, 3), (6, 4), (6, 5)\}$ (d) $E_1 \cap E_2 = \emptyset$
    (e) $E_1 \cup E_3' = \{(3, 1), (3, 2), (3, 4), (3, 5), (3, 6), (1, 2), (1, 3), (1, 4), (1, 6), (2, 1), (2, 3), (2, 4), (2, 6), (4, 1), (4, 2), (4, 3), (4, 6), (5, 1),$
        $(5, 2), (5, 3), (5, 4), (5, 6), (6, 1), (6, 2), (6, 3), (6, 4)\}$
29. (a) $\{BBB, BWB\}$ (b) $\{WBB, WWB, WBW, WWW, BBB, BBW\}$ (c) $\{BBW, WBW, BWW, WWW\}$
    (d) E_1: The adult male is black, E_2: The child is black; $E_1 \cap E_2$
    (e) E_1: The adult male is white, E_2: The adult female is black; $E_1 \cup E_2$ (f) E'
31. A sample space for an experiment is a set that describes all the possible outcomes of the experiment.
33. E and F occur.   35. E does not occur or F occurs or both E does not occur and F occurs.

SECTION 5.2

1. $\frac{1}{2}$ 3. $\frac{5}{6}$ 5. $\frac{1}{2}$ 7. $\frac{3}{5}$ 9. $\frac{2}{5}$ 11. $\frac{3}{13}$ 13. $\frac{1}{9}$ 15. $\frac{1}{6}$ 17. $\frac{1}{2}$ 19. 0.60   21. $\frac{8}{11}$ 23. 0.75
25. $\frac{1}{5}$ 27. 0.9   29. 0.6   31. 0   33. 0.7   35. 0   37. 0.8   39. 0.4
41. The role of a probability assignment is to assign to each element in the sample space a positive number in such a way that the sum of all the numbers assigned is 1.
43. The probability of an event is the sum of the probabilities of the elements in the event.
45. (1) Physical situation: The drawing of a card from a deck of ordinary playing cards; physical question: What is the likelihood that a picture card is drawn?
    (2) Mathematical model: A sample space of equally likely elements is constructed.
    $S = \{C2, C3, C4, C5, C6, C7, C8, C9, C10, CJ, CQ, CK, CA, D2, D3, D4, D5, D6, D7, D8, D9, D10, DJ, DQ, DK, DA,$
    $H2, H3, H4, H5, H6, H7, H8, H9, H10, HJ, HQ, HK, HA, S2, S3, S4, S5, S6, S7, S8, S9, S10, SJ, SQ, SK, SA\}$
    with event $E = \{CJ, CQ, CK, DJ, DQ, DK, HJ, HQ, HK, SJ, SQ, SK\}$
    Mathematical question: What is $P(E)$?
    (3) Mathematical solution: $P(E) = \frac{3}{13}$
    (4) Interpretation: In a large number of trials, it would be expected that $\frac{3}{13}$ of the time a picture card will be drawn when 1 card is selected from a deck of ordinary playing cards.

SECTION 5.3

1. (a) 18   (b)

3. 132   5. (a) $(24)(23)(22) = 12,144$ (b) $(24)(24)(24) = 13,824$ 7. (a) 375   (b) $(3)(4)(3)(2) = 72$ 9. 8
11. (a) 1,296   (b) 216   (c) $\frac{1}{6}$ (d) 36   (e) $\frac{1}{36}$ 13. (a) 120   (b) 6   (c) $\frac{1}{5}$ (d) $\frac{1}{3}$

SECTION 5.4

1. $3! = 6$ 3. $11! = 39,916,800$ 5. $0! = 1$ 7. $C(7, 7) = 1$ 9. $C(12, 3) = 220$ 11. $C(5, 3) = 10$ 13. $C(12, 4) = 495$
15. $C(3, 2) \cdot C(4, 2) = 18$ 17. (a) $C(7, 2) \cdot C(5, 3) = 210$ (b) $\frac{3}{35}$ 19. $C(100, 3) \cdot C(435, 3) = 161,700 \times 13,624,345$
21. $7!7! = 25,401,600$ 23. (a) $C(12, 3) = 210$ (b) $\frac{18}{35}$ (c) $\frac{2}{5}$ 25. (a) 210   (b) $\frac{3}{7}$ 27. $\frac{6}{55}$
29. (a) $P(12, 2) = 132$ (b) $P(24, 3) = 12,144$ 31. A combination is an unordered subset of a given set.

SECTION 5.5

1. $\frac{1}{2}$ 3. $\frac{7}{25}$ 5. $\frac{5}{6}$ 7. $P(E') = P(b) + P(c) = 0.4$; $P(E') = 1 - P(E) = 1 - (0.5 + 0.1) = 0.4$
9. $P(\text{Under 7 or exactly 7}) = P(\text{Under 7}) + P(\text{Exactly 7}) = \frac{7}{12}$; $P(\text{Not over 7}) = 1 - P(\text{Over 7}) = 1 - \frac{15}{36} = \frac{7}{12}$;
   $P(\text{Under 7 or exactly 7}) = P(\text{Under 7}) + P(\text{Exactly 7}) - P(\text{Under and exactly 7}) = \frac{15}{36} + \frac{6}{36} - 0 = \frac{7}{12}$
11. (a) 0.52   (b) 0.85   13. $\frac{7}{10}$ 15. 0.15   17. $\frac{17}{50}$ 19. 0.6   21. 1   23. $\frac{1}{8}$ 25. 1   27. 2 to 3
29. 3 to 7   31. $\frac{3}{10}$
33. Events E and F are mutually exclusive if $E \cap F = \varnothing$; they represent outcomes of an experiment that cannot occur simultaneously.

SECTION 5.6

1. (a) $\frac{1}{169}$ (b) $\frac{4}{663}$ 3. (a) $\frac{49}{144}$ (b) $\frac{35}{144}$ (c) $\frac{35}{144}$ (d) $\frac{35}{72}$ 5. Independent   7. Independent
9. (a) and (b) $\frac{3}{10}$ 11. (a) $\frac{3}{4}$ (b) $\frac{1}{4}$ 13. (a) $\frac{1}{8}$ (b) $\frac{7}{8}$ 15. (a) 0.0005   (b) 0.0595   (c) 0.9405   17. $\frac{2}{9}$
19. $(\frac{5}{36})^2 = \frac{25}{1,296}$ 21. $\frac{125}{1,296}$ 23. $(0.38)(0.67) = 0.2546$ 25. 0.0117
27. Events E and F are independent if $P(E|F) = P(E)$; generally they represent outcomes where the occurrence of one does not affect the occurrence of the other.
29. The probability that each of two events occur   31. If E and F are independent events, $P(E \cap F) = P(F)P(E)$.

REVIEW EXERCISES

1. $\frac{1}{4}, \frac{1}{8}, \frac{1}{8}$ 2. $\frac{1}{2}$ 3. $\frac{1}{2}$ 4. $S = \{A, B, C, D\}$; $P(A) = \frac{1}{2}$, $P(B) = \frac{1}{4}$, $P(C) = \frac{1}{8}$, $P(D) = \frac{1}{8}$ 5. $\frac{3}{8}$ 6. $\frac{3}{8}$
7. $\frac{7}{8}$ 8. (a) $\frac{1}{4}$ (b) $\frac{9}{64}$ 9. $C(5, 2) = 10$ 10. {LH, LT, LJ, LB, HT, HJ, HB, TJ, TB, JB}
11. $E_3 = E_1 \cap E_2$, $E_4 = E_1 \cup E_2$, $E_5 = E'_2$ 12. $P(E_1) = \frac{2}{5}$, $P(E_2) = \frac{2}{5}$, $P(E_3) = \frac{1}{10}$, $P(E_4) = \frac{7}{10}$, $P(E_5) = \frac{3}{5}$ 13. 16
14. $S = \{HHHH, THHH, HTHH, HHTH, HHHT, TTHH, THTH, THHT, HTTH, HTHT, HHTT, HTTT, THTT, TTHT, TTTH, TTTT\}$
15. $E_3 = E_1 \cup E_2$, $E_4 = E_1 \cap E_2$, $E_5 = E'_2$ 16. $P(E_1) = \frac{5}{16}$, $P(E_2) = \frac{1}{4}$, $P(E_3) = \frac{9}{16}$, $P(E_4) = 0$, $P(E_5) = \frac{3}{4}$ 17. 0.6
18. 0.6   19. 0.4   20. 0.4   21. 0.8   22. 0.6   23. 0.3   24. 0.35   25. 0.55   26. 0.1   27. 0.7   28. 0.9
29. 0   30. 0.65   31. 1   32. $\frac{1}{5}$ 33. $\frac{19}{20}$ 34. $\frac{3}{4}$ 35. No, $P(E \cap F) \neq 0$ 36. Yes, $P(E \cap F) = P(E) \cdot P(F)$
37. 0.05   38. 0.85   39. No, $P(E \cap F) \neq 0$ 40. No, $P(E \cap F) \neq P(E) \cdot P(F)$ 41. (a) $\frac{11}{40}$ (b) $\frac{7}{10}$ (c) $\frac{23}{40}$
42. 4   43. (a) $C(12, 5) = 792$ (b) $C(7, 5) = 21$ (c) $C(7, 3) \cdot C(5, 2) = 350$ 44. (a) $6! = 720$ (b) $C(6, 3) = 20$
45. (a) $C(10, 3) = 120$ (b) $[C(6, 2) \cdot C(4, 1)]/120 = \frac{1}{2}$ (c) $\frac{2}{3}$ 46. (a) $\frac{43}{88}$ (b) $\frac{43}{66}$ 47. $\frac{17}{20}$
48. (a) 9 to 11   (b) 3 to 2   49. (a) It is correct.   (b) $\frac{1}{3}$ 50. (a) $\frac{4}{663}$ (b) $\frac{4}{663}$ (c) $\frac{8}{663}$ (d) $\frac{1}{52}$
51. (a) $\frac{1}{169}$ (b) $\frac{1}{169}$ (c) $\frac{2}{169}$ (d) $\frac{1}{52}$

CHAPTER 6   SECTION 6.1

1. $C(6, 4)(0.5)^4(0.5)^2 = 0.234375$ 3. $C(6, 6)(0.5)^6 = 0.015625$ 5. $C(6, 2)(0.5)^2(0.5)^4 = 0.234375$ 7. $C(4, 2)(\frac{1}{6})^2(\frac{5}{6})^2 = 0.11574$
9. $C(4, 0)(\frac{5}{6})^4 = 0.48225$ 11. $C(4, 2)(\frac{1}{6})^2(\frac{5}{6})^2 = 0.11574$ 13. $C(8, 4)(\frac{1}{3})^4(\frac{2}{3})^4 = 0.17071$ 15. $C(8, 8)(\frac{1}{3})^8 = 0.00015$
17. $C(3, 1)(0.5)^1(0.5)^2 = 0.375$ 19. $C(3, 3)(0.5)^3 = 0.125$ 21. $C(6, 2)(0.2)^2(0.8)^4 = 0.24576$ 23. $C(12, 7)(\frac{1}{5})^7(\frac{4}{5})^5 = 0.003322$
25. $C(6, 4)(0.5)^4(0.5)^2 + C(6, 5)(0.5)^5(0.5)^1 + C(6, 6)(0.5)^6 = 0.34375$ 27. $P(0) + P(1) + P(2) + P(3) + P(4) = 0.890625$
29. $C(15, 12)(0.6)^{12}(0.4)^3 = 0.0634$ 31. $1 - P(12 \text{ or more}) = 0.9095$ 33. $P(0) + P(1) + P(11) + P(12 \text{ or more}) = 0.2173$
35. 0.9163   37. $P(3) + P(2) + P(1) + P(0) = 0.2897$ 39. 0.250   41. 0.836   43. (a) 0.317   (b) 0.455   (c) 8
45. (1) The trials are independent. (2) On each trial interest is in only two outcomes, success or failures.
    (3) The probability of a success is the same for each trial.

SECTION 6.2

1. 0.24   3. 0.14   5. 0.06   7. (a) $0.2\overline{2}$ (b) 0.40   9. 0.08   11. 0.06   13. (a) 0.91   (b) 0.994   15. 0.49
17. (a) 0.02   (b) 0.979

SECTION 6.3

1. $P^2 = \begin{bmatrix} 0.44 & 0.56 \\ 0.40 & 0.60 \end{bmatrix}$, $P^3 = \begin{bmatrix} 0.412 & 0.588 \\ 0.42 & 0.58 \end{bmatrix}$ 3. (a) 0.5   (b) 0.40   (c) 0.58   5. (a) 0.8   (b) 0.32   (c) 0.698

7. $P^2 = \begin{bmatrix} 0.09 & 0.35 & 0.56 \\ 0.10 & 0.59 & 0.31 \\ 0.05 & 0.60 & 0.35 \end{bmatrix}$, $P^3 = \begin{bmatrix} 0.062 & 0.525 & 0.413 \\ 0.089 & 0.568 & 0.343 \\ 0.075 & 0.595 & 0.33 \end{bmatrix}$

9. (a) 0.5   (b) 0.60   (c) 0.595   11. (a) 0.4   (b) 0.3   (c) 0.72

13. (a) $P = \begin{bmatrix} 0.70 & 0.30 \\ 0.60 & 0.40 \end{bmatrix}$, $P^2 = \begin{bmatrix} 0.67 & 0.33 \\ 0.66 & 0.34 \end{bmatrix}$, $P^3 = \begin{bmatrix} 0.667 & 0.333 \\ 0.666 & 0.334 \end{bmatrix}$, $P^4 = \begin{bmatrix} 0.6667 & 0.3333 \\ 0.6666 & 0.3334 \end{bmatrix}$

(b) 0.30, 0.333, 0.3333   (c) 0.40, 0.34, 0.334

15.
	AA	Aa	aA	aa
AA	$\frac{1}{2}$	$\frac{1}{2}$	0	0
Aa	$\frac{1}{4}$	$\frac{1}{4}$	$\frac{1}{4}$	$\frac{1}{4}$
aA	$\frac{1}{4}$	$\frac{1}{4}$	$\frac{1}{4}$	$\frac{1}{4}$
aa	0	0	$\frac{1}{2}$	$\frac{1}{2}$

; $P(\text{AA grandchild given Aa grandmother}) = \frac{1}{4}$

17. $(0.01)^7 = 0.00000000000001$ 19. 0.729   21. 0.97
23. A Markov chain is a sequence of repeated experiments having the following three properties: (1) the sample space remains the same for each experiment; (2) the probability of obtaining any particular element in the sample space possibly depends upon the outcome of the preceding trial but upon no other previous trial; and (3) for each i and j, the probability p_{ij}, that a_j will occur given that a_i occurred on the preceding trial remains constant throughout the sequence.
25. Transition probabilities are the probabilities that the system will change from one state to another.

SECTION 6.4

1. Regular, $[\frac{4}{7} \ \frac{3}{7}]$ 3. Regular, $[\frac{13}{28} \ \frac{15}{28}]$ 5. Not regular   7. Regular, $[\frac{35}{141} \ \frac{58}{141} \ \frac{48}{141}]$
9. (a) $[\frac{2}{3} \ \frac{1}{3}]$ (b) $\frac{2}{3}$ of the stock should be frozen and $\frac{1}{3}$ of the stock should be unfrozen orange juice.
11. (a) $[\frac{2}{5} \ \frac{2}{5} \ \frac{1}{5}]$ (b) Either city A or B   13. [0.938 0.062], give up the first machine.
17. A probability vector is a vector with nonnegative entries whose sum is one.
19. A transition matrix is regular if the matrix or some power of it contains only positive numbers.
21. (1) Physical situation: The sale of orange juice in a store; physical problem: Determine the proportion of frozen and unfrozen orange juice to be carried in the store.
(2) The mathematical model is the transition matrix. Mathematical question: What is the fixed probability vector of the transition matrix?
(3) Mathematical solution: $[\frac{2}{3} \ \frac{1}{3}]$
(4) Interpretation: $\frac{2}{3}$ of the stock should be frozen orange juice and $\frac{1}{3}$ should be unfrozen orange juice.

SECTION 6.5

1. 3.1   3. 1.5   5. (a) $\frac{35}{18}$ (b) $4.47\overline{2}$ 7. (a) $-\$3.00$ (b) $-\$4.99$ 9. -5.2¢
11. (a) Not fair   (b) Not fair   (c) Not fair   (d) Not fair   13. $2,000   15. 1.85 errors
17. A random variable on a sample space S is a function (or rule) that assigns a unique number to each element in S.
19. If X is a random variable with values $x_1, x_2, \ldots, x_n$, the expected value of the random variable is the sum $x_1 P(x_1) + x_2 P(x_2) + \cdots + x_n P(x_n)$.

SECTION 6.6

1. 6 is a saddle point; r_1 and c_2; value of 6   3. No saddle point   5. 4 is a saddle point; r_3 and c_2; value of 4
7. 3 is a saddle point; r_2 and c_2; value of 3   9. 0 is a saddle point; r_3 and c_4; value of 0   11. Not strictly determined

13. Strictly determined; not fair   15. Strictly determined; fair   17. Not strictly determined
19. Best strategy: r_2 and c_3, or P for the union and O for management; wage increase is 35¢ per hour.

21. Player I $\begin{array}{c} \text{Player II} \\ \begin{array}{c} 1 2 3 \end{array} \\ \begin{array}{c} 1 \\ 2 \\ 3 \end{array} \begin{bmatrix} 0 & 1 & -1 \\ -1 & 0 & 1 \\ 1 & -1 & 0 \end{bmatrix} \end{array}$; not strictly determined

23. Company $\begin{array}{c} \text{Nature} \\ \text{B} \text{NB} \\ \begin{array}{c} \text{G} \\ \text{NG} \end{array} \begin{bmatrix} -20 & -5 \\ -21 & -3 \end{bmatrix} \end{array}$; use the guaranteed transistors.

25. The payoff matrix for a two-person game is a matrix that gives the payoffs for the various combinations of plays of two players.
27. Any combination of appropriate plays for a player is called a strategy for that player.
29. The value of a strictly determined game is the value of the saddle point.

SECTION 6.7

1. $-\frac{7}{12}$

3. $[\frac{2}{3} \ \frac{1}{3}], \begin{bmatrix} \frac{4}{9} \\ \frac{5}{9} \end{bmatrix}$; expected payoff is $\frac{1}{3}$. The row player should play row 1 with probability $\frac{2}{3}$ and row 2 with probability $\frac{1}{3}$. The column player should play column 1 with probability $\frac{4}{9}$ and column 2 with probability $\frac{5}{9}$.

5. $[\frac{11}{26} \ \frac{15}{26}], \begin{bmatrix} \frac{6}{13} \\ \frac{7}{13} \end{bmatrix}$; expected payoff is $\frac{80}{169}$. The row player should play row 1 with probability $\frac{11}{26}$ and row 2 with probability $\frac{15}{26}$. The column player should play column 1 with probability $\frac{6}{13}$ and column 2 with probability $\frac{7}{13}$.

7. $[\frac{5}{53} \ \frac{48}{53}], \begin{bmatrix} \frac{15}{53} \\ \frac{38}{53} \end{bmatrix}$; expected payoff is $\frac{870}{53}$. The row player should play row 1 with probability $\frac{5}{53}$ and row 2 with probability $\frac{48}{53}$. The column player should play column 1 with probability $\frac{15}{53}$ and column 2 with probability $\frac{38}{53}$.

9. $\begin{bmatrix} 3 & 2 \\ -1 & 3 \end{bmatrix}$ 11. $\begin{bmatrix} 2 & -1 \\ -3 & 2 \end{bmatrix}$

13. $[\frac{1}{2} \ \frac{1}{2}] \begin{bmatrix} \frac{1}{2} \\ \frac{1}{2} \end{bmatrix}$; value is $-\frac{1}{2}$. The row player should play row 1 with probability $\frac{1}{2}$ and row 3 with probability $\frac{1}{2}$. The column player should play column 1 with probability $\frac{1}{2}$ and column 2 with probability $\frac{1}{2}$.

15. The row player should play row 1 and the column player should play column 1. The value is 0.

17. The row player should play row 2 $\frac{7}{10}$ of the time and row 3 $\frac{3}{10}$ of the time; the column player should play column 1 $\frac{3}{5}$ of the time and column 2 $\frac{2}{5}$ of the time. The value of the game is $-\frac{1}{5}$.

19. $\begin{bmatrix} 5{,}100 & 2{,}700 \\ 3{,}300 & 3{,}600 \end{bmatrix}$; she should invest $3,333.33 in money market funds and $26,666.67 in stocks, with an expected return of $3,500.

21. The value of a game that is not strictly determined is the expected payoff corresponding to the probability vectors P and Q, which give the best strategies for the players.

23. Column c_i is dominated by column c_j if each entry in c_j is less than the corresponding entry in c_i.

25. (1) Physical situation: Investing $30,000 for 1 year; physical question: How much should be invested in each type of investment and what return can be expected for the year?

   (2) Mathematical model: The payoff matrix $\begin{bmatrix} 0.17 & 0.09 \\ 0.11 & 0.12 \end{bmatrix}$.

   Mathematical question: What is the probability vector giving the best strategy for the row player, and what is the value of the game?

   (3) Mathematical solution: $P = [\frac{1}{9} \ \frac{8}{9}]$; the value of the game is $11.\overline{6}\%$.

   (4) Interpretation: $\frac{1}{9}$ or $3,333.33 should be invested in the money market fund and $\frac{8}{9}$ or $26,666.67 should be invested in stocks for an expected return of $11.\overline{6}\%$ or $3,500.00.

APPENDIX C ANSWERS TO ODD-NUMBERED AND REVIEW EXERCISES

REVIEW EXERCISES

1. $C(5, 3)(\frac{1}{2})^3(\frac{1}{2})^2 = 0.3125$ 2. $C(5, 4)(\frac{1}{2})^4(\frac{1}{2}) + C(5, 5)(\frac{1}{2})^5 = 0.1875$ 3. $1 - (0.3125 + 0.1875) = 0.5$ 4. $0.5 + 0.3125 = 0.8125$
5. $0.5 + 0.1875 = 0.6875$, or $1 - 0.3125 = 0.6875$ 6. (a) $C(10, 7)(0.6)^7(0.4)^3 = 0.21498$ (b) $C(10, 3)(0.6)^3(0.4)^7 = 0.0425$
7. 0.7462   8. 0.6367   9. $n = 7$ 10. (a) 0.85   (b) 0.15   11. 0.3085   12. (a) 0.0787   (b) 0.4331
13. (a) 0.5   (b) 0.45   (c) 0.555   14. P is regular since it contains only positive entries.   15. $[\frac{4}{9} \quad \frac{5}{9}]$
16. (a) $\begin{bmatrix} 0.7 & 0.3 \\ 0.1 & 0.9 \end{bmatrix}$ (b) 0.84   (c) $[\frac{1}{4} \quad \frac{3}{4}]$ 17. Lower level, $\frac{1}{7}$; middle level, $\frac{2}{7}$; upper level, $\frac{4}{7}$ 18. 5.65   19. $185
20. 6.24   21. The row player should play row 2 and the column player should play column 2; value is 0.
22. The row player should play row 1 with probability $\frac{6}{7}$ and row 2 with probability $\frac{1}{7}$. The column player should play column 1 with probability $\frac{6}{7}$ and column 2 with probability $\frac{1}{7}$. The value is $\frac{8}{7}$.
23. The row player should play row 2 with probability $\frac{3}{10}$ and row 3 with probability $\frac{7}{10}$. The column player should play column 2 with probability $\frac{2}{5}$ and column 3 with probability $\frac{3}{5}$. The value is $\frac{1}{5}$.
24. The row player should play row 1 with probability 0.3 and row 2 with probability 0.7. The column player should play column 1 with probability 0.5 and column 2 with probability 0.5. The value is 0.5.
25. Company R's advertising should be 50% on television and 50% in magazines.
26. (1) Physical situation: Two companies and their advertising strategy; physical question: What percent of company R's advertising should be on TV and what percent in magazines?
    (2) Mathematical model: The given payoff matrix in Exercise 25. Mathematical question: What is the probability vector that gives the best strategy for the row player?
    (3) Mathematical solution: [0.5 0.5]
    (4) Interpretation: 50% of company R's advertising should be on television and 50% should be in magazines.

CHAPTER 7 SECTION 7.1

1. | Measurement | Frequency |
    |---|---|
    | 10 | 7 |
    | 11 | 6 |
    | 12 | 4 |
    | 13 | 3 |
    | 14 | 1 |
    | 15 | 5 |

3. | Measurement | Frequency |
    |---|---|
    | 9.0 | 2 |
    | 9.2 | 4 |
    | 9.4 | 7 |
    | 9.6 | 8 |
    | 9.8 | 4 |
    | 10.0 | 2 |

5. | Intervals | Frequency |
    |---|---|
    | 15–19 | 8 |
    | 20–24 | 7 |
    | 25–29 | 16 |
    | 30–34 | 13 |
    | 35–39 | 6 |

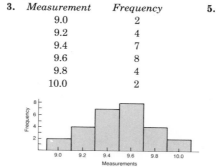

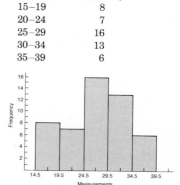

7. | Intervals | Frequency |
    |---|---|
    | 40–49 | 2 |
    | 50–59 | 0 |
    | 60–69 | 7 |
    | 70–79 | 14 |
    | 80–89 | 10 |
    | 90–99 | 4 |

9. | Intervals | Frequency |
    |---|---|
    | 125.0–126.9 | 4 |
    | 127.0–128.9 | 5 |
    | 129.0–130.9 | 10 |
    | 131.0–132.9 | 11 |
    | 133.0–134.9 | 10 |
    | 135.0–136.9 | 5 |

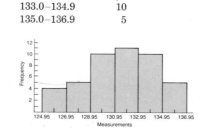

SECTION 7.2

1. Mean = 27; median = 26
3. Mean = 206; median = 207
5. (a) Mean$_A$ = 42; mean$_B$ = 41.2 (b) Type A is better. (c) Median$_A$ = 41.9; median$_B$ = 41.9 (d) No difference
7. Mean = 12; median = 11.5
9. Mean = 9.5; median = 9.6
11. Mean = 27.2; median = 27.545
13. Mean = 75.85; median = 76.1$\overline{6}$
15. Mean = 131.42; median = 131.62
17. (a) One-half of the families in the United States had incomes less than $19,684 and one-half had incomes greater than $19,684.
    (b) Too vague to interpret. What is "typical equipment"? Is mean based on total number of cars manufactured or on one car per model?
19. Models that serve as measures of central tendency are called averages.
21. The median of a set of numbers is the middle measurement when the numbers are arranged in order of magnitude.
23. The midrange of a set of numbers is the mean of the largest and smallest measurements.

SECTION 7.3

1. 5.497; $Q_3 - Q_1 = 14$
3. 9.46; $Q_3 - Q_1 = 19$
5. 7; $Q_3 - Q_1 = 16$; range = 20
7. (a) Yes, no (b) First
9. 1.82; $Q_3 - Q_1 = 3.5$; range = 5
11. 0.263; $Q_3 - Q_1 = 0.2$; range = 1.0
13. 6.2; $Q_3 - Q_1 = 9.9$
15. 11.66; $Q_3 - Q_1 = 16.7$
17. 2.872; $Q_3 - Q_1 = 4.545$
21. 5.497
23. The standard deviation of a set of data $\{x_1, x_2, \ldots, x_n\}$ is the square root of $\left[\sum_{i=1}^{n}(x_i - \bar{x})^2\right]/n$.
25. The third quartile of a set of data is a number such that one-fourth of the measurements are greater than or equal to this number.
27. The difference between the largest and smallest measurement of a set of data is called the *range*.
29. Measures of dispersion are used to indicate how closely data center about an average.

SECTION 7.4

1. Type A plug is superior.
3. Neither fertilizer is superior.
5. High octane gasoline gives better mileage.
7. The diet is effective.
9. The Republican candidate has the majority of the votes.
11. Use either flavoring.

SECTION 7.5

1. 0.3925
3. 0.2357
5. 0.0628
7. 0.2730
9. 0.8185
11. 0.0359
13. 0.9332
15. 0.1894
17. 0.0456
19. 0.4772
21. 0.4772
23. 0.1359
25. 0.1540
27. 0.8964
29. 0.0013
31. 0.8849
33. 0.9987
35. 0.2743
37. 0.0947
39. 0.1331
41. 0.4986
43. 0.0726
45. 0.4701
47. 0.4172
49. 0.6554
51. 0.0808
53. 0.2420
55. 0.9251
57. (a) 0.3830 (b) 0.6678
59. (a) 0.1587 (b) 6.68%
61. 0.13%
63. 1.96
65. −1.645
67. 73.72
69. 68.96
71. 34%
73. 48%
75. (1) Physical situation: Grade-point averages at a college; physical question: What is the likelihood that a student randomly chosen has a grade-point average greater than 3.0?
    (2) Mathematical model: The mathematical description of the grade-point averages is the normal distribution with mean 2.4 and standard deviation of 0.6; mathematical question: What is the probability of obtaining a value of x greater than 3.0 if x has the given normal distribution?
    (3) Mathematical solution: 0.1587
    (4) Interpretation: If checking grade-point averages is done in this manner a large number of times, about 15.87% of the averages checked will be greater than 3.0; to this extent the result indicates the likelihood of a randomly chosen student having a grade-point average greater than 3.0.

SECTION 7.6

1. Reject the claim.
3. Accept the claim.
5. The batch does not meet specifications.
7. The mileage is significantly different than advertised.
9. Accept the claim.
11. Accept the claim.
13. Reject the claim.
15. Central limit theorem: If samples of size n are randomly chosen and if the population has mean μ and standard deviation σ', then the sample means are approximately normally distributed with mean μ and standard deviation $\sigma = \sigma'/\sqrt{n}$.

REVIEW EXERCISES

1. Mean = 17; median = 18; standard deviation = 4.3; $Q_3 - Q_1 = 9$
2. Mean = 99; median = 97; standard deviation = 7.7; $Q_3 - Q_1 = 15$
3. Mean = 54.4; median = 54.25; standard deviation = 4.2; $Q_3 - Q_1 = 7.95$
4. Mean = 100.5; median = 100.5; standard deviation = 0.33; $Q_3 - Q_1 = 0.8$
5.
Measurement	Frequency
23	4
24	5
25	3
26	0
27	2
28	3

6.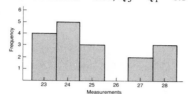

7. Mean = 25; median = 24

8. Standard deviation = 1.81; $Q_3 - Q_1 = 4$

9.
Measurement	Frequency
110.0	4
110.1	3
110.2	0
110.3	1
110.4	5

10.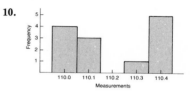

11. Mean = 110.2; median = 110.1
12. Standard deviation = 0.18; $Q_3 - Q_1 = 0.4$
13.
Intervals	Frequency
0–4	7
5–9	5
10–14	6
15–19	9
20–24	7

14.

15. Mean = 12.6; median = 13.41
16. Standard deviation = 7.2; $Q_3 - Q_1 = 13$

17.
Intervals	Frequency
10.0–12.9	5
13.0–15.9	6
16.0–18.9	8
19.0–21.9	6

18.

19. Mean = 16.25; median = 16.61

20. Standard deviation = 3.2; $Q_3 - Q_1 = 6$
21. Mean = $37,480
22. Mean = 2.195 seconds
23. Mean = $97.88\overline{3}$°F
24. A: 0; B: 0; C: $25; D: $50; E: 0; F: 0; G: $50; H: 0; I: $50; J: $25; K: $25; L: 0
25. (a) Mean$_X$ = 3.94; mean$_Y$ = 3.87; brand X   (b) Standard deviation$_X$ = 0.34, standard deviation$_Y$ = 0.20; brand Y   (c) Neither brand is superior.
26. (a) Mean$_A$ = $1.2\overline{6}$; mean$_B$ = $0.9\overline{3}$; brand B   (b) Standard deviation$_A$ = 1.84; standard deviation$_B$ = 1.95; brand A   (c) Both brands are about the same.
27. Not effective
28. Drinking has an effect on reaction times.
29. 0.4906   30. 0.4192   31. 0.0913   32. 0.8873   33. 0.8413   34. 0.1587   35. 0.8413   36. 0.1587
37. 0.4332   38. 0.4938   39. 0.1587   40. 0.6826   41. 0.8413   42. 0.3085   43. 0.1359   44. 0.9772
45. 0.2934   46. 0   47. 0.6826   48. 0.1747   49. 0.0228   50. 0.9332   51. 0.3944   52. 0.5987
53. 0.3753   54. 0.6215   55. 0.0228   56. 0.6826   57. 0.0228   58. 0.4772   59. 0.5   60. 0.9544
61. 0.0013   62. 0.0013   63. Accept the claim.   64. Accept the claim.   65. Switch to reservoir.
66. Reject the claim.   67. Accept the claim.

CHAPTER 8 SECTION 8.1

1.

x	y
3	27
4	81
−1	$\frac{1}{3}$
0	1

3.

x	y
−2	−0.08
$\frac{2}{3}$	−20
0	−5
$\sqrt{2}$	−94.6

5.

x	y
−0.5	1.5
−2	40.5
0.75	0.096
0.68	0.112

7.

x	y
−1	0.0183
−1.3	0.0055
0	1
2.4	14,765

9. $y = 3(2^x)$ **11.** $y = -2^{-x}$ **13.** $y = 3^{-x}$ **15.** $y = e^x$ **17.** $y = 2^x + 2$

 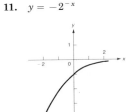 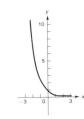  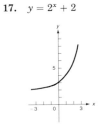

19. $y = 1 - e^{-x}$ **21.** **23.** (a) $y = 1,000(1.09)^n$ (b) $n \geq 0$, n an integer
(c) About $2,000
(d) $y = 1,000(1.09)^8 = \$1,992.56$
(e) $2,367.36

 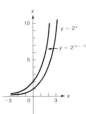

25. (a) $y = 1,200(1.01)^n$ (b) About $1,350 **27.** (a) $y = 2,400\, e^{0.1x}$ (b) About 6,500
(c) $1,352.19
(d) $1,435.38

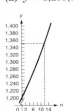 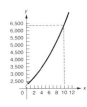

29. (a) $q = 500(2^p) - 500$ (b) $p \geq 0$ **31.** (a) $y = 6 - 5e^{-n}$, $n \geq 1$ (b) About 5,300 units **33.**
(c) About 5,100 units (c) 5,323 units
(d) 5,908 units

SECTION 8.2

1. 2 **3.** 3 **5.** 4 **7.** 2 **9.** 2 **11.** 0 **13.** −1 **15.** 1 **17.** 2 **19.** 5 **21.** 17 **23.** 51
25. e^2 **27.** x **29.** 1.13783 **31.** 2.24071 **33.** 0.4518 **35.** 0.6021 **37.** 0.8779 **39.** $x = -2$ **41.** $x = 16$
43. $\frac{1}{4}$ **45.** $x = 2$ **47.** $x = 10$ **49.** $x = \sqrt{5}/5$ **51.** $x = 5$ **53.** $x = 4$ **55.** $x = \frac{1}{5}$ **57.** $x = \frac{1}{2}$

59. $x = e^{-2}$ **61.** 7.4 **63.** 1,000
65. (a) (i) $2 \cdot 5$ (ii) 5^2 (iii) $\frac{5}{2}$ (iv) $2^2 \div 5$ (b) (i) 2.30259 (ii) 3.21888 (iii) 0.91629 (iv) -0.22314
67. 2.4849 **69.** 0.28768 **71.** -0.405465 **73.** 0.47712 **75.** 1.68124 **77.** -0.69315 **79.** -0.28768 **81.** $x = 20$
83. $x = 2$ **85.** $x = 125$ **87.** $x = 2,000$ **89.** (a) Approximately 1 year (b) Approximately 7 years, 8 months
91. (a) Approximately 1 year, 4 months (b) Approximately 2 years, 3 months
93. The logarithm of x to the base b has the value y if $b^y = x$.
95. $\log_2(-4) = x$ requires that $2^x = -4$, which is true for no value of x. **97.** False, $\log_b a - \log_b c = \log_b(a/c)$
99. The logarithm to some base of the product of two numbers is equal to the sum of the logarithm to the same base of the first number and the logarithm to the same base of the second number.
101. The logarithm to some base of a number raised to a power is equal to the power times the logarithm to the same base of the number.

SECTION 8.3

1. $y = \log x$ **3.** $y = 3 \log(2x)$ **5.** $y = -2 \log_4 x$ **7.** $y = 3 - \log_{1/3}(-x)$ **9.**

11. (a) $q = \log_{1/3}(\frac{1}{10}p)$ (b) As the price is lowered (gets closer to zero), the quantity demanded increases very quickly.
(c) Demand for the product ceases when the price reaches $10.

13. (a) 40 decibels (b) 60 decibels (c) 140 decibels (d) $I = 10^{5.3} I_0$ watts

REVIEW EXERCISES

1.
x	y
2	48
$\frac{1}{2}$	6
0	3
-1	$\frac{3}{4}$

2.
x	y
3	$-3,645$
1	-45
-1	$-\frac{5}{9}$
-2	$-\frac{5}{81}$

3.
x	y
4	24
2	4
-1	0.27
-2	$\frac{1}{9}$

4.
x	y
3	-0.003
2	-0.03
-2	-300
-1.5	-94.9

5.
x	y
2	0.010
0	4
-0.5	17.93
-2.3	3,969.09

6. $y = 3^x$ **7.** $y = 4(2^{-x})$ **8.** $y = -4(2^x)$ **9.** $y = -\frac{1}{2}(3^{2x})$ **10.** $y = 3(4^{-x/2})$

   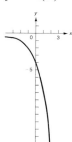 

11. $y = -\frac{3}{4}e^{-x}$ **12.** $y = 1.5e^x$ **13.** $y = 2^x - 2$ **14.** (a) $y = 1,200(1.06)^n$ (b) About $1,600 (c) $1,605.87 (d) $1,804.36

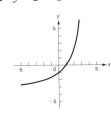
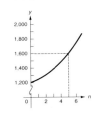

15. (a) $y = 500e^{0.08x}$ (b) About $575 (c) $575.15 (d) $688.55

16. (a) $y = 1,000(0.94)^n$ (b) About $780 (c) $780.75 (d) $538.61

17. 2 **18.** 0 **19.** -2 **20.** 3 **21.** 3 **22.** -2 **23.** 1 **24.** 3 **25.** 0 **26.** -2 **27.** -3 **28.** 6
29. 5 **30.** -3 **31.** -3 **32.** 3 **33.** 1.5 **34.** 47 **35.** 3.5 **36.** 50 **37.** $\frac{1}{2}$ **38.** x^2 **39.** 16
40. e^3 **41.** $9^{1/3}$ **42.** $5^{1/2}$ **43.** 36 **44.** 64 **45.** 1 **46.** 5 **47.** $\frac{1}{25}$ **48.** $\frac{1}{2}$ **49.** 2 **50.** 3 **51.** e^2
52. 8.31 **53.** 6.10 **54.** 0.01 **55.** 7.05 **56.** 5 **57.** 4 **58.** 3 **59.** 8 **60.** $\frac{1}{2}$ **61.** $\frac{1}{4}$ **62.** $\frac{2}{3}$
63. $x > 0$ and $x \neq 1$ **64.** 20 **65.** 42 **66.** 2 **67.** 3 **68.** 4 **69.** 3.58351 **70.** 0.81093 **71.** 2.77258
72. 1.09861 **73.** 3.00815 **74.** 0.60208 **75.** 0.95424 **76.** 1.55632 **77.** 0.77816 **78.** 1.38024
79. Approximately 2 years, 3 months **80.** Approximately 4 years, 2 months **81.** Approximately 1 year, 10 months
82. Approximately 6 years, 11 months
83. $y = \log_3 x$ **84.** $y = -3\log_2 x$ **85.** $y = -3\log_2(4x)$ **86.** $y = \log_3(-2x)$ **87.** $y = 2\log 3x$

88. $y = 1 - 2\ln x$

CHAPTER 9 SECTION 9.1

1. Yes **3.** No **5.** Yes **7.** Yes **9.** Yes **11.** No **13.** No **15.** No **17.** $a_3 = 100$ **19.** $a_4 = -24$
21. $a_7 = 2$ **23.** $a_6 = -\frac{5}{32}$ **25.** $a_4 = 0.007$ **27.** $a_{27} = 0$ **29.** 3, 12, 48, 192, 768 **31.** $-16, -4, -1, -\frac{1}{4}, -\frac{1}{16}$
33. 1, 0.01, 0.0001, 0.000001, 0.00000001 **35.** $-2, 4, -8, 16, -32$ **37.** $\frac{1}{27}, \frac{1}{9}, \frac{1}{3}, 1, 3$
39. 15, -75, 375, $-1,875$, 9,375 or 15, 75, 375, 1,875, 9,375 **41.** 31 **43.** $\frac{40}{9}$ **45.** $-\frac{31}{8}$ **47.** $-\frac{13}{4}$ **49.** 0
51. $2,051.45 **53.** About 1,778 **55.** 9 days **57.** $1,127.42
59. A geometric progression is a sequence of numbers such that each number in the sequence, after the first, can be obtained from the preceding number by multiplying by some constant.

APPENDIX C ANSWERS TO ODD-NUMBERED AND REVIEW EXERCISES   683

61. Quantities that are positive and form a geometric progression with a common ratio greater than 1 are said to increase geometrically.

SECTION 9.2

1. $1,000   **3.** $106.88   **5.** $192   **7.** $204.33   **9.** $1,050   **11.** Yes, for a monthly savings of about $1.   **13.** 7%
15. 9.5%   **17.** 9%   **19.** (a) $470   (b) 12.77%   **21.** 12.37%   **23.** 19.46%
25. (a) $7,578   (b) $1,578   (c) Monthly rate = 1.25%; yearly = 15%   **27.** $339.81   **29.** $800   **31.** Pay $200 now.

SECTION 9.3

1. Average daily balance = $82.70; interest = $1.24   **3.** Average daily balance = $54.70; interest = $0.82
5. Average daily balance = $591.23; interest = $2.71   **7.** Average daily balance = $559.73; interest = $2.57

SECTION 9.4

1. $140 + $149.80 + $160.29 + $171.51 + $2,000 = $2,621.60   **3.** $2,621.59   **5.** $3,693.64   **7.** $1,793.43
9. $S = \$634.87, I = \$134.87$ **11.** $S = \$511.71, I = \$36.71$ **13.** $2,392.83   **15.** $168.24   **17.** $1,305.88   **19.** Wait
21. (1) Physical situation: holding of a $2,000 savings certificate; physical question: What will it be worth in 2 years at 9% compounded monthly?
(2) Mathematical model: $S = P(1 + i)^n$; mathematical question: What is the value of S when $P = 2,000$, $i = 0.0075$, and $n = 24$?
(3) Mathematical solution: 2,392.82
(4) Interpretation: In 2 years, the certificate will be worth $2,392.82.
23. The difference between compound and simple interest is that (after the first period) compound interest is based on the interest from previous periods as well as on the original principal; simple interest is based only on the original principal.

SECTION 9.5

1. $4,345.97   **3.** $1,387.42   **5.** $11,275.24   **7.** No ($9,249.47)   **9.** $250,012.37   **11.** $2,408.65   **13.** $64.38
15. $66,311.89   **17.** $432.46
19. (1) Physical situation: accumulating funds for warehouse expansion; physical question: What payments must be made to accumulate the desired amount?
(2) Mathematical model: $R = S\left[\dfrac{i}{(1+i)^n - 1}\right]$; mathematical question: What is the value of R when $S = 700,000$, $i = 0.015$, and $n = 36$?
(3) Mathematical solution: $R = 14,806.68$
(4) Interpretation: Monthly payments of $14,806.68 must be made to accumulate the funds for the warehouse expansion.

SECTION 9.6

1. $9,277.73   **3.** $3,186.51   **5.** $18,589.97   **7.** $10,783.25   **9.** $1,067.45   **11.** $864.29   **13.** $3,027.20
15. $228.58   **17.** $2,967.60   **19.** $83.04; $2,989.44   **21.** $758.31; $208,991.60
23. Amortization, the process of paying off an interest-bearing debt by a series of equal payments, forms an annuity whose present value is the amount borrowed.

REVIEW EXERCISES

1. $-3, -6, -12, -24$ **2.** $15, -3, \frac{3}{5}, -\frac{3}{25}$ **3.** $6, 3, \frac{3}{2}, \frac{3}{4}$ **4.** $5, -10, 20, -40$ **5.** $-1, 0.4, -0.16, 0.064$ **6.** $\frac{4}{5}, \frac{2}{5}, \frac{1}{5}, \frac{1}{10}$
7. $2, 6, 18, 54$ or $2, -6, 18, -54$ **8.** $\frac{4}{3}, \frac{2}{3}, \frac{1}{3}, \frac{1}{6}$ or $\frac{4}{3}, -\frac{2}{3}, \frac{1}{3}, -\frac{1}{6}$ **9.** 135   **10.** 112   **11.** $-\frac{4}{9}$ **12.** 375   **13.** 0.0125
14. $\frac{27}{16}$ **15.** No   **16.** Yes   **17.** Yes   **18.** Yes   **19.** No   **20.** No   **21.** 160   **22.** -441 **23.** $\frac{488}{81}$

24. 1,995 **25.** $\frac{4}{9}(3 + \sqrt{3})$ **26.** 5.5555 **27.** 0 **28.** (a) $25.60 (b) $204.75 **29.** 13.17 **30.** $366,025
31. (a) $7,830.09 (b) $40,629.47 **32.** $225 **33.** $22.46 **34.** $936 **35.** $638.75 **36.** 12% **37.** 16%
38. 13% **39.** 11% **40.** $200 **41.** (a) $539.25 (b) 15% **42.** 17% **43.** (a) $1,226.42 (b) $756.76
44. $1,071.43 **45.** $500 now **46.** Average daily balance = $198.49; interest = $3.47
47. Average daily balance = $131.52; interest = $2.30 **48.** Average daily balance = $691.87; interest = $3.46
49. Average daily balance = $744.12; interest = $3.72 **50.** $937.67 **51.** $1,468.68 **52.** $S = \$1,166.50; I = \$191.50$
53. $S = \$2,077.15; I = \$527.15$ **54.** $2,326.99 **55.** $593.80 **56.** $599.89 **57.** $1,030.11 **58.** $600 now
59. $1,953.16 **60.** $972.58 **61.** $327.51 **62.** $321.49 **63.** No **64.** $1,570.83 **65.** $66,495.99 **66.** $2,326.83
67. $1,731.79 **68.** $174.73 **69.** $3,184.25 **70.** $19,727.55 **71.** $44,859.19 **72.** $77,910.04 **73.** $720.47; $159,891

CHAPTER 10 SECTION 10.1

1. 14 **3.** 0 **5.** $3x^2 + 6xh + 3h^2 + x + h$ **7.** $\frac{2}{81}$ **9.** $2(3^b)$ **11.** 9 **13.** $\frac{25}{4}$ **15.** $t^2 + 10t + 25$ **17.** $\frac{5}{14}$
19. $\frac{-a+1}{a^2-2}$ **21.** 48 **23.** $x^2 + 2xh + h^2 - 1$ **25.** $2x + h$ **27.** 2.2 **29.** $6t + 1$ **31.** 2 **33.** $2xh + h^2 - 2h$
35. $x^3 + 3x^2h + 3xh^2 + h^3$ **37.** $3x^2 + 3xh + h^2$ **39.** 0 **41.** 9 **43.** 121 **45.** 49 **47.** 9
49. (a) $f(x) = 500 + 0.15x, x \geq 0$ (b) 1,625 (c) 2,000
51. (a) For $x > 0$

$$f(x) = \begin{cases} 0.03x & 0 < x \leq 2{,}000 \\ 60 + 0.04(x - 2{,}000) & 2{,}000 < x \leq 4{,}000 \\ 140 + 0.05(x - 4{,}000) & 4{,}000 < x \leq 6{,}000 \\ 240 + 0.06(x - 6{,}000) & 6{,}000 < x \leq 10{,}000 \\ 480 + 0.07(x - 10{,}000) & 10{,}000 < x \end{cases}$$

(b) 450 (c) 2,282.50

53. The set of all real numbers **55.** The set of all real numbers **57.** The set of all positive real numbers
59. The set of all real numbers except -1 and 2 **61.** The set of all positive real numbers
63. The set of all real numbers greater than or equal to 1 **65.** The set of all real numbers greater than or equal to -1
67. The set of all real numbers
69. A function is a rule that associates a unique number to each member of some set D. The set D is the domain of the function.

SECTION 10.2

1. $f(x) = 2x + 3$ **3.** $y = 4 - 2x, x \geq 0$ **5.** $f(x) = 3$ **7.** $h(x) = x^2 + 2x - 1, x \leq 0$ **9.** $y = 5 \ln x$

11. $y = 3^{-x}$ **13.** $f(x) = \begin{cases} -x & x \leq 1 \\ x - 2 & x > 1 \end{cases}$ **15.** $g(x) = \begin{cases} x - 3 & x \leq 0 \\ x^2 + 2 & x > 0 \end{cases}$ **17.** $y = \frac{x^2 - 4}{x - 2}$

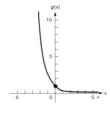

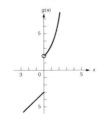

19. $y = \sqrt{x}$ **21.** Linear or polynomial   **23.** Linear or polynomial   **25.** Linear or polynomial

27. Quadratic or polynomial   **29.** Logarithmic   **31.** Exponential   **33.** $h(x) = 6x^2, x > 0$

35.

37. The graph of a function f is the graph of the equation $y = f(x)$.

39. A function f is linear if $f(x)$ can be written in the form $f(x) = ax + b$, where a and b are constants.
41. A function f is exponential if $f(x)$ can be written in the form $f(x) = Ab^{kx}$, where A, b, and k are constants, $b > 0$.
43. The function f is polynomial if $f(x)$ can be written in the form $f(x) = a_n x^n + a_{n-1} x^{n-1} + \cdots + a_1 x + a_0$, where the a_i's are constants.

SECTION 10.3

1. 7   **3.** 3   **5.** 1   **7.** 85   **9.** $-\frac{17}{9}$ **11.** $\frac{5}{4}$ **13.** 2   **15.** No limit exists.   **17.** 25   **19.** -5 **21.** 11
23. 1   **25.** 0   **27.** $\frac{2}{9}$ **29.** -1 **31.** No limit exists.   **33.** $\frac{1}{20}$ **35.** 8   **37.** $\frac{3}{7}$ **39.** No limit exists.
41. $\frac{1}{2}$ **43.** 4   **45.** 1,250   **47.** 3.25   **49.** 5.75   **51.** 165   **53.** 620

SECTION 10.4

1. $f(3) = 9 = \lim_{x \to 3} f(x)$ **3.** $g(-1) = 5 = \lim_{x \to -1} g(x)$ **5.** $f(4) = 81 = \lim_{x \to 4} f(x)$ **7.** $f(1) = 5 = \lim_{t \to 1} f(t)$
9. $g(3) = \frac{4}{5} = \lim_{t \to 3} \frac{4}{5} = \lim_{t \to 3} \frac{t^2 - 2t - 3}{t^2 - t - 6}$ **11.** Not continuous   **13.** Not continuous   **15.** Continuous   **17.** Continuous
19. y is not defined in the real number system for $x \le 4$.   **21.** f is not defined for $x = -1$.
23. $-3 < x < -2, -2 < x < -1, -1 < x < 0, 0 < x < 1, 1 < x < 2, 2 < x < 3$ **25.** $-\infty < x < 1, 1 < x < \infty$
27. $-\infty < t < -2, -2 < t < \infty$ **29.** $-\infty < t < 2, 2 < t < 3, 3 < t < \infty$ **31.** $x > 0$ **33.** $x < 0$ **35.** $x > 0$
37. A function f is continuous on an interval if it is continuous at each point in the interval.

SECTION 10.5

1. $x = 4, x = 1$ **3.** $x = 1$ **5.** $x = \frac{2}{3}, x = -\frac{1}{2}$ **7.** (a) $y = 3$
9. (a) No horizontal asymptote   (b) $\lim_{x \to \infty} f(x) = \infty$ (c) $\lim_{x \to -\infty} f(x) = \infty$

11. (a) $y = 0$ 13. (a) $x = 1$ (b) $y = 3$
15. (a) $x = -1$ (b) No horizontal asymptote (c) $\lim\limits_{x \to \infty} f(x) = \infty$ (d) $\lim\limits_{x \to -\infty} f(x) = -\infty$
17. (a) $x = 5$ (b) No horizontal asymptote (c) $\lim\limits_{x \to \infty} f(x) = -\infty$ (d) $\lim\limits_{x \to -\infty} f(x) = \infty$

19. $f(x) = \dfrac{x+3}{x-1}$ 21. $f(x) = \dfrac{-x}{x^2-1}$ 23. $g(x) = \dfrac{x^2-4}{x}$ 25. $y = \dfrac{8}{x-4}$

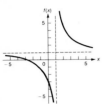 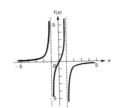 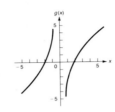 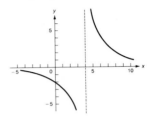

27. $y = \dfrac{2-x^2}{x}$ 29. $y = \dfrac{3x+3}{x^2+x-2}$ 31. (a) $y = \dfrac{5x}{2+x}$ (b) $x \geq 0$ (c) 5

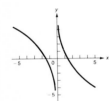 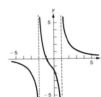 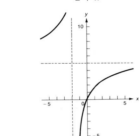

33. 0% 35. $\tfrac{3}{4}$ 37. $-\infty$ 39. 0
41. The vertical lines corresponding to values of x for which the denominator of a rational function is zero and at which the values of f become unbounded are called *vertical asymptotes* for the graph of a rational function.

REVIEW EXERCISES

1. 19 2. 4 3. 4 4. $t^2 + 3$ 5. All real numbers 6. -1 7. 0 8. -0.27 9. 2.86
10. All real numbers except $x = -1$ and $x = 2$ 11. 1 12. 0.1 13. 3 14. 11.27
15. All real numbers greater than or equal to -2 16. $\tfrac{3}{16}$ 17. $\tfrac{3}{2}$ 18. 3 19. 192 20. All real numbers
21. 3 22. 1 23. 0 24. -2 25. All positive real numbers 26. $\tfrac{3}{4}$ 27. $\tfrac{1}{2}$ 28. 0 29. $\tfrac{7}{8}$
30. All real numbers 31. 8 32. 199 33. 8 34. 289 35. All real numbers 36. 3 37. $4x + 2h - 1$
38. $-2x - h$ 39. $-2x - 4 - h$
40. $y = 4 - 2x$ 41. $f(x) = 2$ 42. $y = -x, x \leq 0$ 43. $g(x) = x^2 - 6x + 7$

44. $f(x) = -x^2 + 4x - 3$ **45.** $y = x^2 - 4$, $-2 \leq x \leq 2$ **46.** $g(x) = \log(x/2)$ **47.** $y = 3^x$

  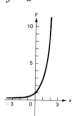

48. $y = \begin{cases} 2x & x \geq 2 \\ x - 2 & x < 2 \end{cases}$ **49.** $f(x) = \begin{cases} x^2 & x \geq 1 \\ e^{x-1} & x < 1 \end{cases}$ **50.** $y = \dfrac{x^2 - 25}{x + 5}$ **51.** $g(x) = \sqrt{x - 2}$

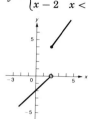  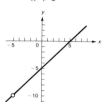

52. -15 **53.** 0 **54.** -8 **55.** $\frac{13}{4}$ **56.** 5 **57.** 11 **58.** -1 **59.** No limit exists   **60.** $\ln 2$ **61.** 0
62. -2 **63.** -2 **64.** $\frac{6}{7}$ **65.** $\frac{5}{6}$ **66.** No limit exists   **67.** 0 **68.** No limit exists   **69.** $\frac{5}{2}$ **70.** 6
71. -5 **72.** Continuous   **73.** Continuous   **74.** Continuous   **75.** Continuous   **76.** Continuous
77. Continuous   **78.** Continuous   **79.** Not continuous   **80.** Continuous   **81.** Continuous   **82.** Continuous
83. Continuous   **84.** Continuous   **85.** Not continuous   **86.** Not continuous   **87.** Continuous
88. Not continuous   **89.** Not continuous   **90.** $\lim\limits_{x \to -3} (7) = 7 \neq \lim\limits_{x \to -3} \dfrac{x^2 - 9}{x + 3} = -6$ **91.** $-\infty < x < \infty$
92. $-\infty < x < \infty$ **93.** $-\infty < t < 1, 1 < t < \infty$ **94.** $-\infty < x < \infty$ **95.** $-\infty < x < -4, -4 < x < 2, 2 < x < \infty$
96. $-\infty < x < -3, -3 < x < 1, 1 < x < \infty$ **97.** $x > 3, x < -3$
98. (a) $f(x) = \begin{cases} 200 + 0.05x & 0 \leq x \leq 1{,}000 \\ 250 + 0.10(x - 1{,}000) & x > 1{,}000 \end{cases}$ (b) $f(575) = 228.75$; $f(1{,}500) = 300$

(c)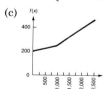

(d) $\lim\limits_{x \to 575} f(x) = 228.75$; $\lim\limits_{x \to 1{,}000} f(x) = 250$ (e) Continuous at $x = 500$; continuous at $x = 1{,}000$ (f) $0 \leq x < \infty$

99. (a) $g(x) = \begin{cases} 0.005x & 0 < x \leq 5{,}000 \\ 25 + 0.01(x - 5{,}000) & 5{,}000 < x \leq 10{,}000 \\ 75 + 0.02(x - 10{,}000) & 10{,}000 < x \leq 15{,}000 \\ 175 + 0.025(x - 15{,}000) & 15{,}000 < x \leq 20{,}000 \\ 300 + 0.03(x - 20{,}000) & 20{,}000 < x \leq 40{,}000 \\ 900 + 0.035(x - 40{,}000) & x > 40{,}000 \end{cases}$

(b) $g(7{,}500) = 50$; $g(12{,}750) = 130$; $g(48{,}500) = 1{,}197.50$ (c) $\lim\limits_{x \to 5{,}000} g(x) = 25$; $\lim\limits_{x \to 17{,}000} g(x) = 225$; $\lim\limits_{x \to 40{,}000} g(x) = 900$

(d) Continuous at $x = 5{,}000$; continuous at $x = 17{,}000$; continuous at $x = 40{,}000$ (e) $0 \leq x < \infty$
100. (a) $x = -3$ (b) $y = 0$ **101.** (a) No vertical asymptote   (b) $y = -3$ **102.** (a) $x = 1, x = -5$ (b) $y = 0$
103. (a) $x = -\frac{3}{4}$ (b) No horizontal asymptote   (c) $\lim\limits_{x \to \infty} f(x) = \infty$.   (d) $\lim\limits_{x \to -\infty} f(x) = -\infty$

104. (a) $x = -2$, $x = -3$ (b) No horizontal asymptote (c) $\lim_{x \to \infty} f(x) = \infty$ (d) $\lim_{x \to -\infty} f(x) = -\infty$

105. $x = -\frac{1}{2}$ (b) No horizontal asymptote (c) $\lim_{x \to \infty} f(x) = \infty$ (d) $\lim_{x \to -\infty} f(x) = \infty$

106. $f(x) = \dfrac{2x}{x+3}$ 107. $g(x) = \dfrac{x^2+9}{x^2-9}$ 108. $y = \dfrac{4}{x-2}$

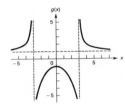

109. $y = \dfrac{x^2}{x+3}$ 110. $f(x) = \dfrac{x^2+1}{x}$ 111. $y = \dfrac{x^2}{2-x}$ 112. $\frac{1}{2}$

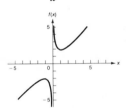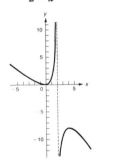

CHAPTER 11 SECTION 11.1

1. (a) 4 (b) $2x + h$ 3. (a) 24 (b) $23 + h$ 5. (a) 3 (b) 3   7. (a) -4 (b) $4x + 2h$
9. (a) $8x + 4h - 3$ (b) -3 (c) 1   11. (a) $-1/[x(x+h)]$ (b) $-\frac{1}{4}$ 13. (a) $-2x - h + 1$ (b) -2
15. The average rate of change of a function f on an interval of length h beginning at x is $\dfrac{f(x+h) - f(x)}{h}$.

SECTION 11.2

1. (a) $2x$ (b) 2 (c) -4 3. (a) $2x - 2$ (b) -2 (c) -10 5. (a) $1 - 6x$ (b) -11 (c) 1
7. (a) -2 (b) -2 (c) -2 9. (a) $4x$ (b) -8 (c) 5.2   11. (a) $2x - 3$ (b) -9 (c) -1.6
13. (a) 0 (b) 0 (c) 0   15. (a) $-2/(x+1)^2$ (b) $-\frac{2}{25}$ (c) -2 (d) $g(x)$ has no value at $x = -1$.
17. The derivative of a function f is the function f' where $f'(x) = \lim_{h \to 0} \dfrac{f(x+h) - f(x)}{h}$.

SECTION 11.3

1. $y = 4x - 3$ 3. (a) $y = -3x + 2$ (b) $y = 5x - 14$ 5. $y = -2x - 2$ 7. (a) $C'(x) = 4x + 16$ (b) $C'(20) = 96$
9. -0.6 pound per minute
11. (a) 3 seconds (b) 96 feet per second (c) The ball is moving away from the reference point (building top).
13. 4 words per hour

15. (1) Physical situation—Intravenous administration of an experimental drug to a monkey; physical question—What is the rate of change of the weight of the monkey 3 minutes after the administration of the drug is started?
    (2) Mathematical model—$f(x) = 70 - 0.1x^2$, $0 \leq x \leq 5$, the weight of the monkey, in pounds, x minutes after the drug is started; mathematical question—What is the value of $f'(3)$?
    (3) Mathematical solution—$f'(3) = -0.6$
    (4) Interpretation—The monkey's weight is changing at a rate of -0.6 pound per minute after 3 minutes.
17. Marginal change is the instantaneous rate of change of a quantity.

SECTION 11.4

1. $6x^5$ 3. $\dfrac{3\sqrt[4]{x^3}}{4x}$ 5. $\dfrac{-2\sqrt[3]{t}}{3t^2}$ 7. $-\dfrac{2}{27}$ 9. 1 11. $8x^3$ 13. $u^{1/2}$ 15. $\dfrac{3}{4}$ 17. 0 19. 0
21. $20x^3 + 15x^4$ 23. -11 25. $6x - \dfrac{2\sqrt[3]{x}}{x}$ 27. $\dfrac{-1}{u^2} + 1$ 29. $\dfrac{29}{6}$ 31. $f'(x) = 6x^2 - 3$ 33. $g'(64) = \dfrac{19}{16}$
35. $y = -11x - 25$ 37. $y = 6x + 1$ 39. (a) $f'(5) = 226$ feet per second (b) $f'(1.5) = 21.25$ feet per second
41. (a) $S'(x) = \frac{1}{2}x^2 - 6x + 16$ (b) $S'(10) = 6$ (c) $S'(11) = 10.5$ 43. 324π cubic centimeters per centimeter
45. $f'(x) = \lim\limits_{h \to 0} \dfrac{u(x+h) - v(x+h) - u(x) + v(x)}{h}$

$= \lim\limits_{h \to 0} \dfrac{u(x+h) - u(x) - [v(x+h) - v(x)]}{h}$

$= \lim\limits_{h \to 0} \dfrac{u(x+h) - u(x)}{h} - \lim\limits_{h \to 0} \dfrac{v(x+h) - v(x)}{h} = u'(x) - v'(x)$

47. The power rule states that for any constant n,
$\dfrac{d}{dx}(x^n) = nx^{n-1}$

49. The derivative of any constant function is zero.

SECTION 11.5

1. $\dfrac{dy}{dx} = 2(x^3 + 2)(3x^2) = 6x^2(x^3 + 2)$ 3. $\dfrac{dy}{dx} = -4(2x+1)^{-5}(2) = \dfrac{-8}{(2x+1)^5}$
5. $g'(t) = 3(4t^2 + t^{-1})^2(8t - t^{-2}) = \dfrac{3(4t^3 + 1)^2(8t^3 - 1)}{t^4}$ 7. $f'(x) = \dfrac{1}{2}\left(x + \dfrac{1}{x}\right)^{-1/2}\left(1 - \dfrac{1}{x^2}\right) = \dfrac{1}{2}\left[\dfrac{x^2 - 1}{x(x^2 + 1)}\right]\left(\dfrac{x^2 + 1}{x}\right)^{1/2}$
9. $f'(x) = 4(2x^{-2} - 3)^3(-4x^{-3}) = \dfrac{-16(2 - 3x^2)^3}{x^9}$ 11. $g'(x) = -\dfrac{1}{2}(x^2 - x)^{-3/2}(2x - 1) = \dfrac{\sqrt{x^2 - x}(2x - 1)}{-2(x^2 - x)^2}$
13. $\dfrac{dy}{dx} = 6(3x^4 - x^2)^2(12x^3 - 2x) = (72x^3 - 12x)(3x^4 - x^2)^2$ 15. $g'(x) = (x^{-1} + 3x)(-x^{-2} + 3) = \dfrac{9x^4 - 1}{x^3}$ 17. $f'(3) = -24$
19. $g'(\frac{1}{2}) = \dfrac{81}{16}$ 21. $6x^2 - 6$ 23. $\dfrac{-5}{x^2}$ 25. 0 27. $\dfrac{6}{(x+6)^3}$
29. $\dfrac{d}{dx}(3x - 6x^2)^{2/3} = \dfrac{2}{3}(3x - 6x^2)^{-1/3}(3 - 12x) = \dfrac{(2 - 8x)(3x - 6x^2)^{2/3}}{3x - 6x^2}$ 31. -4 33. $27x^2 + 3(x + 2)^2$ 35. 2
37. (a) 8 (b) -48

39. Chain rule: If y can be written as a function of u where u is a function of x, then $\dfrac{dy}{dx} = \dfrac{dy}{du} \cdot \dfrac{du}{dx}$.

Power rule: If u is a function of x having a derivative du/dx, then $\dfrac{d}{dx}(u^n) = (nu^{n-1}) \cdot \dfrac{du}{dx}$.

SECTION 11.6

1. $f'(x) = 2x(2x^3 - 4) + x^2(6x^2) = 10x^4 - 8x$ 3. $f'(x) = 2(4x + 6) + 4(2x - 1) = 16x + 8$ 5. -9

7. $\dfrac{dy}{dt} = 2(2t^2 + t)(4t + 1)(-4t^3) + (2t^2 + t)^2(-12t^2) = (2t^2 + t)(-4t^2)(14t^2 + 5t)$ 9. $f'(x) = \dfrac{(1)(x^2 - 4) - (2x)x}{(x^2 - 4)^2} = \dfrac{-x^2 - 4}{(x^2 - 4)^2}$

11. $y' = \dfrac{5(x - 7x^2) - (1 - 14x)(5x + 1)}{(x - 7x^2)^2} = \dfrac{35x^2 + 14x - 1}{x^2(1 - 7x)^2}$ 13. $f'(t) = \dfrac{-(3t^2)(-4)}{(t^3 - 1)^2} = \dfrac{12t^2}{(t^3 - 1)^2}$

15. $f'(x) = 2\left(\dfrac{x}{2x - 3}\right)\left[\dfrac{(1)(2x - 3) - 2x}{(2x - 3)^2}\right] = \dfrac{-6x}{(2x - 3)^3}$ 17. $g'(x) = 6(-1)\left(\dfrac{5x}{x + 3}\right)^{-2}\left[\dfrac{5(x + 3) - 1(5x)}{(x + 3)^2}\right] = \dfrac{-18}{5x}$

19. $y' = \dfrac{1}{2}\left(\dfrac{2x}{x^2 + 1}\right)^{-1/2}\left[\dfrac{2(x^2 + 1) - 2x(2x)}{(x^2 + 1)^2}\right] = \dfrac{1}{2}\left(\dfrac{x^2 + 1}{2x}\right)^{1/2}\left[\dfrac{-2x^2 + 2}{(x^2 + 1)^2}\right]$

21. $f'(x) = 4[(2)(1 - x^2) + (-2x)(2x - 3)] = 4(2 + 6x - 6x^2)$

23. $y' = \dfrac{50}{(5 + 2x)^2}; y'\big|_{x=5} = \tfrac{2}{9}$ item per hour   25. $y' = \dfrac{30 - 12x^2}{(4x^2 + 10)^2}; y'\big|_{x=2} = -0.027\%$ per hour   27. $y = -\tfrac{9}{2}x + 4$

29. (1) Physical situation—Concentration of a drug in the bloodstream; physical question—What is the rate at which the concentration is decreasing 2 hours after the drug is administered?
   (2) Mathematical model—$y = \dfrac{3x}{4x^2 + 10}$, $x \geq 0$; mathematical question—What is the value of $y'\big|_{x=2}$?
   (3) Mathematical solution—$y'\big|_{x=2} = -0.027$
   (4) Interpretation—The concentration is decreasing at a rate of 0.027% per hour.

31. The derivative of a quotient is the denominator times the derivative of the numerator minus the numerator times the derivative of the denominator, all divided by the denominator squared.

SECTION 11.7

1. Product rule, Theorem 9(a) (chain rule and Theorem 9), power rule   3. $3(7^x) \ln 7$ 5. $e^x(1 + x)$ 7. $3e^x + 2$

9. $-1/t$ 11. $\ln x + 1$ 13. $2e^{2x}$ 15. $\dfrac{3^x(x \ln 3 - 2)}{x^3}$ 17. $e^{-x-1}[2x - (x^2 + 1)]$ 19. $3/t$ 21. $\dfrac{2 \log x}{x \ln 10}$

23. $\dfrac{2x + 1}{x^2 + x}$ 25. $4\dfrac{(\ln x)^3(1 - \ln x)}{x^5}$ 27. 0

29. (a) $y'\big|_{x=3} = -\tfrac{3}{8} \ln 2 = -0.26$ error per day
   (b) y gets very close to zero. The longer the employee is on the job, the fewer errors he/she makes (approaches zero).

31. $y'\big|_{t=5,000} = -0.0010$ gram per year   33. $d'\big|_{t=2} = \tfrac{12}{13}$ foot per second   35. $y'\big|_{x=9} = 0.1$; $y = 0.1x + 1.1$

SECTION 11.8

1. b   3. a   5. b   7. a   9. $-\infty < x < 0, 0 < x < \infty$ 11. $-\infty < x < -4, -4 < x < \infty$
13. $-\infty < x < -2, -2 < x < \infty$ 15. $0 < x < \infty$ 17. (a) $-1 < x < 0, 0 < x < 1, 1 < x < 2$ (b) $f'(x) = 0$ over each interval
19. A function is said to be differentiable at a value of x if the function has a derivative at that value of x.

REVIEW EXERCISES

1. (a) $2x + h$ (b) 8 2. (a) $2 - 2x - h$ (b) 2 3. (a) 3 (b) 3 4. (a) $\dfrac{-1}{(x + h - 1)(x - 1)}$ (b) -1

5. (a) 6 (b) $2x$ 6. (a) $-2x$ (b) 2 7. (a) $f'(6) = 14$ (b) $f'(x) = 2 + 2x$ 8. (a) $g'(0) = 0$ (b) $g'(x) = -4x$

9. (a) $f'(t) = 3$ (b) $f'(4) = 3$ 10. (a) $f'(x) = \dfrac{-1}{(x - 1)^2}$ (b) $f'(0.5) = -4$ 11. (a) $g'(x) = 2x + 1$ (b) $g'(-2) = -3$

12. The instantaneous rate of change and the derivative are equal; $f'(x) = x + \tfrac{1}{2}$ 13. $f'(x) = 12x^3$; 2 14. $f'(x) = 4x - 6$; 4

15. $g'(x) = \dfrac{-x - 32}{4x^3}$; 4 16. $y' = \dfrac{\sqrt{x}}{2x} + \dfrac{2\sqrt[3]{x^2}}{3x}$; 4 17. $f'(t) = 2t + 2$; 4 18. $y' = -5x^4$; 2 19. $y' = -5(-x)^4$; 5

20. $g'(x) = 8(2x + 3)^3$; 5   21. $g'(x) = \dfrac{-6x}{(x^2 + 1)^4}$; 5   22. $f'(t) = \left(\dfrac{4}{t}\right)(\ln t + 2)^3$; 5   23. $y' = -8(3x + x^2)(3 + 2x)$; 2

24. $g'(x) = 0$; 3   25. $h'(t) = 1$; 1   26. $f'(x) = \dfrac{(2x - 1 - \sqrt{x})\sqrt{x + \sqrt{x}}}{4(x^2 - x)}$; 5   27. $f'(x) = \dfrac{e^{\sqrt{x}}\sqrt[3]{x}}{3x}$; 8(a)

28. $g'(x) = \dfrac{\sqrt{x}}{2x}e^{\sqrt{x}} - \dfrac{\sqrt{x}}{2x}e^{-\sqrt{x}}$; 4   29. $y' = \dfrac{7\sqrt{x}}{2x}e^{\sqrt{x}}$; 2   30. $y' = \dfrac{6}{x}$; 9(a)   31. $y' = \dfrac{3(\ln x)^2}{x}$; 5   32. $f'(x) = \dfrac{-5}{(2x - 3)^2}$; 7

33. $y' = \dfrac{-2}{(x + 1)(x - 1)}$; 9(a)   34. $f'(x) = 3x^2(\ln x) + x^2$; 6   35. $g'(x) = (3x^2 - x^3)e^{-x}$; 6   36. $g'(x) = -4e^x(1 + x)$; 6

37. $h'(x) = \dfrac{\sqrt{x + 3}}{2(x + 3)}e^{\sqrt{x+3}}$; 8(a)   38. $g'(t) = \dfrac{\sqrt{e^{t+3}}}{2}$; 5   39. $h'(t) = \dfrac{2 - 2\ln t}{t^2}$; 7   40. $g'(r) = \dfrac{4r^3(7 - r^2)}{(r^2 + 7)^5}$; 5

41. $f'(x) = \dfrac{-2}{(x - 4)^2}$; 7   42. $g'(x) = \dfrac{1 - 2\log x \ln 10}{(\ln 10)x^3}$; 7   43. $y' = (\log x)^2\left(\log x + \dfrac{3}{\ln 10}\right)$; 6   44. $h'(x) = (4^x - 4^{-x})\ln 4$; 4

45. $f'(x) = \dfrac{-e^x}{(1 + e^x)^2}$; 7   46. $y' = 12x^2 + 2x - 12$; 6   47. $f'(x) = \dfrac{x(1 + 2x)(4x^2 - 1) + 2(x + x^2)}{x^3}$; 7

48. $y' = 6(3x^4 + 1)(2x - 7)^2(11x^4 - 28x^3 + 1)$; 6   49. $g'(t) = \dfrac{-12t^2 - 40t}{(3t + 5)^2}$; 7   50. $f'(x) = \dfrac{-x - 3}{(x - 1)^3}$; 7   51. $f'(x) = \dfrac{4x^7(x - 2)}{(x - 1)^5}$; 5

52. $f'(t) = \dfrac{-24t^2}{(3t - 4)^4}$; 2   53. $y = 10x - 16$ 54. $y = -6x - 17$ 55. $y = -2x - 2$ 56. $y = 0.64x + 0.08$ 57. $f'(1) = -8$
58. $g'(-1) = -\tfrac{2}{9}$ 59. $f'(0) = 0.69315$ 60. $g'(3) = 18$ 61. 791 people per year   62. Leaking 1 gallon per second
63. $g'(2) = -\tfrac{7}{2}$ or $3,500 per month   64. 6.9 million per second   65. $C'(3) = 76$ or $76,000 per month
66. $R'(10,000) = \$5.50$ per week   67. $P'(50) = \$17$ 68. 48 feet per second
69. $s'(10) = -64$ feet per second; toward the ground
70. (1) Physical situation—A leaking 20-gallon gasoline tank; physical question—How fast is the gasoline tank leaking 10 seconds after the leak occurs?
    (2) Mathematical model—$V = 20\left(1 - \dfrac{t^2}{400}\right)$; mathematical question—What is the value of $V'|_{t=10}$?
    (3) Mathematical solution—$V'|_{t=10} = -1$
    (4) Interpretation—The tank is leaking at a rate of 1 gallon per second.
71. (1) Physical situation—The profit of a manufacturer of electric toasters; physical question—What is the approximate profit from producing the 51st toaster?
    (2) Mathematical model—$P(x) = 18x - 0.01x^2 - 4,000$; mathematical question—What is the value of $P'(50)$?
    (3) Mathematical solution—$P'(50) = 17$
    (4) Interpretation—The approximate profit from producing the 51st unit will be $17.
72. A function is a rule that assigns a unique number to each element of some set D.
73. The derivative of a function f is the function f' where $f'(x) = \lim\limits_{h \to 0} \dfrac{f(x + h) - f(x)}{h}$.

CHAPTER 12   SECTION 12.1

1. $f''(x) = 12x^2 - 12x$; $f'''(x) = 24x - 12$ 3. $f''(x) = 6 - 40x^3$; $f'''(x) = -120x^2$ 5. $y'' = 12x^2 + 6x + 2$; $y''' = 24x + 6$
7. $g''(x) = 0$; $g'''(x) = 0$ 9. $y'' = -\dfrac{1}{4}\left(\dfrac{\sqrt{x}}{x^2}\right)$; $y''' = \dfrac{3}{8}\left(\dfrac{\sqrt{x}}{x^3}\right)$ 11. $f''(x) = \dfrac{15}{4}\sqrt{x} + \dfrac{3\sqrt{x}}{4x}$; $f'''(x) = \dfrac{15\sqrt{x}}{8x} - \dfrac{3\sqrt{x}}{8x^2}$
13. $f''(t) = \dfrac{4}{t^3}$; $f'''(t) = \dfrac{-12}{t^4}$ 15. $y'' = 6(x - 3)$; $y''' = 6$ 17. $f''(x) = -\dfrac{25\sqrt{3 - 5x}}{4(3 - 5x)^2}$; $f'''(x) = -\dfrac{375\sqrt{3 - 5x}}{8(3 - 5x)^3}$

19. $f''(t) = -e^t$; $f'''(t) = -e^t$ 21. $g''(x) = 2e^x + xe^x$; $g'''(x) = 3e^x + xe^x$ 23. $f''(x) = \dfrac{-1}{x^2}$; $f'''(x) = \dfrac{2}{x^3}$

25. $f''(x) = \dfrac{-2}{x^2}$; $f'''(x) = \dfrac{4}{x^3}$ 27. $y'' = \dfrac{2}{x^3}$; $y''' = \dfrac{-6}{x^4}$

29. The second derivative of a function f is the derivative of the first derivative of f.
31. The fourth derivative of a function f is the derivative of the third derivative of f.

SECTION 12.2

1. $y = \frac{1}{3}x^3 - x^2 + 2$

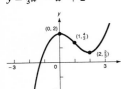

3. $y = x^3 - 3x + 1$

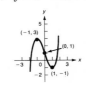

5. $f(x) = -2x^3 + 6x^2 - 4$

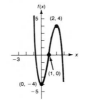

7. $y = x^2 + 4x + 6$

9. $y = -x^2 + x - 3$

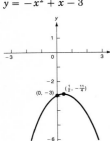

11. $f(x) = -x^4$

13. $y = 1 + x^{1/3}$

15. $y = (1 + x)^{1/3}$

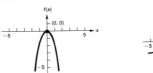

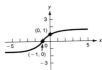

17. $f(x) = \frac{1}{2}x^2 - \ln x$

19. (a) The supply will increase for $x \geq 0$.   (b) The supply will increase for $x \geq 0$.

21. (a) $C'(x) = 0.06x^2 - 0.3x + 2$ (b) $R'(x) = 5$ (c) $P(x) = R(x) - C(x) = 5x - 0.02x^3 + 0.15x^2 - 2x - 10$, $x \geq 0$
    (d) $P'(x) = 5 - 0.06x^2 + 0.3x - 2$; profit is decreasing for sales levels greater than 10 units; profit is increasing for sales levels greater than or equal to 0 but less than 10 units.
    (e) Profit is maximum for $x = 10$. The maximum profit is $1,500. There is no minimal profit.

23. (1) Physical situation—Monthly sales profits; physical question—Over what sales levels are profits increasing? Decreasing?
    (2) Mathematical model—$P(x) = 100x - x^2$, $x \geq 0$; mathematical question—Over what intervals is P' positive? Negative?
    (3) Mathematical solution—$P'(x) > 0$ for $0 < x < 50$, $P'(x) < 0$ for $x > 50$
    (4) Interpretation—Profits are increasing for sales levels between 0 and 50 units. Profits are decreasing for sales levels above 50 units.

SECTION 12.3

1. y has a relative minimum of -5 at $x = 3$.   3. f has a relative maximum of $\frac{35}{4}$ at $x = \frac{3}{2}$.
5. f has a relative maximum of $\frac{13}{6}$ at $x = -1$ and a relative minimum of $-\frac{7}{3}$ at $x = 2$.
7. y has a relative maximum of 46 at $x = -4$ and a relative minimum of $-\frac{130}{27}$ at $x = \frac{2}{3}$.
9. g has neither a relative maximum nor a relative minimum.
11. f has relative minima of 0 at $x = 1$ and $x = -1$, and a relative maximum of 1 at $x = 0$.
13. y has a relative minimum of 0 at $x = 0$.   15. f has a relative minimum of 1 at $x = 0$.
17. f has a relative minimum of -7 at $x = -2$.   19. f has a relative maximum of 9 at $x = 2$.
21. y has a relative maximum of $\frac{25}{3}$ at $x = -2$ and a relative minimum of $-\frac{25}{2}$ at $x = 3$.
23. y has neither a relative maximum nor a relative minimum.
25. f has a relative maximum of 0 at $x = 0$ and a relative minimum of $-\frac{1}{24}$ at $x = \frac{1}{2}$.

27. f has a relative minimum of -48 at $x = -2$.   29. y has a relative minimum of 0 at $x = 3$.
31. f has a relative maximum of -10 at $x = -5$ and a relative minimum of 10 at $x = 5$.
33. y has a relative minimum of 0 at $x = -2$.   35. y has a relative minimum of 1 at $x = 0$.
37.

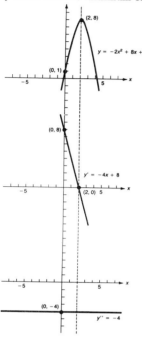

41. If $x = a$ is a critical value of the function f, and the first derivative is positive for all values of x just to the left of a and is negative for all values of x just to the right of a, then f has a relative maximum at $x = a$. If the first derivative is negative for all values of x just to the left of a and positive for all values of x just to the right of a, then f has a relative minimum at $x = a$.
43. A function f has a relative maximum [minimum] at $x = a$ if $f(a)$ is greater than $f(x)$ [$f(a)$ is less than $f(x)$] for all x in some interval containing a in its interior.

SECTION 12.4

1. f has an absolute maximum of 7 at $x = 4$ and an absolute minimum of 3 at $x = 2$.
3. y has an absolute maximum of 7 at $x = 2$ and an absolute minimum of 3 at $x = 0$.
5. f has an absolute maximum of 17 at $x = 4$ and an absolute minimum of -27 at $x = 2$.
7. f has an absolute maximum of 81 at $x = -4$ and an absolute minimum of -27 at $x = 2$.
9. g has an absolute maximum of $\frac{230}{27}$ at $x = \frac{4}{3}$ and an absolute minimum of 2 at $x = 0$.
11. y has absolute maxima of 9 at $x = 2$ and $x = -2$, and absolute minima of 0 at $x = 1$ and $x = -1$.
13. f has an absolute minimum of 0 at $x = 3$ and an absolute maximum of 3 at $x = 0$.
15. g has an absolute maximum of 1 at $x = 1$ and an absolute minimum of -1.55 at $x = 2$.
17. A function f has an absolute maximum [minimum] at $x = a$ if $f(x)$ is less than or equal to $f(a)$ [$f(x)$ is greater than or equal to $f(a)$] for all values of x in the domain of f.

SECTION 12.5

1. Maximum volume is approximately 303 cubic inches.   3. $50   5. $4,225   7. 225 square feet
9. Maximum profit of $6,250,000 is reached with sales of 250 units.
11. Average cost per unit is minimal when 400 units are produced.
13. $P(x) = 3x^2 - 60x + 310$; profit is minimal when 1,000 units are sold; profit at this sales level is $10,000.
15. (a) Statement is true for Exercise 14 but false for Exercise 13.
    (b) No, marginal cost equals marginal revenue at the critical point(s) of the profit function, but these point(s) are not always maxima.

19. After 2.5 hours the maximum number of words will be memorized; the maximum number is 25.
21. The population has a relative maximum of 494.4 million at $t = \frac{10}{3}$ seconds. This is not an absolute maximum, since y has another critical value in the interval $[0, \infty)$.
23. 66,000 miles
25. (1) Physical situation—A psychologist's learning experiment where new words are memorized; physical question—After how many hours of study will an individual have memorized the most words? What is this maximum number?
    (2) Mathematical model—$f(x) = 20(x - 0.2x^2)$, $0 \le x \le 4$; mathematical question—What is the absolute maximum of $f(x)$? For what value of x does this maximum occur?
    (3) Mathematical solution—$f(x)$ has an absolute maximum of 25 at $x = 2.5$.
    (4) Interpretation—After 2.5 hours a person will have memorized a maximum number of words. That maximum number is 25.

SECTION 12.6

1. $(1, -5)$ 3. No inflection points   5. $(3, -\frac{297}{2})$ and $(-2, -56)$ 7. No inflection points   9. $(-2, -0.27)$

(For Exercises 11–27, see graphs in Section 12.2 for the location of the inflection points.)
11. $(1, \frac{4}{3})$ 13. $(0, 1)$ 15. $(1, 0)$ 17. No inflection points   19. No inflection points   21. No inflection points
23. $(0, 1)$ 25. $(-1, 0)$ 27. No inflection points

REVIEW EXERCISES

1. $f''(x) = 24x^2 - 10$; $f'''(x) = 48x$ 2. $f''(x) = 4x^3 + 6x$; $f'''(x) = 12x^2 + 6$ 3. $y'' = \frac{-\sqrt{x}(x+3)}{4x^3}$; $y''' = \frac{3\sqrt{x}(x+5)}{8x^4}$

4. $g''(x) = \frac{-3\sqrt[4]{x^3}}{2x^2} + \frac{4\sqrt[3]{(3x)^2}}{(3x)^3}$; $g'''(x) = \frac{15\sqrt[4]{x^3}}{8x^3} - \frac{28\sqrt[3]{(3x)^2}}{(3x)^4}$

5. $g''(t) = -\frac{9\sqrt{2-3t}}{4(2-3t)^2}$; $g'''(t) = -\frac{81\sqrt{2-3t}}{8(2-3t)^3}$ 6. $y'' = 24(2x - 1)$; $y''' = 48$

7. $f''(x) = 2e^{x^2} + 4x^2 e^{x^2}$; $f'''(x) = 4xe^{x^2} + 8xe^{x^2} + 8x^3 e^{x^2}$ 8. $g''(x) = 2 \ln x + 3$; $g'''(x) = \frac{2}{x}$

9. $g''(x) = \frac{-1}{(2-x)^2}$; $g'''(x) = \frac{-2}{(2-x)^3}$ 10. $f''(x) = \frac{2}{(x-1)^3}$; $f'''(x) = \frac{-6}{(x-1)^4}$

11. (a) $x = 0$ and $x = 2$
    (b) Increasing: $x < 0$ and $x > 2$
        Decreasing: $0 < x < 2$
    (c) $(1, 0)$
    (d)

12. (a) $x = 3$ and $x = -1$
    (b) Increasing: $-1 < x < 3$
        Decreasing: $x < -1$ and $x > 3$
    (c) $(1, \frac{2}{3})$
    (d)

13. (a) $x = -\frac{3}{2}$
    (b) Increasing: $x < -\frac{3}{2}$
        Decreasing: $x > -\frac{3}{2}$
    (c) No inflection points
    (d)

14. (a) $x = \frac{1}{2}$
    (b) Increasing: $x > \frac{1}{2}$
        Decreasing: $x < \frac{1}{2}$
    (c) No inflection points
    (d)

15. (a) $x = 0$ and $x = \frac{3}{4}$
    (b) Increasing: $x > \frac{3}{4}$
        Decreasing: $x < 0$, $0 < x < \frac{3}{4}$
    (c) $(0, 0)$ and $(\frac{1}{2}, -\frac{1}{16})$
    (d)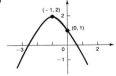

16. (a) $x = -1$
    (b) Increasing: $x < -1$
        Decreasing: $x > -1$
    (c) No inflection points
    (d)

17. (a) $x = 2$
    (b) Increasing: $x > 2$
        Decreasing: $x < 2$
    (c) No inflection points
    (d)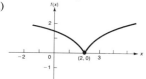

18. (a) $x = 1$
    (b) Increasing: $x < 1$
        Decreasing: $x > 1$
    (c) $(2, 0.28)$
    (d)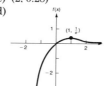

19. (a) None
    (b) Increasing: $x > 0$
    (c) No inflection points
    (d)

20. (a) $x = 1$
    (b) Increasing: $x > 1$
        Decreasing: $0 < x < 1$
    (c) No inflection points
    (d)

21. (a) Increasing: $0 \le x < 3$; decreasing: $x > 3$ (b) Increasing: $x > \frac{5}{2}$; decreasing: $0 \le x < \frac{5}{2}$

22. (a) Increasing: $0 < x < 2$; decreasing: $2 < x < 22$ (b) Increasing: $0 < x < 1$; decreasing: $x > 1$
23. Cost is decreasing for $0 < x < 12$ and increasing for $x > 12$. 24. y has a relative minimum of -17 at $x = -3$.
25. y has a relative maximum of $\frac{25}{4}$ at $x = \frac{5}{4}$. 26. y has a relative minimum of 0 at $x = 0$ and a relative maximum of 1 at $x = 1$.
27. f has a relative maximum of 2.22 at $x = 1 - \sqrt{2}$ and a relative minimum of -1.55 at $x = 1 + \sqrt{2}$.
28. f has no relative maxima or minima.
29. g has a relative maximum of 3 at $x = 0$, and relative minima of 2 at $x = 1$ and $x = -1$.
30. y has no relative maxima or minima. 31. y has a relative maximum of 1 at $x = 0$.
32. y has a relative minimum of 8 at $x = 2$.
33. f has a relative maximum of -3 at $x = -3$ and a relative minimum of 0 at $x = 0$.
34. f has a relative minimum of $-\frac{5}{4}$ at $x = \frac{3}{2}$. 35. y has a relative maximum of 13 at $x = -3$.
36. y has no relative maxima or minima.
37. f has a relative maximum of 1 at $x = 0$, and relative minima of -15 at $x = 2$ and $x = -2$.
38. y has a relative minimum of 3 at $x = 0$. 39. y has no relative maxima or minima
40. f has no relative maxima or minima. 41. g has a relative minimum of -9 at $x = -1$.
42. y has a relative minimum of -81 at $x = -3$.
43. f has a relative minimum of 2.83 at $x = \sqrt{2}$ and a relative maximum of -2.83 at $x = -\sqrt{2}$.
44. f has a relative minimum of -0.39 at $x = 1$. 45. y has a relative minimum of 0 at $x = -5$.
46. y has an absolute minimum of $\frac{14}{3}$ at $x = \frac{2}{3}$ and an absolute maximum of 38 at $x = 4$.
47. y has an absolute maximum of 70 at $x = -4$ and an absolute minimum of 6 at $x = 0$.
48. f has an absolute maximum of 6 at $x = -1$ and an absolute minimum of -21 at $x = 2$.
49. f has an absolute minimum of -16 at $x = 1$ and an absolute maximum of 11 at $x = -2$.
50. f has an absolute minimum of -21 at $x = 2$ and an absolute maximum of 11 at $x = -2$.
51. y has an absolute maximum of 1 at $x = 0$ and an absolute minimum of -5.75 at $x = 3$.
52. y has an absolute maximum of 1.396 at $x = -0.5$ and an absolute minimum of -5.75 at $x = 3$.
53. y has an absolute minimum of 1 at $x = 0$ and an absolute maximum of 2 at $x = \ln 2$.
54. f has an absolute minimum of 0 at $x = 0$ and an absolute maximum of 0.3679 at $x = 1$.
55. g has an absolute minimum of -3 at $x = 1$. 56. y has an absolute maximum of 81 at $x = 3$.
57. 12 inches by 8 inches 58. 9 inches by 6 inches 59. 720,000 square feet 60. 4 feet by 4 feet by $\frac{25}{8}$ feet
61. 687 units 62. 200 units 63. Daily production should be 150 hand calculators to reach a maximum profit of $457.50.
64. The concentration is greatest at 5 centimeters.
65. (1) Physical situation—The designing of a shipping carton; physical question—What are the dimensions of the box with minimal cost?

    (2) Mathematical model—$C(x) = 0.5x^2 + \dfrac{64}{x}$, $x > 0$; mathematical question—What is the value of x at the absolute minimum of $C(x)$?

(3) Mathematical solution—$x = 4$
(4) Interpretation—The dimensions of the box with minimal cost are 4 feet by 4 feet by $\frac{25}{8}$ feet.
66. (1) Physical situation—Profit from sales of a product; physical question—What is the sales level when profit is maximum?
(2) Mathematical model—$P(x) = -x^3 + 9x^2 + 18x - 90$; mathematical question—What is the value of x at the absolute maximum of $P(x)$?
(3) Mathematical solution—$x = 6.87$
(4) Interpretation—Profit will be maximum when sales are 687 units.

CHAPTER 13 SECTION 13.1

1. $\frac{1}{3}x^3 + C$ 3. $\frac{-1}{3t^3} + C$ 5. $-3t + C$ 7. $x + C$ 9. $\frac{2}{3}t^{3/2} + C$ 11. $\frac{3}{8}x^8 + C$ 13. $\frac{2(3^x)}{\ln 3} + C$ 15. $\frac{1}{4}e^x + C$
17. $2\sqrt{x} - e^x + C$ 19. $t^2 + 5e^t + C$ 21. $\frac{2}{3}x^{3/2} - \frac{3}{4}x^{4/3} + \frac{8}{5}x^{5/4} + C$ 23. $2t^2 + \ln t + C$ 25. $\frac{1}{3}x^3 - \frac{1}{2}x^2 - x + C$
27. $\frac{1}{2}x^2 + 3x + C$ 29.

x	-3	-1	0	4
$\|x+1\|$	1	0	1	5
$\|x-1\|$	2			3

31. The indefinite integral of a function $f(x)$ is the function $F(x) + C$, where $F'(x) = f(x)$ and C is an arbitrary constant.

SECTION 13.2

1. $3 \ln x + C$ 3. $\frac{4^t}{\ln 4} + 2 \ln t + C$ 5. $\frac{1}{x^2} + 5x + C$ 7. $\frac{2}{3}x^{3/2} + 8x^{1/2} + C$ 9. $\frac{1}{3}\ln x - \frac{1}{3}x + C$ 11. $t - \frac{1}{2t^2} + C$
13. $z^3 + 3 \ln z + C$ 15. $x + C$ 17. $f(x) = x^2 - x + 3$ 19. $f(x) = 3e^x$ 21. $g(t) = \frac{1}{2}t^4 - 3t^2 + 1$ 23. $C(x) = 5x + 15$
25. $C(x) = 4x + 3$ 27. $D(x) = -x^2 + 100$ 29. $R(x) = 100x$
31. (a) $C(x) = x^2 + 70x + 1.5$ (b) $R(x) = 4x^2 + 10x$ (c) $P(x) = 3x^2 - 60x - 1.5$ 33. 10 seconds 35. 6.1 minutes

SECTION 13.3

1. (a) $\frac{33}{4}$ square units; error $= \frac{19}{12}$ square units (b) $\frac{119}{16}$ square units; error $= \frac{37}{38}$ square unit
3. (a) 3 square units; error $= \frac{7}{3}$ square units (b) $\frac{17}{4}$ square units; error $= -\frac{13}{12}$ square units
5. (a) 24 square units; error $= 4$ square units (b) 22 square units; error $= 2$ square units
7. (a) 9 square units; error $= 0$ (b) 9 square units; error $= 0$
9. (a) $1.0\overline{6}$ square units; error $= -0.54$ square unit (b) 1.28 square units; error $= -0.32$ square unit

SECTION 13.4

1. 8 3. 1.082 5. $\frac{4}{5}$ 7. $\frac{5}{18}$ 9. 1 11. $\frac{6}{\ln 2} = 8.66$ 13. 2 15. 7.125 17. -32
19. $2 - \frac{3}{2}(\ln 2)^2 = 1.28$ 21. $\frac{20}{3}$ 23. $-\frac{64}{15}$ 25. $-\frac{456}{5}$ 27. $-\frac{4}{3}$ 29. $4(\ln 3 + 1) = 8.39$ 31. 75
33. $F'(x) = x^2 - 2x$
35. If the function $f(x)$ is continuous on the interval $[a, b]$, then $\int_a^b f(x)\, dx = F(b) - F(a)$, where $F(x)$ is an indefinite integral of $f(x)$.

SECTION 13.5

1. 69 square units 3. $\frac{243}{4}$ square units 5. $\frac{59}{2}$ square units 7. $\frac{69}{2}$ square units 9. $\frac{20}{3}$ square units
11. $\frac{16}{3}$ square units 13. 20 square units 15. 9 square units 17. 1.6 square units 19. 8 square units

21. 1.02 square units **23.** 36 square units **25.** 4.5 square units **27.** 0 **29.** 0

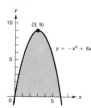
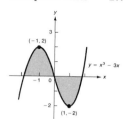

31. $3\int_{-2}^{2}(2x^2 - x)\,dx = 32$ **33.** $4\int_{5}^{7}2\,dx - 4\int_{5}^{7}3x\,dx = -128$ **35.** $\int_{1}^{3}(x^{-2} + x^{-3})\,dx + \int_{3}^{5}(x^{-2} + x^{-3})\,dx = \frac{32}{25}$
37. The integral of the sum (difference) of two functions is the integral of the first function plus (minus) the integral of the second function.

SECTION 13.6

1. $\frac{45}{4}$ square units **3.** $\frac{25}{6}$ square units **5.** 22.5 square units **7.** $\frac{32}{3}$ square units **9.** $\frac{32}{3}$ square units
13. 48 square units **15.** $\frac{1}{2}$ square unit **17.** $75 **19.** $21.33 **21.** $42.35 **23.** $24,900 **25.** $140,724.31
27. 80 feet **29.** 8 words
31. (1) Physical situation—Supply and demand for a product; physical question—What is the consumer surplus?
(2) Mathematical model—The price/demand function $P(x) = -\frac{1}{4}x + 20$ and the equilibrium point (4, 16); mathematical question—What is the value of the integral $\int_{0}^{4}[(-\frac{1}{4}x + 20) - 16]\,dx$?
(3) Mathematical solution—The value of the definite integral is 14.
(4) Interpretation—The consumer surplus is $14, that is, the consumers who would be willing to buy at a higher price would save $14.
33. The definite integral of the instantaneous rate of change of a function on the given interval is the total change in the values of the function over the given interval.

REVIEW EXERCISES

1. $2x^{3/2} + C$ **2.** $-5e^x + C$ **3.** $\frac{5^x}{\ln 5} + C$ **4.** $\ln|y| + C$ **5.** $x^3 - 2x + C$ **6.** $2x^2 - 2x^4 + C$ **7.** $\frac{2}{3}x^{3/2} + \frac{3}{2}x^{4/3} + C$
8. $\frac{4}{5}x^{5/4} + \frac{4}{3}x^{3/4} + C$ **9.** $\frac{-2}{x} + \ln x + C$ **10.** $\frac{5}{2}x^2 + \frac{5^x}{\ln 5} + C$ **11.** $\frac{1}{6}t^6 + 5\ln|t| + C$ **12.** $3e^x - 4x + C$
13. $x^3 - x^2 + x + C$ **14.** $\frac{1}{2}x^2 + 2e^x - 3x + C$ **15.** $y^3 + \ln|y| + C$ **16.** $\frac{2}{5}t^{5/2} + t + C$ **17.** $\frac{1}{2}x^2 + 3x + C$
18. $\frac{1}{2}x^2 - 2x + C$ **19.** 32 **20.** $\frac{7}{6}$ **21.** 5.41 **22.** 0 **23.** $\frac{7}{6}$ **24.** 38 **25.** 0 **26.** 18 **27.** 2.69315
28. $-\frac{11}{2}$ **29.** 5 **30.** 40 **31.** $g(x) = 4x^3 - x + 2$ **32.** $f(x) = \frac{-2}{x} + x$ **33.** $C(x) = \frac{1}{4}x^4 - \frac{2}{3}x^3 + 760$
34. $D(x) = \frac{1}{2x} + 2.5$ **35.** $R(x) = 480x - 0.02x^2$
36. (a) $C(x) = 1,800 + 144x - 0.012x^2$ (b) $R(x) = 24x$ (c) $P(x) = 0.012x^2 - 120x - 1,800$
37. (a) $s(t) = -16t^2 + 160t$ (b) The object will return to the ground at $t = 10$ seconds. **38.** $\frac{40}{3}$ square units
39. 384.43 square units **40.** 51 square units **41.** $\frac{8}{3}$ square units **42.** $\frac{4}{3}$ square units **43.** 6.93 square units
44. 16 square units **45.** 6.63 square units **46.** $\frac{32}{3}$ square units **47.** $\frac{125}{6}$ square units **48.** $10 **49.** $122.50
50. Consumer surplus = $166.67; producer surplus = $125 **51.** $16,000 **52.** $2,280 **53.** 336 feet **54.** −3,495.19
55. (1) Physical situation—Production costs; physical question—What is the total change in cost if production is changed from 10 to 30 units?
(2) Mathematical model—The marginal cost function $C'(x) = 0.03x^2 - 0.4x + 3$; mathematical question—What is the value of the integral $\int_{10}^{30}(0.03x^2 - 0.4x + 3)\,dx$?
(3) Mathematical solution—160
(4) Interpretation—The total change in cost is $16,000.
56. (1) Physical situation—A killing agent is introduced into a bacteria colony; physical question—What is the change in the population from the second through the fifth seconds after the killing agent is introduced?
(2) Mathematical model—The rate of decrease $2e^{-0.2t}$; mathematical question—What is the value of the integral $\int_{2}^{5}(2e^{-0.2t})\,dt$?
(3) Mathematical solution— −3.49519
(4) Interpretation—The bacteria colony population decreases by 3,495.19 during the given time.

CHAPTER 14 SECTION 14.1

1. $df = 2x\, dx$
3. $df = 4(3x^2 + 1)\, dx$
5. $dg = \dfrac{-2}{t^2}\, dt$
7. $dy = 8(2x + 3)^3\, dx$
9. $df = \dfrac{1}{(4-x)^2}\, dx$
11. $dy = (6x + 2)\, dx$
13. $df = (9x^2 - 8x)\, dx$
15. $\dfrac{(x^2+5)^4}{8} + C$
17. $\dfrac{(2x^3+1)^3}{18} + C$
19. $\dfrac{-1}{12(3-6x^2)^2} + C$
21. $\dfrac{(t^2+1)^{3/2}}{3} + C$
23. $\dfrac{-1}{3(3x+1)} + C$
25. $\tfrac{1}{3}e^{x^3} + C$
27. $-\tfrac{1}{3}e^{-3t} + C$
29. $\tfrac{1}{3}e^{3x^2} + C$
31. $\tfrac{1}{2}e^{x^2-2} + C$
33. $\dfrac{4^{x^3}}{\ln 4} + C$
35. $\dfrac{-6^{1-4x^2}}{2\ln 6} + C$
37. $\ln|2x+1| + C$
39. $-\tfrac{1}{3}\ln|4-3x| + C$
41. $\tfrac{1}{2}\ln|5+x^2| + C$
43. $\dfrac{(x^2+2x)^4}{8} + C$
45. $\tfrac{1}{2}e^{t^2-2t} + C$
47. $\tfrac{1}{3}\ln|3x^2+9x| + C$
49. $(t^2+3)^4 + C$
51. $\tfrac{4}{3}(2x^2-1)^{3/2} + C$
53. $2e^{5t^2} + C$
55. $\dfrac{4(7^{2x^3+2})}{\ln 7} + C$
57. $\tfrac{1}{2}(3^{x^2}) + C$
59. $5\ln|2t-1| + C$
61. $\tfrac{5}{2}\ln|2x+1| + C$
63. $-\tfrac{5}{6}e^{-3t^2} + C$
65. $\tfrac{2}{3}\ln|2x^3+6x| + C$
67. $\tfrac{1}{2}\ln|x| - \tfrac{1}{3}\ln|3x+1| + C$
69. $\tfrac{1}{3}(2x)^{3/2} + \tfrac{5}{2}x^2 + C$
71. $2x + \tfrac{1}{2}\ln|x^2+2| + C$
73. 20
75. $\tfrac{1}{2}$
77. $\dfrac{120}{\ln 4} = 86.6$
79. 120
81. $\tfrac{1}{2}\ln|e^2+1|$
83. 24
85. $\tfrac{1}{6}$ square unit
87. $\tfrac{1}{6}$ square unit

SECTION 14.2

1. $\dfrac{1}{12}\ln\left|\dfrac{3+2x}{3-2x}\right| + C$
3. $\dfrac{25x^3}{3}\left[\ln(5x) - \dfrac{1}{3}\right] + C$
5. $\tfrac{1}{3}\ln\left|3x + \sqrt{9x^2-16}\right| + C$
7. $-\tfrac{1}{3}e^{-3x} + C$
9. $\ln\left|\dfrac{x}{1+3x}\right| + C$
11. $2 + \tfrac{2}{3}t - 2\ln|9+3t| + C$
13. $\ln|x+3+\sqrt{6x+x^2}| + C$
15. $\tfrac{3}{4}te^{4t} - \tfrac{3}{16}e^{4t} + C$
17. $\ln|e^x + \sqrt{e^{2x}+4}| + C$
19. $\dfrac{1}{12}\ln\left|\dfrac{2x-3}{2x+3}\right| + C$
21. $\tfrac{1}{12}\ln|2+6x^2| + C$
23. $e^x(x^2 - 2x + 2) + C$
25. $\tfrac{1}{2}\ln 3 = 0.55$
27. $\ln 1.83 = 0.60$

SECTION 14.3

1. $\dfrac{2x}{3}(x+3)^{3/2} - \dfrac{4}{15}(x+3)^{5/2} + C$
3. $\dfrac{x}{3}(2x-1)^{3/2} - \dfrac{1}{15}(2x-1)^{5/2} + C$
5. $\dfrac{x}{3}(x+4)^3 - \dfrac{1}{12}(x+4)^4 + C$
7. $\dfrac{2x}{3}(x+4)^3 - \dfrac{1}{6}(x+4)^4 + C$
9. $\dfrac{-2x}{9}(-3x+4)^3 - \dfrac{1}{54}(-3x+4)^4 + C$
11. $xe^{3x} - \tfrac{1}{3}e^{3x} + C$
13. $\dfrac{x}{3}e^{3x} - \tfrac{1}{9}e^{3x} + C$
15. $-xe^{-x} - e^{-x} + C$
17. $-3xe^{-x} - 3e^{-x} + C$
19. 37.2
21. $x\ln x - x + C$
23. We would have $\tfrac{1}{2}\int ue^u\, du$ and not $\int e^u\, du$.

SECTION 14.4

7. $y = x^2 + 4x + C$
9. $y = 2x + (1/x) + C$
11. General solution: $y = x^3 - x^2 + C$; particular solution: $y = x^3 - x^2 + 2$
13. General solution: $y = \ln|t| + C$; particular solution: $y = \ln|t| - 10$
15. $y = \pm\sqrt{x^2 + C}$
17. $y = Ae^{x^2}$, where $A = e^c$
19. General solution: $y = \pm\sqrt{x^3 + C}$; particular solution: $y = -\sqrt{x^3+1}$
21. General solution: $y = \sqrt[3]{t^2 + t + C}$; particular solution: $y = \sqrt[3]{t^2 + t + 15}$
23. $5,637.50
25. 1,353 bacteria
27. (a) $P(t) = 5{,}000e^{-0.09t}$ (b) $P(2) = \$4{,}176.35$
29. $N(t) = Pe^{-kt}$
33. 8.7 years
35. 3.5 seconds
37. 7.7 years
39. (1) Physical situation—The investment of $5,000 at 9% compounded continuously; physical question—What will the investment be worth in 16 months?
    (2) Mathematical model—$dQ/dt = 0.09Q$; mathematical question—What is the value of $Q(\tfrac{4}{3})$?
    (3) Mathematical solution—$Q(\tfrac{4}{3}) = 5{,}637.5$
    (4) Interpretation—The investment will be worth $5,637.50 in 16 months.
41. A differential equation is an equation containing the derivative of an unknown function.

43. A general solution of a differential equation is a solution containing an arbitrary constant.

SECTION 14.5

1. Convergent; $\frac{1}{2}$ **3.** Divergent **5.** Convergent; 3 **7.** Convergent; $-\frac{5}{2}$. **9.** Divergent **11.** Convergent; $\frac{1}{3}$
13. Divergent **15.** Divergent **17.** Convergent; 0
19. The value of the improper integral $\int_{-\infty}^{0} e^x \, dx$ is the area between the curve $y = e^x$ and the x-axis, to the left of the y-axis.
21. A divergent improper integral is an integral that has no value according to the given definitions.

SECTION 14.6

1. (b) $\frac{1}{2}$ (c) The area under the density function $f(x) = \frac{1}{4}$ over the interval [3, 5]
3. (b) $\frac{1}{8}$ (c) The area under the density function $f(x) = \frac{3}{64}x^2$ over the interval [0, 2]
5. (b) $P(0 \le X \le 5) = 0.993$; $P(X \ge 4) = 0.018$; $P(X \le 2) = 0.86$
   (c) $P(0 \le X \le 5)$ is the area under the density function $f(x) = e^{-x}$ over the interval [0, 5]. $P(X \ge 4)$ is the area under the density function $f(x) = e^{-x}$ to the right of $x = 4$. $P(X \le 2)$ is the area under the density function $f(x) = e^{-x}$ over the interval [0, 2].
7. $k = \frac{1}{32}$ **9.** (a) 0.23 (b) 0.37 **11.** 0.86 **13.** $\frac{1}{6}$ **15.** $\frac{14}{15}$
17. (1) Physical situation—The height at maturity of a particular plant; physical question—What is the probability that one of the plants, arbitrarily chosen, will grow to a height of less than 20 inches?
   (2) Mathematical model—The density function $f(x) = \frac{1}{6}$, $18 \le x \le 24$; mathematical question—What is the value of the integral $\int_{18}^{20} \frac{1}{6} \, dx$?
   (3) Mathematical solution—The value of the given integral is $\frac{1}{3}$.
   (4) Interpretation—The probability that one of the plants, arbitrarily chosen, will grow to a height of less than 20 inches is $\frac{1}{3}$.

REVIEW EXERCISES

1. $\frac{1}{9}(x^3 - 4)^3 + C$ **2.** $\frac{-5}{3(x^3 - 4)} + C$ **3.** $\frac{1}{6}\ln|3x^2 - 4| + C$ **4.** $\frac{7}{4}\ln|5 + 4y| + C$ **5.** $\frac{1}{4}e^{2t^2} + C$ **6.** $\frac{1}{3}e^{x^3 - 3x} + C$
7. 36 **8.** $\frac{1,024}{3}$ **9.** $\frac{1}{4}\left[\ln|1 + 2x| + \frac{1}{1 + 2x}\right] + C$ **10.** $\frac{t^4}{4}\left(\ln t - \frac{1}{4}\right) + C$ **11.** $\frac{1}{4}\ln\left|\frac{3 + 2x}{3 - 2x}\right| + C$
12. $\ln|x + 3 + \sqrt{6x + x^2}| + C$ **13.** $4e^x(x^2 - 2x + 2) + C$ **14.** $3(1 + y^2)^{1/2} + C$ **15.** $\frac{1}{3} - \frac{1}{9}\ln 4 = 0.18$
16. $\frac{1}{4}\ln\frac{7}{15} = -0.19$ **17.** $\frac{x}{3}(x - 7)^3 - \frac{1}{12}(x - 7)^4 + C$ **18.** $\frac{4x}{9}(3x - 1)^{3/2} - \frac{8}{135}(3x - 1)^{5/2} + C$ **19.** $-6xe^{-x} - 6e^{-x} + C$
20. $\frac{x^2}{2}\ln x - \frac{1}{4}x^2 + C$ **21.** 271.9 **22.** 14.1 **23.** $\frac{1}{6}(2x^2 + 3)^{3/2} + C$ **24.** $-x(x + 5)^3 + \frac{1}{4}(x + 5)^4 + C$
25. $\ln|x + 2 + \sqrt{4x + x^2}| + C$ **26.** $2e^{2x^2 + 3x} + C$ **27.** $2e^x - xe^x + C$ **28.** $\frac{1}{25}[2 + 5x - 2\ln|2 + 5x|] + C$ **29.** $1\frac{1}{3}$
30. $\frac{3}{4}$ **31.** Convergent; $\frac{5}{3}$ **32.** Divergent **33.** Divergent **34.** Convergent; $-\frac{1}{10}$ **35.** Convergent; -1.82
36. Divergent **37.** 28.53 square units **38.** 10.1 square units **39.** 8.4 square units **40.** 0.26 square unit
41. Solution **42.** Not a solution **43.** Not a solution **44.** Not a solution **45.** $y = \frac{1}{2}x^2 + \frac{1}{x} + C$
46. $y = 3e^x + x + C$ **47.** General solution: $y = (x + 1)^2 + C$; particular solution: $y = (x + 1)^2 - 3$
48. General solution: $y = \frac{4}{3}(t + 1)^3 + C$; particular solution: $y = \frac{4}{3}(t + 1)^3 + \frac{11}{3}$ **49.** $y = \sqrt[3]{x^2 + C}$ **50.** $y = Ce^{x^2/2}$
51. General solution: $y = \pm\sqrt{\frac{2t^2 + ct + 1}{t}}$; particular solution: $y = -\sqrt{\frac{2t^2 + 1}{t}}$
52. General solution: $y = \frac{-3(x^{3/2} - C)}{2(x^3 - C^2)}$; particular solution: $y = \frac{-3(x^{3/2} + \frac{3}{2})}{2(x^3 - \frac{9}{4})}$ **53.** (a) $Q(t) = 6{,}000e^{0.08t}$ (b) $6{,}765

54. (a) $P(t) = 6{,}000e^{-0.08t}$ (b) $5,112.86 **55.** $1,491.80 **56.** (a) 8.7 years (b) 8.7 years
57. (a) $P(t) = 150e^{0.013t}$ (b) 183.21 million **58.** (b) $(-\ln 2)/K$ (c) 168.733 grams (d) 4,077.4 years
59. (a) $y = (0.5t + 20)^2$ (b) 576 fish **60.** (b) $\frac{2}{5}$ **61.** (b) 0.39 (c) 0.61 **62.** (a) $k = \frac{1}{12}$ (b) $\frac{2}{3}$
63. No, at $x = 0$, $4/x$ is not defined. **64.** 0.15625 **65.** (a) 0.49 (b) 0.37
66. (1) Physical situation—The investing of $6,000 at 8% compounded continuously; physical question—What will be the amount 18 months later?
(2) Mathematical model—$Q(t) = 6{,}000e^{0.08t}$, the amount at any time t; mathematical question—What is the value of $Q(1.5)$?
(3) Mathematical solution—$Q(1.5) = 6{,}756$
(4) Interpretation—The amount 18 months later will be $6,756.
67. (1) Physical situation—The time that an experimental treatment for immunizing rabbits is effective; physical question—What is the probability that the treatment on one of the rabbits will be effective less than 6 months?
(2) Mathematical model—the given density function $f(t) = \frac{3}{4}(2t - t^2)$, $t \in [0, 2]$; mathematical question—What is the value of $\int_0^{0.5} f(t)\,dt$?
(3) Mathematical solution—$\int_0^{0.5} f(t)\,dt = 0.15625$
(4) Interpretation—The probability that the treatment on one of the rabbits will be effective less than 6 months is 0.15625.

CHAPTER 15 SECTION 15.1

1. -7 **3.** 1 **5.** Does not exist **7.** 2 **9.** $\sqrt{30}/6$ **11.** 0 **13.** 9 **15.** 20 **17.** -15 **19.** $z = xy$
21. $I = PRT$ **23.** $8xy$ **25.** $2x$

SECTION 15.2

1. 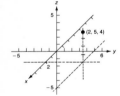 **3.** 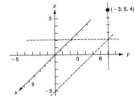 **5.**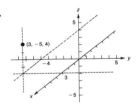

7. 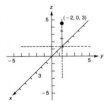 **9.** 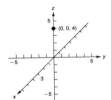 **11.**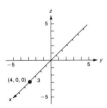

13. $(0, 4, 0)$ **15.** $(5, 0, 0)$ **17.** $(5, 0, -2)$ **19.** $(5, 4, -2)$ **21.** $(5, 4, 3)$
23. 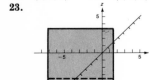 **25.** 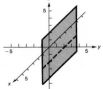 **27.**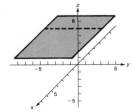

29. 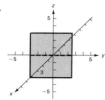 31. 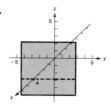 33.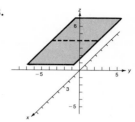

SECTION 15.3

1. (a) $z_x = 2x + 3y$ (b) $z_y = -2y + 3x$ 3. (a) $f_x(x, y) = 6x^2y^2 - \dfrac{y^4}{x^2}$ (b) $f_y(x, y) = 4x^3y + \dfrac{4y^3}{x}$

5. (a) $z_x = \dfrac{y\sqrt{xy}}{2xy}$ (b) $z_y = \dfrac{x\sqrt{xy}}{2xy}$ 7. (a) $f_x(3, 1) = 96$ (b) $f_y(3, 1) = 96$

9. (a) $\dfrac{\partial z}{\partial x} = 24xy(3x^2 - y)$ (b) $\dfrac{\partial z}{\partial y} = 2(3x^2 - y)^2 - 4y(3x^2 - y)$ 11. (a) $f_x(x, y) = 2xe^{x+y} + x^2 e^{x+y}$ (b) $f_y(x, y) = x^2 e^{x+y}$

13. (a) $z_x = ye^{xy}$ (b) $z_y = xe^{xy}$ 15. (a) $f_x(3, -4) = \tfrac{8}{49}$ (b) $f_y(0, 2) = 0$ 17. (a) $z_x = \dfrac{2xy^2}{(2x + y^2)^3}$ (b) $z_y = \dfrac{-4x^2 y}{(2x + y^2)^3}$

19. (a) $f_x(x, y, z) = y + z$ (b) $f_y(x, y, z) = x + z$ (c) $f_z(x, y, z) = x + y$
21. (a) $f_x(1, 2, 3) = \tfrac{5}{6}$ (b) $f_y(-1, 0, 2) = -2$ (c) $f_z(0, 1, -2) = 0$
23. If f is a function of the two variables x and y, the partial derivative of f with respect to x is the derivative of f with y held at a fixed value.

SECTION 15.4

1. 2   3. $2e$ 5. Slope is zero at $(2, -4, 12)$.   7. The function is increasing at the rate of $\tfrac{1}{3}$ unit per unit change in y.
9. (a) $P_y(200, y) = 10y - 400$ (b) 2,100
11. (a) $C_x(x, 6) = 6 - x$; $C_x(x, 6)$ is the marginal cost of labor when material cost is held fixed at 6.
    (b) $C_x(4, 6) = 2$; $C_x(4, 6)$ is the marginal cost of 4 units of labor when material cost is held fixed at 6.
13. (a) $\tfrac{3}{20}$ pound per square inch   (b) $-\tfrac{9}{400}$ pound per square inch   (c) $-\tfrac{405}{98}$ cubic inches per pound

SECTION 15.5

1. $f_x = 3x^2 + 7y$; $f_y = 3y^2 + 7x$; $f_{xx} = 6x$; $f_{yy} = 6y$; $f_{xy} = 7$; $f_{yx} = 7$
3. $z_x = 6xy - 4y^3$; $z_y = 3x^2 - 12xy^2$; $z_{xx} = 6y$; $z_{yy} = -24xy$; $z_{xy} = 6x - 12y^2$; $z_{yx} = 6x - 12y^2$
5. $f_x = \dfrac{8x^3}{y} + 3y$; $f_y = -\dfrac{2x^4}{y^2} + 3x$; $f_{xx} = \dfrac{24x^2}{y}$; $f_{yy} = \dfrac{4x^4}{y^3}$; $f_{xy} = -\dfrac{8x^3}{y^2} + 3$; $f_{yx} = -\dfrac{8x^3}{y^2} + 3$
7. $z_x = 4x^3 y^2 - 4xy$; $z_y = 2x^4 y - 2x^2$; $z_{xx} = 12x^2 y^2 - 4y$; $z_{yy} = 2x^4$; $z_{xy} = 8x^3 y - 4x$; $z_{yx} = 8x^3 y - 4x$
9. $z_x = \dfrac{\sqrt{xy}}{2x}$; $z_y = \dfrac{\sqrt{xy}}{2y}$; $z_{xx} = -\dfrac{\sqrt{xy}}{4x^2}$; $z_{yy} = -\dfrac{\sqrt{xy}}{4y^2}$; $z_{xy} = \dfrac{\sqrt{xy}}{4xy}$; $z_{yx} = \dfrac{\sqrt{xy}}{4xy}$
11. $f_x = ye^{xy}$; $f_y = xe^{xy}$; $f_{xx} = y^2 e^{xy}$; $f_{yy} = x^2 e^{xy}$; $f_{xy} = e^{xy} + xye^{xy}$; $f_{yx} = e^{xy} + xye^{xy}$ 13. (a) $18x$ (b) 0   (c) $-24y^2$
15. (a) $6x$ (b) 6   (c) 18   17. (a) $-\tfrac{1}{2}$ (b) $\tfrac{1}{16}$ (c) $\tfrac{1}{16}$ 19. (a) $3x^2$ (b) $6xy$ (c) 12
21. (a) $-1/x^2$ (b) 0   (c) $\tfrac{1}{4}$

SECTION 15.6

1. f has a relative minimum of -9 at $(3, 0)$.   3. f has a relative maximum of 4 at $(-1, -2)$.
5. z has a relative maximum of 7 at $(0, 1)$.   7. f has no relative maximum or minimum.

9. z has no relative maximum or minimum.   11. f has a relative minimum of 3 at $(1, 1)$.
13. The company should spend $5 million on research and $15 million on labor to maximize its profits at $90 million.
15. A maximum number of additional votes of 16,875 will be picked up by spending $22,500 on television advertising and $15,000 on radio advertising.
17. A minimum inventory cost of $2,500 occurs when 500 units are on hand and 1,000 units are ordered at a time.
19. (1) Physical situation—The manufacturing of two items, A and B; physical question—What is the number of units of each item the company should produce each year to maximize its profit?
    (2) Mathematical model—$z = 16xy - 17x^2 - 4y^2 + 3x$; mathematical question—For what values of x and y is z maximum?
    (3) Mathematical solution—$x = 1.5$, $y = 3$
    (4) Interpretation—The company will obtain maximum profit if it manufactures 1,500 units of item A and 3,000 units of item B annually.
21. (1) Physical situation—The use of advertising funds to best influence voters; physical question—How much money should be spent on radio advertising and how much on television advertising in order to maximize the number of additional votes picked up by the candidate? What is the maximum number of additional votes?
    (2) Mathematical model—$z = 3xy - y^3 - x^2$; mathematical question—For what values of x and y is z maximum and what is the maximum value of z?
    (3) Mathematical solution—$x = \frac{9}{4}$, $y = \frac{3}{2}$ for a maximum of $z = 1.6875$
    (4) Interpretation—A maximum number of additional votes of 16,875 will be picked up by spending $22,500 on television advertising and $15,000 on radio advertising.

SECTION 15.7

1. f has a minimum of 5 at $(2, 1)$.   3. f has a maximum of $-\frac{432}{61}$ at $(\frac{156}{61}, \frac{84}{61})$.   5. z has a minimum of $\frac{2}{19}$ at $(\frac{5}{19}, -\frac{1}{19})$.
7. z has a minimum of $\frac{14}{27}$ at $(\frac{2}{3}, -\frac{1}{3})$ and a maximum of 10 at $(-2, -3)$.   9. f has a minimum of $\frac{4}{9}$ at $(\frac{2}{9}, \frac{4}{9}, -\frac{4}{9})$.
11. A maximum profit of $105,000 is achieved when 550 units of item A and 450 units of item B are manufactured.
13. When 156 units of labor and 130 units of capital are used, maximum production of 5,960 units is achieved.
15. A box with dimensions 3 feet by 3 feet by 6 feet will have a minimal cost of $162.
17. (1) Physical situation—Production; physical question—How many units of labor and how many units of capital will maximize production? What is maximum production?
    (2) Mathematical model—$f(x, y) = 40x^{3/4}y^{1/4}$; mathematical question—What values of x and y give a maximum for f? What is the maximum?
    (3) Mathematical solution—The maximum of the given function occurs when $x = 156$ and $y = 130$. The maximum is 5,960.
    (4) Interpretation—When 156 units of labor and 130 units of capital are used, maximum production of 5,960 is achieved.

REVIEW EXERCISES

1. $-\frac{4}{9}$ 2. 0   3. -0.25 4. 2   5. 2   6. 0   7. 63   8. 0.004   9. 64   10. 32   11. $a = \frac{1}{2}xy$
12. 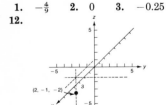   13. 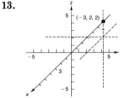   14.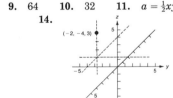
15. 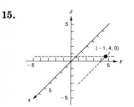   16. 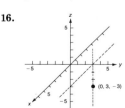   17.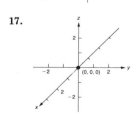

18. 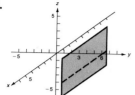 19. 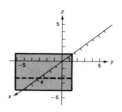 20.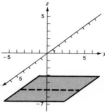

21. $6xy - 2$ 22. $3x^2 + 2y$ 23. $6y$ 24. $6x$ 25. 2 26. 6 27. -18 28. 118 29. 0 30. $6x(x^2 - y^2)^2$
31. $-6y(x^2 - y^2)^2$ 32. $6(x^2 - y^2)(5x^2 - y^2)$ 33. $-24xy(x^2 - y^2)$ 34. $-6(x^2 - y^2)(x^2 - 5y^2)$ 35. 108 36. -480
37. $-\frac{27}{16}$ 38. 0 39. $\dfrac{-y}{(x-y)^2}$ 40. $\dfrac{x}{(x-y)^2}$ 41. $\dfrac{2y}{(x-y)^3}$ 42. $\dfrac{-(x+y)}{(x-y)^3}$ 43. $\dfrac{2x}{(x-y)^3}$ 44. $\dfrac{-(x+y)}{(x-y)^3}$
45. $e^{x^2}(1 + 2x^2 + 2xy)$ 46. e^{x^2} 47. $e^{x^2}(4x^3 + 4x^2y + 6x + 2y)$ 48. 0 49. $2xe^{x^2}$ 50. $2xe^{x^2}$ 51. $2xz$
52. $z^2 - 9y^2$ 53. $x^2 + 2zy$ 54. $2x$ 55. $2z$ 56. $2y$ 57. -32 58. 18 59. -9 60. $y^2 - z$
61. $2y(x + z^2)$ 62. $2y^2z - 3z^2 - x$ 63. -1 64. $4yz$ 65. $2(x + z^2)$ 66. -1 67. 0 68. 5 69. 2
70. The slope is zero at $(2, 2, -8)$. 71. The slope is greatest at $(1, 3, 0)$. 72. (a) $P_y(20, y) = 40y - 120$ (b) $P_y(20, 30) = \$1,080$
73. (a) $P_x(x, 30) = 4x + 720$ (b) $P_x(20, 30) = \$800$ 74. f has a relative minimum of -16 at $(0, 4)$.
75. f has a relative maximum of 9 at $(2, 2)$. 76. z has a relative minimum of $-\frac{33}{4}$ at $(-\frac{1}{2}, -\frac{3}{2})$.
77. z has no relative maximum or minimum. 78. f has no relative maximum or minimum.
79. f has a relative minimum of 4 at $(1, 1)$.
80. The company should spend $8 million on labor and $6 million on advertising to achieve a maximum profit of $178 million.
81. The company should manufacture 300 units of item A and 200 units of item B to minimize the daily cost at $5,000.
82. f has a maximum of $\frac{3}{13}$ at $(\frac{7}{13}, -\frac{14}{13})$. 83. f has a minimum of -24.6 at $(0.23, 0.21)$ and a maximum of -23.7 at $(-0.36, 0.52)$.
84. w has a maximum of $-\frac{23}{34}$ at $(\frac{99}{34}, \frac{33}{34}, \frac{11}{34})$. 85. 360 units of labor and 257 units of capital will maximize production at 19,300 units.
86. A box of dimensions 5 feet by 5 feet by 4 feet will have a minimal cost of $300.
87. (1) Physical situation—The manufacturing of two items, A and B; physical question—How many units of each item should be produced to minimize the daily cost? What is the minimum daily cost?
   (2) Mathematical model—$C(x, y) = 4x + 9y + 36\left(\dfrac{x+y}{xy}\right) - 10$; mathematical question—What is the minimum value of the given function and for what values of x and y is the minimum obtained?
   (3) Mathematical solution—The function has a minimum value of 50 at $x = 3$, $y = 2$.
   (4) Interpretation—The company should produce 300 units of item A and 200 units of item B to minimize the daily cost at $5,000.
88. (1) Physical situation—Making a rectangular box; physical question—If the box is to have no top and a volume of 100 cubic feet, what dimensions of the box will give minimal cost? What is the minimal cost?
   (2) Mathematical model—$a(x, y, z) = 4xy + 5yz + 5xz$ and the constraint $xyz = 100$; mathematical question—What is the minimum of $a(x, y, z)$ subject to $xyz = 100$? For what values of x, y, and z is the function minimal?
   (3) Mathematical solution—The function has a minimum of 300 at $x = 5$, $y = 5$, $z = 4$.
   (4) Interpretation—A box of dimensions 5 feet long by 5 feet wide by 4 feet deep will have a minimal cost of $300.

APPENDIX A   SECTION A.1

1. 18 3. 1 5. -3 7. 2 9. 2 11. 192 13. 6 15. 6 17. $\frac{52}{3}$ 19. 192 21. 24 23. -2
25. Evaluation of exponents, multiplication and division (proceeding left to right), then addition and subtraction (proceeding left to right)

SECTION A.2

1. 2 and x 3. $(a + b)$ and $(a - b)$ 5. $(a - 2)$ and $(a^2 + 3)$ 7. $(x + y)$ and $(x + y)$ 9. $(2x + 5)$ and $\frac{1}{3}$
11. $2x$ and 3 13. x^2y, $4y^2x$, and -9 15. a, $2b$, and $-6a^2b^2$ 17. x, $-3y$, and 2; z, y, and $-2x$ 19. 4
21. -26 23. 14 25. -6 27. 2 29. $9x^2 - 31x$ 31. $2 - 5x + 2x^2$ 33. $-2x^2 - 12x$ 35. $2a + 17$
37. $-6x^2 - x - 3$ 39. $28x^2 + 7x$ 41. (a) -2 (b) $10 - 4x$ (c) -2 43. (a) 0 (b) $7t + 7$ (c) 0
45. The letters in algebra may stand for any one in a group of numbers (in which case, they are called variables) or they may stand for a single number (in which case, they are called constants).

SECTION A.3

1. $(x + 3)(x - 3)$ 3. $(x - 2)(x^2 + 2x + 4)$ 5. $(t + 3)(t^2 - 3t + 9)$ 7. $(x + 3)^2$ 9. $(x - 3)^2$ 11. $(y + 4)(y + 5)$
13. $(y + 5)(y - 4)$ 15. $(x - 6)(x + 2)$ 17. $(x + y - 2)(x + y + 2)$ 19. $(y + 3)(y - 2)$ 21. Not factorable
23. $(x - 3)^2$, or $(3 - x)^2$ 25. Not factorable   27. $(x - 1)(x + 1)(x^2 + 1)$ 29. $3(x - 2)(x + 2)$ 31. $4(y - 2)(y^2 + 2y + 4)$
33. $2(x - 3)^2$ 35. $-3(x - 3)^2$ 37. $-x(x + 4)^2$ 39. $x^2(x + 2)$ 41. $25(4 - x)$ 43. $2(x - 3)$
45. $3(x - 2)(x + 2)(x - 4)$ 47. $(x + 4)(8 - x)$ 49. $(x - 4)(x + 1)$ 51. $(x - 4)(x + 1)$ 53. Not factorable
55. $(x + 2)(x^2 - 2x + 4)$ 57. $(x + 2)(x + 1)$ 59. $2(x - 3)$ 61. $(8 - x)(4 + x)$

SECTION A.4

1. $x = 4$ 3. $x = -\frac{9}{2}$ 5. $x = 3$ 7. $y = 7$ 9. $x = \frac{6}{5}$ 11. $x = -12$ 13. $x = -\frac{10}{11}$ 15. $t = 5$ 17. $y = 15$
19. $x = -1$ 21. $x = 3, -4$ 23. $x = 3, -1$ 25. $x = 3, -5$ 27. $x = -3, -2$ 29. $y = -2$ 31. $x = 1, -5$
33. $x = 0, -7$ 35. $x = 1, -\frac{5}{2}$ 37. $x = 1, -\frac{3}{2}$ 39. $x = \frac{4}{3}, -\frac{1}{2}$ 41. No real solution   43. $y = \frac{1}{2}$
45. $x = -3, -2$ 47. $x = 1, -5$ 49. $x = 0, -7$
51. An equation is a statement that two (or more) mathematical expressions are equal.
53. By factoring and by use of the quadratic formula

SECTION A.5

1. $\dfrac{2}{x - 1}$ 3. -1 5. $1 - t$ 7. $\frac{1}{2}$ 9. $\dfrac{1}{2x^3}$ 11. 1 13. $\dfrac{x + 1}{3}$ 15. $\dfrac{x + 1}{2(x - 2)}$ 17. $\dfrac{x^2 - 1}{3}$
19. (1) Factor each numerator and denominator, if possible. (2) Cancel any factor appearing in both numerator and denominator. (3) Multiply the remaining factors in the numerator and in the denominator.
21. $\dfrac{4x + 1}{x^2}$ 23. $\dfrac{x^2 - 3x + 3}{x^2 - x}$ 25. $\dfrac{3x + 2}{4x}$ 27. $\dfrac{2}{x - 1}$ 29. $\frac{2}{5}$ 31. $\dfrac{x - 1}{x(x + 1)}$ 33. $\dfrac{2}{(x + 1)(x + 3)}$
35. $\dfrac{2(10 + 9t - 15t^2)}{15t^2}$ 37. $\dfrac{2}{y - 1}$ 39. $\dfrac{-x - 2}{x(x + 1)}$ 41. $\dfrac{3x - 6}{x - 7}$ 43. $\dfrac{2(2 + y)}{y^2}$ 45. $2x - 4$ 47. $\dfrac{x + 3}{2}$ 49. $\dfrac{-(x + 4)}{x - 3}$
51. (1) Factor each denominator (constant factors should be written as products of prime numbers). (2) Form the L.C.D. by multiplying together the factors occurring in the different denominators. The highest power to which a factor appears in any denominator is used. (3) Rewrite each fraction with the L.C.D. as its denominator; multiply the numerator and denominator of each fraction by the factors of the L.C.D. not in the original denominator.

SECTION A.6

1. $x = 0, -5, 3$ 3. $x = 0, 1, -\frac{5}{3}$ 5. $x = 1, 2, 3$ 7. $x = \frac{5}{3}, -\frac{7}{4}, -\frac{1}{2}$ 9. $x = -\frac{1}{2}$ 11. $y = 2$ 13. $x = 11$
15. $x = \frac{3}{2}$ 17. No solution   19. $x = -1 + \sqrt{3}, -1 - \sqrt{3}$ 21. $x = 0, -3$ 23. $x = -\frac{3}{2}$
25. $x = 2 + \sqrt{6}, 2 - \sqrt{6}$

SECTION A.7

1. 9 3. 125 5. $\frac{4}{9}$ 7. -4 9. 1 11. $-\frac{1}{125}$ 13. 3 15. -2 17. 3^5 19. $\frac{2}{3}$ 21. $\frac{1}{36}$ 23. 1
25. $\frac{1}{64}$ 27. 49 29. $-\frac{27}{125}$ 31. $\frac{4}{27}$ 33. $\frac{5}{4}$ 35. $\dfrac{x^6}{y^{12}}$ 37. $\dfrac{a^2\sqrt{b}}{b}$ 39. $x = 2$ 41. $x = -2$
43. $x = \frac{1}{3}$ 45. $x = \frac{1}{2}$ 47. $x = 0$ 49. $x = 3$ 51. $x = 1$ 53. $x = 2$ 55. $x = 5$ 57. $x = 2$ 59. $x = 8$
61. False, a^{m+n} 63. True   65. False, x^{2m} 67. False, 2^{x+y}

SECTION A.8

1. $\{-6, -4, -2, 0, 1, 2, 3, 4, 5, 6\}$ 3. $\{-6, -3, 0, 2, 3, 4, 6\}$ 5. $\{-5, -4, -2, -1, 1, 2, 4, 5\}$ 7. $\{-5, -1, 1, 5\}$
9. $\{-4, -1, 0, 1, 2, 4\}$ 11. $\{-4, -3, -1, 0, 1, 3\}$ 13. $\{0, 2, 3\}$ 15. $\varnothing$ 17. $\{p, a\}$ 19. $\{j, i, m, t, o\}$
21. $\{d, l, n, o, u\}$ 23. $D = \{d, a, n\}$ 25. $D \subset A$ 27. $\{3\} \subset C$ 29. $U \not\subset B$ 31. $m \in (J \cap T)$ 33. $D \subset J'$

35. $j \in A'$ **37.** $P \not\subseteq D$ **39.** The set of all male students
41. The set of all students taking a mathematics or a business course
43. The set of all male students taking a mathematics or a business course
45. $\varnothing$, the set of all students who are both male and female   **47.** $R' = \{\text{May, June, July, August}\}$
49. {February, March, April, September, October, November, December}   **51.** {January}

53.    **55.** 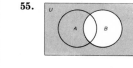   **57.** 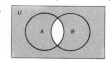   **59.**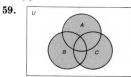

61. 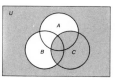   **63.** 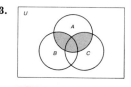   **65.** A set is a collection of objects.

67. The universal set is the set of all elements that enter into a particular discussion.
69. The complement of a set A is the set of all elements of the universal set that are not in A.
71. The intersection of two sets A and B is the set of all those elements that are in both A and B.

INDEX

Pages shown in bold italics indicate the beginning of a chapter or section.

Absolute maximum, **472**
   alternate test, 477
Absolute minimum, **472**
   alternate test, 477
Absolute value, 497
Addition of matrices, **80**
   associative law, 84
   commutative law, 84
Addition rule
   derivatives, 427
   probability, **186**
Algebra, **599**
Algebraic fractions, **618**
Alternate test, 477
Amortization, 362
Amount
   annuity, 356
   compound interest, 352
   simple interest, 344
Annuities, ordinary, **355**
   amount, 356
   payment size, 358, 362
   present value, **360**
Antiderivative, 496
Application of mathematics, **11**
Associative law
   matrix addition, 84
   matrix multiplication, 89
Asymptotes, 396
Augmented coefficient matrix, 69
Average cost, 477
Average daily balance, **346**
Average rate of change, **410**
Averages, 266
Axes, coordinate, 26

Basic feasible solutions, 133
Basic solutions, **127**
Basic variables, 128
Bayes, Rev. Thomas, 216
Bayes' rule, **215**
Bernoulli, Jacob, 212
Bernoulli trials, 212

Binomial probabilities, **208**
   table, 642
Break-even point, 64

Cardan, Jerome, 159
Cartesian plane, **24**
Central limit theorem, 299
Central tendency, measures of, **266**
Chain rule, **432**
Cobb-Douglas production function, 592
Coefficient matrix, 91
Combinations, **179**
Common logarithms, 319
   table, 652
Commutative law
   matrix addition, 84
Complement
   of a set, 636
   of an event, 162
Completing the square, 46
Composition of functions, 432
Compound interest, **350**
   amount, 352
   continuous, 315, 322
   conversion period, 352
   present value, 354
Concave up/down, 481
Conditional probability, 193
Connected graph, 3
Constant functions, 427
Constraint, 586
Consumer surplus, 516
Continuity, **390**
   at a point, 390
   on an interval, 393
Continuous compound interest, 315, 322
   present value, 546
Convergent improper integral, 550
Conversion period, 352
Coordinate(s), 25, 26, 567
Coordinate axes, 26
Coordinate system, 27

INDEX

Counting techniques, **173**
  combinations, **179**
  fundamental counting principle, 174, 175
  permutations, 180, 186
Critical value(s), 285, 466

Data organization, **262**
de Fermat, Pierre, 159
Decay factor, 544
Decreasing function, **458**
Definite integral(s), **502, 507**
  definition, 505
Demand vector, 101
Departing variable, 136
Dependent variable, 381, 564
Depreciation
  linear, 41
  straight line, 41
Derivative(s), **414, 455**
  addition rule, 427
  chain rule, **432**
  definition, 417
  higher-order, **456**
  interpretation of, **419**
  of a constant function, 427
  of a constant times a function, 426
  of a logarithmic function, 442, 444
  of an exponential function, 440, 441
  partial, **570, 576**
  power rule, 425, 434
  product rule, 436
  quotient rule, 437
  table, 451
Descartes, Rene, 17
Difference quotient, 411
Differentiable function(s), **446**
  at a point, 448
  on an interval, 448
Differential(s), 496, 526
Differential equation(s), **540**
  general solution, 541
  particular solution, 541
  solution, 540
Discounted loan, 343
Dispersion, measures of, **272**
Distributive laws
  matrices, 89
Divergent improper integral, 550
Domain, 373, 565
Dominated column, 253
Dominated row, 253
Dual problem, 145

Echelon form, 71
Edge, 2
Elements of a set, 634
Elimination, Gauss-Jordan, 75
Empty set, 636

Entering variable, 135
Equal matrices, 84
Equal sets, 635
Equation(s)
  as mathematical models, **18, 40, 49**
  differential, **540**
  graphs of, **28**
  linear, **37, 613**
  quadratic, **46, 613**
  solution of, **613, 627**
  systems of, **57**
Equilibrium point, 65, 516
Equilibrium price, 65, 516
Equilibrium quantity, 65
Equivalent systems of equations, 59
Euler, Leonhard, 6
Euler path, 6
  existence, 7
Even vertex, 2
Event(s), **160**
  independent, 196
  mutually exclusive, 188
  probability of, 166
Expected value, **236**
Exponent(s), **630**
  properties, 631
Exponential equations, 310
  graphs of, **310**
Exponential function, 381
  derivative of, 440, 441
Exponential increase/decrease, 544

Factor(s), 603
Factorials, 180
Factoring, **608**
Fair game, 241, 245, 251
Feasible solution, 119, 133
Finance, **333**
First derivative test, 466
Fixed cost, 65
Fixed probability vector, 231
Fraction(s)
  algebraic, **618**
  complex, 621
Frequency table, 262
Function(s), **372**
  Cobb-Douglas production, 592
  composition of, 432
  constant, 427
  continuous, 390, 393
  decreasing, **458**
  differentiable, **446**
  domain of, 373, 565
  exponential, 381
  graphs of, **378,** 458
  increasing, **458**
  integrable, 505
  linear, 381
  logarithmic, 381

Function(s), *continued*
  objective, 113
  of several variables, **564**
  of two variables, 564
  polynomial, 381
  probability density, **551**
  quadratic, 381
  range of, 377
  rational, **395**
Fundamental counting principle, 174, 175
Fundamental theorem of calculus, 507

Game
  fair, 241, 245, 251
  strictly determined, 245
  theory, **243, 248**
  two-person, 243
  value of, 245, 251
  zero sum, 254
Gauss, Carl F., 77
Gauss-Jordan elimination, **75**
General solution, 541
Genes, 197
Genotypes, 197
Geometric progression(s), **334**
  common ratio, 334
  nth term, 336
  sum of terms, 338
  terms, 335
Graph, 2
  connected, 3
Graph of an equation, **28**
  linear equation, 38
  quadratic equation, 46
Graph of an inequality, **114**
  system of inequalities, 116
Graphs of functions, **378,** 458, 568
Grouped data, 263
Growth factor, 544

Higher-order derivatives, **456**
Histogram, 262

Identity matrices, **95**
Improper integrals, **548**
  convergent, 550
  divergent, 550
Increasing function, **458**
Indefinite integral(s), **492**
  definition, 492
Independent events, 196
Independent variable(s), 381, 564
Inequality, linear, 114
  graph of, **114**
  solution of, 114
  system of, 116
Infinite limits, 399
Inflection points, **481**

Initial condition, 541
Input-output matrix, 102
Instantaneous rate of change, 415
Integrable function, 505
Integral(s)
   definite, **502**
   improper, **548**
   indefinite, **492**
   table, 533
Integrand, 496
Integration, **491, 525**
   by parts, **537**
   by substitution, **526**
   limits of, 508
   use of tables, **532**
   variable of, 496
Interest
   compound, **350**
   continuous compound, 315, 322
   simple, **341**
Interquartile range, 275
Intersection
   of events, 161
   of sets, 637
Interval, 25
Inverse of a matrix, **95**

Jordan, Camille, 77

Lagrange, Joseph L., 586
Lagrange multipliers, **586**
Leibniz, Gottfried, 410
Leontief, Wassily, 104
Leontief input-output models, **101**
Limits, **383**
   at infinity, 397
   infinite, 399
   of integration, 508
Line
   equation of, **31**
   point-slope form, 34
   slope of, 31
   slope-intercept form, 36
Linear depreciation, 41
Linear equation(s), **37, 613**
   as mathematical models, **40**
   graph of, 38
   systems of, **57**
Linear function, 381
Linear inequality, 114
   graph of, **114**
   solution of, 114
   system of, 116
Linear programming, **113**
   geometric method, **119**
   minimization, **145**
   simplex method, **132**
Linear programming problem, 113
   feasible solution(s), 119, 133

Logarithm(s), **317**
   common, 319
      table, 652
   natural, 319
      table, 650
   properties, 320
Logarithmic equations, 325
   graphs of, **325**
Logarithmic function, 381
   derivative of, 442, 444

Main diagonal, 96
Marginal change, 421
Markov, Andrei A., 222
Markov chains, **221**
   regular, **230**
Mathematical model(s), 11, **18, 40, 49**
Matrices, **57**
   addition of, **80**
   augmented coefficient, 69
   coefficient, 91
   difference of, 82
   dimension, 81
   echelon form, 71
   equal, 84
   identity, **95**
   input-output, 102
   inverse, **95**
   main diagonal of, 96
   multiplication of, **87**
   payoff, 243
   regular transition, 231
   row operations, 71
   scalar multiplication, **80**
   square, 96
   transition, 222
   transpose, 145
Mean, 266
   normal distribution, 290
   test, **299**
Measures of central tendency, **266**
   mean, 266, 290
   median, 266
   midrange, 270
   mode, 270
Measures of dispersion, **272**
   interquartile range, 275
   range, 278
   standard deviation, 273, 290
Median, 266
Mendel, Gregor, 197
Midrange, 270
Mixed strategies, **248**
Mode, 270
Model(s)
   Leontief input-output, **101**
   mathematical (see Mathematical model)
Multiplication of matrices, **87**
   associative law, 89

Multiplication rule, **193**
Mutually exclusive events, 188

Natural logarithms, 319
   table, 650
Newton, Isaac, 410
Nonbasic variable, 128
Normal curve, **289**
   standard, 291
   table, 644
Normal distribution, 290
   mean, 290
   standard deviation, 290

Objective function, 113
Odd vertex, 2
Odds, 192
Ordinary annuity (see Annuities, ordinary)
Origin
   Cartesian plane, 27
   number line, 25

Parabola, 46
   vertex, 46
Partial derivatives, **570**
   second-order, **576**
Particular solution, 541
Pascal, Blaise, 159
Path, 3
   Euler, 6
Payoff matrix, 243
Permutations, 180, 186
Pivot, 136
Pivoting, 136
Point-slope form, 34
Polynomial, 381
   degree of, 382
Polynomial function, 381
Power rule, 425, 434
Present value
   annuities, **360**
   compound interest, 354
   continuous compound interest, 546
   simple interest, 344
Primal problem, 146
Probability, **159, 207**
   a posteriori, 220
   a priori, 220
   addition rule, **186**
   assignment, 166
   binomial, **208**
   conditional, 193
   multiplication rule, **193**
   normal distribution, 290
   of an event, 166
   vector, 230
Probability density functions, **551**

Probability vector, 230
   fixed, 231
Producer surplus, 516
Product rule, 436
Production vector, 102

Quadratic equation(s), **46, 613**
   as mathematical models, **49**
   graph of, 46
Quadratic formula, 616
Quadratic function, 381
Quartiles, 275
Quotient rule, 437

Random variable, 237
Range, 278, 377
Rate of change
   average, **410**
   instantaneous, 415
Rational functions, **395**
   asymptotes of, 396
Real number line, **24**
Regular Markov chains, **230**
Regular transition matrix, 231
Relative maximum
   first derivative test, 466
   function of one variable, **466**
   function of two variables, 580
   second derivative test, 468
Relative minimum
   first derivative test, 466
   function of one variable, **466**
   function of two variables, 580
   second derivative test, 468
Row operations, 59

Saddle point, 244
Sample space, **160**
Scalar multiplication, **80**
Second derivative test, 468
Second-order partial derivatives, **576**
Set(s), **634**
   complement, 636
   elements of, 634
   empty, 636
   equal, 635
   intersection, 637
   subset, 635
   union, 637

Set(s), *continued*
   universal, 636
   Venn diagrams for, 636
Set-builder notation, 635
Sign test, **281**
Silk, Leonard, 350
Simple interest, **341**
   amount, 344
   present value, 344
Simplex method, **132**
Sinking fund, 358
Slack variables, 133
Slope of a line, 31
Slope-intercept form, 36
Solution(s)
   basic, **127**
   basic feasible, 133
   feasible, 119, 133
   of a differential equation, 540, 541
   of a system of equations, 59, 67
   of an equation, **613, 627**
   of an inequality, 114
Solving equations, **613, 627**
Square matrix, 96
Standard deviation, 273
   normal distribution, 290
Standard normal curve, 291
Statistics, **261**
Straight line depreciation, 41
Strategy, 244
   best, 244
   mixed, **248**
Strictly determined game, 245
Subset, 635
Subtraction of matrices, 82
System of inequalities
   graph of, 116
Systems of linear equations, **57**, 59
   basic solutions, **127**
   equivalent, 59
   solution, 59, 67

Tableau, simplex, 135
Tables
   areas under the normal curve, 644
   binomial probabilities, 642
   common logarithms, 652
   derivatives, 451
   integrals, 533

Tables, *continued*
   natural logarithms, 650
   values of e^x and e^{-x}, 645
   values of $(1 + i)^n$, 654
   values of $(1 + i)^{-n}$, 655
Terms
   algebraic, 604
   geometric progression, 335
Theorem, 7
Transition matrix, 222
   regular, 231
Transition probabilities, 222
Transpose of a matrix, 145
Two-person game, 243

Union
   of events, 161
   of sets, 637
Unit of distance, 25
Universal set, 636

Variable(s)
   basic, 128
   departing, 136
   dependent, 381, 564
   entering, 135
   independent, 381, 564
   nonbasic, 128
   of integration, 496
   random, 237
   slack, 133
Variable cost, 65
Vectors, 91
   demand, 101
   fixed probability, 231
   probability, 230
   production, 102
Velocity, 422
Venn diagrams, 636
Vertex, 2
   degree of, 2
   even, 2
   odd, 2
   parabola, 46

Zero-sum game, 254

DATE DUE

OCT 26 '89			
OCT 1 5 '91			
JUL 22 '~~~			

DEMCO 38-297

QA37.2.K47 1985
  KERTZ, GEORGE J.
    APPLIED FINITE MATHEMATICS AND
CALCULUS.

11223108

PENN STATE
DELAWARE COUNTY
CAMPUS LIBRARY

CALCULUS

DERIVATIVES

(b, c, and n are constants, $b > 0$, $b \neq 1$; u is a function of x)

1. $\dfrac{d}{dx} x^n = nx^{n-1}$

1(a). $\dfrac{d}{dx} u^n = nu^{n-1} \cdot \dfrac{du}{dx}$ General Power Rule

2. $\dfrac{d}{dx}[cf(x)] = c \cdot f'(x)$

3. $\dfrac{d}{dx} c = 0$

4. $\dfrac{d}{dx}[u(x) \pm v(x)] = u'(x) \pm v'(x)$ Addition Rule

5. $\dfrac{d}{dx}[f(u)] = \dfrac{d}{du} f(u) \cdot \dfrac{du}{dx}$ Chain Rule

6. $\dfrac{d}{dx}[u(x) \cdot v(x)] = u(x) \cdot v'(x) + v(x) \cdot u'(x)$ Product Rule

7. $\dfrac{d}{dx}\left[\dfrac{u(x)}{v(x)}\right] = \dfrac{v(x) \cdot u'(x) - u(x) \cdot v'(x)}{v(x)^2}$ Quotient Rule

8. $\dfrac{d}{dx} b^x = b^x (\ln b)$ and $\dfrac{d}{dx} e^x = e^x$

8(a). $\dfrac{d}{dx} b^u = b^u (\ln b) \cdot \dfrac{du}{dx}$ and $\dfrac{d}{dx} e^u = e^u \cdot \dfrac{du}{dx}$

9. $\dfrac{d}{dx} \log_b x = \dfrac{1}{(\ln b)x}$ and $\dfrac{d}{dx} \ln x = \dfrac{1}{x}$

9(a). $\dfrac{d}{dx} \log_b u = \dfrac{1}{(\ln b)\, u} \cdot \dfrac{du}{dx}$ and $\dfrac{d}{dx} \ln u = \dfrac{1}{u} \cdot \dfrac{du}{dx}$